Springer Tracts in Advanced Robotics

Volume 60

Editors: Bruno Siciliano · Oussama Khatib · Frans Groen

Wen-Hong Zhu

Virtual Decomposition Control

Toward Hyper Degrees of Freedom Robots

Springer

Author

Wen-Hong Zhu
Canadian Space Agency
6767 route de l'Aeroport
Saint-Hubert, QC J3Y 8Y9
Canada
E-mail: Wen-Hong.Zhu@asc-csa.gc.ca

ISBN 978-3-642-10723-8 e-ISBN 978-3-642-10724-5

DOI 10.1007/978-3-642-10724-5

Springer Tracts in Advanced Robotics ISSN 1610-7438

Library of Congress Control Number: 2009944183

Typeset & Cover Design: Scientific Publishing Services Pvt. Ltd., Chennai, India.

Printed on acid-free paper

5 4 3 2 1 0

springer.com

STAR (Springer Tracts in Advanced Robotics) has been promoted under the auspices of EURON (European Robotics Research Network)

This book is dedicated to

George Vukovich
Septimiu E. (Tim) Salcudean
Joris De Schutter
Zeungnam Bien
Zhongjun Zhang
Yugeng Xi
Huitang Chen

for their academic inspiration, essential help, and financial support

Foreword

By the dawn of the new millennium, robotics has undergone a major transformation in scope and dimensions. This expansion has been brought about by the maturity of the field and the advances in its related technologies. From a largely dominant industrial focus, robotics has been rapidly expanding into the challenges of the human world. The new generation of robots is expected to safely and dependably co-habitat with humans in homes, workplaces, and communities, providing support in services, entertainment, education, healthcare, manufacturing, and assistance.

Beyond its impact on physical robots, the body of knowledge robotics has produced is revealing a much wider range of applications reaching across diverse research areas and scientific disciplines, such as: biomechanics, haptics, neurosciences, virtual simulation, animation, surgery, and sensor networks among others. In return, the challenges of the new emerging areas are proving an abundant source of stimulation and insights for the field of robotics. It is indeed at the intersection of disciplines that the most striking advances happen.

The goal of the series of *Springer Tracts in Advanced Robotics (STAR)* is to bring, in a timely fashion, the latest advances and developments in robotics on the basis of their significance and quality. It is our hope that the wider dissemination of research developments will stimulate more exchanges and collaborations among the research community and contribute to further advancement of this rapidly growing field.

The monograph written by Wen Hong Zhu is the outcome of several years of work by the author, stemming from his doctoral thesis. The main contribution of this book is the presentation of the so-called Virtual Decomposition Control (VDC) to dynamics and control of complex robotic systems. The pursued approach can be adopted for a wide range of robots, including the actuation system. Constrained and multi-arm robots, space robots, humanoid robots, teleoperated robots, flexible robots and more generally hyper degree-of-freedom modular systems can be keenly dealt with VDC while ensuring their stability. Remarkably, the theoretical analysis is supported by several useful examples and practical case studies.

This volume is a very fine addition to our STAR series!

Naples, Italy
September 2009

Bruno Siciliano
STAR Editor

Preface

Robot control, a subject aimed at making robots behave as desired, has been extensively developed for more than two decades. Among many books being published on this subject, a common feature is to treat a robot as a single system that is to be controlled by a variety of control algorithms depending on different scenarios and control objectives. However, when a robot becomes more complex and its degrees of freedom of motion increase substantially, the needed control computation can easily go beyond the scope a modern computer can handle within a pre-specified sampling period. A solution is to base the control on subsystem dynamics.

This book entitled *Virtual Decomposition Control* (VDC) is the first book that talks about subsystem based control of robots without compromising control performances. The VDC approach uses subsystem dynamics to conduct control computation, while rigorously guaranteeing the stability of the entire robot without imposing additional approximations. Unlike decentralized control, the dynamic interactions among subsystems are mathematically handled through the definition of *virtual power flows*. A unique property with this novel approach is that the sum of all *virtual power flows* will be or must be zero at the end.

This book comprises two parts with fourteen chapters. The first part comprised of Chapters 1 to 4 presents all theoretical contents necessarily needed to conduct the VDC for any complex robot. Specifically, Chapter 1 gives an introductory description about the background that motivates the creation of this novel control approach. Chapter 2 collects all mathematical foundations to be used later on by the VDC. Most importantly, the concept of the *virtual stability* is introduced. The unique characteristic of the *virtual stability* is the inclusion of the *virtual power flows* in the time derivative of the *non-negative accompanying function* assigned to each subsystem. The importance of defining the *virtual stability* lies in the fact that the (L_2 and L_∞) stability of the entire complex robot can be guaranteed as long as *every* subsystem combined with its respective control equations qualifies to be *virtually stable*. Chapter 3 takes a two degrees of freedom planar robot as a simple example to demonstrate how the VDC works. Finally in Chapter 4, a general formulation of the VDC when being applied to a generalized complex robot is given. This chapter ends with discussions on its relationships to the passivity theory and to PID control.

The second part comprised of Chapters 5 to 14 is dedicated to specific applications. Chapter 5 is a *must-read* chapter that applies the VDC to a class of the most popular six-joint robot manipulators in both free motion and constrained motion. Three joint control modes suitable for different electronic control interfaces are presented. Starting from Chapter 6, each chapter becomes independent.

Prospective readers are suggested to read Chapters 1 to 3 and 5 to fully understand the basic concepts of the VDC approach. From this point, readers might have several options. A reader might decide to go back to Chapter 4, if a general formulation of the VDC is a concern. Chapter 6 is recommended for readers who are interested in motor-transmission mechanisms including the commonly used harmonic drives. Readers working on hydraulic robots might find it interesting to go to Chapter 7. Chapter 8 is particularly dedicated to the coordinated control of multiple robot manipulators holding a common object. Readers in the space robotics society might have interest in Chapter 9. Furthermore, a possible application of the VDC to humanoid robots is initialized in Chapter 10. Readers interested in force-reflected bilateral teleoperation are suggested to examine Chapter 11 for an alternative solution based on adaptive control to automatically accommodate the system uncertainties, leading to superior operational performances. Chapter 12 applies the modularity of the VDC to modular robot manipulators, aimed at giving the modular robot manipulators the same dynamics based control performances as these integrated robot manipulators. Readers who have been working on control of flexible link robots are encouraged to investigate Chapter 13 in which the VDC is extended to distributed parameter systems. Finally in Chapter 14, the extension of the VDC to a current control problem of electrical circuits is briefly discussed for the first time, making use of the duality between mechanical and electrical systems.

The author is grateful to Dr. George Vukovich for his invaluable helps and efforts on editing a draft version of the first four chapters to enhance the readability of this book. His comments on the presentation certainly help improve the quality of the book further. The author would also like to thank Dr. Bruno Siciliano, STAR Editor, for his rigorous handling of the review process, leading to the acceptance of the book proposal. Finally, special thanks are due to Dr. Thomas Ditzinger, Senior Editor of Engineering, Springer-Verlag, for his consistent support of this book project right from the beginning, seeing it to the successful completion with great patience.

Brossard, Québec
Canada

Wen-Hong Zhu
September 2009

Contents

Part I: Virtual Decomposition Control Theory

Part II: Virtual Decomposition Control Applications

List of Figures

List of Tables

Part I

Virtual Decomposition Control Theory

1 Introduction

1.1 History of Robotic Systems

Robots, with different definitions and unfinished scope, have substantially influenced our lives. As is by now well known, the term "robot" was initiated in Karel Capek's 1921 play Rossum's Universal Robots (RUR), and the concept quickly found its way into the popular imagination, becoming the subject of various dramas, books, and movies. The term "robotics" and the famous three laws of robotics were first introduced in Isaac Asimov's science fiction. Real robots could be said to date from a collaboration between inventor George Devol and entrepreneur Joseph Engelberger in 1956 which established Unimation—the first company started robot business which characterizes the birth of robot-centered flexible automation.

Common industrial robots are generally heavy rigid devices. They operate in precisely structured environments and perform highly repetitive tasks under preprogrammed control. Today, over one million industrial robots are operating worldwide.

In the 1950s and 1960s, the robot industry was in its infancy, producing robots which performed simple operations in transportation with basic mechanical functions under simple computer-assisted hydraulic control, such as the first Unimate industrial robot (made by Unimation in 1961) operating in a General Motors automobile factory in New Jersey. The early version of the hydraulically driven Unimate robots was developed with joint servo control and simple computer-assisted task programming. The Unimation hydraulic robot design was licensed to Kawasaki Heavy Industries in 1968, which set-off robot productions in Japan. During this time period, the well-known homogeneous transformation technique was developed to compute the robot kinematics, which still remains the foundation of robotic kinematics nowadays. From the late 1960s to the 1970s, a transition phase for industrial robots, the demands for small to medium payload robots motivated the development of electrically driven robots, initialized by Vicarm using the design from Victor Scheinman. In 1977, after Vicarm was sold to Unimation, the Programmable Universal Machine for Assembly (PUMA)

W.-H. Zhu: Virtual Decomposition Control, STAR 60, pp. 3–13.
springerlink.com

robots were created. In Europe, ASEA—the previous company of ABB Robotics of Sweden developed its first fully electrically driven robot in 1974. The German company KUKA launched its first six-joint electrically driven industrial robot in 1977. Introductions of electric motors and artificial intelligence allowed the vision and contact technologies to be incorporated.

The major developments in modern robotics began in the late 1970s when interest in industrial robotics swelled and many companies entered the field. A contract between Fanuc Robotics and General Motors in 1982 made a landmark role for robots performing massive assembly and welding tasks in the automotive industry. Cincinnati Milacron produced hydraulic robots T3 (The Tomorrow Tools) in 1974 for spot-welding and this robotic technology was transferred to ABB in 1990. In 1984, the first robot company Unimation was acquired by Westinghouse and then sold to Stäubli 4 years later. Today, the major market players include the German company KUKA, the Swedish company ABB Robotics with its key product series IRB, the US companies Adept, Barrett Technology Inc., and MobileRobots Inc. (formerly ActivMedia Robotics), and the Japanese companies Fanuc, Kawasaki Heavy Industries, Yaskawa Electric Corp., Mitsubishi Heavy Industries, and Matsushita (Panasonic). Industrial robots are now used in machine industry, electronics industry, automotive industry, and food industry. The main tasks performed by industrial robots have been extended from painting, die-casting, material handling, and spot welding to arc welding, assembly, grinding, deburring, polishing, and packing, where precise motion control, sensor-based contact control, effective programming, and integration of robots with complementary systems are needed.

While industrial robots are mainly being used in manufacturing, special purpose robots and research robots have gone far beyond. Joseph Engelberger started the Transition Robotics company producing HelpMate robots engaged in the service robotics field. Several prototypes have been developed to demonstrate their potential for helping disabled and elderly and for cleaning households. Possessing anthropomorphic appearance, the Honda showcase Asimo and the Sony Dream robot are able to perform several motion patterns with a certain level of artificial intelligence. Built by MD Robotics of Canada, the Space Station Remote Manipulator System (SSRMS) was successfully launched into orbit in 2001 and began operations to complete the assembly of the International Space Station. Underwater robots such as the Aqua Explorer are used to inspect the sea bed for the purpose of laying optical-fiber cables. Field robots are bing tested in the agriculture industry and security robots are being used in dangerous environments to clean up hazardous objects and to perform anti-terrorism activities.

1.2 Overview of Robot Control

Control systems are essential to robots. They produce desired behaviours, maximize potential performances, and deliver achievable accuracy, subject to robustness requirements.

1.2.1 Classical Control

The most widely used control scheme for industrial robots is based on the joint Proportional, Integral, and Derivative (PID) servo control. In this scheme, inverse kinematics is used to convert a desired robot tip (end-effector) position into the desired joint positions. Then, a joint PID controller migrated from industrial servoing systems is used to make the actual joint positions track their desired values. While this scheme is easy to implement and is able to move a robot to its desired position in the steady state, the dynamic properties of the robot can vitiate this approach, since the dynamic coupling among the joints and the mass matrix variations can invalidate the PID control.[1] In order to prevent overshoots in contact motion (a potentially dangerous situation with a possibility of having undesirable collisions with the environment), the joint PID servo controllers are designed to be over-damped in most circumstances. As a result, these robots are capable only of performing regulation tasks that require accuracies only in the steady state. Such tasks include spot welding, static material holding, and object transportation in a static environment. For tasks requiring tracking accuracies, such as arc welding, material cutting, grinding, deburring, polishing, assembly, and motion/force coordination in a dynamic environment, the joint PID control is generally inadequate.

1.2.2 Limitations of PID Control

A typical second-order system, which can represent a one degree of freedom (DOF) robot $G(s)$, subject to a PID control $C(s)$, is illustrated in Figure 1.1. Without loss of generality, the system model is represented by a linear second-order differential equation in the Laplace domain as

$$G(s) = \frac{1}{ms^2 + bs + k} \tag{1.1}$$

where m, b, k are three positive parameters characterizing the mass, the damping coefficient, and the stiffness, respectively. On the other hand, the PID controller takes the form of

$$C(s) = k_d s + k_p + \frac{k_I}{s} \tag{1.2}$$

where k_d, k_p, and k_I are three control parameters. If the parameters are chosen such that

$$k_d = m\lambda \tag{1.3}$$

$$k_p = b\lambda \tag{1.4}$$

$$k_I = k\lambda \tag{1.5}$$

[1] The PID type of control is originally designed for second-order mechanical systems without subject to significant coupling among the axes (such as the machine tools). However, this is not the case for robots that exhibit significant variations in mass properties (when robots move to different configurations and with different payloads) and possess inherent high dynamic coupling among the axes.

hold with $\lambda > 0$, then the closed loop transfer function becomes

$$\frac{x(s)}{r(s)} = \frac{\lambda}{s + \lambda} . \tag{1.6}$$

This example leads to an observation that a PID control with a special set of gains can have the closed-loop transfer function reduced to a first order system. The cut-off frequency is dependent on a factor in the control gains. However, due to the limitations such as sampling delays and the use of anti-aliasing low-pass filters, the control gains are always limited by stability conditions. As a result, the PID control is able to deliver acceptable control performances at low frequencies, but it is unable to produce adequate control performances at high frequencies beyond the cut-off frequency.

Unlike process control, an amount of 1% error is very significant and unacceptable in general.[2] In robot applications, control errors usually must be kept to $< 0.1\%$. While being able to achieve this control accuracy for tasks involving regulations, PID control is insufficient for tasks involving tracking operations. Therefore, researchers have long been seeking advanced control approaches that are able to achieve higher control performances. One solution is to use dynamics based control.

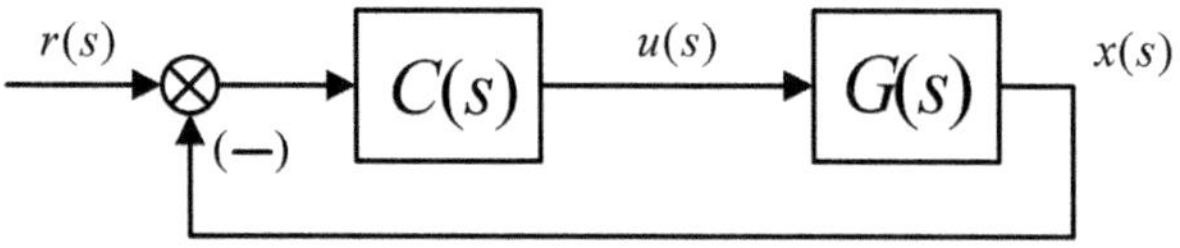

Fig. 1.1. A PID control system.

1.2.3 Dynamics Based Control

A typical dynamics based feedforward plus PID feedback control (simply called dynamics based control) is illustrated in Figure 1.2, in which a new feedforward block $F(s)$ is added.

If $F(s) = 1/G(s)$ holds indefinitely, that is, the feedforward transfer function takes exactly the inverse dynamics of the system to be controlled, then the ideal transfer function $\frac{x(j\omega)}{r(j\omega)} = 1$, $\forall \omega \in [0, \infty)$, can be achieved. This result implies that an "infinite" control bandwidth is possible as long as a proper feedforward control is designed. In this approach, the feedforward term mainly contributes to high control accuracies, whereas the feedback term is primarily used to overcome uncertainties, to maintain stability, and to address transition issues.

An alternative way to achieve the ideal transfer function $x(s)/r(s) = 1$ is to build an open-loop compensator shown as $D(s)$ in Figure 1.3. If $D(s) = \frac{s+\lambda}{\lambda}$ is chosen, then $x(s)/r(s) = 1$ can be achieved in the absence of uncertainties.

[2] For a robot having an arm length of 1 m, an amount of 1% error indicates an error of 1 cm at the tip.

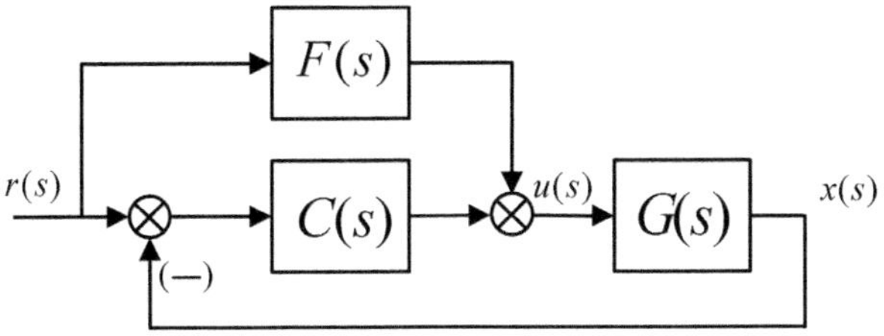

Fig. 1.2. A dynamics based feedforward and PID feedback control system.

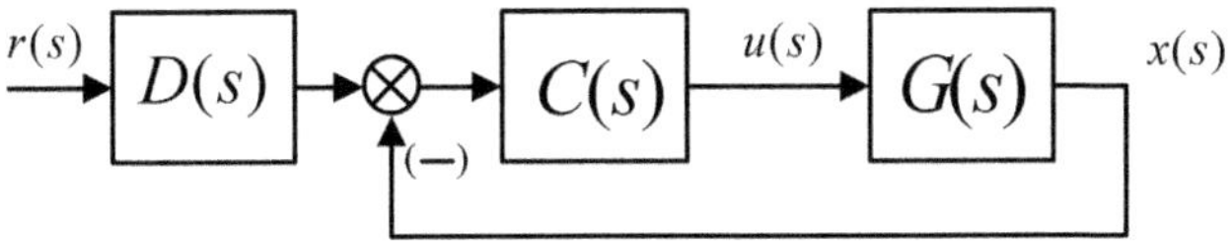

Fig. 1.3. An open-loop compensator and PID control system.

However, these two approaches are not equivalent. The advantages of the dynamics based control approach as illustrated in Figure 1.2 over the open-loop compensator approach as illustrated in Figure 1.3 can be summarized as follows. First, the design of $F(s)$ in Figure 1.2 is completely based on the robot inverse dynamics and can be obtained independently from the feedback controller $C(s)$. These inverse dynamics are available from both well-known Lagrangian and Newton–Euler formulations. Second, unlike the open-loop compensator approach which is limited to linear control systems, the dynamics based control approach is generally applicable to highly coupled nonlinear systems such as robots [155].

In conclusion, the dynamics based control approach (see Figure 1.2) is able to achieve high bandwidth control without relying on the feedback gains. Having high bandwidth control leads to at least two positive impacts on the applications. First, it allows accurate executions of some dynamically challenging tasks, such as high speed cutting, grinding, deburring, and polishing, which would be otherwise impossible by using PID controlled robots. Second, it allows rapid executions of tasks which were previously executed by PID controlled robots with a much lower speed. Thus, the control approach illustrated in Figure 1.2 will be the baseline approach from which this book evolves.

1.2.4 Related Work

Set-Point Control

This type of control targets at regulation tasks in which only the steady state accuracy is required. Examples include spot welding and pick-and-place operations. Generally, no dynamic model of the robots is used, which sacrifices the performances when performing high-speed trajectory tracking control. The control law can be either a joint PID control or a joint PD control plus gravity compensation [166]. Passivity theory is used to guarantee the stability [10].

Nonlinear Feedback Linearization

This type of control has been widely adopted in the robot control community [4, 11, 39, 145, 149, 161, 200]. The fundamental concept is to use a specially chosen feedback which ideally cancels out system nonlinearities, leaving a linearized system which can then be handled by conventional linear control methods. While being useful in many circumstances, this approach is hampered by the need for accurate models, access to certain state variables, and subject to restricted regions of applicability.

Model-Based Adaptive Control

An early version of this type of control for robotic systems was proposed by Slotine and Li [153, 154]. The fundamental control structure consists of a feedforward term and a feedback term, similar to Figure 1.2. The feedforward term plays a major role to guarantee motion tracking performances under a parameter adaptation law, whereas the feedback term is used only to overcome the uncertainties and external disturbances. The feedforward part is directly based on the inverse robot dynamics. Unlike the nonlinear feedback linearization approaches, no inverse of the mass matrix is used. This feature allows the robot dynamics based feedforward term to be expressed in a linear parametrization form so that a direct adaptive control can be applied to yield the asymptotic stability of the motion with bounded parameter errors [155].

1.2.5 Technical Challenges in Control of Complex Robots

In many books on control of robots, such as [4, 7, 8, 11, 21, 39, 54, 60, 63, 92, 100, 138, 140, 145, 149, 161, 189, 200], the control designs are based on the complete dynamic models of robots. Usually, robots with two or three degrees of freedom were taken as examples in simulations or experiments in order to make control implementations manageable. However, substantial technical challenges in control implementations arise when the number of degrees of freedom of motion is well over six. This is due to the fact that the complexity (computational burden) of robot dynamics is proportional to the fourth power of the number of degrees of freedom of motion. For a system with more than thirty-DOF of motion, such as a humanoid robot, it is very difficult, if not impossible, to implement the complete dynamics based control on a single computer.

In contrast, the *virtual decomposition control* approach, as will be outlined below and presented in this book, is a general, efficient, and powerful tool particularly suitable to handle the full-dynamics-based control problem of complex robots.

1.3 Virtual Decomposition Control: Toward Hyper Degrees of Freedom Robots

1.3.1 Subsystem Dynamics Based Control

With respect to the technical challenges arisen from the application of dynamics based control to complex robots, one possible solution is to base control directly on subsystem dynamics, rather than on the complete dynamics of a robot with hyper degrees of freedom. In view of the fact that no matter how complicated a robotic system is, the dynamics of the subsystems (components) remain relatively simple with fixed dynamic structures invariant to target systems. Thus, control computations are then proportional to the number of subsystems and can be performed even by locally embedded hardware/software.

Whereas the subsystem dynamics based control can be viewed as an efficient way to handle the control problems of systems with hyper degrees of freedom, the remaining issue is to address the dynamic interactions among the subsystems.

1.3.2 Virtual Decomposition Control

The *virtual decomposition control* (VDC) uses the dynamics of subsystems (rigid links and joints) to conduct control design, while maintaining the L_2 and L_∞ stability and convergence of the entire robotic system. A feature of this approach is the introduction of a scalar term, namely *virtual power flow* (defined as an inner product of the velocity error and the force error in a common frame), at every virtual *cutting point* at which a virtual "disconnection" is placed. The *virtual power flows* uniquely define the dynamic interactions among the subsystems and play a vital role in the definition of the *virtual stability*[3] of the subsystems, leading to the L_2 and L_∞ stability of the entire system. Furthermore, the VDC is capable of accomplishing a variety of control objectives, such as motion control, internal force control, and optimizations, without restrictions on target systems.

One of the advantages of using the VDC approach is that the change of the dynamics of a subsystem (such as substituting a hydraulic actuator for an electrical motor) only affects the respective local control equations associated with this subsystem, while keeping the control equations associated with the rest of the system unchanged. For example, when an electrical motor is replaced by a hydraulic actuator, only the control equations for this particular joint change, whereas the control equations for the rest of the system keep unchanged.

In summary, the VDC approach provides an insight into the internal dynamic structure of each subsystem with an unlimited ability to exploit systems in detail, since any subsystem can be further virtually decomposed into several sub-subsystems without foreseeable boundaries.

[3] The formal definition of the *virtual stability* will be given in Chapter 2.

1.3.3 An Illustrated Example

A very simple system with one degree of freedom is taken as an example to demonstrate the basic procedure of applying the VDC approach.

An illustrated system is given in Figure 1.4. It is an object with (point) mass m moving in a guided rail subject to a friction force $g(x, \dot{x}, t)$. Without loss of generality, it is assumed that this system is a subsystem of a complex system and is subject to two exerting forces, $f_1(t)$ and $f_2(t)$, being exerted to and from this subsystem, respectively. This system can be viewed as a representative of a robot link moving in a one-dimensional space. The two exerting forces represent the dynamic interactions between "the link" and the rest of the system, when "the link" has two mechanical connection points (or virtual *cutting points*) with the rest of the system.

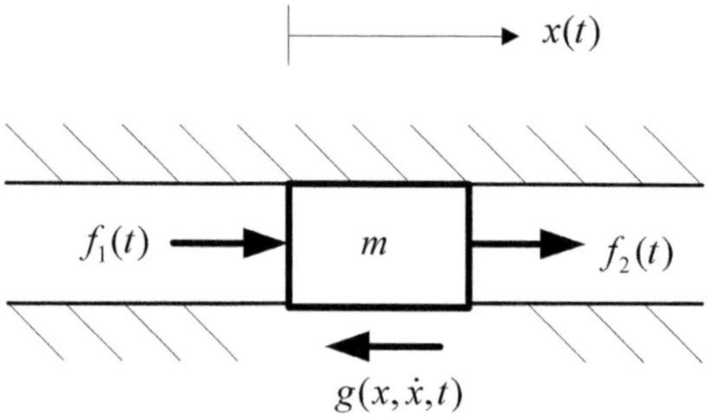

Fig. 1.4. An illustrated system.

Let

$$m\ddot{x}(t) + g(x, \dot{x}, t) = f^*(t) \tag{1.7}$$

be the dynamics of the subsystem, where $x(t)$ is the position and f^* is the net force subject to

$$f^*(t) = f_1(t) - f_2(t). \tag{1.8}$$

In the control design, required (design) variables are being widely used. The meaning of the required variables is that control objectives are fulfilled once the actual variables approach their respective required variables. In other words, the required variables are (actual) variables that are required to fulfil the control objectives.

According to (1.7), the model based required net force is designed as

$$f_r^*(t) = m\ddot{x}_r(t) + g(x, \dot{x}, t) + k_s(\dot{x}_r(t) - \dot{x}(t)) \tag{1.9}$$

where $k_s > 0$ is a feedback gain and $\ddot{x}_r(t)$ is the time-derivative of the required velocity $\dot{x}_r$ defined by

$$\dot{x}_r(t) = \dot{x}_d(t) + \lambda(x_d(t) - x(t)) \tag{1.10}$$

with $x_d(t)$ being the desired position and $\lambda > 0$. Meanwhile, the required net force can be expressed as

$$f_r^*(t) = f_{1r}(t) - f_{2r}(t) \tag{1.11}$$

where f_{ir} denotes the required exerting force for $i = 1, 2$. Note that (1.11) yields f_{1r} from given f_{2r} as

$$f_{1r}(t) = f_r^*(t) + f_{2r}(t). \tag{1.12}$$

Remark 1.1. In (1.9), the first two terms in the right hand side are model based feedforward terms which are based on (1.7), whereas the last term is a feedback control term. Unlike conventional feedforward terms, the required velocity and acceleration are used in lieu of the desired velocity and acceleration.

Subtracting (1.7) from (1.9) yields the error dynamics

$$m(\ddot{x}_r(t) - \ddot{x}(t)) + k_s(\dot{x}_r(t) - \dot{x}(t)) = f_r^*(t) - f^*(t) \tag{1.13}$$

which will be used in the stability analysis.

Let

$$\nu(t) = \frac{1}{2}m(\dot{x}_r(t) - \dot{x}(t))^2 \tag{1.14}$$

be a non-negative accompanying function assigned to this subsystem. In view of (1.8), (1.12), and (1.13), the time-derivative of (1.14) can be written as

$$\begin{aligned}\dot{\nu}(t) &= m(\dot{x}_r(t) - \dot{x}(t))(\ddot{x}_r(t) - \ddot{x}(t)) \\ &= -k_s(\dot{x}_r(t) - \dot{x}(t))^2 \\ &\quad +(\dot{x}_r(t) - \dot{x}(t))(f_r^*(t) - f^*(t)) \\ &= -k_s(\dot{x}_r(t) - \dot{x}(t))^2 \\ &\quad +(\dot{x}_r(t) - \dot{x}(t))(f_{1r}(t) - f_1(t)) \\ &\quad -(\dot{x}_r(t) - \dot{x}(t))(f_{2r}(t) - f_2(t)).\end{aligned} \tag{1.15}$$

The last two terms in the right hand side of (1.15) are named the *virtual power flows* which are the products of velocity errors and force errors. As will be seen in the next chapter, the appearance of *virtual power flows* in the time-derivative of a non-negative accompanying function assigned to a subsystem is required to ensure that this subsystem is *virtually stable*. If every subsystem is *virtually stable*, then the entire system is stable. This is due to the fact that at any given *cutting point*, the *virtual power flows* of the two adjacent subsystems take the same magnitude with opposite signs (determined by the exerting force reference direction). Therefore, all *virtual power flows* cancel out when being added up to ensure the stability of the entire system.

The essence of the VDC approach is that the controller uses the dynamics of each subsystem rather than the dynamics of the entire system (which are very complex) to ensure the *virtual stability* of each subsystem in the first place. Then, when all subsystems are *virtually stable*, the stability of the entire system is guaranteed.

When parameter uncertainties are concerned, (1.9) should be re-written as

$$f_r^*(t) = \hat{m}\ddot{x}_r(t) + \hat{g}(x, \dot{x}, t) + k_s(\dot{x}_r(t) - \dot{x}(t)) \tag{1.16}$$

and (1.13) changes to

$$\begin{aligned} f_r^*(t) - f^*(t) = {} & m(\ddot{x}_r(t) - \ddot{x}(t)) + \lambda(\ddot{x}_r(t) - \ddot{x}(t)) \\ & -(m - \hat{m})\ddot{x}_r - [g(x, \dot{x}, t) - \hat{g}(x, \dot{x}, t)]. \end{aligned} \tag{1.17}$$

Accordingly, a non-negative term

$$\nu_\theta(t) \geqslant 0 \tag{1.18}$$

characterizing the parameter uncertainties should be added to (1.14) to form

$$\nu(t) = \frac{1}{2}m(\dot{x}_r(t) - \dot{x}(t))^2 + \nu_\theta(t). \tag{1.19}$$

It follows from (1.8), (1.12), and (1.17) that

$$\begin{aligned} \dot{\nu}(t) &= m(\dot{x}_r(t) - \dot{x}(t))(\ddot{x}_r(t) - \ddot{x}(t)) + \dot{\nu}_\theta(t) \\ &= -k_s(\dot{x}_r(t) - \dot{x}(t))^2 \\ &\quad +(\dot{x}_r(t) - \dot{x}(t))(f_r^*(t) - f^*(t)) \\ &\quad +(\dot{x}_r(t) - \dot{x}(t))[(m - \hat{m})\ddot{x}_r + (g(x, \dot{x}, t) - \hat{g}(x, \dot{x}, t))] \\ &\quad +\dot{\nu}_\theta(t) \\ &= -k_s(\dot{x}_r(t) - \dot{x}(t))^2 \\ &\quad +(\dot{x}_r(t) - \dot{x}(t))(f_{1r}(t) - f_1(t)) \\ &\quad -(\dot{x}_r(t) - \dot{x}(t))(f_{2r}(t) - f_2(t)) \\ &\quad +(\dot{x}_r(t) - \dot{x}(t))[(m - \hat{m})\ddot{x}_r + (g(x, \dot{x}, t) - \hat{g}(x, \dot{x}, t))] \\ &\quad +\dot{\nu}_\theta(t) \end{aligned} \tag{1.20}$$

holds. Parameter adaptation laws are to be designed to make sure that

$$(\dot{x}_r(t) - \dot{x}(t))[(m - \hat{m})\ddot{x}_r + (g(x, \dot{x}, t) - \hat{g}(x, \dot{x}, t))] + \dot{\nu}_\theta \leqslant 0 \tag{1.21}$$

holds to guarantee the *virtual stability* of the subsystem.

1.4 Experimental Demonstrations

The VDC approach has been successfully applied to several real robotic systems, after it was first published in [211]. The first set of experiments includes two industrial robots KUMA 361 and KUMA 160 performing rigid contact control and holding a raw egg rigidly [215, 218]. In these experiments, hybrid motions with rigid contacts were realized. In [217], the VDC was successfully applied to force-reflected bilateral teleoperation systems enabling arbitrary motion/force scaling in both position and rate modes. It was the first time that the force-reflected bilateral teleoperation control can be based directly on the complete nonlinear dynamic models of both master and slave robots, while permitting a rigid contact with the environment. Asymptotic position and velocity tracking between the master and slave robots in both free and contact motions is another important result of this approach.

Other two workable examples include the control of hydraulic robots and the control of robots with harmonic drives [221, 222]. In these two examples, the feature that the dynamics and control of the joints can be conducted independently from the rest of the system was experimentally demonstrated. The VDC approach allows both hydraulic actuators and harmonic drives to be controlled independently regardless of the dynamics of the remaining subsystems. This feature allowed emulations of an electrically driven space robot by a hydraulically driven ground robot [219] to be possible. This is due to the fact that the only difference between the control of an electrically-driven robot and the control of a hydraulically driven robot situates in the local control equations affiliated with the actuators.

1.5 Concluding Remarks

It is the difficulty arisen from the implementation of a complete dynamics based control on robots with hyper degrees of freedom that gives rise to the creation of the VDC. This approach allows us to conduct control design by solely using the subsystem dynamics, while guaranteeing the stability and convergence of the complete robot. Starting next chapter, it will be shown that the technical merit of the VDC approach is the introduction of the *virtual power flows* that completely represent the dynamic interactions among the subsystems. The introduction of the *virtual power flows* further leads to the definition of the *virtual stability*. Finally, the equivalence between the *virtual stability* of every subsystem and the stability of the complete robot is established, based on the fact that the summation of all *virtual power flows* gives zero.

Chapter 2 collects all mathematical materials necessarily to be used by the VDC approach. Among them, the concept of the *virtual stability* together with the terminology of the *virtual power flow* is introduced and defined. Chapters 3–5 form the technical core of the VDC approach. Chapter 3 shows how the VDC approach works with respect to a two degrees of freedom planar robot. Chapter 5 applies the VDC approach to a class of the most popular six-joint robot manipulators in both the free motion and the constrained motion, whereas Chapter 4 presents a general formulation of the VDC approach when being applied to a generalized complex robot. Chapters 6 and 7 deal with actuator-transmission and hydraulic actuation, respectively. Chapters 8–12 handle five different types of robots, namely coordinated multiple robot manipulators, space robots, humanoid robots, force-reflected bilateral teleoperation, and modular robot manipulators. Finally in Chapters 13 and 14, the VDC approach is extended to distributed parameter systems and to electrical circuits.

2

Mathematical Preliminaries

In this chapter, mathematical tools to be used throughout this book are given.

Starting from notation, this chapter presents spaces and groups first. After that, vectors and coordinate systems are introduced, particularly the body-frame referenced vector expressions are emphasized. Orientations expressed by quaternions are given. Then, linear/angular velocity vectors and force/moment vectors are defined, leading to the rigid body dynamics expressed in a body frame. Two functions to be used in the parameter adaptation are subsequently defined. Furthermore, the concept of the virtual decomposition is presented and the terminology of the *virtual power flow* is introduced, giving rise to the definition of the *virtual stability* in conjunction with a simple oriented graph. This chapter ends with a formal definition of algebraic loops.

2.1 Notation

The following rules apply to the notation throughout this book unless otherwise specified:

- An italic letter in lower case represents a scalar.
- A bold letter in lower case with an overhead right arrow, such as $\vec{\mathbf{v}}$, represents a directed (physical) vector; a bold letter in lower case, such as $\mathbf{v}$, represents a valued vector.
- A bold letter in upper case, such as $\mathbf{U}$, represents a matrix.
- A bold letter in upper case surrounded by a pair of braces, such as $\{\mathbf{A}\}$, represents a coordinate frame.
- A *mathcal* letter, such as $\mathcal{V}$, represents an augmented vector or matrix or represents a special function such as the projection function $\mathcal{P}$.
- A bold zero represents either a zero vector or a zero matrix.

W.-H. Zhu: Virtual Decomposition Control, STAR 60, pp. 15–43.
springerlink.com

2.2 Spaces and Groups

In this section, definitions on the Euclidean n-space, the Lebesgue space, the special orthogonal group, and the special Euclidean group are given first, followed by several lemmas that state the properties. The following definitions are cited from [61, 142].

Definition 2.1. *Euclidean n-space is the space of all n-tuples of real numbers, denoted as $\mathbb{R}^n$, such that $\mathbf{x} = [x_1, x_2, \cdots, x_n]^T \in \mathbb{R}^n$ is defined. The Euclidean norm, denoted as $\|\mathbf{x}\|$, is defined as $\|\mathbf{x}\| = \sqrt{\sum_{i=1}^{n} x_i^2}$.*

Definition 2.2. *Special orthogonal group of degree 3, denoted as $SO(3)$, is the group of 3×3 orthogonal matrices.*

Definition 2.3. *Special Euclidean group, denoted as $SE(3)$, is the group of 4×4 matrices having the form of*

$$\begin{bmatrix} \mathbf{R} & \mathbf{v} \\ 0 & 1 \end{bmatrix}$$

with $\mathbf{R} \in SO(3) \bigcap \mathbb{R}^{3\times 3}$ and $\mathbf{v} \in \mathbb{R}^3$. Space $SE(3)$ is homomorphic to $\mathbb{R}^3 \times SO(3)$.

Definition 2.4. *Lebesgue space, denoted as L_p with p being a positive integer, contains all Lebesgue measurable and integrable functions $f(t)$ subject to*

$$\|f\|_p = \lim_{T\to\infty} \left[\int_0^T |f(t)|^p \, d\tau \right]^{\frac{1}{p}} < +\infty. \tag{2.1}$$

Two particular cases are given:

(a) A Lebesgue measurable function $f(t)$ belongs to L_2 if and only if $\lim_{T\to\infty} \int_0^T |f(t)|^2 \, d\tau < +\infty$.

(b) A Lebesgue measurable function $f(t)$ belongs to L_∞ if and only if $\max_{t\in[0,\infty)} |f(t)| < +\infty$.

Definition 2.5. *A vectored Lebesgue measurable function*[1] *$\mathbf{f}(t) = [f_1(t), f_2(t), \cdots, f_n(t)]^T \in L_p$, $p = 1, 2, \cdots, \infty$, implies $f_i(t) \in L_p$ for all $i \in \{1, n\}$.*

The following definition is cited from [87, 178].

Definition 2.6. *A system with input $u(t)$ and output $y(t)$ is said to have a finite L_2 gain of γ_0 if there exist $\gamma_0 > 0$ and $\gamma_g > 0$ such that*

[1] In some literature, such as [5], the space containing vectored Lebesgue measurable functions is denoted as L_p^n with n representing the number of dimensions. For simplicity, L_p^n is written as L_p throughout this book.

$$\left(\int_0^T |y(t)|^2 \, d\tau\right)^{\frac{1}{2}} \leqslant \gamma_0 \left(\int_0^T |u(t)|^2 \, d\tau\right)^{\frac{1}{2}} + \gamma_g, \quad \forall T > 0 \tag{2.2}$$

holds.

The following three lemmas will be frequently used to prove the L_2 and L_∞ stability of target systems.

Lemma 2.1. *For a given non-negative piecewise continuous function $\xi(t) \geqslant 0$, if its time derivative is governed by*

$$\dot{\xi}(t) \leqslant -\mathbf{y}(t)^T \mathbf{Q} \mathbf{y}(t) - s(t) \tag{2.3}$$

almost everywhere, where $\mathbf{y} \in \mathbb{R}^m$, $m \geqslant 1$, $\mathbf{Q} \in \mathbb{R}^{m \times m}$ is a symmetric positive-definite matrix, and $s(t)$ is subject to

$$\int_0^\infty s(t)dt \geqslant -\gamma_0 \tag{2.4}$$

with $0 \leqslant \gamma_0 < \infty$, then it follows that

$$\mathbf{y}(t) \in L_2$$

holds.

Proof: Re-write (2.3) as

$$\mathbf{y}(t)^T \mathbf{Q} \mathbf{y}(t) \leqslant -\dot{\xi}(t) - s(t).$$

Integrating both sides over time and using (2.4) yields

$$\begin{aligned} \int_0^\infty \mathbf{y}(t)^T \mathbf{Q} \mathbf{y}(t) dt &\leqslant -\int_0^\infty \dot{\xi}(t) dt - \int_0^\infty s(t) dt \\ &\leqslant \xi(0) - \xi(\infty) + \gamma_0 \\ &\leqslant \xi(0) + \gamma_0 \end{aligned} \tag{2.5}$$

which implies $\mathbf{y}(t) \in L_2$. ■

Lemma 2.2. *Consider a non-negative piecewise continuous function $\xi(t)$ defined as*

$$\xi(t) \geqslant \frac{1}{2} \mathbf{x}^T(t) \mathbf{P} \mathbf{x}(t) \tag{2.6}$$

with $\mathbf{x}(t) \in \mathbb{R}^n$, $n \geqslant 1$, and $\mathbf{P} \in \mathbb{R}^{n \times n}$ being a symmetric positive-definite matrix. If the time derivative of $\xi(t)$ is governed by

$$\dot{\xi}(t) \leqslant -s(t) \tag{2.7}$$

where $s(t)$ is subject to (2.4) with $0 \leqslant \gamma_0 < \infty$, then it follows that

$$\begin{aligned} \xi(t) &\in L_\infty \\ \mathbf{x}(t) &\in L_\infty \end{aligned}$$

hold.

Proof: Integrating (2.7) over time and using (2.4) yields $\xi(t) \leqslant \xi(0)+\gamma_0$, $\forall t > 0$, which implies that $\xi(t) \in L_\infty$ holds. Define $a^- = \lambda_{min}(\mathbf{P})$. It follows from (2.6) that

$$\|\mathbf{x}(t)\|^2 \leqslant 2\frac{\xi(t)}{a^-} < \infty \tag{2.8}$$

holds for $t > 0$, leading to $\mathbf{x}(t) \in L_\infty$. ∎

Lemma 2.3. *Consider a non-negative piecewise continuous function $\xi(t)$ defined as*

$$\xi(t) \geqslant \frac{1}{2}\mathbf{x}(t)^T\mathbf{P}\mathbf{x}(t) \tag{2.9}$$

with $\mathbf{x}(t) \in \mathbb{R}^n$, $n \geqslant 1$, and $\mathbf{P} \in \mathbb{R}^{n\times n}$ being a symmetric positive-definite matrix. If the time derivative of $\xi(t)$ is governed by

$$\dot{\xi}(t) \leqslant -\mathbf{y}(t)^T\mathbf{Q}\mathbf{y}(t) - s(t) \tag{2.10}$$

where $\mathbf{y}(t) \in \mathbb{R}^m$, $m \geqslant 1$, $\mathbf{Q} \in \mathbb{R}^{m\times m}$ is a symmetric positive-definite matrix, and $s(t)$ is subject to

$$\int_0^\infty s(t)dt \geqslant -\gamma_0 \tag{2.11}$$

with $0 \leqslant \gamma_0 < \infty$, then it follows that

$$\begin{aligned} \xi(t) &\in L_\infty \\ \mathbf{x}(t) &\in L_\infty \\ \mathbf{y}(t) &\in L_2 \end{aligned}$$

hold.

Proof: The proof follows from Lemmas 2.1 and 2.2 straightforwardly. ∎

Remark 2.1. Note that $s(t) = 0$ is a special case that validates (2.4) with $\gamma_0 = 0$. Therefore, Lemmas 2.1, 2.2, and 2.3 are all valid for $s(t) = 0$.

The following lemma shows that a L_2 and L_∞ signal retains its properties after passing through a first-order filter. This lemma will be used later in the text for stability analysis.

Lemma 2.4. *Consider a first-order system described by*

$$\dot{x}(t) + cx(t) = u(t) \tag{2.12}$$

with $c > 0$. If $u(t) \in L_p$ holds, then $x(t) \in L_p$ and $\dot{x}(t) \in L_p$ hold for $p = 2, \infty$.

Proof: Define

$$v_1(t) = \frac{1}{2}x(t)^2. \tag{2.13}$$

It follows from (2.12) that

$$\begin{aligned}\dot{v}_1(t) &= x(t)\dot{x}(t)\\ &= -cx(t)^2 + x(t)u(t)\\ &\leqslant -cx(t)^2 + x(t)u(t) + \frac{c}{2}\left[x(t) - \frac{1}{c}u(t)\right]^2\\ &= -cx(t)^2 + \frac{c}{2}\left(x(t)^2 + \frac{1}{c^2}u(t)^2\right)\\ &= -\frac{c}{2}x(t)^2 + \frac{1}{2c}u(t)^2 \end{aligned} \tag{2.14}$$

holds. Re-write (2.14) as

$$x(t)^2 \leqslant \frac{1}{c^2}u(t)^2 - \frac{2}{c}\dot{v}_1(t).$$

Integrating both sides over time yields

$$\begin{aligned}\int_0^\infty x(t)^2 dt &\leqslant \frac{1}{c^2}\int_0^\infty u(t)^2 dt - \frac{2}{c}\left[v_1(\infty) - v_1(0)\right]\\ &\leqslant \frac{1}{c^2}\int_0^\infty u(t)^2 dt + \frac{2}{c}v_1(0).\end{aligned} \tag{2.15}$$

Thus, $u(t) \in L_2$ implies $x(t) \in L_2$. Furthermore, it follows from (2.12) that

$$\begin{aligned}\int_0^\infty u(t)^2 dt &= \int_0^\infty [\dot{x}(t) + cx(t)]^2 dt\\ &= \int_0^\infty \dot{x}(t)^2 dt + c^2\int_0^\infty x(t)^2 dt + 2c\int_0^\infty \dot{x}(t)x(t)dt\\ &= \int_0^\infty \dot{x}(t)^2 dt + c^2\int_0^\infty x(t)^2 dt + cx(\infty)^2 - cx(0)^2\end{aligned} \tag{2.16}$$

holds. Equation (2.16) implies

$$\int_0^\infty \dot{x}(t)^2 dt \leqslant \int_0^\infty u(t)^2 dt + cx(0)^2. \tag{2.17}$$

Thus, $u(t) \in L_2$ implies $\dot{x}(t) \in L_2$.

In view of (2.14), it follows that $\dot{v}_1(t) \leqslant 0$ holds for $x(t) \in \left\{x(t) \,\middle|\, |x(t)| \geqslant \frac{1}{c}|u(t)|\right\}$, $\forall t > 0$. It further leads to

$$\max_{t>0}\left(|x(t)|\right) \leqslant \max\left(|x(0)|, \frac{1}{c}\max_{t>0}|u(t)|\right).$$

Thus, $u(t) \in L_\infty$ implies $x(t) \in L_\infty$ and furthermore implies $\dot{x}(t) \in L_\infty$ from (2.12). ■

Above lemma can be extended to time-varying systems and MIMO systems.

Lemma 2.5. *Consider a first-order time-varying system described by*

$$a(t)\dot{x}(t) + b(t)x(t) = u(t) \tag{2.18}$$

with $0 < a_0 \leqslant a(t) < +\infty$, $0 < b_0 \leqslant b(t) < +\infty$, *and* $u(t) \in L_\infty$. *It follows that* $x(t) \in L_\infty$ *and* $\dot{x}(t) \in L_\infty$ *hold.*

Proof: Re-write (2.18) as

$$\dot{x}(t) + c(t)x(t) = y(t) \tag{2.19}$$

with $c(t) = \frac{b(t)}{a(t)} > 0$ and $y(t) = \frac{u(t)}{a(t)}$. Since $a(t) \geqslant a_0 > 0$, $u(t) \in L_\infty$ implies $y(t) = \frac{u(t)}{a(t)} \in L_\infty$. With

$$v_1(t) = \frac{1}{2}x(t)^2 \tag{2.20}$$

following the same procedure leading to (2.14) by substituting $c(t)$ for c yields

$$\dot{v}_1(t) \leqslant -\frac{c(t)}{2}x(t)^2 + \frac{1}{2c(t)}y(t)^2. \tag{2.21}$$

In view of (2.21), it yields $\dot{v}_1(t) \leqslant 0$ for $x(t) \in \left\{ x(t) \,\middle|\, |x(t)| \geqslant \frac{1}{c(t)}|y(t)| \right\}, \forall t > 0$. It further leads to

$$\max_{t>0}(|x(t)|) \leqslant \max\left(|x(0)|, \max_{t>0} \frac{1}{c(t)}|y(t)| \right). \tag{2.22}$$

Note that $b(t) \geqslant b_0 > 0$ and $a(t) < +\infty$ imply $\frac{1}{c(t)} = \frac{a(t)}{b(t)} < +\infty$. Thus, $y(t) \in L_\infty$ implies $x(t) \in L_\infty$ and further $\dot{x}(t) \in L_\infty$ from (2.22) and (2.19). ■

Lemma 2.6. *Consider a MIMO first-order system described by*

$$\dot{\mathbf{x}}(t) + \mathbf{K}\mathbf{x}(t) = \mathbf{u}(t) \tag{2.23}$$

with $\mathbf{x}(t) \in \mathbb{R}^n$, $\mathbf{u}(t) \in \mathbb{R}^n$, *and* $\mathbf{K} \in \mathbb{R}^{n\times n}$ *being symmetrical and positive-definite. If* $\mathbf{u}(t) \in L_2 \bigcap L_\infty$ *holds, then* $\mathbf{x}(t) \in L_2 \bigcap L_\infty$ *and* $\dot{\mathbf{x}}(t) \in L_2 \bigcap L_\infty$ *hold.*

Proof: The proof follows the similar procedure as for Lemma 2.4.

Let $\mathbf{K}$ be expressed as $\mathbf{U}^T\mathbf{U}$, where $\mathbf{U} \in \mathbb{R}^{n\times n}$ is invertible. Define

$$v_2(t) = \frac{1}{2}\mathbf{x}(t)^T\mathbf{x}(t). \tag{2.24}$$

It follows from (2.23) that

$$\begin{aligned}
\dot{v}_2(t) &= \mathbf{x}(t)^T\dot{\mathbf{x}}(t) \\
&= -\mathbf{x}(t)^T\mathbf{K}\mathbf{x}(t) + \mathbf{x}(t)^T\mathbf{u}(t) \\
&\leqslant -\mathbf{x}(t)^T\mathbf{K}\mathbf{x}(t) + \mathbf{x}(t)^T\mathbf{u}(t) \\
&\quad + \frac{1}{2}\left[\mathbf{U}\mathbf{x}(t) - \mathbf{U}^{-T}\mathbf{u}(t)\right]^T\left[\mathbf{U}\mathbf{x}(t) - \mathbf{U}^{-T}\mathbf{u}(t)\right] \\
&= -\mathbf{x}(t)^T\mathbf{K}\mathbf{x}(t) + \frac{1}{2}\mathbf{x}(t)^T\mathbf{U}^T\mathbf{U}\mathbf{x}(t) + \frac{1}{2}\mathbf{u}(t)^T\mathbf{U}^{-1}\mathbf{U}^{-T}\mathbf{u}(t) \\
&= -\frac{1}{2}\mathbf{x}(t)^T\mathbf{K}\mathbf{x}(t) + \frac{1}{2}\mathbf{u}(t)^T\mathbf{K}^{-1}\mathbf{u}(t) \qquad (2.25)
\end{aligned}$$

holds. Note that $\mathbf{u}(t) \in L_2$ implies

$$-\int_0^\infty \mathbf{u}(t)^T\mathbf{K}^{-1}\mathbf{u}(t)dt \geqslant -\gamma_0$$

with $0 \leqslant \gamma_0 < \infty$. It follows from Lemma 2.3 that $\mathbf{x}(t) \in L_2 \bigcap L_\infty$ holds.

Furthermore, $\mathbf{x}(t) \in L_\infty$ implies $\dot{\mathbf{x}}(t) \in L_\infty$ from (2.23) with $\mathbf{u}(t) \in L_\infty$.

Finally, express $\mathbf{K}$ as either $\mathbf{U}^T\mathbf{U}$ or $\mathbf{R}^T\mathbf{\Lambda}\mathbf{R}$, where $\mathbf{\Lambda}$ is a diagonal positive-definite matrix and $\mathbf{R}^T = \mathbf{R}^{-1}$ holds. It follows from (2.23) that

$$\begin{aligned}
\int_0^\infty \mathbf{u}(t)^T\mathbf{u}(t)dt &= \int_0^\infty [\dot{\mathbf{x}}(t) + \mathbf{K}\mathbf{x}(t)]^T[\dot{\mathbf{x}}(t) + \mathbf{K}\mathbf{x}(t)]dt \\
&= \int_0^\infty \dot{\mathbf{x}}(t)^T\dot{\mathbf{x}}(t)dt + \int_0^\infty \mathbf{x}(t)^T\mathbf{K}\mathbf{K}\mathbf{x}(t)dt \\
&\quad + \int_0^\infty \left[\dot{\mathbf{x}}(t)^T\mathbf{K}\mathbf{x}(t) + \mathbf{x}(t)^T\mathbf{K}\dot{\mathbf{x}}(t)\right]dt \\
&= \int_0^\infty \dot{\mathbf{x}}(t)^T\dot{\mathbf{x}}(t)dt + \int_0^\infty \mathbf{x}(t)^T\mathbf{R}^T\mathbf{\Lambda}\mathbf{\Lambda}\mathbf{R}\mathbf{x}(t)dt \\
&\quad + \mathbf{x}(\infty)^T\mathbf{K}\mathbf{x}(\infty) - \mathbf{x}(0)^T\mathbf{K}\mathbf{x}(0) \qquad (2.26)
\end{aligned}$$

holds. Equation (2.26) implies

$$\int_0^\infty \dot{\mathbf{x}}(t)^T\dot{\mathbf{x}}(t)dt \leqslant \int_0^\infty \mathbf{u}(t)^T\mathbf{u}(t)dt + \mathbf{x}(0)^T\mathbf{K}\mathbf{x}(0). \qquad (2.27)$$

Thus, $\mathbf{u}(t) \in L_2$ implies $\dot{\mathbf{x}}(t) \in L_2$. ∎

The following lemma shows that a L_2 and L_∞ vectored signal retains its properties after passing through a MIMO second-order filter.

Lemma 2.7. *Consider a second-order MIMO system described by*

$$\mathbf{M}\ddot{\mathbf{x}}(t) + \mathbf{D}\dot{\mathbf{x}}(t) + \mathbf{K}\mathbf{x}(t) = \mathbf{u}(t) \qquad (2.28)$$

where $\mathbf{M}$, $\mathbf{D}$, *and* $\mathbf{K}$ *are symmetric positive-definite matrices. If* $\mathbf{u}(t) \in L_p$ *holds, then* $\mathbf{x}(t) \in L_p$ *and* $\dot{\mathbf{x}}(t) \in L_p$ *hold for* $p = 2, \infty$.

Proof: Express (2.28) in a state space form as

$$\begin{bmatrix} \dot{\mathbf{x}}(t) \\ \ddot{\mathbf{x}}(t) \end{bmatrix} = \mathbf{A} \begin{bmatrix} \mathbf{x}(t) \\ \dot{\mathbf{x}}(t) \end{bmatrix} + \mathbf{B}\mathbf{u}(t) \tag{2.29}$$

where

$$\mathbf{A} = \begin{bmatrix} \mathbf{0} & \mathbf{I} \\ -\mathbf{M}^{-1}\mathbf{K} & -\mathbf{M}^{-1}\mathbf{D} \end{bmatrix}, \qquad \mathbf{B} = \begin{bmatrix} \mathbf{0} \\ \mathbf{M}^{-1} \end{bmatrix}.$$

Construct a matrix

$$\mathbf{P} = \begin{bmatrix} \mathbf{D} + \delta_m \mathbf{K} & \mathbf{M} \\ \mathbf{M} & \delta_m \mathbf{M} \end{bmatrix} \tag{2.30}$$

with $\delta_m > 0$ such that

$$-\mathbf{Q} \stackrel{def}{=} \mathbf{P}\mathbf{A} + \mathbf{A}^T\mathbf{P} = \begin{bmatrix} -2\mathbf{K} & \mathbf{0} \\ \mathbf{0} & -2(\delta_m \mathbf{D} - \mathbf{M}) \end{bmatrix} \tag{2.31}$$

holds. It needs to prove that both $\mathbf{P}$ and $\mathbf{Q}$ are positive-definite. By defining

$$\delta_m \geqslant \frac{\lambda_{\max}(\mathbf{M})}{\lambda_{\min}(\mathbf{D})} + \delta_m^* \tag{2.32}$$

with $\delta_m^* > 0$, where $\lambda_{\max}(\bullet)$ and $\lambda_{\min}(\bullet)$ denote the maximum and minimum eigenvalues of the corresponding matrices, it follows from (2.30) and (2.31) that

$$\begin{aligned} \begin{bmatrix} \mathbf{x}(t) \\ \dot{\mathbf{x}}(t) \end{bmatrix}^T \mathbf{P} \begin{bmatrix} \mathbf{x}(t) \\ \dot{\mathbf{x}}(t) \end{bmatrix} &= \mathbf{x}(t)^T \left(\delta_m \mathbf{K} + \mathbf{D} - \frac{1}{\delta_m} \mathbf{M} \right) \mathbf{x}(t) \\ &\quad + \left(\frac{1}{\sqrt{\delta_m}} \mathbf{x}(t) + \sqrt{\delta_m} \dot{\mathbf{x}}(t) \right)^T \mathbf{M} \left(\frac{1}{\sqrt{\delta_m}} \mathbf{x}(t) + \sqrt{\delta_m} \dot{\mathbf{x}}(t) \right) \\ &> 0, \qquad \forall \left\| \begin{matrix} \mathbf{x}(t) \\ \dot{\mathbf{x}}(t) \end{matrix} \right\| \neq 0 \end{aligned}$$

and

$$\begin{aligned} -\begin{bmatrix} \mathbf{x}(t) \\ \dot{\mathbf{x}}(t) \end{bmatrix}^T \mathbf{Q} \begin{bmatrix} \mathbf{x}(t) \\ \dot{\mathbf{x}}(t) \end{bmatrix} &= -2\mathbf{x}(t)^T \mathbf{K} \mathbf{x}(t) - 2\dot{\mathbf{x}}(t)^T (\delta_m \mathbf{D} - \mathbf{M}) \dot{\mathbf{x}}(t) \\ &< 0, \qquad \forall \left\| \begin{matrix} \mathbf{x}(t) \\ \dot{\mathbf{x}}(t) \end{matrix} \right\| \neq 0 \end{aligned}$$

hold. Therefore, both $\mathbf{P}$ and $\mathbf{Q}$ defined by (2.30) and (2.31) are positive-definite, subject to (2.32). Return to the quadratic form above and define

$$v_3(t) = \begin{bmatrix} \mathbf{x}(t) \\ \dot{\mathbf{x}}(t) \end{bmatrix}^T \mathbf{P} \begin{bmatrix} \mathbf{x}(t) \\ \dot{\mathbf{x}}(t) \end{bmatrix} \geqslant 0. \tag{2.33}$$

It follows from (2.29) and (2.31) that

$$
\begin{aligned}
\dot{v}_3(t) &= -\begin{bmatrix}\mathbf{x}(t)\\ \dot{\mathbf{x}}(t)\end{bmatrix}^T \mathbf{Q} \begin{bmatrix}\mathbf{x}(t)\\ \dot{\mathbf{x}}(t)\end{bmatrix} + 2\begin{bmatrix}\mathbf{x}(t)\\ \dot{\mathbf{x}}(t)\end{bmatrix}^T \begin{bmatrix}\mathbf{I}\\ \delta_m \mathbf{I}\end{bmatrix} \mathbf{u}(t) \\
&\leqslant -\lambda_{min}(\mathbf{Q}) \left\| \begin{matrix}\mathbf{x}(t)\\ \dot{\mathbf{x}}(t)\end{matrix} \right\|^2 + 2\left[\mathbf{x}(t)^T\mathbf{u}(t) + \delta_m \dot{\mathbf{x}}(t)^T \mathbf{u}(t)\right] \\
&\leqslant -\lambda_{min}(\mathbf{Q}) \left\| \begin{matrix}\mathbf{x}(t)\\ \dot{\mathbf{x}}(t)\end{matrix} \right\|^2 + 2\overline{\delta}_m \left(\|\mathbf{x}(t)\|\|\mathbf{u}(t)\| + \|\dot{\mathbf{x}}(t)\|\|\mathbf{u}(t)\|\right) \\
&\quad + \frac{\lambda_{min}(\mathbf{Q})}{2}\left(\|\mathbf{x}(t)\| - \frac{2\overline{\delta}_m}{\lambda_{min}(\mathbf{Q})}\|\mathbf{u}(t)\|\right)^2 \\
&\quad + \frac{\lambda_{min}(\mathbf{Q})}{2}\left(\|\dot{\mathbf{x}}(t)\| - \frac{2\overline{\delta}_m}{\lambda_{min}(\mathbf{Q})}\|\mathbf{u}(t)\|\right)^2 \\
&= -\lambda_{min}(\mathbf{Q}) \left\| \begin{matrix}\mathbf{x}(t)\\ \dot{\mathbf{x}}(t)\end{matrix} \right\|^2 + \frac{\lambda_{min}(\mathbf{Q})}{2}\left(\|\mathbf{x}(t)\|^2 + \frac{4\overline{\delta}_m^2}{\lambda_{min}(\mathbf{Q})^2}\|\mathbf{u}(t)\|^2\right) \\
&\quad + \frac{\lambda_{min}(\mathbf{Q})}{2}\left(\|\dot{\mathbf{x}}(t)\|^2 + \frac{4\overline{\delta}_m^2}{\lambda_{min}(\mathbf{Q})^2}\|\mathbf{u}(t)\|^2\right) \\
&= -\frac{\lambda_{min}(\mathbf{Q})}{2} \left\| \begin{matrix}\mathbf{x}(t)\\ \dot{\mathbf{x}}(t)\end{matrix} \right\|^2 + \frac{4\overline{\delta}_m^2}{\lambda_{min}(\mathbf{Q})}\|\mathbf{u}(t)\|^2
\end{aligned} \tag{2.34}
$$

holds, where $\mathbf{I}$ denotes an identity matrix and

$$\overline{\delta}_m = \max(1, \delta_m). \tag{2.35}$$

Re-write (2.34) as

$$\left\| \begin{matrix}\mathbf{x}(t)\\ \dot{\mathbf{x}}(t)\end{matrix} \right\|^2 \leqslant \frac{8\overline{\delta}_m^2}{\lambda_{min}(\mathbf{Q})^2}\|\mathbf{u}(t)\|^2 - \frac{2}{\lambda_{min}(\mathbf{Q})}\dot{v}_3(t).$$

Integrating both sides over time yields

$$
\begin{aligned}
\int_0^\infty \left\| \begin{matrix}\mathbf{x}(t)\\ \dot{\mathbf{x}}(t)\end{matrix} \right\|^2 dt &\leqslant \frac{8\overline{\delta}_m^2}{\lambda_{min}(\mathbf{Q})^2}\int_0^\infty \|\mathbf{u}(t)\|^2 dt - \frac{2}{\lambda_{min}(\mathbf{Q})}\left[v_3(\infty) - v_3(0)\right] \\
&\leqslant \frac{8\overline{\delta}_m^2}{\lambda_{min}(\mathbf{Q})^2}\int_0^\infty \|\mathbf{u}(t)\|^2 dt + \frac{2}{\lambda_{min}(\mathbf{Q})}v_3(0).
\end{aligned} \tag{2.36}
$$

Thus, $\mathbf{u}(t) \in L_2$ implies $[\mathbf{x}(t)^T, \dot{\mathbf{x}}(t)^T]^T \in L_2$.

Furthermore, it follows from (2.34) that

$$\dot{v}_3(t) \leqslant 0 \text{ holds for } \begin{bmatrix}\mathbf{x}(t)\\ \dot{\mathbf{x}}(t)\end{bmatrix} \in \left\{ \begin{bmatrix}\mathbf{x}(t)\\ \dot{\mathbf{x}}(t)\end{bmatrix} \middle| \left\| \begin{matrix}\mathbf{x}(t)\\ \dot{\mathbf{x}}(t)\end{matrix} \right\| \geqslant \frac{2\sqrt{2}\,\overline{\delta}_m}{\lambda_{min}(\mathbf{Q})}\|\mathbf{u}(t)\| \right\}, \ \forall t > 0$$

which implies

$$\max_{t>0}(v_3(t)) \leqslant \max\left(v_3(0), \lambda_{max}(\mathbf{P})\frac{8\bar{\delta}_m^2}{\lambda_{min}(\mathbf{Q})^2}\max_{t>0}\|\mathbf{u}(t)\|^2\right)$$

such that

$$\max_{t>0}\left(\left\|\begin{matrix}\mathbf{x}(t)\\ \dot{\mathbf{x}}(t)\end{matrix}\right\|\right) \leqslant \sqrt{\frac{1}{\lambda_{min}(\mathbf{P})}\max_{t>0}(v_3(t))}$$
$$\leqslant \sqrt{\frac{\lambda_{max}(\mathbf{P})}{\lambda_{min}(\mathbf{P})}}\max\left(\left\|\begin{matrix}\mathbf{x}(0)\\ \dot{\mathbf{x}}(0)\end{matrix}\right\|, \frac{2\sqrt{2}\,\bar{\delta}_m}{\lambda_{min}(\mathbf{Q})}\max_{t>0}\|\mathbf{u}(t)\|\right)$$

holds. Thus, $\mathbf{u}(t) \in L_\infty$ implies $[\mathbf{x}(t)^T, \dot{\mathbf{x}}(t)^T]^T \in L_\infty$. ■

The following lemma will be used to prove the asymptotic stability later in the text [168]:

Lemma 2.8. *If* $e(t) \in L_2$ *and* $\dot{e}(t) \in L_\infty$, *then* $\lim_{t\to\infty} e(t) = 0$. ■

2.3 Vectors and Coordinate Systems

Vectors can be classified as *free vectors*, *line vectors*, and *point vectors* [39]. The *free vectors* are vectors characterized by their dimensions, magnitudes, and directions, regardless of their points of action. Two *free vectors* with different points of action are considered to be the same as long as they have the same dimensions, magnitude, and direction. Examples of these *free vectors* include moments, angular velocities, and relative positions. A *line vector* possesses the same properties as a *free vector*, but its point of action is limited to a line oriented by the direction of the *line vector*. An example of a *line vector* is an acting force. A *point vector* possesses the same properties as a *free vector*, but its point of action is fixed.

Throughout this book, a vector refers to a *free vector* or directed vector unless otherwise specified, since a *line vector* can be replaced by a *point vector* and a *point vector* can be expressed by two *free vectors*—one specifying the dimensions, magnitude, and direction and the other specifying the point of action with respect to a reference point.

Two types of coordinate systems are used throughout this book. A first type coordinate system, called a *frame* for simplicity, is constructed by three mutually orthogonal three-dimensional unit vectors as bases. A second type coordinate system simply consists of a single three-dimensional unit vector.

Let $\{\mathbf{A}\}$ be a frame with its bases being denoted as $\{\mathbf{A}\} = [\vec{\mathbf{a}}_1, \vec{\mathbf{a}}_2, \vec{\mathbf{a}}_3]$, and let $\{\mathbf{B}\}$ be a frame with its bases being denoted as $\{\mathbf{B}\} = [\vec{\mathbf{b}}_1, \vec{\mathbf{b}}_2, \vec{\mathbf{b}}_3]$.

Definition 2.7. *A three-dimensional vector* $\vec{\mathbf{x}}$ *can be expressed in an arbitrary frame, such as* $\{\mathbf{A}\}$, *as* ${}^{\mathbf{A}}\mathbf{x} = [\vec{\mathbf{a}}_1 \cdot \vec{\mathbf{x}}, \vec{\mathbf{a}}_2 \cdot \vec{\mathbf{x}}, \vec{\mathbf{a}}_3 \cdot \vec{\mathbf{x}}]^T \in \mathbb{R}^3$ *subject to*

$$\vec{\mathbf{x}} = \{\mathbf{A}\}{}^{\mathbf{A}}\mathbf{x}. \tag{2.37}$$

Note that ${}^{\mathbf{A}}\mathbf{x} \in \mathbb{R}^3$ is the expression of the directed (physical) vector $\vec{\mathbf{x}}$ in frame $\{\mathbf{A}\}$. The directed (physical) vector $\vec{\mathbf{x}}$ is resolved into three components forming the valued vector ${}^{\mathbf{A}}\mathbf{x} \in \mathbb{R}^3$. For a given $\vec{\mathbf{x}}$, ${}^{\mathbf{A}}\mathbf{x}$ can be different, depending on what frame $\{\mathbf{A}\}$ is.

In view of Definition 2.7, any one of the three vectors forming the bases of frame $\{\mathbf{A}\}$ can be expressed in frame $\{\mathbf{B}\}$ as a valued vector, denoted as ${}^{\mathbf{B}}\mathbf{a}_i \in \mathbb{R}^3$, subject to

$$\vec{\mathbf{a}}_i = \{\mathbf{B}\}{}^{\mathbf{B}}\mathbf{a}_i, \quad i = 1, 2, 3. \tag{2.38}$$

Grouping all three base vectors ($\vec{\mathbf{a}}_i$ for $i = 1, 2, 3$) of frame $\{\mathbf{A}\}$ yields

$$\{\mathbf{A}\} = \{\mathbf{B}\}{}^{\mathbf{B}}\mathbf{R}_{\mathbf{A}} \tag{2.39}$$

where ${}^{\mathbf{B}}\mathbf{R}_{\mathbf{A}} = [{}^{\mathbf{B}}\mathbf{a}_1, {}^{\mathbf{B}}\mathbf{a}_2, {}^{\mathbf{B}}\mathbf{a}_3] \in \mathbb{R}^{3\times 3}$ denotes a rotation (direction cosine) matrix connecting frame $\{\mathbf{A}\}$ with frame $\{\mathbf{B}\}$ in two meanings. First, it transforms a physical vector expressed in frame $\{\mathbf{A}\}$ to the same physical vector expressed in frame $\{\mathbf{B}\}$. Second, it rotates frame $\{\mathbf{B}\}$ to coincide with frame $\{\mathbf{A}\}$.

2.4 Orientation Expressions by Quaternions

Orientations can be described in many ways. Among these are Euler angles, rotation (direction cosine) matrices, and quaternions.

Although physically appealing, the Euler angles suffer from singularity problem,[2] computation inefficiency, and the possibility of losing isomorphism (univalence and bijection property) for large angle rotations. On the other hand, direction cosine matrices, in which nine parameters subject to six constraints are used, can be computationally efficient and singularity-free, but they are not well suited for control purposes. Thus, the quaternion, which uses a scalar and a vector subject to one constraint and limits the singularity uniquely to any 180-degree rotation, turns out to be the most useful approach for control purposes. Therefore, the quaternion approach will be used in orientation control throughout this book.

2.4.1 Conversions between a Quaternion and a Rotation Matrix

Given frames $\{\mathbf{A}\}$ and $\{\mathbf{B}\}$, there exist a unit axis $\vec{\mathbf{k}}$ and a scalar ϕ such that rotating frame $\{\mathbf{B}\}$ about $\vec{\mathbf{k}}$ through ϕ takes the orientation of frame $\{\mathbf{A}\}$.

A quaternion specifying the orientation difference from frame $\{\mathbf{B}\}$ to frame $\{\mathbf{A}\}$ is defined as $[q_0, \mathbf{q}^T]^T \in \mathbb{R}^4$ with

[2] For an orientation formed by rotating along axis x through α, followed by rotating along axis y through β, and finally by rotating along axis z through γ, the singularity occurs at $\beta = 90°$.

$$q_0 = \cos\left(\frac{\phi}{2}\right) \tag{2.40}$$

$$\mathbf{q} = \sin\left(\frac{\phi}{2}\right) {}^{\mathbf{B}}\mathbf{k} \tag{2.41}$$

where ${}^{\mathbf{B}}\mathbf{k} \in \mathbb{R}^3$ denotes the unit vector $\vec{\mathbf{k}}$ expressed in frame $\mathbf{B}$, in view of Definition 2.7. Note that ${}^{\mathbf{B}}\mathbf{k} = {}^{\mathbf{A}}\mathbf{k}$ holds.

The quaternion $[q_0, \mathbf{q}^T]^T \in \mathbb{R}^4$ defined by (2.40) and (2.41) can be related to the rotation matrix ${}^{\mathbf{B}}\mathbf{R}_{\mathbf{A}} \in \mathbb{R}^{3\times3}$ in $SO(3)$ by [185]

$$\begin{aligned} {}^{\mathbf{B}}\mathbf{R}_{\mathbf{A}} &= \mathbf{I}_3 + 2q_0(\mathbf{q}\times) + 2(\mathbf{q}\times)^2 \\ &= (2q_0^2 - 1)\mathbf{I}_3 + 2\left[\mathbf{q}\mathbf{q}^T + q_0(\mathbf{q}\times)\right] \end{aligned} \tag{2.42}$$

with $\mathbf{I}_3$ being a 3×3 identity matrix and

$$(\mathbf{q}\times) = \begin{bmatrix} 0 & -q_3 & q_2 \\ q_3 & 0 & -q_1 \\ -q_2 & q_1 & 0 \end{bmatrix}.$$

Note that (2.42) converts either $[q_0, \mathbf{q}^T]$ or $[-q_0, -\mathbf{q}^T]$ to the same ${}^{\mathbf{B}}\mathbf{R}_{\mathbf{A}}$.

Alternatively, for a given matrix ${}^{\mathbf{B}}\mathbf{R}_{\mathbf{A}}$, the quaternion $[q_0, \mathbf{q}^T]^T \in \mathbb{R}^4$ can be obtained as

$$q_0 = \frac{\sqrt{1 + r_{11} + r_{22} + r_{33}}}{2} \tag{2.43}$$

$$q_1 = \operatorname{sgn}(r_{32} - r_{23})\sqrt{r_{11}/2 + (1 - r_{11} - r_{22} - r_{33})/4} \tag{2.44}$$

$$q_2 = \operatorname{sgn}(r_{13} - r_{31})\sqrt{r_{22}/2 + (1 - r_{11} - r_{22} - r_{33})/4} \tag{2.45}$$

$$q_3 = \operatorname{sgn}(r_{21} - r_{12})\sqrt{r_{33}/2 + (1 - r_{11} - r_{22} - r_{33})/4} \tag{2.46}$$

where q_i, $i \in \{1,3\}$, is the ith element of $\mathbf{q} \in \mathbb{R}^3$and r_{ij} is an element of ${}^{\mathbf{B}}\mathbf{R}_{\mathbf{A}}$ at row i and column j for all $i \in \{1,3\}$ and $j \in \{1,3\}$.

2.4.2 Quaternion Multiplications

For three given frames $\{\mathbf{A}\}$, $\{\mathbf{B}\}$, and $\{\mathbf{C}\}$, let $[q_0, \mathbf{q}^T]^T \in \mathbb{R}^4$, $[a_0, \mathbf{a}^T]^T \in \mathbb{R}^4$, and $[b_0, \mathbf{b}^T]^T \in \mathbb{R}^4$ be three quaternions corresponding to ${}^{\mathbf{B}}\mathbf{R}_{\mathbf{A}}$, ${}^{\mathbf{B}}\mathbf{R}_{\mathbf{C}}$, and ${}^{\mathbf{C}}\mathbf{R}_{\mathbf{A}}$, respectively. It follows that

$$\begin{bmatrix} q_0 \\ \mathbf{q} \end{bmatrix} = \begin{bmatrix} a_0 \\ \mathbf{a} \end{bmatrix} \otimes \begin{bmatrix} b_0 \\ \mathbf{b} \end{bmatrix} = \begin{bmatrix} a_0 b_0 - \mathbf{a}^T\mathbf{b} \\ a_0\mathbf{b} + b_0\mathbf{a} + (\mathbf{a}\times)\mathbf{b} \end{bmatrix} \tag{2.47}$$

holds [32]. Note that this formulation makes it clear that quaternion multiplications are not commutative, since they involve vector cross products that are not commutative.

2.4.3 Quaternion Derivatives and Angular Velocities

In this subsection, the relationship between the time derivative of a quaternion and the corresponding angular velocity difference is given.

Let ${}^{\mathbf{B}}\tilde{\boldsymbol{\omega}} \in \mathbb{R}^3$ be the angular velocity difference between frame $\{\mathbf{A}\}$ and frame $\{\mathbf{B}\}$, expressed in frame $\{\mathbf{B}\}$, subject to ${}^{\mathbf{B}}\tilde{\boldsymbol{\omega}} \overset{def}{=} {}^{\mathbf{B}}\mathbf{R}_{\mathbf{I}}(\omega_{\mathbf{A}} - \omega_{\mathbf{B}})$, where frame $\{\mathbf{I}\}$ represents the inertial frame. The time derivative of $[q_0, \mathbf{q}^T]^T \in \mathbb{R}^4$ can be written as

$$\dot{q}_0 = -\frac{1}{2} {}^{\mathbf{B}}\tilde{\boldsymbol{\omega}}^T \mathbf{q} \tag{2.48}$$

$$\dot{\mathbf{q}} = \frac{1}{2}\left[q_0 {}^{\mathbf{B}}\tilde{\boldsymbol{\omega}} + ({}^{\mathbf{B}}\tilde{\boldsymbol{\omega}}\times)\mathbf{q}\right]. \tag{2.49}$$

Conversely, the angular velocity difference between frame $\{\mathbf{A}\}$ and frame $\{\mathbf{B}\}$, expressed in frame $\{\mathbf{B}\}$, can be obtained from the time derivative of the corresponding quaternion as

$${}^{\mathbf{B}}\tilde{\boldsymbol{\omega}} = 2\left[q_0\dot{\mathbf{q}} - \dot{q}_0\mathbf{q} + (\mathbf{q}\times)\dot{\mathbf{q}}\right]. \tag{2.50}$$

The following relationship is important to quaternion-based orientation control.

The integral of the inner product of the angular velocity difference ${}^{\mathbf{B}}\tilde{\boldsymbol{\omega}}$ and the vector part of the quaternion $\mathbf{q}$ can be written as

$$\begin{aligned}
\int_0^t \mathbf{q}^{T\mathbf{B}}\tilde{\boldsymbol{\omega}} d\tau &= \int_0^t \mathbf{q}^T\left[q_0 {}^{\mathbf{B}}\tilde{\boldsymbol{\omega}} + ({}^{\mathbf{B}}\tilde{\boldsymbol{\omega}}\times)\mathbf{q}\right] + (1-q_0)\mathbf{q}^{T\mathbf{B}}\tilde{\boldsymbol{\omega}} d\tau \\
&= \int_0^t \left[2\mathbf{q}^T\dot{\mathbf{q}} - 2(1-q_0)\dot{q}_0\right] d\tau \\
&= \left[\mathbf{q}(t)^T\mathbf{q}(t) + (1-q_0(t))^2\right] \\
&\quad - \left[\mathbf{q}(0)^T\mathbf{q}(0) + (1-q_0(0))^2\right].
\end{aligned} \tag{2.51}$$

In view of (2.40) and (2.41), it yields

$$\begin{aligned}
\mathbf{q}^T\mathbf{q} + (1-q_0)^2 &= \left(\sin\frac{\phi}{2}\right)^2 + \left(1 - \cos\frac{\phi}{2}\right)^2 \\
&= 4\left(\sin\frac{\phi}{4}\right)^2.
\end{aligned} \tag{2.52}$$

Therefore, the following inequality holds indefinitely

$$-4 \leqslant -4\left(\sin\frac{\phi(0)}{4}\right)^2 \leqslant \int_0^t \mathbf{q}^{T\mathbf{B}}\tilde{\boldsymbol{\omega}} d\tau \leqslant 4\left(\sin\frac{\phi(t)}{4}\right)^2 \leqslant 4. \tag{2.53}$$

2.5 Expressions of Velocities and Forces in Body Frames

2.5.1 Velocities and Forces in Body Frames

Let $\{\mathbf{A}\}$ be a frame attached to a rigid body. Let $\vec{\mathbf{f}}_{\mathbf{A}}$ and $\vec{\mathbf{m}}_{\mathbf{A}}$ be two directed (physical) vectors representing the force and moment applied to the origin of frame $\{\mathbf{A}\}$, and let $\vec{\mathbf{v}}_{\mathbf{A}}$ and $\vec{\omega}_{\mathbf{A}}$ be two directed (physical) vectors representing the linear and angular velocities[3] of frame $\{\mathbf{A}\}$, referenced to the inertial frame $\{\mathbf{I}\}$.

In view of Definition 2.7, the directed vector $\vec{\mathbf{v}}_{\mathbf{A}}$, which represents the linear velocity of the origin of frame $\{\mathbf{A}\}$, can be expressed either in the inertial frame $\{\mathbf{I}\}$ or in body frame $\{\mathbf{A}\}$ as

$$\vec{\mathbf{v}}_{\mathbf{A}} = \{\mathbf{I}\}^{\mathbf{I}}\mathbf{v}_{\mathbf{A}} = \{\mathbf{A}\}^{\mathbf{A}}\mathbf{v}_{\mathbf{A}}. \tag{2.54}$$

Applying (2.39) yields $\{\mathbf{A}\} = \{\mathbf{I}\}^{\mathbf{I}}\mathbf{R}_{\mathbf{A}}$, where ${}^{\mathbf{I}}\mathbf{R}_{\mathbf{A}} \in \mathbb{R}^{3\times3}$ is a rotation matrix. Substituting it into (2.54) yields

$${}^{\mathbf{I}}\mathbf{v}_{\mathbf{A}} = {}^{\mathbf{I}}\mathbf{R}_{\mathbf{A}}{}^{\mathbf{A}}\mathbf{v}_{\mathbf{A}} \tag{2.55}$$

or

$${}^{\mathbf{A}}\mathbf{v}_{\mathbf{A}} = {}^{\mathbf{A}}\mathbf{R}_{\mathbf{I}}{}^{\mathbf{I}}\mathbf{v}_{\mathbf{A}} \tag{2.56}$$

with ${}^{\mathbf{A}}\mathbf{R}_{\mathbf{I}} = {}^{\mathbf{I}}\mathbf{R}_{\mathbf{A}}^{-1} = {}^{\mathbf{I}}\mathbf{R}_{\mathbf{A}}^{T}$.

Note that ${}^{\mathbf{I}}\mathbf{v}_{\mathbf{A}} \in \mathbb{R}^3$ represents the velocity of the origin of frame $\{\mathbf{A}\}$ in a conventional sense. Therefore, the linear (point) velocity of the origin of frame $\{\mathbf{A}\}$, expressed in frame $\{\mathbf{A}\}$, is obtained by premultiplying the conventional velocity, denoted as ${}^{\mathbf{I}}\mathbf{v}_{\mathbf{A}} \in \mathbb{R}^3$, by a rotation matrix ${}^{\mathbf{A}}\mathbf{R}_{\mathbf{I}} \in \mathbb{R}^{3\times3}$.

For simplicity, ${}^{\mathbf{A}}\mathbf{v}_{\mathbf{A}}$ will be written as ${}^{\mathbf{A}}\mathbf{v}$ throughout this book, unless otherwise specified.

Similarly, let ${}^{\mathbf{A}}\mathbf{v} \in \mathbb{R}^3$ and ${}^{\mathbf{A}}\omega \in \mathbb{R}^3$ be the linear and angular velocity vectors of frame $\{\mathbf{A}\}$, expressed in frame $\{\mathbf{A}\}$; let ${}^{\mathbf{A}}\mathbf{f} \in \mathbb{R}^3$ and ${}^{\mathbf{A}}\mathbf{m} \in \mathbb{R}^3$ be the force and moment vectors applied to the origin of frame $\{\mathbf{A}\}$, expressed in frame $\{\mathbf{A}\}$; subject to

$$\vec{\mathbf{v}}_{\mathbf{A}} = \{\mathbf{A}\}^{\mathbf{A}}\mathbf{v} \tag{2.57}$$

$$\vec{\omega}_{\mathbf{A}} = \{\mathbf{A}\}^{\mathbf{A}}\omega \tag{2.58}$$

$$\vec{\mathbf{f}}_{\mathbf{A}} = \{\mathbf{A}\}^{\mathbf{A}}\mathbf{f} \tag{2.59}$$

$$\vec{\mathbf{m}}_{\mathbf{A}} = \{\mathbf{A}\}^{\mathbf{A}}\mathbf{m}. \tag{2.60}$$

Remark 2.2. Expressing velocities in a body frame, such as using ${}^{\mathbf{A}}\mathbf{v} = {}^{\mathbf{A}}\mathbf{v}_{\mathbf{A}}$ instead of ${}^{\mathbf{I}}\mathbf{v}_{\mathbf{A}}$, will not simplify the kinematics, since $\frac{d}{dt}\left({}^{\mathbf{A}}\mathbf{v}\right) = \frac{d}{dt}\left({}^{\mathbf{A}}\mathbf{R}_{\mathbf{I}}{}^{\mathbf{I}}\mathbf{v}_{\mathbf{A}}\right) = \frac{d}{dt}\left({}^{\mathbf{A}}\mathbf{R}_{\mathbf{I}}\right){}^{\mathbf{I}}\mathbf{v}_{\mathbf{A}} + {}^{\mathbf{A}}\mathbf{R}_{\mathbf{I}}\frac{d}{dt}\left({}^{\mathbf{I}}\mathbf{v}_{\mathbf{A}}\right) \neq {}^{\mathbf{A}}\mathbf{R}_{\mathbf{I}}\frac{d}{dt}\left({}^{\mathbf{I}}\mathbf{v}_{\mathbf{A}}\right)$. However, as will be seen below, the body-frame-expressed velocities and accelerations make the dynamics expressions much more efficient, since the inertial matrix of a rigid body becomes a time-invariant and symmetric positive-definite matrix.

---

[3] The linear velocity of frame $\{\mathbf{A}\}$ refers to the point velocity of the frame origin.

2.5.2 Expressions of Linear/Angular Velocity Vectors and Force/Moment Vectors

In this book, both linear/angular velocities and forces/moments are properly integrated to facilitate the transformations of velocities and forces/moments among different frames.

Definition 2.8. *Let* ${}^{\mathbf{A}}\mathbf{v} \in \mathbb{R}^3$ *defined in (2.57) and* ${}^{\mathbf{A}}\boldsymbol{\omega} \in \mathbb{R}^3$ *defined in (2.58) be the linear and angular velocity vectors of frame* $\{\mathbf{A}\}$*, expressed in frame* $\{\mathbf{A}\}$*. The linear/angular velocity vector of frame* $\{\mathbf{A}\}$ *is defined by*

$$ {}^{\mathbf{A}}V \overset{def}{=} \begin{bmatrix} {}^{\mathbf{A}}\mathbf{v} \\ {}^{\mathbf{A}}\boldsymbol{\omega} \end{bmatrix} \in \mathbb{R}^6. \tag{2.61} $$

Definition 2.9. *Let* ${}^{\mathbf{A}}\mathbf{f} \in \mathbb{R}^3$ *defined in (2.59) and* ${}^{\mathbf{A}}\mathbf{m} \in \mathbb{R}^3$ *defined in (2.60) be the force and moment vectors that are being measured and expressed in frame* $\{\mathbf{A}\}$*. The force/moment vector in frame* $\{\mathbf{A}\}$ *is defined by*

$$ {}^{\mathbf{A}}F \overset{def}{=} \begin{bmatrix} {}^{\mathbf{A}}\mathbf{f} \\ {}^{\mathbf{A}}\mathbf{m} \end{bmatrix} \in \mathbb{R}^6. \tag{2.62} $$

2.5.3 The Duality of Linear/Angular Velocity and Force/Moment Transformations

Consider two given frames, denoted as $\{\mathbf{A}\}$ and $\{\mathbf{B}\}$, being fixed to a common rigid body moving freely and subject to a pair of physical force and moment vectors. The following relations hold

$$ {}^{\mathbf{B}}V = {}^{\mathbf{A}}\mathbf{U}_{\mathbf{B}}^{T}\,{}^{\mathbf{A}}V \tag{2.63} $$
$$ {}^{\mathbf{A}}F = {}^{\mathbf{A}}\mathbf{U}_{\mathbf{B}}\,{}^{\mathbf{B}}F \tag{2.64} $$

where

$$ {}^{\mathbf{A}}\mathbf{U}_{\mathbf{B}} = \begin{bmatrix} {}^{\mathbf{A}}\mathbf{R}_{\mathbf{B}} & \mathbf{0}_{3\times 3} \\ ({}^{\mathbf{A}}\mathbf{r}_{\mathbf{AB}}\times){}^{\mathbf{A}}\mathbf{R}_{\mathbf{B}} & {}^{\mathbf{A}}\mathbf{R}_{\mathbf{B}} \end{bmatrix} \in \mathbb{R}^{6\times 6} \tag{2.65} $$

denotes a force/moment transformation matrix that transforms the force/moment vector measured and expressed in frame $\{\mathbf{B}\}$ to the same force/moment vector measured and expressed in frame $\{\mathbf{A}\}$. In (2.65), ${}^{\mathbf{A}}\mathbf{R}_{\mathbf{B}} = {}^{\mathbf{B}}\mathbf{R}_{\mathbf{A}}^{T}$ holds with ${}^{\mathbf{B}}\mathbf{R}_{\mathbf{A}} \in \mathbb{R}^{3\times 3}$ being defined in (2.39), and ${}^{\mathbf{A}}\mathbf{r}_{\mathbf{AB}} \in \mathbb{R}^3$ denotes a vector from the origin of frame $\{\mathbf{A}\}$ to the origin of frame $\{\mathbf{B}\}$, expressed in frame $\{\mathbf{A}\}$. It is obvious that ${}^{\mathbf{B}}\mathbf{U}_{\mathbf{A}} \in \mathbb{R}^{6\times 6}$ is constant.

In view of both ${}^{\mathbf{A}}V \in \mathbb{R}^6$ and ${}^{\mathbf{A}}F \in \mathbb{R}^6$ being defined by (2.61) and (2.62) and further by (2.57)–(2.60), both ${}^{\mathbf{B}}V \in \mathbb{R}^6$ and ${}^{\mathbf{B}}F \in \mathbb{R}^6$ are having exactly the same definitions by substituting frame $\{\mathbf{B}\}$ for frame $\{\mathbf{A}\}$.

Remark 2.3. Equations (2.63) and (2.64) indicate the duality between the linear/angular velocity transformations and the force/moment transformations.

Remark 2.4. Note that (2.64) is valid only when both ${}^{\mathbf{A}}F \in \mathbb{R}^6$ and ${}^{\mathbf{B}}F \in \mathbb{R}^6$ represent a pair of the same physical force and moment vectors being measured and expressed in different ($\{\mathbf{A}\}$ and $\{\mathbf{B}\}$) frames.

2.6 Rigid Body Dynamics in a Body Frame

In this section, both net force and moment vectors applied to a rigid body are defined first. Then, rigid body dynamics expressed in a body frame are derived. Finally, the linear parametrization expression to be used by parameter adaptation is given.

2.6.1 Net Forces and Moments

Consider a rigid body to which frame $\{\mathbf{A}\}$ is attached. The net (summation) force and moment vectors applied to this rigid body are written as

$$\vec{\mathbf{f}}^*_{\mathbf{A}} \stackrel{def}{=} \{\mathbf{A}\}^{\mathbf{A}}\mathbf{f}^* \tag{2.66}$$

$$\vec{\mathbf{m}}^*_{\mathbf{A}} \stackrel{def}{=} \{\mathbf{A}\}^{\mathbf{A}}\mathbf{m}^* \tag{2.67}$$

where $\vec{\mathbf{f}}^*_{\mathbf{A}}$ denotes the sum of all force vectors applied to this rigid body, $\vec{\mathbf{m}}^*_{\mathbf{A}}$ denotes the sum of all moment vectors and all force-induced moment vectors applied to this rigid body, and ${}^{\mathbf{A}}\mathbf{f}^* \in \mathbb{R}^3$ and ${}^{\mathbf{A}}\mathbf{m}^* \in \mathbb{R}^3$ denote the net (summation) force and moment vectors expressed in frame $\{\mathbf{A}\}$, respectively.

Definition 2.10. *Let ${}^{\mathbf{A}}\mathbf{f}^* \in \mathbb{R}^3$ defined in (2.66) and ${}^{\mathbf{A}}\mathbf{m}^* \in \mathbb{R}^3$ defined in (2.67) be the net force and moment vectors that are being applied to a rigid body and being measured and expressed in a body frame $\{\mathbf{A}\}$. The net force/moment vector of the rigid body in frame $\{\mathbf{A}\}$ is defined by*

$${}^{\mathbf{A}}F^* \stackrel{def}{=} \begin{bmatrix} {}^{\mathbf{A}}\mathbf{f}^* \\ {}^{\mathbf{A}}\mathbf{m}^* \end{bmatrix}. \tag{2.68}$$

2.6.2 Rigid Body Dynamics

Two frames, denoted as $\{\mathbf{A}\}$ and $\{\mathbf{B}\}$, are fixed to a rigid object. Frame $\{\mathbf{A}\}$ is used to express the rigid body dynamics, and frame $\{\mathbf{B}\}$ is assumed to be located at the center of mass.

The dynamic equation of the rigid body in free motion, expressed in the inertial frame $\{\mathbf{I}\}$, is

$$\begin{bmatrix} m_{\mathbf{A}}\mathbf{I}_3 & \\ & \mathbf{I}_o(t) \end{bmatrix} \begin{bmatrix} \dot{\mathbf{v}} \\ \dot{\boldsymbol{\omega}} \end{bmatrix} + \begin{bmatrix} m_{\mathbf{A}}\mathbf{g} \\ (\boldsymbol{\omega}\times)\mathbf{I}_o(t)\boldsymbol{\omega} \end{bmatrix} = \begin{bmatrix} \mathbf{f}^* \\ \mathbf{m}^* \end{bmatrix} \tag{2.69}$$

where $\mathbf{I}_3$ is a 3×3 identity matrix, $m_\mathbf{A} \in \mathbb{R}$ denotes the mass of the rigid body, $\mathbf{I}_o(t) \in \mathbb{R}^{3\times3}$ denotes the moment of inertia matrix around the center of mass, $\mathbf{v} \in \mathbb{R}^3$ and $\boldsymbol{\omega} \in \mathbb{R}^3$ denote the linear velocity vector of the center of mass and the angular velocity vector, respectively, $\mathbf{g} = [0, 0, 9.8]^T \in \mathbb{R}^3$ denotes the gravitational vector, and $\mathbf{f}^* \in \mathbb{R}^3$ and $\mathbf{m}^* \in \mathbb{R}^3$ denote the net force and moment vectors applied to the center of mass, respectively.

The net force/moment vector of this rigid body in frame $\{\mathbf{A}\}$ can be written as

$$\begin{aligned} {}^{\mathbf{A}}F^* &= {}^{\mathbf{A}}\mathbf{U}_\mathbf{B}{}^{\mathbf{B}}F^* \\ &= {}^{\mathbf{A}}\mathbf{U}_\mathbf{B}\begin{bmatrix} {}^{\mathbf{B}}\mathbf{R}_\mathbf{I} & 0 \\ 0 & {}^{\mathbf{B}}\mathbf{R}_\mathbf{I} \end{bmatrix}\begin{bmatrix} \mathbf{f}^* \\ \mathbf{m}^* \end{bmatrix} \end{aligned} \tag{2.70}$$

with ${}^{\mathbf{A}}\mathbf{U}_\mathbf{B} \in \mathbb{R}^{6\times6}$ being defined by (2.65).

Meanwhile, the linear and angular velocity vectors of the rigid body can be re-written as

$$\begin{bmatrix} \mathbf{v} \\ \boldsymbol{\omega} \end{bmatrix} = \begin{bmatrix} {}^{\mathbf{I}}\mathbf{R}_\mathbf{B} & 0 \\ 0 & {}^{\mathbf{I}}\mathbf{R}_\mathbf{B} \end{bmatrix} {}^{\mathbf{A}}\mathbf{U}_\mathbf{B}^T{}^{\mathbf{A}}V. \tag{2.71}$$

Using

$$\frac{d}{dt}\left({}^{\mathbf{I}}\mathbf{R}_\mathbf{B}\right) = (\boldsymbol{\omega}\times){}^{\mathbf{I}}\mathbf{R}_\mathbf{B} \tag{2.72}$$

and differentiating (2.71) with respect to time yields

$$\begin{aligned} \begin{bmatrix} \dot{\mathbf{v}} \\ \dot{\boldsymbol{\omega}} \end{bmatrix} &= \begin{bmatrix} (\boldsymbol{\omega}\times){}^{\mathbf{I}}\mathbf{R}_\mathbf{B} & 0 \\ 0 & (\boldsymbol{\omega}\times){}^{\mathbf{I}}\mathbf{R}_\mathbf{B} \end{bmatrix} {}^{\mathbf{A}}\mathbf{U}_\mathbf{B}^T{}^{\mathbf{A}}V \\ &+ \begin{bmatrix} {}^{\mathbf{I}}\mathbf{R}_\mathbf{B} & 0 \\ 0 & {}^{\mathbf{I}}\mathbf{R}_\mathbf{B} \end{bmatrix} {}^{\mathbf{A}}\mathbf{U}_\mathbf{B}^T\frac{d}{dt}({}^{\mathbf{A}}V). \end{aligned} \tag{2.73}$$

Premultiply both sides of (2.69) by ${}^{\mathbf{A}}\mathbf{U}_\mathbf{B}\begin{bmatrix} {}^{\mathbf{B}}\mathbf{R}_\mathbf{I} & 0 \\ 0 & {}^{\mathbf{B}}\mathbf{R}_\mathbf{I} \end{bmatrix}$. Then applying (2.70) and (2.73) into it and using ${}^{\mathbf{A}}\mathbf{R}_\mathbf{I}(\boldsymbol{\omega}\times) = ({}^{\mathbf{A}}\boldsymbol{\omega}\times){}^{\mathbf{A}}\mathbf{R}_\mathbf{I}$ yields

$$\mathbf{M}_\mathbf{A}\frac{d}{dt}({}^{\mathbf{A}}V) + \mathbf{C}_\mathbf{A}({}^{\mathbf{A}}\omega){}^{\mathbf{A}}V + \mathbf{G}_\mathbf{A} = {}^{\mathbf{A}}F^* \tag{2.74}$$

where

$$\mathbf{M}_\mathbf{A} = \begin{bmatrix} m_\mathbf{A}\mathbf{I}_3 & -m_\mathbf{A}({}^{\mathbf{A}}\mathbf{r}_{\mathbf{AB}}\times) \\ m_\mathbf{A}({}^{\mathbf{A}}\mathbf{r}_{\mathbf{AB}}\times) & \mathbf{I}_\mathbf{A} - m_\mathbf{A}({}^{\mathbf{A}}\mathbf{r}_{\mathbf{AB}}\times)^2 \end{bmatrix} \tag{2.75}$$

$$\mathbf{C}_\mathbf{A}({}^{\mathbf{A}}\omega) = \begin{bmatrix} m_\mathbf{A}({}^{\mathbf{A}}\boldsymbol{\omega}\times) & -m_\mathbf{A}({}^{\mathbf{A}}\boldsymbol{\omega}\times)({}^{\mathbf{A}}\mathbf{r}_{\mathbf{AB}}\times) \\ m_\mathbf{A}({}^{\mathbf{A}}\mathbf{r}_{\mathbf{AB}}\times)({}^{\mathbf{A}}\boldsymbol{\omega}\times) & ({}^{\mathbf{A}}\boldsymbol{\omega}\times)\mathbf{I}_\mathbf{A} + \mathbf{I}_\mathbf{A}({}^{\mathbf{A}}\boldsymbol{\omega}\times) - m_\mathbf{A}({}^{\mathbf{A}}\mathbf{r}_{\mathbf{AB}}\times)({}^{\mathbf{A}}\boldsymbol{\omega}\times)({}^{\mathbf{A}}\mathbf{r}_{\mathbf{AB}}\times) \end{bmatrix} \tag{2.76}$$

$$\mathbf{G}_\mathbf{A} = \begin{bmatrix} m_\mathbf{A}{}^{\mathbf{A}}\mathbf{R}_\mathbf{I}\mathbf{g} \\ m_\mathbf{A}({}^{\mathbf{A}}\mathbf{r}_{\mathbf{AB}}\times){}^{\mathbf{A}}\mathbf{R}_\mathbf{I}\mathbf{g} \end{bmatrix} \tag{2.77}$$

and $\mathbf{I}_\mathbf{A} = {}^{\mathbf{A}}\mathbf{R}_\mathbf{I}\mathbf{I}_o(t){}^{\mathbf{I}}\mathbf{R}_\mathbf{A}$ is time-invariant.

2.6.3 Linear Parametrization Expression

Let ${}^{\mathbf{A}}V_r \in \mathbb{R}^6$ be the required vector (a design vector) of ${}^{\mathbf{A}}V \in \mathbb{R}^6$. The following equation is defined

$$\mathbf{Y_A}\boldsymbol{\theta}_\mathbf{A} \overset{def}{=} \mathbf{M_A}\frac{d}{dt}({}^{\mathbf{A}}V_r) + \mathbf{C_A}({}^{\mathbf{A}}\omega){}^{\mathbf{A}}V_r + \mathbf{G_A}. \tag{2.78}$$

Equation (2.78) is based on the rigid body dynamics (2.74) with $\mathbf{M_A}$, $\mathbf{C_A}({}^{\mathbf{A}}\omega)$, and $\mathbf{G_A}$ being given by (2.75)–(2.77). The detail expressions of the regressor matrix $\mathbf{Y_A} \in \mathbb{R}^{6\times 13}$ and the parameter vector $\boldsymbol{\theta}_\mathbf{A} \in \mathbb{R}^{13}$ are given in Appendix A.

2.7 Projection Functions

Two projection functions defined below will be used for parameter adaptation.

Definition 2.11. *A projection function $\mathcal{P}(s(t), k, a(t), b(t), t) \in \mathbb{R}$ is a differentiable scalar function defined for $t \geqslant 0$ such that its time derivative is governed by*

$$\dot{\mathcal{P}} = ks(t)\kappa \tag{2.79}$$

with

$$\kappa = \begin{cases} 0 & \text{if } \mathcal{P} \leqslant a(t) \text{ and } s(t) \leqslant 0 \\ 0 & \text{if } \mathcal{P} \geqslant b(t) \text{ and } s(t) \geqslant 0 \\ 1 & \text{otherwise} \end{cases}$$

where $s(t) \in \mathbb{R}$ is a scalar variable, $k > 0$ is a constant, and $a(t) \leqslant b(t)$ holds.

Lemma 2.9. *Consider a $\mathcal{P}$ function defined by (2.79). For any constant $\mathcal{P}_c$ subject to $a(t) \leqslant \mathcal{P}_c \leqslant b(t)$, it follows that*

$$(\mathcal{P}_c - \mathcal{P})\left(s(t) - \frac{1}{k}\dot{\mathcal{P}}\right) \leqslant 0 \tag{2.80}$$

holds.

Proof: Substituting (2.79) into (2.80) yields

$$(\mathcal{P}_c - \mathcal{P})\left(s(t) - \frac{1}{k}\dot{\mathcal{P}}\right) = (\mathcal{P}_c - \mathcal{P})\, s(t)\,(1-\kappa). \tag{2.81}$$

When $\mathcal{P} \leqslant a(t) \leqslant \mathcal{P}_c$ and $s(t) \leqslant 0$ hold, it yields

$$(\mathcal{P}_c - \mathcal{P})\, s(t)\,(1-\kappa) = (\mathcal{P}_c - \mathcal{P})\, s(t) \leqslant 0.$$

When $\mathcal{P} \geqslant b(t) \geqslant \mathcal{P}_c$ and $s(t) \geqslant 0$ hold, it yields

$$(\mathcal{P}_c - \mathcal{P})\, s(t)\,(1-\kappa) = (\mathcal{P}_c - \mathcal{P})\, s(t) \leqslant 0.$$

Otherwise

$$(\mathcal{P}_c - \mathcal{P})\, s(t)\,(1-\kappa) = 0$$

holds. ■

Definition 2.12. *A projection function $\mathcal{P}_2(s(t), k, a(t), b(t), c, t) \in \mathbb{R}$ is a second-order differentiable scalar function defined for $t \geqslant 0$ such that its time derivative is governed by*

$$\dot{\mathcal{P}}_2 = k\,(s(t) + c\kappa_2) \tag{2.82}$$

with

$$\kappa_2 = \begin{cases} a(t) - \mathcal{P}_2 & \text{if } \mathcal{P}_2 \leqslant a(t) \\ b(t) - \mathcal{P}_2 & \text{if } \mathcal{P}_2 \geqslant b(t) \\ 0 & \text{otherwise} \end{cases}$$

where $s(t) \in \mathbb{R}$ is a scalar variable, $k > 0$ and $c > 0$ are two constants, and $a(t) \leqslant b(t)$ holds.

Lemma 2.10. *Consider a $\mathcal{P}_2$ function defined by (2.82). For any constant $\mathcal{P}_{c2}$ subject to $a(t) \leqslant \mathcal{P}_{c2} \leqslant b(t)$, it follows that*

$$(\mathcal{P}_{c2} - \mathcal{P}_2)\left(s(t) - \frac{1}{k}\dot{\mathcal{P}}_2\right) \leqslant -c\kappa_2^2 \leqslant 0 \tag{2.83}$$

holds.

Proof: Substituting (2.82) into (2.83) yields

$$(\mathcal{P}_{c2} - \mathcal{P}_2)\left(s(t) - \frac{1}{k}\dot{\mathcal{P}}_2\right) = -c\kappa_2\,(\mathcal{P}_{c2} - \mathcal{P}_2)\,. \tag{2.84}$$

When $\mathcal{P}_2 \leqslant a(t) \leqslant \mathcal{P}_{c2}$ holds, it yields

$$-c\kappa_2\,(\mathcal{P}_{c2} - \mathcal{P}_2) \leqslant -c\kappa_2\,(a(t) - \mathcal{P}_2) = -c\kappa_2^2.$$

When $\mathcal{P}_2 \geqslant b(t) \geqslant \mathcal{P}_{c2}$ holds, it yields

$$\begin{aligned} -c\kappa_2\,(\mathcal{P}_{c2} - \mathcal{P}_2) &= -c(-\kappa_2)\,(\mathcal{P}_2 - \mathcal{P}_{c2}) \\ &\leqslant -c(-\kappa_2)\,(\mathcal{P}_2 - b(t)) \\ &= -c\kappa_2\,(b(t) - \mathcal{P}_2) \\ &= -c\kappa_2^2. \end{aligned}$$

Otherwise

$$-c\kappa_2\,(\mathcal{P}_{c2} - \mathcal{P}_2) = 0$$

holds. ■

Both $\mathcal{P}$ and $\mathcal{P}_2$ functions are assigned with lower and upper bounds denoted by $a(t)$ and $b(t)$. Within these bounds, the first order time-derivatives of the two functions are driven by $s(t)$ with gain k. Outside the bounds, the $\mathcal{P}$ function is only allowed to move toward the interval $[a(t), b(t)]$, while a corrective term brings $\mathcal{P}_2$ back to the interval $[a(t), b(t)]$. The purpose of introducing the $\mathcal{P}$ and $\mathcal{P}_2$ functions is to prevent parameter estimates from drifting. Consequently, a parameter adaptation by using the $\mathcal{P}$ or $\mathcal{P}_2$ function is robust against unmodeled disturbances. The major difference between the $\mathcal{P}$ and $\mathcal{P}_2$ functions is that $\ddot{\mathcal{P}}_2$ exists if $|\dot{s}(t)| < +\infty$, while $\ddot{\mathcal{P}}$ does not exist. The $\mathcal{P}$ function is thus preferable when there is no $\ddot{\mathcal{P}}$ required in the control algorithm. Otherwise, the $\mathcal{P}_2$ function should be considered.

The definitions of the $\mathcal{P}$ and $\mathcal{P}_2$ functions can be found in [214] where $\underline{\mathcal{P}}$ was substituted for $\mathcal{P}_2$ and the lower and upper bounds $a(t)$ and $b(t)$ were constant. In this book, both $a(t)$ and $b(t)$ are allowed to be time-variant.

2.8 Virtual Cutting Points and Oriented Graphs

The concept of the virtual *cutting point* is central to the VDC approach. It allows a complex robotic system to be conceptually *broken down* into subsystems. The *cutting points* are virtual in the sense that a robotic system is cut conceptually rather than physically. After the virtual "cutting" process is performed, a simple oriented graph is to be used to represent the entire robot with each node representing a corresponding subsystem.

2.8.1 Virtual Cutting Points

A *cutting point* forms a virtual cutting surface at which three-dimensional force vectors and three-dimensional moment vectors can be exerted from one part to another.

Definition 2.13. *A cutting point is a directed separation interface that conceptually cuts through a rigid body. At the cutting point, the two parts resulting from the virtual cut maintain equal position and orientation. The cutting point is interpreted as a driving cutting point by one part and is simultaneously interpreted as a driven cutting point by another part.*[4] *A force vector* $\mathbf{f} \in \mathbb{R}^3$ *and a moment vector* $\mathbf{m} \in \mathbb{R}^3$ *are exerted from one part to which the cutting point is interpreted as a driving cutting point to the other part to which the cutting point is interpreted as a driven cutting point.*

2.8.2 Oriented Graphs

Simple oriented graphs will be used to represent the topological structure and control relations of a complex robot.

Definition 2.14. *A graph consists of nodes and edges. A directed graph is a graph in which all the edges have directions. An oriented graph is a directed graph in which each edge has a unique direction. A simple oriented graph is an oriented graph in which no loop is formed [24].*

[4] In [211], a *driven cutting point* was named as a "controlled point" and a *driving cutting point* was named as a "controlling point."

In a simple oriented graph, each subsystem is represented by a node, while each *cutting point* is represented by a directed edge defining the reference direction of these forces and moments passing through this *cutting point*. The forces and moments at a *cutting point* are exerted from the subsystem to which the *cutting point* is interpreted as a *driving cutting point* to the adjacent subsystem to which the *cutting point* is interpreted as a *driven cutting point*. Some nodes are called *source nodes* that have pointing-away edges only, and some nodes are called *sink nodes* that have pointing-to edges only.

2.9 Virtual Stability

After a complex robot is virtually decomposed into subsystems, a natural concern is on the properties each subsystem should have in order to maintain the stability of the entire complex robot. These properties are being addressed in this section around the central concept of the *virtual stability* to be defined below.

2.9.1 Non-negative Accompanying Functions

Definition 2.15. *A non-negative accompanying function $\nu(t) \in \mathbb{R}$ is a piecewise differentiable function possessing the following properties:*

(i) $\nu(t) \geqslant 0$ for $t > 0$, and
(ii) $\dot{\nu}(t)$ exists almost everywhere.

In this book, every subsystem (or a node in a simple oriented graph) is assigned a non-negative accompanying function to conduct the stability and convergence analysis.

2.9.2 Virtual Power Flow

Virtual power flows (VPFs) will be defined and used to characterize the dynamic interactions among subsystems. The introduction of this terminology is an important step leading to the definition of the *virtual stability* to be outlined in the next subsection.

Definition 2.16. *With respect to a frame $\{\mathbf{A}\}$, the virtual power flow (VPF) is defined as the inner product of the linear/angular velocity vector error and the force/moment vector error, that is*

$$p_{\mathbf{A}} \stackrel{def}{=} ({}^{\mathbf{A}}V_r - {}^{\mathbf{A}}V)^T({}^{\mathbf{A}}F_r - {}^{\mathbf{A}}F) \tag{2.85}$$

where ${}^{\mathbf{A}}V_r \in \mathbb{R}^6$ and ${}^{\mathbf{A}}F_r \in \mathbb{R}^6$ represent the required (design) vectors of ${}^{\mathbf{A}}V \in \mathbb{R}^6$ and ${}^{\mathbf{A}}F \in \mathbb{R}^6$, respectively.

Let two frames, denoted as $\{\mathbf{A}\}$ and $\{\mathbf{B}\}$, be attached to a common rigid body. If the required linear/angular velocity vectors and the required force/moment vectors are subject to the same constraints as that imposed on the linear/angular velocity vectors and on the force/moment vectors, that is

$$ {}^{\mathbf{B}}V_r = {}^{\mathbf{A}}\mathbf{U}_{\mathbf{B}}^{T}\,{}^{\mathbf{A}}V_r \tag{2.86} $$

$$ {}^{\mathbf{A}}F_r = {}^{\mathbf{A}}\mathbf{U}_{\mathbf{B}}\,{}^{\mathbf{B}}F_r \tag{2.87} $$

then it follows from (2.63), (2.64), and (2.85)–(2.87) that

$$ p_{\mathbf{A}} = p_{\mathbf{B}} \tag{2.88} $$

holds.[5]

Remark 2.5. Equation (2.88) indicates that the VPF defined by (2.85) is invariant to frames fixed to a common rigid body. This feature is similar to the power flow within a rigid body. The difference is that a *virtual power flow* represents an inner product of a velocity vector error and a force vector error in $\mathbb{R}^6$, rather than an inner product of a velocity vector and a force vector themselves.

Note that the conditions (2.86) and (2.87) being used to validate (2.88) need to be always satisfied in control designs.

2.9.3 Virtual Stability

Definition 2.17. *A subsystem that is virtually decomposed from a complex robot is said to be virtually stable with its affiliated vector* $\mathbf{x}(t)$ *being a virtual function in* L_∞ *and its affiliated vector* $\mathbf{y}(t)$ *being a virtual function in* L_2*, if and only if there exists a non-negative accompanying function*

$$ \nu(t) \geqslant \frac{1}{2}\mathbf{x}(t)^T\mathbf{P}\mathbf{x}(t) \tag{2.89} $$

such that

$$ \dot{\nu}(t) \leqslant -\mathbf{y}(t)^T\mathbf{Q}\mathbf{y}(t) - s(t) + \sum_{\{\mathbf{A}\}\in\Phi} p_{\mathbf{A}} - \sum_{\{\mathbf{C}\}\in\Psi} p_{\mathbf{C}} \tag{2.90} $$

holds, subject to

$$ \int_0^\infty s(t)d\tau \geqslant -\gamma_s \tag{2.91} $$

with $0 \leqslant \gamma_s < \infty$*, where* $\mathbf{P}$ *and* $\mathbf{Q}$ *are two block-diagonal positive-definite matrices, set* Φ *contains frames being placed at the driven cutting points of this subsystem and set* Ψ *contains frames being placed at the driving cutting points of this subsystem, and* $p_{\mathbf{A}}$ *and* $p_{\mathbf{C}}$ *denote the virtual power flows in Definition 2.16.*

[5] Note that the validity of (2.88) is linked to Remark 2.4. When two different pairs of physical force and moment vectors apply respectively to frames $\{\mathbf{A}\}$ and $\{\mathbf{B}\}$, (2.88) becomes invalid.

Remark 2.6. For a given subsystem, the *virtual stability* in Definition 2.17 requires the appearance of the VPFs in the time derivative of the non-negative accompanying function assigned to this subsystem. The *virtual power flows* are with a positive sign at the *driven cutting points* and with a negative sign at the *driving cutting points*. This is the unique feature of the *virtual stability*.

Remark 2.7. Note that $s(t) = 0$ is a special case that satisfies (2.91).

Remark 2.8. In Definition 2.17, a *virtual function* in L_p is a candidate function in L_p for $p = 2, \infty$. As will be seen below, a *virtual function* in L_p become a function in L_p, when every subsystem of a complex robot qualifies to be *virtually stable*.

The following lemma shows that two adjacent subsystems that are *virtually stable* are equivalent to a single subsystem that is *virtually stable*.

Lemma 2.11. *Every two adjacent subsystems that are virtually stable can be equivalent to a single subsystem that is virtually stable in the sense of Definition 2.17. Every virtual function in L_p affiliated with any one of the two adjacent subsystems remains to be a virtual function in L_p affiliated with the equivalent subsystem for $p = 2, \infty$.*

Proof: Let $\mathbf{x}_i(t)$ be a *virtual function* in L_∞, and let $\mathbf{y}_i(t)$ be a *virtual function* in L_2, affiliated with the ith subsystem for $i = 1, 2$. Let Φ_i be a set containing frames being placed at the *driven cutting points* of the ith subsystem, and let Ψ_i be a set containing frames being placed at the *driving cutting points* of the ith subsystem, for $i = 1, 2$. It follows from Definition 2.17 that there exist two non-negative accompanying functions

$$\nu_i(t) \geqslant \frac{1}{2}\mathbf{x}_i(t)^T \mathbf{P}_i \mathbf{x}_i(t) \tag{2.92}$$

for $i = 1, 2$, such that

$$\dot{\nu}_i(t) \leqslant -\mathbf{y}_i(t)^T \mathbf{Q}_i \mathbf{y}_i(t) - s_i(t) + \sum_{\{\mathbf{A}\}\in\Phi_i} p_{\mathbf{A}} - \sum_{\{\mathbf{C}\}\in\Psi_i} p_{\mathbf{C}} \tag{2.93}$$

holds for $i = 1, 2$, subject to

$$\int_0^\infty s_i(t)dt \geqslant -\gamma_{si} \tag{2.94}$$

with $0 \leqslant \gamma_{si} < \infty$, where $\mathbf{P}_i$ and $\mathbf{Q}_i$ are block-diagonal positive-definite matrices.

Let

$$\Xi = \left(\Phi_1 \bigcap \Psi_2\right) \bigcup \left(\Phi_2 \bigcap \Psi_1\right)$$

be a set[6] containing frames being placed at the *cutting points* between the two adjacent subsystems.

[6] In fact, either $\Phi_1 \bigcap \Psi_2$ or $\Phi_2 \bigcap \Psi_1$ must be a null set, since there is no circular path in any simple oriented graph in terms of Definition 2.14.

Define

$$\begin{aligned}\nu(t) &= \nu_1(t) + \nu_2(t) \\ &\geqslant \sum_{i=1}^{2} \frac{1}{2}\mathbf{x}_i(t)^T \mathbf{P}_i \mathbf{x}_i(t)\end{aligned} \tag{2.95}$$

as the non-negative accompanying function assigned to the equivalent subsystem. It follows from (2.93) and the definition of set $\varXi$ that

$$\begin{aligned}\dot{\nu}(t) &= \dot{\nu}_1(t) + \dot{\nu}_2(t) \\ &\leqslant -\sum_{i=1}^{2} \mathbf{y}_i(t)^T \mathbf{Q}_i \mathbf{y}_i(t) - \sum_{i=1}^{2} s_i(t) + \sum_{i=1}^{2} \sum_{\{\mathbf{A}\}\in\varPhi_i} p_{\mathbf{A}} - \sum_{i=1}^{2} \sum_{\{\mathbf{C}\}\in\varPsi_i} p_{\mathbf{C}} \\ &= -\sum_{i=1}^{2} \mathbf{y}_i(t)^T \mathbf{Q}_i \mathbf{y}_i(t) - \sum_{i=1}^{2} s_i(t) + \sum_{\{\mathbf{A}\}\in\varPhi} p_{\mathbf{A}} - \sum_{\{\mathbf{C}\}\in\varPsi} p_{\mathbf{C}}\end{aligned} \tag{2.96}$$

holds with

$$\varPhi = \varPhi_1 \bigcup \varPhi_2 - \varXi \tag{2.97}$$

$$\varPsi = \varPsi_1 \bigcup \varPsi_2 - \varXi. \tag{2.98}$$

Consider the fact that set $\varPhi$ contains frames being placed at the *driven cutting points* of the equivalent subsystem and set $\varPsi$ contains frames being placed at the *driving cutting points* of the equivalent subsystem. Using (2.95) and (2.96) completes the proof, in terms of Definition 2.17. ■

Remark 2.9. The unique characteristic of the *virtual stability* is the appearance of the VPFs in the time derivative of the non-negative accompanying function assigned to each subsystem. These *virtual power flows* completely represent the dynamic interactions among subsystems.

When every subsystem is *virtually stable* in the sense of Definition 2.17, the following theorem ensures that the L_2 and L_∞ stability of the entire complex robot can be guaranteed.

Theorem 2.1. *Consider a complex robot that is virtually decomposed into subsystems and is represented by a simple oriented graph in Definition 2.14. If every subsystem is virtually stable in the sense of Definition 2.17, then all virtual functions in L_2 are functions in L_2 and all virtual functions in L_∞ are functions in L_∞.*

Proof: Without loss of generality, it is assumed that a complex robot is virtually decomposed into $n > 1$ subsystems. Every subsystem is *virtually stable* in the sense of Definition 2.17. For the ith subsystem, $i \in \{1, n\}$, its associated non-negative accompanying function and the time derivative can be written as

$$\nu_i(t) \geqslant \frac{1}{2}\mathbf{x}_i(t)^T\mathbf{P}_i\mathbf{x}_i(t) \tag{2.99}$$

$$\dot{\nu}_i(t) \leqslant -\mathbf{y}_i(t)^T\mathbf{Q}_i\mathbf{y}_i(t) - s_i(t) + \sum_{\{\mathbf{A}\}\in\Phi_i} p_\mathbf{A} - \sum_{\{\mathbf{C}\}\in\Psi_i} p_\mathbf{C} \tag{2.100}$$

subject to

$$\int_0^\infty s_i(t)dt \geqslant -\gamma_{si} \tag{2.101}$$

with $0 \leqslant \gamma_{si} < \infty$, where $\mathbf{P}_i$ and $\mathbf{Q}_i$ are two block-diagonal positive-definite matrices, set Φ_i contains frames being placed at the *driven cutting points* of the ith subsystem, and set Ψ_i contains frames being placed at the *driving cutting points* of the ith subsystem.

Define a non-negative function for the entire complex robot as

$$\nu(t) = \sum_{i=1}^{n} \nu_i(t) \geqslant \sum_{i=1}^{n} \frac{1}{2}\mathbf{x}_i(t)^T\mathbf{P}_i\mathbf{x}_i(t). \tag{2.102}$$

It follows from (2.99)–(2.102) that

$$\begin{aligned}\dot{\nu}(t) \leqslant & -\sum_{i=1}^{n} \mathbf{y}_i(t)^T\mathbf{Q}_i\mathbf{y}_i(t) - \sum_{i=1}^{n} s_i(t) \\ & + \sum_{i=1}^{n}\left[\sum_{\{\mathbf{A}\}\in\Phi_i} p_\mathbf{A} - \sum_{\{\mathbf{C}\}\in\Psi_i} p_\mathbf{C}\right]\end{aligned} \tag{2.103}$$

holds, subject to

$$\int_0^\infty \sum_{i=1}^{n} s_i(t)dt \geqslant -\sum_{i=1}^{n} \gamma_{si} \tag{2.104}$$

with $0 \leqslant \gamma_{si} < \infty$ for all $i \in \{1, n\}$.

Consider two adjacent subsystems with a *cutting point* in between. This *cutting point* is interpreted as a *driving cutting point* by one subsystem and is also interpreted as a *driven cutting point* by the other subsystem. In other words, for any given *driving cutting point* there exists a corresponding *driven cutting point* and *vice versa*. Being a scalar, the VPF by Definition 2.16 at a *driving cutting point* takes exactly the same magnitude as the VPF at the corresponding *driven cutting point*, but the sign. Therefore, the sum of VPFs at all *driving cutting points* equals the sum of VPFs at all *driven cutting points*, or mathematically

$$\sum_{i=1}^{n} \sum_{\{\mathbf{A}\}\in\Phi_i} p_\mathbf{A} = \sum_{i=1}^{n} \sum_{\{\mathbf{C}\}\in\Psi_i} p_\mathbf{C} \tag{2.105}$$

holds, leading to

$$\sum_{i=1}^{n}\left[\sum_{\{\mathbf{A}\}\in\Phi_i} p_\mathbf{A} - \sum_{\{\mathbf{C}\}\in\Psi_i} p_\mathbf{C}\right] = 0. \tag{2.106}$$

Substituting (2.106) into (2.103) and using Lemma 2.3 yields

$$\mathbf{x}_i(t) \in L_\infty, \ \forall i \in \{1, n\} \tag{2.107}$$

$$\mathbf{y}_i(t) \in L_2, \ \forall i \in \{1, n\}. \tag{2.108}$$

■

Remark 2.10. Theorem 2.1 is the most important theorem in this book. It establishes the equivalence between the *virtual stability* of every subsystem and the stability of the entire complex robot. Consequently, it allows us to concentrate on ensuring the *virtual stability* of every subsystem, in lieu of the stability of the entire complex robot. Therefore, Theorem 2.1 can be considered as the theoretical foundation of the *virtual decomposition control.*

2.10 Algebraic Loop Issue

The algebraic loop issue is widely encountered in control of robots, particularly in robot force control. Algebraic loops are formed by control laws that use the time derivatives of system states, such as forces or accelerations. In a graphic representation, algebraic loops represent direct circular paths passing through gain blocks without encountering any dynamic element, such as an integrator.

Definition 2.18. *Given a control system, if there exists a vectored (or scalar) variable* $\mathbf{f}(t) \in \mathbb{R}^n$, $n \geqslant 1$, *satisfying*

$$\mathbf{f}(t) = \varsigma \left[\mathbf{K}(t)\mathbf{f}(t) + \mathbf{u}_a(t)\right] + \mathbf{u}_b(t) \tag{2.109}$$

with

$$\varsigma \overset{def}{=} \begin{cases} 1 & \text{control enabled} \\ 0 & \text{control disabled} \end{cases}$$

where $\mathbf{K}(t) \in \mathbb{R}^{n \times n}$ *is a non-zero matrix and* $\mathbf{u}_a(k) \in \mathbb{R}^n$ *and* $\mathbf{u}_b(k) \in \mathbb{R}^n$ *are two residual vectored (or scalar) variables, then an algebraic loop about* $\mathbf{f}(t) \in \mathbb{R}^n$ *with gain* $\mathbf{K}(t) \neq 0$ *is formed.*

The term $\mathbf{f}(t)$ in the left hand side of (2.109) is a physical variable, while the first term in the right hand side of (2.109) is induced by control. Considering the sampling delay in digital implementations, it is realistic to re-write (2.109) as

$$\mathbf{f}(t) = \varsigma \left[\mathbf{K}(kT)\mathbf{f}(kT) + \mathbf{u}_a(kT)\right] + \mathbf{u}_b(t) \quad \text{for } kT < t \leqslant (k+1)T \tag{2.110}$$

where k is a positive integer representing the sampling time index and T is the sampling period.

The discrete-time representation of (2.110) becomes

$$\mathbf{f}(k+1) = \varsigma \mathbf{K}(k)\mathbf{f}(k) + \mathbf{u}(k+1) \tag{2.111}$$

with

$$\mathbf{u}(k+1) \overset{def}{=} \varsigma \mathbf{u}_a(k) + \mathbf{u}_b(k+1). \tag{2.112}$$

The stability of the algebraic loop is governed by one lemma and one theorem to be presented below.

Lemma 2.12. *Consider two positive discrete-time signals $x(k) \geqslant 0$ and $u(k) \geqslant 0$ subject to*

$$x(k+1) \leqslant c_k x(k) + u(k+1) \tag{2.113}$$

with $0 < c_k < 1$ being a positive constant. If $u(t) \in L_\infty$ holds, then it yields

$$x(k) \leqslant \max\left\{x(0), \frac{1}{1-c_k} \max_k(u(k))\right\}. \tag{2.114}$$

Proof: When $u(t) \in L_\infty$ holds, the existence of $\max_k(u(k)) < \infty$ is ensured.

In case of $x(0) \geqslant \frac{1}{1-c_k} \max_k(u(k))$, it follows from (2.113) that $x(1) \leqslant c_k x(0) + u(1) \leqslant c_k x(0) + \max_k(u(k)) \leqslant c_k x(0) + (1-c_k)x(0) = x(0)$ holds. Consequently, if $x(0) \geqslant x(k) \geqslant \frac{1}{1-c_k} \max_k(u(k))$ holds for a $k \geqslant 1$, it then leads to $x(k+1) \leqslant c_k x(k) + u(k+1) \leqslant c_k x(k) + \max_k(u(k)) \leqslant c_k x(k) + (1-c_k)x(k) = x(k)$. Thus, $x(k) \leqslant x(0)$ holds indefinitely for all integer $k \geqslant 1$.

In case of $x(0) \leqslant \frac{1}{1-c_k} \max_k(u(k))$, it follows from (2.113) that $x(1) \leqslant c_k x(0) + u(1) \leqslant \frac{c_k}{1-c_k} \max_k(u(k)) + \max_k(u(k)) = \frac{1}{1-c_k} \max_k(u(k))$ holds. Consequently, if $x(k) \leqslant \frac{1}{1-c_k} \max_k(u(k))$ holds for a $k \geqslant 1$, it then leads to $x(k+1) \leqslant c_k x(k) + u(k+1) \leqslant \frac{c_k}{1-c_k} \max_k(u(k)) + \max_k(u(k)) = \frac{1}{1-c_k} \max_k(u(k))$. Thus, $x(k) \leqslant \frac{1}{1-c_k} \max_k(u(k))$ holds indefinitely for all integer $k \geqslant 1$. ■

Theorem 2.2. *Consider an algebraic loop characterized by (2.109) or (2.110) in continuous-time and by (2.111) in discrete-time with $\mathbf{u}_a(t) \in L_\infty$ and $\mathbf{u}_b(t) \in L_\infty$. Let $0 < \delta_\epsilon \ll 1$ be a small positive constant. If the maximum singular value of $\mathbf{K}(t) \in \mathbb{R}^n$ over time satisfies*

$$\max_{t>0} (\overline{\sigma}(\mathbf{K}(t))) \leqslant \sqrt{\frac{1}{1+\delta_\epsilon}} \tag{2.115}$$

then $\mathbf{f}(t) \in L_\infty$ holds for $t > 0$.

Proof: Let $\delta_a > 0$ and $\delta_b > 0$ be two small positive constants subject to

$$\delta_a + \delta_b = \delta_\epsilon. \tag{2.116}$$

With $\varsigma = 1$, it follows from (2.111) that

$$\begin{aligned}
\|\mathbf{f}(k+1)\|^2 &= \mathbf{f}(k+1)^T \mathbf{f}(k+1) \\
&= \mathbf{f}(k)^T \mathbf{K}(k)^T \mathbf{K}(k) \mathbf{f}(k) \\
&\quad + \mathbf{u}(k+1)^T \mathbf{u}(k+1) + 2\mathbf{f}(k)^T \mathbf{K}(k)^T \mathbf{u}(k+1) \\
&\leqslant \mathbf{f}(k)^T \mathbf{K}(k)^T \mathbf{K}(k) \mathbf{f}(k) \\
&\quad + \mathbf{u}(k+1)^T \mathbf{u}(k+1) + 2\mathbf{f}(k)^T \mathbf{K}(k)^T \mathbf{u}(k+1) \\
&\quad + \delta_a \left[\mathbf{K}(k)\mathbf{f}(k) - \frac{1}{\delta_a}\mathbf{u}(k+1)\right]^T
\end{aligned}$$

$$\begin{aligned}&\left[\mathbf{K}(k)\mathbf{f}(k)-\frac{1}{\delta_a}\mathbf{u}(k+1)\right]\\
&=(1+\delta_a)\,\mathbf{f}(k)^T\mathbf{K}(k)^T\mathbf{K}(k)\mathbf{f}(k)\\
&\quad+\left(1+\frac{1}{\delta_a}\right)\mathbf{u}(k+1)^T\mathbf{u}(k+1)\\
&=(1+\delta_a)\,\|\mathbf{K}(k)\mathbf{f}(k)\|^2+\left(1+\frac{1}{\delta_a}\right)\|\mathbf{u}(k+1)\|^2\\
&\leqslant(1+\delta_a)\,\overline{\sigma}(\mathbf{K}(k))^2\|\mathbf{f}(k)\|^2\\
&\quad+\left(1+\frac{1}{\delta_a}\right)\|\mathbf{u}(k+1)\|^2\end{aligned} \tag{2.117}$$

holds, where $\overline{\sigma}(\mathbf{K}(k))$ denotes the maximum singular value of $\mathbf{K}(k)$. Using (2.115) yields

$$(1+\delta_a)\,\overline{\sigma}(\mathbf{K}(k))^2\leqslant\frac{1+\delta_a}{1+\delta_\epsilon}. \tag{2.118}$$

Furthermore, using (2.116)-(2.118) and Lemma 2.12 yields

$$\max_k(\|\mathbf{f}(k)\|^2)\leqslant\max\left\{\|\mathbf{f}(0)\|^2,\frac{1+\delta_\epsilon}{\delta_b}\left(1+\frac{1}{\delta_a}\right)\max_k(\|\mathbf{u}(k)\|^2)\right\}. \tag{2.119}$$

Note that $\mathbf{u}_a(t)\in L_\infty$ and $\mathbf{u}_b(t)\in L_\infty$ imply $\max_k\|\mathbf{u}(k)\|<\infty$. Consequently, it follows from (2.119) that $\max_k(\|\mathbf{f}(k)\|)<\infty$ holds. Thus, the boundedness of the control action represented by $\mathbf{K}(kT)\mathbf{f}(kT)+\mathbf{u}_a(kT)$ is ensured. Finally, it follows from (2.110) that $\mathbf{f}(t)\in L_\infty$ holds for $t>0$. ■

Remark 2.11. To avoid having algebraic loops in the control design should be considered in the first place. If this attempt fails, attentions ought to be paid to ensure the stability of the algebraic loops induced by control. In Theorem 2.2, the sufficient condition to ensure the stability of an algebraic loop is to limit the algebraic loop gain matrix by (2.115) under bounded input signals. The maximum singular value of the algebraic loop gain matrix must be smaller than *one* to guarantee the stability.

2.11 Concluding Remarks

This chapter prepares readers with necessary mathematical backgrounds and fundamental technical elements that will be extensively used in this book. After the notation was presented in Section 2.1, Lebesgue integrable functions with basic properties were presented in Section 2.2. This section forms the basis for theoretical results to develop.

In Section 2.3, vectors and coordinate systems were introduced. Orientation expressions by using quaternions were discussed in Section 2.4. Then, the body-frame referenced linear/angular velocity vectors and force/moment vectors were defined and their duality was discussed in Section 2.5. Furthermore, the body-frame referenced rigid body dynamics were presented in Section 2.6. In Section 2.7, the two projection functions to be used in parameter adaptation were given. Unlike conventional parameter adaptation that is persistently driven by control errors, the projection functions used in this book impose limits on parameter estimates to prevent them from drifting. Corresponding lemmas guarantee that parameter estimates with limits can be smoothly incorporated into respective controllers without jeopardizing the stability properties. These five sections constitute the basic contents of the *virtual decomposition control* (VDC) computations which extensively use body-frame referenced rigid body dynamics and kinematics-based velocity/force transformations.

Sections 2.8 and 2.9 presented fundamental concepts for the stability analysis particularly associated with the VDC approach. The presentation started with the definition of *cutting points*, followed by system representations through the use of simple oriented graphs. Then, the concept of the *virtual power flow* was introduced, leading to the definition of the *virtual stability* which is vital to the stability analysis of the entire system. The importance of Theorem 2.1 is that the stability issue of a complex robot can be converted to the *virtual stability* issue of individual subsystems. The design objective is, therefore, to make every individual subsystem qualified to be *virtually stable* in the sense of Definition 2.17, through designing respective control equations.

The physical interpretation of Theorem 2.1 has been given. Due to the force/moment reference directions, the two VPFs (associated with two adjacent subsystems) at a given *cutting point* take the same magnitude with different signs. This fact ensures that all *virtual power flows* at the *cutting points* cancel out in the time derivative of the overall non-negative function to guarantee the stability and convergence of the entire complex robot.

At the end, the algebraic loop issue for control systems arisen from the use of the time-derivatives of system states has been formally defined in Section 2.10. The sufficient condition to ensure the stability in digital implementations has also been given under bounded input signals. The algebraic loop issue is very important, since it has been deeply involved in many robot control strategies including force control.

3

Virtual Decomposition Control - A Two DOF Example

In this chapter, the *virtual decomposition control* (VDC) approach is being presented with respect to a simple robot having two degrees of freedom (DOFs) in free motion. The purpose of this chapter is to give readers an intuitive impression and a detailed understanding on how this approach works.

3.1 Original System

A two degrees of freedom robot moving in a horizontal plane without subject to the gravity is illustrated in Figure 3.1.

In planar motion, the x and y axes of each frame are aligned with the motion plane to describe all two-dimensional positions and forces, and the z axis of each frame is perpendicular to the motion plane to describe all one-dimensional rotations and torques. Thus, the rotation matrices are 2×2 matrices.

As shown in Figure 3.1, the robot has two joints and two links. The ith link, $i = 1, 2$, has a length of l_i, a mass of m_i with the center of mass being located at a distance of d_i to the ith joint, and a moment of inertia around the center of mass of I_{oi}. The joint angles are represented by q_1 and q_2, respectively. A frame $\{\mathbf{O}\}$ is attached to the end-effector (end of link 2) of the robot to describe its motion. Without loss of generality, the x axis of frame $\{\mathbf{O}\}$ aligns with the axis of link 2.

3.2 Virtual Decomposition

The original system, as illustrated in Figure 3.1, is virtually decomposed into four subsystems (two joints and two links) by placing three *cutting points* into the system, as illustrated in Figure 3.2.

A *cutting point* is a separation interface that virtually cuts through a link. The two surfaces resulting form the virtual cut maintain the same position and orientation. Three coordinate frames, namely $\{\mathbf{B}_1\}$, $\{\mathbf{B}_2\}$, and $\{\mathbf{T}_2\}$, are assigned to

W.-H. Zhu: Virtual Decomposition Control, STAR 60, pp. 45–62.
springerlink.com

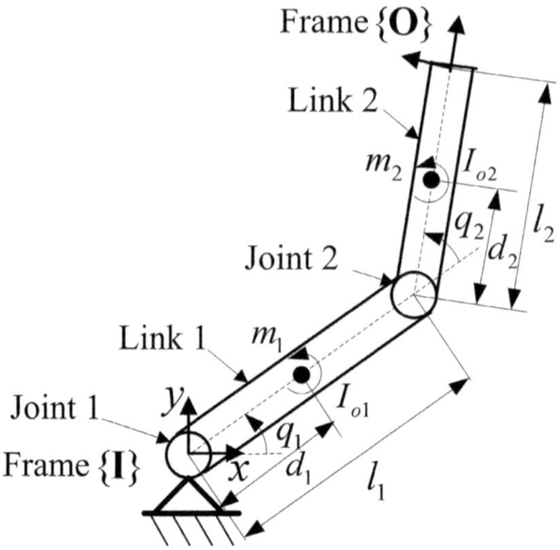

Fig. 3.1. A two-DOF robot.

the three *cutting points*, respectively. Each frame is assigned to a *cutting point* in such a way that it is simultaneously fixed to the two adjacent subsystems resulting form the virtual cut. At frames $\{\mathbf{B}_i\}$, $i = 1, 2$, the exerting forces/moment are referenced from joint i to link i. At frame $\{\mathbf{T}_2\}$, the exerting forces/moment are referenced from link 1 to joint 2.

Similar to frame $\{\mathbf{O}\}$, the x axes of frames $\{\mathbf{B}_1\}$ and $\{\mathbf{T}_2\}$ align with the axis of link 1, and the x axis of frame $\{\mathbf{B}_2\}$ aligns with the axis of link 2.

The *cutting point* associated with frame $\{\mathbf{B}_2\}$ is called the *driven cutting point* of link 2 and is simultaneously called the *driving cutting point* of joint 2. The *cutting point* associated with frame $\{\mathbf{T}_2\}$ is called the *driven cutting point* of joint 2 and is simultaneously called the *driving cutting point* of link 1. The *cutting point* associated with frame $\{\mathbf{B}_1\}$ is called the *driven cutting point* of link 1 and is simultaneously called the *driving cutting point* of joint 1. It follows that

- Link 2 has only one *driven cutting point* associated with frame $\{\mathbf{B}_2\}$.
- Joint 2 has one *driving cutting point* associated with frame $\{\mathbf{B}_2\}$ and one *driven cutting point* associated with frame $\{\mathbf{T}_2\}$.
- Link 1 has one *driving cutting point* associated with frame $\{\mathbf{T}_2\}$ and one *driven cutting point* associated with frame $\{\mathbf{B}_1\}$.
- Joint 1 has one *driving cutting point* associated with frame $\{\mathbf{B}_1\}$.

3.3 Kinematics

In this section, all kinematics computations including velocity transformations among different body-attached frames will be given.

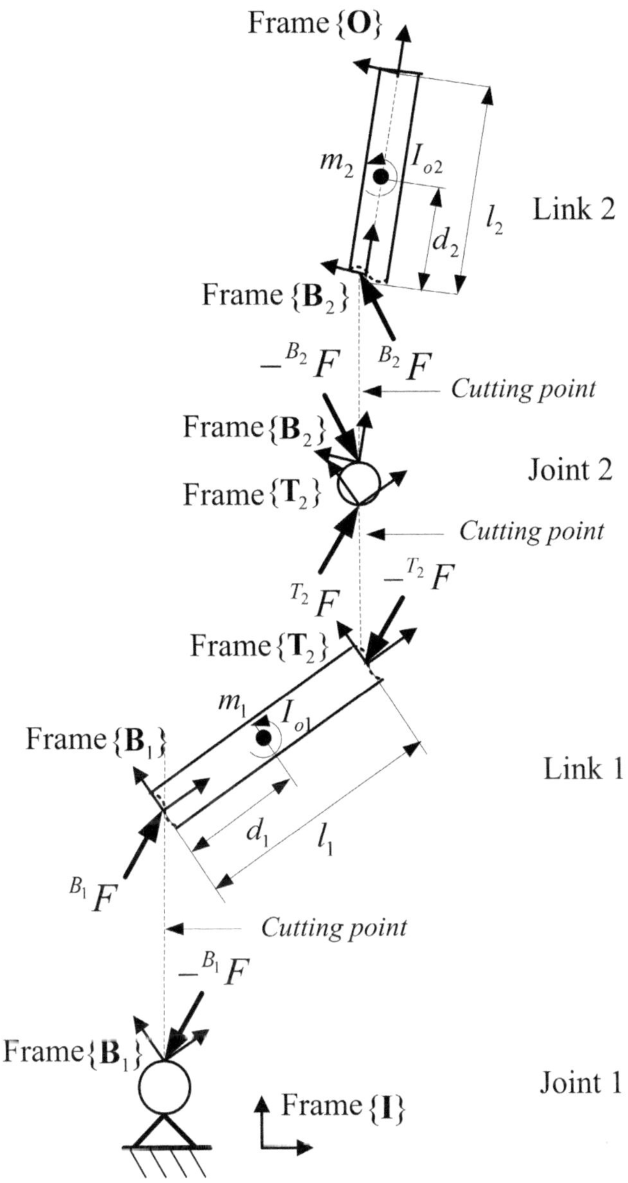

Fig. 3.2. The robot being virtually decomposed.

3.3.1 End-Effector Positions and Velocities

In view of Figure 3.1 and the fact that the x axis of frame $\{\mathbf{O}\}$ aligns with the axis of link 2, the rotation matrix that transforms a two-dimensional vector expressed in frame $\{\mathbf{O}\}$ to the same vector expressed in inertia frame $\{\mathbf{I}\}$ can be written as

$$ {}^{\mathbf{I}}\mathbf{R}_{\mathbf{O}} = \begin{bmatrix} \cos(q_1+q_2) & -\sin(q_1+q_2) \\ \sin(q_1+q_2) & \cos(q_1+q_2) \end{bmatrix} \in \mathbb{R}^{2\times 2} \tag{3.1} $$

with ${}^{\mathbf{O}}\mathbf{R}_{\mathbf{I}} = {}^{\mathbf{I}}\mathbf{R}_{\mathbf{O}}^T$.

Meanwhile, the forward kinematics from the joint angles to the position of the origin of frame $\{\mathbf{O}\}$, denoted as $[x_o, y_o]^T \in \mathbb{R}^2$, can be expressed as

$$ x_o = l_1\cos(q_1) + l_2\cos(q_1+q_2) \tag{3.2} $$

$$ y_o = l_1\sin(q_1) + l_2\sin(q_1+q_2). \tag{3.3} $$

Consequently, the time derivatives of these two equations can be further written as

$$ \begin{bmatrix} \dot{x}_o \\ \dot{y}_o \end{bmatrix} = \begin{bmatrix} -l_1\sin(q_1) - l_2\sin(q_1+q_2) & -l_2\sin(q_1+q_2) \\ l_1\cos(q_1) + l_2\cos(q_1+q_2) & l_2\cos(q_1+q_2) \end{bmatrix} \begin{bmatrix} \dot{q}_1 \\ \dot{q}_2 \end{bmatrix}. \tag{3.4} $$

Then, the end-effector velocity vector expressed in the body frame $\{\mathbf{O}\}$, denoted as ${}^{\mathbf{O}}\mathbf{v} \in \mathbb{R}^{2\times 2}$, can be obtained by premultiplying both sides of (3.4) by ${}^{\mathbf{O}}\mathbf{R}_{\mathbf{I}} = {}^{\mathbf{I}}\mathbf{R}_{\mathbf{O}}^T$, that is

$$ {}^{\mathbf{O}}\mathbf{v} = \mathbf{J}\dot{\mathbf{q}} \tag{3.5} $$

where

$$ {}^{\mathbf{O}}\mathbf{v} = {}^{\mathbf{O}}\mathbf{R}_{\mathbf{I}} \begin{bmatrix} \dot{x}_o \\ \dot{y}_o \end{bmatrix} \in \mathbb{R}^2 \tag{3.6} $$

$$ \mathbf{J} = {}^{\mathbf{O}}\mathbf{R}_{\mathbf{I}} \begin{bmatrix} -l_1\sin(q_1) - l_2\sin(q_1+q_2) & -l_2\sin(q_1+q_2) \\ l_1\cos(q_1) + l_2\cos(q_1+q_2) & l_2\cos(q_1+q_2) \end{bmatrix} \in \mathbb{R}^{2\times 2} \tag{3.7} $$

$$ \dot{\mathbf{q}} = \begin{bmatrix} \dot{q}_1 \\ \dot{q}_2 \end{bmatrix} \in \mathbb{R}^2. \tag{3.8} $$

3.3.2 Rotation and Transformation Matrices

In view of Figure 3.1, the rotation (direction cosine) matrices associated with frames $\{\mathbf{I}\}$, $\{\mathbf{B}_1\}$, and $\{\mathbf{B}_2\}$ can be expressed by

$$ {}^{\mathbf{I}}\mathbf{R}_{\mathbf{B}_1} = \begin{bmatrix} \cos(q_1) & -\sin(q_1) \\ \sin(q_1) & \cos(q_1) \end{bmatrix}; \quad {}^{\mathbf{B}_1}\mathbf{R}_{\mathbf{I}} = \begin{bmatrix} \cos(q_1) & \sin(q_1) \\ -\sin(q_1) & \cos(q_1) \end{bmatrix} \tag{3.9} $$

$$ {}^{\mathbf{B}_1}\mathbf{R}_{\mathbf{B}_2} = \begin{bmatrix} \cos(q_2) & -\sin(q_2) \\ \sin(q_2) & \cos(q_2) \end{bmatrix}; \quad {}^{\mathbf{B}_2}\mathbf{R}_{\mathbf{B}_1} = \begin{bmatrix} \cos(q_2) & \sin(q_2) \\ -\sin(q_2) & \cos(q_2) \end{bmatrix} \tag{3.10} $$

with ${}^{\mathbf{I}}\mathbf{R}_{\mathbf{B}_2} = {}^{\mathbf{I}}\mathbf{R}_{\mathbf{B}_1}{}^{\mathbf{B}_1}\mathbf{R}_{\mathbf{B}_2} = {}^{\mathbf{I}}\mathbf{R}_{\mathbf{O}}$.

A force/moment transformation matrix, such as ${}^{\mathbf{A}}\mathbf{U}_{\mathbf{B}}$ in (2.65), becomes a 3×3 matrix in planar motion. As a result, three 3×3 transformation matrices are defined as

$$ {}^{\mathbf{B}_1}\mathbf{U}_{\mathbf{T}_2} = \begin{bmatrix} 1 & 0 & 0 \\ 0 & 1 & 0 \\ 0 & l_1 & 1 \end{bmatrix} \tag{3.11} $$

$$ {}^{\mathbf{T}_2}\mathbf{U}_{\mathbf{B}_2} = \begin{bmatrix} \cos(q_2) & -\sin(q_2) & 0 \\ \sin(q_2) & \cos(q_2) & 0 \\ 0 & 0 & 1 \end{bmatrix} \tag{3.12} $$

$$ \begin{aligned} {}^{\mathbf{B}_1}\mathbf{U}_{\mathbf{B}_2} &= {}^{\mathbf{B}_1}\mathbf{U}_{\mathbf{T}_2}\,{}^{\mathbf{T}_2}\mathbf{U}_{\mathbf{B}_2} \\ &= \begin{bmatrix} \cos(q_2) & -\sin(q_2) & 0 \\ \sin(q_2) & \cos(q_2) & 0 \\ l_1\sin(q_2) & l_1\cos(q_2) & 1 \end{bmatrix}. \end{aligned} \tag{3.13} $$

Note that these three matrices will play an important role in transforming linear/angular velocity vectors and force/moment vectors, as outlined below.

3.3.3 Linear/Angular Velocity Vectors

A linear/angular velocity vector given by Definition 2.8 becomes a three-dimensional vector in planar motion. As a result, the linear/angular velocity vector of frame $\{\mathbf{B}_1\}$ is assembled as

$$ {}^{\mathbf{B}_1}V = \begin{bmatrix} {}^{\mathbf{B}_1}\mathbf{R}_{\mathbf{I}} \begin{bmatrix} \dot{x}_{\mathbf{B}_1} \\ \dot{y}_{\mathbf{B}_1} \end{bmatrix} \\ \omega_{\mathbf{B}_1} \end{bmatrix} \tag{3.14} $$

with ${}^{\mathbf{B}_1}\mathbf{R}_{\mathbf{I}} \in \mathbb{R}^{2\times 2}$ being defined in (3.9), where $[\dot{x}_{\mathbf{B}_1}, \dot{y}_{\mathbf{B}_1}]^T \in \mathbb{R}^2$ denotes the linear velocity vector of the origin of frame $\{\mathbf{B}_1\}$, expressed in the inertial frame, while $\omega_{\mathbf{B}_1} \in \mathbb{R}$ denotes the angular velocity of frame $\{\mathbf{B}_1\}$ or the angular velocity of link 1.

The data structure of (3.14) also applies to the linear/angular velocity vectors of frames $\{\mathbf{T}_2\}$ and $\{\mathbf{B}_2\}$, denoted as ${}^{\mathbf{T}_2}V \in \mathbb{R}^3$ and ${}^{\mathbf{B}_2}V \in \mathbb{R}^3$, respectively, where frames $\{\mathbf{T}_2\}$ and $\{\mathbf{B}_2\}$ are substituted for frame $\{\mathbf{B}_1\}$ in (3.14).

3.3.4 Force/Moment Vectors

A force/moment vector given by Definition 2.9 becomes a three-dimensional vector in planar motion. As a result, the force/moment vector in frame $\{\mathbf{B}_1\}$ is assembled as

$$ {}^{\mathbf{B}_1}F = \begin{bmatrix} {}^{\mathbf{B}_1}\mathbf{R}_{\mathbf{I}} \begin{bmatrix} f_{\mathbf{B}_1 x} \\ f_{\mathbf{B}_1 y} \end{bmatrix} \\ m_{\mathbf{B}_1} \end{bmatrix} \tag{3.15} $$

with ${}^{\mathbf{B}_1}\mathbf{R}_{\mathbf{I}} \in \mathbb{R}^{2\times 2}$ being defined in (3.9), where $[f_{\mathbf{B}_1 x}, f_{\mathbf{B}_1 y}]^T \in \mathbb{R}^2$ denotes the force vector applied from joint 1 to link 1 and expressed in the inertial frame, while $m_{\mathbf{B}_1} \in \mathbb{R}$ denotes the moment applied from joint 1 to link 1 and measured by frame $\{\mathbf{B}_1\}$, in view of Figure 3.2.

The data structure of (3.15) also applies to the force/moment vectors in frames $\{\mathbf{T}_2\}$ and $\{\mathbf{B}_2\}$, denoted as ${}^{\mathbf{T}_2}F \in \mathbb{R}^3$ and ${}^{\mathbf{B}_2}F \in \mathbb{R}^3$, respectively, where frames $\{\mathbf{T}_2\}$ and $\{\mathbf{B}_2\}$ are substituted for frame $\{\mathbf{B}_1\}$ in (3.15).

In view of Figure 3.2, ${}^{\mathbf{T}_2}F \in \mathbb{R}^3$ denotes the forces and moment applied from link 1 to joint 2 and expressed in frame $\{\mathbf{T}_2\}$, and ${}^{\mathbf{B}_2}F \in \mathbb{R}^3$ denotes the forces and moment applied from joint 2 to link 2 and expressed in frame $\{\mathbf{B}_2\}$.

3.3.5 Velocity Mapping

Define $\mathbf{z} = [0, 0, 1]^T \in \mathbb{R}^3$ and refer to Figure 3.1. The linear/angular velocity vector of frame $\{\mathbf{B}_1\}$ can be written as

$$ {}^{\mathbf{B}_1}V = \mathbf{z}\dot{q}_1. \tag{3.16} $$

Then, the linear/angular velocity vector of frame $\{\mathbf{T}_2\}$ can be written as

$$ {}^{\mathbf{T}_2}V = {}^{\mathbf{B}_1}\mathbf{U}_{\mathbf{T}_2}^T \, {}^{\mathbf{B}_1}V. \tag{3.17} $$

Furthermore, the linear/angular velocity vector of frame $\{\mathbf{B}_2\}$ can be written as

$$ \begin{aligned} {}^{\mathbf{B}_2}V &= \mathbf{z}\dot{q}_2 + {}^{\mathbf{T}_2}\mathbf{U}_{\mathbf{B}_2}^T \, {}^{\mathbf{T}_2}V \\ &= \mathbf{z}\dot{q}_2 + {}^{\mathbf{B}_1}\mathbf{U}_{\mathbf{B}_2}^T \, {}^{\mathbf{B}_1}V. \end{aligned} \tag{3.18} $$

From (3.16) and (3.18), an augmented velocity vector can be defined, that is

$$ \mathcal{V}_2 \overset{def}{=} \begin{bmatrix} \dot{q}_1 \\ \dot{q}_2 \\ {}^{\mathbf{B}_1}V \\ {}^{\mathbf{B}_2}V \end{bmatrix} = \begin{bmatrix} 1 & 0 \\ 0 & 1 \\ \mathbf{z} & \mathbf{0} \\ {}^{\mathbf{B}_1}\mathbf{U}_{\mathbf{B}_2}^T\mathbf{z} & \mathbf{z} \end{bmatrix} \begin{bmatrix} \dot{q}_1 \\ \dot{q}_2 \end{bmatrix} \overset{def}{=} \mathcal{J}_2 \begin{bmatrix} \dot{q}_1 \\ \dot{q}_2 \end{bmatrix}. \tag{3.19} $$

The velocity mapping is an important step in the VDC approach. Unlike the approach in [110], both linear velocities and angular velocities are properly integrated to allow easy transformations.

3.3.6 Required Velocities

The terminology of *required velocity* gives rise to an important concept in the VDC approach. A required velocity is different from a desired velocity which usually serves as the reference trajectory of a velocity with respect to time. The meaning of a required velocity is that if the true velocity tracks the required velocity, then the control objectives including position control and force control can be achieved. Therefore, the general format of a required velocity includes both the desired velocity and one or more terms that are related to control errors, such as position errors and/or force errors.

The control objective in this chapter is to make the robot position trajectory track its desired position trajectory. It will be shown below that the position control can be performed through a velocity controller by incorporating a position error term into the required velocity.

If the position control is performed in the joint space, then the required joint velocities can be designed as

$$\begin{bmatrix} \dot{q}_{1r} \\ \dot{q}_{2r} \end{bmatrix} = \begin{bmatrix} \dot{q}_{1d} \\ \dot{q}_{2d} \end{bmatrix} + \lambda \begin{bmatrix} q_{1d} - q_1 \\ q_{2d} - q_2 \end{bmatrix} \tag{3.20}$$

where q_{1d} and q_{2d} denote the desired joint angles and $\lambda > 0$ is a control parameter.

Alternatively, if the Jacobian matrix $\mathbf{J}$ defined by (3.7) is of full rank, then the position control can be performed in the Cartesian space to make the end-effector position vector $[x_o, y_o]^T \in \mathbb{R}^2$ track its desired position vector $[x_{od}, y_{od}]^T \in \mathbb{R}^2$ in the Cartesian space. Consequently, the required Cartesian velocity vector expressed in frame $\{\mathbf{O}\}$ is obtained by

$$^{\mathbf{O}}\mathbf{v}_r = {^{\mathbf{O}}\mathbf{R}_{\mathbf{I}}} \left(\begin{bmatrix} \dot{x}_{od} \\ \dot{y}_{od} \end{bmatrix} + \lambda \begin{bmatrix} x_{od} - x_o \\ y_{od} - y_o \end{bmatrix} \right) \tag{3.21}$$

and the corresponding required joint velocities are obtained as

$$\begin{bmatrix} \dot{q}_{1r} \\ \dot{q}_{2r} \end{bmatrix} = \mathbf{J}^{-1}\,{^{\mathbf{O}}\mathbf{v}_r}. \tag{3.22}$$

Once the required joint velocities are obtained, the required linear/angular velocity vectors of the three body-fixed frames are computed as

$$^{\mathbf{B}_1}V_r = \mathbf{z}\dot{q}_{1r} \tag{3.23}$$

$$^{\mathbf{T}_2}V_r = {^{\mathbf{B}_1}\mathbf{U}_{\mathbf{T}_2}^T}\,{^{\mathbf{B}_1}V_r} \tag{3.24}$$

$$\begin{aligned} ^{\mathbf{B}_2}V_r &= \mathbf{z}\dot{q}_{2r} + {^{\mathbf{T}_2}\mathbf{U}_{\mathbf{B}_2}^T}\,{^{\mathbf{T}_2}V_r} \\ &= \mathbf{z}\dot{q}_{2r} + {^{\mathbf{B}_1}\mathbf{U}_{\mathbf{B}_2}^T}\,{^{\mathbf{B}_1}V_r} \end{aligned} \tag{3.25}$$

which are equivalent to

$$\mathcal{V}_r \overset{def}{=} \begin{bmatrix} \dot{q}_{1r} \\ \dot{q}_{2r} \\ ^{\mathbf{B}_1}V_r \\ ^{\mathbf{B}_2}V_r \end{bmatrix} = \mathcal{J}_2 \begin{bmatrix} \dot{q}_{1r} \\ \dot{q}_{2r} \end{bmatrix} \tag{3.26}$$

with $\mathcal{J}_2$ being defined in (3.19).

Computing the required velocity vectors of all three frames is a necessary step toward the computations of the subsystem dynamics based control, as will be seen in the next section.

3.4 Dynamics and Control of the Two Links

In this section, the dynamics and control problem of the two links will be handled independently. The presentations start with the kinematics and dynamics of the two links, followed by the respective control equations. The *virtual stability* in the sense of Definition 2.17 will be ensured at the end.

3.4.1 Kinematics and Dynamics

This subsection comprises three parts, namely link kinematics, rigid link dynamics, and resultant force/moment equations which relate the net forces/moment of a link to all exerting forces/moments on this link.

Link Kinematics

The link kinematics establishes the linear/angular velocity mapping relationships among all frames. Equation (3.17) or (3.19) in general serves the purpose.

Rigid Link Dynamics

The rigid link dynamics in the two-dimensional planar motion can be obtained from the three-dimensional dynamics developed in Section 2.6 with following modifications:

- Substituting frame $\{\mathbf{B}_i\}$, $i = 1, 2$, for frame $\{\mathbf{A}\}$.
- Substituting mass m_i, $i = 1, 2$, for mass $m_{\mathbf{A}}$.
- Substituting $[d_i, 0, 0]^T$, $i = 1, 2$, for ${}^{\mathbf{A}}\mathbf{r}_{\mathbf{AB}}$, since the center of mass is located on the x axis of frame $\{\mathbf{B}_i\}$ with a distance d_i from the frame origin.
- Substituting $[0, 0, \omega_{zi}]^T$, $i = 1, 2$, for ${}^{\mathbf{A}}\boldsymbol{\omega}$, since only the orientation in z axis is permitted.
- Substituting $\mathrm{diag}(0, 0, I_{oi})$ for $\mathbf{I}_{\mathbf{A}}$.
- Remove the third to fifth rows and the third to fifth columns of $\mathbf{M}_{\mathbf{B}_i}$ and $\mathbf{C}_{\mathbf{B}_i}({}^{\mathbf{B}_i}\omega)$.

Thus, the dynamic equation of the two rigid links in the two-dimensional motion without subject to the gravity can be expressed as

$$\mathbf{M}_{\mathbf{B}_i}\frac{d}{dt}\left({}^{\mathbf{B}_i}V\right) + \mathbf{C}_{\mathbf{B}_i}{}^{\mathbf{B}_i}V = {}^{\mathbf{B}_i}F^*, \quad i = 1, 2 \tag{3.27}$$

where

$$\mathbf{M}_{\mathbf{B}_i} = \begin{bmatrix} m_i & 0 & 0 \\ 0 & m_i & m_i d_i \\ 0 & m_i d_i & I_{oi} + m_i d_i^2 \end{bmatrix} \tag{3.28}$$

$$\mathbf{C}_{\mathbf{B}_i} = \begin{bmatrix} 0 & -m_i & -m_i d_i \\ m_i & 0 & 0 \\ m_i d_i & 0 & 0 \end{bmatrix} \omega_{zi} \tag{3.29}$$

with $\omega_{z1} = \dot{q}_1$ and $\omega_{z2} = \dot{q}_1 + \dot{q}_2$.

Resultant Forces/Moments

The net force/moment vectors in (3.27) are governed by

$$^{\mathbf{B}_2}F^* = {}^{\mathbf{B}_2}F \tag{3.30}$$

$$^{\mathbf{B}_1}F^* = {}^{\mathbf{B}_1}F - {}^{\mathbf{B}_1}\mathbf{U}_{\mathbf{T}_2}{}^{\mathbf{T}_2}F \tag{3.31}$$

in terms of Definition 2.10, where

$$^{\mathbf{T}_2}F = {}^{\mathbf{T}_2}\mathbf{U}_{\mathbf{B}_2}{}^{\mathbf{B}_2}F \tag{3.32}$$

holds with ${}^{\mathbf{B}_1}\mathbf{U}_{\mathbf{T}_2}$ and ${}^{\mathbf{T}_2}\mathbf{U}_{\mathbf{B}_2}$ being defined by (3.11) and (3.12), respectively.

3.4.2 Link Control Equations

The link control equations are fully model-based. Therefore, these control equations have the same structure as that in the previous subsection.

Required Velocities

The transformations of the required linear/angular velocities are defined by (3.24) or (3.26) in general.

Required Net Force/Moment Vectors

Two design vectors, namely required net force/moment vectors, will be designed so that the link dynamics combined with the control equations are qualified to be *virtually stable* in the sense of Definition 2.17.

Based on the dynamic model (3.27), the feedforward term with perfect parameters is formed by substituting the required linear/angular velocity vector and its time-derivative for the actual linear/angular velocity vector and its time-derivative, resulting in

$$\mathbf{M}_{\mathbf{B}_i}\frac{d}{dt}\left({}^{\mathbf{B}_i}V_r\right) + \mathbf{C}_{\mathbf{B}_i}{}^{\mathbf{B}_i}V_r, \quad i = 1, 2.$$

Then, express this term in a linear parametrization form

$$\mathbf{Y}_i\boldsymbol{\theta}_i = \mathbf{M}_{\mathbf{B}_i}\frac{d}{dt}\left({}^{\mathbf{B}_i}V_r\right) + \mathbf{C}_{\mathbf{B}_i}{}^{\mathbf{B}_i}V_r, \quad i - 1, 2 \tag{3.33}$$

where

$$\boldsymbol{\theta}_i \overset{def}{=} [m_i, m_i d_i, I_{oi} + m_i d_i^2]^T \in \mathbb{R}^3 \tag{3.34}$$

is a parameter vector and

$$\mathbf{Y}_i = \begin{bmatrix} \mathbf{y}_{i1} & \mathbf{y}_{i2} & \mathbf{y}_{i3} \end{bmatrix} \tag{3.35}$$

is a 3×3 regressor matrix with

$$\mathbf{y}_{i1} = \begin{bmatrix} \frac{d}{dt}\left({}^{\mathbf{B}_i}v_r\right)(1) - \omega_{zi}{}^{\mathbf{B}_i}v_r(2) \\ \frac{d}{dt}\left({}^{\mathbf{B}_i}v_r\right)(2) + \omega_{zi}{}^{\mathbf{B}_i}v_r(1) \\ 0 \end{bmatrix}$$

$$\mathbf{y}_{i2} = \begin{bmatrix} -\omega_{zi}{}^{\mathbf{B}_i}v_r(3) \\ \frac{d}{dt}\left({}^{\mathbf{B}_i}v_r\right)(3) \\ \frac{d}{dt}\left({}^{\mathbf{B}_i}v_r\right)(2) + \omega_{zi}{}^{\mathbf{B}_i}v_r(1) \end{bmatrix}$$

$$\mathbf{y}_{i3} = \begin{bmatrix} 0 \\ 0 \\ \frac{d}{dt}\left({}^{\mathbf{B}_i}v_r\right)(3) \end{bmatrix}$$

in which $\omega_{z1} = \dot{q}_1$ and $\omega_{z2} = \dot{q}_1 + \dot{q}_2$ hold, whereas $\frac{d}{dt}\left({}^{\mathbf{B}_i}v_r\right)(j) \in \mathbb{R}$ denotes the jth element of $\frac{d}{dt}\left({}^{\mathbf{B}_i}V_r\right) \in \mathbb{R}^3$ and ${}^{\mathbf{B}_i}v_r(j) \in \mathbb{R}$ denotes the jth element of ${}^{\mathbf{B}_i}V_r \in \mathbb{R}^3$ for all $j \in \{1, 3\}$ and $i \in \{1, 2\}$.

Furthermore, by using the parameter estimates rather than the unavailable perfect parameters, the required net force/moment vectors of the two links are designed as

$$ {}^{\mathbf{B}_i}F_r^* = \mathbf{Y}_i\hat{\boldsymbol{\theta}}_i + \mathbf{K}_{si}\left({}^{\mathbf{B}_i}V_r - {}^{\mathbf{B}_i}V\right), \quad i = 1, 2 \tag{3.36}$$

where $\hat{\boldsymbol{\theta}}_i \in \mathbb{R}^3$ denotes the estimate of $\boldsymbol{\theta}_i \in \mathbb{R}^3$ and $\mathbf{K}_{si} \in \mathbb{R}^{3\times 3}$ is a symmetric positive-definite matrix.

Equations (3.33) and (3.36) show how the model based control is designed. First, the model based term with perfect parameters is formed by substituting the required linear/angular velocity vector and its time-derivative for the actual linear/angular velocity vector and its time-derivative in (3.27). Second, the model based term is expressed in a linear parametrization form as appeared in (3.33). Then, the perfect parameter vector is replaced by its estimate to form the model based feedforward compensation term that is added to the velocity feedback term in the control equation (3.36). Finally, these parameter estimates are updated by using the $\mathcal{P}$ function in Definition 2.11, as will be seen below.

Define

$$\mathbf{s}_i = \mathbf{Y}_i^T\left({}^{\mathbf{B}_i}V_r - {}^{\mathbf{B}_i}V\right). \tag{3.37}$$

The parameter estimates in $\hat{\boldsymbol{\theta}}_i \in \mathbb{R}^3$ are updated by using the $\mathcal{P}$ function in Definition 2.11 as

$$\hat{\theta}_{i\gamma} = \mathcal{P}\left(s_{i\gamma}, \rho_{i\gamma}, \underline{\theta}_{i\gamma}, \overline{\theta}_{i\gamma}, t\right) \tag{3.38}$$

for $i = 1, 2$ and $\gamma = 1, 2, 3$, where $\hat{\theta}_{i\gamma}$ denotes the γth element of $\hat{\boldsymbol{\theta}}_i \in \mathbb{R}^3$, $s_{i\gamma}$ denotes the γth element of $\mathbf{s}_i$ defined by (3.37), $\rho_{i\gamma} > 0$ is a parameter update gain, and $\underline{\theta}_{i\gamma} \in \mathbb{R}$ and $\overline{\theta}_{i\gamma} \in \mathbb{R}$ denote the lower and upper bounds of $\theta_{i\gamma}$—the γth element of $\boldsymbol{\theta}_i \in \mathbb{R}^3$ defined by (3.34).

Required Force/Moment Vectors

After the required net force/moment vector ${}^{\mathbf{B}_i}F_r^* \in \mathbb{R}^3$ being obtained from (3.36) for all $i \in \{1, 2\}$, three new design vectors, namely the required force/moment vectors in frames $\{\mathbf{B}_1\}$, $\{\mathbf{T}_2\}$, and $\{\mathbf{B}_2\}$ (at the three *cutting points*), are obtained in line with (3.30)–(3.32) as

$$ {}^{\mathbf{B}_2}F_r = {}^{\mathbf{B}_2}F_r^* \tag{3.39}$$

$$ {}^{\mathbf{T}_2}F_r = {}^{\mathbf{T}_2}\mathbf{U}_{\mathbf{B}_2}\,{}^{\mathbf{B}_2}F_r \tag{3.40}$$

$$ {}^{\mathbf{B}_1}F_r = {}^{\mathbf{B}_1}F_r^* + {}^{\mathbf{B}_1}\mathbf{U}_{\mathbf{T}_2}\,{}^{\mathbf{T}_2}F_r \tag{3.41}$$

with ${}^{\mathbf{B}_1}\mathbf{U}_{\mathbf{T}_2}$ and ${}^{\mathbf{T}_2}\mathbf{U}_{\mathbf{B}_2}$ being defined by (3.11) and (3.12), respectively.

3.4.3 Virtual Stability

In this subsection, it will be shown that each of the two links combined with its control equations qualifies to be *virtually stable*. The results are presented by a lemma followed by a theorem.

Lemma 3.1. *Consider the ith link, $i = 1, 2$, described by (3.27) and combined with its respective control equation (3.36) and with the parameter adaptation (3.37) and (3.38). If a non-negative accompanying function for this link is chosen as*

$$\nu_i = \frac{1}{2}\left[\left({}^{\mathbf{B}_i}V_r - {}^{\mathbf{B}_i}V\right)^T \mathbf{M}_{\mathbf{B}_i}\left({}^{\mathbf{B}_i}V_r - {}^{\mathbf{B}_i}V\right) + \sum_{\gamma=1}^{3}\left(\theta_{i\gamma} - \hat{\theta}_{i\gamma}\right)^2 / \rho_{i\gamma}\right] \geqslant 0 \tag{3.42}$$

then it follows that

$$\begin{aligned}\dot{\nu}_i \leqslant &-\left({}^{\mathbf{B}_i}V_r - {}^{\mathbf{B}_i}V\right)^T \mathbf{K}_{si}\left({}^{\mathbf{B}_i}V_r - {}^{\mathbf{B}_i}V\right) \\ &+\left({}^{\mathbf{B}_i}V_r - {}^{\mathbf{B}_i}V\right)^T\left({}^{\mathbf{B}_i}F_r^* - {}^{\mathbf{B}_i}F^*\right)\end{aligned} \tag{3.43}$$

holds.

Proof: Subtracting (3.27) from (3.36) yields

$$\begin{aligned}{}^{\mathbf{B}_i}F_r^* - {}^{\mathbf{B}_i}F^* = &\mathbf{M}_{\mathbf{B}_i}\left[\frac{d}{dt}\left({}^{\mathbf{B}_i}V_r\right) - \frac{d}{dt}\left({}^{\mathbf{B}_i}V\right)\right] + \mathbf{C}_{\mathbf{B}_i}\left({}^{\mathbf{B}_i}V_r - {}^{\mathbf{B}_i}V\right) \\ &+\mathbf{K}_{si}\left({}^{\mathbf{B}_i}V_r - {}^{\mathbf{B}_i}V\right) - \mathbf{Y}_i(t)\left(\boldsymbol{\theta}_i - \hat{\boldsymbol{\theta}}_i\right), \quad i = 1, 2.\end{aligned} \tag{3.44}$$

In view of (3.44), the time derivative of (3.42) is

$$\begin{aligned}\dot{\nu}_i = &\left({}^{\mathbf{B}_i}V_r - {}^{\mathbf{B}_i}V\right)^T \mathbf{M}_{\mathbf{B}_i}\left[\frac{d}{dt}\left({}^{\mathbf{B}_i}V_r\right) \quad \frac{d}{dt}\left({}^{\mathbf{B}_i}V\right)\right] \\ &-\sum_{\gamma=1}^{3}\left(\theta_{i\gamma} - \hat{\theta}_{i\gamma}\right)\frac{\dot{\hat{\theta}}_{i\gamma}}{\rho_{i\gamma}} \\ = &-\left({}^{\mathbf{B}_i}V_r - {}^{\mathbf{B}_i}V\right)^T \mathbf{K}_{si}\left({}^{\mathbf{B}_i}V_r - {}^{\mathbf{B}_i}V\right) \\ &+\left({}^{\mathbf{B}_i}V_r - {}^{\mathbf{B}_i}V\right)^T\left({}^{\mathbf{B}_i}F_r^* - {}^{\mathbf{B}_i}F^*\right) \\ &+\sum_{\gamma=1}^{3}\left(\theta_{i\gamma} - \hat{\theta}_{i\gamma}\right)\left[\mathbf{y}_{i\gamma}(t)^T\left({}^{\mathbf{B}_i}V_r - {}^{\mathbf{B}_i}V\right) - \frac{\dot{\hat{\theta}}_{i\gamma}}{\rho_{i\gamma}}\right]\end{aligned} \tag{3.45}$$

where $\mathbf{y}_{i\gamma} \in \mathbb{R}^{3\times 1}$ denotes the γth column of $\mathbf{Y}_i \in \mathbb{R}^{3\times 3}$ defined by (3.35). In (3.45), $\left({}^{\mathbf{B}_i}V_r - {}^{\mathbf{B}_i}V\right)^T \mathbf{C}_{\mathbf{B}_i} \left({}^{\mathbf{B}_i}V_r - {}^{\mathbf{B}_i}V\right) = 0$ is used, since $\mathbf{C}_{\mathbf{B}_i}$ defined by (3.29) is skew-symmetric.

In view of (3.37), (3.38), and Lemma 2.9, it follows that

$$\left(\theta_{i\gamma} - \hat{\theta}_{i\gamma}\right) \left[\mathbf{y}_{i\gamma}(t)^T \left({}^{\mathbf{B}_i}V_r - {}^{\mathbf{B}_i}V\right) - \frac{\dot{\hat{\theta}}_{i\gamma}}{\rho_{i\gamma}}\right] \leqslant 0 \tag{3.46}$$

holds for all $i \in \{1, 2\}$ and $\gamma \in \{1, 3\}$. Substituting (3.46) into (3.45) yields (3.43). ■

Theorem 3.1. *The ith link, $i = 1, 2$, described by (3.17), (3.27), (3.30), and (3.31), combined with its respective control equations (3.24), (3.36), (3.39), and (3.41) and with the parameter adaptation (3.37) and (3.38), is virtually stable with its affiliated vector ${}^{\mathbf{B}_i}V_r - {}^{\mathbf{B}_i}V$ being a virtual function in both L_2 and L_∞, in the sense of Definition 2.17.*

Proof: It follows from (3.17), (3.24), (3.30), (3.31), (3.39), (3.41), and (2.85) in Definition 2.16 that

$$\begin{aligned}
&\left({}^{\mathbf{B}_1}V_r - {}^{\mathbf{B}_1}V\right)^T \left({}^{\mathbf{B}_1}F_r^* - {}^{\mathbf{B}_1}F^*\right) \\
&\quad = \left({}^{\mathbf{B}_1}V_r - {}^{\mathbf{B}_1}V\right)^T \left({}^{\mathbf{B}_1}F_r - {}^{\mathbf{B}_1}F\right) \\
&\qquad - \left({}^{\mathbf{B}_1}V_r - {}^{\mathbf{B}_1}V\right)^T {}^{\mathbf{B}_1}\mathbf{U}_{\mathbf{T}_2} \left({}^{\mathbf{T}_2}F_r - {}^{\mathbf{T}_2}F\right) \\
&\quad = \left({}^{\mathbf{B}_1}V_r - {}^{\mathbf{B}_1}V\right)^T \left({}^{\mathbf{B}_1}F_r - {}^{\mathbf{B}_1}F\right) - \left({}^{\mathbf{T}_2}V_r - {}^{\mathbf{T}_2}V\right)^T \left({}^{\mathbf{T}_2}F_r - {}^{\mathbf{T}_2}F\right) \\
&\quad = p_{\mathbf{B}_1} - p_{\mathbf{T}_2}
\end{aligned} \tag{3.47}$$

$$\begin{aligned}
&\left({}^{\mathbf{B}_2}V_r - {}^{\mathbf{B}_2}V\right)^T \left({}^{\mathbf{B}_2}F_r^* - {}^{\mathbf{B}_2}F^*\right) \\
&\quad = \left({}^{\mathbf{B}_2}V_r - {}^{\mathbf{B}_2}V\right)^T \left({}^{\mathbf{B}_2}F_r - {}^{\mathbf{B}_2}F\right) \\
&\quad = p_{\mathbf{B}_2}
\end{aligned} \tag{3.48}$$

hold, where $p_{\mathbf{B}_1}$, $p_{\mathbf{T}_1}$, and $p_{\mathbf{B}_1}$ denote three *virtual power flows* at the three *cutting points*.

Substituting (3.47) and (3.48) into (3.43) yields

$$\dot{\nu}_1 = -\left({}^{\mathbf{B}_1}V_r - {}^{\mathbf{B}_1}V\right)^T \mathbf{K}_{s1} \left({}^{\mathbf{B}_1}V_r - {}^{\mathbf{B}_1}V\right) + p_{\mathbf{B}_1} - p_{\mathbf{T}_2} \tag{3.49}$$

$$\dot{\nu}_2 = -\left({}^{\mathbf{B}_2}V_r - {}^{\mathbf{B}_2}V\right)^T \mathbf{K}_{s2} \left({}^{\mathbf{B}_2}V_r - {}^{\mathbf{B}_2}V\right) + p_{\mathbf{B}_2}. \tag{3.50}$$

Consider the fact that link 1 has one *driving cutting point* associated with frame $\{\mathbf{T}_2\}$ and one *driven cutting point* associated with frame $\{\mathbf{B}_1\}$. Also, link 2 has only one *driven cutting point* associated with frame $\{\mathbf{B}_2\}$. Thus, the proof is completed, in view of (3.42), (3.49), (3.50), and Definition 2.17. ■

3.5 Dynamics and Control of the Two Joints

With the same structure as that of the previous section, three subsections, namely joint kinematics and dynamic, model-based joint control, and *virtual stability*, are presented in this section.

3.5.1 Joint Kinematics and Dynamics

Joint Kinematics

The joint kinematics is defined by (3.16) and (3.18) or by (3.19) in general.

Joint Dynamics

The dynamics of the two joints can be written as

$$I_{mi}\ddot{q}_i + k_{ci}\text{sign}(\dot{q}_i) \overset{def}{=} \tau_i^* = \tau_i - \tau_{ai} \tag{3.51}$$

for $i = 1, 2$, where $I_{mi} \in \mathbb{R}$ denotes the moment of inertia of the ith joint, $k_{ci} \in \mathbb{R}$ denotes the Coulomb friction coefficient, $\tau_i^* \in \mathbb{R}$ denotes the net torque applied to the ith joint, $\tau_i \in \mathbb{R}$ denotes the control torque of the ith joint, and

$$\tau_{ai} = \mathbf{z}^T {}^{\mathbf{B}_i}F, \quad i = 1, 2 \tag{3.52}$$

denotes the output torque of the ith joint toward the links, which equals in magnitude the torque projected from the links.

3.5.2 Joint Control Equations

Required Velocities

The relationships between the required velocities of the joints and the required velocities of the links are defined by (3.23) and (3.25) or by (3.26) in general.

Joint Control Equations

Two new design variables, namely the required net torques, for the two joints are specified by

$$\tau_{ir}^* = \mathbf{Y}_{ai}(t)\hat{\boldsymbol{\theta}}_{ai} + k_{ai}\left(\dot{q}_{ir} - \dot{q}_i\right), \quad i = 1, 2 \tag{3.53}$$

with

$$\mathbf{Y}_{ai}(t) = \begin{bmatrix} y_{ai1}(t) & y_{ai2}(t) \end{bmatrix} = \begin{bmatrix} \ddot{q}_{ir} & \text{sign}(\dot{q}_{ir}) \end{bmatrix} \in \mathbb{R}^{1\times 2} \tag{3.54}$$

$$\boldsymbol{\theta}_{ai} = \begin{bmatrix} \theta_{ai1} \\ \theta_{ai2} \end{bmatrix} = \begin{bmatrix} I_{mi} \\ k_{ci} \end{bmatrix} \in \mathbb{R}^2. \tag{3.55}$$

In (3.53), $\hat{\boldsymbol{\theta}}_{ai} \in \mathbb{R}^2$ denotes the estimate of $\boldsymbol{\theta}_{ai} \in \mathbb{R}^2$ defined by (3.55), and $k_{ai} > 0$ denotes a feedback gain.

Two other design variables, namely the required output torques at the two joints, are then specified by

$$\tau_{air} = \mathbf{z}^T {}^{\mathbf{B}_i}F_r, \quad i = 1, 2. \tag{3.56}$$

Finally, the two control torques are designed as

$$\tau_i = \tau_{ir}^* + \tau_{air}, \quad i = 1, 2 \tag{3.57}$$

with τ_{ir}^* and τ_{air} being specified by (3.53) and (3.56), respectively.

The estimates of the two parameters in $\boldsymbol{\theta}_{ai} \in \mathbb{R}^2$ need to be updated. Define

$$\mathbf{s}_{ai} = \mathbf{Y}_{ai}^T (\dot{q}_{1r} - \dot{q}_1). \tag{3.58}$$

The parameter estimates in $\hat{\boldsymbol{\theta}}_{ai} \in \mathbb{R}^2$ are updated by using the $\mathcal{P}$ function in Definition 2.11 as

$$\hat{\theta}_{ai\gamma} = \mathcal{P}\left(s_{ai\gamma}, \rho_{ai\gamma}, \underline{\theta}_{ai\gamma}, \overline{\theta}_{ai\gamma}, t\right) \tag{3.59}$$

for $i = 1, 2$ and $\gamma = 1, 2$, where $s_{ai\gamma}$ denotes the γth element of $\mathbf{s}_{ai}$ defined by (3.58), $\rho_{ai\gamma} > 0$ is a parameter update gain, and $\underline{\theta}_{ai\gamma} \in \mathbb{R}$ and $\overline{\theta}_{ai\gamma} \in \mathbb{R}$ denote the lower and upper bounds of $\theta_{ai\gamma}$ ($\theta_{ai1} = I_{mi}$ and $\theta_{ai2} = k_{ci}$).

3.5.3 Virtual Stability

In this subsection, it will be shown that each of the two joints combined with its respective control equations qualifies to be *virtually stable*.

Theorem 3.2. *The* i*th joint,* $i = 1, 2$*, described by (3.16), (3.18), (3.32), (3.51), and (3.52), combined with the control equations (3.23), (3.25), (3.40), and (3.53)–(3.57) and with the parameter adaptation (3.58) and (3.59), is virtually stable with its affiliated variable* $\dot{q}_{ir} - \dot{q}_i$ *being a virtual function in both* L_2 *and* L_∞*, in the sense of Definition 2.17.*

Proof: Subtracting (3.51) from (3.53) and using (3.54) and (3.55) yields

$$\begin{aligned} \tau_{ir}^* - \tau_i^* &= I_{mi}(\ddot{q}_{ir} - \ddot{q}_i) + k_{ci}(\mathrm{sign}(\dot{q}_{ir}) - \mathrm{sign}(\dot{q}_i)) \\ &\quad + k_{ai}(\dot{q}_{ir} - \dot{q}_i) - \mathbf{Y}_{ai}(t)\left(\boldsymbol{\theta}_{ai} - \hat{\boldsymbol{\theta}}_{ai}\right), \quad i = 1, 2. \end{aligned} \tag{3.60}$$

A non-negative accompanying function for the ith joint is chosen as

$$\nu_{ai} = \frac{1}{2}\left[I_{mi}(\dot{q}_{ir} - \dot{q}_i)^2 + \sum_{\gamma=1}^{2}\left(\theta_{ai\gamma} - \hat{\theta}_{ai\gamma}\right)^2 / \rho_{ai\gamma}\right] \geqslant 0, \quad i = 1, 2. \tag{3.61}$$

In view of (3.60), the time derivative of (3.61) becomes

$$\begin{aligned}\dot{\nu}_{ai} &= (\dot{q}_{ir} - \dot{q}_i) I_{mi} (\ddot{q}_{ir} - \ddot{q}_i) - \sum_{\gamma=1}^{2} \left(\theta_{ai\gamma} - \hat{\theta}_{ai\gamma}\right) \frac{\dot{\hat{\theta}}_{ai\gamma}}{\rho_{ai\gamma}} \\ &= -k_{ai} (\dot{q}_{ir} - \dot{q}_i)^2 - k_{ci} [\mathrm{sign}(\dot{q}_{ir}) - \mathrm{sign}(\dot{q}_i)] (\dot{q}_{ir} - \dot{q}_i) \\ &\quad + \sum_{\gamma=1}^{2} \left(\theta_{ai\gamma} - \hat{\theta}_{ai\gamma}\right) \left[y_{ai\gamma}(t) - \frac{\dot{\hat{\theta}}_{ai\gamma}}{\rho_{ai\gamma}}\right] \\ &\quad + (\dot{q}_{ir} - \dot{q}_i) (\tau_{ir}^* - \tau_i^*) \end{aligned} \tag{3.62}$$

for all $i \in \{1,2\}$ and $\gamma \in \{1,2\}$.

From (3.58), (3.59), and Lemma 2.9, it follows that

$$\left(\theta_{ai\gamma} - \hat{\theta}_{ai\gamma}\right) \left[y_{ai\gamma}(t) - \frac{\dot{\hat{\theta}}_{ai\gamma}}{\rho_{ai\gamma}}\right] \leqslant 0 \tag{3.63}$$

holds for all $i \in \{1,2\}$ and $\gamma \in \{1,2\}$. Meanwhile, the following relationship

$$- k_{ci} [\mathrm{sign}(\dot{q}_{ir}) - \mathrm{sign}(\dot{q}_i)] (\dot{q}_{ir} - \dot{q}_i) \leqslant 0 \tag{3.64}$$

holds for $k_{ci} > 0$ with $i \in \{1,2\}$.

Substituting (3.57) into (3.51) yields

$$(\dot{q}_{ir} - \dot{q}_i) (\tau_{ir}^* - \tau_i^*) = - (\dot{q}_{ir} - \dot{q}_i) (\tau_{air} - \tau_{ai}) \tag{3.65}$$

for all $i \in \{1,2\}$. Furthermore, it follows from (3.16), (3.18), (3.23), (3.25), (3.32), (3.40), (3.52), (3.56), and Definition 2.16 that

$$\begin{aligned} &- (\dot{q}_{1r} - \dot{q}_1) (\tau_{a1r} - \tau_{a1}) \\ &\quad = - (\dot{q}_{1r} - \dot{q}_1) \mathbf{z}^T \left({}^{\mathbf{B}_1}F_r - {}^{\mathbf{B}_1}F\right) \\ &\quad = - \left({}^{\mathbf{B}_1}V_r - {}^{\mathbf{B}_1}V\right)^T \left({}^{\mathbf{B}_1}F_r - {}^{\mathbf{B}_1}F\right) \\ &\quad = -p_{\mathbf{B}_1} \end{aligned} \tag{3.66}$$

$$\begin{aligned} &- (\dot{q}_{2r} - \dot{q}_2) (\tau_{a2r} - \tau_{a2}) \\ &\quad = - (\dot{q}_{2r} - \dot{q}_2) \mathbf{z}^T \left({}^{\mathbf{B}_2}F_r - {}^{\mathbf{B}_2}F\right) \\ &\quad = - \left[\left({}^{\mathbf{B}_2}V_r - {}^{\mathbf{B}_2}V\right) - {}^{\mathbf{T}_2}\mathbf{U}_{\mathbf{B}_2}^T \left({}^{\mathbf{T}_2}V_r - {}^{\mathbf{T}_2}V\right)\right]^T \left({}^{\mathbf{B}_2}F_r - {}^{\mathbf{B}_2}F\right) \\ &\quad = - \left({}^{\mathbf{B}_2}V_r - {}^{\mathbf{B}_2}V\right)^T \left({}^{\mathbf{B}_2}F_r - {}^{\mathbf{B}_2}F\right) \\ &\qquad + \left({}^{\mathbf{T}_2}V_r - {}^{\mathbf{T}_2}V\right)^T {}^{\mathbf{T}_2}\mathbf{U}_{\mathbf{B}_2} \left({}^{\mathbf{B}_2}F_r - {}^{\mathbf{B}_2}F\right) \\ &\quad = -p_{\mathbf{B}_2} + p_{\mathbf{T}_2} \end{aligned} \tag{3.67}$$

hold. Finally, substituting (3.63)–(3.67) into (3.62) yields

$$\dot{\nu}_{a1} \leqslant -k_{s1} (\dot{q}_{1r} - \dot{q}_1)^2 - p_{\mathbf{B}_1} \tag{3.68}$$

$$\dot{\nu}_{a2} \leqslant -k_{s2} (\dot{q}_{2r} - \dot{q}_2)^2 + p_{\mathbf{T}_2} - p_{\mathbf{B}_2}. \tag{3.69}$$

Consider that fact that joint 1 has only one *driving cutting point* associated with frame $\{\mathbf{B}_1\}$. Also, joint 2 has one *driving cutting point* associated with frame $\{\mathbf{B}_2\}$ and one *driven cutting point* associated with frame $\{\mathbf{T}_2\}$. Thus, the proof is completed, in view of (3.61), (3.68), (3.69), and Definition 2.17. ■

3.6 L_2 and L_∞ Stability of the Entire System

Following the same procedure leading to Theorem 2.1, a non-negative function of the entire system is chosen as

$$\nu = \nu_1 + \nu_{a1} + \nu_2 + \nu_{a2}. \tag{3.70}$$

In view of (3.49), (3.50), (3.68), and (3.69), it follows that

$$\dot{\nu} \leqslant -\sum_{i=1}^{2} \left({}^{\mathbf{B}_i}V_r - {}^{\mathbf{B}_i}V\right)^T \mathbf{K}_{si} \left({}^{\mathbf{B}_i}V_r - {}^{\mathbf{B}_i}V\right) - \sum_{i=1}^{2} k_{ai} \left(\dot{q}_{ir} - \dot{q}_i\right)^2 \tag{3.71}$$

holds. Note that all *virtual power flows* are cancelled out due to the fact that the two *virtual power flows* associated with a common frame at a common *cutting point* take the same magnitude with opposite signs.

Consequently, it follows from (3.42), (3.61), (3.70), (3.71), and Lemma 2.3 that

$$ {}^{\mathbf{B}_i}V_r - {}^{\mathbf{B}_i}V \in L_2 \bigcap L_\infty \tag{3.72}$$

$$\dot{q}_{ir} - \dot{q}_i \in L_2 \bigcap L_\infty \tag{3.73}$$

hold for all $i \in \{1, 2\}$, leading to

$$\dot{q}_{id} - \dot{q}_i \in L_2 \bigcap L_\infty \tag{3.74}$$

$$q_{id} - q_i \in L_2 \bigcap L_\infty \tag{3.75}$$

for all $i \in \{1, 2\}$, in view of (3.20), (3.73), and Lemma 2.4.

3.7 Forward Dynamics

Forward dynamics define the relationship between the joint control torques and the joint accelerations.

In terms of the definition of $\mathcal{J}_2$ in (3.19), it follows from (3.30)–(3.32), (3.51), and (3.52) that

$$\begin{bmatrix} \tau_1 \\ \tau_2 \end{bmatrix} = \mathcal{J}_2^T \begin{bmatrix} \tau_1^* \\ \tau_2^* \\ {}^{\mathbf{B}_1}F^* \\ {}^{\mathbf{B}_2}F^* \end{bmatrix} \tag{3.76}$$

holds.

In view of (3.19), (3.27), (3.51), and (3.76), the forward dynamics can be expressed as

$$\mathcal{J}_2^T \mathcal{M}_e \mathcal{J}_2 \begin{bmatrix} \ddot{q}_1 \\ \ddot{q}_2 \end{bmatrix} + \mathcal{J}_2^T \mathcal{M}_e \dot{\mathcal{J}}_2 \begin{bmatrix} \dot{q}_1 \\ \dot{q}_2 \end{bmatrix} + \mathcal{J}_2^T \mathcal{C}_e = \begin{bmatrix} \tau_1 \\ \tau_2 \end{bmatrix} \tag{3.77}$$

where

$$\mathcal{M}_e = \text{diag}\left(I_{m1}, I_{m2}, \mathbf{M}_{\mathbf{B}_1}, \mathbf{M}_{\mathbf{B}_2}\right)$$
$$\mathcal{C}_e = \begin{bmatrix} k_{c1}\text{sign}(\dot{q}_1) \\ k_{c2}\text{sign}(\dot{q}_2) \\ \mathbf{C}_{\mathbf{B}_1}{}^{\mathbf{B}_1}V \\ \mathbf{C}_{\mathbf{B}_2}{}^{\mathbf{B}_2}V \end{bmatrix}.$$

3.8 Asymptotic Stability

Given bounded reference signals q_{id}, $\dot{q}_{id}$, and $\ddot{q}_{id}$ for all $i \in \{1, 2\}$, it follows from (3.36), (3.39)–(3.41), (3.53)–(3.57), and (3.73) that

$$\tau_i \in L_\infty \tag{3.78}$$

holds for all $i \in \{1, 2\}$. In view of (3.78) and the forward dynamics (3.77), the boundedness of $\ddot{q}_i$ can be ensured for all $i \in \{1, 2\}$, leading to the boundedness of $\ddot{q}_{id} - \ddot{q}_i$ for all $i \in \{1, 2\}$. Therefore, it follows from Lemma 2.8 that the asymptotic stability in the sense of

$$\dot{q}_{id} - \dot{q}_i \to 0 \tag{3.79}$$
$$q_{id} - q_i \to 0 \tag{3.80}$$

can be ensured for all $i \in \{1, 2\}$.

Remark 3.1. For direct drive robots, where $I_{mi} = 0$ is used, the result of (3.72) is still valid, which ensures (3.73) and all other stability results developed in this chapter.

3.9 Discussions and Concluding Remarks

In this chapter, the *virtual decomposition control* (VDC) approach has been demonstrated in detail with respect to a typical robot having two degrees of freedom. After the original system is introduced and the *virtual decomposition* is applied, the robot kinematics that governs the transformations from the independent joint velocities to the link velocities is presented. The same kinematic transformations apply also to the required velocities—the velocities that are required to accomplish control objectives.

Sections 3.4 and 3.5 constitute the central contents of this chapter. Section 3.4 handles the dynamics and control issue of the two links, and Section 3.5 handles the dynamics and control issue of the two joints. Particularly, the required net force/moment vectors of all links and the required net torques of all joints are computed, along with the computations of the required force/moment vectors in all frames. Parameter adaptations with lower and upper bounds are used to accommodate the parameter uncertainties. Both sections are relatively independent, provided that each link or each joint combined with its respective control equations qualifies to be *virtually stable* in the sense of Definition 2.17. When every subsystem (links and joints) is *virtually stable*, the L_2 and L_∞ stability of the entire robot can be guaranteed by Theorem 2.1.

Whereas the subsystem dynamics based control computation is one feature of the VDC approach, the stability analysis is another feature of the approach. Each subsystem (a rigid link or a joint) is assigned a non-negative accompanying function. Given a subsystem, the time-derivative of the corresponding non-negative accompanying function reveals that the dynamic interactions between the addressed subsystem and the rest of the robot are uniquely represented by the *virtual power flows* at its *cutting points*. At a given *cutting point*, the two *virtual power flows* associated with the two adjacent subsystems take the same magnitude with opposite signs. This fact enables the cancellation of all *virtual power flows* in the time-derivative of the overall non-negative function which is formed by adding all the non-negative accompanying functions assigned to the subsystems.

While not being evident in this chapter, the true advantages of using the VDC approach will be demonstrated in the following chapters. In the next chapter, the VDC for a class of general complex robots will be presented in detail. It will be shown that no matter how complex a robotic system is, the subsystem dynamics based control merely uses the dynamics of rigid bodies and the dynamics of (single-DOF or three-DOF) joints. It will also be shown that the complete control design procedure is very similar to the one presented in this chapter.

4

Virtual Decomposition Control - General Formulation

4.1 Overview

In this section, the general formulation of the *virtual decomposition control* (VDC) approach will be presented, aimed at achieving a full-dynamics-based control on robots with hyper degrees of freedom. At the beginning, a complex robot is virtually "broken" into "pieces" (subsystems) and is represented by a simple oriented graph. Then, the subsystem dynamics based control is applied to make each subsystem qualified to be *virtually stable*, subject to certain geometric and force constraints. The *virtual stability* of every subsystem results in the stability and convergence of the entire robot, in which the concept of *virtual power flow* plays a vital role in representing the dynamic interactions among the subsystems.

The VDC approach guarantees velocity convergences which ensure position control when position errors are incorporated into the required independent velocities.

After these main results are presented, this chapter follows with discussions on the relations of the VDC approach to the passivity theory and to the PID control. This chapter ends with a discussion on the implementation issues.

4.2 Problem Formulation

The definition of a general complex robot to be addressed in this chapter is first given.

Definition 4.1. *A complex robotic system, such as the system illustrated in Figure 4.1, is a base-fixed robot*[1] *comprised of links connected by joints, having possible contacts with the environment.*

[1] A space robot in which the base is floating can be converted to a base-fixed robot by placing a virtual zero-mass six-degree-of-freedom (six-DOF) manipulator between the base of the space robot and the absolute base on the ground [211].

W.-H. Zhu: Virtual Decomposition Control, STAR 60, pp. 63–110.
springerlink.com

The control objective is to find appropriate control actions (forces or torques) in the actuated joints so that the system's independent states asymptotically track their desired trajectories over time, while maintaining the control of the general constraint forces.[2]

A fundamental assumption is used throughout this book:

Assumption 1. *The kinematics of a complex robotic system is known, and there is no singularity in the kinematics.*

4.3 Virtual Decomposition of a Complex Robot

In this chapter, a complex robotic system, such as the system illustrated in Figure 4.1, is being virtually decomposed into n_o *objects* and n_c *open chains* by placing these conceptual *cutting points* defined by Definition 2.13. Both *objects* and *open chains* are to be defined.

4.3.1 The Graph Representation of a Complex Robot

After being virtually decomposed, the original complex robot, as in Figure 4.1, can be represented by a simple oriented graph as in Figure 4.2. In this simple oriented graph, the n_o *objects* and the n_c *open chains* are connected alternatively. An *object* connects with several *open chains* and may or may not be in contact with the environment. An *open chain* connects with at most two *objects* without in touch with the environment. An *object* or an *open chain* is represented by a node in the corresponding simple oriented graph. Each *cutting point* is represented by a directed edge that defines the reference direction of the force/moment vector at the *cutting point* (from the *driving cutting point* to the *driven cutting point*). The reference directions are assigned in such a way that no circular path is formed in an oriented graph. Thus, some nodes become *source nodes* from which the directed edges emanate, and some nodes become *sink nodes* at which the directed edges terminate.

4.3.2 *Open Chains*

The jth *open chain*, $j \in \{1, n_c\}$, comprises $l_j + 1$ rigid links connected by l_j joints, $l_j \geqslant 1$, with two *cutting points* (or one *cutting point* if the jth *open chain* is a *source node*) located at both ends, see Figure 4.3. The $l_j + 1$ rigid links are numbered sequentially from 0 to l_j with joint k, $k \in \{1, l_j\}$, connecting link $k-1$ with link k. Frames $\{\mathbf{B}_j\}$ and $\{\mathbf{T}_j\}$ are located at the two *cutting points* and, therefore, are fixed to link 0 and link l_j, respectively. The force/moment vector in frame $\{\mathbf{B}_j\}$ is exerted (directed) toward the jth *open chain*, and the force/moment vector in frame $\{\mathbf{T}_j\}$ is exerted (directed) away from the jth *open*

[2] The general constraint forces refer to all the forces that do not affect the robot motion.

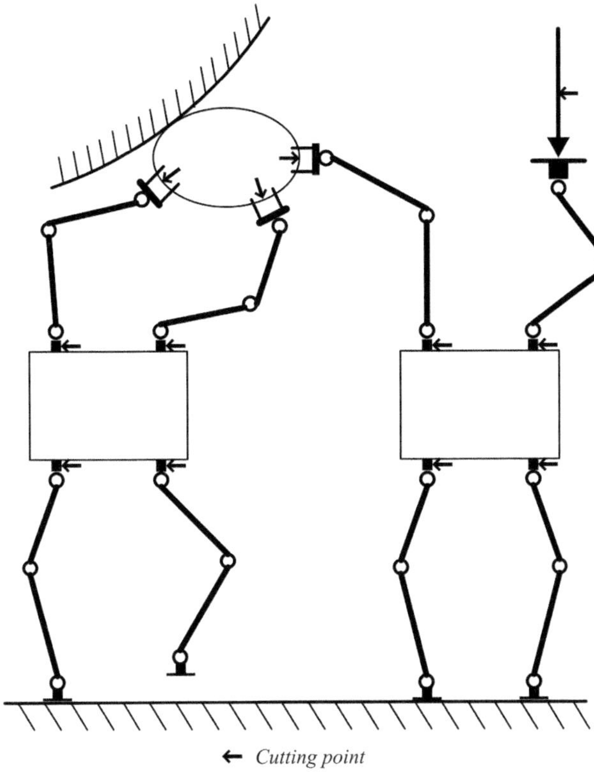

Fig. 4.1. A complex robotic system.

chain. Among the l_j joints, l_{1j} joints are single-degree-of-freedom (single-DOF) prismatic or revolute joints, and l_{3j} joints are three-degree-of-freedom (three-DOF) spherical (ball) joints,[3] subject to

$$l_j = l_{1j} + l_{3j}. \tag{4.1}$$

Three sets are defined:

$\mathcal{C}_{1j}$ consists of the ordinal numbers of the l_{1j} single-DOF joints.
$\mathcal{C}_{3j}$ consists of the ordinal numbers of the l_{3j} three-DOF joints.
$\mathcal{U}_j$ consists of the ordinal numbers of the unactuated joints.

In addition, $2l_j$ subsidiary *cutting points* are placed to the jth *open chain* aimed at virtually isolating the l_j joints from the $l_j{+}1$ rigid links. Accordingly, $2l_j$ frames are assigned to the $2l_j$ subsidiary *cutting points*. For joint k, $k \in \{1, l_j\}$, a pair of frames, denoted as $\{\mathbf{B}_{jk}\}$ and $\{\mathbf{T}_{jk}\}$, is located to the joint with frame $\{\mathbf{B}_{jk}\}$ being fixed to link k and frame $\{\mathbf{T}_{jk}\}$ being fixed to link $k-1$. If the kth joint is a single-DOF joint, then the z axes of both frames $\{\mathbf{B}_{jk}\}$ and $\{\mathbf{T}_{jk}\}$ align with the joint axis.

[3] A universal two-DOF joint can be expressed as two single-DOF revolute joints connected serially.

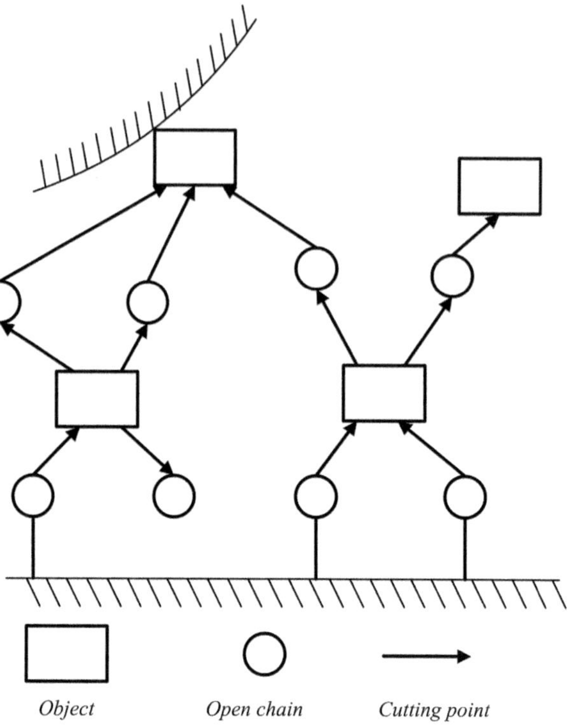

Fig. 4.2. Graphic expression of Figure 4.1.

In view of Definition 2.13, the *cutting point* associated with frame $\{\mathbf{B}_j\}$ is called the *driven cutting point* of link 0. The *cutting point* associated with frame $\{\mathbf{B}_{jk}\}$, $k \in \{1, l_j\}$, is called the *driving cutting point* of joint k and is simultaneously called the *driven cutting point* of link k. The *cutting point* associated with frame $\{\mathbf{T}_{jk}\}$, $k \in \{1, l_j\}$, is called the *driving cutting point* of link $k-1$ and is simultaneously called the *driven cutting point* of joint k. Finally, the *cutting point* associated with frame $\{\mathbf{T}_j\}$ is called the *driving cutting point* of link l_j. It follows that

- The l_jth link has one *driving cutting point* associated with frame $\{\mathbf{T}_j\}$ and one *driven cutting point* associated with frame $\{\mathbf{B}_{jl_j}\}$.
- The kth link has one *driving cutting point* associated with frame $\{\mathbf{T}_{j(k+1)}\}$ and one *driven cutting point* associated with frame $\{\mathbf{B}_{jk}\}$ for all $k \in \{1, l_j - 1\}$.
- The 0th link has one *driving cutting point* associated with frame $\{\mathbf{T}_{j1}\}$ and one *driven cutting point* associated with frame $\{\mathbf{B}_j\}$.
- The kth joint has one *driving cutting point* associated with frame $\{\mathbf{B}_{jk}\}$ and one *driven cutting point* associated with frame $\{\mathbf{T}_{jk}\}$ for all $k \in \{1, l_j\}$.

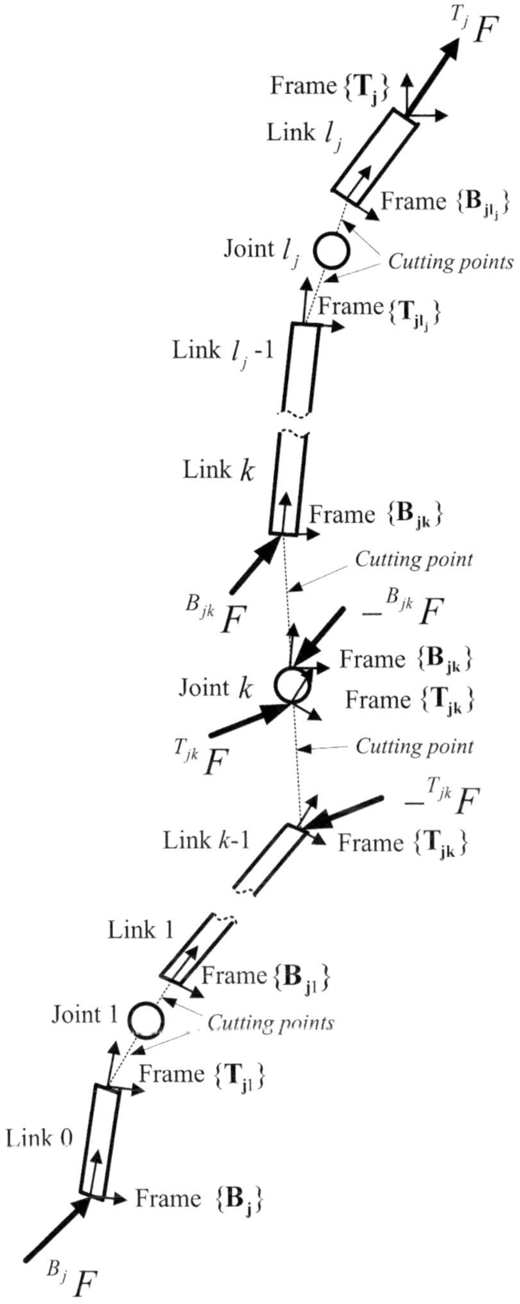

Fig. 4.3. The jth *open chain*.

Two additional assumptions are made for modelling the joints:

Assumption 2. *The motion of the motor rotor is a pure rotation with respect to an inertial frame, or equivalently, the inertial force/torque acts along the joint axis with its magnitude proportional to the acceleration of the relative joint motion.*

Assumption 3. *The mass properties of each motor rotor are neglected to linkage dynamics.*

Remark 4.1. The above two assumptions are mainly made for these robots driven by electrical motors through transmissions with high gear ratios. Due to the high gear ratio, a motor rotor moves much faster along its axis than any other directions. This fact makes Assumption 2, which was used first in [159], valid by extracting the inertial force/torque along the joint axis and neglecting inertial forces/torques in other directions. On the other hand, a high gear ratio implies that the motor torque is much smaller (gear ratio times smaller) than the joint driving torque. Thus, the mass properties of the motor rotor become negligible, which validates Assumption 3.

Remark 4.2. A procedure for the rigorous modelling of high gear ratio transmissions without using Assumptions 2 and 3 will be given in Chapter 6.

Remark 4.3. For direct drive robots, the rotor and stator of each motor are included in the corresponding links. Consequently, Assumptions 2 and 3 are not needed.

4.3.3 *Objects*

An *object* is a rigid body on which the motion and force control specifications are given. An *object* may or may not be in contact with the environment. Figure 4.4 shows the ith *object*, $i \in \{1, n_o\}$, to which frame $\{\mathbf{O}_i\}$ is fixed to describe the motion and force specifications. Frame $\{\mathbf{S}_i\}$ is located at the contact point with the environment. The force/moment vector in frame $\{\mathbf{S}_i\}$ is referenced toward the environment.

Remark 4.4. In case the contact point with the environment is changing, a cluster of body-fixed frames is used so that each frame is located at a possible contact point. Frame $\{\mathbf{S}_i\}$ is/takes the frame located at the contact point at time being.

There are several *cutting points* associated with the ith *object*, $i \in \{1, n_o\}$. If the jth *open chain*, $j \in \{1, n_c\}$, is adjacent to the ith *object* with frame $\{\mathbf{B}_j\}$ being placed at the *cutting point* between them, then this *cutting point* is called a *driving cutting point* of the ith *object*. The force/moment vector in frame $\{\mathbf{B}_j\}$ is exerted from the ith *object* toward the jth *open chain*. Alternatively, if the jth *open chain*, $j \in \{1, n_c\}$, is adjacent to the ith *object* with frame $\{\mathbf{T}_j\}$ being placed at the *cutting point* between them, then this *cutting point* is called a

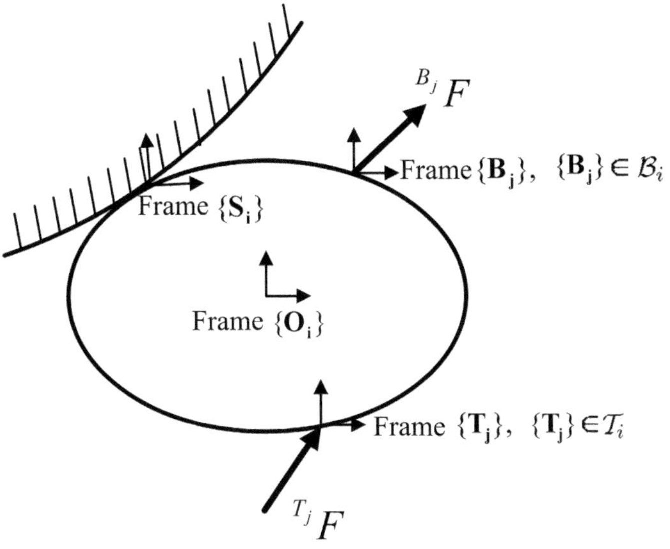

Fig. 4.4. The ith *object*.

driven cutting point of the ith *object*. The force/moment vector in frame $\{\mathbf{T}_j\}$ is exerted from the jth *open chain* toward the ith *object*.

Two sets are therefore defined:

$\mathcal{B}_i$ contains frame $\{\mathbf{B}_j\}$, when frame $\{\mathbf{B}_j\}$ is associated with a *driving cutting point* of the ith *object* for all $j \in \{1, n_c\}$.

$\mathcal{T}_i$ contains frame $\{\mathbf{T}_j\}$, when frame $\{\mathbf{T}_j\}$ is associated with a *driven cutting point* of the ith *object* for all $j \in \{1, n_c\}$.

Let $n_i \geqslant 1$ be the number of elements in set $\mathcal{T}_i$.

Note that $\mathcal{B}_i \bigcap \mathcal{T}_i = \emptyset$ holds, since no circular edge appears in any simple oriented graph.

4.4 Kinematics

In this section, the independent velocity coordinates[4] are given first, followed by the use of velocity mapping matrices that map the independent velocity coordinates to the extended velocity coordinates of all the *objects* and *open chains*. Then, the corresponding required independent velocity coordinates are designed, giving rise to the determination of the required extended velocity coordinates of all subsystems.

[4] The terminology terms "coordinates" and "coordinate vector" are equivalent throughout this book.

4.4.1 Velocity Mapping Matrices

Without loss of generality, it is assumed that a complex robotic system possesses $m > 0$ degrees of freedom of motion in the velocity configuration space. Thus, an independent velocity vector, denoted as $\mathcal{V} \in \mathbb{R}^m$, is defined. This vector expressed in an m-dimensional Euclidean (configuration) space characterizes the velocity profile of the entire system. Generally, this vector can be partitioned into three components after being pre-multiplied by a reordering matrix $\mathcal{R} \in \mathbb{R}^{m \times m}$ as

$$\mathcal{R}\mathcal{V} = [\mathcal{V}_p^T, \mathcal{V}_o^T, \mathcal{V}_u^T]^T \tag{4.2}$$

where $\mathcal{V}_p$ is for position control, $\mathcal{V}_o$ is for optimizations aimed at minimizing certain operational velocity measures, and $\mathcal{V}_u$ is to deal with the force/torque constraints imposed on unactuated joints. Note that $\mathcal{R}^{-1} = \mathcal{R}^T$ holds indefinitely.

Let $\dot{\mathbf{q}}_j \in \mathbb{R}^{(l_{1j}+3l_{3j})}$ be the joint velocity vector of the jth *open chain*. Thus, both ${}^{\mathbf{B}_j}V \in \mathbb{R}^6$ (linear/angular velocity vector of frame $\{\mathbf{B}_j\}$) and $\dot{\mathbf{q}}_j \in \mathbb{R}^{(l_{1j}+3l_{3j})}$ specify the velocity profile of the jth *open chain*.

For the ith *object*, ${}^{\mathbf{O}_i}V \in \mathbb{R}^6$ (linear/angular velocity vector of frame $\{\mathbf{O}_i\}$) specifies the velocity profile of the *object*.

The velocities of all the *objects* and *open chains* do not actually represent the independent velocity coordinates of the original system due to the physical constraints imposed on the *cutting points*. Thus, there exists a velocity mapping matrix, denoted as $\mathcal{J}_e$, which maps the independent velocity coordinate vector $\mathcal{V} \in \mathbb{R}^m$ to the extended velocity coordinates of all the *objects* and *open chains*. The mathematical expression is

$$\mathcal{V}_e = \mathcal{J}_e \mathcal{V} \tag{4.3}$$

where

$$\mathcal{V}_e \stackrel{def}{=} \left[{}^{\mathbf{O}_1}V^T, \cdots, {}^{\mathbf{O}_i}V^T, \cdots, {}^{\mathbf{O}_{n_o}}V^T, \mathcal{V}_{e1}^T, \cdots, \mathcal{V}_{ej}^T, \cdots, \mathcal{V}_{en_c}^T \right]^T \tag{4.4}$$

denotes the extended velocity coordinate vector of the entire robot and

$$\mathcal{V}_{ej} \stackrel{def}{=} \left[\dot{\mathbf{q}}_j^T, {}^{\mathbf{B}_j}V^T, {}^{\mathbf{B}_{j1}}V^T, \cdots, {}^{\mathbf{B}_{jk}}V^T, \cdots, {}^{\mathbf{B}_{jl_j}}V^T \right]^T \tag{4.5}$$

denotes the extended velocity coordinate vector of the jth *open chain* for all $j \in \{1, n_c\}$.

Equation (4.3) represents a conceptual mapping for velocities. Practically, the velocity mappings are computed in two steps. The first step is to map the independent velocity coordinate vector $\mathcal{V} \in \mathbb{R}^m$ to the joint velocity coordinates of the robot by

$$\mathcal{V}_a = \mathcal{J}_a \mathcal{V} \tag{4.6}$$

where

$$\mathcal{V}_a \stackrel{def}{=} \left[\dot{\mathbf{q}}_1^T, \cdots, \dot{\mathbf{q}}_j^T, \cdots, \dot{\mathbf{q}}_{n_c}^T \right]^T. \tag{4.7}$$

The second step is to map $\mathcal{V}_a$ to the velocity coordinates of all rigid bodies as

$$\mathcal{V}_b = \mathcal{J}_{ba}\mathcal{V}_a = \mathcal{J}_b\mathcal{V} \tag{4.8}$$

where

$$\mathcal{V}_b \stackrel{def}{=} \left[{}^{\mathbf{O}_1}V^T, \cdots, {}^{\mathbf{O}_i}V^T, \cdots, {}^{\mathbf{O}_{n_o}}V^T, \mathcal{V}_{b1}^T, \cdots, \mathcal{V}_{bj}^T, \cdots, \mathcal{V}_{bn_c}^T\right]^T \tag{4.9}$$

with

$$\mathcal{V}_{bj} \stackrel{def}{=} \left[{}^{\mathbf{B}_j}V^T, {}^{\mathbf{B}_{j1}}V^T, \cdots, {}^{\mathbf{B}_{jk}}V^T, \cdots, {}^{\mathbf{B}_{jl_j}}V^T\right]^T, \ j \in \{1, n_c\}. \tag{4.10}$$

It follows from (4.6) and (4.8) that

$$\mathcal{J}_b = \mathcal{J}_{ba}\mathcal{J}_a \tag{4.11}$$

holds.

Remark 4.5. Matrix $\mathcal{J}_a$, which transforms the independent velocity coordinates to the joint velocity coordinates, is analogous to the inverse of the Jacobian matrix of a conventional robot manipulator in Cartesian space operations. For a given complex robot, $\mathcal{J}_a$ is formed first to compute $\mathcal{V}_a$. Then, $\mathcal{J}_{ba}$ is formed to compute $\mathcal{V}_b$, starting from *source nodes* and ending with *sink nodes* in a simple oriented graph.

Remark 4.6. The number of dimensions of $\mathcal{V}_a$ plus the number of dimensions of $\mathcal{V}_b$ equals the number of dimensions of $\mathcal{V}_e$.

4.4.2 Required Velocity Mapping

Let $\mathcal{V}_r \in \mathbb{R}^m$ be a design vector, namely the *required independent velocity vector*, subject to

$$\mathcal{R}\mathcal{V}, \stackrel{def}{=} [\mathcal{V}_{pr}^T, \mathcal{V}_{or}^T, \mathcal{V}_{ur}^T]^T \tag{4.12}$$

where $\mathcal{V}_{pr}$ is for position control, $\mathcal{V}_{or}$ is for optimizations, and $\mathcal{V}_{ur}$ is used to handle the force/torque constraints imposed on unactuated joints, in accordance with (4.2).

Consequently, the required vector of $\mathcal{V}_e$ is computed as

$$\mathcal{V}_{er} = \mathcal{J}_e\mathcal{V}_r \tag{4.13}$$

where

$$\mathcal{V}_{er} \stackrel{def}{=} [{}^{\mathbf{O}_1}V_r^T, \cdots, {}^{\mathbf{O}_i}V_r^T, \cdots, {}^{\mathbf{O}_{n_o}}V_r^T, \mathcal{V}_{e1r}^T, \cdots, \mathcal{V}_{ejr}^T, \cdots, \mathcal{V}_{en_cr}^T]^T \tag{4.14}$$

with

$$\mathcal{V}_{ejr} \stackrel{def}{=} \left[\dot{\mathbf{q}}_{jr}^T, {}^{\mathbf{B}_j}V_r^T, {}^{\mathbf{B}_{j1}}V_r^T, \cdots, {}^{\mathbf{B}_{jk}}V_r^T, \cdots, {}^{\mathbf{B}_{jl_j}}V_r^T\right]^T \tag{4.15}$$

for all $j \in \{1, n_c\}$. Moreover, the time derivative of (4.13) can be written as

$$\dot{\mathcal{V}}_{er} = \mathcal{J}_e \dot{\mathcal{V}}_r + \dot{\mathcal{J}}_e \mathcal{V}_r. \tag{4.16}$$

Alternatively, the required vectors of $\mathcal{V}_a$ and $\mathcal{V}_b$ are computed as

$$\mathcal{V}_{ar} = \mathcal{J}_a \mathcal{V}_r \tag{4.17}$$

$$\mathcal{V}_{br} = \mathcal{J}_{ba} \mathcal{V}_{ar} = \mathcal{J}_b \mathcal{V}_r \tag{4.18}$$

where

$$\mathcal{V}_{ar} \overset{def}{=} \left[\dot{\mathbf{q}}_{1r}^T, \cdots, \dot{\mathbf{q}}_{jr}^T, \cdots, \dot{\mathbf{q}}_{n_c r}^T\right]^T \tag{4.19}$$

$$\mathcal{V}_{br} \overset{def}{=} \left[{}^{\mathbf{O}_1}V_r^{\ T}, \cdots, {}^{\mathbf{O}_i}V_r^{\ T}, \cdots, {}^{\mathbf{O}_{n_o}}V_r^{\ T}, \mathcal{V}_{b1r}^T, \cdots, \mathcal{V}_{bjr}^T, \cdots, \mathcal{V}_{bn_c r}^T\right]^T \tag{4.20}$$

with

$$\mathcal{V}_{bjr} \overset{def}{=} \left[{}^{\mathbf{B}_j}V_r^{\ T}, {}^{\mathbf{B}_{j1}}V_r^{\ T}, \cdots, {}^{\mathbf{B}_{jk}}V_r^{\ T}, \cdots, {}^{\mathbf{B}_{jl_j}}V_r^{\ T}\right]^T \tag{4.21}$$

for all $j \in \{1, n_c\}$.

Remark 4.7. It can be found from (4.13) that the required velocity vectors are subject to the same constraints being applied to the velocity vectors themselves. This observation validates (2.86).

4.5 Dynamics and Control of *Objects*

The *virtual decomposition control* (VDC) approach allows the dynamics and control issue of each *object* to be handled independently, provided that each *object* combined with its respective control equations qualifies to be *virtually stable* in the sense of Definition 2.17. This section comprises three parts, namely kinematics and dynamics, control equations, and *virtual stability*.

4.5.1 Kinematics and Dynamics

In this subsection, both kinematics and dynamics of each *object* are presented.

Velocity Transformations

Consider the ith *object* in Figure 4.4. Let $0 \leqslant n_{fi} < 6$ be the number of independent constraints applied to the *object* through its contact with the environment.

Besides all frames that are defined in Figure 4.4, a new frame, denoted as $\{\mathbf{S}_{bi}\}$, is defined by having the same origin as frame $\{\mathbf{S}_i\}$ with some axes describing the motion and others the constraints. Thus, the linear/angular velocity vector of frame $\{\mathbf{S}_i\}$ with the ith *object* can be expressed as

$${}^{\mathbf{S}_i}V = \mathrm{diag}({}^{\mathbf{S}_i}\mathbf{R}_{\mathbf{S}_{bi}}, {}^{\mathbf{S}_i}\mathbf{R}_{\mathbf{S}_{bi}})\mathbf{T}_{si}\boldsymbol{\chi}_i \tag{4.22}$$

where $\mathbf{T}_{si} \in \mathbb{R}^{6\times(6-n_{fi})}$ is of full column-rank with each column containing a single *one* and five *zeros* and $\chi_i \in \mathbb{R}^{6-n_{fi}}$ denotes the velocity coordinate vector of frame $\{\mathbf{S}_{bi}\}$.

In view of (2.63), the following relationships hold for velocity transformations

$$\begin{aligned} {}^{\mathbf{O}_i}V &= {}^{\mathbf{S}_i}\mathbf{U}_{\mathbf{O}_i}^T\, {}^{\mathbf{S}_i}V \\ &= {}^{\mathbf{T}_j}\mathbf{U}_{\mathbf{O}_i}^T\, {}^{\mathbf{T}_j}V \\ &= {}^{\mathbf{B}_j}\mathbf{U}_{\mathbf{O}_i}^T\, {}^{\mathbf{B}_j}V \end{aligned} \tag{4.23}$$

with $\{\mathbf{B}_j\} \in \mathcal{B}_i$ and $\{\mathbf{T}_j\} \in \mathcal{T}_i$.

Remark 4.8. Note that $\chi_i \in \mathbb{R}^{6-n_{fi}}$ in (4.22) may not necessarily be the independent velocity coordinates of the entire robot, since the number of dimensions for the independent motion can be further reduced by these constraints from the adjacent *open chains*. It is clear that χ_i can be obtained from the independent velocity coordinate vector $\mathcal{V} \in \mathbb{R}^m$ through (4.3), (4.22), and (4.23).

Remark 4.9. According to Remark 4.4, if a cluster of body-fixed frames is used, from which frame $\{\mathbf{S}_i\}$ is selected, then matrix ${}^{\mathbf{S}_i}\mathbf{U}_{\mathbf{O}_i} \in \mathbb{R}^{6\times6}$ in (4.23) is no longer time-invariant.

Rigid Body Dynamics

Referring back to (2.74), the dynamics of the ith *object* are given by

$$\mathbf{M}_{\mathbf{O}_i}\frac{d}{dt}({}^{\mathbf{O}_i}V) + \mathbf{C}_{\mathbf{O}_i}({}^{\mathbf{O}_i}\omega){}^{\mathbf{O}_i}V + \mathbf{G}_{\mathbf{O}_i} = {}^{\mathbf{O}_i}F^* \tag{4.24}$$

where frame $\{\mathbf{O}_i\}$ is substituted for frame $\{\mathbf{A}\}$ and ${}^{\mathbf{O}_i}V \in \mathbb{R}^6$ is obtained from (4.3) or (4.8). On the other hand, the net force/moment vector in frame $\{\mathbf{O}_i\}$ is governed by

$${}^{\mathbf{O}_i}F^* = \sum_{\{\mathbf{T}_j\}\in\mathcal{T}_i} {}^{\mathbf{O}_i}\mathbf{U}_{\mathbf{T}_j}\, {}^{\mathbf{T}_j}F - \sum_{\{\mathbf{B}_j\}\in\mathcal{B}_i} {}^{\mathbf{O}_i}\mathbf{U}_{\mathbf{B}_j}\, {}^{\mathbf{B}_j}F - {}^{\mathbf{O}_i}\mathbf{U}_{\mathbf{S}_i}\, {}^{\mathbf{S}_i}F \tag{4.25}$$

in view of Figure 4.4, where sets $\mathcal{T}_i$ and $\mathcal{B}_i$ are defined in Subsection 4.3.3 and the force/moment transformation matrices are defined in Subsection 2.5.3.

Constraint Forces

In (4.25), the contact/constraint force/moment vector in frame $\{\mathbf{S}_i\}$, denoted as ${}^{\mathbf{S}_i}F \in \mathbb{R}^6$, can be generally written as

$${}^{\mathbf{S}_i}F = \mathrm{diag}({}^{\mathbf{S}_i}\mathbf{R}_{\mathbf{S}_{bi}}, {}^{\mathbf{S}_i}\mathbf{R}_{\mathbf{S}_{bi}})\,(\mathbf{T}_{si}\psi_i + \mathbf{T}_{fi}\varphi_i) \tag{4.26}$$

where $\mathbf{T}_{si} \in \mathbb{R}^{6\times(6-n_{fi})}$ is defined in (4.22) and $\mathbf{T}_{fi} \in \mathbb{R}^{6\times n_{fi}}$ is of full column-rank with each column containing a single *one* and five *zeros*. The selection of $\mathbf{T}_{si} \in \mathbb{R}^{6\times(6-n_{fi})}$ and $\mathbf{T}_{fi} \in \mathbb{R}^{6\times n_{fi}}$ makes sure that $\mathbf{T}_{si}^T\mathbf{T}_{si} = \mathbf{I}_{6-n_{fi}}$, $\mathbf{T}_{fi}^T\mathbf{T}_{fi} = \mathbf{I}_{n_{fi}}$, and $\mathbf{T}_{si}^T\mathbf{T}_{fi} = \mathbf{0}$ hold. In (4.26), $\boldsymbol{\psi}_i \in \mathbb{R}^{6-n_{fi}}$ and $\boldsymbol{\varphi}_i \in \mathbb{R}^{n_{fi}}$ denote two coordinate vectors in the motion and constraint configuration spaces, respectively.

Let $\boldsymbol{\psi}_i \in \mathbb{R}^{6-n_{fi}}$ in (4.26) (the independent tangential force coordinate vector) be expressed in a linear parametrization form as

$$\boldsymbol{\psi}_i = \mathbf{Y}_{si}\boldsymbol{\theta}_{si} \tag{4.27}$$

with $\boldsymbol{\theta}_{si}$ being a parameter vector.

Force/Moment Vectors at the Driven Cutting Points

Let $n_i \geqslant 1$ be the number of elements in set $\mathcal{T}_i$, which equals the number of *driven cutting points* of the ith *object*. Let

$$\mathrm{col}\left({}^{\mathbf{T}_j}F \mid \{\mathbf{T}_j\} \in \mathcal{T}_i\right) \in \mathbb{R}^{6n_i}$$

be an extended vector of the exerting forces/moments at all *driven cutting points* of the ith *object*. Besides the resultant force equation (4.25) which gives six constraints, there must exist a $6(n_i - 1)$ dimensional internal force vector to have $\mathrm{col}\left({}^{\mathbf{T}_j}F \mid \{\mathbf{T}_j\} \in \mathcal{T}_i\right) \in \mathbb{R}^{6n_i}$ fully resolved. Let $\boldsymbol{\eta}_i \in \mathbb{R}^{6(n_i-1)}$ be an internal force vector. It follows from (4.25) that

$$\begin{aligned}\mathrm{col}\left({}^{\mathbf{T}_j}F \mid \{\mathbf{T}_j\} \in \mathcal{T}_i\right) = {} & \mathrm{diag}\left({}^{\mathbf{T}_j}\mathbf{U}_{\mathbf{O}_i} \mid \{\mathbf{T}_j\} \in \mathcal{T}_i\right)\left[\boldsymbol{\Omega}_{mi} \quad \boldsymbol{\Omega}_{fi}\right] \\ & \times \begin{bmatrix} {}^{\mathbf{O}_i}F^* + \sum_{\{\mathbf{B}_j\}\in\mathcal{B}_i} {}^{\mathbf{O}_i}\mathbf{U}_{\mathbf{B}_j}\,{}^{\mathbf{B}_j}F + {}^{\mathbf{O}_i}\mathbf{U}_{\mathbf{S}_i}\,{}^{\mathbf{S}_i}F \\ \boldsymbol{\eta}_i \end{bmatrix}\end{aligned} \tag{4.28}$$

holds, where $\mathrm{diag}\left({}^{\mathbf{T}_j}\mathbf{U}_{\mathbf{O}_i} \mid \{\mathbf{T}_j\} \in \mathcal{T}_i\right) \in \mathbb{R}^{6n_i\times 6n_i}$ is a block diagonal matrix with its diagonal elements being ${}^{\mathbf{T}_j}\mathbf{U}_{\mathbf{O}_i} \in \mathbb{R}^{6\times 6}$ for all $\{\mathbf{T}_j\} \in \mathcal{T}_i$. In (4.28), $\boldsymbol{\Omega}_{mi} \in \mathbb{R}^{6n_i\times 6}$ serves as a load distribution matrix [96, 134], and $\boldsymbol{\Omega}_{fi} \in \mathbb{R}^{6n_i\times 6(n_i-1)}$ serves as an internal force mapping matrix. The two matrices are of full column-rank such that the inverse of $\left[\boldsymbol{\Omega}_{mi}\ \boldsymbol{\Omega}_{fi}\right]$ exists and satisfies

$$\begin{bmatrix} \mathcal{H}_i \\ \boldsymbol{\Psi}_i \end{bmatrix} \stackrel{def}{=} \left[\boldsymbol{\Omega}_{mi}\ \boldsymbol{\Omega}_{fi}\right]^{-1} \tag{4.29}$$

with $\mathcal{H}_i = \underbrace{[\mathbf{I}_6, \cdots, \mathbf{I}_6, \cdots]}_{n_i} \in \mathbb{R}^{6\times 6n_i}$ and $\boldsymbol{\Psi}_i \in \mathbb{R}^{6(n_i-1)\times 6n_i}$. A detailed expression of these matrices will be given in Chapter 8.

Note that premultiplying (4.28) by $\mathcal{H}_i\mathrm{diag}\left({}^{\mathbf{T}_j}\mathbf{U}_{\mathbf{O}_i} \mid \{\mathbf{T}_j\} \in \mathcal{T}_i\right)^{-1}$ yields (4.25).

4.5.2 Control Equations

In this subsection, the model-based control equations are to be presented.

Required Velocity Transformations

In accordance with (4.22), the required linear/angular velocity vector at the contact point can be designed as

$$ {}^{\mathbf{S}_i}V_r = \mathrm{diag}({}^{\mathbf{S}_i}\mathbf{R}_{\mathbf{S}_{bi}}, {}^{\mathbf{S}_i}\mathbf{R}_{\mathbf{S}_{bi}})\mathbf{T}_{si}\boldsymbol{\chi}_{ir} \tag{4.30} $$

where $\boldsymbol{\chi}_{ir} \in \mathbb{R}^{6-n_{fi}}$ denotes the required vector of $\boldsymbol{\chi}_i \in \mathbb{R}^{6-n_{fi}}$.

On the other hand, in order to validate (2.86), the following relationships hold for the required linear/angular velocity vectors as well

$$ \begin{aligned} {}^{\mathbf{O}_i}V_r &= {}^{\mathbf{S}_i}\mathbf{U}_{\mathbf{O}_i}^T\,{}^{\mathbf{S}_i}V_r \\ &= {}^{\mathbf{T}_j}\mathbf{U}_{\mathbf{O}_i}^T\,{}^{\mathbf{T}_j}V_r \\ &= {}^{\mathbf{B}_j}\mathbf{U}_{\mathbf{O}_i}^T\,{}^{\mathbf{B}_j}V_r \end{aligned} \tag{4.31} $$

with $\{\mathbf{B}_j\} \in \mathcal{B}_i$ and $\{\mathbf{T}_j\} \in \mathcal{T}_i$.

Remark 4.10. The required velocity vector $\boldsymbol{\chi}_{ir} \in \mathbb{R}^{6-n_{fi}}$ may not necessarily be an independent design vector. It can be computed from the required independent velocity vector $\mathcal{V}_r \in \mathbb{R}^m$ through (4.13), (4.30), and (4.31).

Required Net Force/Moment Vector

After ${}^{\mathbf{O}_i}V_r$ being obtained from (4.13) and (4.14), a new design vector, namely the required net force/moment vector, is specified as

$$ {}^{\mathbf{O}_i}F_r^* = \mathbf{Y}_{\mathbf{O}_i}\hat{\boldsymbol{\theta}}_{\mathbf{O}_i} + \mathbf{K}_{\mathbf{O}_i}\left({}^{\mathbf{O}_i}V_r - {}^{\mathbf{O}_i}V\right) \tag{4.32} $$

where $\mathbf{K}_{\mathbf{O}_i} \in \mathbb{R}^{6\times 6}$ is a positive-definite gain matrix characterizing the velocity feedback control. The term $\mathbf{Y}_{\mathbf{O}_i}\hat{\boldsymbol{\theta}}_{\mathbf{O}_i}$ denotes the model based feedforward compensation term by using the required velocities, their time derivatives, and the estimated parameters. The regressor matrix $\mathbf{Y}_{\mathbf{O}_i} \in \mathbb{R}^{6\times 13}$ and the parameter vector $\boldsymbol{\theta}_{\mathbf{O}_i} \in \mathbb{R}^{13}$ are defined in (2.78) and further in Appendix A by substituting frame $\{\mathbf{O}_i\}$ for frame $\{\mathbf{A}\}$.

With respect to (4.32), define

$$ \mathbf{s}_{\mathbf{O}_i} = \mathbf{Y}_{\mathbf{O}_i}^T\left({}^{\mathbf{O}_i}V_r - {}^{\mathbf{O}_i}V\right). \tag{4.33} $$

The $\mathcal{P}$ function defined by (2.79) is used to update each of the 13 elements of $\hat{\boldsymbol{\theta}}_{\mathbf{O}_i} \in \mathbb{R}^{13}$ as

$$ \hat{\theta}_{\mathbf{O}_i\gamma} = \mathcal{P}\left(s_{\mathbf{O}_i\gamma}, \rho_{\mathbf{O}_i\gamma}, \underline{\theta}_{\mathbf{O}_i\gamma}, \overline{\theta}_{\mathbf{O}_i\gamma}, t\right), \ \gamma \in \{1, 13\} \tag{4.34} $$

where $\hat{\theta}_{\mathbf{O}_i\gamma}$ denotes the γth element of $\hat{\boldsymbol{\theta}}_{\mathbf{O}_i}$ and $s_{\mathbf{O}_i\gamma}$ denotes the γth element of $\mathbf{s}_{\mathbf{O}_i}$ defined by (4.33), $\rho_{\mathbf{O}_i\gamma} > 0$ is an update gain, and $\underline{\theta}_{\mathbf{O}_i\gamma}$ and $\overline{\theta}_{\mathbf{O}_i\gamma}$ denote the lower and upper bounds of $\theta_{\mathbf{O}_i\gamma}$, the γth element of $\boldsymbol{\theta}_{\mathbf{O}_i} \in \mathbb{R}^{13}$.

The force resultant equation (4.25) is also applied to the required forces. It follows that

$$ {}^{\mathbf{O}_i}F_r^* = \sum_{\{\mathbf{T}_j\}\in\mathcal{T}_i} {}^{\mathbf{O}_i}\mathbf{U}_{\mathbf{T}_j}\,{}^{\mathbf{T}_j}F_r - \sum_{\{\mathbf{B}_j\}\in\mathcal{B}_i} {}^{\mathbf{O}_i}\mathbf{U}_{\mathbf{B}_j}\,{}^{\mathbf{B}_j}F_r - {}^{\mathbf{O}_i}\mathbf{U}_{\mathbf{S}_i}\,{}^{\mathbf{S}_i}F_r \tag{4.35} $$

holds.

Required Constraint Forces

In accordance with (4.26), the required constraint force/moment vector in frame $\{\mathbf{S}_i\}$ is designed as

$$^{\mathbf{S}_i}F_r = \mathrm{diag}(^{\mathbf{S}_i}\mathbf{R}_{\mathbf{S}_{bi}}, {}^{\mathbf{S}_i}\mathbf{R}_{\mathbf{S}_{bi}}) \left(\mathbf{T}_{si}\mathbf{Y}_{si}\hat{\boldsymbol{\theta}}_{si} + \mathbf{T}_{fi}\varphi_{id}\right) \tag{4.36}$$

where $\varphi_{id} \in \mathbb{R}^{n_{fi}}$ denotes the desired (design) constraint force coordinate vector and $\hat{\boldsymbol{\theta}}_{si}$ denotes the estimate of $\boldsymbol{\theta}_{si}$.

With respect to (4.36), define

$$\mathbf{s}_{si} = \mathbf{Y}_{si}^T \left(\boldsymbol{\chi}_{ir} - \boldsymbol{\chi}_i\right). \tag{4.37}$$

The γth element of $\boldsymbol{\theta}_{si}$ is updated by

$$\hat{\theta}_{si\gamma} = \mathcal{P}\left(s_{si\gamma}, \rho_{si\gamma}, \underline{\theta}_{si\gamma}, \overline{\theta}_{si\gamma}, t\right) \tag{4.38}$$

where $\hat{\theta}_{si\gamma}$ is the γth element of $\hat{\boldsymbol{\theta}}_{si}$, $s_{si\gamma}$ is the γth element of $\mathbf{s}_{si}$, $\rho_{si\gamma} > 0$ is an update gain, and $\underline{\theta}_{si\gamma}$ and $\overline{\theta}_{si\gamma}$ denote the lower and upper bounds of $\theta_{si\gamma}$—the γth element of $\boldsymbol{\theta}_{si}$ in (4.27).

Required Force/Moment Vectors at the Driven Cutting Points

Given the required force/moment vectors at the *driving cutting points*, denoted as $^{\mathbf{B}_j}F_r \in \mathbb{R}^6$ for all $\{\mathbf{B}_j\} \in \mathcal{B}_i$, and the desired (design) internal force vector $\boldsymbol{\eta}_{id} \in \mathbb{R}^{6(n_i-1)}$, the required force/moment vectors at the *driven cutting points*, denoted as $^{\mathbf{T}_j}F_r \in \mathbb{R}^6$ for all $\{\mathbf{T}_j\} \in \mathcal{T}_i$, are obtained as

$$\begin{aligned}\mathrm{col}\left(^{\mathbf{T}_j}F_r \mid \{\mathbf{T}_j\} \in \mathcal{T}_i\right) &= \mathrm{diag}\left(^{\mathbf{T}_j}\mathbf{U}_{\mathbf{O}_i} \mid \{\mathbf{T}_j\} \in \mathcal{T}_i\right)\left[\boldsymbol{\Omega}_{mi} \quad \boldsymbol{\Omega}_{fi}\right] \\ &\times \begin{bmatrix} ^{\mathbf{O}_i}F_r^* + \sum_{\{\mathbf{B}_j\}\in\mathcal{B}_i} {}^{\mathbf{O}_i}\mathbf{U}_{\mathbf{B}_j}{}^{\mathbf{B}_j}F_r + {}^{\mathbf{O}_i}\mathbf{U}_{\mathbf{S}_i}{}^{\mathbf{S}_i}F_r \\ \boldsymbol{\eta}_{id} \end{bmatrix}.\end{aligned} \tag{4.39}$$

Equation (4.39) is in line with (4.28).

Remark 4.11. The solution (4.39) can be viewed as a computational transformation from $^{\mathbf{B}_j}F_r \in \mathbb{R}^6$ for all $\{\mathbf{B}_j\} \in \mathcal{B}_i$ to $^{\mathbf{T}_j}F_r \in \mathbb{R}^6$ for all $\{\mathbf{T}_j\} \in \mathcal{T}_i$ by incorporating $^{\mathbf{S}_i}F_r \in \mathbb{R}^6$ and $\boldsymbol{\eta}_{id} \in \mathbb{R}^{6(n_i-1)}$.

4.5.3 Virtual Stability

In this subsection, it will be shown that each *object* combined with its respective control equations qualifies to be *virtually stable* in the sense of Definition 2.17.

Lemma 4.1. *Consider the ith object described by (4.24) and combined with its respective control equation (4.32) and with the parameter adaptation (4.33) and (4.34). If a non-negative accompanying function for this object is chosen as*

$$\nu_{\mathbf{O}_i} = \frac{1}{2}\left({}^{\mathbf{O}_i}V_r - {}^{\mathbf{O}_i}V\right)^T \mathbf{M}_{\mathbf{O}_i}\left({}^{\mathbf{O}_i}V_r - {}^{\mathbf{O}_i}V\right) + \frac{1}{2}\sum_{\gamma=1}^{13}\left(\theta_{\mathbf{O}_i\gamma} - \hat{\theta}_{\mathbf{O}_i\gamma}\right)^2 / \rho_{\mathbf{O}_i\gamma} \tag{4.40}$$

then it follows that

$$\dot{\nu}_{\mathbf{O}_i} \leqslant -\left({}^{\mathbf{O}_i}V_r - {}^{\mathbf{O}_i}V\right)^T \mathbf{K}_{\mathbf{O}_i}\left({}^{\mathbf{O}_i}V_r - {}^{\mathbf{O}_i}V\right) + \left({}^{\mathbf{O}_i}V_r - {}^{\mathbf{O}_i}V\right)^T\left({}^{\mathbf{O}_i}F_r^* - {}^{\mathbf{O}_i}F^*\right) \tag{4.41}$$

holds.

The proof is given in Appendix B.1. ■

Theorem 4.1. *The ith object described by (4.22)-(4.27), combined with its respective control equations (4.30)-(4.32), (4.36), and (4.39) and with the parameter adaptation (4.33), (4.34), (4.37), and (4.38), is virtually stable with its affiliated vector* ${}^{\mathbf{O}_i}V_r - {}^{\mathbf{O}_i}V$ *being a virtual function in both* L_2 *and* L_∞*, in the sense of Definition 2.17.*

Proof: Premultiplying (4.39) by $\mathcal{H}_i\mathrm{diag}\left({}^{\mathbf{T}_j}\mathbf{U}_{\mathbf{O}_i} \mid \{\mathbf{T}_j\} \in \mathcal{T}_i\right)^{-1}$ and using $\mathcal{H}_i\mathbf{\Omega}_{mi} = \mathbf{I}_6$ and $\mathcal{H}_i\mathbf{\Omega}_{fi} = \mathbf{0}$ yields

$$\sum_{\{\mathbf{T}_j\}\in\mathcal{T}_i} {}^{\mathbf{O}_i}\mathbf{U}_{\mathbf{T}_j}\,{}^{\mathbf{T}_j}F_r = {}^{\mathbf{O}_i}F_r^* + \sum_{\{\mathbf{B}_j\}\in\mathcal{B}_i} {}^{\mathbf{O}_i}\mathbf{U}_{\mathbf{B}_j}\,{}^{\mathbf{B}_j}F_r + {}^{\mathbf{O}_i}\mathbf{U}_{\mathbf{S}_i}\,{}^{\mathbf{S}_i}F_r. \tag{4.42}$$

Furthermore, it follows from (4.23), (4.25), (4.31), (4.42), and (2.85) that

$$\begin{aligned}
&\left({}^{\mathbf{O}_i}V_r - {}^{\mathbf{O}_i}V\right)^T\left({}^{\mathbf{O}_i}F_r^* - {}^{\mathbf{O}_i}F^*\right) \\
&= \left({}^{\mathbf{O}_i}V_r - {}^{\mathbf{O}_i}V\right)^T \sum_{\{\mathbf{T}_j\}\in\mathcal{T}_i} {}^{\mathbf{O}_i}\mathbf{U}_{\mathbf{T}_j}\left({}^{\mathbf{T}_j}F_r - {}^{\mathbf{T}_j}F\right) \\
&\quad - \left({}^{\mathbf{O}_i}V_r - {}^{\mathbf{O}_i}V\right)^T \sum_{\{\mathbf{B}_j\}\in\mathcal{B}_i} {}^{\mathbf{O}_i}\mathbf{U}_{\mathbf{B}_j}\left({}^{\mathbf{B}_j}F_r - {}^{\mathbf{B}_j}F\right) \\
&\quad - \left({}^{\mathbf{O}_i}V_r - {}^{\mathbf{O}_i}V\right)^T {}^{\mathbf{O}_i}\mathbf{U}_{\mathbf{S}_i}\left({}^{\mathbf{S}_i}F_r - {}^{\mathbf{S}_i}F\right) \\
&= \sum_{\{\mathbf{T}_j\}\in\mathcal{T}_i} p_{\mathbf{T}_j} - \sum_{\{\mathbf{B}_j\}\in\mathcal{B}_i} p_{\mathbf{B}_j} - p_{\mathbf{S}_i}
\end{aligned} \tag{4.43}$$

holds.

Due to the orthogonality between $\mathbf{T}_{si}$ and $\mathbf{T}_{fi}$ in (4.26), it follows from (4.22), (4.26), (4.27), (4.30), (4.36)-(4.38), and Lemma 2.9 that

$$\begin{aligned}
p_{\mathbf{S}_i} &= \left(\boldsymbol{\chi}_{ir} - \boldsymbol{\chi}_i\right)^T \mathbf{Y}_{si}\left(\hat{\boldsymbol{\theta}}_{si} - \boldsymbol{\theta}_{si}\right) \\
&= -\sum_{\gamma}\left(\theta_{si\gamma} - \hat{\theta}_{si\gamma}\right)\frac{\dot{\hat{\theta}}_{si\gamma}}{\rho_{si\gamma}}
\end{aligned}$$

$$-\sum_{\gamma}\left\{\left(\theta_{si\gamma}-\hat{\theta}_{si\gamma}\right)\left[s_{si\gamma}-\frac{\dot{\hat{\theta}}_{si\gamma}}{\rho_{si\gamma}}\right]\right\}$$

$$\geqslant -\sum_{\gamma}\left(\theta_{si\gamma}-\hat{\theta}_{si\gamma}\right)\frac{\dot{\hat{\theta}}_{si\gamma}}{\rho_{si\gamma}} \tag{4.44}$$

holds. Integrating (4.44) over time yields

$$\int_0^t p_{\mathbf{S}_i} d\tau \geqslant \sum_{\gamma}\frac{1}{2\rho_{si\gamma}}\left(\theta_{si\gamma}(t)-\hat{\theta}_{si\gamma}(t)\right)^2-\sum_{\gamma}\frac{1}{2\rho_{si\gamma}}\left(\theta_{si\gamma}-\hat{\theta}_{si\gamma}(0)\right)^2$$

$$\geqslant -\sum_{\gamma}\frac{1}{2\rho_{si\gamma}}\left(\theta_{si\gamma}-\hat{\theta}_{si\gamma}(0)\right)^2 . \tag{4.45}$$

Consider the fact that the ith *object* has all its *driving cutting points* associated with frame $\{\mathbf{B}_j\}$ in set $\mathcal{B}_i$ and all its *driven cutting points* associated with frame $\{\mathbf{T}_j\}$ in set $\mathcal{T}_i$. Substituting (4.43) into (4.41) and using (4.40) and (4.45) proves the theorem, in view of Definition 2.17. ■

Remark 4.12. Equations (4.41) and (4.43) indicate that the dynamic interactions of the ith *object* with its adjacent *open chains* are uniquely defined by a scalar term $\left({}^{\mathbf{O}_i}V_r - {}^{\mathbf{O}_i}V\right)^T\left({}^{\mathbf{O}_i}F_r^* - {}^{\mathbf{O}_i}F^*\right)$ which equals the sum of all the *virtual power flows* at the *cutting points* and at the contact point with the environment. The sign of a *virtual power flow* is positive at a *driven cutting point* at which the force/moment vector is referenced toward the *object* and is negative at a *driving cutting point* or at the contact point with the environment at which the force/moment vector is referenced away from the *object*.

The following lemma gives the sufficient conditions to achieve (4.45).

Lemma 4.2. *If one of the following conditions holds*

(i) ${}^{\mathbf{S}_i}V_r = {}^{\mathbf{S}_i}V$

(ii) ${}^{\mathbf{S}_i}F_r = {}^{\mathbf{S}_i}F$

(iii) $\left({}^{\mathbf{S}_i}V_r - {}^{\mathbf{S}_i}V\right) \perp \left({}^{\mathbf{S}_i}F_r - {}^{\mathbf{S}_i}F\right)$

then $p_{\mathbf{S}_i} = 0$ *holds, which satisfies (4.45).* ■

4.6 Dynamics and Control of Rigid Links

With a simplified structure of the last section, the dynamics and control issue of all rigid links of the jth *open chain* is to be addressed in this section. The objective is to show that every link of the jth *open chain* combined with its respective control equations qualifies to be *virtually stable* in the sense of Definition 2.17.

4.6.1 Kinematics and Dynamics

In this subsection, both kinematics and dynamics of all rigid links of the jth *open chain* are presented.

Velocity Transformations

Consider the fact that a link is a rigid body. It follows from (2.63) that

$$
\begin{aligned}
{}^{\mathbf{T}_j}V &= {}^{\mathbf{B}_{jl_j}}\mathbf{U}_{\mathbf{T}_j}^T \, {}^{\mathbf{B}_{jl_j}}V && (4.46)\\
{}^{\mathbf{T}_{j(k+1)}}V &= {}^{\mathbf{B}_{jk}}\mathbf{U}_{\mathbf{T}_{j(k+1)}}^T \, {}^{\mathbf{B}_{jk}}V, \quad \forall k \in \{1, l_j - 1\} && (4.47)\\
{}^{\mathbf{T}_{j1}}V &= {}^{\mathbf{B}_j}\mathbf{U}_{\mathbf{T}_{j1}}^T \, {}^{\mathbf{B}_j}V && (4.48)
\end{aligned}
$$

hold with the force/moment transformation matrices being defined by (2.65).

Link Dynamics

Note that the dynamics of an *object* described by (4.24) are also applicable to each link of an *open chain*. In fact, the dynamics of all l_j+1 links can be obtained from (4.24) by substituting frames $\{\mathbf{B}_j\}$ and $\{\mathbf{B}_{jk}\}$ for all $k \in \{1, l_j\}$ for frame $\{\mathbf{O}_i\}$, respectively. It follows that

$$
\mathbf{M}_{\mathbf{B}_j} \frac{d}{dt}({}^{\mathbf{B}_j}V) + \mathbf{C}_{\mathbf{B}_j}({}^{\mathbf{B}_j}\omega){}^{\mathbf{B}_j}V + \mathbf{G}_{\mathbf{B}_j} = {}^{\mathbf{B}_j}F^* \tag{4.49}
$$

$$
\mathbf{M}_{\mathbf{B}_{jk}} \frac{d}{dt}({}^{\mathbf{B}_{jk}}V) + \mathbf{C}_{\mathbf{B}_{jk}}({}^{\mathbf{B}_{jk}}\omega){}^{\mathbf{B}_{jk}}V + \mathbf{G}_{\mathbf{B}_{jk}} = {}^{\mathbf{B}_{jk}}F^*, \quad \forall k \in \{1, l_j\} \tag{4.50}
$$

hold. These net force/moment vectors can be further expressed by

$$
\begin{aligned}
{}^{\mathbf{B}_{jl_j}}F^* &= {}^{\mathbf{B}_{jl_j}}F - {}^{\mathbf{B}_{jl_j}}\mathbf{U}_{\mathbf{T}_j} \, {}^{\mathbf{T}_j}F && (4.51)\\
{}^{\mathbf{B}_{jk}}F^* &= {}^{\mathbf{B}_{jk}}F - {}^{\mathbf{B}_{jk}}\mathbf{U}_{\mathbf{T}_{j(k+1)}} \, {}^{\mathbf{T}_{j(k+1)}}F \\
&= {}^{\mathbf{B}_{jk}}F - {}^{\mathbf{B}_{jk}}\mathbf{U}_{\mathbf{B}_{j(k+1)}} \, {}^{\mathbf{B}_{j(k+1)}}F, \quad k = l_j - 1, \cdots, 1 && (4.52)\\
{}^{\mathbf{B}_j}F^* &= {}^{\mathbf{B}_j}F - {}^{\mathbf{B}_j}\mathbf{U}_{\mathbf{T}_{j1}} \, {}^{\mathbf{T}_{j1}}F \\
&= {}^{\mathbf{B}_j}F - {}^{\mathbf{B}_j}\mathbf{U}_{\mathbf{B}_{j1}} \, {}^{\mathbf{B}_{j1}}F && (4.53)
\end{aligned}
$$

in view of (2.64) and Figure 4.3.

Remark 4.13. Note that the relationship ${}^{\mathbf{T}_{jk}}F = {}^{\mathbf{T}_{jk}}\mathbf{U}_{\mathbf{B}_{jk}} \, {}^{\mathbf{B}_{jk}}F$ used in (4.52) and (4.53) is derived from Assumption 3.

4.6.2 Control Equations

The control equations are model-based, aimed at achieving the *virtual stability* of every link.

Required Velocity Transformations

In order to validate (2.86), the following relationships are imposed on the required linear/angular velocity vectors of all links

$$
\begin{aligned}
{}^{\mathbf{T}_j}V_r &= {}^{\mathbf{B}_{jl_j}}\mathbf{U}_{\mathbf{T}_j}^T \, {}^{\mathbf{B}_{jl_j}}V_r && (4.54)\\
{}^{\mathbf{T}_{j(k+1)}}V_r &= {}^{\mathbf{B}_{jk}}\mathbf{U}_{\mathbf{T}_{j(k+1)}}^T \, {}^{\mathbf{B}_{jk}}V_r, \quad \forall k \in \{1, l_j - 1\} && (4.55)\\
{}^{\mathbf{T}_{j1}}V_r &= {}^{\mathbf{B}_j}\mathbf{U}_{\mathbf{T}_{j1}}^T \, {}^{\mathbf{B}_j}V_r. && (4.56)
\end{aligned}
$$

Required Net Force/Moment Vectors

Accordingly, the required net force/moment vectors of the $l_j + 1$ links can also be obtained in a similar way leading to (4.32). It follows that

$$
\begin{aligned}
{}^{\mathbf{B}_j}F_r^* &= \mathbf{Y}_{\mathbf{B}_j}\hat{\boldsymbol{\theta}}_{\mathbf{B}_j} + \mathbf{K}_{\mathbf{B}_j}\left({}^{\mathbf{B}_j}V_r - {}^{\mathbf{B}_j}V\right) && (4.57)\\
{}^{\mathbf{B}_{jk}}F_r^* &= \mathbf{Y}_{\mathbf{B}_{jk}}\hat{\boldsymbol{\theta}}_{\mathbf{B}_{jk}} + \mathbf{K}_{\mathbf{B}_{jk}}\left({}^{\mathbf{B}_{jk}}V_r - {}^{\mathbf{B}_{jk}}V\right), \quad \forall k \in \{1, l_j\} && (4.58)
\end{aligned}
$$

hold, where $\mathbf{K}_{\mathbf{B}_j} \in \mathbb{R}^{6\times 6}$ and $\mathbf{K}_{\mathbf{B}_{jk}} \in \mathbb{R}^{6\times 6}$ for all $k \in \{1, l_j\}$ are symmetric positive-definite matrices representing the velocity feedback control gain matrices. The terms $\mathbf{Y}_{\mathbf{B}_j}\hat{\boldsymbol{\theta}}_{\mathbf{B}_j}$ and $\mathbf{Y}_{\mathbf{B}_{jk}}\hat{\boldsymbol{\theta}}_{\mathbf{B}_{jk}}$ for all $k \in \{1, l_j\}$ denote the model-based feedforward compensation terms with the regressor matrices $\mathbf{Y}_{\mathbf{B}_j} \in \mathbb{R}^{6\times 13}$ and $\mathbf{Y}_{\mathbf{B}_{jk}} \in \mathbb{R}^{6\times 13}$ and the parameter vectors $\boldsymbol{\theta}_{\mathbf{B}_j}$ and $\boldsymbol{\theta}_{\mathbf{B}_{jk}}$ being defined in (2.78) and further in Appendix A by substituting frames $\{\mathbf{B}_j\}$ and $\{\mathbf{B}_{jk}\}$ for frame $\{\mathbf{A}\}$, respectively.

The parameter estimates need to be updated. Define

$$
\begin{aligned}
\mathbf{s}_{\mathbf{B}_j} &= \mathbf{Y}_{\mathbf{B}_j}^T\left({}^{\mathbf{B}_j}V_r - {}^{\mathbf{B}_j}V\right) && (4.59)\\
\mathbf{s}_{\mathbf{B}_{jk}} &= \mathbf{Y}_{\mathbf{B}_{jk}}^T\left({}^{\mathbf{B}_{jk}}V_r - {}^{\mathbf{B}_{jk}}V\right), \quad \forall k \in \{1, l_j\}. && (4.60)
\end{aligned}
$$

The $\mathcal{P}$ function defined by (2.79) is used to update each element of $\hat{\boldsymbol{\theta}}_{\mathbf{B}_j} \in \mathbb{R}^{13}$ and of $\hat{\boldsymbol{\theta}}_{\mathbf{B}_{jk}} \in \mathbb{R}^{13}$ for all $k \in \{1, l_j\}$ as

$$
\begin{aligned}
\hat{\theta}_{\mathbf{B}_j\gamma} &= \mathcal{P}\left(s_{\mathbf{B}_j\gamma}, \rho_{\mathbf{B}_j\gamma}, \underline{\theta}_{\mathbf{B}_j\gamma}, \overline{\theta}_{\mathbf{B}_j\gamma}, t\right), \quad \gamma \in \{1, 13\} && (4.61)\\
\hat{\theta}_{\mathbf{B}_{jk}\gamma} &= \mathcal{P}\left(s_{\mathbf{B}_{jk}\gamma}, \rho_{\mathbf{B}_{jk}\gamma}, \underline{\theta}_{\mathbf{B}_{jk}\gamma}, \overline{\theta}_{\mathbf{B}_{jk}\gamma}, t\right), \quad \gamma \in \{1, 13\} && (4.62)
\end{aligned}
$$

where

$\hat{\theta}_{\mathbf{B}_j\gamma}$	γth element of $\hat{\boldsymbol{\theta}}_{\mathbf{B}_j}$.
$\hat{\theta}_{\mathbf{B}_{jk}\gamma}$	γth element of $\hat{\boldsymbol{\theta}}_{\mathbf{B}_{jk}}$.
$s_{\mathbf{B}_j\gamma}$	γth element of $\mathbf{s}_{\mathbf{B}_j}$ defined by (4.59).
$s_{\mathbf{B}_{jk}\gamma}$	γth element of $\mathbf{s}_{\mathbf{B}_{jk}}$ defined by (4.60).
$\rho_{\mathbf{B}_j\gamma}$	Parameter update gain.
$\rho_{\mathbf{B}_{jk}\gamma}$	Parameter update gain.
$\underline{\theta}_{\mathbf{B}_j\gamma}$	Lower bound of $\theta_{\mathbf{B}_j\gamma}$.

$\overline{\theta}_{\mathbf{B}_j\gamma}$	Upper bound of $\theta_{\mathbf{B}_j\gamma}$.
$\underline{\theta}_{\mathbf{B}_{jk}\gamma}$	Lower bound of $\theta_{\mathbf{B}_{jk}\gamma}$.
$\overline{\theta}_{\mathbf{B}_{jk}\gamma}$	Upper bound of $\theta_{\mathbf{B}_{jk}\gamma}$.
$\theta_{\mathbf{B}_j\gamma}$	γth element of $\boldsymbol{\theta}_{\mathbf{B}_j}$.
$\theta_{\mathbf{B}_{jk}\gamma}$	γth element of $\boldsymbol{\theta}_{\mathbf{B}_{jk}}$.

Required Force/Moment Vectors

After ${}^{\mathbf{B}_j}F_r^*$ and ${}^{\mathbf{B}_{jk}}F_r^*$ being obtained from (4.57) and (4.58), the required force/moment vectors in frames $\{\mathbf{B}_j\}$ and $\{\mathbf{B}_{jk}\}$ for all $k \in \{1, l_j\}$ can be obtained from given ${}^{\mathbf{T}_j}F_r$ as

$$
{}^{\mathbf{B}_{jl_j}}F_r = {}^{\mathbf{B}_{jl_j}}F_r^* + {}^{\mathbf{B}_{jl_j}}\mathbf{U}_{\mathbf{T}_j}\,{}^{\mathbf{T}_j}F_r \tag{4.63}
$$

$$
\begin{aligned}
{}^{\mathbf{B}_{jk}}F_r &= {}^{\mathbf{B}_{jk}}F_r^* + {}^{\mathbf{B}_{jk}}\mathbf{U}_{\mathbf{T}_{j(k+1)}}\,{}^{\mathbf{T}_{j(k+1)}}F_r \\
&= {}^{\mathbf{B}_{jk}}F_r^* + {}^{\mathbf{B}_{jk}}\mathbf{U}_{\mathbf{B}_{j(k+1)}}\,{}^{\mathbf{B}_{j(k+1)}}F_r, \quad k = l_j - 1, \cdots, 1
\end{aligned} \tag{4.64}
$$

$$
\begin{aligned}
{}^{\mathbf{B}_j}F_r &= {}^{\mathbf{B}_j}F_r^* + {}^{\mathbf{B}_j}\mathbf{U}_{\mathbf{T}_{j1}}\,{}^{\mathbf{T}_{j1}}F_r \\
&= {}^{\mathbf{B}_j}F_r^* + {}^{\mathbf{B}_j}\mathbf{U}_{\mathbf{B}_{j1}}\,{}^{\mathbf{B}_{j1}}F_r
\end{aligned} \tag{4.65}
$$

in which ${}^{\mathbf{T}_{jk}}F_r = {}^{\mathbf{T}_{jk}}\mathbf{U}_{\mathbf{B}_{jk}}\,{}^{\mathbf{B}_{jk}}F_r$ is used for all $k \in \{1, l_j\}$ and $j \in \{1, n_c\}$.

Remark 4.14. Equations (4.63)–(4.65) compute ${}^{\mathbf{B}_j}F_r \in \mathbb{R}^6$ from given ${}^{\mathbf{T}_j}F_r \in \mathbb{R}^6$. Like (4.39), the computation is executed from the required force/moment vector at the *driving cutting point* to the required force/moment vector at the *driven cutting point*.

4.6.3 Virtual Stability

In this subsection, it will be shown that each link combined with its respective control equations qualifies to be *virtually stable* in the sense of Definition 2.17.

Theorem 4.2. *Each rigid link of the jth open chain described by (4.49) or (4.50) subject to (4.46)–(4.48) and (4.51)–(4.53), combined with its respective control equations (4.54)–(4.56), (4.57) or (4.58), and (4.63)–(4.65) and with the parameter adaptation described by (4.59) and (4.61) or by (4.60) and (4.62), is virtually stable with its affiliated vector* ${}^{\mathbf{B}_j}V_r - {}^{\mathbf{B}_j}V$ *or* ${}^{\mathbf{B}_{jk}}V_r - {}^{\mathbf{B}_{jk}}V$, $k \in \{1, l_j\}$, *being a virtual function in both* L_2 *and* L_∞, *in the sense of Definition 2.17.*

Proof: Let

$$
\begin{aligned}
\nu_{\mathbf{B}_j} &= \frac{1}{2}\left({}^{\mathbf{B}_j}V_r - {}^{\mathbf{B}_j}V\right)^T \mathbf{M}_{\mathbf{B}_j}\left({}^{\mathbf{B}_j}V_r - {}^{\mathbf{B}_j}V\right) \\
&\quad + \frac{1}{2}\sum_{\gamma=1}^{13}\left(\theta_{\mathbf{B}_j\gamma} - \hat{\theta}_{\mathbf{B}_j\gamma}\right)^2 / \rho_{\mathbf{B}_j\gamma}
\end{aligned} \tag{4.66}
$$

$$
\begin{aligned}
\nu_{\mathbf{B}_{jk}} &= \frac{1}{2}\left({}^{\mathbf{B}_{jk}}V_r - {}^{\mathbf{B}_{jk}}V\right)^T \mathbf{M}_{\mathbf{B}_{jk}}\left({}^{\mathbf{B}_{jk}}V_r - {}^{\mathbf{B}_{jk}}V\right) \\
&\quad + \frac{1}{2}\sum_{\gamma=1}^{13}\left(\theta_{\mathbf{B}_{jk}\gamma} - \hat{\theta}_{\mathbf{B}_{jk}\gamma}\right)^2 / \rho_{\mathbf{B}_{jk}\gamma}
\end{aligned} \tag{4.67}
$$

be two non-negative accompanying functions respectively assigned to the base link to which frame $\{\mathbf{B}_j\}$ is attached and to the kth link to which frame $\{\mathbf{B}_{jk}\}$ is attached. It follows from (4.49), (4.50), (4.57)–(4.62), and Lemma 4.1 that

$$\begin{aligned} \dot{\nu}_{\mathbf{B}_j} \leqslant & -\left({}^{\mathbf{B}_j}V_r - {}^{\mathbf{B}_j}V\right)^T \mathbf{K}_{\mathbf{B}_j}\left({}^{\mathbf{B}_j}V_r - {}^{\mathbf{B}_j}V\right) \\ & + \left({}^{\mathbf{B}_j}V_r - {}^{\mathbf{B}_j}V\right)^T \left({}^{\mathbf{B}_j}F_r^* - {}^{\mathbf{B}_j}F^*\right) \end{aligned} \quad (4.68)$$

$$\begin{aligned} \dot{\nu}_{\mathbf{B}_{jk}} \leqslant & -\left({}^{\mathbf{B}_{jk}}V_r - {}^{\mathbf{B}_{jk}}V\right)^T \mathbf{K}_{\mathbf{B}_{jk}}\left({}^{\mathbf{B}_{jk}}V_r - {}^{\mathbf{B}_{jk}}V\right) \\ & + \left({}^{\mathbf{B}_{jk}}V_r - {}^{\mathbf{B}_{jk}}V\right)^T \left({}^{\mathbf{B}_{jk}}F_r^* - {}^{\mathbf{B}_{jk}}F^*\right), \ \forall k \in \{1, l_j\} \end{aligned} \quad (4.69)$$

hold.

In view of (2.85), (4.46)–(4.48), (4.51)–(4.56), and (4.63)–(4.65), it yields

$$\begin{aligned} & \left({}^{\mathbf{B}_j}V_r - {}^{\mathbf{B}_j}V\right)^T \left({}^{\mathbf{B}_j}F_r^* - {}^{\mathbf{B}_j}F^*\right) \\ & = \left({}^{\mathbf{B}_j}V_r - {}^{\mathbf{B}_j}V\right)^T \left[\left({}^{\mathbf{B}_j}F_r - {}^{\mathbf{B}_j}F\right) - {}^{\mathbf{B}_j}\mathbf{U}_{\mathbf{T}_{j1}}\left({}^{\mathbf{T}_{j1}}F_r - {}^{\mathbf{T}_{j1}}F\right)\right] \\ & = p_{\mathbf{B}_j} - p_{\mathbf{T}_{j1}} \end{aligned} \quad (4.70)$$

$$\begin{aligned} & \left({}^{\mathbf{B}_{jk}}V_r - {}^{\mathbf{B}_{jk}}V\right)^T \left({}^{\mathbf{B}_{jk}}F_r^* - {}^{\mathbf{B}_{jk}}F^*\right) \\ & = \left({}^{\mathbf{B}_{jk}}V_r - {}^{\mathbf{B}_{jk}}V\right)^T \\ & \quad \times \left[\left({}^{\mathbf{B}_{jk}}F_r - {}^{\mathbf{B}_{jk}}F\right) - {}^{\mathbf{B}_{jk}}\mathbf{U}_{\mathbf{T}_{j(k+1)}}\left({}^{\mathbf{T}_{j(k+1)}}F_r - {}^{\mathbf{T}_{j(k+1)}}F\right)\right] \\ & = p_{\mathbf{B}_{jk}} - p_{\mathbf{T}_{j(k+1)}}, \quad \forall k \in \{1, l_j - 1\} \end{aligned} \quad (4.71)$$

$$\begin{aligned} & \left({}^{\mathbf{B}_{jl_j}}V_r - {}^{\mathbf{B}_{jl_j}}V\right)^T \left({}^{\mathbf{B}_{jl_j}}F_r^* - {}^{\mathbf{B}_{jl_j}}F^*\right) \\ & = \left({}^{\mathbf{B}_{jl_j}}V_r - {}^{\mathbf{B}_{jl_j}}V\right)^T \left[\left({}^{\mathbf{B}_{jl_j}}F_r - {}^{\mathbf{B}_{jl_j}}F\right) - {}^{\mathbf{B}_{jl_j}}\mathbf{U}_{\mathbf{T}_j}\left({}^{\mathbf{T}_j}F_r - {}^{\mathbf{T}_j}F\right)\right] \\ & = p_{\mathbf{B}_{jl_j}} - p_{\mathbf{T}_j}. \end{aligned} \quad (4.72)$$

Consider the fact that the kth link has one *driving cutting point* associated with frame $\{\mathbf{T}_{j(k+1)}\}$ for all $k \in \{0, l_j - 1\}$ or associated with frame $\{\mathbf{T}_j\}$ for $k = l_j$ and one *driven cutting point* associated with frame $\{\mathbf{B}_j\}$ for $k = 0$ or associated with frame $\{\mathbf{B}_{jk}\}$ for all $k \in \{1, l_j\}$. Substituting (4.70)–(4.72) into (4.68) and (4.69) and using (4.66), (4.67), and Definition 2.17 completes the proof. ∎

4.7 Required Force/Moment Vector Computations of the Entire Complex Robot

By using (4.39) and (4.63)–(4.65), the computations of the required force/moment vectors of the entire complex robot are executed along the opposite directions of the simple oriented graph, starting from the *sink nodes* toward the *source nodes*. Consider the fact that a node represents either an *object* or an *open chain*. Equation (4.39) is used for a node representing an *object*, and (4.63)–(4.65) are used for a node representing an *open chain*. The alternative connection of *objects* and *open chains* allows the alternative use of (4.39) and (4.63)–(4.65), leading to the completion of the computations.

4.8 Dynamics and Control of Joints

After the dynamics and control issue of rigid bodies that represent both the *objects* and the links of *open chains* has been addressed in the previous sections, the dynamics and control issue of joints is to be addressed in this section.

Two subsections concerning both single-DOF joints and three-DOF joints are developed.

4.8.1 Dynamics and Control of Single-DOF Joints

This subsection comprises three parts, namely kinematics and dynamics, joint control equations, and *virtual stability*. The objective is to show that a single-DOF joint combined with its respective control equations qualifies to be *virtually stable* in the sense of Definition 2.17.

Kinematics and Dynamics

Assume that the kth joint of the jth *open chain* is a single-DOF prismatic or revolute joint with $k \in \mathcal{C}_{1j}$ and $j \in \{1, n_c\}$. In view of (2.63) and (4.3), the relationship between the velocity of this joint and the linear/angular velocity vectors of the adjacent links can be expressed by

$$\begin{aligned} {}^{\mathbf{B}_{jk}}V &= \mathbf{z}\dot{q}_{jk} + {}^{\mathbf{T}_{jk}}\mathbf{U}_{\mathbf{B}_{jk}}^{T}\,{}^{\mathbf{T}_{jk}}V \\ &= \mathbf{z}\dot{q}_{jk} + {}^{\mathbf{B}_{j(k-1)}}\mathbf{U}_{\mathbf{B}_{jk}}^{T}\,{}^{\mathbf{B}_{j(k-1)}}V \end{aligned} \tag{4.73}$$

for all $k \in \mathcal{C}_{1j}$ with ${}^{\mathbf{B}_{j0}}V = {}^{\mathbf{B}_j}V$ and $j \in \{1, n_c\}$, where $\mathbf{z} = [0,0,1,0,0,0]^T$ holds for a prismatic joint and $\mathbf{z} = [0,0,0,0,0,1]^T$ holds for a revolute joint.

The dynamics of this joint under Assumption 2 can be written as

$$\tau_{jk}^{*} \overset{def}{=} J_{mjk}\ddot{q}_{jk} + \xi_{jk}(t) = \tau_{jk} - \mathbf{z}^{T}\,{}^{\mathbf{B}_{jk}}F \tag{4.74}$$

where $J_{mjk} \in \mathbb{R}$ is the equivalent mass or moment of inertia under Assumption 2, $q_{jk} \in \mathbb{R}$ is the joint displacement, $\tau_{jk} \in \mathbb{R}$ is the joint control force/torque,[5] $\tau_{jk}^{*} \in \mathbb{R}$ represents the net force/torque devoted to the joint dynamics, ${}^{\mathbf{B}_{jk}}F \in \mathbb{R}^6$ denotes the force/moment vector in frame $\{\mathbf{B}_{jk}\}$ exerted from link $k-1$ to link k, and $\xi_{jk}(t) \in \mathbb{R}$ is the friction force/torque having a form of

$$\xi_{jk}(t) = k_{cjk}\mathrm{sign}(\dot{q}_{jk}) + k_{vjk}\dot{q}_{jk} + c_{jk} \tag{4.75}$$

where $k_{cjk} > 0$ and $k_{vjk} > 0$ denote the Coulomb and viscous friction coefficients, respectively, and $c_{jk} \in \mathbb{R}$ denotes an offset that accommodates asymmetric Coulomb frictions.

[5] Note that $\tau_{jk} = 0$ holds for an unactuated joint.

Remark 4.15. If a direct drive joint is concerned, $J_{mjk} = 0$ should be used in (4.74) and in all corresponding equations afterwards.

Control Equations

Derived from (4.13), the relationship between the required velocity of the kth joint of the jth *open chain* and the required linear/angular velocity vectors of its adjacent links can be expressed by

$$\begin{aligned} {}^{\mathbf{B}_{jk}}V_r &= \mathbf{z}\dot{q}_{jkr} + {}^{\mathbf{T}_{jk}}\mathbf{U}_{\mathbf{B}_{jk}}^T \, {}^{\mathbf{T}_{jk}}V_r \\ &= \mathbf{z}\dot{q}_{jkr} + {}^{\mathbf{B}_{j(k-1)}}\mathbf{U}_{\mathbf{B}_{jk}}^T \, {}^{\mathbf{B}_{j(k-1)}}V_r \end{aligned} \tag{4.76}$$

for all $k \in \mathcal{C}_{1j}$ with ${}^{\mathbf{B}_{j0}}V_r = {}^{\mathbf{B}_j}V_r$ and $j \in \{1, n_c\}$.

After $\dot{q}_{jkr} \in \mathbb{R}$ being obtained from (4.13), the joint control equations are designed as

$$\tau_{jkr}^* = \mathbf{Y}_{jk}\hat{\boldsymbol{\theta}}_{jk} + k_{jk}\left(\dot{q}_{jkr} - \dot{q}_{jk}\right) \tag{4.77}$$

$$\tau_{jk} = \tau_{jkr}^* + \mathbf{z}^T \, {}^{\mathbf{B}_{jk}}F_r \tag{4.78}$$

where $k_{jk} > 0$ denotes the velocity feedback gain of the joint, ${}^{\mathbf{B}_{jk}}F_r \in \mathbb{R}^6$ (the required force/moment vector from link $k-1$ to link k) is obtained from (4.63) and (4.64), and $\hat{\boldsymbol{\theta}}_{jk} \in \mathbb{R}^4$ denotes the estimate of $\boldsymbol{\theta}_{jk} \in \mathbb{R}^4$. Both $\mathbf{Y}_{jk} \in \mathbb{R}^{1\times 4}$ and $\boldsymbol{\theta}_{jk} \in \mathbb{R}^4$ are expressed by

$$\mathbf{Y}_{jk} = \left[\ddot{q}_{jkr} \quad \mathrm{sign}(\dot{q}_{jkr}) \quad \dot{q}_{jkr} \quad 1\right] \tag{4.79}$$

$$\boldsymbol{\theta}_{jk} = \left[J_{mjk} \quad k_{cjk} \quad k_{vjk} \quad c_{jk}\right]^T. \tag{4.80}$$

The parameter estimates need to be updated. Define

$$\mathbf{s}_{jk} = \mathbf{Y}_{jk}^T\left(\dot{q}_{jkr} - \dot{q}_{jk}\right). \tag{4.81}$$

The $\mathcal{P}$ function defined by (2.79) is used to update each of the four parameters of $\hat{\boldsymbol{\theta}}_{jk} \in \mathbb{R}^4$ as

$$\hat{\theta}_{jk\gamma} = \mathcal{P}\left(s_{jk\gamma}, \rho_{jk\gamma}, \underline{\theta}_{jk\gamma}, \overline{\theta}_{jk\gamma}, t\right), \ \gamma \in \{1, 4\} \tag{4.82}$$

where $\hat{\theta}_{jk\gamma}$ denotes the γth element of $\hat{\boldsymbol{\theta}}_{jk}$ and $s_{jk\gamma}$ denotes the γth element of $\mathbf{s}_{jk}$ defined by (4.81), $\rho_{jk\gamma} > 0$ is an update gain, and $\underline{\theta}_{jk\gamma}$ and $\overline{\theta}_{jk\gamma}$ denote the lower and upper bounds of $\theta_{jk\gamma}$, the γth element of $\boldsymbol{\theta}_{jk} \in \mathbb{R}^4$.

The joint control equation (4.78) mainly consists of two parts. The first part, comprised of a model based feedforward term $\mathbf{Y}_{jk}\hat{\boldsymbol{\theta}}_{jk}$ and a joint velocity feedback term as formed by (4.77), contributes to the dynamics based control of the joint itself. The second part, denoted as $\mathbf{z}^T \, {}^{\mathbf{B}_{jk}}F_r$, contributes to the control of robot links.

Virtual Stability

The following theorem ensures that a single-DOF joint combined with its respective control equations qualifies to be *virtually stable* in the sense of Definition 2.17.

Theorem 4.3. *Given $k \in \mathcal{C}_{1j}$ and $j \in \{1, n_c\}$, the kth joint of the jth open chain described by (4.73)–(4.75), combined with its respective control equations (4.76)–(4.80) and with the parameter adaptation (4.81) and (4.82), is virtually stable with its affiliated variable $\dot{q}_{jkr} - \dot{q}_{jk}$ being a virtual function in both L_2 and L_∞, in the sense of Definition 2.17.*

Proof: Subtracting (4.74) from (4.77) and using (4.75), (4.79), and (4.80) yields

$$\begin{aligned} \tau_{jkr}^* - \tau_{jk}^* = & J_{mjk}\left(\ddot{q}_{jkr} - \ddot{q}_{jk}\right) + k_{cjk}\left[\mathrm{sign}(\dot{q}_{jkr}) - \mathrm{sign}(\dot{q}_{jk})\right] \\ & + k_{vjk}\left(\dot{q}_{jkr} - \dot{q}_{jk}\right) - \mathbf{Y}_{jk}\left(\boldsymbol{\theta}_{jk} - \hat{\boldsymbol{\theta}}_{jk}\right) \\ & + k_{jk}\left(\dot{q}_{jkr} - \dot{q}_{jk}\right). \end{aligned} \tag{4.83}$$

A non-negative accompanying function is chosen as

$$\nu_{1jk} = \frac{1}{2} J_{mjk}\left(\dot{q}_{jkr} - \dot{q}_{jk}\right)^2 + \frac{1}{2}\sum_{\gamma=1}^{4}\left(\theta_{jk\gamma} - \hat{\theta}_{jk\gamma}\right)^2 / \rho_{jk\gamma}. \tag{4.84}$$

By using

$$\left[\mathrm{sign}(\dot{q}_{jkr}) - \mathrm{sign}(\dot{q}_{jk})\right]\left(\dot{q}_{jkr} - \dot{q}_{jk}\right) \geqslant 0 \tag{4.85}$$

together with (4.81)–(4.83) and Lemma 2.9, the time derivative of (4.84) becomes

$$\begin{aligned} \dot{\nu}_{1jk} = & J_{mjk}\left(\dot{q}_{jkr} - \dot{q}_{jk}\right)\left(\ddot{q}_{jkr} - \ddot{q}_{jk}\right) \\ & - \sum_{\gamma=1}^{4}\left(\theta_{jk\gamma} - \hat{\theta}_{jk\gamma}\right)\frac{\dot{\hat{\theta}}_{jk\gamma}}{\rho_{jk\gamma}} \\ = & -k_{cjk}\left(\dot{q}_{jkr} - \dot{q}_{jk}\right)\left[\mathrm{sign}(\dot{q}_{jkr}) - \mathrm{sign}(\dot{q}_{jk})\right] - k_{vjk}\left(\dot{q}_{jkr} - \dot{q}_{jk}\right)^2 \\ & + \sum_{\gamma=1}^{4}\left(\theta_{jk\gamma} - \hat{\theta}_{jk\gamma}\right)\left[s_{jk\gamma} - \frac{\dot{\hat{\theta}}_{jk\gamma}}{\rho_{jk\gamma}}\right] \\ & - k_{jk}\left(\dot{q}_{jkr} - \dot{q}_{jk}\right)^2 + \left(\dot{q}_{jkr} - \dot{q}_{jk}\right)\left(\tau_{jkr}^* - \tau_{jk}^*\right) \\ \leqslant & -\left(k_{vjk} + k_{jk}\right)\left(\dot{q}_{jkr} - \dot{q}_{jk}\right)^2 + \left(\dot{q}_{jkr} - \dot{q}_{jk}\right)\left(\tau_{jkr}^* - \tau_{jk}^*\right). \end{aligned} \tag{4.86}$$

Substituting (4.78) into (4.74) yields

$$\tau_{jkr}^* - \tau_{jk}^* = -\mathbf{z}^T\left({}^{\mathbf{B}_{jk}}F_r - {}^{\mathbf{B}_{jk}}F\right). \tag{4.87}$$

It then follows from (2.85), (4.73), (4.76), and (4.87) that

$$\begin{aligned}
&\left(\dot{q}_{jkr} - \dot{q}_{jk}\right)\left(\tau^*_{jkr} - \tau^*_{jk}\right) \\
&\quad = -\left(\dot{q}_{jkr} - \dot{q}_{jk}\right)\mathbf{z}^T\left({}^{\mathbf{B}_{jk}}F_r - {}^{\mathbf{B}_{jk}}F\right) \\
&\quad = -\left[\left({}^{\mathbf{B}_{jk}}V_r - {}^{\mathbf{B}_{jk}}V\right) - {}^{\mathbf{T}_{jk}}\mathbf{U}^T_{\mathbf{B}_{jk}}\left({}^{\mathbf{T}_{jk}}V_r - {}^{\mathbf{T}_{jk}}V\right)\right]^T \\
&\qquad\qquad \times\left({}^{\mathbf{B}_{jk}}F_r - {}^{\mathbf{B}_{jk}}F\right) \\
&\quad = -\left({}^{\mathbf{B}_{jk}}V_r - {}^{\mathbf{B}_{jk}}V\right)^T\left({}^{\mathbf{B}_{jk}}F_r - {}^{\mathbf{B}_{jk}}F\right) \\
&\qquad + \left({}^{\mathbf{T}_{jk}}V_r - {}^{\mathbf{T}_{jk}}V\right)^T {}^{\mathbf{T}_{jk}}\mathbf{U}_{\mathbf{B}_{jk}}\left({}^{\mathbf{B}_{jk}}F_r - {}^{\mathbf{B}_{jk}}F\right) \\
&\quad = -p_{\mathbf{B}_{jk}} + p_{\mathbf{T}_{jk}}
\end{aligned} \tag{4.88}$$

holds, in which ${}^{\mathbf{T}_{jk}}F = {}^{\mathbf{T}_{jk}}\mathbf{U}_{\mathbf{B}_{jk}}\,{}^{\mathbf{B}_{jk}}F$ and ${}^{\mathbf{T}_{jk}}F_r = {}^{\mathbf{T}_{jk}}\mathbf{U}_{\mathbf{B}_{jk}}\,{}^{\mathbf{B}_{jk}}F_r$ are used.

Finally, substituting (4.88) into (4.86) yields

$$\dot{\nu}_{1jk} \leqslant -\left(k_{vjk} + k_{jk}\right)\left(\dot{q}_{jkr} - \dot{q}_{jk}\right)^2 - p_{\mathbf{B}_{jk}} + p_{\mathbf{T}_{jk}}. \tag{4.89}$$

Using (4.84) and (4.89) completes the proof, since frame $\{\mathbf{B}_{jk}\}$ is placed at the sole *driving cutting point* of the joint and frame $\{\mathbf{T}_{jk}\}$ is placed at the sole *driven cutting point* of the joint. ■

Remark 4.16. Inequality (4.89) indicates that the dynamic interactions of the kth joint with its two adjacent links are uniquely represented by the two *virtual power flows* at the two subsidiary *cutting points*. The signs of the two *virtual power flows* are positive at the *driven cutting point* and negative at the *driving cutting point*.

4.8.2 Dynamics and Control of Three-DOF Spherical Joints

With a similar structure, the dynamics and control issue of a spherical joint is to be presented in this subsection. The objective is to show that a spherical joint combined with its respective control equations qualifies to be *virtually stable* in the sense of Definition 2.17.

Kinematics and Dynamics

Assume that the kth joint of the jth *open chain* is a three-DOF spherical joint for $k \in \mathcal{C}_{3j}$ and $j \in \{1, n_c\}$. The velocity mapping equation that describes the relationship between the velocity vector of the joint and the linear/angular velocity vectors of its adjacent links is similar to (4.73) and is expressed by

$$\begin{aligned}
{}^{\mathbf{B}_{jk}}V &= \mathbf{Z}\dot{\mathbf{q}}_{jk} + {}^{\mathbf{T}_{jk}}\mathbf{U}^T_{\mathbf{B}_{jk}}\,{}^{\mathbf{T}_{jk}}V \\
&= \mathbf{Z}\dot{\mathbf{q}}_{jk} + {}^{\mathbf{B}_{j(k-1)}}\mathbf{U}^T_{\mathbf{B}_{jk}}\,{}^{\mathbf{B}_{j(k-1)}}V
\end{aligned} \tag{4.90}$$

with ${}^{\mathbf{B}_{j0}}V = {}^{\mathbf{B}_j}V$ and $j \in \{1, n_c\}$, where $\mathbf{Z} = [\mathbf{0}_{3\times3}, \mathbf{I}_3]^T \in \mathbb{R}^{6\times3}$ is a matrix and $\dot{\mathbf{q}}_{jk} \in \mathbb{R}^3$ denotes the joint velocity vector.

Meanwhile, the dynamics of the joint can be expressed as

$$\boldsymbol{\tau}^*_{jk} \overset{def}{=} \boldsymbol{\xi}_{jk}(t) = \boldsymbol{\tau}_{jk} - \mathbf{Z}^T \, {}^{\mathbf{B}_{jk}}F \tag{4.91}$$

where $\boldsymbol{\tau}_{jk} \in \mathbb{R}^3$ denotes the joint control torque vector and $\boldsymbol{\tau}^*_{jk} \in \mathbb{R}^3$ represents the net torque vector of the joint dynamics that includes the joint friction torque vector $\boldsymbol{\xi}_{jk}(t) \in \mathbb{R}^3$ only,[6] while ${}^{\mathbf{B}_{jk}}F \in \mathbb{R}^6$ denotes the force/moment vector in frame $\{\mathbf{B}_{jk}\}$, exerted from link $k-1$ to link k.

From now on, the control equations and the *virtual stability* can be written straightforwardly.

Control Equations

Similar to (4.76), the relationship between the required velocity vector of the joint and the required linear/angular velocity vectors of its adjacent links can be expressed by

$$\begin{aligned} {}^{\mathbf{B}_{jk}}V_r &= \mathbf{Z}\dot{\mathbf{q}}_{jkr} + {}^{\mathbf{T}_{jk}}\mathbf{U}^T_{\mathbf{B}_{jk}} \, {}^{\mathbf{T}_{jk}}V_r \\ &= \mathbf{Z}\dot{\mathbf{q}}_{jkr} + {}^{\mathbf{B}_{j(k-1)}}\mathbf{U}^T_{\mathbf{B}_{jk}} \, {}^{\mathbf{B}_{j(k-1)}}V_r \end{aligned} \tag{4.92}$$

for all $k \in \mathcal{C}_{3j}$ with ${}^{\mathbf{B}_{j0}}V_r = {}^{\mathbf{B}_j}V_r$ and $j \in \{1, n_c\}$.

The joint control equations are designed as

$$\boldsymbol{\tau}^*_{jkr} = \mathbf{Y}_{jk}\hat{\boldsymbol{\theta}}_{jk} + \mathbf{K}_{jk}\left(\dot{\mathbf{q}}_{jkr} - \dot{\mathbf{q}}_{jk}\right) \tag{4.93}$$

$$\boldsymbol{\tau}_{jk} = \boldsymbol{\tau}^*_{jkr} + \mathbf{Z}^T \, {}^{\mathbf{B}_{jk}}F_r \tag{4.94}$$

where $\mathbf{K}_{jk} \in \mathbb{R}^{3\times 3}$ denotes a symmetric positive-definite matrix for the joint velocity feedback and $\hat{\boldsymbol{\theta}}_{jk}$ denotes the estimate of $\boldsymbol{\theta}_{jk}$ subject to

$$\mathbf{Y}_{jk}\boldsymbol{\theta}_{jk} = \boldsymbol{\xi}_{jk}(t). \tag{4.95}$$

The parameter estimates need to be updated. Define

$$\mathbf{s}_{jk} = \mathbf{Y}^T_{jk}\left(\dot{\mathbf{q}}_{jkr} - \dot{\mathbf{q}}_{jk}\right). \tag{4.96}$$

The $\mathcal{P}$ function defined by (2.79) is used to update each of the parameters of $\hat{\boldsymbol{\theta}}_{jk}$. It follows that

$$\dot{\hat{\theta}}_{jk\gamma} = \mathcal{P}\left(s_{jk\gamma}, \rho_{jk\gamma}, \underline{\theta}_{jk\gamma}, \overline{\theta}_{jk\gamma}, t\right) \tag{4.97}$$

holds, where $\hat{\theta}_{jk\gamma}$ denotes the γth element of $\hat{\boldsymbol{\theta}}_{jk}$, $s_{jk\gamma}$ denotes the γth element of $\mathbf{s}_{jk}$ defined by (4.96), $\rho_{jk\gamma} > 0$ is an update gain, and $\underline{\theta}_{jk\gamma}$ and $\overline{\theta}_{jk\gamma}$ denote the lower and upper bounds of $\theta_{jk\gamma}$, the γth element of $\boldsymbol{\theta}_{jk}$.

[6] For a spherical joint the inertia properties are included in the corresponding links.

Virtual Stability

It will be shown that a spherical joint combined with its respective control equations qualifies to be *virtually stable* in the sense of Definition 2.17.

Theorem 4.4. *Given $k \in \mathcal{C}_{3j}$ and $j \in \{1, n_c\}$, the kth joint of the jth open chain described by (4.90) and (4.91), combined with its control equations (4.92)-(4.95) and with the parameter adaptation (4.96) and (4.97), is virtually stable with its affiliated vector $\dot{\mathbf{q}}_{jkr} - \dot{\mathbf{q}}_{jk}$ being a virtual function in L_2, in the sense of Definition 2.17.*

Proof: Subtracting (4.91) from (4.93) and using (4.95) yields

$$\boldsymbol{\tau}_{jkr}^* - \boldsymbol{\tau}_{jk}^* = -\mathbf{Y}_{jk}\left(\boldsymbol{\theta}_{jk} - \hat{\boldsymbol{\theta}}_{jk}\right) + \mathbf{K}_{jk}\left(\dot{\mathbf{q}}_{jkr} - \dot{\mathbf{q}}_{jk}\right). \tag{4.98}$$

Premultiplying (4.98) by $\left(\dot{\mathbf{q}}_{jkr} - \dot{\mathbf{q}}_{jk}\right)^T$ and using (4.96) yields

$$\begin{aligned} -\mathbf{s}_{jk}^T\left(\boldsymbol{\theta}_{jk} - \hat{\boldsymbol{\theta}}_{jk}\right) &= \left(\dot{\mathbf{q}}_{jkr} - \dot{\mathbf{q}}_{jk}\right)^T\left(\boldsymbol{\tau}_{jkr}^* - \boldsymbol{\tau}_{jk}^*\right) \\ &\quad - \left(\dot{\mathbf{q}}_{jkr} - \dot{\mathbf{q}}_{jk}\right)^T \mathbf{K}_{jk}\left(\dot{\mathbf{q}}_{jkr} - \dot{\mathbf{q}}_{jk}\right). \end{aligned} \tag{4.99}$$

A non-negative accompanying function can now be chosen as

$$\nu_{3jk} = \frac{1}{2}\sum_{\gamma}\left(\theta_{jk\gamma} - \hat{\theta}_{jk\gamma}\right)^2/\rho_{jk\gamma}. \tag{4.100}$$

It follows from (4.96), (4.97), (4.99), and Lemma 2.9 that

$$\begin{aligned} \dot{\nu}_{3jk} &= -\sum_{\gamma}\left(\theta_{jk\gamma} - \hat{\theta}_{jk\gamma}\right)\frac{\dot{\hat{\theta}}_{jk\gamma}}{\rho_{jk\gamma}} \\ &= \sum_{\gamma}\left(\theta_{jk\gamma} - \hat{\theta}_{jk\gamma}\right)\left(s_{jk\gamma} - \frac{\dot{\hat{\theta}}_{jk\gamma}}{\rho_{jk\gamma}}\right) - \mathbf{s}_{jk}^T\left(\boldsymbol{\theta}_{jk} - \hat{\boldsymbol{\theta}}_{jk}\right) \\ &\leqslant -\left(\dot{\mathbf{q}}_{jkr} - \dot{\mathbf{q}}_{jk}\right)^T \mathbf{K}_{jk}\left(\dot{\mathbf{q}}_{jkr} - \dot{\mathbf{q}}_{jk}\right) \\ &\quad + \left(\dot{\mathbf{q}}_{jkr} - \dot{\mathbf{q}}_{jk}\right)^T\left(\boldsymbol{\tau}_{jkr}^* - \boldsymbol{\tau}_{jk}^*\right) \end{aligned} \tag{4.101}$$

holds.

Substituting (4.94) into (4.91) yields

$$\boldsymbol{\tau}_{jkr}^* - \boldsymbol{\tau}_{jk}^* = -\mathbf{Z}^T\left({}^{\mathbf{B}_{jk}}F_r - {}^{\mathbf{B}_{jk}}F\right) \tag{4.102}$$

and subtracting (4.90) from (4.92) gives

$$\mathbf{Z}^T\left(\dot{\mathbf{q}}_{jkr} - \dot{\mathbf{q}}_{jk}\right) = \left({}^{\mathbf{B}_{jk}}V_r - {}^{\mathbf{B}_{jk}}V\right) - {}^{\mathbf{T}_{jk}}\mathbf{U}_{\mathbf{B}_{jk}}^T\left({}^{\mathbf{T}_{jk}}V_r - {}^{\mathbf{T}_{jk}}V\right). \tag{4.103}$$

Consequently, it follows from (2.85), (4.102) and (4.103) that

$$\begin{aligned}
&\left(\dot{\mathbf{q}}_{jkr}-\dot{\mathbf{q}}_{jk}\right)^{T}\left(\boldsymbol{\tau}^{*}_{jkr}-\boldsymbol{\tau}^{*}_{jk}\right)\\
&\quad=-\left(\dot{\mathbf{q}}_{jkr}-\dot{\mathbf{q}}_{jk}\right)^{T}\mathbf{Z}^{T}\left({}^{\mathbf{B}_{jk}}F_{r}-{}^{\mathbf{B}_{jk}}F\right)\\
&\quad=-\left[\left({}^{\mathbf{B}_{jk}}V_{r}-{}^{\mathbf{B}_{jk}}V\right)-{}^{\mathbf{T}_{jk}}\mathbf{U}^{T}_{\mathbf{B}_{jk}}\left({}^{\mathbf{T}_{jk}}V_{r}-{}^{\mathbf{T}_{jk}}V\right)\right]^{T}\\
&\qquad\qquad\times\left({}^{\mathbf{B}_{jk}}F_{r}-{}^{\mathbf{B}_{jk}}F\right)\\
&\quad=-\left({}^{\mathbf{B}_{jk}}V_{r}-{}^{\mathbf{B}_{jk}}V\right)^{T}\left({}^{\mathbf{B}_{jk}}F_{r}-{}^{\mathbf{B}_{jk}}F\right)\\
&\qquad+\left({}^{\mathbf{T}_{jk}}V_{r}-{}^{\mathbf{T}_{jk}}V\right)^{T}{}^{\mathbf{T}_{jk}}\mathbf{U}_{\mathbf{B}_{jk}}\left({}^{\mathbf{B}_{jk}}F_{r}-{}^{\mathbf{B}_{jk}}F\right)\\
&\quad=-p_{\mathbf{B}_{jk}}+p_{\mathbf{T}_{jk}}
\end{aligned}\tag{4.104}$$

holds, in which ${}^{\mathbf{T}_{jk}}F={}^{\mathbf{T}_{jk}}\mathbf{U}_{\mathbf{B}_{jk}}{}^{\mathbf{B}_{jk}}F$ and ${}^{\mathbf{T}_{jk}}F_{r}={}^{\mathbf{T}_{jk}}\mathbf{U}_{\mathbf{B}_{jk}}{}^{\mathbf{B}_{jk}}F_{r}$ are used.

Finally, substituting (4.104) into (4.101) and using Definition 2.17 completes the proof, since frame $\{\mathbf{B}_{jk}\}$ is placed at the sole *driving cutting point* of the joint and frame $\{\mathbf{T}_{jk}\}$ is placed at the sole *driven cutting point* of the joint. ■

Remark 4.17. In Theorem 4.4, only $\dot{\mathbf{q}}_{jkr}-\dot{\mathbf{q}}_{jk}\in L_2$ is ensured. However, $\dot{\mathbf{q}}_{jkr}-\dot{\mathbf{q}}_{jk}\in L_\infty$ can be ensured from (4.103) when both ${}^{\mathbf{B}_{jk}}V_{r}-{}^{\mathbf{B}_{jk}}V\in L_\infty$ and ${}^{\mathbf{B}_{j(k-1)}}V_{r}-{}^{\mathbf{B}_{j(k-1)}}V\in L_\infty$ hold, as guaranteed by Theorem 4.2.

4.8.3 Unactuated Joints with Force/Torque Constraints

In view of the definition of an *open chain*, either a single-DOF prismatic or revolute joint or a three-DOF spherical joint can be unactuated, given $k\in\mathcal{U}_j$. Thus, the following constraints apply to the overall control designs

$$\tau_{jk}=0,\quad \forall k\in\mathcal{U}_j\bigcap\mathcal{C}_{1j}\tag{4.105}$$

and

$$\boldsymbol{\tau}_{jk}=\mathbf{0},\quad \forall k\in\mathcal{U}_j\bigcap\mathcal{C}_{3j}.\tag{4.106}$$

Let

$$\boldsymbol{\tau}_u=\begin{bmatrix}\mathrm{col}(\tau_{jk}\mid k\in\mathcal{U}_j\bigcap\mathcal{C}_{1j}, j\in\{1,n_c\})\\ \mathrm{col}(\boldsymbol{\tau}_{jk}\mid k\in\mathcal{U}_j\bigcap\mathcal{C}_{3j}, j\in\{1,n_c\})\end{bmatrix}\tag{4.107}$$

be a vector for all unactuated joints. Then, the alternative use of (4.39) and (4.63)–(4.65) allows $\boldsymbol{\tau}_u$ to be expressed as an affine function of the desired internal force vector $\boldsymbol{\eta}_{id}$ for all $i\in\{1,n\}$, the required constraint force/moment vector ${}^{\mathbf{S}_i}F_r$ for all $i\in\{1,n\}$, and the time derivative of the required independent velocity vector $\dot{\mathcal{V}}_r$. It follows that

$$\boldsymbol{\tau}_u=\mathcal{J}_\eta\mathrm{col}(\boldsymbol{\eta}_{id}\mid i\in\{1,n_o\})+\mathcal{J}_s\mathrm{col}({}^{\mathbf{S}_i}F_r\mid i\in\{1,n_o\})+\mathcal{J}_v\dot{\mathcal{V}}_r+\boldsymbol{\tau}_{ua}=\mathbf{0}\tag{4.108}$$

holds, where $\mathcal{J}_\eta$, $\mathcal{J}_s$, and $\mathcal{J}_v$ are three matrices and $\boldsymbol{\tau}_{ua}$ is a remaining vector.

In view of (4.12) and (4.36), re-write (4.108) as

$$\boldsymbol{\tau}_u = \mathcal{J}_\eta \mathrm{col}(\boldsymbol{\eta}_{id} \mid i \in \{1, n_o\}) + \mathcal{J}_s \mathrm{col}(\boldsymbol{\varphi}_{id} \mid i \in \{1, n_o\}) + \mathcal{J}_{vu} \dot{\mathcal{V}}_{ur} + \boldsymbol{\tau}_{ub} = \mathbf{0} \tag{4.109}$$

with $\boldsymbol{\tau}_{ub}$ being a remaining term.

Then, design two re-ordering matrices $\mathcal{R}_\eta$ and $\mathcal{R}_\varphi$ to make

$$\begin{aligned}
\mathcal{R}_\eta \mathrm{col}(\boldsymbol{\eta}_{id} \mid i \in \{1, n_o\}) &= \begin{bmatrix} \boldsymbol{\eta}_{cr} \\ \boldsymbol{\eta}_{ur} \end{bmatrix} \\
\mathcal{R}_\varphi \mathrm{col}(\boldsymbol{\varphi}_{id} \mid i \in \{1, n_o\}) &= \begin{bmatrix} \boldsymbol{\varphi}_{cr} \\ \boldsymbol{\varphi}_{ur} \end{bmatrix} \\
\mathcal{J}_\eta \mathcal{R}_\eta^{-1} &= \begin{bmatrix} \mathcal{J}_{\eta c} & \mathcal{J}_{\eta u} \end{bmatrix} \\
\mathcal{J}_s \mathcal{R}_\varphi^{-1} &= \begin{bmatrix} \mathcal{J}_{sc} & \mathcal{J}_{su} \end{bmatrix}
\end{aligned}$$

such that matrix

$$\begin{bmatrix} \mathcal{J}_{\eta u} & \mathcal{J}_{su} & \mathcal{J}_{vu} \end{bmatrix}$$

is of full rank. Finally, re-express (4.109) as

$$\begin{aligned}
\boldsymbol{\tau}_u &= \begin{bmatrix} \mathcal{J}_{\eta c} \; \mathcal{J}_{\eta u} \end{bmatrix} \begin{bmatrix} \boldsymbol{\eta}_{cr} \\ \boldsymbol{\eta}_{ur} \end{bmatrix} + \begin{bmatrix} \mathcal{J}_{sc} \; \mathcal{J}_{su} \end{bmatrix} \begin{bmatrix} \boldsymbol{\varphi}_{cr} \\ \boldsymbol{\varphi}_{ur} \end{bmatrix} + \mathcal{J}_{vu} \dot{\mathcal{V}}_{ur} + \boldsymbol{\tau}_{ub} \\
&= \begin{bmatrix} \mathcal{J}_{\eta u} & \mathcal{J}_{su} & \mathcal{J}_{vu} \end{bmatrix} \begin{bmatrix} \boldsymbol{\eta}_{ur} \\ \boldsymbol{\varphi}_{ur} \\ \dot{\mathcal{V}}_{ur} \end{bmatrix} + \mathcal{J}_{\eta c} \boldsymbol{\eta}_{cr} + \mathcal{J}_{sc} \boldsymbol{\varphi}_{cr} + \boldsymbol{\tau}_{ub} \\
&= \mathbf{0}
\end{aligned} \tag{4.110}$$

such that $\boldsymbol{\eta}_{ur}$, $\boldsymbol{\varphi}_{ur}$, and $\dot{\mathcal{V}}_{ur}$ become affine functions of $\boldsymbol{\eta}_{cr}$ and $\boldsymbol{\varphi}_{cr}$.

As a result, only $\boldsymbol{\eta}_{cr}$ and $\boldsymbol{\varphi}_{cr}$ are subject to independent internal force control and constraint force control, while $\boldsymbol{\eta}_{ur}$, $\boldsymbol{\varphi}_{ur}$, and $\dot{\mathcal{V}}_{ur}$ are released to satisfy (4.110).

4.9 Forward Dynamics

The computations in Sections 4.4–4.8 perform the function of inverse dynamics that yield the control forces/torques from given (desired) accelerations and velocities. The forward dynamics that map the joint control forces/torques to the acceleration coordinates in motion space are to be addressed in this section.

According to (4.7), (4.9), and (4.10), three augmented vectors are defined as

$$\mathcal{F}_a \overset{def}{=} \left[\boldsymbol{\tau}_1^{*T}, \cdots, \boldsymbol{\tau}_j^{*T}, \cdots, \boldsymbol{\tau}_{n_c}^{*T}\right]^T \tag{4.111}$$

$$\mathcal{F}_b \overset{def}{=} \left[{}^{\mathbf{O}_1}F^{*T}, \cdots, {}^{\mathbf{O}_i}F^{*T}, \cdots, {}^{\mathbf{O}_{n_o}}F^{*T}, \mathcal{F}_{b1}^T, \cdots, \mathcal{F}_{bj}^T, \cdots, \mathcal{F}_{bn_c}^T\right]^T \tag{4.112}$$

$$\mathcal{F}_{bj} = \left[{}^{\mathbf{B}_j}F^{*T}, {}^{\mathbf{B}_{j1}}F^{*T}, \cdots, {}^{\mathbf{B}_{jk}}F^{*T}, \cdots, {}^{\mathbf{B}_{jl_j}}F^{*T}\right]^T, \; j \in \{1, n_c\} \tag{4.113}$$

where $\boldsymbol{\tau}_j^* \in \mathbb{R}^{(l_{1j}+3l_{3j})}$, $j \in \{1, n_c\}$, denotes the net joint force/torque vector of the jth *open chain* with elements being defined by (4.74) and (4.91).

Partition the velocity mapping matrix $\mathcal{J}_a$ defined by (4.6) into rows as

$$\mathcal{J}_a = \left[\mathcal{J}_{a1}^T, \cdots, \mathcal{J}_{aj}^T, \cdots, \mathcal{J}_{an_c}^T\right]^T \tag{4.114}$$

where $\mathcal{J}_{aj}$, which maps the independent velocity coordinate vector $\mathcal{V}$ to the joint velocity vector of the jth *open chain* $\dot{\mathbf{q}}_j \in \mathbb{R}^{(l_{1j}+3l_{3j})}$, can be further partitioned to

$$\mathcal{J}_{aj} = \left[\mathcal{J}_{j1}^T, \cdots, \mathcal{J}_{jk}^T, \cdots, \mathcal{J}_{jl_j}^T\right]^T \tag{4.115}$$

where $\mathcal{J}_{jk}$ maps the independent velocity coordinate vector $\mathcal{V}$ to the joint velocity of the kth joint of the jth *open chain*, denoted as $\dot{q}_{jk} \in \mathbb{R}$ if $k \in \mathcal{C}_{1j}$ or $\dot{\mathbf{q}}_{jk} \in \mathbb{R}^3$ if $k \in \mathcal{C}_{3j}$.

Meanwhile, partition the velocity mapping matrix $\mathcal{J}_b$ defined by (4.8) into rows as

$$\mathcal{J}_b = \left[\mathcal{J}_{\mathbf{O}_1}^T, \cdots, \mathcal{J}_{\mathbf{O}_i}^T, \cdots, \mathcal{J}_{\mathbf{O}_{n_o}}^T, \mathcal{J}_{b1}^T, \cdots, \mathcal{J}_{bj}^T, \cdots, \mathcal{J}_{bn_c}^T\right]^T \tag{4.116}$$

where $\mathcal{J}_{\mathbf{O}_i}$ maps the independent velocity coordinate vector $\mathcal{V}$ to ${}^{\mathbf{O}_i}V$ and $\mathcal{J}_{bj}$ maps $\mathcal{V}$ to $\mathcal{V}_{bj}$. Matrix $\mathcal{J}_{bj}$ can be further partitioned to

$$\mathcal{J}_{bj} = \left[\mathcal{J}_{\mathbf{B}_j}^T, \mathcal{J}_{\mathbf{B}_{j1}}^T, \cdots, \mathcal{J}_{\mathbf{B}_{jk}}^T, \cdots, \mathcal{J}_{\mathbf{B}_{jl_j}}^T\right]^T \tag{4.117}$$

according to (4.10), where $\mathcal{J}_{\mathbf{B}_j}$ and $\mathcal{J}_{\mathbf{B}_{jk}}$ map $\mathcal{V}$ to ${}^{\mathbf{B}_j}V$ and ${}^{\mathbf{B}_{jk}}V$, respectively.

Note the fact that the n_o *objects* are all connected to the n_c *open chains*, which makes

$$\bigcup_{i \in \{1, n_o\}} \mathcal{T}_i = \{\{\mathbf{T}_1\}, \cdots, \{\mathbf{T}_j\}, \cdots, \{\mathbf{T}_{n_c}\}\} \tag{4.118}$$

$$\bigcup_{i \in \{1, n_o\}} \mathcal{B}_i \subseteq \{\{\mathbf{B}_1\}, \cdots, \{\mathbf{B}_j\}, \cdots, \{\mathbf{B}_{n_c}\}\}\,. \tag{4.119}$$

Expressions (4.118) and (4.119) mean that the n_c *driving cutting points* of all *open chains* are connected to the *driven cutting points* of the n_o *objects*, while some *open chains* are connected to the base (as *source nodes*) with $\mathcal{J}_{\mathbf{B}_j} = \mathbf{0}$.

Premultiplying $\mathcal{F}_a$ in (4.111) by $\mathcal{J}_a^T$ in (4.114) and using (4.74) and (4.91) yields

$$\begin{aligned}
\mathcal{J}_a^T \mathcal{F}_a &= \sum_{j=1}^{n_c} \mathcal{J}_{aj}^T \boldsymbol{\tau}_j^* \\
&= \sum_{j=1}^{n_c} \mathcal{J}_{aj}^T \boldsymbol{\tau}_j - \sum_{j=1}^{n_c} \left(\sum_{k \in \mathcal{C}_{1j}} \mathcal{J}_{jk}^T \mathbf{z}^{T\,\mathbf{B}_{jk}}F + \sum_{k \in \mathcal{C}_{3j}} \mathcal{J}_{jk}^T \mathbf{Z}^{T\,\mathbf{B}_{jk}}F \right)
\end{aligned} \tag{4.120}$$

where $\mathcal{J}_{aj}$ is defined by (4.115) and $\boldsymbol{\tau}_j \in \mathbb{R}^{(l_{1j}+3l_{3j})}$ denotes the joint control force/torque vector of the jth *open chain*.

In view of (4.118) and (4.119), premultiplying $\mathcal{F}_b$ in (4.112) by $\mathcal{J}_b^T$ in (4.116) and using (4.25), (4.51)–(4.53), (4.73), (4.112), (4.113), and (4.117) yields

$$\mathcal{J}_b^T \mathcal{F}_b = -\sum_{i=1}^{n_o} \mathcal{J}_{\mathbf{O}_i}^T {}^{\mathbf{O}_i}\mathbf{U}_{\mathbf{S}_i} {}^{\mathbf{S}_i}F + \sum_{j=1}^{n_c} \left(\sum_{k\in\mathcal{C}_{1j}} \mathcal{J}_{jk}^T \mathbf{z}^T {}^{\mathbf{B}_{jk}}F + \sum_{k\in\mathcal{C}_{3j}} \mathcal{J}_{jk}^T \mathbf{Z}^T {}^{\mathbf{B}_{jk}}F \right). \tag{4.121}$$

The detailed derivation is given in Appendix B.2.

On the other hand, $\mathcal{F}_a$ in (4.111) can be re-written as

$$\mathcal{F}_a = \mathcal{M}_a \mathcal{J}_a \dot{\mathcal{V}} + \mathcal{M}_a \dot{\mathcal{J}}_a \mathcal{V} + \mathcal{G}_a \tag{4.122}$$

by using (4.6), (4.7), (4.74), and (4.91), where

$$\mathcal{M}_a = \mathrm{diag}(\mathbf{M}_{a1}, \cdots, \mathbf{M}_{aj}, \cdots, \mathbf{M}_{an_c}) \tag{4.123}$$

$$\mathcal{G}_a = \left[\mathcal{G}_{a1}^T, \cdots, \mathcal{G}_{aj}^T, \cdots, \mathcal{G}_{an_c}^T\right]^T \tag{4.124}$$

with

$$\mathbf{M}_{aj} = \mathrm{diag}(I_{j1}, \cdots, I_{jk}, \cdots, I_{jl_j}) \tag{4.125}$$

$$\mathcal{G}_{aj} = \left[\boldsymbol{\alpha}_{j1}^T, \cdots, \boldsymbol{\alpha}_{jk}^T, \cdots, \boldsymbol{\alpha}_{jl_j}^T\right]^T \tag{4.126}$$

subject to

$$I_{jk} = \begin{cases} J_{mjk} & \text{if } k \in \mathcal{C}_{1j} \\ \mathbf{0} & \text{if } k \in \mathcal{C}_{3j} \end{cases} \tag{4.127}$$

$$\boldsymbol{\alpha}_{jk} = \begin{cases} \xi_{jk} & \text{if } k \in \mathcal{C}_{1j} \\ \boldsymbol{\xi}_{jk} & \text{if } k \in \mathcal{C}_{3j} \end{cases}. \tag{4.128}$$

Similarly, $\mathcal{F}_b$ can be written as

$$\mathcal{F}_b = \mathcal{M}_b \mathcal{J}_b \dot{\mathcal{V}} + \mathcal{M}_b \dot{\mathcal{J}}_b \mathcal{V} + \mathcal{C}_b \mathcal{J}_b \mathcal{V} + \mathcal{G}_b \tag{4.129}$$

by using (4.8), (4.24), (4.49), and (4.50), where

$$\mathcal{M}_b = \mathrm{diag}(\mathbf{M}_{\mathbf{O}_1}, \cdots, \mathbf{M}_{\mathbf{O}_i}, \cdots, \mathbf{M}_{\mathbf{O}_{n_o}}, \mathcal{M}_{b1}, \cdots, \mathcal{M}_{bj}, \cdots, \mathcal{M}_{bn_c}) \tag{4.130}$$

$$\mathcal{C}_b = \mathrm{diag}(\mathbf{C}_{\mathbf{O}_1}, \cdots, \mathbf{C}_{\mathbf{O}_i}, \cdots, \mathbf{C}_{\mathbf{O}_{n_o}}, \mathcal{C}_{b1}, \cdots, \mathcal{C}_{bj}, \cdots, \mathcal{C}_{bn_c}) \tag{4.131}$$

$$\mathcal{G}_b = \left[\mathbf{G}_{\mathbf{O}_1}^T, \cdots, \mathbf{G}_{\mathbf{O}_i}^T, \cdots, \mathbf{G}_{\mathbf{O}_{n_o}}^T, \mathcal{G}_{b1}^T, \cdots, \mathcal{G}_{bj}^T, \cdots, \mathcal{G}_{bn_c}^T\right]^T \tag{4.132}$$

with

$$\mathcal{M}_{bj} = \mathrm{diag}(\mathbf{M}_{\mathbf{B}_j}, \mathbf{M}_{\mathbf{B}_{j1}}, \cdots, \mathbf{M}_{\mathbf{B}_{jk}}, \cdots, \mathbf{M}_{\mathbf{B}_{jl_j}}) \tag{4.133}$$

$$\mathcal{C}_{bj} = \mathrm{diag}(\mathbf{C}_{\mathbf{B}_j}, \mathbf{C}_{\mathbf{B}_{j1}}, \cdots, \mathbf{C}_{\mathbf{B}_{jk}}, \cdots, \mathbf{C}_{\mathbf{B}_{jl_j}}) \tag{4.134}$$

$$\mathcal{G}_{bj} = \left[\mathbf{G}_{\mathbf{B}_j}^T, \mathbf{G}_{\mathbf{B}_{j1}}^T, \cdots, \mathbf{G}_{\mathbf{B}_{jk}}^T \cdots, \mathbf{G}_{\mathbf{B}_{jl_j}}^T\right]^T. \tag{4.135}$$

Finally, substituting (4.122) and (4.129) into (4.120) and (4.121) yields

$$\mathcal{M}\dot{\mathcal{V}} + \mathcal{C}\mathcal{V} + \mathcal{G} + \mathcal{F}_s = \mathcal{J}_a^T \boldsymbol{\tau} \tag{4.136}$$

where

$$\mathcal{M} = \mathcal{J}_b^T \mathcal{M}_b \mathcal{J}_b + \mathcal{J}_a^T \mathcal{M}_a \mathcal{J}_a \tag{4.137}$$

$$\mathcal{C} = \mathcal{J}_b^T \mathcal{M}_b \dot{\mathcal{J}}_b + \mathcal{J}_a^T \mathcal{M}_a \dot{\mathcal{J}}_a + \mathcal{J}_b^T \mathcal{C}_b \mathcal{J}_b \tag{4.138}$$

$$\mathcal{G} = \mathcal{J}_b^T \mathcal{G}_b + \mathcal{J}_a^T \mathcal{G}_a \tag{4.139}$$

$$\mathcal{F}_s = \sum_{i=1}^{n_o} \mathcal{J}_{\mathbf{O}_i}^T \, {}^{\mathbf{O}_i}\mathbf{U}_{\mathbf{S}_i} \, {}^{\mathbf{S}_i}F \tag{4.140}$$

$$\boldsymbol{\tau} = \left[\boldsymbol{\tau}_1^T, \cdots, \boldsymbol{\tau}_j^T, \cdots, \boldsymbol{\tau}_{n_c}^T\right]^T. \tag{4.141}$$

In view of the definition of $\mathcal{J}_{\mathbf{O}_i}$ in (4.116), it follows from (4.22) and (4.23) that

$$\begin{aligned} {}^{\mathbf{S}_i}V &= \mathrm{diag}\left({}^{\mathbf{S}_i}\mathbf{R}_{\mathbf{S}_{bi}}, {}^{\mathbf{S}_i}\mathbf{R}_{\mathbf{S}_{bi}}\right) \mathbf{T}_{si} \boldsymbol{\chi}_i \\ &= {}^{\mathbf{O}_i}\mathbf{U}_{\mathbf{S}_i}^T \, {}^{\mathbf{O}_i}V \\ &= {}^{\mathbf{O}_i}\mathbf{U}_{\mathbf{S}_i}^T \mathcal{J}_{\mathbf{O}_i} \mathcal{V} \end{aligned} \tag{4.142}$$

holds. By using $\mathbf{T}_{si}^T \mathbf{T}_{si} = \mathbf{I}$ and $\mathbf{T}_{fi}^T \mathbf{T}_{si} = \mathbf{0}$, it yields

$$\mathbf{T}_{fi}^T \mathrm{diag}\left({}^{\mathbf{S}_{bi}}\mathbf{R}_{\mathbf{S}_i}, {}^{\mathbf{S}_{bi}}\mathbf{R}_{\mathbf{S}_i}\right) {}^{\mathbf{O}_i}\mathbf{U}_{\mathbf{S}_i}^T \mathcal{J}_{\mathbf{O}_i} = \mathbf{0} \tag{4.143}$$

from (4.142). It then follows from (4.26) and (4.143) that

$$\mathcal{J}_{\mathbf{O}_i}^T \, {}^{\mathbf{O}_i}\mathbf{U}_{\mathbf{S}_i} \, {}^{\mathbf{S}_i}F = \mathcal{J}_{\mathbf{O}_i}^T \, {}^{\mathbf{O}_i}\mathbf{U}_{\mathbf{S}_i} \mathrm{diag}\left({}^{\mathbf{S}_i}\mathbf{R}_{\mathbf{S}_{bi}}, {}^{\mathbf{S}_i}\mathbf{R}_{\mathbf{S}_{bi}}\right) \mathbf{T}_{si} \boldsymbol{\psi}_i \tag{4.144}$$

holds.

Remark 4.18. Equations (4.140) and (4.144) suggest that the term $\mathcal{F}_s$ in (4.136) denotes the sum of the projections of the contact force/moment vectors of the n_o *objects* onto the independent velocity configuration space. When this term solely includes friction forces and moments and can be expressed by

$$\|\boldsymbol{\psi}_i\| \leqslant k_{\psi 1} \|\boldsymbol{\chi}_i\| + k_{\psi 2}, \quad \forall i \in \{1, n_o\} \tag{4.145}$$

with $0 < k_{\psi 1} < +\infty$ and $0 < k_{\psi 2} < +\infty$, the boundedness of $\mathcal{F}_s$ is completely dependent on the boundedness of $\mathcal{V}$.

Remark 4.19. In (4.137), $\mathcal{M}_b$ is a symmetric positive-definite matrix and $\mathcal{M}_a$ is a symmetric semi-positive-definite matrix. Note that $\mathcal{J}_b$ is of full column-rank. Therefore, $\mathcal{M}$ is of full rank.

Equation (4.136) gives the forward dynamics of the entire complex robot comprised of n_o *objects* and n_c *open chains*. Given the joint control force/torque vectors of all the *open chains*, denoted as $\boldsymbol{\tau}_j$ for all $j \in \{1, n_c\}$, the time derivative of the independent velocity coordinate vector $\dot{\mathcal{V}}$ can be obtained from (4.136), in view of Remark 4.19.

Theorem 4.5. *The forward dynamics of a complex robot given by Definition 4.1 can be expressed by the form of (4.136), together with (4.137)–(4.141). The time derivative of the independent velocity coordinate vector, denoted as* $\dot{\mathcal{V}}$, *is bounded, if* $\boldsymbol{\tau}_j \in L_\infty$ *for all* $j \in \{1, n_c\}$, $\mathcal{V} \in L_\infty$, *and* $\mathcal{F}_s \in L_\infty$ *hold.*

Proof: The first statement is valid by following the development in this subsection. The second statement is valid, based on the fact that a bounded $\mathcal{V}$ yields a bounded matrix $\mathcal{C}$, a bounded vector $\mathcal{G}$, and a bounded $\mathcal{F}_s$ in view of Remark 4.18. Then, the bounded control force/torque vector $\boldsymbol{\tau}_j \in L_\infty$ for all $j \in \{1, n_c\}$ results in a bounded $\dot{\mathcal{V}} \in L_\infty$, in view of Remark 4.19. ■

4.10 Stability and Convergence

In this section, the stability and convergence results of the entire complex robot are to be presented.

4.10.1 L_2 and L_∞ Stability

In view of Theorems 4.1–4.4, it follows from Theorem 2.1 that

$$ {}^{\mathbf{O}_i}V_r - {}^{\mathbf{O}_i}V \in L_2 \bigcap L_\infty, \ \forall i \in \{1, n_o\} \tag{4.146} $$

$$ {}^{\mathbf{B}_j}V_r - {}^{\mathbf{B}_j}V \in L_2 \bigcap L_\infty, \ \forall j \in \{1, n_c\} \tag{4.147} $$

$$ {}^{\mathbf{B}_{jk}}V_r - {}^{\mathbf{B}_{jk}}V \in L_2 \bigcap L_\infty, \ \forall j \in \{1, n_c\}, \ \forall k \in \{1, l_j\} \tag{4.148} $$

$$ \dot{q}_{jkr} - \dot{q}_{jk} \in L_2 \bigcap L_\infty, \ \forall j \in \{1, n_c\}, \ \forall k \in \mathcal{C}_{1j} \tag{4.149} $$

$$ \dot{\mathbf{q}}_{jkr} - \dot{\mathbf{q}}_{jk} \in L_2 \bigcap L_\infty, \ \forall j \in \{1, n_c\}, \ \forall k \in \mathcal{C}_{3j} \tag{4.150} $$

hold. Expressions (4.146)–(4.150) imply

$$ \mathcal{V}_r - \mathcal{V} \in L_2 \bigcap L_\infty \tag{4.151} $$

from (4.3) and (4.13).

4.10.2 Asymptotic Stability

In view of (4.151) and Lemma 2.8, the asymptotic stability represented by $\mathcal{V}_r - \mathcal{V} \to \mathbf{0}$ depends on the boundedness of $\dot{\mathcal{V}}_r - \dot{\mathcal{V}}$.

Expression (4.151) indicates that $\mathcal{V}$ is bounded for a bounded $\mathcal{V}_r$. If the time derivative of the required independent velocity vector $\dot{\mathcal{V}}_r$ is bounded through a proper motion planning, then $\dot{\mathcal{V}}_{er}$ defined by (4.16) is bounded, since $\dot{\mathcal{J}}_e$ is bounded for a bounded $\mathcal{V}$. The boundedness of $\dot{\mathcal{V}}_{er}$ and $\mathcal{V}_{er}$ further ensures the boundedness of ${}^{\mathbf{O}_i}F_r^*$ for all $i \in \{1, n_o\}$, the boundedness of ${}^{\mathbf{B}_j}F_r^*$ for all $j \in \{1, n_c\}$, and the boundedness of ${}^{\mathbf{B}_{jk}}F_r^*$ for all $k \in \{1, l_j\}$ and $j \in \{1, n_c\}$, in view of (4.32), (4.57), (4.58), and (2.78). It also ensures the boundedness of τ_{jkr}^* from (4.77) and the boundedness of $\boldsymbol{\tau}_{jkr}^*$ from (4.93). This, in turn, implies the boundedness of

$$\mathcal{F}_{er} = \Big[{}^{\mathbf{O}_1}F_r^{*T}, \cdots, {}^{\mathbf{O}_i}F_r^{*T}, \cdots, {}^{\mathbf{O}_{n_o}}F_r^{*T}, \\ \mathcal{F}_{e1r}^T, \cdots, \mathcal{F}_{ejr}^T, \cdots, \mathcal{F}_{en_cr}^T\Big]^T \tag{4.152}$$

with

$$\mathcal{F}_{ejr} = \Big[\boldsymbol{\tau}_{jr}^{*T}, {}^{\mathbf{B}_j}F_r^{*T}, {}^{\mathbf{B}_{j1}}F_r^{*T}, \cdots, {}^{\mathbf{B}_{jk}}F_r^{*T}, \cdots, {}^{\mathbf{B}_{jl_j}}F_r^{*T}\Big]^T \tag{4.153}$$

where $\boldsymbol{\tau}_{jr}^*$ is a vector with its kth element being either τ_{jkr}^* for $k \in \mathcal{C}_{1j}$ or $\boldsymbol{\tau}_{jkr}^*$ for $k \in \mathcal{C}_{3j}$. If ${}^{\mathbf{S}_i}F_r$ and $\boldsymbol{\eta}_{id}$ are bounded for all $i \in \{1, n_o\}$, then the boundedness of ${}^{\mathbf{T}_j}F_r$ for all $\{\mathbf{T}_j\} \in \mathcal{T}_i$ and $i \in \{1, n_o\}$ can be ensured, in view of (4.39). It further yields the boundedness of ${}^{\mathbf{B}_{jk}}F_r$ and ${}^{\mathbf{B}_j}F_r$ for all $k \in \{1, l_j\}$ and $j \in \{1, n_c\}$, in view of (4.63)–(4.65). Finally, it yields the boundedness of τ_{jk} for all $k \in \mathcal{C}_{1j}$ and the boundedness of $\boldsymbol{\tau}_{jk}$ for all $k \in \mathcal{C}_{3j}$, in view of (4.78) and (4.94).

On the other hand, the boundedness of $\mathcal{V}$ implies the boundedness of $\mathcal{J}_{\mathbf{O}_i}^T {}^{\mathbf{O}_i}\mathbf{U}_{\mathbf{S}_i} {}^{\mathbf{S}_i}F$ for all $i \in \{1, n_o\}$ from Remark 4.18. Thus, the boundedness of $\dot{\mathcal{V}}$ can be guaranteed from Theorem 4.5. Eventually, the boundedness of $\dot{\mathcal{V}}_r - \dot{\mathcal{V}}$ ensures the asymptotic velocity tracking of

$$\mathcal{V}_r - \mathcal{V} \to \mathbf{0} \tag{4.154}$$

from (4.151) and Lemma 2.8.

Expression (4.154) implies

$$ {}^{\mathbf{O}_i}V_r - {}^{\mathbf{O}_i}V \to \mathbf{0}, \ \forall i \in \{1, n_o\} \tag{4.155}$$

$$ {}^{\mathbf{B}_j}V_r - {}^{\mathbf{B}_j}V \to \mathbf{0}, \ \forall j \in \{1, n_c\} \tag{4.156}$$

$$ \dot{q}_{jkr} - \dot{q}_{jk} \to 0, \ \forall j \in \{1, n_c\}, \forall k \in \mathcal{C}_{1j} \tag{4.157}$$

$$ \dot{\mathbf{q}}_{jkr} - \dot{\mathbf{q}}_{jk} \to \mathbf{0}, \ \forall j \in \{1, n_c\}, \forall k \in \mathcal{C}_{3j}. \tag{4.158}$$

4.10.3 Force Errors

After the asymptotic stability of motion has been studied in the previous subsection, both constraint and internal force errors are to be studied in this subsection.

From (4.25) and (4.35), it follows that

$$\begin{aligned}{}^{\mathbf{O}_i}\mathbf{U}_{\mathbf{S}_i}\left({}^{\mathbf{S}_i}F_r - {}^{\mathbf{S}_i}F\right) = & \sum_{\{\mathbf{T}_j\}\in\mathcal{T}_i} {}^{\mathbf{O}_i}\mathbf{U}_{\mathbf{T}_j}\left({}^{\mathbf{T}_j}F_r - {}^{\mathbf{T}_j}F\right) \\ & - \sum_{\{\mathbf{B}_j\}\in\mathcal{B}_i} {}^{\mathbf{O}_i}\mathbf{U}_{\mathbf{B}_j}\left({}^{\mathbf{B}_j}F_r - {}^{\mathbf{B}_j}F\right) \\ & - \left({}^{\mathbf{O}_i}F_r^* - {}^{\mathbf{O}_i}F^*\right) \end{aligned} \quad (4.159)$$

holds. Subtracting (4.28) from (4.39) and then premultiplying it by $\mathbf{\Psi}_i \text{diag}\left({}^{\mathbf{O}_i}\mathbf{U}_{\mathbf{T}_j} \mid \{\mathbf{T}_j\} \in \mathcal{T}_i\right)$ yields

$$\boldsymbol{\eta}_{id} - \boldsymbol{\eta}_i = \mathbf{\Psi}_i \text{diag}\left({}^{\mathbf{O}_i}\mathbf{U}_{\mathbf{T}_j} \mid \{\mathbf{T}_j\} \in \mathcal{T}_i\right) \text{col}\left({}^{\mathbf{T}_j}F_r - {}^{\mathbf{T}_j}F \mid \{\mathbf{T}_j\} \in \mathcal{T}_i\right). \quad (4.160)$$

Equations (4.159) and (4.160) indicate that the internal force error $\boldsymbol{\eta}_{id} - \boldsymbol{\eta}_i$ depends on the force/moment vector error ${}^{\mathbf{T}_j}F_r - {}^{\mathbf{T}_j}F$ for all $\{\mathbf{T}_j\} \in \mathcal{T}_i$, whereas the constraint force error ${}^{\mathbf{S}_i}F_r - {}^{\mathbf{S}_i}F$ depends on the force/moment vector errors ${}^{\mathbf{T}_j}F_r - {}^{\mathbf{T}_j}F$ for all $\{\mathbf{T}_j\} \in \mathcal{T}_i$ and ${}^{\mathbf{B}_j}F_r - {}^{\mathbf{B}_j}F$ for all $\{\mathbf{B}_j\} \in \mathcal{B}_i$ and also depends on the net force/moment vector error ${}^{\mathbf{O}_i}F_r^* - {}^{\mathbf{O}_i}F^*$ described by (B.1).

Furthermore, (4.51)–(4.53) and (4.63)–(4.65) imply that

$${}^{\mathbf{B}_{jk}}F_r - {}^{\mathbf{B}_{jk}}F = \sum_{\gamma=k}^{l_j} {}^{\mathbf{B}_{jk}}\mathbf{U}_{\mathbf{B}_{j\gamma}}\left({}^{\mathbf{B}_{j\gamma}}F_r^* - {}^{\mathbf{B}_{j\gamma}}F^*\right) + {}^{\mathbf{B}_{jk}}\mathbf{U}_{\mathbf{T}_j}\left({}^{\mathbf{T}_j}F_r - {}^{\mathbf{T}_j}F\right) \quad (4.161)$$

and

$${}^{\mathbf{B}_{jk}}F_r - {}^{\mathbf{B}_{jk}}F = -\sum_{\gamma=0}^{k-1} {}^{\mathbf{B}_{jk}}\mathbf{U}_{\mathbf{B}_{j\gamma}}\left({}^{\mathbf{B}_{j\gamma}}F_r^* - {}^{\mathbf{B}_{j\gamma}}F^*\right) + {}^{\mathbf{B}_{jk}}\mathbf{U}_{\mathbf{B}_j}\left({}^{\mathbf{B}_j}F_r - {}^{\mathbf{B}_j}F\right) \quad (4.162)$$

hold with $\mathbf{B}_{j0} = \mathbf{B}_j$.

Premultiplying (4.161) and (4.162) by $\mathbf{z}^T$ if $k \in \mathcal{C}_{1j}$ or by $\mathbf{Z}^T$ if $k \in \mathcal{C}_{3j}$ and listing them all together for all $k \in \{1, l_j\}$ yields

$$\begin{aligned} & \begin{bmatrix} \text{col}\left(\mathbf{z}^T\left({}^{\mathbf{B}_{jk}}F_r - {}^{\mathbf{B}_{jk}}F\right) \mid k \in \mathcal{C}_{1j}\right) \\ \text{col}\left(\mathbf{Z}^T\left({}^{\mathbf{B}_{jk}}F_r - {}^{\mathbf{B}_{jk}}F\right) \mid k \in \mathcal{C}_{3j}\right) \end{bmatrix} \\ & = \mathcal{A}_{Tj}\left({}^{\mathbf{T}_j}F_r - {}^{\mathbf{T}_j}F\right) + \mathcal{B}_{Tj} && (4.163) \\ & = \mathcal{A}_{Bj}\left({}^{\mathbf{B}_j}F_r - {}^{\mathbf{B}_j}F\right) - \mathcal{B}_{Bj} && (4.164) \end{aligned}$$

where

$$\mathcal{A}_{Tj} = \begin{bmatrix} \text{col}\left(\mathbf{z}^T {}^{\mathbf{B}_{jk}}\mathbf{U}_{\mathbf{T}_j} \mid k \in \mathcal{C}_{1j}\right) \\ \text{col}\left(\mathbf{Z}^T {}^{\mathbf{B}_{jk}}\mathbf{U}_{\mathbf{T}_j} \mid k \in \mathcal{C}_{3j}\right) \end{bmatrix} \quad (4.165)$$

$$\mathcal{B}_{Tj} = \begin{bmatrix} \text{col}\left(\mathbf{z}^T \sum_{\gamma=k}^{l_j} {}^{\mathbf{B}_{jk}}\mathbf{U}_{\mathbf{B}_{j\gamma}}\left({}^{\mathbf{B}_{j\gamma}}F_r^* - {}^{\mathbf{B}_{j\gamma}}F^*\right) \mid k \in \mathcal{C}_{1j}\right) \\ \text{col}\left(\mathbf{Z}^T \sum_{\gamma=k}^{l_j} {}^{\mathbf{B}_{jk}}\mathbf{U}_{\mathbf{B}_{j\gamma}}\left({}^{\mathbf{B}_{j\gamma}}F_r^* - {}^{\mathbf{B}_{j\gamma}}F^*\right) \mid k \in \mathcal{C}_{3j}\right) \end{bmatrix} \quad (4.166)$$

$$\mathcal{A}_{Bj} = \begin{bmatrix} \text{col}\left(\mathbf{z}^{T\mathbf{B}_{jk}}\mathbf{U}_{\mathbf{B}_j} \mid k \in \mathcal{C}_{1j}\right) \\ \text{col}\left(\mathbf{Z}^{T\mathbf{B}_{jk}}\mathbf{U}_{\mathbf{B}_j} \mid k \in \mathcal{C}_{3j}\right) \end{bmatrix} \tag{4.167}$$

$$\mathcal{B}_{Bj} = \begin{bmatrix} \text{col}\left(\mathbf{z}^T \sum_{\gamma=0}^{k-1} {}^{\mathbf{B}_{jk}}\mathbf{U}_{\mathbf{B}_{j\gamma}} \left({}^{\mathbf{B}_{j\gamma}}F_r^* - {}^{\mathbf{B}_{j\gamma}}F^*\right) \mid k \in \mathcal{C}_{1j}\right) \\ \text{col}\left(\mathbf{Z}^T \sum_{\gamma=0}^{k-1} {}^{\mathbf{B}_{jk}}\mathbf{U}_{\mathbf{B}_{j\gamma}} \left({}^{\mathbf{B}_{j\gamma}}F_r^* - {}^{\mathbf{B}_{j\gamma}}F^*\right) \mid k \in \mathcal{C}_{3j}\right) \end{bmatrix} \tag{4.168}$$

with $\mathbf{B}_{j0} = \mathbf{B}_j$.

Using (4.83) and (4.87) results in

$$\begin{aligned} \mathbf{z}^{T\mathbf{B}_{jk}}F_r - \mathbf{z}^{T\mathbf{B}_{jk}}F &= -\left(\tau_{jkr}^* - \tau_{jk}^*\right) \\ &= -J_{mjk}\left(\ddot{q}_{jkr} - \ddot{q}_{jk}\right) - k_{cjk}\left[\text{sign}(\dot{q}_{jkr}) - \text{sign}(\dot{q}_{jk})\right] \\ &\quad -k_{vjk}\left(\dot{q}_{jkr} - \dot{q}_{jk}\right) + \mathbf{Y}_{jk}\left(\boldsymbol{\theta}_{jk} - \hat{\boldsymbol{\theta}}_{jk}\right) \\ &\quad -k_{jk}\left(\dot{q}_{jkr} - \dot{q}_{jk}\right). \end{aligned} \tag{4.169}$$

Similarly, substituting (4.94) into (4.91) and using (4.93) and (4.95) yields

$$\begin{aligned} \mathbf{Z}^{T\mathbf{B}_{jk}}F_r - \mathbf{Z}^{T\mathbf{B}_{jk}}F &= -\left(\boldsymbol{\tau}_{jkr}^* - \boldsymbol{\tau}_{jk}^*\right) \\ &= \mathbf{Y}_{jk}\left(\boldsymbol{\theta}_{jk} - \hat{\boldsymbol{\theta}}_{jk}\right) - \mathbf{K}_{jk}\left(\dot{\mathbf{q}}_{jkr} - \dot{\mathbf{q}}_{jk}\right). \end{aligned} \tag{4.170}$$

Equations (4.169) and (4.170) imply that the term in the left-hand side of (4.163) is a function of the dynamic uncertainties and the feedback control terms of the l_j joints of the jth *open chain*. Accordingly, $\mathcal{B}_{Tj}$ and $\mathcal{B}_{Bj}$ in (4.163) and (4.164) are functions of the dynamic uncertainties and the feedback control terms of the $l_j + 1$ links of the jth *open chain*. Consequently, as long as $\mathcal{A}_{Tj}$ and $\mathcal{A}_{Bj}$ in (4.163) and (4.164) are of full column-rank, ${}^{\mathbf{T}_j}F_r - {}^{\mathbf{T}_j}F$ and ${}^{\mathbf{B}_j}F_r - {}^{\mathbf{B}_j}F$ can be determined as functions of the dynamic uncertainties and the feedback control terms of the jth *open chain*. Thus, $\boldsymbol{\eta}_{id} - \boldsymbol{\eta}_i$ and ${}^{\mathbf{S}_i}F_r - {}^{\mathbf{S}_i}F$ are functions of the dynamic uncertainties and the feedback control terms of the entire complex robot.

The feedback control terms comprise control gains and state errors. Large control gains will produce large constraint/internal force errors for given state errors. When Assumption 1 is valid, the asymptotic stability of the entire system eventually overcomes the transient effects caused by the feedback control action. Thus, the constraint/internal force errors eventually depend only on the dynamic uncertainties which are completely determined by the parameter uncertainties. Therefore, a dynamically well identified system produces small constraint/internal force errors. For a system with both known kinematics and known dynamics, the asymptotic stability of the constraint/internal force error dynamics can be ensured ultimately.

4.11 Position Control Based on Velocity Convergence

The essence of the VDC approach provides a velocity control interface, subject to constraint forces and internal forces, in the sense of

$$\mathcal{V}_r - \mathcal{V} \in L_2 \bigcap L_\infty \tag{4.171}$$

and

$$\mathcal{V}_r - \mathcal{V} \to \mathbf{0} \tag{4.172}$$

as described by (4.151) and (4.154). In this subsection, it will be shown that the position and orientation control can be easily accomplished based on an asymptotically stable velocity controller.

Expressions (4.171) and (4.172) imply

$$\mathcal{V}_{pr} - \mathcal{V}_p \in L_2 \bigcap L_\infty \tag{4.173}$$

$$\mathcal{V}_{pr} - \mathcal{V}_p \to \mathbf{0} \tag{4.174}$$

based on the definitions in (4.2) and (4.12). Taking the same form as (3.20), the required independent velocity coordinate vector for position control $\mathcal{V}_{pr}$ can be designed as

$$\mathcal{V}_{pr} = \mathcal{V}_{pd} + \mathcal{E} \tag{4.175}$$

where $\mathcal{V}_{pd}$ denotes the desired independent velocity coordinate vector for position control and $\mathcal{E}$ is a vector representing the scaled position and orientation errors and must satisfy the following two conditions:

$$\text{i)} \int_0^\infty (\mathcal{V}_{pd} - \mathcal{V}_p)^T \mathcal{E} dt \geqslant -\gamma_p \tag{4.176}$$

$$\text{ii)} \begin{cases} \mathcal{V}_{pd} - \mathcal{V}_p \in L_\infty \\ \mathcal{E} \in L_\infty \end{cases} \quad \text{for given} \quad \mathcal{V}_{pr} - \mathcal{V}_p \in L_\infty \tag{4.177}$$

with $\gamma_p \geqslant 0$ being a constant.

Three lemmas below are to give specific designs satisfying the two conditions.

Lemma 4.3. *Let* $\mathbf{e}(t)$ *be a linear position error vector. If* $\mathcal{E}(t) = \lambda \mathbf{e}(t)$ *and* $\mathcal{V}_{pd}(t) - \mathcal{V}_p(t) = \dot{\mathbf{e}}(t)$ *hold with* $\lambda > 0$*, then (4.176) and (4.177) are satisfied.*

Proof: It follows that

$$\int_0^\infty (\mathcal{V}_{pd}(t) - \mathcal{V}_p(t))^T \mathcal{E}(t) dt = \lambda \int_0^\infty \dot{\mathbf{e}}(t)^T \mathbf{e}(t) dt \geqslant -\frac{\lambda}{2} \mathbf{e}(0)^T \mathbf{e}(0) \tag{4.178}$$

holds, which satisfies (4.176). Meanwhile, given $\dot{\mathbf{e}} + \lambda \mathbf{e} \in L_\infty$, it yields $\dot{\mathbf{e}} \in L_\infty$ and $\mathbf{e} \in L_\infty$ from Lemma 2.4. Thus, (4.177) is satisfied. ∎

Lemma 4.4. *Let* $\mathbf{e}(t)$ *be a linear position error vector and* $\mathbf{R}(t)$ *be a rotation matrix satisfying* $\mathbf{R}(t)^T = \mathbf{R}(t)^{-1}$*. If* $\mathcal{E}(t) = \lambda \mathbf{R}(t)\mathbf{e}(t)$ *and* $\mathcal{V}_{pd}(t) - \mathcal{V}_p(t) = \mathbf{R}(t)\dot{\mathbf{e}}(t)$ *hold with* $\lambda > 0$*, then (4.176) and (4.177) are satisfied.*

Proof: Note that (4.178) is still valid, which satisfies (4.176). Meanwhile, $\mathbf{R}(t)\dot{\mathbf{e}}(t) + \lambda \mathbf{R}(t)\mathbf{e}(t) \in L_\infty$ implies $\dot{\mathbf{e}}(t) + \lambda \mathbf{e}(t) \in L_\infty$ and further implies $\dot{\mathbf{e}}(t) \in L_\infty$ and $\mathbf{e}(t) \in L_\infty$ from Lemma 2.4, which satisfies (4.177). ∎

Lemma 4.5. *Let* $\mathbf{q}(t)$ *be the vector part of a quaternion. If* $\mathcal{E}(t) = \lambda\mathbf{q}(t)$ *and* $\mathcal{V}_{pd}(t) - \mathcal{V}_p(t) = {}^{\mathbf{B}}\tilde{\boldsymbol{\omega}}(t)$ *hold, where* $\lambda > 0$ *is a constant and* ${}^{\mathbf{B}}\tilde{\boldsymbol{\omega}}(t)$ *denotes an angular velocity error vector as defined by (2.50), then (4.176) and (4.177) are satisfied.*

Proof: Inequality (2.53) implies (4.176), and the validity of (4.177) is straightforward from the definition of quaternions. ■

Remark 4.20. In a particular application, a combination of the three cases defined in Lemmas 4.3 to 4.5 can be used. Normally, Lemmas 4.3 and 4.4 would be used to handle the position control, and Lemma 4.5 would be used to handle the orientation control.

Theorem 4.6. *Consider (4.173) with* $\mathcal{V}_{pr}$ *being defined by (4.175), subject to (4.176) and (4.177). It follows that*

$$\mathcal{V}_{pd} - \mathcal{V}_p \in L_2 \bigcap L_\infty \tag{4.179}$$

$$\mathcal{E} \in L_2 \bigcap L_\infty \tag{4.180}$$

hold.

Proof: Note the fact that

$$\int_0^\infty (\mathcal{V}_{pr} - \mathcal{V}_p)^T(\mathcal{V}_{pr} - \mathcal{V}_p)dt = \int_0^\infty (\mathcal{V}_{pd} - \mathcal{V}_p)^T(\mathcal{V}_{pd} - \mathcal{V}_p)dt + \int_0^\infty \mathcal{E}^T\mathcal{E}dt + 2\int_0^\infty (\mathcal{V}_{pd} - \mathcal{V}_p)^T\mathcal{E}dt$$

holds. Therefore, it yields

$$\mathcal{V}_{pd} - \mathcal{V}_p \in L_2 \tag{4.181}$$

$$\mathcal{E} \in L_2 \tag{4.182}$$

for given $\mathcal{V}_{pr} - \mathcal{V}_p \in L_2$ and (4.176).

On the other hand, (4.177) implies

$$\mathcal{V}_{pd} - \mathcal{V}_p \in L_\infty \tag{4.183}$$

$$\mathcal{E} \in L_\infty. \tag{4.184}$$

Combining (4.181)-(4.184) completes the proof. ■

When $\mathcal{V}_{pd} - \mathcal{V}_p \in L_\infty$ implies $\dot{\mathcal{E}} \in L_\infty$, it follows from (4.180) and Lemma 2.8 that

$$\mathcal{E} \to \mathbf{0} \tag{4.185}$$

holds, leading to

$$\mathcal{V}_{pd} - \mathcal{V}_p \to \mathbf{0} \tag{4.186}$$

from (4.174) and (4.175).

4.12 Optimization

Once $\mathcal{V}_{pr}$ is specified for position control, the second part of the required independent velocity coordinate vector, denoted as $\mathcal{V}_{or}$, is used for the optimization purposes. Let $\mathbf{p}$ be a cost function vector governed by

$$\mathbf{p} = \mathbf{A}_o \mathcal{V}_o + \mathbf{b}_o \tag{4.187}$$

where $\mathbf{A}_o$ is a matrix of full column-rank and $\mathbf{b}_o$ is a remaining vector.

With a pseudo-inverse design, the required velocity coordinate vector for the optimization is designed as

$$\mathcal{V}_{or} = -\left(\mathbf{A}_o^T \mathbf{W} \mathbf{A}_o\right)^{-1} \mathbf{A}_o^T \mathbf{W} \mathbf{b}_o \tag{4.188}$$

to achieve

$$\min\left(\mathbf{p}^T \mathbf{W} \mathbf{p}\right)$$

under $\mathcal{V}_{or} - \mathcal{V}_o \to \mathbf{0}$, where $\mathbf{W}$ is a weighting matrix.

4.13 Design Procedure

In this section, the design procedure of the *virtual decomposition control* approach, when being applied to a complex robotic system, is summarized.

4.13.1 Forward Dynamics

The forward dynamics equation (4.136) that determines the time derivative of the independent velocity coordinate vector $\dot{\mathcal{V}}$ from the given joint control forces/torques consists of two parts. The first part deals with the dynamics of individual subsystems—the rigid bodies and joints, characterized by (4.122) and (4.129), which can be computed in parallel. The second part deals with two velocity mapping matrices $\mathcal{J}_a$ and $\mathcal{J}_b$ formed from the system kinematics. Note that the last term in the left-hand side of (4.136) is a summation of the projected contact forces/moments of all *objects* onto the motion configuration space. This term is a sole function of $\mathcal{V}$.

4.13.2 Inverse Dynamics and Control

The inverse dynamics and control problem is essentially a process that determines the joint control forces/torques from the given (required) accelerations and velocities in the task space. The general computations consist of the following steps:

Step 1: Decompose a complex robotic system into n_o *objects* and n_c *open chains*, establish the corresponding simple oriented graph described in Section 4.3, and assign necessary frames.

Step 2: Specify the m-dimensional independent velocity coordinate vector $\mathcal{V}$ according to the number of the degrees of freedom of motion, and partition $\mathcal{V}$ into $[\mathcal{V}_p^T, \mathcal{V}_o^T, \mathcal{V}_u^T]^T$ according to (4.2). Then, form the velocity mapping matrix $\mathcal{J}_e$ defined by (4.3) and obtain the extended velocity vector of the entire robotic system $\mathcal{V}_e$, or alternatively, form the velocity mapping matrices $\mathcal{J}_a$ and $\mathcal{J}_{ba}$ defined by (4.6) and (4.8) and obtain the extended velocity vectors of the entire robotic system $\mathcal{V}_a$ and $\mathcal{V}_b$.

Step 3: Compute $\mathcal{V}_{pr}$ and $\dot{\mathcal{V}}_{pr}$ in terms of (4.175), and compute $\mathcal{V}_{or}$ and $\dot{\mathcal{V}}_{or}$ in terms of (4.188).

Step 4: Compute the required extended velocity vector $\mathcal{V}_{er}$ and its time derivative $\dot{\mathcal{V}}_{er}$ in terms of (4.13) and (4.16). In this step, $\mathcal{V}_{er}$ and $\dot{\mathcal{V}}_{er}$ are affine functions of $\mathcal{V}_{ur}$ and $\dot{\mathcal{V}}_{ur}$, respectively.

Step 5: Calculate the required net force/moment vector for each rigid body (an *object* or a rigid link of the *open chains*, according to (4.32), (4.57), and (4.58)). Perform parameter adaptation (4.34), (4.61), and (4.62). In this step, ${}^{\mathbf{O}_i}F_r^*$, ${}^{\mathbf{B}_j}F_r^*$, and ${}^{\mathbf{B}_{jk}}F_r^*$ are affine functions of $\dot{\mathcal{V}}_{ur}$ as well.

Step 6: Compute $\mathrm{col}\left({}^{\mathbf{T}_j}F_r \mid \{\mathbf{T}_j\} \in \mathcal{T}_i\right)$ for all $i \in \{1, n_o\}$ from (4.39), and compute ${}^{\mathbf{B}_j}F_r$ and ${}^{\mathbf{B}_{jk}}F_r$ for all $k \in \{1, l_j\}$ and $j \in \{1, n_c\}$ from (4.63)-(4.65). This procedure starts from the *sink nodes* and goes all the way back to the *source nodes*. In this step, ${}^{\mathbf{B}_{jk}}F_r$, $k \in \{1, l_j\}$ and $j \in \{1, n_c\}$, remains to be an affine function of $\boldsymbol{\eta}_{id}$ and $\boldsymbol{\varphi}_{id}$ for all $i \in \{1, n_o\}$ and $\dot{\mathcal{V}}_{ur}$.

Step 7: For all the unactuated joints with $k \in \mathcal{U}_j$, $j \in \{1, n_c\}$, express (4.108) as (4.109). Then, re-order $\mathrm{col}(\boldsymbol{\eta}_{id} \mid i \in \{1, n_o\})$ and $\mathrm{col}(\boldsymbol{\varphi}_{id} \mid i \in \{1, n_o\})$ and re-express (4.109) as (4.110). Finally, solving (4.110) makes $\boldsymbol{\eta}_{ur}$, $\boldsymbol{\varphi}_{ur}$, and $\dot{\mathcal{V}}_{ur}$ affine functions of $\boldsymbol{\eta}_{cr}$ and $\boldsymbol{\varphi}_{cr}$.

Step 8: Specify $\boldsymbol{\eta}_{cr}$ and $\boldsymbol{\varphi}_{cr}$ in terms of the force control specifications.

Step 9: After specifying $\dot{\mathcal{V}}_{ur}$ and $\boldsymbol{\eta}_{id}$ and ${}^{\mathbf{S}_i}F_r$ for all $i \in \{1, n_o\}$, use the results from step 6 to compute ${}^{\mathbf{B}_{jk}}F_r$ for all $k = \{1, l_j\}$ and $j \in \{1, n_c\}$. Furthermore, compute τ_{jk} from (4.77) and (4.78) for $k \in \mathcal{C}_{1j}$, and compute $\boldsymbol{\tau}_{jk}$ from (4.93) and (4.94) for $k \in \mathcal{C}_{3j}$. Finally, perform parameter adaptation (4.82) and (4.97). Return to step 3.

Steps 1 and 2 perform the initialization and steps 3–9 perform a loop computation with step 7 particularly being devoted to handle the force/torque constraints on the unactuated joints. Specifically, steps 5 and 9 deal with the parallel computations of the subsystem dynamics based control and the parameter adaptation. Steps 3 and 8 give the motion and force control specifications, respectively. Steps 4 and 6 deal with iterative computations for obtaining the linear/angular velocity vectors and the force/moment vectors at all *cutting points*.

When there is no unactuated joint in the system, that is $\mathcal{U}_j = \emptyset$, the control algorithm can be substantially simplified. Apart from eliminating step 7, the terms $\mathcal{V}_{ur}$ and $\dot{\mathcal{V}}_{ur}$ do not appear from step 4 to step 6. Meanwhile, step 8 can precede step 6 to specify ${}^{\mathbf{S}_i}F_r$ and $\boldsymbol{\eta}_{id}$ for all $i \in \{1, n_o\}$.

When unactuated joints are involved in the system, the main idea in step 7 is to find certain releasable variables to satisfy the force/torque constraints imposed on all unactuated joints. These releasable variables are selected among

$\dot{\mathcal{V}}_r$ and ${}^{\mathbf{S}_i}F_r$ and $\boldsymbol{\eta}_{id}$ for all $i \in \{1, n_o\}$ so that the number of total released variables equals the number of the independent force/torque constraints on the unactuated joints.

4.14 Relation to Passivity Theory

Passivity theory has been widely used to guarantee the system stability [10], particularly in teleoperation systems [5]. The book by Van der Schaft defines a close relationship between the passivity theory and the L_2 stability [178]. Therefore, it would be interesting to view the main results of the VDC approach developed in this chapter from the point of view of the passivity theory.

Definition 4.2. *For a single-input-single-output (SISO) system with $f(t)$ as the input and $v(t)$ as the output, the system is said to be passive if*

$$\int_0^t v(\tau) f(\tau) d\tau \geqslant -\gamma_1 \tag{4.189}$$

holds with $\gamma_1 \geqslant 0$.

Let $f(t)$ be the force applied to a mechanical system and $v(t)$ be the velocity. A physical interpretation of (4.189) is that the addressed mechanical system is energy dissipative and γ_1 characterizes the initial energy stored in the system.

This interpretation of passivity can be generalized by specifying different inputs and outputs in multiple-input-multiple-output (MIMO) systems.

4.14.1 Passivity of *Objects*

For an *object* defined in this chapter, design a standard feedback system, as illustrated in Figure 4.5, with

$$\begin{aligned}
\mathbf{r}_1 &= \sum_{\{\mathbf{T}_j\} \in \mathcal{T}_i} {}^{\mathbf{O}_i}\mathbf{U}_{\mathbf{T}_j} \left({}^{\mathbf{T}_j}F_r - {}^{\mathbf{T}_j}F \right) - \sum_{\{\mathbf{B}_j\} \in \mathcal{B}_i} {}^{\mathbf{O}_i}\mathbf{U}_{\mathbf{B}_j} \left({}^{\mathbf{B}_j}F_r - {}^{\mathbf{B}_j}F \right) \\
\mathbf{u}_1 &= {}^{\mathbf{O}_i}F_r^* - {}^{\mathbf{O}_i}F^* \\
\mathbf{y}_1 &= {}^{\mathbf{O}_i}V_r - {}^{\mathbf{O}_i}V \\
\mathbf{r}_2 &= \mathbf{0} \\
\mathbf{u}_2 &= \mathbf{y}_1 \\
\mathbf{y}_2 &= {}^{\mathbf{O}_i}\mathbf{U}_{\mathbf{S}_i} \left({}^{\mathbf{S}_i}F_r - {}^{\mathbf{S}_i}F \right)
\end{aligned}$$

in which

$$\begin{aligned}
{}^{\mathbf{O}_i}F_r^* - {}^{\mathbf{O}_i}F^* = & \sum_{\{\mathbf{T}_j\} \in \mathcal{T}_i} {}^{\mathbf{O}_i}\mathbf{U}_{\mathbf{T}_j} \left({}^{\mathbf{T}_j}F_r - {}^{\mathbf{T}_j}F \right) - \sum_{\{\mathbf{B}_j\} \in \mathcal{B}_i} {}^{\mathbf{O}_i}\mathbf{U}_{\mathbf{B}_j} \left({}^{\mathbf{B}_j}F_r - {}^{\mathbf{B}_j}F \right) \\
& - {}^{\mathbf{O}_i}\mathbf{U}_{\mathbf{S}_i} \left({}^{\mathbf{S}_i}F_r - {}^{\mathbf{S}_i}F \right)
\end{aligned} \tag{4.190}$$

derived from (4.25) and (4.42) is used.

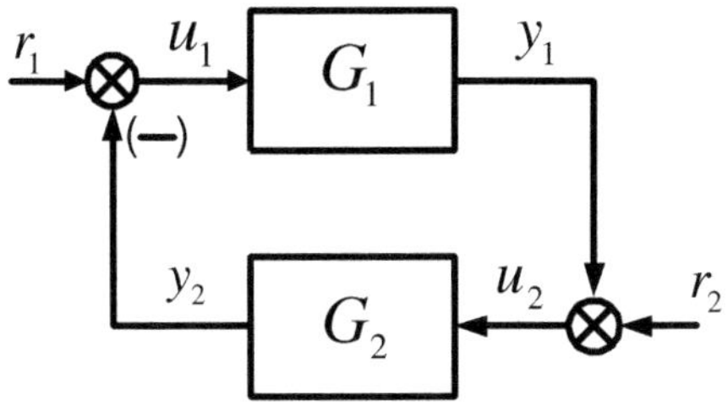

Fig. 4.5. Standard feedback configuration.

Integrating (4.41) over time yields

$$\begin{aligned}\int_0^t \mathbf{u}_1^T \mathbf{y}_1 dt &= \int_0^t \left({}^{\mathbf{O}_i}V_r - {}^{\mathbf{O}_i}V\right)^T \left({}^{\mathbf{O}_i}F_r^* - {}^{\mathbf{O}_i}F^*\right) dt \\ &\geqslant \int_0^t \left({}^{\mathbf{O}_i}V_r - {}^{\mathbf{O}_i}V\right)^T \mathbf{K}_{\mathbf{O}_i} \left({}^{\mathbf{O}_i}V_r - {}^{\mathbf{O}_i}V\right) dt + \int_0^t \dot{\nu}_{\mathbf{O}_i} dt \\ &\geqslant \int_0^t \left({}^{\mathbf{O}_i}V_r - {}^{\mathbf{O}_i}V\right)^T \mathbf{K}_{\mathbf{O}_i} \left({}^{\mathbf{O}_i}V_r - {}^{\mathbf{O}_i}V\right) dt - \nu_{\mathbf{O}_i}(0). \quad (4.191)\end{aligned}$$

This inequality indicates that the transfer function G_1 qualifies to be *strictly output passive*, in the sense of Definition 2.2.8 in [178, page 17].

Meanwhile, (4.45) guarantees the passivity of G_2 with $\mathbf{u}_2^T \mathbf{y}_2 = p_{\mathbf{S}_i}$. Thus, Theorem 2.2.11 in [178, pages 17-18] ensures that the complete system in Figure 4.5 is *strictly output passive* with input $\mathbf{r}_1$ and output $\mathbf{y}_1$.

On the other hand, the *strictly output passivity* from input $\mathbf{r}_1$ to output $\mathbf{y}_1$ can be verified from the results achieved in this chapter. It follows from (4.41), (4.45), (4.190), and (4.191) that

$$\begin{aligned}\int_0^t \mathbf{r}_1^T \mathbf{y}_1 dt &= \int_0^t \left({}^{\mathbf{O}_i}V_r - {}^{\mathbf{O}_i}V\right)^T \left({}^{\mathbf{O}_i}F_r^* - {}^{\mathbf{O}_i}F^*\right) dt + \int_0^t p_{\mathbf{S}_i} dt \\ &\geqslant \int_0^t \left({}^{\mathbf{O}_i}V_r - {}^{\mathbf{O}_i}V\right)^T \mathbf{K}_{\mathbf{O}_i} \left({}^{\mathbf{O}_i}V_r - {}^{\mathbf{O}_i}V\right) dt - \nu_{\mathbf{O}_i}(0) \\ &\quad - \sum_\gamma \frac{1}{2\rho_{si\gamma}} \left(\theta_{si\gamma} - \hat{\theta}_{si\gamma}(0)\right)^2 \quad (4.192)\end{aligned}$$

holds.

Therefore, the result derived from the passivity theory is consistent with what has been achieved in this chapter.

4.14.2 Passivity of Rigid Links

Figure 4.5 can also represent the passivity of a rigid link. Let

$$\begin{aligned}
\mathbf{r}_1 &= \left({}^{\mathbf{B}_{jk}}F_r - {}^{\mathbf{B}_{jk}}F\right) - {}^{\mathbf{B}_{jk}}\mathbf{U}_{\mathbf{T}_{j(k+1)}}\left({}^{\mathbf{T}_{j(k+1)}}F_r - {}^{\mathbf{T}_{j(k+1)}}F\right) \\
\mathbf{u}_1 &= {}^{\mathbf{B}_{jk}}F_r^* - {}^{\mathbf{B}_{jk}}F^* \\
\mathbf{y}_1 &= {}^{\mathbf{B}_{jk}}V_r - {}^{\mathbf{B}_{jk}}V \\
\mathbf{r}_2 &= \mathbf{0} \\
\mathbf{u}_2 &= \mathbf{y}_1 \\
\mathbf{y}_2 &= \mathbf{0}.
\end{aligned}$$

Integrating (4.69) over time gives

$$\begin{aligned}
\int_0^t \mathbf{u}_1^T \mathbf{y}_1 dt &= \int_0^t \left({}^{\mathbf{B}_{jk}}V_r - {}^{\mathbf{B}_{jk}}V\right)^T \left({}^{\mathbf{B}_{jk}}F_r^* - {}^{\mathbf{B}_{jk}}F^*\right) dt \\
&\geqslant \int_0^t \left({}^{\mathbf{B}_{jk}}V_r - {}^{\mathbf{B}_{jk}}V\right)^T \mathbf{K}_{\mathbf{B}_{jk}} \left({}^{\mathbf{B}_{jk}}V_r - {}^{\mathbf{B}_{jk}}V\right) dt + \int_0^t \dot{\nu}_{\mathbf{B}_{jk}} dt \\
&\geqslant \int_0^t \left({}^{\mathbf{B}_{jk}}V_r - {}^{\mathbf{B}_{jk}}V\right)^T \mathbf{K}_{\mathbf{B}_{jk}} \left({}^{\mathbf{B}_{jk}}V_r - {}^{\mathbf{B}_{jk}}V\right) dt - \nu_{\mathbf{B}_{jk}}(0)
\end{aligned} \tag{4.193}$$

which ensures that G_1 is *strictly output passive*. Meanwhile, $\mathbf{y}_2 = \mathbf{0}$ ensures that G_2 is passive. Thus, the system with input $\mathbf{r}_1$ and output $\mathbf{y}_1$ is *strictly output passive*, in the sense of Definition 2.2.8 in [178, page 17].

On the other hand, the *strictly output passivity* from input $\mathbf{r}_1$ to output $\mathbf{y}_1$ can be verified from the results achieved in this chapter. It follows from (4.52) and (4.64) that

$$\begin{aligned}
\int_0^t \mathbf{r}_1^T \mathbf{y}_1 dt &= \int_0^t \left({}^{\mathbf{B}_{jk}}V_r - {}^{\mathbf{B}_{jk}}V\right)^T \left({}^{\mathbf{B}_{jk}}F_r^* - {}^{\mathbf{B}_{jk}}F^*\right) dt \\
&\geqslant \int_0^t \left({}^{\mathbf{B}_{jk}}V_r - {}^{\mathbf{B}_{jk}}V\right)^T \mathbf{K}_{\mathbf{B}_{jk}} \left({}^{\mathbf{B}_{jk}}V_r - {}^{\mathbf{B}_{jk}}V\right) dt - \nu_{\mathbf{B}_{jk}}(0)
\end{aligned} \tag{4.194}$$

holds.

Therefore, the result derived from the passivity theory is consistent with what has been achieved in this chapter.

4.14.3 Passivity of Single-DOF Joints

Note that the passivity holds for each individual joint as well. Let

$$\begin{aligned}
r_1 &= \tau_{jkr}^* - \tau_{jk}^* \\
u_1 &= r_1 \\
y_1 &= \dot{q}_{jkr} - \dot{q}_{jk} \\
r_2 &= 0 \\
u_2 &= y_1 \\
y_2 &= 0
\end{aligned}$$

for all $k \in \mathcal{C}_{1j}$ and $j \in \{1, n_c\}$. Integrating (4.86) over time yields

$$\begin{aligned} \int_0^t r_1 y_1 dt &= \int_0^t (\dot{q}_{jkr} - \dot{q}_{jk}) \left(\tau^*_{jkr} - \tau^*_{jk}\right) dt \\ &\geqslant \int_0^t (k_{vjk} + k_{jk}) (\dot{q}_{jkr} - \dot{q}_{jk})^2 \, dt + \int_0^t \dot{\nu}_{1jk} dt \\ &\geqslant \int_0^t (k_{vjk} + k_{jk}) (\dot{q}_{jkr} - \dot{q}_{jk})^2 \, dt - \nu_{1jk}(0) \end{aligned} \tag{4.195}$$

which ensures that the system with input r_1 and output y_1 is *strictly output passive.*

4.14.4 Passivity of Three-DOF Joints

The same property holds for all $k \in \mathcal{C}_{3j}$ and $j \in \{1, n_c\}$. Let

$$\begin{aligned} \mathbf{r}_1 &= \boldsymbol{\tau}^*_{jkr} - \boldsymbol{\tau}^*_{jk} \\ \mathbf{u}_1 &= \mathbf{r}_1 \\ \mathbf{y}_1 &= \dot{\mathbf{q}}_{jkr} - \dot{\mathbf{q}}_{jk} \\ \mathbf{r}_2 &= \mathbf{0} \\ \mathbf{u}_2 &= \mathbf{y}_1 \\ \mathbf{y}_2 &= \mathbf{0} \end{aligned}$$

it then follows from (4.101) that

$$\int_0^t \mathbf{r}_1^T \mathbf{y}_1 dt \geqslant \int_0^t \left(\dot{\mathbf{q}}_{jkr} - \dot{\mathbf{q}}_{jk}\right)^T \mathbf{K}_{jk} \left(\dot{\mathbf{q}}_{jkr} - \dot{\mathbf{q}}_{jk}\right) dt - \nu_{3jk}(0) \tag{4.196}$$

holds.

4.14.5 Passivity and L_2 Stability

Inequalities (4.192), (4.194), (4.195), and (4.196) ensure that each subsystem (an *object*, a rigid link, or a joint) incorporating with its respective control equations is *strictly output passive*. Below, it will be shown that the properties of the *strictly output passive* subsystems lead to the L_2 stability of the entire system.

The left hand-side of (4.192) can be re-written as

$$\int_0^t \mathbf{r}_1^T \mathbf{y}_1 dt = \int_0^t \left(\sum_{\{\mathbf{T}_j\} \in \mathcal{T}_i} p_{\mathbf{T}_j} - \sum_{\{\mathbf{B}_j\} \in \mathcal{B}_i} p_{\mathbf{B}_j} \right) dt \overset{def}{=} w_{\mathbf{O}_i}. \tag{4.197}$$

Similarly, the left-hand side of (4.194) can be re-written as

$$\int_0^t \mathbf{r}_1^T \mathbf{y}_1 dt = \int_0^t \left(p_{\mathbf{B}_{jk}} - p_{\mathbf{T}_{j(k+1)}}\right) dt \overset{def}{=} w_{\mathbf{B}_{jk}} \tag{4.198}$$

and the left hand-side of (4.195) or (4.196) can be re-written as

$$\int_0^t r_1 y_1 dt = \int_0^t \left(-p_{\mathbf{B}_{jk}} + p_{\mathbf{T}_{jk}}\right) dt \stackrel{def}{=} w_{jk} \tag{4.199}$$

or

$$\int_0^t \mathbf{r}_1^T \mathbf{y}_1 dt = \int_0^t \left(-p_{\mathbf{B}_{jk}} + p_{\mathbf{T}_{jk}}\right) dt \stackrel{def}{=} w_{jk}. \tag{4.200}$$

Define

$$w_j = \sum_{k=0}^{l_j} w_{\mathbf{B}_{jk}} + \sum_{k=1}^{l_j} w_{jk} \tag{4.201}$$

with $\mathbf{B}_j = \mathbf{B}_{j0}$ and $\mathbf{T}_{j(l_j+1)} = \mathbf{T}_j$. It follows that

$$w_j = \int_0^t \left(p_{\mathbf{B}_j} - p_{\mathbf{T}_j}\right) dt \tag{4.202}$$

holds. In view of

$$\sum_{i=1}^{n_o} \left(\sum_{\{\mathbf{T}_j\} \in \mathcal{T}_i} p_{\mathbf{T}_j} - \sum_{\{\mathbf{B}_j\} \in \mathcal{B}_i} p_{\mathbf{B}_j} \right) + \sum_{j=1}^{n_c} \left(p_{\mathbf{B}_j} - p_{\mathbf{T}_j}\right) = 0 \tag{4.203}$$

it yields

$$\begin{aligned} & \sum_{i=1}^{n_o} w_{\mathbf{O}_i} + \sum_{j=1}^{n_c} w_j \\ & = \int_0^t \left[\sum_{i=1}^{n_o} \left(\sum_{\{\mathbf{T}_j\} \in \mathcal{T}_i} p_{\mathbf{T}_j} - \sum_{\{\mathbf{B}_j\} \in \mathcal{B}_i} p_{\mathbf{B}_j} \right) + \sum_{j=1}^{n_c} \left(p_{\mathbf{B}_j} - p_{\mathbf{T}_j}\right) \right] dt \\ & = 0. \end{aligned} \tag{4.204}$$

Equation (4.203) is the mathematical expression of the fact that the sum of the *virtual power flows* associated with all the subsystems of a complex robotic system equals zero,[7] which validates (4.204). In view of (4.197)–(4.201) and (4.204), it can be concluded that the sum of the left hand side terms of (4.192), (4.194), (4.195), and (4.196) associated with all the subsystems (*objects*, rigid links, and joints) equals zero. Thus, by adding the right hand terms of (4.192), (4.194), (4.195), and (4.196) altogether, the *strict output passivity* of the subsystems leads to the L_2 stability of the entire system in the sense of (4.146)–(4.150).

[7] This is mathematically equivalent to (2.106). The fact is that at each given *cutting point* the two *virtual power flows* associated with the two adjacent subsystems take the same magnitude with different signs. Therefore, the summation of all the *virtual power flows* gives a zero.

4.15 Relation to PID Control

As aforementioned in Section 1.2, the VDC approach described in this book generally consists of a model based feedforward control part and a feedback control part. For an actuated joint, the feedback control part plus the estimated force/torque bias can be equivalent to a joint PID control. This fact builds a close relationship between the VDC approach and the PID robot control that is widely used in industry. It implies that a joint PID controller constitutes a special case of the VDC approach. It can also be seen in principle that the VDC approach is capable of delivering all the performances a joint PID controller is able to deliver. Following paragraphs give the details.

In the joint position control of a robot, let q_{jk} be an element of $\mathcal{V} = \mathcal{V}_p \in \mathbb{R}^m$, and let $\lambda_{jk}(q_{jkd} - q_{jk})$ be the corresponding element in $\mathcal{E}$ such that (4.176) is valid, where $\lambda_{jk} > 0$ is a control parameter and q_{jkd} is the desired variable of q_{jk}. It follows from (4.175) that

$$\dot{q}_{jkr} = \dot{q}_{jkd} + \lambda_{jk}(q_{jkd} - q_{jk}) \tag{4.205}$$

holds.

Let ${}^{\mathbf{O}_i}F_r^* = \mathbf{0}$, ${}^{\mathbf{S}_i}F_r = \mathbf{0}$ and $\boldsymbol{\eta}_{id} = \mathbf{0}$ for all $i \in \{1, n_o\}$, and ${}^{\mathbf{B}_j}F_r^* = \mathbf{0}$ and ${}^{\mathbf{B}_{jk}}F_r^* = \mathbf{0}$ for all $j \in \{1, n_c\}$ and $k \in \{1, l_j\}$. It results in ${}^{\mathbf{B}_{jk}}F_r = \mathbf{0}$ from (4.39) and (4.63)–(4.65). Then, it follows from (4.77) and (4.78) that

$$\tau_{jk} = \tau_{jkr}^* = \mathbf{Y}_{jk}\hat{\boldsymbol{\theta}}_{jk} + k_{jk}(\dot{q}_{jkr} - \dot{q}_{jk}) \tag{4.206}$$

holds. Furthermore, let

$$\boldsymbol{\theta}_{jk} = \begin{bmatrix} 0 & 0 & 0 & c_{jk} \end{bmatrix}^T$$

be used with $\underline{c}_{jk} \rightarrow -\infty$ and $\overline{c}_{jk} \rightarrow +\infty$ such that

$$\hat{c}_{jk} = \rho_{jk4} \int_0^t (\dot{q}_{jkr} - \dot{q}_{jk})\, dt \tag{4.207}$$

holds. Finally, it follows from (4.205)-(4.207) that

$$\begin{aligned} \tau_{jk} &= \hat{c}_{jk} + k_{jk}(\dot{q}_{jkr} - \dot{q}_{jk}) \\ &= k_{jk}(\dot{q}_{jkr} - \dot{q}_{jk}) + \rho_{jk4} \int_0^t (\dot{q}_{jkr} - \dot{q}_{jk})\, dt \\ &= k_{jk}(\dot{q}_{jkd} - \dot{q}_{jk}) + (k_{jk}\lambda_{jk} + \rho_{jk4})(q_{jkd} - q_{jk}) \\ &\quad + \rho_{jk4}\lambda_{jk} \int_0^t (q_{jkd} - q_{jk})\, dt \end{aligned} \tag{4.208}$$

holds. Thus, the three positive parameters k_{jk}, λ_{jk}, and ρ_{jk4} uniquely determine the equivalent derivative, proportional, and integral gains of a PID controller described in (1.2) as

$$k_d = k_{jk} \tag{4.209}$$

$$k_p = k_{jk}\lambda_{jk} + \rho_{jk4} \tag{4.210}$$

$$k_I = \rho_{jk4}\lambda_{jk}. \tag{4.211}$$

4.16 Implementation Issues

This section particularly focuses on the implementation issues associated with the VDC approach. Some guidelines on tuning the control parameters and the parameter update gains will be presented.

For a given complex robot with known kinematics, the design parameters required for the control implementations appear in (4.32) and (4.34) for an *object*, in (4.57), (4.58), (4.61), and (4.62) for a rigid link, in (4.77) and (4.82) for a single-DOF joint, and in (4.93) and (4.97) for a three-DOF joint. These design parameters can be grouped in Table 4.1. The first group contains feedback gains for all rigid bodies (*objects* and rigid links) and for all (single-DOF and three-DOF) joints. The relationship between a typical single-DOF joint control design and the PID control, as outlined in the previous section, allows us to use any existing PID control design tool to specify the values for k_{jk}, λ_{jk}, and ρ_{jk4} based on (4.209)–(4.211). The values of $\mathbf{K}_{\boldsymbol{\alpha}}$ and $\mathbf{K}_{jk}$ should be gradually increased from zero to a point at which the instability occurs, then the matrices should be halved. This step might involve some "work of art."

Unlike conventional direct adaptive control approaches, the parameter update gains should be made as large as possible, since no parameter drift beyond the given lower and upper bounds is possible. It follows from (4.40), (4.66), (4.67), (4.84), and (4.100) that large parameter update gains reduce the portion of the parameter uncertainties contributing to the corresponding non-negative accompanying functions. In the tuning process, a parameter update gain is gradually increased from zero to a point at which the parameter estimate starts oscillating, then the gain should be halved.

The parameter bounds are very important to achieve good control results. In principle, a known parameter implies that its lower bound equals its upper bound and further equals its true value. In contrast, a completely unknown parameter implies that its lower bound tends to minus infinity and its upper bound tends to plus infinity. Lemma 2.9 requires that the interval spanned by the lower and upper bounds has to include the true value. In practice, the lower and upper bounds should be set as close to the true value as possible. Some off-line identification methods or theoretical computations can be conducted to determine the lower and upper bounds for a particular parameter. As permitted, the lower and upper bounds can be time-varying functions.

Table 4.1. Parameters in control implementations.

	Feedback gains	Adaptation gains	Parameter bounds
Rigid bodies	$\mathbf{K}_{\boldsymbol{\alpha}}$	$\rho_{\boldsymbol{\alpha}\gamma}$	$\underline{\theta}_{\boldsymbol{\alpha}\gamma}$, $\overline{\theta}_{\boldsymbol{\alpha}\gamma}$, $\gamma = 1, \cdots, 13$
Single-DOF joints	k_{jk}, λ_{jk}, ρ_{jk4}	$\rho_{jk1}, \rho_{jk2}, \rho_{jk3}$	$\underline{\theta}_{jk\gamma}$, $\overline{\theta}_{jk\gamma}$, $\gamma = 1, \cdots, 4$
Three-DOF joints	$\mathbf{K}_{jk}$	$\rho_{jk\gamma}$	$\underline{\theta}_{jk\gamma}$, $\overline{\theta}_{jk\gamma}$

where $\boldsymbol{\alpha} \in \{\mathbf{O}_i, \mathbf{B}_j, \mathbf{B}_{jk}\}$, $i \in \{1, n_o\}$, $j \in \{1, n_c\}$, $k \in \{1, l_j\}$.

4.17 Concluding Remarks

This chapter presents the essence of the VDC approach applicable to a class of general complex robotic systems containing multiple closed chains with combined actuated/unactuated joints. The VDC approach aims at using the dynamics of subsystems (rigid bodies and joints) to conduct model-based control, while guaranteeing the stability and convergence of the entire robotic system.

The virtual decomposition process virtually "cuts" a complex robot into "pieces," namely *objects* and *open chains*. An *open chain* can then be virtually "cut" into individual rigid links and joints. In fact, this cutting process can even be extended further to reach the basic components or elements of a subsystem without foreseeable limitations.

The control computations of the VDC approach mainly comprise four parts as developed in Sections 4.4–4.8. The first part computes the actual velocity vectors and the required velocity vectors for all *objects* and *open chains* from the independent velocity coordinate vector and the required independent velocity coordinate vector, as described in Section 4.4. The second part computes the required net force/moment vectors for all rigid bodies including all the *objects* and all the links of the *open chains*, as described in Sections 4.5 and 4.6. The third part calculates the required force/moment vectors at all *cutting points*, as described in Section 4.7. This part is done in a fashion essentially dual to that of the first part. The fourth part computes the joint dynamics based control equations for all the joints, as described in Section 4.8.

The concept of the *virtual stability* plays the central role in ensuring the stability of a complex robot. The *virtual stability* of a subsystem requires the appearance of *virtual power flows* in the time derivative of the non-negative accompanying function assigned to this subsystem. These *virtual power flows* uniquely represent the dynamic interactions this very subsystem has with the rest of the system. The *virtual stability* of every subsystem is mathematically equivalent to the stability of the entire complex robot. This stability equivalence allows us to focus on the *virtual stability* of each subsystem by designing respective subsystem-dynamics-based control equations—a relatively simple task, rather than to focus on the stability of the original system—a relatively complex task.

An important feature of the VDC approach is that the changes of the dynamics of a particular subsystem (for instance, replacing a hydraulic motor with an electrical motor) affect only the respective local control equations associated with this very subsystem to maintain its *virtual stability*, while keeping the control equations associated with the rest of the complex robot unchanged.

The VDC approach also allows each dynamic parameter to be updated independently, provided that each true parameter is indeed within the lower and upper bounds known to the controller. High gains for parameter adaptation are recommended, since no parameter drifting is possible.

The VDC approach guarantees the asymptotic stability of the velocity control. Based on this velocity controller, the position/orientation control can be automatically achieved by including a position/orientation error term into the corresponding required velocity coordinate vector. Due to the nature of the model-based control, the VDC allows a controlled complex robot to possess a much higher trajectory tracking control bandwidth than the one determined by feedback gains alone. In addition, both constraint and internal force errors are bounded by the system state errors and the dynamic parameter errors. For a dynamically known robot, the asymptotic convergence of both constraint and internal force errors may be achieved.

Furthermore, the relations of the VDC approach to the passivity theory and to the joint PID control have been examined. It turns out that the joint PID control is a particular case of the VDC approach. This scenario ensures that the VDC approach can always achieve what a PID controller can deliver.

Finally, careful considerations need to be taken to apply the VDC approach successfully. First, the entire approach is based on the assumption that the kinematics is known. Second, the order of the dynamics describing each subsystem must be known exactly for the stability results to hold. Third, the number of control parameters and the number of parameter update gains remain large. The tuning process involves some "work of art."

Part II

Virtual Decomposition Control Applications

5

Control of Electrically Driven Robots

5.1 Introduction

The control of single-arm robot manipulators is one of the mostly studied topics in robotics community. With respect to this very popular type of systems, the application of the *virtual decomposition control* (VDC) approach will be demonstrated in this chapter. In the first part, the control of single-arm robot manipulators in free motion is to be mainly focused by showing how a robot manipulator is virtually decomposed into several rigid links and joints and how the control design for each link or joint can be conducted independently from the rest of the manipulator, while ensuring the *virtual stability* of every subsystem necessarily needed to guarantee the stability of the entire robot. Three joint control modes, namely *torque control mode, current control mode*, and *voltage control mode*, are to be presented in detail. In the second part of the chapter, the force control issues will be exploited.

Most technical contents developed in this chapter can be re-used in the following chapters.

5.2 Prior Work

This section lists some of the remarkable references prior to the development of the *virtual decomposition control* (VDC). Readers are suggested to refer to [150] for a timely and up-to-date reference.

Since the installation of the first Unimate industrial robot in 1961 and the creations of the electrically driven robots, namely the Vicarm in 1972, the first ASEA robot in 1974, and the first KUKA robot in 1977, the control of single-arm robot manipulators has drawn extraordinary attentions of both academia and industry, motivated by the need to precisely control the motion and even the contact forces. Various types of control algorithms have been developed, ranging from joint PID control—transformed from industrial servo systems [39, 138], Cartesian space control [88], hybrid control—utilizing the orthogonality between motion spaces and

W.-H. Zhu: Virtual Decomposition Control, STAR 60, pp. 113–147.
springerlink.com

contact force spaces [141], impedance control—trading off between position errors and force errors [71], compliance control—incorporating contact environment dynamics [46, 147], nonlinear feedback linearization—converting nonlinear systems into linear systems [117], parallel position/force control—performing symmetric position control and force control [31], to adaptive control—accommodating parameter uncertainties [22, 79, 105, 154, 182, 203, 206].

In the meantime, the impact of the actuator dynamics on robot control performances was first studied in [59] and then in [41, 102, 163, 170].

Whereas robot motion control keeps the end-effector (or joint angles) moving along its planned trajectory, robot force control becomes necessary when the end-effector contacts with the environment. Among early works on robot force control, hybrid control [141] and impedance control [71] were proposed first. Then, a general framework for dynamic modeling and control of robots in Cartesian space was suggested [88]. In [117], a non-linear hybrid control scheme was proposed for constrained robots, in which the robot dynamics are projected into both the free motion space and the constraint force space so that the constraint force can be controlled independently from the motion control. An interesting finding in this research is that the constraint force itself does not affect the motion dynamics, but the motion dynamics do affect the constraint force. When the environment stiffness is concerned, a compliant control approach was developed in [46]. To make position control and force control symmetric, an approach called parallel position/force control was proposed [31]. Furthermore, to accommodate parameter uncertainties associated with the contact environment, an adaptive force control approach was proposed [22]. In [149], both theoretical and experimental treatments were given to robot interaction control that incorporates stiffness control, impedance control, direct force control, and passivity-based adaptive control with output feedback. For early industrial applications of force control on a six-degree-of-freedom (six-DOF) industrial robot equipped with a force sensor, readers should refer to [128]. Research results developed in East Europe on robot force control can be found in [60].

5.3 Virtual Decomposition

Without loss of generality, a robot manipulator grasping a rigid object, as illustrated in Figure 5.1, is studied. This robot manipulator has six single-degree-of-freedom (single-DOF) joints allowing its end-effector to possess *six* degrees of freedom (DOFs) of motion without encountering kinematics singularity.

The complete system is virtually decomposed into the held object and the manipulator by placing one *cutting point* at the end-effector. Thus, it results in one *object* that represents the held object and one *open chain* that represents the manipulator with $n_o = 1$ and $n_c = 1$, see Figure 5.2. The manipulator or the unique *open chain* is further virtually decomposed into six rigid links and six joints by placing 12 subsidiary *cutting points*. Fourteen coordinate frames are defined. Frame $\{\mathbf{O}\}$ is attached to the held object. Frame $\{\mathbf{C}\}$ is located at the *cutting point* between the held object and the manipulator and is fixed to both of them. Frames $\{\mathbf{B}_k\}$ and $\{\mathbf{T}_k\}$, $k = 1, 2, \cdots, 6$, are located at the two subsidiary *cutting points* of the kth joint and are fixed to the kth link and to the $(k-1)$th link, respectively, with their z axes being coincident with the kth joint axis.

In view of Definition 2.13, the *cutting point* associated with frame $\{\mathbf{C}\}$ is called the *driven cutting point* of the *object* and is simultaneously called the *driving cutting point* of link 6. The *cutting point* associated with frame $\{\mathbf{B}_k\}$ is called the *driven cutting point* of link k and is simultaneously called the *driving cutting point* of joint k for all $k \in \{1, 6\}$. The *cutting point* associated with frame $\{\mathbf{T}_k\}$ is called the *driven cutting point* of joint k and is simultaneously called the *driving cutting point* of link $k-1$ for all $k \in \{1, 6\}$. It follows that

- The held object has only one *driven cutting point* associated with frame $\{\mathbf{C}\}$.
- The 6th link has one *driving cutting point* associated with frame $\{\mathbf{C}\}$ and one *driven cutting point* associated with frame $\{\mathbf{B}_6\}$.
- The kth link has one *driving cutting point* associated with frame $\{\mathbf{T}_{k+1}\}$ and one *driven cutting point* associated with frame $\{\mathbf{B}_k\}$ for all $k \in \{1, 5\}$.
- The kth joint has one *driving cutting point* associated with frame $\{\mathbf{B}_k\}$ and one *driven cutting point* associated with frame $\{\mathbf{T}_k\}$ for all $k \in \{1, 6\}$.

5.4 Kinematics

In this section, all kinematics computations are presented. After the independent velocity coordinates are chosen, the velocity mappings to all linear/angular velocity vectors of the thirteen frames associated with the *cutting points* are given. It ends with the design of the required velocities.

5.4.1 Independent Velocities and the Jacobian Matrix

Since there is neither unactuated joint nor redundant joint in the system, it yields

$$\mathcal{V} - \mathcal{V}_p = {}^{\mathbf{O}}V. \tag{5.1}$$

This equation indicates that the independent velocity coordinate vector $\mathcal{V} \in \mathbb{R}^6$ takes the linear/angular velocity vector of the held object ${}^{\mathbf{O}}V \in \mathbb{R}^6$, subject to a position/orientation control in all six dimensions.

The Cartesian velocity vector ${}^{\mathbf{O}}V \in \mathbb{R}^6$ and the joint velocity vector $\dot{\mathbf{q}} = [\dot{q}_1, \cdots, \dot{q}_6]^T \in \mathbb{R}^6$ are governed by a Jacobian matrix $\mathbf{J}_q \in \mathbb{R}^{6\times 6}$ as follows

$${}^{\mathbf{O}}V = \mathbf{J}_q \dot{\mathbf{q}} \tag{5.2}$$

with

$$\mathbf{J}_q \overset{def}{=} \left[{}^{\mathbf{B}_1}\mathbf{U}_{\mathbf{O}}^T\mathbf{z}, {}^{\mathbf{B}_2}\mathbf{U}_{\mathbf{O}}^T\mathbf{z}, \cdots, {}^{\mathbf{B}_6}\mathbf{U}_{\mathbf{O}}^T\mathbf{z} \right] \in \mathbb{R}^{6\times 6} \tag{5.3}$$

being invertible as permitted by Assumption 1, where $\mathbf{z} = [0, 0, 1, 0, 0, 0]^T$ holds for a prismatic joint and $\mathbf{z} = [0, 0, 0, 0, 0, 1]^T$ holds for a revolute joint.

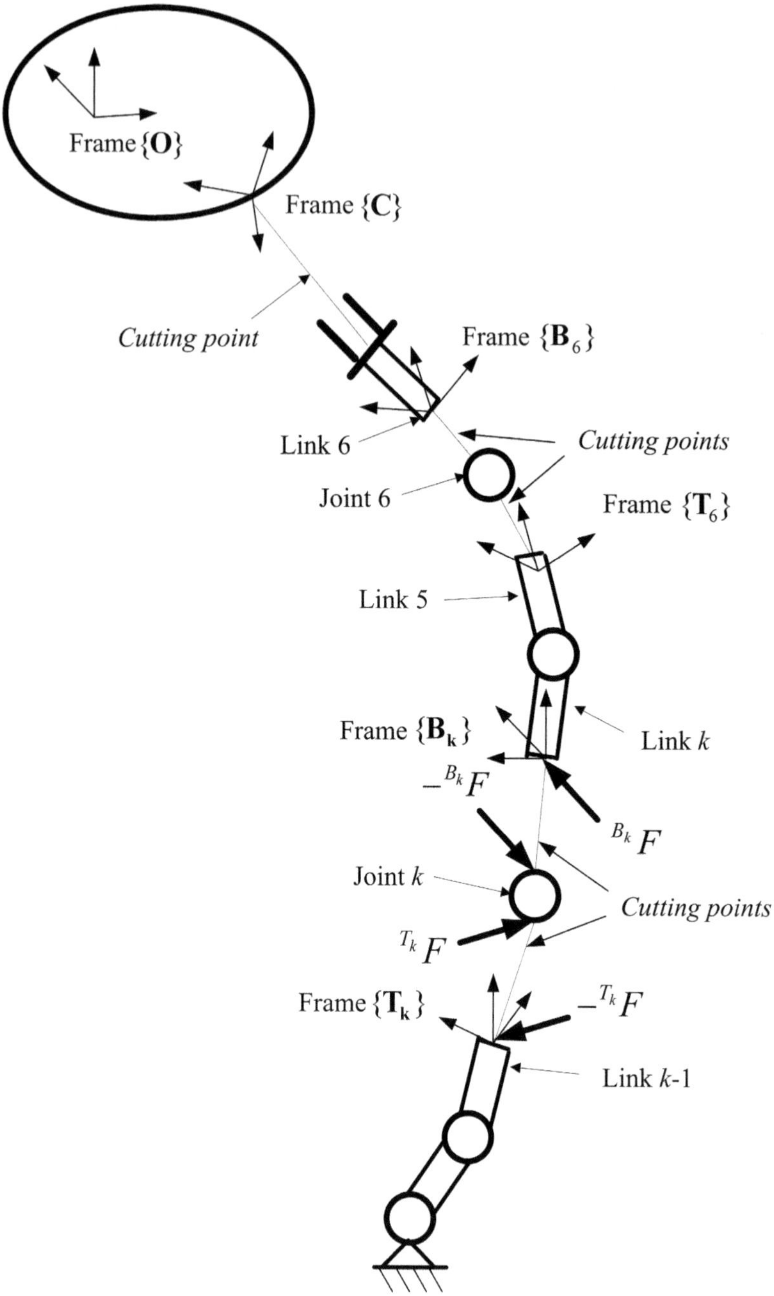

Fig. 5.1. A six degrees of freedom robot manipulator.

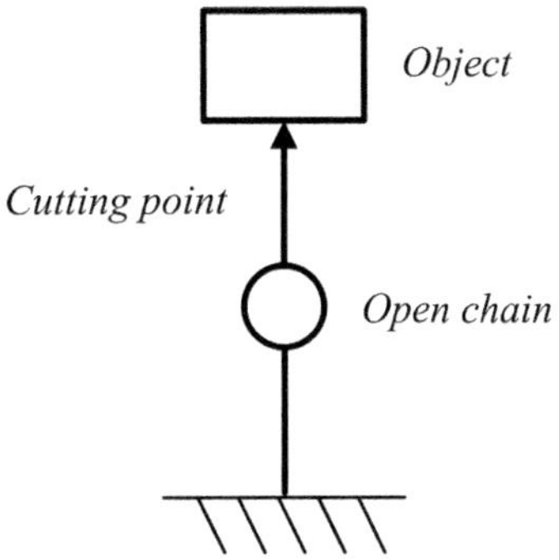

Fig. 5.2. Graphic representation of the system.

Alternatively, the diffeomorphism between the Cartesian velocity vector ${}^{\mathbf{O}}V$ and the joint velocity vector $\dot{\mathbf{q}}$ makes it possible to use the joint velocity vector $\dot{\mathbf{q}}$ as the independent velocity coordinate vector $\mathcal{V}$. However, as will be seen below, the use of the Cartesian velocity vector as the independent velocity coordinate vector is very convenient when contact tasks are involved. Therefore, (5.1) will be used throughout this chapter.

5.4.2 Extended Velocities

In this subsection, the velocity mappings used to create all linear/angular velocity vectors at the *cutting points* are to be presented.

The extended velocity vector of the robot can be written as

$$\mathcal{V}_e \stackrel{def}{=} \left[{}^{\mathbf{O}}V^T, \dot{\mathbf{q}}^T, {}^{\mathbf{B}_1}V^T, \cdots, {}^{\mathbf{B}_k}V^T, \cdots, {}^{\mathbf{B}_6}V^T\right]^T \in \mathbb{R}^{48}$$

subject to

$$\mathcal{V}_e = \mathcal{J}_e {}^{\mathbf{O}}V \tag{5.4}$$

with

$$\mathcal{J}_e = \begin{bmatrix} \mathbf{J}_q \\ \mathbf{I}_6 \\ \begin{bmatrix} \mathbf{z} & \mathbf{0} & \cdots \mathbf{0} \\ {}^{\mathbf{B}_1}\mathbf{U}_{\mathbf{B}_2}^T\mathbf{z} & \mathbf{z} & \cdots \mathbf{0} \\ \vdots & \vdots & \ddots \vdots \\ {}^{\mathbf{B}_1}\mathbf{U}_{\mathbf{B}_6}^T\mathbf{z} & {}^{\mathbf{B}_2}\mathbf{U}_{\mathbf{B}_6}^T\mathbf{z} & \cdots \mathbf{z} \end{bmatrix} \end{bmatrix} \mathbf{J}_q^{-1} \in \mathbb{R}^{48\times 6} \tag{5.5}$$

where $\mathbf{I}_6$ is a 6×6 identity matrix.

Alternatively, with respect to (4.7), (4.9) and (4.10), define

$$\mathcal{V}_a = \dot{\mathbf{q}} \in \mathbb{R}^6 \tag{5.6}$$

$$\mathcal{V}_b = \left[{}^{\mathbf{O}}V^T, {}^{\mathbf{B}_1}V^T, \cdots, {}^{\mathbf{B}_k}V^T, \cdots, {}^{\mathbf{B}_6}V^T\right]^T \in \mathbb{R}^{42} \tag{5.7}$$

such that

$$\mathcal{V}_a = \mathcal{J}_a \mathcal{V} \tag{5.8}$$
$$\mathcal{V}_b = \mathcal{J}_{ba} \mathcal{V}_a = \mathcal{J}_b \mathcal{V} \tag{5.9}$$

hold with

$$\mathcal{J}_a = \mathbf{J}_q^{-1} \tag{5.10}$$
$$\mathcal{J}_{ba} = \begin{bmatrix} \mathbf{J}_q \\ \begin{bmatrix} \mathbf{z} & \mathbf{0} & \cdots & \mathbf{0} \\ {}^{\mathbf{B}_1}\mathbf{U}_{\mathbf{B}_2}^T \mathbf{z} & \mathbf{z} & \cdots & \mathbf{0} \\ \vdots & \vdots & \ddots & \vdots \\ {}^{\mathbf{B}_1}\mathbf{U}_{\mathbf{B}_6}^T \mathbf{z} & {}^{\mathbf{B}_2}\mathbf{U}_{\mathbf{B}_6}^T \mathbf{z} & \cdots & \mathbf{z} \end{bmatrix} \end{bmatrix} \tag{5.11}$$
$$\mathcal{J}_b = \mathcal{J}_{ba} \mathcal{J}_a. \tag{5.12}$$

5.4.3 Required Velocities

Define a new frame $\{\mathbf{D}\}$, whose origin and orientation represent the desired position and orientation of frame $\{\mathbf{O}\}$. The motion control objective is to make frame $\{\mathbf{O}\}$ track frame $\{\mathbf{D}\}$.

The required independent velocity vector is designed as

$$\mathcal{V}_r = {}^{\mathbf{O}}V_r = \begin{bmatrix} {}^{\mathbf{O}}\mathbf{R}_{\mathbf{I}} \dot{\mathbf{x}}_d + k_x {}^{\mathbf{O}}\mathbf{R}_{\mathbf{I}} (\mathbf{x}_d - \mathbf{x}) \\ {}^{\mathbf{O}}\mathbf{R}_{\mathbf{I}} \boldsymbol{\omega}_d + k_\omega {}^{\mathbf{O}}\boldsymbol{\mu} \end{bmatrix} \tag{5.13}$$

where $k_x > 0$ and $k_\omega > 0$ are two positive constants, $\mathbf{x}_d \in \mathbb{R}^3$ denotes the desired position vector, $\boldsymbol{\omega}_d \in \mathbb{R}^3$ denotes the desired angular velocity vector, and ${}^{\mathbf{O}}\boldsymbol{\mu} \in SO^3$ denotes the vector part of a quaternion representing the rotation from frame $\{\mathbf{O}\}$ to frame $\{\mathbf{D}\}$.

Accordingly, the required extended velocity vector is written as

$$\mathcal{V}_{er} = \mathcal{J}_e \mathcal{V}_r \tag{5.14}$$

with

$$\mathcal{V}_{er} \overset{def}{=} \left[{}^{\mathbf{O}}V_r^T, \dot{\mathbf{q}}_r^T, {}^{\mathbf{B}_1}V_r^T, \cdots, {}^{\mathbf{B}_k}V_r^T, \cdots, {}^{\mathbf{B}_6}V_r^T \right]^T \in \mathbb{R}^{48}$$

or alternatively

$$\mathcal{V}_{ar} = \mathcal{J}_a \mathcal{V}_r \tag{5.15}$$
$$\mathcal{V}_{br} = \mathcal{J}_{ba} \mathcal{J}_a \mathcal{V}_r = \mathcal{J}_b \mathcal{V}_r \tag{5.16}$$

with

$$\mathcal{V}_{ar} \overset{def}{=} \dot{\mathbf{q}}_r \in \mathbb{R}^6 \tag{5.17}$$
$$\mathcal{V}_{br} \overset{def}{=} \left[{}^{\mathbf{O}}V_r^T, {}^{\mathbf{B}_1}V_r^T, \cdots, {}^{\mathbf{B}_k}V_r^T, \cdots, {}^{\mathbf{B}_6}V_r^T \right]^T \in \mathbb{R}^{42}. \tag{5.18}$$

Equation (5.14) validates (2.86), so do (5.15) and (5.16).

5.5 Dynamics and Control of the Object

With a similar structure as that of Sections 4.5 and 4.6, three parts will be presented in this section. The first part gives the kinematics and dynamics of the *object*, the second part presents the model-based control, and the third part guarantees the *virtual stability*.

5.5.1 Kinematics and Dynamics

Equation (5.4) implies

$$^{\mathbf{O}}V = {}^{\mathbf{C}}\mathbf{U}_{\mathbf{O}}^{T}\,{}^{\mathbf{C}}V \tag{5.19}$$

with ${}^{\mathbf{C}}\mathbf{U}_{\mathbf{O}} \in \mathbb{R}^{6\times 6}$ being defined by (2.65) with appropriate frame substitutions.

Recall (2.74) to obtain the dynamics of the *object* with frame $\{\mathbf{O}\}$ being substituted for frame $\{\mathbf{A}\}$. It yields

$$\mathbf{M_O}\frac{d}{dt}({}^{\mathbf{O}}V) + \mathbf{C_O}({}^{\mathbf{O}}\omega){}^{\mathbf{O}}V + \mathbf{G_O} = {}^{\mathbf{O}}F^{*}. \tag{5.20}$$

In view of (4.25), the force resultant equation of the *object* is then written as

$$^{\mathbf{O}}F^{*} = {}^{\mathbf{O}}\mathbf{U_C}\,{}^{\mathbf{C}}F. \tag{5.21}$$

5.5.2 Control Equations

Equation (5.14) implies

$$^{\mathbf{O}}V_r = {}^{\mathbf{C}}\mathbf{U}_{\mathbf{O}}^{T}\,{}^{\mathbf{C}}V_r \tag{5.22}$$

which validates (2.86).

Accordingly, recall (4.32) to compute the required net force/moment vector of the *object* with an appropriate frame substitution. It follows that

$$^{\mathbf{O}}F_r^{*} = \mathbf{Y_O}\hat{\boldsymbol{\theta}}_{\mathbf{O}} + \mathbf{K_O}\left({}^{\mathbf{O}}V_r - {}^{\mathbf{O}}V\right) \tag{5.23}$$

holds, where $\mathbf{K_O} \in \mathbb{R}^{6\times 6}$ is a positive-definite gain matrix and $\hat{\boldsymbol{\theta}}_{\mathbf{O}} \in \mathbb{R}^{13}$ denotes the estimate of $\boldsymbol{\theta}_{\mathbf{O}} \in \mathbb{R}^{13}$ with $\mathbf{Y_O}\boldsymbol{\theta}_{\mathbf{O}}$ being defined by (2.78) and expressed in Appendix A by substituting frames $\{\mathbf{O}\}$ for frame $\{\mathbf{A}\}$.

With respect to (5.23), define

$$\mathbf{s_O} = \mathbf{Y}_{\mathbf{O}}^{T}\left({}^{\mathbf{O}}V_r - {}^{\mathbf{O}}V\right). \tag{5.24}$$

The $\mathcal{P}$ function defined by (2.79) is used to update each of the 13 parameters of $\hat{\boldsymbol{\theta}}_{\mathbf{O}} \in \mathbb{R}^{13}$ as

$$\hat{\theta}_{\mathbf{O}\gamma} = \mathcal{P}\left(s_{\mathbf{O}\gamma}, \rho_{\mathbf{O}\gamma}, \underline{\theta}_{\mathbf{O}\gamma}, \overline{\theta}_{\mathbf{O}\gamma}, t\right) \tag{5.25}$$

for all $\gamma \in \{1, 13\}$, where

$\theta_{\mathbf{O}\gamma}$	γth element of $\boldsymbol{\theta}_{\mathbf{O}}$.
$\hat{\theta}_{\mathbf{O}\gamma}$	Estimate of $\theta_{\mathbf{O}\gamma}$.
$s_{\mathbf{O}\gamma}$	γth element of $\mathbf{s}_{\mathbf{O}}$.
$\rho_{\mathbf{O}\gamma}$	Parameter update gain.
$\underline{\theta}_{\mathbf{O}\gamma}$	Lower bound of $\theta_{\mathbf{O}\gamma}$.
$\overline{\theta}_{\mathbf{O}\gamma}$	Upper bound of $\theta_{\mathbf{O}\gamma}$.

Finally, the required force/moment vector in frame $\{\mathbf{C}\}$ is computed as

$$ {}^{\mathbf{C}}F_r = {}^{\mathbf{C}}\mathbf{U}_{\mathbf{O}}\,{}^{\mathbf{O}}F_r^*. \tag{5.26}$$

5.5.3 Virtual Stability

In this subsection, it will be shown that the held object combined with its respective control equations qualifies to be *virtually stable* in the sense of Definition 2.17.

Theorem 5.1. *The held object described by (5.20) subject to (5.19) and (5.21), combined with its respective control equations (5.22), (5.23), and (5.26) and with the parameter adaptation (5.24) and (5.25), is virtually stable with its affiliated vector* ${}^{\mathbf{O}}V_r - {}^{\mathbf{O}}V$ *being a virtual function in both* L_2 *and* L_∞*, in the sense of Definition 2.17.*

Proof: In view of (4.40), the non-negative accompanying function for the held object is chosen to be

$$\begin{aligned}\nu_{\mathbf{O}} = &\frac{1}{2}\left({}^{\mathbf{O}}V_r - {}^{\mathbf{O}}V\right)^T \mathbf{M}_{\mathbf{O}}\left({}^{\mathbf{O}}V_r - {}^{\mathbf{O}}V\right)\\ &+\frac{1}{2}\sum_{\gamma=1}^{13}\left(\theta_{\mathbf{O}\gamma} - \hat{\theta}_{\mathbf{O}\gamma}\right)^2/\rho_{\mathbf{O}\gamma}.\end{aligned} \tag{5.27}$$

With an appropriate frame substitution, it follows from (5.20), (5.23)–(5.25), and Lemma 4.1 that

$$\begin{aligned}\dot{\nu}_{\mathbf{O}} \leqslant &-\left({}^{\mathbf{O}}V_r - {}^{\mathbf{O}}V\right)^T \mathbf{K}_{\mathbf{O}}\left({}^{\mathbf{O}}V_r - {}^{\mathbf{O}}V\right)\\ &+\left({}^{\mathbf{O}}V_r - {}^{\mathbf{O}}V\right)^T\left({}^{\mathbf{O}}F_r^* - {}^{\mathbf{O}}F^*\right)\end{aligned} \tag{5.28}$$

holds.

In view of (2.85), (5.19), (5.21), (5.22), and (5.26), it yields

$$\begin{aligned}&\left({}^{\mathbf{O}}V_r - {}^{\mathbf{O}}V\right)^T\left({}^{\mathbf{O}}F_r^* - {}^{\mathbf{O}}F^*\right)\\ &= \left({}^{\mathbf{O}}V_r - {}^{\mathbf{O}}V\right)^T {}^{\mathbf{O}}\mathbf{U}_{\mathbf{C}}\left({}^{\mathbf{C}}F_r - {}^{\mathbf{C}}F\right)\\ &= \left[{}^{\mathbf{O}}\mathbf{U}_{\mathbf{C}}^T\left({}^{\mathbf{O}}V_r - {}^{\mathbf{O}}V\right)\right]^T\left({}^{\mathbf{C}}F_r - {}^{\mathbf{C}}F\right)\\ &= p_{\mathbf{C}}.\end{aligned} \tag{5.29}$$

Consider the fact that the held object has only one *driven cutting point* associated with frame $\{\mathbf{C}\}$. Substituting (5.29) into (5.28) and using (5.27) and Definition 2.17 completes the proof. ■

5.6 Dynamics and Control of Rigid Links

With a similar structure as that of the last section, three parts, namely kinematics and dynamics, model-based control, and *virtual stability*, are to be presented for each of the six rigid links in this section.

5.6.1 Kinematics and Dynamics

Equation (5.4) implies the following relationships

$$
\begin{aligned}
{}^{\mathbf{C}}V &= {}^{\mathbf{B}_6}\mathbf{U}_{\mathbf{C}}^{T}\,{}^{\mathbf{B}_6}V && (5.30)\\
{}^{\mathbf{T}_{k+1}}V &= {}^{\mathbf{B}_k}\mathbf{U}_{\mathbf{T}_{k+1}}^{T}\,{}^{\mathbf{B}_k}V, \ \forall k \in \{1,5\} && (5.31)
\end{aligned}
$$

with the two force/moment transformation matrices being defined by (2.65) with appropriate frame substitutions.

Recall (2.74) to obtain the dynamics of each rigid link with frame $\{\mathbf{B}_k\}$ being substituted for frame $\{\mathbf{A}\}$ for all $k \in \{1,6\}$. It follows that

$$
\mathbf{M}_{\mathbf{B}_k}\frac{d}{dt}({}^{\mathbf{B}_k}V) + \mathbf{C}_{\mathbf{B}_k}({}^{\mathbf{B}_k}\omega){}^{\mathbf{B}_k}V + \mathbf{G}_{\mathbf{B}_k} = {}^{\mathbf{B}_k}F^*, \quad \forall k \in \{1,6\} \qquad (5.32)
$$

holds.

In view of (4.51)–(4.53), the force resultant equations of the six rigid links can be written as

$$
\begin{aligned}
{}^{\mathbf{B}_6}F^* &= {}^{\mathbf{B}_6}F - {}^{\mathbf{B}_6}\mathbf{U}_{\mathbf{C}}\,{}^{\mathbf{C}}F && (5.33)\\
{}^{\mathbf{B}_k}F^* &= {}^{\mathbf{B}_k}F - {}^{\mathbf{B}_k}\mathbf{U}_{\mathbf{T}_{k+1}}\,{}^{\mathbf{T}_{j+1}}F \\
&= {}^{\mathbf{B}_k}F - {}^{\mathbf{B}_k}\mathbf{U}_{\mathbf{B}_{j+1}}\,{}^{\mathbf{B}_{j+1}}F, \ \forall k \in \{1,5\} && (5.34)
\end{aligned}
$$

in which ${}^{\mathbf{T}_k}F = {}^{\mathbf{T}_k}\mathbf{U}_{\mathbf{B}_k}\,{}^{\mathbf{B}_k}F$ is used for all $k \in \{1,6\}$.

5.6.2 Control Equations

Equation (5.14) implies the following relationships for the required linear/angular velocity vectors

$$
\begin{aligned}
{}^{\mathbf{C}}V_r &= {}^{\mathbf{B}_6}\mathbf{U}_{\mathbf{C}}^{T}\,{}^{\mathbf{B}_6}V_r && (5.35)\\
{}^{\mathbf{T}_{k+1}}V_r &= {}^{\mathbf{B}_k}\mathbf{U}_{\mathbf{T}_{k+1}}^{T}\,{}^{\mathbf{B}_k}V_r, \ \forall k \in \{1,5\}. && (5.36)
\end{aligned}
$$

These equations validate (2.86).

Accordingly, recall (4.32) to compute the required net force/moment vectors of the six rigid links with appropriate frame substitutions. It follows that

$$
{}^{\mathbf{B}_k}F_r^* = \mathbf{Y}_{\mathbf{B}_k}\hat{\boldsymbol{\theta}}_{\mathbf{B}_k} + \mathbf{K}_{\mathbf{B}_k}\left({}^{\mathbf{B}_k}V_r - {}^{\mathbf{B}_k}V\right), \quad \forall k \in \{1,6\} \qquad (5.37)
$$

holds, where $\mathbf{K}_{\mathbf{B}_k} \in \mathbb{R}^{6\times 6}$ is a positive-definite gain matrix and $\hat{\boldsymbol{\theta}}_{\mathbf{B}_k} \in \mathbb{R}^{13}$ denotes the estimate of $\boldsymbol{\theta}_{\mathbf{B}_k} \in \mathbb{R}^{13}$ with $\mathbf{Y}_{\mathbf{B}_k}\boldsymbol{\theta}_{\mathbf{B}_k}$ being defined by (2.78) and expressed in Appendix A by substituting frame $\{\mathbf{B}_k\}$ for frame $\{\mathbf{A}\}$.

With respect to (5.37), define

$$\mathbf{s}_{\mathbf{B}_k} = \mathbf{Y}_{\mathbf{B}_k}^T \left({}^{\mathbf{B}_k}V_r - {}^{\mathbf{B}_k}V \right), \quad \forall k \in \{1, 6\}. \tag{5.38}$$

The $\mathcal{P}$ function defined by (2.79) is used to update each of the 13 parameters of each rigid link as

$$\hat{\theta}_{\mathbf{B}_k\gamma} = \mathcal{P}\left(s_{\mathbf{B}_k\gamma}, \rho_{\mathbf{B}_k\gamma}, \underline{\theta}_{\mathbf{B}_k\gamma}, \overline{\theta}_{\mathbf{B}_k\gamma}, t \right) \tag{5.39}$$

for all $\gamma \in \{1, 13\}$ and $k \in \{1, 6\}$, where

$\theta_{\mathbf{B}_k\gamma}$	γth element of $\boldsymbol{\theta}_{\mathbf{B}_k}$.
$\hat{\theta}_{\mathbf{B}_k\gamma}$	Estimate of $\theta_{\mathbf{B}_k\gamma}$.
$s_{\mathbf{B}_k\gamma}$	γth element of $\mathbf{s}_{\mathbf{B}_k}$.
$\rho_{\mathbf{B}_k\gamma}$	Parameter update gain.
$\underline{\theta}_{\mathbf{B}_k\gamma}$	Lower bound of $\theta_{\mathbf{B}_k\gamma}$.
$\overline{\theta}_{\mathbf{B}_k\gamma}$	Upper bound of $\theta_{\mathbf{B}_k\gamma}$.

Finally, the required force/moment vectors in frame $\{\mathbf{B}_k\}$ for all $k \in \{1, 6\}$ are computed as

$${}^{\mathbf{B}_6}F_r = {}^{\mathbf{B}_6}F_r^* + {}^{\mathbf{B}_6}\mathbf{U}_{\mathbf{C}}\,{}^{\mathbf{C}}F_r \tag{5.40}$$

$$\begin{aligned} {}^{\mathbf{B}_k}F_r &= {}^{\mathbf{B}_k}F_r^* + {}^{\mathbf{B}_k}\mathbf{U}_{\mathbf{T}_{k+1}}\,{}^{\mathbf{T}_{j+1}}F_r \\ &= {}^{\mathbf{B}_k}F_r^* + {}^{\mathbf{B}_k}\mathbf{U}_{\mathbf{B}_{j+1}}\,{}^{\mathbf{B}_{j+1}}F_r, \; k = 5, \cdots, 1 \end{aligned} \tag{5.41}$$

in which ${}^{\mathbf{T}_k}F_r = {}^{\mathbf{T}_k}\mathbf{U}_{\mathbf{B}_k}\,{}^{\mathbf{B}_k}F_r$ is used for all $k \in \{1, 6\}$.

5.6.3 Virtual Stability

In this subsection, it will be shown that each rigid link combined with its respective control equations qualifies to be *virtually stable* in the sense of Definition 2.17.

Theorem 5.2. *The kth rigid link, $k \in \{1, 6\}$, described by (5.32) subject to (5.30), (5.31), (5.33), and (5.34), combined with its respective control equations (5.35)–(5.37), (5.40), and (5.41) and with the parameter adaptation (5.38) and (5.39), is virtually stable with its affiliated vector ${}^{\mathbf{B}_k}V_r - {}^{\mathbf{B}_k}V$ being a virtual function in both L_2 and L_∞, in the sense of Definition 2.17.*

Proof: In view of (4.40), the non-negative accompanying function for the kth rigid link is chosen to be

$$\begin{aligned} \nu_{\mathbf{B}_k} &= \frac{1}{2}\left({}^{\mathbf{B}_k}V_r - {}^{\mathbf{B}_k}V \right)^T \mathbf{M}_{\mathbf{B}_k} \left({}^{\mathbf{B}_k}V_r - {}^{\mathbf{B}_k}V \right) \\ &\quad + \frac{1}{2}\sum_{\gamma=1}^{13} \left(\theta_{\mathbf{B}_k\gamma} - \hat{\theta}_{\mathbf{B}_k\gamma} \right)^2 / \rho_{\mathbf{B}_k\gamma} \end{aligned} \tag{5.42}$$

for all $k \in \{1, 6\}$.

With an appropriate frame substitution, it follows from (5.32), (5.37)–(5.39), and Lemma 4.1 that

$$\begin{aligned} \dot{\nu}_{\mathbf{B}_k} \leqslant & -\left({}^{\mathbf{B}_k}V_r - {}^{\mathbf{B}_k}V\right)^T \mathbf{K}_{\mathbf{B}_k} \left({}^{\mathbf{B}_k}V_r - {}^{\mathbf{B}_k}V\right) \\ & + \left({}^{\mathbf{B}_k}V_r - {}^{\mathbf{B}_k}V\right)^T \left({}^{\mathbf{B}_k}F_r^* - {}^{\mathbf{B}_k}F^*\right) \end{aligned} \tag{5.43}$$

holds for all $k \in \{1, 6\}$.

In view of (2.85), (5.30), (5.31), (5.33)–(5.36), (5.40), and (5.41), it yields

$$\begin{aligned} & \left({}^{\mathbf{B}_6}V_r - {}^{\mathbf{B}_6}V\right)^T \left({}^{\mathbf{B}_6}F_r^* - {}^{\mathbf{B}_6}F^*\right) \\ & = \left({}^{\mathbf{B}_6}V_r - {}^{\mathbf{B}_6}V\right)^T \left[\left({}^{\mathbf{B}_6}F_r - {}^{\mathbf{B}_6}F\right) - {}^{\mathbf{B}_6}\mathbf{U}_{\mathbf{C}} \left({}^{\mathbf{C}}F_r - {}^{\mathbf{C}}F\right)\right] \\ & = p_{\mathbf{B}_6} - \left[{}^{\mathbf{B}_6}\mathbf{U}_{\mathbf{C}}^T \left({}^{\mathbf{B}_6}V_r - {}^{\mathbf{B}_6}V\right)\right]^T \left({}^{\mathbf{C}}F_r - {}^{\mathbf{C}}F\right) \\ & = p_{\mathbf{B}_6} - p_{\mathbf{C}} \end{aligned} \tag{5.44}$$

$$\begin{aligned} & \left({}^{\mathbf{B}_k}V_r - {}^{\mathbf{B}_k}V\right)^T \left({}^{\mathbf{B}_k}F^*{}_r - {}^{\mathbf{B}_k}F^*\right) \\ & = \left({}^{\mathbf{B}_k}V_r - {}^{\mathbf{B}_k}V\right)^T \left[\left({}^{\mathbf{B}_k}F_r - {}^{\mathbf{B}_k}F\right) - {}^{\mathbf{B}_k}\mathbf{U}_{\mathbf{T}_{k+1}} \left({}^{\mathbf{T}_{k+1}}F_r - {}^{\mathbf{T}_{k+1}}F\right)\right] \\ & = p_{\mathbf{B}_k} - \left[{}^{\mathbf{B}_k}\mathbf{U}_{\mathbf{T}_{k+1}}^T \left({}^{\mathbf{B}_k}V_r - {}^{\mathbf{B}_k}V\right)\right]^T \left({}^{\mathbf{T}_{k+1}}F_r - {}^{\mathbf{T}_{k+1}}F\right) \\ & = p_{\mathbf{B}_k} - p_{\mathbf{T}_{k+1}}, \quad k = 1, \cdots, 5. \end{aligned} \tag{5.45}$$

Consider the fact that the kth link has one *driving cutting point* associated with frame $\{\mathbf{T}_{k+1}\}$ for all $k \in \{1, 5\}$ or associated with frame $\{\mathbf{C}\}$ for $k = 6$ and one *driven cutting point* associated with frame $\{\mathbf{B}_k\}$ for all $k \in \{1, 6\}$. Substituting (5.44) and (5.45) into (5.43) and using (5.42) and Definition 2.17 completes the proof. ∎

5.7 Dynamics and Control of Joints

Electrically driven robots use electric motors to drive their joints. Electric motors generate forces/torques through armature currents resulting from the voltages applied to the motors. An electric motor can be in the *motor torque control* mode when the armature current is well controlled through a current servo amplifier and the motor torque/current constant is known. Otherwise, an electric motor should be in the *motor current control* mode when only the armature current is well controlled but the torque/current constant is unknown. Finally, an electric motor must be in the *motor voltage control* mode when no current servo control is available. Below, all the three joint control modes are presented progressively.

5.7.1 Motor Torque Control

Similar to Subsection 4.8.1, three parts, namely kinematics and dynamics, joint control equations, and *virtual stability*, are presented in this subsection.

Kinematics and Dynamics

The velocity mapping equation (5.4) results in a relationship between the joint velocity and the linear/angular velocity vectors of the adjacent links, that is

$$^{\mathbf{B}_k}V = \mathbf{z}\dot{q}_k + {}^{\mathbf{T}_k}\mathbf{U}_{\mathbf{B}_k}^T \, {}^{\mathbf{T}_k}V. \tag{5.46}$$

Recall (4.74). The dynamics of the kth joint including drive mechanisms can be written as

$$J_{mk}\ddot{q}_k + \xi_k(q_k, \dot{q}_k) = \tau_k - \mathbf{z}^T {}^{\mathbf{B}_k}F \tag{5.47}$$

where $J_{mk} \in \mathbb{R}$ is the equivalent mass or moment of inertia, $\xi_k(q_k, \dot{q}_k) \in \mathbb{R}$ is a term representing the friction and gravitation force/torque, $\tau_k \in \mathbb{R}$ is the control force/torque, and $\mathbf{z}^T {}^{\mathbf{B}_k}F$ is the projection of the force/moment vector onto the kth joint axis.

Joint Control Equations

Equation (5.14) implies

$$^{\mathbf{B}_{jk}}V_r = \mathbf{z}\dot{q}_{jkr} + {}^{\mathbf{T}_{jk}}\mathbf{U}_{\mathbf{B}_{jk}}^T \, {}^{\mathbf{T}_{jk}}V_r. \tag{5.48}$$

Recall (4.77) and (4.78). The desired joint force/torque of the kth joint is designed as

$$\tau_{kd}^* = \mathbf{Y}_{\tau k}\hat{\boldsymbol{\theta}}_{\tau k} + k_{\tau k}(\dot{q}_{kr} - \dot{q}_k) \tag{5.49}$$

$$\begin{aligned} \tau_{kd} &= \tau_{kd}^* + \mathbf{z}^T {}^{\mathbf{B}_k}F_r \\ &= \mathbf{Y}_{\tau k}\hat{\boldsymbol{\theta}}_{\tau k} + \mathbf{z}^T {}^{\mathbf{B}_k}F_r + k_{\tau k}(\dot{q}_{kr} - \dot{q}_k) \end{aligned} \tag{5.50}$$

with $k_{\tau k} > 0$, subject to

$$\mathbf{Y}_{\tau k}\boldsymbol{\theta}_{\tau k} = J_{mk}\ddot{q}_{kr} + \xi_k(q_k, \dot{q}_k) \tag{5.51}$$

where $\dot{q}_{kr}$ is the kth element of $\dot{\mathbf{q}}_r \in \mathbb{R}^6$ obtained from (5.14).

Define

$$\mathbf{s}_{\tau k} = \mathbf{Y}_{\tau k}^T (\dot{q}_{kr} - \dot{q}_k) \, . \tag{5.52}$$

The estimate of $\boldsymbol{\theta}_{\tau k}$ is updated by using the $\mathcal{P}$ function defined by (2.79) as

$$\hat{\theta}_{\tau k\gamma} = \mathcal{P}\left(s_{\tau k\gamma}, \rho_{\tau k\gamma}, \underline{\theta}_{\tau k\gamma}, \overline{\theta}_{\tau k\gamma}, t\right) \tag{5.53}$$

where $\hat{\theta}_{\tau k\gamma}$ denotes the estimate of $\theta_{\tau k\gamma}$—the γth element of $\boldsymbol{\theta}_{\tau k}$, $s_{\tau k\gamma}$ denotes the γth element of $\mathbf{s}_{\tau k}$ defined by (5.52), $\rho_{\tau k\gamma} > 0$ denotes an update gain, and $\underline{\theta}_{\tau k\gamma}$ and $\overline{\theta}_{\tau k\gamma}$ denote the lower and upper bounds of $\theta_{\tau k\gamma}$.

Virtual Stability

The following theorem ensures that each joint combined with its respective control equations in the *motor torque control* mode qualifies to be *virtually stable* in the sense of Definition 2.17.

Theorem 5.3. *The kth joint of the manipulator, $k \in \{1, 6\}$, described by (5.46) and (5.47), combined with the joint motor torque control equations (5.48)–(5.51) and with the parameter adaptation (5.52) and (5.53), subject to*

$$\tau_k = \tau_{kd} \tag{5.54}$$

is virtually stable with its affiliated variable $\dot{q}_{kr} - \dot{q}_k$ being a virtual function in both L_2 and L_∞, in the sense of Definition 2.17.

Proof: Substituting (5.50) into (5.47) and using (5.49) and (5.51) yields

$$\begin{aligned} J_{mk}(\ddot{q}_{kr} - \ddot{q}_k) = \tau_{kd} - \tau_k + \mathbf{Y}_{\tau k}(\boldsymbol{\theta}_{\tau k} - \hat{\boldsymbol{\theta}}_{\tau k}) \\ -\mathbf{z}^T({}^{\mathbf{B}_k}F_r - {}^{\mathbf{B}_k}F) - k_{\tau k}(\dot{q}_{kr} - \dot{q}_k). \end{aligned} \tag{5.55}$$

The first non-negative accompanying function is chosen as

$$\nu_{\tau k} = \frac{1}{2} J_{mk}(\dot{q}_{kr} - \dot{q}_k)^2 + \frac{1}{2} \sum_\gamma \left(\theta_{\tau k\gamma} - \hat{\theta}_{\tau k\gamma}\right)^2 / \rho_{\tau k\gamma}. \tag{5.56}$$

It follows from (5.52), (5.53), (5.55), and Lemma 2.9 that

$$\begin{aligned} \dot{\nu}_{\tau k} &= (\dot{q}_{kr} - \dot{q}_k) J_{mk}(\ddot{q}_{kr} - \ddot{q}_k) - \sum_\gamma \left(\theta_{\tau k\gamma} - \hat{\theta}_{\tau k\gamma}\right) \dot{\hat{\theta}}_{\tau k\gamma} / \rho_{\tau k\gamma} \\ &= -k_{\tau k}(\dot{q}_{kr} - \dot{q}_k)^2 + (\dot{q}_{kr} - \dot{q}_k)(\tau_{kd} - \tau_k) \\ &\quad -(\dot{q}_{kr} - \dot{q}_k)\mathbf{z}^T({}^{\mathbf{B}_k}F_r - {}^{\mathbf{B}_k}F) + (\dot{q}_{kr} - \dot{q}_k)\mathbf{Y}_{\tau k}(\boldsymbol{\theta}_{\tau k} - \hat{\boldsymbol{\theta}}_{\tau k}) \\ &\quad - \sum_\gamma \left(\theta_{\tau k\gamma} - \hat{\theta}_{\tau k\gamma}\right) \dot{\hat{\theta}}_{\tau k\gamma} / \rho_{\tau k\gamma} \\ &= -k_{\tau k}(\dot{q}_{kr} - \dot{q}_k)^2 + (\dot{q}_{kr} - \dot{q}_k)(\tau_{kd} - \tau_k) \\ &\quad -(\dot{q}_{kr} - \dot{q}_k)\mathbf{z}^T({}^{\mathbf{B}_k}F_r - {}^{\mathbf{B}_k}F) \\ &\quad + \sum_\gamma \left(\theta_{\tau k\gamma} - \hat{\theta}_{\tau k\gamma}\right) \left(s_{\tau k\gamma} - \dot{\hat{\theta}}_{\tau k\gamma} / \rho_{\tau k\gamma}\right) \\ &\leqslant -k_{\tau k}(\dot{q}_{kr} - \dot{q}_k)^2 + (\dot{q}_{kr} - \dot{q}_k)(\tau_{kd} - \tau_k) \\ &\quad -(\dot{q}_{kr} - \dot{q}_k)\mathbf{z}^T({}^{\mathbf{B}_k}F_r - {}^{\mathbf{B}_k}F) \end{aligned} \tag{5.57}$$

holds.

In view of (2.85), (5.46), and (5.48), the last term in the right-hand side of (5.57) can be re-written as

$$
\begin{aligned}
&-(\dot{q}_{kr} - \dot{q}_k)\mathbf{z}^T({}^{\mathbf{B}_k}F_r - {}^{\mathbf{B}_k}F) \\
&\quad = -\left[({}^{\mathbf{B}_k}V_r - {}^{\mathbf{B}_k}V) - {}^{\mathbf{T}_k}\mathbf{U}_{\mathbf{B}_k}^T\left({}^{\mathbf{T}_k}V_r - {}^{\mathbf{T}_k}V\right)\right]^T({}^{\mathbf{B}_k}F_r - {}^{\mathbf{B}_k}F) \\
&\quad = -p_{\mathbf{B}_k} + p_{\mathbf{T}_k}
\end{aligned}
\tag{5.58}
$$

in which ${}^{\mathbf{T}_k}F_r = {}^{\mathbf{T}_k}\mathbf{U}_{\mathbf{B}_k}{}^{\mathbf{B}_k}F_r$ and ${}^{\mathbf{T}_k}F = {}^{\mathbf{T}_k}\mathbf{U}_{\mathbf{B}_k}{}^{\mathbf{B}_k}F$ are used for all $k \in \{1,6\}$.

In the *motor torque control* mode, the joint control torque is commanded to validate (5.54).

Consider the fact that the kth joint has one *driving cutting point* associated with frame $\{\mathbf{B}_k\}$ and one *driven cutting point* associated with frame $\{\mathbf{T}_k\}$ for all $k \in \{1,6\}$. Substituting (5.54) and (5.58) into (5.57) and using (5.56) and Definition 2.17 completes the proof. ■

5.7.2 Motor Current Control

Motor current control mode is used when the motor armature current can be directly controlled, but the torque/current constant is uncertain. The design is based on the results achieved in the previous subsection.

In addition to the dynamics of the kth joint described by (5.47), the armature current is directly related to the electromagnetic force/torque by

$$\tau_k = k_{ck} I_{ck} \tag{5.59}$$

where I_{ck} denotes the motor armature current and $k_{ck} > 0$ denotes the unknown torque/current constant.

The corresponding desired armature current is designed as

$$I_{ckd} = \tau_{kd}/\hat{k}_{ck} \tag{5.60}$$

with $\hat{k}_{ck}$ being updated through the $\mathcal{P}$ function defined by (2.79) as

$$\hat{k}_{ck} = \mathcal{P}\left(s_{ck}, \rho_{ck}, \underline{k}_{ck}, \overline{k}_{ck}, t\right) \tag{5.61}$$

where

$$s_{ck} \overset{def}{=} -I_{ckd}(\dot{q}_{kr} - \dot{q}_k) \tag{5.62}$$

$\rho_{ck} > 0$ is an update gain, and $\underline{k}_{ck} > 0$ and $\overline{k}_{ck} > 0$ represent the lower and upper bounds of k_{ck}.

The following theorem shows that each joint combined with its respective control equations in the *motor current control* mode qualifies to be *virtually stable* in the sense of Definition 2.17.

Theorem 5.4. *The kth joint of the manipulator, $k \in \{1,6\}$, described by (5.46), (5.47), and (5.59), combined with the joint motor current control equations (5.48)–(5.51) and (5.60) and with the parameter adaptation (5.52), (5.53), (5.61), and (5.62), subject to*

$$I_{ck} = I_{ckd} \tag{5.63}$$

is virtually stable with its affiliated variable $\dot{q}_{kr} - \dot{q}_k$ being a virtual function in both L_2 and L_∞, in the sense of Definition 2.17.

Proof: Combining (5.59) with (5.60) gives

$$\tau_{kd} - \tau_k = k_{ck}(I_{ckd} - I_{ck}) - (k_{ck} - \hat{k}_{ck})I_{ckd}. \tag{5.64}$$

The second non-negative accompanying function is constructed as

$$\nu_{ck} = \nu_{\tau k} + \frac{1}{2}(k_{ck} - \hat{k}_{ck})^2 \tag{5.65}$$

with $\nu_{\tau k}$ being defined by (5.56).

In view of (2.85), (5.57), (5.58), (5.61), (5.62), (5.64), and Lemma 2.9, it follows that

$$\begin{aligned}
\dot{\nu}_{ck} &\leqslant k_{\tau k}(\dot{q}_{kr} - \dot{q}_k)^2 + (\dot{q}_{kr} - \dot{q}_k)(\tau_{kd} - \tau_k) \\
&\quad - p_{\mathbf{B}_k} + p_{\mathbf{T}_k} - (k_{ck} - \hat{k}_{ck})\dot{\hat{k}}_{ck}/\rho_{ck} \\
&= -k_{\tau k}(\dot{q}_{kr} - \dot{q}_k)^2 + k_{ck}(\dot{q}_{kr} - \dot{q}_k)(I_{ckd} - I_{ck}) \\
&\quad + (k_{ck} - \hat{k}_{ck})\left(s_{ck} - \dot{\hat{k}}_{ck}/\rho_{ck}\right) - p_{\mathbf{B}_k} + p_{\mathbf{T}_k} \\
&\leqslant -k_{\tau k}(\dot{q}_{kr} - \dot{q}_k)^2 + k_{ck}(\dot{q}_{kr} - \dot{q}_k)(I_{ckd} - I_{ck}) \\
&\quad - p_{\mathbf{B}_k} + p_{\mathbf{T}_k}
\end{aligned} \tag{5.66}$$

holds.

In the *motor current control* mode, the motor current is commanded to validate (5.63). Substituting (5.63) into (5.66) yields

$$\dot{\nu}_{ck} \leqslant -k_{\tau k}(\dot{q}_{kr} - \dot{q}_k)^2 - p_{\mathbf{B}_k} + p_{\mathbf{T}_k}. \tag{5.67}$$

Using (5.65), (5.67), and Definition 2.17 completes the proof. ∎

5.7.3 Motor Voltage Control

When the electric motor of the kth joint is operated in the *motor voltage control* mode, the armature current dynamics can be written as

$$a_k \dot{I}_{ck} + b_k I_{ck} + c_k \dot{q}_k = u_k \tag{5.68}$$

where u_k denotes the control voltage, I_{ck} denotes the motor armature current, a_k represents the armature inductance, b_k denotes the armature resistance, and c_k denotes the motor EMF constant.

With I_{ckd} being obtained from (5.60) and a regressor matrix $\mathbf{Y}_{vk}$ and a parameter vector $\boldsymbol{\theta}_{vk}$ being defined by

$$\mathbf{Y}_{vk} = \begin{bmatrix} \dot{I}_{ckd} & I_{ck} & \dot{q}_k & \dot{q}_{kr} - \dot{q}_k \end{bmatrix} \in \mathbb{R}^{1\times 4} \tag{5.69}$$

$$\boldsymbol{\theta}_{vk} = \begin{bmatrix} a_k & b_k & c_k & k_{ck} \end{bmatrix}^T \in \mathbb{R}^4 \tag{5.70}$$

the desired voltage u_{kd} is designed as

$$u_{kd} = \mathbf{Y}_{vk}\hat{\boldsymbol{\theta}}_{vk} + k_{vk}(I_{ckd} - I_{ck}) \tag{5.71}$$

where $k_{vk} > 0$ is a feedback gain and $\hat{\boldsymbol{\theta}}_{vk}$ denotes the estimate of $\boldsymbol{\theta}_{vk}$. The γth element of $\hat{\boldsymbol{\theta}}_{vk}$, denoted as $\hat{\theta}_{vk\gamma}$, is updated by using the $\mathcal{P}$ function defined by (2.79) as

$$\hat{\theta}_{vk\gamma} = \mathcal{P}\left(s_{vk\gamma}, \rho_{vk\gamma}, \underline{\theta}_{vk\gamma}, \overline{\theta}_{vk\gamma}, t\right) \tag{5.72}$$

for all $\gamma \in \{1, 4\}$, where $s_{vk\gamma}$ denotes the γth element of

$$\mathbf{s}_{vk} \overset{def}{=} \mathbf{Y}_{vk}^T(I_{ckd} - I_{ck}) \tag{5.73}$$

$\rho_{vk\gamma} > 0$ denotes an update gain, and $\underline{\theta}_{vk\gamma}$ and $\overline{\theta}_{vk\gamma}$ denote the lower and upper bounds of $\theta_{vk\gamma}$, the γth element of $\boldsymbol{\theta}_{vk}$.

The following theorem shows that each joint combined with its respective control equations in the *motor voltage control* mode qualifies to be *virtually stable* in the sense of Definition 2.17.

Theorem 5.5. *The kth joint of the manipulator, $k \in \{1, 6\}$, described by (5.46), (5.47), (5.59), and (5.68), combined with the joint motor voltage control equations (5.48)–(5.51), (5.60), and (5.71) and with the parameter adaptation (5.52), (5.53), (5.61), (5.62), (5.72), and (5.73), subject to*

$$u_k = u_{kd} \tag{5.74}$$

is virtually stable with its affiliated variables $\dot{q}_{kr} - \dot{q}_k$ and $I_{ckd} - I_{ck}$ being virtual functions in both L_2 and L_∞, in the sense of Definition 2.17.

Proof: Substituting (5.71) into (5.68) yields

$$\begin{aligned} u_{kd} - u_k &= a_k(\dot{I}_{ckd} - \dot{I}_{ck}) + k_{ck}(\dot{q}_{kr} - \dot{q}_k) \\ &\quad + k_{vk}(I_{ckd} - I_{ck}) - \mathbf{Y}_{vk}(\boldsymbol{\theta}_{vk} - \hat{\boldsymbol{\theta}}_{vk}). \end{aligned} \tag{5.75}$$

The third non-negative accompanying function is constructed as

$$\nu_{vk} = \nu_{ck} + \frac{1}{2}a_k(I_{ckd} - I_{ck})^2 + \frac{1}{2}\sum_{\gamma=1}^{4}\left(\theta_{vk\gamma} - \hat{\theta}_{vk\gamma}\right)^2 / \rho_{vk\gamma} \tag{5.76}$$

with ν_{ck} being defined by (5.65).

In view of (5.66), (5.72), (5.73), (5.75), and Lemma 2.9, it follows that

$$\begin{aligned} \dot{\nu}_{vk} &\leqslant -k_{\tau k}(\dot{q}_{kr} - \dot{q}_k)^2 + k_{ck}(\dot{q}_{kr} - \dot{q}_k)(I_{ckd} - I_{ck}) - p_{\mathbf{B}_k} + p_{\mathbf{T}_k} \\ &\quad + (I_{ckd} - I_{ck})a_k(\dot{I}_{ckd} - \dot{I}_{ck}) - \sum_{\gamma=1}^{4}\left(\theta_{vk\gamma} - \hat{\theta}_{vk\gamma}\right)\dot{\hat{\theta}}_{vk\gamma}/\rho_{vk\gamma} \\ &= -k_{\tau k}(\dot{q}_{kr} - \dot{q}_k)^2 - k_{vk}(I_{ckd} - I_{ck})^2 + (I_{ckd} - I_{ck})(u_{kd} - u_k) \\ &\quad + \sum_{\gamma=1}^{4}\left(\theta_{vk\gamma} - \hat{\theta}_{vk\gamma}\right)\left(s_{vk\gamma} - \dot{\hat{\theta}}_{vk\gamma}/\rho_{vk\gamma}\right) - p_{\mathbf{B}_k} + p_{\mathbf{T}_k} \\ &\leqslant -k_{\tau k}(\dot{q}_{kr} - \dot{q}_k)^2 - k_{vk}(I_{ckd} - I_{ck})^2 \\ &\quad + (I_{ckd} - I_{ck})(u_{kd} - u_k) - p_{\mathbf{B}_k} + p_{\mathbf{T}_k} \end{aligned} \tag{5.77}$$

holds.

In the *motor voltage control* mode, the motor voltage is commanded to validate (5.74). Substituting (5.74) into (5.77) yields

$$\dot{\nu}_{vk} \leqslant -k_{\tau k}(\dot{q}_{kr} - \dot{q}_k)^2 - k_{vk}(I_{ckd} - I_{ck})^2 - p_{\mathbf{B}_k} + p_{\mathbf{T}_k}. \tag{5.78}$$

Using (5.76), (5.78), and Definition 2.17 completes the proof. ■

This section provides three options for conducting adaptive control of each joint driven by an electric motor. The *motor torque control* mode can be used only when both $I_{ckd} = I_{ck}$ and $\hat{k}_{ck} = k_{ck}$ hold, that is, the current servo loop has an ideal transfer function of *one* and the torque/current constant is known. Otherwise, the *motor current control* mode should be used when $I_{ckd} = I_{ck}$ holds. Finally, the *motor voltage control* mode is recommended when neither $I_{ckd} = I_{ck}$ nor $\hat{k}_{ck} = k_{ck}$ holds.

The three joint control modes are presented in a progressive order. The *motor current control* mode is based on the *motor torque control* mode, and the *motor voltage control* mode is further based on the *motor current control* mode. Most importantly, the *virtual stability* ensures that the dynamic interactions of a joint with its adjacent links are uniquely represented by the *virtual power flows* at the two *cutting points* of the joint, denoted as $p_{\mathbf{T}_k} - p_{\mathbf{B}_k}$, regardless of the control mode being used.

5.8 Forward Dynamics

Recall the procedure in Section 4.9 leading to (4.136). The forward dynamics can be obtained as

$$\mathcal{M}\dot{\mathcal{V}} + \mathcal{C}\mathcal{V} + \mathcal{G} = \mathbf{J}_q^{-T}\boldsymbol{\tau} \tag{5.79}$$

where $\boldsymbol{\tau} = [\tau_1, \cdots, \tau_k, \cdots, \tau_6]^T \in \mathbb{R}^6$ denotes the joint control force/torque vector and

$$\mathcal{M} = \mathcal{J}_b^T \mathcal{M}_b \mathcal{J}_b + \mathcal{J}_a^T \mathcal{M}_a \mathcal{J}_a \tag{5.80}$$

$$\mathcal{C} = \mathcal{J}_b^T \mathcal{M}_b \dot{\mathcal{J}}_b + \mathcal{J}_a^T \mathcal{M}_a \dot{\mathcal{J}}_a + \mathcal{J}_b^T \mathcal{C}_b \mathcal{J}_b \tag{5.81}$$

$$\mathcal{G} = \mathcal{J}_b^T \mathcal{G}_b + \mathcal{J}_a^T \mathcal{G}_a \tag{5.82}$$

with

$$\mathcal{M}_a = \mathrm{diag}(J_{m1}, \cdots, J_{m6}) \tag{5.83}$$

$$\mathcal{M}_b = \mathrm{diag}(\mathbf{M_O}, \mathbf{M}_{\mathbf{B}_1}, \cdots, \mathbf{M}_{\mathbf{B}_6}) \tag{5.84}$$

$$\mathcal{C}_b = \mathrm{diag}(\mathbf{C_O}, \mathbf{C}_{\mathbf{B}_1}, \cdots, \mathbf{C}_{\mathbf{B}_6}) \tag{5.85}$$

$$\mathcal{G}_a = [\xi_1, \cdots, \xi_6]^T \tag{5.86}$$

$$\mathcal{G}_b = [\mathbf{G}_{\mathbf{O}}^T, \mathbf{G}_{\mathbf{B}_1}^T, \cdots, \mathbf{G}_{\mathbf{B}_6}^T]^T. \tag{5.87}$$

5.9 Control Algorithm

The overall control algorithm for the six-joint manipulator is summarized into four steps as follows:

Step 1: After the required independent velocity vector being formed by (5.13), use (5.4) and (5.14) to compute the extended velocity vector and the required extended velocity vector of the entire robot.

Step 2: Compute the required net force/moment vectors of the held object and the six rigid links from (5.23) and (5.37), and compute the parameter adaptation described by (5.24), (5.25), (5.38), and (5.39).

Step 3: Compute the required force/moment vectors at all *cutting points* described by (5.26), (5.40), and (5.41).

Step 4: Compute the joint control equations (5.49) and (5.50) and the corresponding parameter adaptation (5.52) and (5.53) in the *motor torque control* mode, perform addition computations (5.60), (5.61), and (5.62) in the *motor current control* mode, and perform even more computations (5.71), (5.72), and (5.73) in the *motor voltage control* mode.

The following lemma is going to show that the control algorithm with respect to a six-joint robot manipulator addressed in this chapter makes the joint desired control force/torque vector an affine function of the time-derivative of the required independent velocity vector.

Lemma 5.1. *The virtual decomposition control equations (5.3), (5.10)-(5.12), (5.14) or (5.15)-(5.16), (5.22), (5.23), (5.26), (5.37), (5.40), (5.41), (5.49), and (5.50) make* $\mathbf{J}_q^{-T}\boldsymbol{\tau}_d \in \mathbb{R}^6$ *an affine function of* $\dot{\mathcal{V}}_r = \frac{d}{dt}({}^{\mathbf{O}}V_r) \in \mathbb{R}^6$, *that is*

$$\begin{aligned}\mathbf{J}_q^{-T}\boldsymbol{\tau}_d &= \hat{\mathcal{M}}\frac{d}{dt}({}^{\mathbf{O}}V_r) + \mathbf{b}_\tau \\ &= \hat{\mathcal{M}}{}^{\mathbf{C}}\mathbf{U}_{\mathbf{O}}^T\frac{d}{dt}({}^{\mathbf{C}}V_r) + \mathbf{b}_\tau \end{aligned} \tag{5.88}$$

with

$$\boldsymbol{\tau}_d \stackrel{def}{=} [\tau_{1d}, \cdots, \tau_{6d}]^T \in \mathbb{R}^6 \tag{5.89}$$

$$\hat{\mathcal{M}} \stackrel{def}{=} \mathcal{J}_b^T\hat{\mathcal{M}}_b\mathcal{J}_b + \mathcal{J}_a^T\hat{\mathcal{M}}_a\mathcal{J}_a \tag{5.90}$$

and $\mathbf{b}_\tau \in \mathbb{R}^6$ *being a remaining vector containing state variables and reference variables.*

Proof: In view of (5.14) or (5.15)-(5.16), $\mathcal{V}_{ar}$ and $\mathcal{V}_{br}$ are affine functions of ${}^{\mathbf{O}}V_r$. Define

$$\mathcal{F}_{ar} \stackrel{def}{=} [\tau_{1d}^*, \cdots, \tau_{6d}^*]^T \in \mathbb{R}^6 \tag{5.91}$$

$$\mathcal{F}_{br} \stackrel{def}{=} \left[{}^{\mathbf{O}}F_r^{*T}, {}^{\mathbf{B}_1}F_r^{*T}, \cdots, {}^{\mathbf{B}_6}F_r^{*T}\right]^T \in \mathbb{R}^{42}. \tag{5.92}$$

The two vectors can be expressed as affine functions of $\frac{d}{dt}({}^{\mathbf{O}}V_r) \in \mathbb{R}^6$ as

$$\mathcal{F}_{ar} = \hat{\mathcal{M}}_a \mathcal{J}_a \frac{d}{dt}({}^{\mathbf{O}}V_r) + \mathbf{b}_{af} \tag{5.93}$$

$$\mathcal{F}_{br} = \hat{\mathcal{M}}_b \mathcal{J}_b \frac{d}{dt}({}^{\mathbf{O}}V_r) + \mathbf{b}_{bf} \tag{5.94}$$

from (5.3), (5.10)-(5.12), (5.23), (5.37), and (5.49), where $\hat{\mathcal{M}}_a \in \mathbb{R}^{6\times 6}$ and $\hat{\mathcal{M}}_b \in \mathbb{R}^{42\times 42}$ represent the estimates of $\mathcal{M}_a$ and $\mathcal{M}_b$ defined by (5.83) and (5.84) with estimated inertial parameters being used and $\mathbf{b}_{af} \in \mathbb{R}^6$ and $\mathbf{b}_{bf} \in \mathbb{R}^6$ are two remaining vectors. Then, it follows from (5.11), (5.26), (5.40), (5.41), (5.50), and (5.89) that

$$\boldsymbol{\tau}_d = \mathcal{F}_{ar} + \mathcal{J}_{ba}^T \mathcal{F}_{br} \tag{5.95}$$

holds. Finally, premultiplying (5.95) by $\mathcal{J}_a^T = \mathbf{J}_q^{-T}$ and using (5.10), (5.12), (5.93), and (5.94) yields the first equation of (5.88). The second equation is straightforward from (5.22). ■

Remark 5.1. Lemma 5.1 will be used below to prove the asymptotic stability of the entire robot and to address the algebraic loop issue in force control.

5.10 Stability

The stability and convergence of the entire robot manipulator subject to the VDC approach are being presented in this section.

5.10.1 L_2 and L_∞ Stability

It follows from Theorems 5.1, 5.2, and 5.3 or 5.4 or 5.5 that every subsystem combined with its respective control equations and parameter adaptation is *virtually stable* in the sense of Definition 2.17. Therefore, it follows from Theorem 2.1 that

$${}^{\mathbf{O}}V_r - {}^{\mathbf{O}}V \in L_2 \bigcap L_\infty \tag{5.96}$$

$${}^{\mathbf{B}_k}V_r - {}^{\mathbf{B}_k}V \in L_2 \bigcap L_\infty, \quad \forall k \in \{1, 6\} \tag{5.97}$$

$$\dot{q}_{kr} - \dot{q}_k \in L_2 \bigcap L_\infty, \quad \forall k \in \{1, 6\} \tag{5.98}$$

hold and furthermore

$$I_{ckd} - I_{ck} \in L_2 \bigcap L_\infty, \quad k \in \{1, 6\} \tag{5.99}$$

holds when the kth joint is in the *motor voltage control* mode.

By equalizing

$$\mathcal{V}_p = {}^{\mathbf{O}}\mathcal{V}$$
$$\mathcal{V}_{pd} = \begin{bmatrix} {}^{\mathbf{O}}\mathbf{R}_{\mathbf{I}}\dot{\mathbf{x}}_d \\ {}^{\mathbf{O}}\mathbf{R}_{\mathbf{I}}\boldsymbol{\omega}_d \end{bmatrix} \in \mathbb{R}^6$$
$$\mathcal{E} = \begin{bmatrix} k_x {}^{\mathbf{O}}\mathbf{R}_{\mathbf{I}} (\mathbf{x}_d - \mathbf{x}) \\ k_\omega {}^{\mathbf{O}}\boldsymbol{\mu} \end{bmatrix} \in \mathbb{R}^6$$

it can be verified that (4.176) and (4.177) are satisfied, in view of Lemmas 4.4 and 4.5. Thus, it follows from (5.13), (5.96), and Theorem 4.6 that

$$\dot{\mathbf{x}}_d - \dot{\mathbf{x}} \in L_2 \bigcap L_\infty \tag{5.100}$$
$$\mathbf{x}_d - \mathbf{x} \in L_2 \bigcap L_\infty \tag{5.101}$$
$$\boldsymbol{\omega}_d - \boldsymbol{\omega} \in L_2 \bigcap L_\infty \tag{5.102}$$
$${}^{\mathbf{O}}\boldsymbol{\mu} \in L_2 \bigcap L_\infty \tag{5.103}$$

hold.

5.10.2 Asymptotic Stability

Note that $\boldsymbol{\omega}_d - \boldsymbol{\omega} \in L_\infty$ implies ${}^{\mathbf{O}}\mathbf{R}_{\mathbf{I}}(\boldsymbol{\omega}_d - \boldsymbol{\omega}) \in L_\infty$ and further implies $\frac{d}{dt}\left({}^{\mathbf{O}}\boldsymbol{\mu}\right) \in L_\infty$ from (2.49). Thus, it follows from $\dot{\mathbf{x}}_d - \dot{\mathbf{x}} \in L_\infty$, $\mathbf{x}_d - \mathbf{x} \in L_2$, $\frac{d}{dt}\left({}^{\mathbf{O}}\boldsymbol{\mu}\right) \in L_\infty$, ${}^{\mathbf{O}}\boldsymbol{\mu} \in L_2$, and Lemma 2.8 that

$$\mathbf{x}_d - \mathbf{x} \to \mathbf{0} \tag{5.104}$$
$${}^{\mathbf{O}}\boldsymbol{\mu} \to \mathbf{0} \tag{5.105}$$

hold which yields the asymptotic stability of the position and orientation tracking control.

To prove the asymptotic stability of the velocity control, it is necessary to ensure $\dot{\mathcal{V}}_r - \dot{\mathcal{V}} \in L_\infty$.

Given $\mathbf{x}_d \in L_\infty$, $\dot{\mathbf{x}}_d \in L_\infty$, $\ddot{\mathbf{x}}_d \in L_\infty$, $\boldsymbol{\omega}_d \in L_\infty$, and $\dot{\boldsymbol{\omega}}_d \in L_\infty$, it follows from (5.13) and $\mathbf{x}_d - \mathbf{x} \in L_\infty$ that $\mathcal{V}_r \in L_\infty$ holds, and furthermore $\mathcal{V} \in L_\infty$ holds from (5.96). Differentiating (5.13) over time and using $\dot{\mathbf{x}}_d - \dot{\mathbf{x}} \in L_\infty$, $\boldsymbol{\omega}_d - \boldsymbol{\omega} \in L_\infty$, and $\boldsymbol{\omega} \in L_\infty$ yields $\dot{\mathcal{V}}_r \in L_\infty$.

It then follows from Lemma 5.1 in Section 5.9 that $\tau_{kd} \in L_\infty$ holds for all $k \in \{1, 6\}$. It further yields $I_{ckd} \in L_\infty$ from (5.60) for all $k \in \{1, 6\}$. Consequently, it results in $\tau_k \in L_\infty$ from (5.54) and $I_{ck} \in L_\infty$ from (5.63) or (5.99) for all $k \in \{1, 6\}$. Finally, the invertibility of $\mathcal{M}$ in (5.79) ensures $\dot{\mathcal{V}} = \frac{d}{dt}\left({}^{\mathbf{O}}V\right) \in L_\infty$.

As a result, the boundedness of $\dot{\mathcal{V}}_r - \dot{\mathcal{V}}$ ensures

$$\mathcal{V}_r - \mathcal{V} \to \mathbf{0} \tag{5.106}$$

in view of (5.1), (5.13), (5.96), and Lemma 2.8. It further guarantees the asymptotic convergence of the linear and angular velocity error vectors

$$\dot{\mathbf{x}}_d - \dot{\mathbf{x}} \rightarrow \mathbf{0} \tag{5.107}$$

$$\boldsymbol{\omega}_d - \boldsymbol{\omega} \rightarrow \mathbf{0} \tag{5.108}$$

in addition to (5.104) and (5.105).

5.11 Hybrid Motion/Force Control

In some applications such as assembly, deburring, polishing, and cleaning, the end-effector of a robot manipulator usually contacts with an environment. In such a scenario, force control along the contact directions becomes necessary to prevent excessive contact forces from happening.

As far as a compliant environment is concerned in robot force control, the environment stiffness is usually incorporated into the robot dynamics to derive the control equations. To take the advantage of a model-based force control, the environment resonant mode determined by the equivalent stiffness and mass needs to be placed well within the control bandwidth. This fact limits the maximum allowable stiffness to be taken into account by the controller. A realistic number is about $10^3 \sim 10^4$ (N/m). In practice, however, most contact environments have stiffness ranging from 10^6 (N/m) for woods to 10^9 (N/m) for steel. Therefore, artificial flexible mechanisms are usually added to the end-effector to render the contact stiffness to a range the control system can handle [46]. However, the use of flexible mechanisms at the end-effector can cause undesirable vibrations in the free motion [48] and, consequently, affects positioning accuracy.

In this section, robot force control without using additional flexible mechanisms at the end-effector is addressed. The solution is to replace the hardware compliance by software compliance. To do this, a filtered contact force term (rather than the contact force itself) is added to the required linear/angular velocity vector of the end-effector. This technique allows the robot manipulator to achieve the asymptotic stability of both motion control and force control for known contact geometry, while retaining the properties of the impedance control for unknown contact geometry.

When a contact between the end-effector and the environment happens, the concept of the *virtual stability* allows us to change the control equations solely associated with the end-effector, while keeping the control equations associated with the rest of the robot unchanged.

5.11.1 Kinematics and Dynamics in Constrained Motion

Figure 5.3 illustrates the held object in contact with the environment. Specifically, let $\{\mathbf{S}\}$ be a body-fixed frame located at the contact point with the environment, and let $\{\mathbf{S}_b\}$ be a frame having the same origin as frame $\{\mathbf{S}\}$. In frame $\{\mathbf{S}_b\}$, some of its axes describe the motion and others the constraints. Let $0 < n_f < 6$ be the number of dimensions of the independent constraint forces at the end-effector, and let $\mathbf{T}_s \in \mathbb{R}^{6\times(6-n_f)}$ and $\mathbf{T}_f \in \mathbb{R}^{6\times n_f}$ be two selection

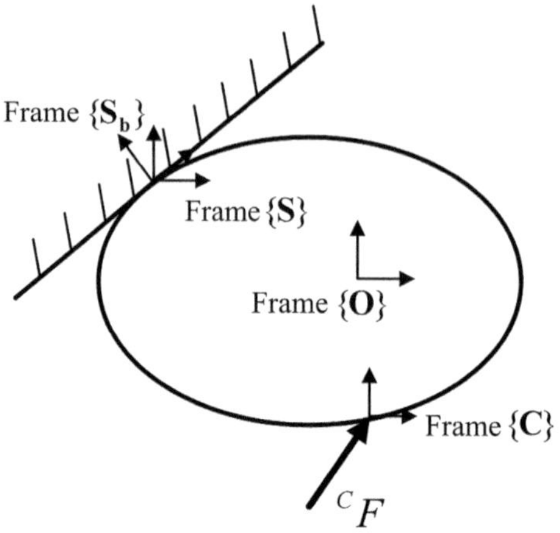

Fig. 5.3. The held object in the constrained motion.

mapping matrices of full column-rank. Each column contains a single *one* and five *zeros*, subject to $\mathbf{T}_s^T\mathbf{T}_s = \mathbf{I}_{(6-n_f)}$, $\mathbf{T}_f^T\mathbf{T}_f = \mathbf{I}_{n_f}$, and $\mathbf{T}_s^T\mathbf{T}_f = \mathbf{0}$.

It follows from (4.22) that

$$^{\mathbf{S}}V = \text{diag}\left(^{\mathbf{S}}\mathbf{R}_{\mathbf{S}_b}, {}^{\mathbf{S}}\mathbf{R}_{\mathbf{S}_b}\right)\mathbf{T}_s\chi \tag{5.109}$$

holds, where $\chi \in \mathbb{R}^{(6-n_f)}$ denotes the independent velocity coordinate vector of the constrained robot manipulator, leading to

$$\mathcal{V} = \chi \in \mathbb{R}^{(6-n_f)} \tag{5.110}$$

with

$$\begin{aligned} ^{\mathbf{O}}V &= {}^{\mathbf{S}}\mathbf{U}_{\mathbf{O}}^T\,{}^{\mathbf{S}}V \\ &= {}^{\mathbf{S}}\mathbf{U}_{\mathbf{O}}^T\text{diag}\left(^{\mathbf{S}}\mathbf{R}_{\mathbf{S}_b}, {}^{\mathbf{S}}\mathbf{R}_{\mathbf{S}_b}\right)\mathbf{T}_s\chi \\ &= {}^{\mathbf{S}}\mathbf{U}_{\mathbf{O}}^T\text{diag}\left(^{\mathbf{S}}\mathbf{R}_{\mathbf{S}_b}, {}^{\mathbf{S}}\mathbf{R}_{\mathbf{S}_b}\right)\mathbf{T}_s\mathcal{V}. \end{aligned} \tag{5.111}$$

In view of (4.26), the contact force/moment vector in frame $\{\mathbf{S}\}$ can thus be written as

$$^{\mathbf{S}}F = \text{diag}\left(^{\mathbf{S}}\mathbf{R}_{\mathbf{S}_b}, {}^{\mathbf{S}}\mathbf{R}_{\mathbf{S}_b}\right)\left(\mathbf{T}_s\psi + \mathbf{T}_f\varphi\right) \tag{5.112}$$

with

$$\psi = \mathbf{Y}_s\boldsymbol{\theta}_s \tag{5.113}$$

where $\psi \in \mathbb{R}^{(6-n_f)}$ and $\varphi \in \mathbb{R}^{n_f}$ denote the independent force coordinate vectors in the motion and constraint configuration spaces, respectively. Equation (5.113) implies that the independent (friction) force coordinate vector in the motion configuration space can be expressed in a linear parametrization form with $\boldsymbol{\theta}_s$ being a parameter vector.

In the presence of the contact force/moment vector $^{\mathbf{S}}F$, the force/moment resultant equation (5.21) changes to

$$ {}^{\mathbf{O}}F^* = {}^{\mathbf{O}}\mathbf{U}_{\mathbf{C}}{}^{\mathbf{C}}F - {}^{\mathbf{O}}\mathbf{U}_{\mathbf{S}}{}^{\mathbf{S}}F \tag{5.114} $$

with ${}^{\mathbf{O}}F^* \in \mathbb{R}^6$ being defined by (5.20).

5.11.2 Hybrid Motion/Force Control Equations

Let $\boldsymbol{\chi}_r \in \mathbb{R}^{(6-n_f)}$ be the required (design) vector of $\boldsymbol{\chi} \in \mathbb{R}^{(6-n_f)}$ and let $\boldsymbol{\varphi}_d \in \mathbb{R}^{n_f}$ be the desired (design) vector of $\boldsymbol{\varphi} \in \mathbb{R}^{n_f}$. Replace (5.13) by

$$ {}^{\mathbf{O}}V_r = {}^{\mathbf{S}}\mathbf{U}_{\mathbf{O}}^T{}^{\mathbf{S}}V_r \tag{5.115} $$

with

$$ {}^{\mathbf{S}}V_r = \text{diag}\left({}^{\mathbf{S}}\mathbf{R}_{\mathbf{S}_b}, {}^{\mathbf{S}}\mathbf{R}_{\mathbf{S}_b}\right)\left[\mathbf{T}_s\boldsymbol{\chi}_r + \mathbf{T}_f\mathbf{A}_f\left(\tilde{\boldsymbol{\varphi}}_d - \tilde{\boldsymbol{\varphi}}\right)\right] \tag{5.116} $$

where $\mathbf{T}_s \in \mathbb{R}^{6\times(6-n_f)}$ and $\mathbf{T}_f \in \mathbb{R}^{6\times n_f}$ are defined in the last subsection, $\mathbf{A}_f \in \mathbb{R}^{n_f\times n_f}$ is a diagonal positive-definite matrix, and $\tilde{\boldsymbol{\varphi}}_d \in \mathbb{R}^{n_f}$ and $\tilde{\boldsymbol{\varphi}} \in \mathbb{R}^{n_f}$ are subject to

$$ \dot{\tilde{\boldsymbol{\varphi}}}_d = -\mathbf{C}_f\tilde{\boldsymbol{\varphi}}_d + \mathbf{C}_f\boldsymbol{\varphi}_d \tag{5.117} $$

$$ \dot{\tilde{\boldsymbol{\varphi}}} = -\mathbf{C}_f\tilde{\boldsymbol{\varphi}} + \mathbf{C}_f\boldsymbol{\varphi} \tag{5.118} $$

with $\mathbf{C}_f \in \mathbb{R}^{n_f\times n_f}$ being a diagonal positive-definite matrix. Equations (5.117) and (5.118) indicate that $\tilde{\boldsymbol{\varphi}}_d \in \mathbb{R}^{n_f}$ and $\tilde{\boldsymbol{\varphi}} \in \mathbb{R}^{n_f}$ are filtered vectors of $\boldsymbol{\varphi}_d \in \mathbb{R}^{n_f}$ and $\boldsymbol{\varphi} \in \mathbb{R}^{n_f}$, respectively.

Comparing (4.30) with (5.116) reveals the difference between the two designs. In (5.116), a second term $\mathbf{T}_f\mathbf{A}_f\left(\tilde{\boldsymbol{\varphi}}_d - \tilde{\boldsymbol{\varphi}}\right)$ is added. As will be seen below, this term plays an important role to ensure the stability and convergence of the filtered force error vector, $\tilde{\boldsymbol{\varphi}}_d - \tilde{\boldsymbol{\varphi}}$.

Recall (4.36)–(4.38) by ignoring the subscript i leading to

$$ {}^{\mathbf{S}}F_r = \text{diag}\left({}^{\mathbf{S}}\mathbf{R}_{\mathbf{S}_b}, {}^{\mathbf{S}}\mathbf{R}_{\mathbf{S}_b}\right)\left(\mathbf{T}_s\hat{\boldsymbol{\psi}} + \mathbf{T}_f\boldsymbol{\varphi}_d\right) \tag{5.119} $$

with

$$ \hat{\boldsymbol{\psi}} = \mathbf{Y}_s\hat{\boldsymbol{\theta}}_s. \tag{5.120} $$

By defining

$$ \mathbf{s}_s = \mathbf{Y}_s^T\left(\boldsymbol{\chi}_r - \boldsymbol{\chi}\right) \tag{5.121} $$

the γth element of $\hat{\boldsymbol{\theta}}_s$ is updated by

$$ \hat{\theta}_{s\gamma} = \mathcal{P}\left(s_{s\gamma}, \rho_{s\gamma}, \underline{\theta}_{s\gamma}, \overline{\theta}_{s\gamma}, t\right) \tag{5.122} $$

where $\hat{\theta}_{s\gamma}$ is the γth element of $\hat{\boldsymbol{\theta}}_s$, $s_{s\gamma}$ is the γth element of $\mathbf{s}_s$, $\rho_{s\gamma} > 0$ is an update gain, and $\underline{\theta}_{s\gamma}$ and $\overline{\theta}_{s\gamma}$ denote the lower and upper bounds of $\theta_{s\gamma}$—the γth element of $\boldsymbol{\theta}_s$.

According to (5.114), the required net force/moment vector in frame $\{\mathbf{O}\}$ is subject to

$$ {}^{\mathbf{O}}F_r^* = {}^{\mathbf{O}}\mathbf{U}_{\mathbf{C}}{}^{\mathbf{C}}F_r - {}^{\mathbf{O}}\mathbf{U}_{\mathbf{S}}{}^{\mathbf{S}}F_r \tag{5.123} $$

or equivalently

$$^{\mathbf{C}}F_r = {}^{\mathbf{C}}\mathbf{U}_{\mathbf{O}}{}^{\mathbf{O}}F_r^* + {}^{\mathbf{C}}\mathbf{U}_{\mathbf{S}}{}^{\mathbf{S}}F_r \tag{5.124}$$

where ${}^{\mathbf{O}}F_r^* \in \mathbb{R}^6$ is defined by (5.23).

The remaining control computations use exactly the same equations as these in the previous sections. Equation (5.14) is used to obtain $\mathcal{V}_{er}$. Then, (5.37) is used to obtain ${}^{\mathbf{B}_k}F_r^*$ for all $k \in \{1, 6\}$. Furthermore, (5.40) and (5.41) are used to obtain ${}^{\mathbf{B}_k}F_r$ for all $k \in \{1, 6\}$. Finally, any one of the three joint control modes presented in Section 5.7 is used.

5.11.3 Virtual Stability

The VDC approach permits the hybrid motion/force control to be handled only with respect to the held object, based on the concept of the *virtual stability*.

Consider the fact that all other subsystems combined with their respective control equations maintain to be *virtually stable* in the sense of Definition 2.17. The purpose of this subsection is to ensure that the constrained held object combined with its respective force control equations developed in this section qualifies to be *virtually stable* in the sense of Definition 2.17. The following lemma and theorem serve the purpose.

Lemma 5.2. *It follows from (5.109)–(5.113) and (5.116)–(5.122) that*

$$\int_0^\infty p_{\mathbf{S}}(t)dt \geqslant -\gamma_s \tag{5.125}$$

holds with $0 \leqslant \gamma_s < \infty$.

Proof: In view of (2.85), (5.109)–(5.113) and (5.116)–(5.122), it follows from Lemma 2.9 that

$$\begin{aligned} p_{\mathbf{S}} &= \left({}^{\mathbf{S}}V_r - {}^{\mathbf{S}}V\right)^T \left({}^{\mathbf{S}}F_r - {}^{\mathbf{S}}F\right) \\ &= \left(\chi_r - \chi\right)^T \left(\hat{\psi} - \psi\right) + \left(\tilde{\varphi}_d - \tilde{\varphi}\right)^T \mathbf{A}_f \left(\varphi_d - \psi\right) \\ &= -\left(\chi_r - \chi\right)^T \mathbf{Y}_s(\boldsymbol{\theta}_s - \hat{\boldsymbol{\theta}}_s) + \left(\tilde{\varphi}_d - \tilde{\varphi}\right)^T \mathbf{A}_f \left(\varphi_d - \varphi\right) \\ &= -\sum_\gamma \left(\theta_{s\gamma} - \hat{\theta}_{s\gamma}\right) \left[s_{s\gamma} - \frac{\dot{\hat{\theta}}_{s\gamma}}{\rho_{s\gamma}} \right] - \sum_\gamma \left(\theta_{s\gamma} - \hat{\theta}_{s\gamma}\right) \frac{\dot{\hat{\theta}}_{s\gamma}}{\rho_{s\gamma}} \\ &\quad + \left(\tilde{\varphi}_d - \tilde{\varphi}\right)^T \mathbf{A}_f \left[\mathbf{C}_f^{-1} \left(\dot{\tilde{\varphi}}_d - \dot{\tilde{\varphi}}\right) + \left(\tilde{\varphi}_d - \tilde{\varphi}\right) \right] \\ &\geqslant -\sum_\gamma \left(\theta_{s\gamma} - \hat{\theta}_{s\gamma}\right) \frac{\dot{\hat{\theta}}_{s\gamma}}{\rho_{s\gamma}} + \left(\tilde{\varphi}_d - \tilde{\varphi}\right)^T \mathbf{A}_f \mathbf{C}_f^{-1} \left(\dot{\tilde{\varphi}}_d - \dot{\tilde{\varphi}}\right) \end{aligned} \tag{5.126}$$

holds. Integrating (5.126) over time yields

$$\begin{aligned}\int_0^t p_{\mathbf{S}}(\tau)d\tau \geqslant & \frac{1}{2}\sum_{\gamma}\left(\theta_{s\gamma}-\hat{\theta}_{s\gamma}(t)\right)^2/\rho_{s\gamma}-\frac{1}{2}\sum_{\gamma}\left(\theta_{s\gamma}-\hat{\theta}_{s\gamma}(0)\right)^2/\rho_{s\gamma} \\ & +\frac{1}{2}\left(\tilde{\varphi}_d(t)-\tilde{\varphi}(t)\right)^T\mathbf{A}_f\mathbf{C}_f^{-1}\left(\tilde{\varphi}_d(t)-\tilde{\varphi}(t)\right) \\ & -\frac{1}{2}\left(\tilde{\varphi}_d(0)-\tilde{\varphi}(0)\right)^T\mathbf{A}_f\mathbf{C}_f^{-1}\left(\tilde{\varphi}_d(0)-\tilde{\varphi}(0)\right) \\ \geqslant & -\frac{1}{2}\sum_{\gamma}\left(\theta_{s\gamma}-\hat{\theta}_{s\gamma}(0)\right)^2/\rho_{s\gamma} \\ & -\frac{1}{2}\left(\tilde{\varphi}_d(0)-\tilde{\varphi}(0)\right)^T\mathbf{A}_f\mathbf{C}_f^{-1}\left(\tilde{\varphi}_d(0)-\tilde{\varphi}(0)\right) \end{aligned} \tag{5.127}$$

which validates (5.125). ■

Theorem 5.6. *The held object described by (5.19), (5.20), (5.22), and (5.109)–(5.114), combined with its control equations (5.23), (5.115)–(5.120), and (5.123) or (5.124) and with the parameter adaptation (5.24), (5.25), (5.121), and (5.122), is virtually stable with its affiliated vector* ${}^{\mathbf{O}}V_r - {}^{\mathbf{O}}V$ *being a virtual function in both* L_2 *and* L_∞*, in the sense of Definition 2.17.*

Proof: Using the same non-negative accompanying function as defined by (5.27) for the held object yields (5.28), since the application of the constraint force to the manipulator does not affect the validity of applying Lemma 4.1 (with an appropriate frame substitution).

In view of (2.85), it follows from (5.19), (5.22), (5.111), (5.114), (5.115), and (5.123) that

$$\left({}^{\mathbf{O}}V_r - {}^{\mathbf{O}}V\right)^T\left({}^{\mathbf{O}}F_r^* - {}^{\mathbf{O}}F^*\right) = p_{\mathbf{C}} - p_{\mathbf{S}} \tag{5.128}$$

holds, where a new term $-p_{\mathbf{S}}$ is added to the right-hand side to characterize the *virtual power flow* between the end-effector and the environment.

Note that frame $\{\mathbf{C}\}$ is located as the sole *driven cutting point* of the held object. The proof immediately follows from (5.27), (5.28), (5.128), and Lemma 5.2, in view of Definition 2.17. ■

5.11.4 Forward Dynamics

The forward dynamics are obtained by projecting the unconstrained dynamics (5.79) into the motion and force configuration spaces of the constrained motion.

Define

$$\mathbf{J}_\Omega = {}^{\mathbf{S}}\mathbf{U}_{\mathbf{O}}^T\mathrm{diag}\left({}^{\mathbf{S}}\mathbf{R}_{\mathbf{S}_b}, {}^{\mathbf{S}}\mathbf{R}_{\mathbf{S}_b}\right)\mathbf{T}_s \tag{5.129}$$

$$\mathbf{J}_\Psi = {}^{\mathbf{S}}\mathbf{U}_{\mathbf{O}}^T\mathrm{diag}\left({}^{\mathbf{S}}\mathbf{R}_{\mathbf{S}_b}, {}^{\mathbf{S}}\mathbf{R}_{\mathbf{S}_b}\right)\mathbf{T}_f. \tag{5.130}$$

Substituting ${}^{\mathbf{O}}V$ for $\mathcal{V}$ in (5.79) and using (5.111) yields

$$\mathcal{M}\mathbf{J}_\Omega\dot{\chi} + \left(\mathcal{C}\mathbf{J}_\Omega + \mathcal{M}\dot{\mathbf{J}}_\Omega\right)\chi + \mathcal{G} = \mathbf{J}_q^{-T}\boldsymbol{\tau}. \tag{5.131}$$

The constraint force term ${}^{\mathbf{O}}\mathbf{U}_{\mathbf{S}}{}^{\mathbf{S}}F$ needs to be added to (5.131) to form

$$\mathcal{M}\mathbf{J}_{\Omega}\dot{\chi} + \left(\mathcal{C}\mathbf{J}_{\Omega} + \mathcal{M}\dot{\mathbf{J}}_{\Omega}\right)\chi + \mathcal{G} = \mathbf{J}_q^{-T}\boldsymbol{\tau} - {}^{\mathbf{O}}\mathbf{U}_{\mathbf{S}}{}^{\mathbf{S}}F. \tag{5.132}$$

Premultiplying (5.132) by

$$\begin{bmatrix} \mathbf{J}_{\Omega}^T \\ \mathbf{J}_{\Psi}^T \end{bmatrix} \in \mathbb{R}^{6\times 6}$$

and using (5.112) yields

$$\mathbf{J}_{\Omega}^T\mathcal{M}\mathbf{J}_{\Omega}\dot{\chi} + \left(\mathbf{J}_{\Omega}^T\mathcal{C}\mathbf{J}_{\Omega} + \mathbf{J}_{\Omega}^T\mathcal{M}\dot{\mathbf{J}}_{\Omega}\right)\chi + \mathbf{J}_{\Omega}^T\mathcal{G} + \boldsymbol{\psi}$$
$$= \mathbf{J}_{\Omega}^T\mathbf{J}_q^{-T}\boldsymbol{\tau} \tag{5.133}$$
$$\mathbf{J}_{\Psi}^T\mathcal{M}\mathbf{J}_{\Omega}\dot{\chi} + \left(\mathbf{J}_{\Psi}^T\mathcal{C}\mathbf{J}_{\Omega} + \mathbf{J}_{\Psi}^T\mathcal{M}\dot{\mathbf{J}}_{\Omega}\right)\chi + \mathbf{J}_{\Psi}^T\mathcal{G}$$
$$= \mathbf{J}_{\Psi}^T\mathbf{J}_q^{-T}\boldsymbol{\tau} - \boldsymbol{\varphi} \tag{5.134}$$

with $\mathcal{M}$, $\mathcal{C}$, and $\mathcal{G}$ being defined by (5.80)–(5.82) and $\boldsymbol{\psi} \in \mathbb{R}^{(6-n_f)}$ and $\boldsymbol{\varphi} \in \mathbb{R}^{n_f}$ being defined in (5.112).

Remark 5.2. The original expressions of (5.133) and (5.134) can be traced back to [117]. The meaning of the two equations is that the constraint force vector $\boldsymbol{\varphi} \in \mathbb{R}^{n_f}$ does not affect the independent (acceleration) motion coordinate vector $\dot{\chi} \in \mathbb{R}^{(6-n_f)}$, but the independent (acceleration) motion coordinate vector does affect the constraint force vector.

Remark 5.3. Both the independent (acceleration) motion coordinate vector and the constraint force vector are linearly related to the control force/torque vector.

5.11.5 L_2 and L_∞ Stability

Theorem 5.6 ensures that the constrained held object combined with its respective control equations qualifies to be *virtually stable* in the sense of Definition 2.17. Meanwhile, all other controlled links and joints also qualify to be *virtually stable* in the sense of Definition 2.17, in view of Theorems 5.1–5.5. Therefore, it follows from Theorem 2.1 that

$${}^{\mathbf{O}}V_r - {}^{\mathbf{O}}V \in L_2 \bigcap L_\infty \tag{5.135}$$

$${}^{\mathbf{B}_k}V_r - {}^{\mathbf{B}_k}V \in L_2 \bigcap L_\infty, \quad \forall k \in \{1, 6\} \tag{5.136}$$

$$\dot{q}_{kr} - \dot{q}_k \in L_2 \bigcap L_\infty, \quad \forall k \in \{1, 6\} \tag{5.137}$$

hold and furthermore

$$I_{ckd} - I_{ck} \in L_2 \bigcap L_\infty, \; k \in \{1, 6\} \tag{5.138}$$

holds when the kth joint is in the *motor voltage control* mode.

Expression (5.135) implies

$$\chi_r - \chi \in L_2 \bigcap L_\infty \tag{5.139}$$

$$\tilde{\varphi}_d - \tilde{\varphi} \in L_2 \bigcap L_\infty \tag{5.140}$$

in view of (5.111), (5.115), and (5.116). Expression (5.139) ensures that the independent velocity errors in the motion configuration space are L_2 and L_∞ signals, and (5.140) ensures that the filtered constraint force errors in the constraint space are L_2 and L_∞ signals as well.

Position control in the motion configuration space can then be accomplished by designing

$$\chi_r = \chi_d + \mathbf{K}_p \mathbf{e} \tag{5.141}$$

where $\chi_d \in \mathbb{R}^{(6-n_f)}$ denotes the desired independent velocity coordinate vector, $\mathbf{K}_p \in \mathbb{R}^{(6-n_f)\times(6-n_f)}$ is a diagonal positive-definite matrix, and $\mathbf{e} \in \mathbb{R}^{(6-n_f)}$ denotes a position error vector subject to (4.176) and (4.177). When $\chi_d - \chi = \dot{\mathbf{e}}$ holds, it results in

$$\chi_d - \chi \in L_2 \bigcap L_\infty \tag{5.142}$$

$$\mathbf{e} \in L_2 \bigcap L_\infty \tag{5.143}$$

from Lemma 2.6.

5.11.6 Algebraic Loop Issue

In this subsection, it will be shown that the developed hybrid motion/force control solution results in an algebraic loop[1] in the system. This algebraic loop involves the constraint forces and the control forces/torques. A sufficient condition to ensure the stability in digital implementations is to be given.

Combining (5.133) and (5.134) yields

$$\begin{bmatrix} \mathcal{M}_s & \mathbf{0} \\ \mathcal{M}_f & \mathbf{I}_{n_f} \end{bmatrix} \begin{bmatrix} \dot{\chi} \\ \varphi \end{bmatrix} = \begin{bmatrix} \mathcal{J}_s \\ \mathcal{J}_f \end{bmatrix} \tau - \begin{bmatrix} \epsilon_s \\ \epsilon_f \end{bmatrix} \tag{5.144}$$

with

$$\mathcal{M}_s = \mathbf{J}_\Omega^T \mathcal{M} \mathbf{J}_\Omega \tag{5.145}$$

$$\mathcal{M}_f = \mathbf{J}_\Psi^T \mathcal{M} \mathbf{J}_\Omega \tag{5.146}$$

$$\mathcal{J}_s = \mathbf{J}_\Omega^T \mathbf{J}_q^{-T} \tag{5.147}$$

$$\mathcal{J}_f = = \mathbf{J}_\Psi^T \mathbf{J}_q^{-T} \tag{5.148}$$

$$\epsilon_s = \left(\mathbf{J}_\Omega^T \mathcal{C} \mathbf{J}_\Omega + \mathbf{J}_\Omega^T \mathcal{M} \dot{\mathbf{J}}_\Omega \right) \chi + \mathbf{J}_\Omega^T \mathcal{G} + \psi \tag{5.149}$$

$$\epsilon_f = \left(\mathbf{J}_\Psi^T \mathcal{C} \mathbf{J}_\Omega + \mathbf{J}_\Psi^T \mathcal{M} \dot{\mathbf{J}}_\Omega \right) \chi + \mathbf{J}_\Psi^T \mathcal{G}. \tag{5.150}$$

---

[1] The definition of *algebraic loops* is given in Section 2.10.

Equation (5.144) indicates that the independent acceleration coordinate vector $\dot{\boldsymbol{\chi}} \in \mathbb{R}^{(6-n_f)}$ and the independent constraint force vector $\boldsymbol{\varphi} \in \mathbb{R}^{n_f}$ are affine functions of the control force/torque vector $\boldsymbol{\tau} \in \mathbb{R}^6$.

The following lemma states that Lemma 5.1 still holds when the end-effector (held object) contacts with the environment.

Lemma 5.3. *Lemma 5.1 is valid when the held object is in contact with the environment such that (5.123) is substituted for (5.26) with* ${}^{\mathbf{S}}F_r \in \mathbb{R}^6$ *being defined by (5.119). Consequently, it follows that*

$$\mathbf{J}_q^{-T}\boldsymbol{\tau}_d = \hat{\mathcal{M}}\frac{d}{dt}({}^{\mathbf{O}}V_r) + {}^{\mathbf{O}}\mathbf{U}_{\mathbf{S}}{}^{\mathbf{S}}F_r + \mathbf{b}_\tau \tag{5.151}$$

holds.

Proof: The difference of (5.123) from (5.26) is the addition of the required contact force/moment vector ${}^{\mathbf{S}}F_r \in \mathbb{R}^6$. Therefore, using (5.88) and the fact that the vector ${}^{\mathbf{S}}F_r \in \mathbb{R}^6$ defined by (5.119) contains only state variables and reference variables completes the proof. ■

In view of (5.115), (5.116), and (5.118), ${}^{\mathbf{O}}V_r$ is an affine function of ${}^{\mathbf{S}}V_r$ and further an affine function of $\tilde{\boldsymbol{\varphi}}$. Thus, it follows from (5.10), (5.12), and (5.151) that

$$\begin{aligned}\boldsymbol{\tau}_d &= \mathbf{J}_q^T\hat{\mathcal{M}}\frac{d}{dt}({}^{\mathbf{O}}V_r) + \mathbf{J}_q^T\mathbf{b}_{\tau s}\\ &= \mathbf{J}_q^T\hat{\mathcal{M}}{}^{\mathbf{S}}\mathbf{U}_{\mathbf{O}}^T\frac{d}{dt}({}^{\mathbf{S}}V_r) + \mathbf{J}_q^T\mathbf{b}_{\tau s}\\ &= -\mathbf{J}_q^T\hat{\mathcal{M}}{}^{\mathbf{S}}\mathbf{U}_{\mathbf{O}}^T\mathrm{diag}({}^{\mathbf{S}}\mathbf{R}_{\mathbf{S}_b}, {}^{\mathbf{S}}\mathbf{R}_{\mathbf{S}_b})\mathbf{T}_f\mathbf{A}_f\dot{\tilde{\boldsymbol{\varphi}}} + \boldsymbol{\epsilon}_{\tau 1}\\ &= -\mathbf{J}_q^T\hat{\mathcal{M}}\mathbf{J}_\Psi\mathbf{A}_f\mathbf{C}_f\boldsymbol{\varphi} + \boldsymbol{\epsilon}_\tau\end{aligned} \tag{5.152}$$

holds with $\hat{\mathcal{M}} \in \mathbb{R}^{6\times 6}$ being defined by (5.90) and $\mathbf{J}_\Psi$ being defined by (5.130), where

$$\mathbf{b}_{\tau s} = \mathbf{b}_\tau + {}^{\mathbf{O}}\mathbf{U}_{\mathbf{S}}{}^{\mathbf{S}}F_r \tag{5.153}$$

and $\boldsymbol{\epsilon}_{\tau 1}$ and $\boldsymbol{\epsilon}_\tau$ are two remaining vectors.

Then, re-expressing (5.60) in a vector form and substituting (5.152) into it yields

$$\begin{aligned}\mathbf{I}_{cd} &= \hat{\mathbf{Z}}_c\boldsymbol{\tau}_d\\ &= -\hat{\mathbf{Z}}_c\mathbf{J}_q^T\hat{\mathcal{M}}\mathbf{J}_\Psi\mathbf{A}_f\mathbf{C}_f\boldsymbol{\varphi} + \hat{\mathbf{Z}}_c\boldsymbol{\epsilon}_\tau\end{aligned} \tag{5.154}$$

with $\mathbf{I}_{cd} = [I_{c1d}, \cdots, I_{c6d}]^T \in \mathbb{R}^6$ and $\hat{\mathbf{Z}}_c = \mathrm{diag}\left(1/\hat{k}_{c1}, \cdots, 1/\hat{k}_{c6}\right) \in \mathbb{R}^{6\times 6}$.

Furthermore, substituting $u_{kd} = u_k$ into (5.75) and using (5.69) and (5.70) yields

$$\begin{aligned}\hat{a}_k \dot{I}_{ckd} + k_{vk} I_{ckd} &= a_k \dot{I}_{ck} + (k_{vk} + b_k - \hat{b}_k) I_{ck} \\ &\quad + (c_k - \hat{c}_k)\dot{q}_k - \hat{k}_{ck}(\dot{q}_{kr} - \dot{q}_k) \end{aligned} \tag{5.155}$$

for all $k \in \{1, 6\}$.

When the *motor torque control* mode is used, $\tau_k = \tau_{kd}$ is valid for all $k \in \{1, 6\}$. It follows from (5.152) that

$$\boldsymbol{\tau} = -\mathbf{J}_q^T \hat{\mathcal{M}} \mathbf{J}_\Psi \mathbf{A}_f \mathbf{C}_f \boldsymbol{\varphi} + \boldsymbol{\epsilon}_\tau \tag{5.156}$$

holds.

In view of (5.144) and (5.156), it is clear that the independent constraint force coordinate vector $\boldsymbol{\varphi} \in \mathbb{R}^{n_f}$ is an affine function of the control force/torque vector $\boldsymbol{\tau} \in \mathbb{R}^6$ through the physical model (5.144) and the control force/torque vector $\boldsymbol{\tau} \in \mathbb{R}^6$ is an affine function of the independent constraint force coordinate vector $\boldsymbol{\varphi} \in \mathbb{R}^{n_f}$ through the control equation (5.156). Thus, an algebraic loop about $\boldsymbol{\varphi} \in \mathbb{R}^{n_f}$ is formed in the sense of Definition 2.18. Its gain matrix is expressed by

$$\begin{aligned}\mathbf{K}(t) &= -\begin{bmatrix} \mathbf{0} & \mathbf{I}_{n_f} \end{bmatrix} \begin{bmatrix} \mathcal{M}_s & \mathbf{0} \\ \mathcal{M}_f & \mathbf{I}_{n_f} \end{bmatrix}^{-1} \begin{bmatrix} \mathcal{J}_s \\ \mathcal{J}_f \end{bmatrix} \mathbf{J}_q^T \hat{\mathcal{M}} \mathbf{J}_\Psi \mathbf{A}_f \mathbf{C}_f \\ &= \left(\mathcal{M}_f \mathcal{M}_s^{-1} \mathbf{J}_\Omega^T - \mathbf{J}_\Psi^T \right) \hat{\mathcal{M}} \mathbf{J}_\Psi \mathbf{A}_f \mathbf{C}_f \end{aligned} \tag{5.157}$$

with $\hat{\mathcal{M}}$, $\mathbf{J}_\Omega$, $\mathbf{J}_\Psi$, $\mathcal{M}_s$, $\mathcal{M}_f$, $\mathcal{J}_s$, and $\mathcal{J}_f$ being defined by (5.90), (5.129), (5.130), and (5.145)–(5.148), respectively.

When the *motor current control* mode is used, $I_{ck} = I_{ckd}$ is valid for all $k \in \{1, 6\}$. It follows from (5.59) and (5.154) that

$$\boldsymbol{\tau} = -\mathbf{K}_c \hat{\mathbf{Z}}_c \mathbf{J}_q^T \hat{\mathcal{M}} \mathbf{J}_\Psi \mathbf{A}_f \mathbf{C}_f \boldsymbol{\varphi} + \mathbf{K}_c \hat{\mathbf{Z}}_c \boldsymbol{\epsilon}_\tau \tag{5.158}$$

holds with $\mathbf{K}_c = \mathrm{diag}\left(k_{c1}, \cdots, k_{c6}\right) \in \mathbb{R}^{6\times 6}$. Thus, an algebraic loop about $\boldsymbol{\varphi} \in \mathbb{R}^{n_f}$ is formed in the sense of Definition 2.18 with its gain matrix being expressed by

$$\mathbf{K}(t) = \left(\mathcal{M}_f \mathcal{M}_s^{-1} \mathbf{J}_\Omega^T - \mathbf{J}_\Psi^T \right) \mathbf{J}_q^{-T} \mathbf{K}_c \hat{\mathbf{Z}}_c \mathbf{J}_q^T \hat{\mathcal{M}} \mathbf{J}_\Psi \mathbf{A}_f \mathbf{C}_f. \tag{5.159}$$

When the *motor voltage control* mode is used, $u_k = u_{kd}$ is valid to yield (5.155) for all $k \in \{1, 6\}$. It follows from (5.59), (5.154), and (5.155) that

$$\begin{aligned}\boldsymbol{\tau} = &-\mathbf{K}_c \left(\mathbf{A}_u s + \mathbf{K}_{\tilde{v}}\right)^{-1} \left(\hat{\mathbf{A}}_u s + \mathbf{K}_v\right) \hat{\mathbf{Z}}_c \mathbf{J}_q^T \hat{\mathcal{M}} \mathbf{J}_\Psi \mathbf{A}_f \mathbf{C}_f \boldsymbol{\varphi} \\ &+ \mathbf{K}_c \left(\mathbf{A}_u s + \mathbf{K}_{\tilde{v}}\right)^{-1} \left(\hat{\mathbf{A}}_u s + \mathbf{K}_v\right) \hat{\mathbf{Z}}_c \boldsymbol{\epsilon}_\tau \\ &+ \mathbf{K}_c \left(\mathbf{A}_u s + \mathbf{K}_{\tilde{v}}\right)^{-1} \boldsymbol{\epsilon}_a \end{aligned} \tag{5.160}$$

holds, where

$$\begin{aligned}\mathbf{A}_u &= \mathrm{diag}\left(a_1, \cdots, a_6\right) \in \mathbb{R}^{6\times 6} \\ \hat{\mathbf{A}}_u &= \mathrm{diag}\left(\hat{a}_1, \cdots, \hat{a}_6\right) \in \mathbb{R}^{6\times 6} \\ \mathbf{K}_v &= \mathrm{diag}(k_{v1}, \cdots, k_{v6}) \in \mathbb{R}^{6\times 6} \\ \mathbf{K}_{\tilde{v}} &= \mathrm{diag}\left((k_{v1} + b_1 - \hat{b}_1), \cdots, (k_{v6} + b_6 - \hat{b}_6)\right) \in \mathbb{R}^{6\times 6}\end{aligned}$$

are four diagonal positive-definite matrices and $\boldsymbol{\epsilon}_a$ is a remaining vector. Thus, an algebraic loop about $\boldsymbol{\varphi} \in \mathbb{R}^{n_f}$ is formed in the sense of Definition 2.18 with its gain matrix being expressed by

$$\mathbf{K}(t) = \left(\mathcal{M}_f \mathcal{M}_s^{-1} \mathbf{J}_\Omega^T - \mathbf{J}_\Psi^T\right) \mathbf{J}_q^{-T} \mathbf{K}_c \mathbf{A}_u^{-1} \hat{\mathbf{A}}_u \hat{\mathbf{Z}}_c \mathbf{J}_q^T \hat{\mathcal{M}} \mathbf{J}_\Psi \mathbf{A}_f \mathbf{C}_f. \tag{5.161}$$

A common factor in (5.157), (5.159), and (5.161) is that the gain matrix $\mathbf{K}(t)$ is proportionally related to $\mathbf{A}_f\mathbf{C}_f$. Therefore, the norm of $\mathbf{A}_f\mathbf{C}_f$ must be limited by

$$\|\mathbf{A}_f \mathbf{C}_f\| \leqslant \gamma_\tau \tag{5.162}$$

where $\gamma_\tau > 0$ is a positive constant, such that the satisfaction of (5.162) implies the validity of (2.115) with $\mathbf{K}(t)$ being defined by (5.157), (5.159), or (5.161) and $\hat{\mathcal{M}}$ being defined by (5.90). The stability of the algebraic loop further leads to

$$\dot{\boldsymbol{\chi}} \in L_\infty \tag{5.163}$$

$$\boldsymbol{\varphi} \in L_\infty \tag{5.164}$$

$$\boldsymbol{\tau} \in L_\infty \tag{5.165}$$

if and only if $\boldsymbol{\epsilon}_s \in L_\infty$, $\boldsymbol{\epsilon}_f \in L_\infty$, $\boldsymbol{\epsilon}_\tau \in L_\infty$, and $\boldsymbol{\epsilon}_a \in L_\infty$ hold, in view of Lemma 2.5 and Theorem 2.2.

Remark 5.4. The upper bound of $\|\mathbf{A}_f\mathbf{C}_f\|$, denoted as γ_τ, is inversely proportional to $\|\hat{\mathcal{M}}\|$—the estimated inertial matrix of a given robot. Therefore, a large and heavy-duty robot is usually associated with a small γ_τ, and a small and light robot is associated with a large γ_τ.

5.11.7 Asymptotic Stability

Given $\boldsymbol{\chi}_r \in L_\infty$, $\dot{\boldsymbol{\chi}}_r \in L_\infty$, and $\boldsymbol{\varphi}_d \in L_\infty$, it follows from $\boldsymbol{\chi}_r - \boldsymbol{\chi} \in L_\infty$, ${}^{\mathbf{B}_k}V_r - {}^{\mathbf{B}_k}V \in L_\infty$, and $\dot{q}_{kr} - \dot{q}_k \in L_\infty$ that $\boldsymbol{\epsilon}_s \in L_\infty$, $\boldsymbol{\epsilon}_f \in L_\infty$, $\boldsymbol{\epsilon}_\tau \in L_\infty$, and $\boldsymbol{\epsilon}_a \in L_\infty$ hold. Therefore, (5.163)–(5.165) become valid. Note that (5.164) also implies $\dot{\tilde{\boldsymbol{\varphi}}} \in L_\infty$ from (5.118). Thus, in view of (5.135)–(5.140) and Lemma 2.8, the asymptotic stability in the sense of

$$\boldsymbol{\chi}_r - \boldsymbol{\chi} \to \mathbf{0} \tag{5.166}$$

$$\tilde{\boldsymbol{\varphi}}_d - \tilde{\boldsymbol{\varphi}} \to \mathbf{0} \tag{5.167}$$

$${}^{\mathbf{O}}V_r - {}^{\mathbf{O}}V \to \mathbf{0} \tag{5.168}$$

$${}^{\mathbf{B}_k}V_r - {}^{\mathbf{B}_k}V \to \mathbf{0}, \quad \forall k \in \{1, 6\} \tag{5.169}$$

$$\dot{q}_{kr} - \dot{q}_k \to 0, \quad \forall k \in \{1, 6\} \tag{5.170}$$

can be ensured. It further ensures

$$I_{ckd} - I_{ck} \to 0, \; k \in \{1, 6\} \tag{5.171}$$

when the *motor voltage control* mode is used for the kth joint.

Remark 5.5. Expression (5.166) ensures that the independent velocity coordinate errors in the motion configuration space converge to zero asymptotically, and (5.167) ensures that the filtered constraint force errors in the constraint space converge to zero asymptotically as well.

If (5.141) is used for the position control with $\dot{\chi}_d \in L_\infty$, it follows from (5.142) and (5.163) that

$$\chi_d - \chi \to \mathbf{0} \tag{5.172}$$

$$\mathbf{e} \to \mathbf{0} \tag{5.173}$$

can be achieved. Expression (5.173) gives the asymptotic stability of the position tracking control in the motion configuration space.

5.12 Motion/Force Control with Unilateral Constraints

The hybrid motion/force control developed in the last section is generally applicable to bilateral constraints.[2] After minor modifications, however, this method can also be applied to unilateral constraints,[3] where both the approach motion and the constrained motion are incorporated.

Define

$$\delta_f = \begin{cases} 0 & \text{approach motion} \\ 1 & \text{constrained motion.} \end{cases} \tag{5.174}$$

To accommodate unilateral constraints, (5.124) is modified to

$$^{\mathbf{C}}F_r = {}^{\mathbf{C}}\mathbf{U}_{\mathbf{O}}{}^{\mathbf{O}}F_r^* + \delta_f{}^{\mathbf{C}}\mathbf{U}_{\mathbf{S}}{}^{\mathbf{S}}F_r \tag{5.175}$$

where δ_f is inserted to enable the required contact force/moment vector when the contact happens.

Accordingly, the linear/angular velocity vector of frame $\{\mathbf{O}\}$ is re-written as

$$^{\mathbf{O}}V = {}^{\mathbf{S}}\mathbf{U}_{\mathbf{O}}^{T}{}^{\mathbf{S}}V = {}^{\mathbf{S}}\mathbf{U}_{\mathbf{O}}^{T}\text{diag}\left({}^{\mathbf{S}}\mathbf{R}_{\mathbf{S}_b}, {}^{\mathbf{S}}\mathbf{R}_{\mathbf{S}_b}\right)\left[\mathbf{T}_s\chi + (1-\delta_f)\mathbf{T}_f\chi_f\right] \tag{5.176}$$

where $\chi_f \in \mathbb{R}^{n_f}$ represents the additional independent velocity coordinates in the approach motion.

In the constrained motion with $\delta_f = 1$, it is clear that (5.175) and (5.176) are equivalent to (5.124) and (5.111), respectively. Thus, all the results achieved in Subsections 5.11.5–5.11.7 hold for the constrained motion.

In the approach motion with $\delta_f = 0$, (5.175) is reduced to (5.26) and it yields (5.29) from (5.21). Therefore, the L_2 and L_∞ stability developed for the free motion maintains and (5.96) is valid. Note that $\tilde{\varphi} = \mathbf{0}$ holds in the approach motion. It follows from (5.96), (5.115), (5.116), and (5.176) that

[2] A bilateral constraint means that the constraint is applied from both directions, such as an object moves in a guided rail.

[3] A unilateral constraint means that the constraint is applied from a single direction, such as a robot tip contacts with the surface of a table.

$$\chi_r - \chi \in L_2 \bigcap L_\infty \tag{5.177}$$

$$\mathbf{A}_f \tilde{\varphi}_d - \chi_f \in L_2 \bigcap L_\infty \tag{5.178}$$

hold.

With a bounded χ_r and a bounded $\tilde{\varphi}_d$, the boundedness of ${}^{\mathbf{O}}V_r$ is ensured, leading to the boundedness of ${}^{\mathbf{O}}V$ from (5.96). With a bounded $\dot{\chi}_r$ and a bounded φ_d, the boundedness of $\dot{\tilde{\varphi}}_d$ is ensured from (5.117), and further the boundedness of $\frac{d}{dt}\left({}^{\mathbf{O}}V_r\right)$ is ensured from (5.115) and (5.116). Consequently, it follows from Lemmas 5.1 and 5.3 and (5.151) that $\tau_{kd} \in L_\infty$ holds, leading to $I_{ckd} \in L_\infty$ from (5.60) for all $k \in \{1, 6\}$. Finally, $\tau_k \in L_\infty$ is ensured from (5.54), or equivalently, $I_{ck} \in L_\infty$ is ensured from (5.63) or (5.99) for all $k \in \{1, 6\}$. Note that the bounded control torques, denoted as $\tau_k \in L_\infty$ for all $k \in \{1, 6\}$, imply $\frac{d}{dt}\left({}^{\mathbf{O}}V\right) \in L_\infty$ from (5.79). It therefore yields the asymptotic stability of the velocity tracking control ${}^{\mathbf{O}}V_r - {}^{\mathbf{O}}V \to \mathbf{0}$, in view of Lemma 2.8, leading to

$$\chi_r - \chi \to \mathbf{0} \tag{5.179}$$

$$\mathbf{A}_f \tilde{\varphi}_d - \chi_f \to \mathbf{0}. \tag{5.180}$$

Remark 5.6. Expression (5.180) implies that the approach velocities are determined by the desired constraint forces scaled by the weighting matrix $\mathbf{A}_f$, subject to (5.162) to maintain the stability of the algebraic loop. Note that this is a feature of this hybrid motion/force control scheme in which the approach velocities for unilateral constraints are proportional to the desired contact forces. Large approach velocities are associated with large contact forces and *vice versa.*

When the manipulator contacts with an inaccurately known environment, the contact forces in the motion configuration space may build up. To address this problem, (5.116) is altered to include a force feedback term in the motion space as

$$ {}^{\mathbf{S}}V_r = \mathrm{diag}\left({}^{\mathbf{S}}\mathbf{R}_{\mathbf{S}_b}, {}^{\mathbf{S}}\mathbf{R}_{\mathbf{S}_b}\right)\left[\mathbf{T}_s\left(\chi_r - \mathbf{A}_m \tilde{\boldsymbol{\psi}}_z\right) + \mathbf{T}_f \mathbf{A}_f\left(\tilde{\varphi}_d - \tilde{\varphi}\right)\right] \tag{5.181}$$

where $\mathbf{A}_m \in \mathbb{R}^{(6-n_f)\times(6-n_f)}$ is a diagonal positive-definite matrix and $\tilde{\boldsymbol{\psi}}_z = [\tilde{\psi}_{z1}, \cdots, \tilde{\psi}_{z(6-n_f)}]^T \in \mathbb{R}^{(6-n_f)}$ is a vector with its components being outputs of a dead-zone filter defined by

$$\tilde{\psi}_{zi} = \begin{cases} \tilde{\psi}_i - z_i & \text{if } \tilde{\psi}_i > z_i \\ 0 & \text{if } |\tilde{\psi}_i| \leqslant z_i \\ \tilde{\psi}_i + z_i & \text{if } \tilde{\psi}_i < -z_i \end{cases} \tag{5.182}$$

for all $i \in \{1, (6 - n_f)\}$. In (5.182), $z_i \geqslant 0$ is a threshold and $\tilde{\psi}_i$ can be further obtained by

$$\dot{\tilde{\psi}}_i = -c_{mi}\tilde{\psi}_i + c_{mi}\psi_i \tag{5.183}$$

for all $i \in \{1, (6 - n_f)\}$, where $c_{mi} > 0$ is a constant and ψ_i is the ith element of $\boldsymbol{\psi} \in \mathbb{R}^{(6-n_f)}$ (the independent contact force coordinates sensed in the motion configuration space) in (5.112).

The threshold z_i in the dead-zone filter needs to be larger than the usual friction force/torque in the corresponding direction so that $\tilde{\psi}_{zi}$ does not show up in normal circumstances. When an unexpected collision or contact with an inaccurately known environment happens, the contact forces in the motion configuration space will be large enough to overcome the thresholds of the dead-zone filter, resulting in a nonzero $\tilde{\boldsymbol{\psi}}_z$. This term is then added to the required linear/angular velocity vector in (5.181) to alter the desired motion, aimed at reducing the contact forces in the motion configuration space.

Remark 5.7. The modification to the required linear/angular velocity vector by introducing a force feedback term with pre-set thresholds in the motion configuration space allows the manipulator to possess hybrid control capability for a known contact environment and also to possess impedance control capability for an inaccurate contact geometry.

5.13 Experiments

Experiments have been conducted to demonstrate the validity of the hybrid motion/force control method when being applied to unilateral constraints [218, 220]. In these experiments, a six-joint industrial robot KUKA 361 was commanded to make a contact with a steel plate put on a steel base representing a rigid environment. A round plate with three ball-bearings is attached to the robot end-effector, which allows a three-DOF of motion after the contact. The three-DOF of motion comprises the linear motion along the two tangential vectors of the steel plate and the rotational motion along the normal vector of the plate. It gives

$$\mathbf{T}_s = \begin{bmatrix} 1 & 0 & 0 \\ 0 & 1 & 0 \\ 0 & 0 & 0 \\ 0 & 0 & 0 \\ 0 & 0 & 0 \\ 0 & 0 & 1 \end{bmatrix}.$$

On the other hand, the linear motion along the normal vector of the plate and the rotational motion along the two tangential vectors of the plate are constrained with

$$\mathbf{T}_f = \begin{bmatrix} 0 & 0 & 0 \\ 0 & 0 & 0 \\ 1 & 0 & 0 \\ 0 & 1 & 0 \\ 0 & 0 & 1 \\ 0 & 0 & 0 \end{bmatrix}.$$

The experiments started by giving a desired contact force perpendicular to the plate. Under this force command, the robot approached the unilateral constraint automatically with the approach velocity being determined by (5.180). Upon the contact, the robot was smoothly transfered from the approach motion to the constrained motion by exerting a desired contact force on the plate. Then, the robot was commanded to move along the plate while keeping the contact. Smooth hybrid motion/force control results have been demonstrated [218, 220].

5.14 Concluding Remarks

In this chapter, the application of the VDC approach to a class of conventional six-joint robot manipulators has been presented in detail. The robot manipulator is first virtually decomposed into one *object* and one *open chain*. The *open chain* is further decomposed into six rigid links and six joints. The seven rigid bodies (one *object* and six rigid links) are subject to their independent control equations being developed in Chapter 4, whereas the six joints are subject to any one of the three joint control modes, namely the *motor torque control* in which the motor torque can be directly commanded, the *motor current control* in which the motor armature current can be directly applied, and the *motor voltage control* in which the motor voltage can be applied.

In the free motion, the six degrees of freedom of motion can be specified either in the joint space or in the Cartesian space. When every subsystem (rigid body or joint) combined with its respective control equations qualifies to be *virtually stable* in the sense of Definition 2.17, the L_2 and L_∞ stability of the entire robot is guaranteed by Theorem 2.1. Given bounded reference signals, the velocity errors converge to zero asymptotically. By adding a position/orientation error term into the required velocity vector, the asymptotic stability for both position/orientation control and velocity control, in the sense of (5.104), (5.105), (5.107), and (5.108), can be achieved.

In the constrained motion, the hybrid motion/force control is designed by incorporating two filtered force terms into the required velocity vector. The filtered force terms comprise one in the constraint space and another in the free motion space. The filtered force term in the constraint space is used to achieve the filtered constraint force control, and the filtered force term in the free motion space is basically used to cope with inaccurate contact geometry and unexpected collisions with unknown obstacles. However, the use of a filtered force term in the constraint space forms an algebraic loop about the constraint force vector in the sense of Definition 2.18. The condition to ensure the stability of the algebraic loop in digital implementations has been given. The L_2 and L_∞ stability, in the sense of (5.139) and (5.140), can be ensured for both the velocity tracking control in the free motion space and the filtered force tracking control in the constraint space. With the stable algebraic loop and with bounded reference signals, both the velocity tracking errors and the filtered constraint force tracking errors converge to zero asymptotically.

By implementing a selective factor to enable/disable the required constraint forces, the hybrid motion/force control scheme can be easily extended to handle both the approach motion and the constrained motion associated with unilateral constraints. In the approach motion, the approach velocity (vector) is determined by the corresponding desired constraint force (vector), which allows a smooth transition from the approach motion to the constrained motion. Experiments of an industrial robot in contact with a rigid environment with unilateral constraints have confirmed the validity of this hybrid motion/force control design.

The contents of this chapter might serve as the basic materials for other applications dealing with more complex robotic systems to be presented in the following chapters.

6

Control of Motor/Transmission Assemblies

6.1 Introduction

The results in the previous chapters reveal that the dynamics of rigid bodies and the dynamics of joints are needed to conduct the *virtual decomposition control* (VDC) design. Consider the fact that a joint motor/transmission assembly consists of merely rigid bodies. A natural question is, therefore, raised about the possibility of replacing the joint dynamics by rigid body dynamics.

In this chapter, the use of rigid body dynamics to model actuator assemblies is presented. This treatment has the advantage of completely eliminating Assumptions 2 and 3. Thus, a rigorous basis for modeling and control is adopted, since no assumption beyond rigid body dynamics is made.

After listing the prior work and introducing the virtual decomposition, this chapter presents the kinematics and dynamics of a motor/transmission assembly as well as its control equations and corresponding parameter adaptation. A non-negative accompanying function for the assembly is then defined, leading to the *virtual stability* when either a rigid or a flexible transmission is used. Finally, the joint torque control of harmonic drives is discussed.

6.2 Prior Work

Control of flexible joint robots is one of the most commonly researched topics. Many contributions have been made during the last two decades. The well-known singular perturbation technique (SPT) was used for the control of robot manipulators with weak joint flexibility (high joint stiffness) [115, 159, 160]. The concept of *integral manifold* was introduced and a three-part control law comprising fast control, corrective control, and rigid control was presented in [160]. A discontinuous control scheme aimed at decoupling the slow subsystems from the fast subsystems was proposed in [122], in which the asymptotic stability for both the slow and fast subsystems was ensured. For the discrete-time control of flexible joint manipulators, an approach using the inverse dynamic control was

W.-H. Zhu: Virtual Decomposition Control, STAR 60, pp. 149–166.
springerlink.com

proposed in [78]. With respect to the set-point control of flexible joint manipulators, a simple joint PD control law was developed in [171], in which the principle of minimum potential energy was used to ensure the asymptotic stability. This control approach was then improved in [85]. On the other hand, robust control approaches for flexible joint manipulators with unmodeled parameters and unknown disturbances were investigated in [15, 172], leading to the guaranteed global uniform ultimate boundedness on all control errors.

Adaptive control approaches for robots with weak joint flexibility were first developed by using SPT [23, 89, 115]. In these approaches, the parameter adaptation laws derived from rigid manipulators were applied directly. Furthermore, two adaptive control approaches for manipulators with arbitrary joint flexibility were reported in [16, 109]. Aimed at eliminating the need of using joint acceleration measurements, an inverse inertial matrix was used in these two approaches, resulting in complex and time-consuming control computations. Meanwhile, the control approaches for flexible joint robots by using only joint position measurements were given by [129, 139]. To achieve accurate modeling and control, two approaches incorporated joint damping effects into the robot models and used the passive nature of flexible joint robots [3, 44]. Survey papers on early research of this topic can be found in [16, 202].

The application of the VDC approach to flexible joint robots had been previously published in [212, 214]. Recently, this approach was applied to robot manipulators equipped with harmonic drives [221, 224, 225]. A common feature is that the VDC approach allows a flexible joint or a harmonic drive to be treated independently from the rest of the robot.

6.3 Virtual Decomposition

A typical single-DOF joint assembly, as illustrated in Figure 6.1, generally comprises four mechanical parts, namely the base, the output shaft, the transmission, and the motor rotor. This assembly is formed by placing two *cutting points* to virtually decompose it from a robotic system. The two *cutting points* are located at the base and at the output shaft, respectively. The *cutting point* located at the base is called the *driven cutting point* of this assembly and the *cutting point* located at the output shaft is called the *driving cutting point* of this assembly.

The base supports the housing of the motor. The motor rotor is connected to the output shaft through a transmission. Three frames, namely $\{\mathbf{T}\}$, $\{\mathbf{A}\}$, and $\{\mathbf{B}\}$, are defined. Frame $\{\mathbf{T}\}$ is located at the *driving cutting point* of the assembly and is fixed to the output shaft with its z axis aligning with the joint axis. Frame $\{\mathbf{A}\}$ is fixed to the motor rotor with its z axis aligning with the motor axis. Frame $\{\mathbf{B}\}$ is located at the *driven cutting point* of the assembly and is fixed to the base.

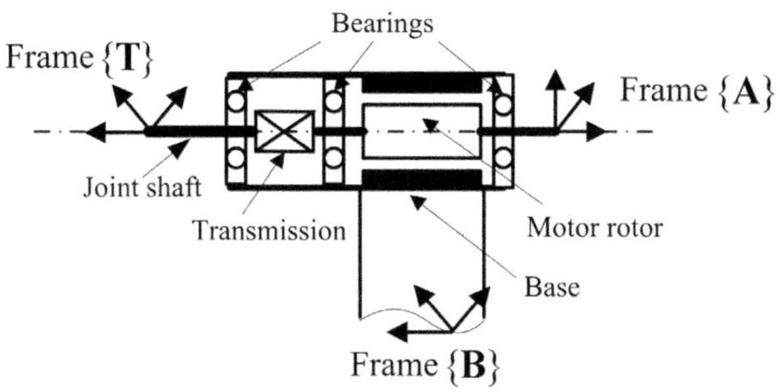

Fig. 6.1. Diagram of a typical joint assembly.

6.4 Kinematics and Dynamics

The kinematics and dynamics of the motor/transmission assembly are presented in this section.

6.4.1 Kinematics

The transmission in Figure 6.1 can be divided into two parts: one part is connected to the joint shaft and the other part is connected to the motor rotor. Thus, the entire joint assembly consists of three rigid bodies:[1] the base, the motor rotor, and the output shaft.

Let $\varrho \geqslant 1$ be the gear ratio of the transmission. Let q be the joint displacement and $\varrho\phi$ be the motor rotor displacement. The velocity relationships among the three rigid bodies can be written as

$$ {}^{\mathbf{A}}V = \mathbf{z}_{\tau}\varrho\dot{\phi} + {}^{\mathbf{B}}\mathbf{U}_{\mathbf{A}}^{T}\,{}^{\mathbf{B}}V \tag{6.1} $$

$$ {}^{\mathbf{T}}V = \mathbf{z}_{\tau}\dot{q} + {}^{\mathbf{B}}\mathbf{U}_{\mathbf{T}}^{T}\,{}^{\mathbf{B}}V \tag{6.2} $$

with $\mathbf{z}_{\tau} = [0, 0, 0, 0, 0, 1]^T \in \mathbb{R}^6$.

6.4.2 Dynamics

Recall (2.74) to obtain the dynamics of each rigid body, denoted as

$$ {}^{\mathbf{T}}F^{*} = \mathbf{M_T}\frac{d}{dt}({}^{\mathbf{T}}V) + \mathbf{C_T}({}^{\mathbf{T}}\omega){}^{\mathbf{T}}V + \mathbf{G_T} \tag{6.3} $$

$$ {}^{\mathbf{A}}F^{*} = \mathbf{M_A}\frac{d}{dt}({}^{\mathbf{A}}V) + \mathbf{C_A}({}^{\mathbf{A}}\omega){}^{\mathbf{A}}V + \mathbf{G_A} \tag{6.4} $$

$$ {}^{\mathbf{B}}F^{*} = \mathbf{M_B}\frac{d}{dt}({}^{\mathbf{B}}V) + \mathbf{C_B}({}^{\mathbf{B}}\omega){}^{\mathbf{B}}V + \mathbf{G_B} \tag{6.5} $$

where the detailed expressions of $\mathbf{M_T}$, $\mathbf{M_B}$, $\mathbf{M_A}$, $\mathbf{C_T}$, $\mathbf{C_B}$, $\mathbf{C_A}$, $\mathbf{G_T}$, $\mathbf{G_B}$, and $\mathbf{G_A}$ are given by (2.75)–(2.77) with appropriate frame substitutions.

[1] When a multiple stage transmission is considered, more than three rigid bodies are formed. However, the treatment follows a similar course.

6.4.3 Force Resultant Equations

Some symbols are defined:

${}^{\mathbf{T}}F_q \in \mathbb{R}^6$	Force/moment vector applied to the output shaft from the base, expressed in frame $\{\mathbf{T}\}$.
${}^{\mathbf{A}}F_\phi \in \mathbb{R}^6$	Force/moment vector applied to the motor rotor from the base, expressed in frame $\{\mathbf{A}\}$.
${}^{\mathbf{T}}F \in \mathbb{R}^6$	Force/moment vector in frame $\{\mathbf{T}\}$ (at the *driving cutting point*), exerted from the output shaft to a neighbour link.
${}^{\mathbf{B}}F \in \mathbb{R}^6$	Force/moment vector in frame $\{\mathbf{B}\}$ (at the *driven cutting point*), exerted from a neighbour link to the base.
$-m_b(\dot{q}) \in \mathbb{R}$	Friction torque applied to the output shaft from the base along with the joint axis (this torque constitutes part of the projection of ${}^{\mathbf{T}}F_q$ onto the joint axis).
$-m_m(\dot{\phi}) \in \mathbb{R}$	Friction torque applied to the motor rotor from the base along with the motor axis (this torque constitutes part of the projection of ${}^{\mathbf{A}}F_\phi$ onto the motor axis).
$\tau_t \in \mathbb{R}$	Effective transmission input torque (this torque can be converted to the output torque at the output shaft by the gear ratio ϱ).
$\tau \in \mathbb{R}$	Motor control torque.

Figure 6.2 illustrates the force relationships inside the actuator assembly.

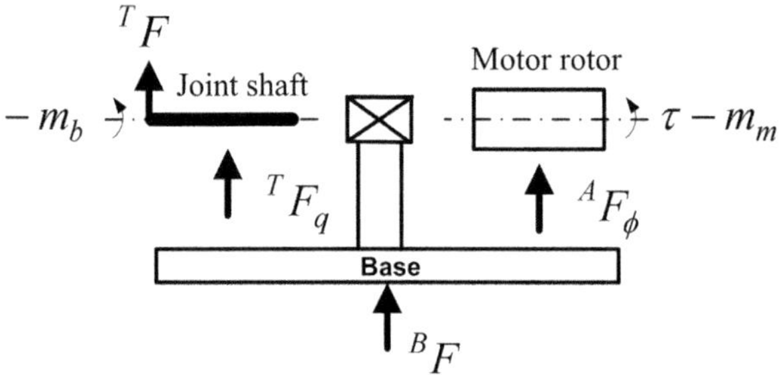

Fig. 6.2. Force relationship inside the assembly.

In view of Figure 6.2 and the definitions of the forces and torques, the force resultant equations of the three rigid bodies are written as

$$ {}^{\mathbf{T}}F^* = -{}^{\mathbf{T}}F + {}^{\mathbf{T}}F_q \tag{6.6} $$

$$ {}^{\mathbf{A}}F^* = {}^{\mathbf{A}}F_\phi \tag{6.7} $$

$$ {}^{\mathbf{B}}F^* = {}^{\mathbf{B}}F - {}^{\mathbf{B}}\mathbf{U}_{\mathbf{T}}{}^{\mathbf{T}}F_q - {}^{\mathbf{B}}\mathbf{U}_{\mathbf{A}}{}^{\mathbf{A}}F_\phi \tag{6.8} $$

with

$$\mathbf{z}_{\tau}^{T\,\mathbf{T}}F_q = \varrho\tau_t - m_b(\dot{q}) \tag{6.9}$$
$$\mathbf{z}_{\tau}^{T\,\mathbf{A}}F_{\phi} = \tau - \tau_t - m_m(\dot{\phi}). \tag{6.10}$$

The joint and motor friction torques can be further expressed as linear parametrization forms as

$$m_b(\dot{q}) = \mathbf{Y}_b(\dot{q})\boldsymbol{\theta}_b \tag{6.11}$$
$$m_m(\dot{\phi}) = \mathbf{Y}_m(\dot{\phi})\boldsymbol{\theta}_m \tag{6.12}$$

where $\boldsymbol{\theta}_b$ and $\boldsymbol{\theta}_m$ are two parameter vectors with $\mathbf{Y}_b(\dot{q})$ and $\mathbf{Y}_m(\dot{\phi})$ being two regressors.

Remark 6.1. Equations (6.6)–(6.8) characterize the resultant force relationships among the three rigid bodies that form the motor/transmission assembly. It would be interesting to view the mathematical meanings of Assumptions 2 and 3 in terms of these equations. Assumption 2 implies that

$$\mathbf{z}_{\tau}^{T\,\mathbf{A}}F^* = I_m\ddot{\phi} \tag{6.13}$$

where I_m represents the (6,6)th element in $\mathbf{M}_{\mathbf{A}}$ multiplied by the gear ratio ϱ. Thus, premultiplying (6.7) by $\mathbf{z}_{\tau}^T$ and using (6.10) and (6.13) yields the dynamics similar to that in (4.74). Assumption 3 simply implies ${}^{\mathbf{T}}F^* = \mathbf{0}$, ${}^{\mathbf{A}}F^* = \mathbf{0}$, and ${}^{\mathbf{B}}F^* = \mathbf{0}$.

6.5 Control Equations

This section includes three subsections, namely required velocities, required net force/moment vectors, and required force/moment vector transformations.

6.5.1 Required Velocities

According to (6.1) and (6.2), the required linear/angular velocity vectors of the three rigid bodies are designed as

$${}^{\mathbf{A}}V_r = \mathbf{z}_{\tau}\varrho\dot{\phi}_r + {}^{\mathbf{B}}\mathbf{U}_{\mathbf{A}}^{T\,\mathbf{B}}V_r \tag{6.14}$$
$${}^{\mathbf{T}}V_r = \mathbf{z}_{\tau}\dot{q}_r + {}^{\mathbf{B}}\mathbf{U}_{\mathbf{T}}^{T\,\mathbf{B}}V_r \tag{6.15}$$

where $\dot{q}_r$ and $\dot{\phi}_r$ denote the required joint velocity and the required motor rotor velocity, respectively. The required joint velocity $\dot{q}_r$ is specified to achieve the overall control objectives of the entire robot, such as by (5.13) and (5.15) in Chapter 5, and $\dot{\phi}_r$ will be specified later for different types of transmissions.

6.5.2 Required Net Force/Moment Vectors

Recall (4.32) to compute the required net force/moment vectors of the three rigid bodies with appropriate frame substitutions. It yields

$$^{\mathbf{T}}F_r^* = \mathbf{Y_T}\hat{\boldsymbol{\theta}}_\mathbf{T} + \mathbf{K_T}\left(^{\mathbf{T}}V_r - {}^{\mathbf{T}}V\right) \tag{6.16}$$
$$^{\mathbf{A}}F_r^* = \mathbf{Y_A}\hat{\boldsymbol{\theta}}_\mathbf{A} + \mathbf{K_A}\left(^{\mathbf{A}}V_r - {}^{\mathbf{A}}V\right) \tag{6.17}$$
$$^{\mathbf{B}}F_r^* = \mathbf{Y_B}\hat{\boldsymbol{\theta}}_\mathbf{B} + \mathbf{K_B}\left(^{\mathbf{B}}V_r - {}^{\mathbf{B}}V\right) \tag{6.18}$$

where $\mathbf{K_T} \in \mathbb{R}^{6\times 6}$, $\mathbf{K_A} \in \mathbb{R}^{6\times 6}$, and $\mathbf{K_B} \in \mathbb{R}^{6\times 6}$ are three positive-definite gain matrices and $\hat{\boldsymbol{\theta}}_\mathbf{T} \in \mathbb{R}^{13}$, $\hat{\boldsymbol{\theta}}_\mathbf{A} \in \mathbb{R}^{13}$, and $\hat{\boldsymbol{\theta}}_\mathbf{B} \in \mathbb{R}^{13}$ denote the estimates of $\boldsymbol{\theta}_\mathbf{T} \in \mathbb{R}^{13}$, $\boldsymbol{\theta}_\mathbf{A} \in \mathbb{R}^{13}$, and $\boldsymbol{\theta}_\mathbf{B} \in \mathbb{R}^{13}$, respectively. The three regressor matrices $\mathbf{Y_T} \in \mathbb{R}^{6\times 13}$, $\mathbf{Y_A} \in \mathbb{R}^{6\times 13}$, and $\mathbf{Y_B} \in \mathbb{R}^{6\times 13}$ and the three parameter vectors $\boldsymbol{\theta}_\mathbf{T} \in \mathbb{R}^{13}$, $\boldsymbol{\theta}_\mathbf{A} \in \mathbb{R}^{13}$, and $\boldsymbol{\theta}_\mathbf{B} \in \mathbb{R}^{13}$ are defined in (2.78) and further expressed in Appendix A with appropriate frame substitutions.

After the required net force/moment vectors of the three rigid bodies are obtained, the parameter estimates need to be updated.

With respect to (6.16)–(6.18), define

$$\mathbf{s_T} = \mathbf{Y_T^{\mathit{T}}}\left(^{\mathbf{T}}V_r - {}^{\mathbf{T}}V\right) \tag{6.19}$$
$$\mathbf{s_A} = \mathbf{Y_A^{\mathit{T}}}\left(^{\mathbf{A}}V_r - {}^{\mathbf{A}}V\right) \tag{6.20}$$
$$\mathbf{s_B} = \mathbf{Y_B^{\mathit{T}}}\left(^{\mathbf{B}}V_r - {}^{\mathbf{B}}V\right). \tag{6.21}$$

The $\mathcal{P}$ function defined by (2.79) is used to update each of the 13 parameters of every rigid body. It follows that

$$\hat{\theta}_{\mathbf{T}\gamma} = \mathcal{P}\left(s_{\mathbf{T}\gamma}, \rho_{\mathbf{T}\gamma}, \underline{\theta}_{\mathbf{T}\gamma}, \overline{\theta}_{\mathbf{T}\gamma}, t\right) \tag{6.22}$$
$$\hat{\theta}_{\mathbf{A}\gamma} = \mathcal{P}\left(s_{\mathbf{A}\gamma}, \rho_{\mathbf{A}\gamma}, \underline{\theta}_{\mathbf{A}\gamma}, \overline{\theta}_{\mathbf{A}\gamma}, t\right) \tag{6.23}$$
$$\hat{\theta}_{\mathbf{B}\gamma} = \mathcal{P}\left(s_{\mathbf{B}\gamma}, \rho_{\mathbf{B}\gamma}, \underline{\theta}_{\mathbf{B}\gamma}, \overline{\theta}_{\mathbf{B}\gamma}, t\right) \tag{6.24}$$

hold for all $\gamma \in \{1, 13\}$, where

$\hat{\theta}_{\mathbf{T}\gamma}$	γth element of $\hat{\boldsymbol{\theta}}_\mathbf{T}$.
$\hat{\theta}_{\mathbf{A}\gamma}$	γth element of $\hat{\boldsymbol{\theta}}_\mathbf{A}$.
$\hat{\theta}_{\mathbf{B}\gamma}$	γth element of $\hat{\boldsymbol{\theta}}_\mathbf{B}$.
$s_{\mathbf{T}\gamma}$	γth element of $\mathbf{s_T}$.
$s_{\mathbf{A}\gamma}$	γth element of $\mathbf{s_A}$.
$s_{\mathbf{B}\gamma}$	γth element of $\mathbf{s_B}$.
$\rho_{\mathbf{T}\gamma}$	Parameter update gain.
$\rho_{\mathbf{A}\gamma}$	Parameter update gain.
$\rho_{\mathbf{B}\gamma}$	Parameter update gain.
$\underline{\theta}_{\mathbf{T}\gamma}$	Lower bound of $\theta_{\mathbf{T}\gamma}$.
$\overline{\theta}_{\mathbf{T}\gamma}$	Upper bound of $\theta_{\mathbf{T}\gamma}$.
$\underline{\theta}_{\mathbf{A}\gamma}$	Lower bound of $\theta_{\mathbf{A}\gamma}$.
$\overline{\theta}_{\mathbf{A}\gamma}$	Upper bound of $\theta_{\mathbf{A}\gamma}$.

$\underline{\theta}_{\mathbf{B}\gamma}$	Lower bound of $\theta_{\mathbf{B}\gamma}$.
$\overline{\theta}_{\mathbf{B}\gamma}$	Upper bound of $\theta_{\mathbf{B}\gamma}$.
$\theta_{\mathbf{T}\gamma}$	γth element of $\boldsymbol{\theta}_{\mathbf{T}}$.
$\theta_{\mathbf{A}\gamma}$	γth element of $\boldsymbol{\theta}_{\mathbf{A}}$.
$\theta_{\mathbf{B}\gamma}$	γth element of $\boldsymbol{\theta}_{\mathbf{B}}$.

6.5.3 Required Force/Moment Vector Transformations

According to (6.6)–(6.12), the following constraints are imposed on the required net force/moment vectors of the three rigid bodies

$$^{\mathbf{T}}F_r^* = -^{\mathbf{T}}F_r + ^{\mathbf{T}}F_{qr} \tag{6.25}$$

$$^{\mathbf{A}}F_r^* = ^{\mathbf{A}}F_{\phi r} \tag{6.26}$$

$$^{\mathbf{B}}F_r^* = ^{\mathbf{B}}F_r - ^{\mathbf{B}}\mathbf{U}_{\mathbf{T}}\,^{\mathbf{T}}F_{qr} - ^{\mathbf{B}}\mathbf{U}_{\mathbf{A}}\,^{\mathbf{A}}F_{\phi r} \tag{6.27}$$

with

$$\mathbf{z}_\tau^{T\,\mathbf{T}}F_{qr} = \varrho\left[\tau_{td} - k_{vq}\left(\dot{q}_r - \dot{q}\right)\right] - \mathbf{Y}_b(\dot{q})\hat{\boldsymbol{\theta}}_b \tag{6.28}$$

$$\mathbf{z}_\tau^{T\,\mathbf{A}}F_{\phi r} = \tau_d - \tau_{td} - k_{v\phi}\left(\dot{\phi}_r - \dot{\phi}\right) - \mathbf{Y}_m(\dot{\phi})\hat{\boldsymbol{\theta}}_m \tag{6.29}$$

where $k_{vq} > 0$ and $k_{v\phi} > 0$ are two control gains, $\mathbf{Y}_b(\dot{q})$ and $\mathbf{Y}_m(\dot{\phi})$ are defined in (6.11) and (6.12), and $\hat{\boldsymbol{\theta}}_b$ and $\hat{\boldsymbol{\theta}}_m$ denote the estimates of $\boldsymbol{\theta}_b$ and $\boldsymbol{\theta}_m$, respectively.

Above five equations can be re-organized to form a mathematically equivalent computational algorithm as follows

$$^{\mathbf{T}}F_{qr} = ^{\mathbf{T}}F_r^* + ^{\mathbf{T}}F_r \tag{6.30}$$

$$^{\mathbf{A}}F_{\phi r} = ^{\mathbf{A}}F_r^* \tag{6.31}$$

$$^{\mathbf{B}}F_r = ^{\mathbf{B}}F_r^* + ^{\mathbf{B}}\mathbf{U}_{\mathbf{T}}\,^{\mathbf{T}}F_{qr} + ^{\mathbf{B}}\mathbf{U}_{\mathbf{A}}\,^{\mathbf{A}}F_{\phi r} \tag{6.32}$$

$$\tau_{td} = \frac{1}{\varrho}\left(\mathbf{z}_\tau^{T\,\mathbf{T}}F_{qr} + \mathbf{Y}_b(\dot{q})\hat{\boldsymbol{\theta}}_b\right) + k_{vq}\left(\dot{q}_r - \dot{q}\right) \tag{6.33}$$

$$\tau_d = \tau_{td} + \mathbf{z}_\tau^{T\,\mathbf{A}}F_{\phi r} + \mathbf{Y}_m(\dot{\phi})\hat{\boldsymbol{\theta}}_m + k_{v\phi}\left(\dot{\phi}_r - \dot{\phi}\right). \tag{6.34}$$

Equations (6.30)–(6.32) transform $^{\mathbf{T}}F_r$ (the required force/moment vector at the *driving cutting point*) to $^{\mathbf{B}}F_r$ (the required force/moment vector at the *driven cutting point*). Then, (6.33) and (6.34) compute τ_d (the desired motor control torque).

Furthermore, the parameter estimates of the friction torques need to be updated. Define

$$\mathbf{s}_b = \mathbf{Y}_b(\dot{q})^T(\dot{q}_r - \dot{q}) \tag{6.35}$$

$$\mathbf{s}_m = \varrho\mathbf{Y}_m(\dot{\phi})^T(\dot{\phi}_r - \dot{\phi}). \tag{6.36}$$

The γth elements of both $\hat{\boldsymbol{\theta}}_b$ in (6.33) and $\hat{\boldsymbol{\theta}}_m$ in (6.34) are updated by using the $\mathcal{P}$ function defined by (2.79) as

$$\hat{\theta}_{b\gamma} = \mathcal{P}\left(s_{b\gamma}, \rho_{b\gamma}, \underline{\theta}_{b\gamma}, \overline{\theta}_{b\gamma}, t\right) \tag{6.37}$$

$$\hat{\theta}_{m\gamma} = \mathcal{P}\left(s_{m\gamma}, \rho_{m\gamma}, \underline{\theta}_{m\gamma}, \overline{\theta}_{m\gamma}, t\right) \tag{6.38}$$

where

$\hat{\theta}_{b\gamma}$	γth element of $\hat{\boldsymbol{\theta}}_b$.
$\hat{\theta}_{m\gamma}$	γth element of $\hat{\boldsymbol{\theta}}_m$.
$s_{b\gamma}$	γth element of $\mathbf{s}_b$.
$s_{m\gamma}$	γth element of $\mathbf{s}_m$.
$\rho_{b\gamma}$	Parameter update gain.
$\rho_{m\gamma}$	Parameter update gain.
$\underline{\theta}_{b\gamma}$	Lower bound of $\theta_{b\gamma}$.
$\overline{\theta}_{b\gamma}$	Upper bound of $\theta_{b\gamma}$.
$\underline{\theta}_{m\gamma}$	Lower bound of $\theta_{m\gamma}$.
$\overline{\theta}_{m\gamma}$	Upper bound of $\theta_{m\gamma}$.
$\theta_{b\gamma}$	γth element of $\boldsymbol{\theta}_b$.
$\theta_{m\gamma}$	γth element of $\boldsymbol{\theta}_m$.

6.6 Non-negative Accompanying Function

The non-negative accompanying function for the motor/transmission assembly is chosen as

$$\nu = \nu_{\mathbf{T}} + \nu_{\mathbf{A}} + \nu_{\mathbf{B}} + \frac{1}{2}\sum_{\gamma}\left(\theta_{b\gamma} - \hat{\theta}_{b\gamma}\right)^2/\rho_{b\gamma} + \frac{1}{2}\sum_{\gamma}\left(\theta_{m\gamma} - \hat{\theta}_{m\gamma}\right)^2/\rho_{m\gamma} \tag{6.39}$$

where

$$\nu_{\mathbf{T}} = \frac{1}{2}\left({}^{\mathbf{T}}V_r - {}^{\mathbf{T}}V\right)^T \mathbf{M}_{\mathbf{T}}\left({}^{\mathbf{T}}V_r - {}^{\mathbf{T}}V\right) + \frac{1}{2}\sum_{\gamma=1}^{13}\left(\theta_{\mathbf{T}\gamma} - \hat{\theta}_{\mathbf{T}\gamma}\right)^2/\rho_{\mathbf{T}\gamma} \tag{6.40}$$

$$\nu_{\mathbf{A}} = \frac{1}{2}\left({}^{\mathbf{A}}V_r - {}^{\mathbf{A}}V\right)^T \mathbf{M}_{\mathbf{A}}\left({}^{\mathbf{A}}V_r - {}^{\mathbf{A}}V\right) + \frac{1}{2}\sum_{\gamma=1}^{13}\left(\theta_{\mathbf{A}\gamma} - \hat{\theta}_{\mathbf{A}\gamma}\right)^2/\rho_{\mathbf{A}\gamma} \tag{6.41}$$

$$\nu_{\mathbf{B}} = \frac{1}{2}\left({}^{\mathbf{B}}V_r - {}^{\mathbf{B}}V\right)^T \mathbf{M}_{\mathbf{B}}\left({}^{\mathbf{B}}V_r - {}^{\mathbf{B}}V\right) + \frac{1}{2}\sum_{\gamma=1}^{13}\left(\theta_{\mathbf{B}\gamma} - \hat{\theta}_{\mathbf{B}\gamma}\right)^2/\rho_{\mathbf{B}\gamma} \tag{6.42}$$

denote the non-negative accompanying functions assigned to the three rigid bodies, respectively.

The following lemma gives the time derivative of the non-negative accompanying function defined by (6.39).

Lemma 6.1. *Consider a motor/transmission assembly described by (6.1)–(6.12) and combined with its control equations (6.14)–(6.18) and (6.25)–(6.29) and with the parameter adaptation (6.19)–(6.24) and (6.35)–(6.38). The time derivative of ν defined by (6.39) can be expressed by*

$$
\begin{aligned}
\dot{\nu} \leqslant & -\left({}^{\mathbf{T}}V_r - {}^{\mathbf{T}}V\right)^T \mathbf{K_T} \left({}^{\mathbf{T}}V_r - {}^{\mathbf{T}}V\right) \\
& -\left({}^{\mathbf{A}}V_r - {}^{\mathbf{A}}V\right)^T \mathbf{K_A} \left({}^{\mathbf{A}}V_r - {}^{\mathbf{A}}V\right) \\
& -\left({}^{\mathbf{B}}V_r - {}^{\mathbf{B}}V\right)^T \mathbf{K_B} \left({}^{\mathbf{B}}V_r - {}^{\mathbf{B}}V\right) \\
& -\varrho k_{vq}(\dot{q}_r - \dot{q})^2 - \varrho k_{v\phi}(\dot{\phi}_r - \dot{\phi})^2 \\
& +\varrho(\dot{q}_r - \dot{q})(\tau_{td} - \tau_t) - \varrho(\dot{\phi}_r - \dot{\phi})(\tau_{td} - \tau_t) + \varrho(\dot{\phi}_r - \dot{\phi})(\tau_d - \tau) \\
& +p_{\mathbf{B}} - p_{\mathbf{T}}
\end{aligned}
\tag{6.43}
$$

where $p_{\mathbf{B}}$ and $p_{\mathbf{T}}$ denote the two virtual power flows by Definition 2.16 at the two cutting points of the assembly. ■

The proof is given in Appendix B.3.

Lemma 6.1 will be used in the next two sections to prove the *virtual stability* of the assembly in the sense of Definition 2.17, when either a rigid or a flexible transmission is used.

6.7 Rigid Transmissions

When a rigid transmission is used, it follows that

$$\phi = q \tag{6.44}$$

holds, which implies that the following relationship

$$\dot{\phi}_r = \dot{q}_r \tag{6.45}$$

is imposed on the required velocities of both the joint and the motor rotor.

The following theorem ensures that the motor-rigid transmission assembly, combined with its respective control equations, qualifies to be *virtually stable* in the sense of Definition 2.17.

Theorem 6.1. *The motor-rigid transmission assembly, described by (6.1)–(6.12) and (6.44), combined with its corresponding control and parameter adaptation described by (6.14)–(6.29), (6.35)–(6.38), and (6.45), in the motor torque control mode*

$$\tau = \tau_d \tag{6.46}$$

is virtually stable with its affiliated vectors ${}^{\mathbf{T}}V_r - {}^{\mathbf{T}}V$, ${}^{\mathbf{A}}V_r - {}^{\mathbf{A}}V$, and ${}^{\mathbf{B}}V_r - {}^{\mathbf{B}}V$ being virtual functions in both L_2 and L_∞, in the sense of Definition 2.17.

Proof: Equations (6.1)–(6.12), (6.14)–(6.29), and (6.35)–(6.38) result in (6.43) from Lemma 6.1.

In view of the non-negative accompanying function ν defined by (6.39), substituting (6.44), (6.45), and (6.46) into (6.43) yields

$$\begin{aligned}\dot{\nu} \leqslant & -\left({}^{\mathbf{T}}V_r - {}^{\mathbf{T}}V\right)^T \mathbf{K}_{\mathbf{T}} \left({}^{\mathbf{T}}V_r - {}^{\mathbf{T}}V\right) \\ & -\left({}^{\mathbf{A}}V_r - {}^{\mathbf{A}}V\right)^T \mathbf{K}_{\mathbf{A}} \left({}^{\mathbf{A}}V_r - {}^{\mathbf{A}}V\right) \\ & -\left({}^{\mathbf{B}}V_r - {}^{\mathbf{B}}V\right)^T \mathbf{K}_{\mathbf{B}} \left({}^{\mathbf{B}}V_r - {}^{\mathbf{B}}V\right) \\ & -\varrho(k_{vq} + k_{v\phi})(\dot{q}_r - \dot{q})^2 \\ & +p_{\mathbf{B}} - p_{\mathbf{T}}\end{aligned} \tag{6.47}$$

where $p_{\mathbf{B}}$ and $p_{\mathbf{T}}$ denote the two *virtual power flows* at the two *cutting points*.

Consider the fact that this assembly has one *driving cutting point* associated with frame $\{\mathbf{T}\}$ and one *driven cutting point* associated with frame $\{\mathbf{B}\}$. Using (6.39)–(6.42), (6.47), and Definition 2.17 completes the proof. ■

When every subsystem of the rest of the robot qualifies to be *virtually stable* in the sense of Definition 2.17, it follows from Theorem 2.1 that

$${}^{\mathbf{T}}V_r - {}^{\mathbf{T}}V \in L_2 \bigcap L_\infty \tag{6.48}$$

$${}^{\mathbf{A}}V_r - {}^{\mathbf{A}}V \in L_2 \bigcap L_\infty \tag{6.49}$$

$${}^{\mathbf{B}}V_r - {}^{\mathbf{B}}V \in L_2 \bigcap L_\infty \tag{6.50}$$

hold. It further follows from (6.1), (6.2), (6.14), and (6.15) that

$$\dot{q}_r - \dot{q} \in L_2 \bigcap L_\infty \tag{6.51}$$

$$\dot{\phi}_r - \dot{\phi} \in L_2 \bigcap L_\infty \tag{6.52}$$

hold.

The asymptotic stability can be achieved once the time derivatives of all L_2 functions are bounded, in view of Lemma 2.8.

6.8 Flexible Transmissions

In the previous section, the transmissions are assumed to be rigid. While being valid for most conventional gear heads including planetary gear heads, this assumption becomes vulnerably fragile when certain sophisticated transmission devices, such as harmonic drives, are used. Since being invented in the late 1950s, harmonic drives have been widely used in robotic applications due to their attractive properties such as high reduction ratio, compact size, lightweight, and coaxial assembly. A typical harmonic drive consists of a wave-generator that is connected to the motor, a circular spline that is connected to the base, and a flexspline that is placed in between and connected to the output shaft [174]. The flexspline is a naturally deformable device and, therefore, demonstrates noticeable torsional deformations when a joint torque is applied. To incorporate harmonic drives into a robot manipulator, a model representing a flexible joint or a flexible transmission is needed.

6.8.1 Dynamics

The dynamics of harmonic drives are generally very complicated, which include specific factors, such as friction, dynamics of the flexspline, nonlinearity, and hysteresis [49, 90, 174]. Experiments in [224, 225] suggested that the frictions and the dynamics of the flexspline constitute the two main factors that significantly affect the control performances.

While the friction torques are modeled as $m_b(\dot{q})$ in (6.11) and $m_m(\dot{\phi})$ in (6.12), the dynamics of the flexspline can be expressed as a pure torsional spring plus a damper to make the model manageable for the control purpose. Thus, (6.44) must be re-written as

$$k_d(\dot{\phi} - \dot{q}) + k_f(\phi - q) = \varrho\tau_t \tag{6.53}$$

where $k_d > 0$ and $k_f > 0$ denote the damping and stiffness coefficients of the flexspline.

6.8.2 Control Equations

Accordingly, $\dot{\phi}_r$ is determined by

$$\hat{k}_d\left(\dot{\phi}_r - \dot{q}_r\right) + \hat{k}_f\left(\phi_r - q_r\right) = \varrho\tau_{td} \tag{6.54}$$

with τ_{td} being specified by (6.33) and

$$\phi_r(t) \stackrel{def}{=} \int_0^t \dot{\phi}_r(t)dt + q(0) \tag{6.55}$$

$$q_r(t) \stackrel{def}{=} \int_0^t \dot{q}_r(t)dt + q(0). \tag{6.56}$$

The two parameter estimates in (6.54) are updated by using the $\mathcal{P}$ function defined by (2.79) as

$$\hat{k}_d = \mathcal{P}\left(s_{kd}, \rho_{kd}, \underline{k}_d, \overline{k}_d, t\right) \tag{6.57}$$

$$\hat{k}_f = \mathcal{P}\left(s_{kf}, \rho_{kf}, \underline{k}_f, \overline{k}_f, t\right) \tag{6.58}$$

with $\rho_{kd} > 0$, $\rho_{kf} > 0$, $0 < \underline{k}_d \leqslant \overline{k}_d$, $0 < \underline{k}_f \leqslant \overline{k}_f$, and

$$s_{kd} \stackrel{def}{=} \left(\dot{\phi}_r - \dot{q}_r\right)\left[(\dot{\phi}_r - \dot{\phi}) - (\dot{q}_r - \dot{q})\right] \tag{6.59}$$

$$s_{kf} \stackrel{def}{=} (\phi_r - q_r)\left[(\dot{\phi}_r - \dot{\phi}) - (\dot{q}_r - \dot{q})\right]. \tag{6.60}$$

6.8.3 Virtual Stability

The following theorem ensures that the motor-flexible transmission assembly, combined with its respective control equations, qualifies to be *virtually stable* in the sense of Definition 2.17.

Theorem 6.2. *The motor-flexible transmission assembly described by (6.1)–(6.12) and (6.53), combined with its respective control equations and parameter adaptation described by (6.14)–(6.29), (6.35)–(6.38), and (6.54)–(6.60), in the motor torque control mode*

$$\tau = \tau_d \tag{6.61}$$

is virtually stable with its affiliated vectors ${}^{\mathbf{T}}V_r - {}^{\mathbf{T}}V$, ${}^{\mathbf{A}}V_r - {}^{\mathbf{A}}V$, *and* ${}^{\mathbf{B}}V_r - {}^{\mathbf{B}}V$ *being virtual functions in both* L_2 *and* L_∞ *and its affiliated variable* $(\phi_r - \phi) - (q_r - q)$ *being a virtual function in* L_∞*, in the sense of Definition 2.17.*

Proof: Subtracting (6.53) from (6.54) yields

$$\begin{aligned}\varrho\left(\tau_{td} - \tau_t\right) = & -(k_d - \hat{k}_d)\left(\dot{\phi}_r - \dot{q}_r\right) \\ & -(k_f - \hat{k}_f)\left(\phi_r - q_r\right) \\ & +k_d\left[(\dot{\phi}_r - \dot{\phi}) - (\dot{q}_r - \dot{q})\right] \\ & +k_f\left[(\phi_r - \phi) - (q_r - q)\right].\end{aligned} \tag{6.62}$$

Equations (6.1)–(6.12), (6.14)–(6.29), and (6.35)–(6.38) result in (6.43) from Lemma 6.1.

Substituting (6.61) and (6.62) into (6.43) and using (6.59)–(6.61) yields

$$\begin{aligned}\dot{\nu} \leqslant & -\left({}^{\mathbf{T}}V_r - {}^{\mathbf{T}}V\right)^T \mathbf{K_T}\left({}^{\mathbf{T}}V_r - {}^{\mathbf{T}}V\right) \\ & -\left({}^{\mathbf{A}}V_r - {}^{\mathbf{A}}V\right)^T \mathbf{K_A}\left({}^{\mathbf{A}}V_r - {}^{\mathbf{A}}V\right) \\ & -\left({}^{\mathbf{B}}V_r - {}^{\mathbf{B}}V\right)^T \mathbf{K_B}\left({}^{\mathbf{B}}V_r - {}^{\mathbf{B}}V\right) \\ & -\varrho k_{vq}(\dot{q}_r - \dot{q})^2 - \varrho k_{v\phi}(\dot{\phi}_r - \dot{\phi})^2 \\ & +\left[(\dot{\phi}_r - \dot{\phi}) - (\dot{q}_r - \dot{q})\right]\Big\{(k_d - \hat{k}_d)\left(\dot{\phi}_r - \dot{q}_r\right) + (k_f - \hat{k}_f)(\phi_r - q_r) \\ & \qquad -k_d\left[(\dot{\phi}_r - \dot{\phi}) - (\dot{q}_r - \dot{q})\right] - k_f\left[(\phi_r - \phi) - (q_r - q)\right]\Big\} \\ & +p_{\mathbf{B}} - p_{\mathbf{T}} \\ = & -\left({}^{\mathbf{T}}V_r - {}^{\mathbf{T}}V\right)^T \mathbf{K_T}\left({}^{\mathbf{T}}V_r - {}^{\mathbf{T}}V\right) \\ & -\left({}^{\mathbf{A}}V_r - {}^{\mathbf{A}}V\right)^T \mathbf{K_A}\left({}^{\mathbf{A}}V_r - {}^{\mathbf{A}}V\right) \\ & -\left({}^{\mathbf{B}}V_r - {}^{\mathbf{B}}V\right)^T \mathbf{K_B}\left({}^{\mathbf{B}}V_r - {}^{\mathbf{B}}V\right) \\ & -\varrho k_{vq}(\dot{q}_r - \dot{q})^2 - \varrho k_{v\phi}(\dot{\phi}_r - \dot{\phi})^2 \\ & -k_d\left[(\dot{\phi}_r - \dot{\phi}) - (\dot{q}_r - \dot{q})\right]^2\end{aligned}$$

$$
\begin{aligned}
&-k_f\left[(\phi_r-\phi)-(q_r-q)\right]\left[(\dot{\phi}_r-\dot{\phi})-(\dot{q}_r-\dot{q})\right]\\
&+(k_d-\hat{k}_d)s_{kd}+(k_f-\hat{k}_f)s_{kf}\\
&+p_{\mathbf{B}}-p_{\mathbf{T}}
\end{aligned} \tag{6.63}
$$

where $p_{\mathbf{B}}$ and $p_{\mathbf{T}}$ denote the two *virtual power flows* at the two *cutting points*.

In view of (6.57) and (6.58), it follows from Lemma 2.9 that

$$
(k_d-\hat{k}_d)\left(s_{kd}-\dot{\hat{k}}_d/\rho_{kd}\right)\leqslant 0 \tag{6.64}
$$

$$
(k_f-\hat{k}_f)\left(s_{kf}-\dot{\hat{k}}_f/\rho_{kf}\right)\leqslant 0 \tag{6.65}
$$

hold.

Finally, let the non-negative accompanying function assigned to the motor-flexible transmission assembly be

$$
\begin{aligned}
\nu_f=\nu&+\frac{1}{2}k_f\left[(\phi_r-\phi)-(q_r-q)\right]^2\\
&+\frac{1}{2}(k_d-\hat{k}_d)^2/\rho_{kd}+\frac{1}{2}(k_f-\hat{k}_f)^2/\rho_{kf}
\end{aligned} \tag{6.66}
$$

with ν being defined by (6.39). It follows from (6.63)–(6.65) that

$$
\begin{aligned}
\dot{\nu}_f\leqslant&-\left({}^{\mathbf{T}}V_r-{}^{\mathbf{T}}V\right)^T\mathbf{K}_{\mathbf{T}}\left({}^{\mathbf{T}}V_r-{}^{\mathbf{T}}V\right)\\
&-\left({}^{\mathbf{A}}V_r-{}^{\mathbf{A}}V\right)^T\mathbf{K}_{\mathbf{A}}\left({}^{\mathbf{A}}V_r-{}^{\mathbf{A}}V\right)\\
&-\left({}^{\mathbf{B}}V_r-{}^{\mathbf{B}}V\right)^T\mathbf{K}_{\mathbf{B}}\left({}^{\mathbf{B}}V_r-{}^{\mathbf{B}}V\right)\\
&-\varrho k_{vq}(\dot{q}_r-\dot{q})^2-\varrho k_{v\phi}(\dot{\phi}_r-\dot{\phi})^2\\
&-k_d\left[(\dot{\phi}_r-\dot{\phi})-(\dot{q}_r-\dot{q})\right]^2\\
&+p_{\mathbf{B}}-p_{\mathbf{T}}
\end{aligned} \tag{6.67}
$$

holds.

Consider the fact that this assembly has one *driving cutting point* associated with frame $\{\mathbf{T}\}$ and one *driven cutting point* associated with frame $\{\mathbf{B}\}$. Using (6.39)–(6.42), (6.66), (6.67), and Definition 2.17 completes the proof. ■

When every subsystem of the rest of the robot qualifies to be *virtually stable* in the sense of Definition 2.17, it follows form Theorem 2.1 that

$$
{}^{\mathbf{T}}V_r-{}^{\mathbf{T}}V\in L_2\bigcap L_\infty \tag{6.68}
$$

$$
{}^{\mathbf{A}}V_r-{}^{\mathbf{A}}V\in L_2\bigcap L_\infty \tag{6.69}
$$

$$
{}^{\mathbf{B}}V_r-{}^{\mathbf{B}}V\in L_2\bigcap L_\infty \tag{6.70}
$$

$$
(\phi_r-q_r)-(\phi-q)\in L_\infty \tag{6.71}
$$

hold. It further follows from (6.1), (6.2), (6.14), and (6.15) that

$$\dot{q}_r - \dot{q} \in L_2 \bigcap L_\infty \tag{6.72}$$

$$\dot{\phi}_r - \dot{\phi} \in L_2 \bigcap L_\infty \tag{6.73}$$

hold.

The asymptotic stability can be achieved once the time derivatives of all L_2 functions are bounded, in view of Lemma 2.8.

6.9 Joint Torque Control of Harmonic Drives

If torque sensors are mounted on the flexspline of a harmonic drives so that τ_t defined in (6.53) is measurable [57, 66, 67, 95], then joint torque control can be conducted by using the dynamics of the motor rotor and the flexspline.

6.9.1 Torque Control

Given a desired torque $\tau_{td} \in L_\infty$, a new variable, namely the desired motor rotor velocity $\dot{\phi}_d$, is governed by

$$\hat{k}_d \left(\dot{\phi}_d - \dot{q}_r \right) + \hat{k}_f \left(\phi_d - q_r \right) = \varrho \tau_{td} \tag{6.74}$$

where $\phi_d(0)$ is determined by $\hat{k}_f(0) \left(\phi_d(0) - q(0) \right) = \varrho \tau_{td}(0)$ and $q_r(t)$ is determined by (6.56).

Then, the required motor rotor velocity $\dot{\phi}_r$ is designed as

$$\dot{\phi}_r = \dot{\phi}_d + \lambda_\tau (\tilde{\tau}_{td} - \tilde{\tau}_t) \tag{6.75}$$

with

$$\phi_r = \phi_d + \lambda_\tau \int_0^t (\tilde{\tau}_{td} - \tilde{\tau}_t) dt \tag{6.76}$$

being integrated from $\dot{\phi}_r$, where $\lambda_\tau > 0$ is a constant and $\tilde{\tau}_{td} - \tilde{\tau}_t$ denotes the filtered joint torque error governed by

$$(\dot{\tilde{\tau}}_{td} - \dot{\tilde{\tau}}_t) + c_\tau (\tilde{\tau}_{td} - \tilde{\tau}_t) = c_\tau (\tau_{td} - \tau_t) \tag{6.77}$$

with $c_\tau > 0$ representing the cut-off frequency of the joint torque filtering.

Furthermore, the two parameter estimates $\hat{k}_d$ and $\hat{k}_f$ in (6.74) are updated by using (6.57) and (6.58) with

$$s_{kd} = \left(\dot{\phi}_d - \dot{q}_r \right) \left[(\dot{\phi}_d - \dot{\phi}) - (\dot{q}_r - \dot{q}) \right] \tag{6.78}$$

$$s_{kf} = \left(\phi_d - q_r \right) \left[(\dot{\phi}_d - \dot{\phi}) - (\dot{q}_r - \dot{q}) \right] \tag{6.79}$$

being substituted for (6.59) and (6.60), respectively.

6.9.2 Virtual Stability

The following theorem ensures that the motor-flexible transmission assembly under the joint torque control qualifies to be *virtually stable* in the sense of Definition 2.17.

Theorem 6.3. *The motor-flexible transmission assembly described by (6.1)–(6.12) and (6.53), combined with its respective control equations and parameter adaptation described by (6.14)–(6.29), (6.35)–(6.38), (6.57), (6.58), (6.74), (6.75), and (6.77)–(6.79), in the motor torque control mode*

$$\tau = \tau_d \tag{6.80}$$

is virtually stable with its affiliated vectors and variables ${}^{\mathbf{T}}V_r - {}^{\mathbf{T}}V$, ${}^{\mathbf{A}}V_r - {}^{\mathbf{A}}V$, ${}^{\mathbf{B}}V_r - {}^{\mathbf{B}}V$, *and* $\tilde{\tau}_{td} - \tilde{\tau}_t$ *being virtual functions in both* L_2 *and* L_∞ *and its affiliated variable* $(\phi_d - \phi) - (q_r - q)$ *being a virtual function in* L_∞, *in the sense of Definition 2.17.*

Proof: Subtracting (6.53) from (6.74) yields

$$\begin{aligned}\varrho\,(\tau_{td} - \tau_t) = &-(k_d - \hat{k}_d)\left(\dot{\phi}_d - \dot{q}_r\right)\\ &-(k_f - \hat{k}_f)\,(\phi_d - q_r)\\ &+k_d\left[(\dot{\phi}_d - \dot{\phi}) - (\dot{q}_r - \dot{q})\right]\\ &+k_f\left[(\phi_d - \phi) - (q_r - q)\right].\end{aligned} \tag{6.81}$$

It follows from (6.75), (6.77)–(6.79), and (6.81) that

$$\begin{aligned}&\varrho(\dot{q}_r - \dot{q})(\tau_{td} - \tau_t) - \varrho(\dot{\phi}_r - \dot{\phi})(\tau_{td} - \tau_t)\\ &= -\varrho(\tau_{td} - \tau_t)\left[(\dot{\phi}_r - \dot{\phi}) - (\dot{q}_r - \dot{q})\right]\\ &= -\varrho\lambda_\tau(\tau_{td} - \tau_t)(\tilde{\tau}_{td} - \tilde{\tau}_t)\\ &\quad -\varrho(\tau_{td} - \tau_t)\left[(\dot{\phi}_d - \dot{\phi}) - (q_r - q)\right]\\ &= -\varrho\lambda_\tau(\tilde{\tau}_{td} - \tilde{\tau}_t)^2 - \frac{\varrho\lambda_\tau}{c_\tau}(\dot{\tilde{\tau}}_{td} - \dot{\tilde{\tau}}_t)(\tilde{\tau}_{td} - \tilde{\tau}_t)\\ &\quad +(k_d - \hat{k}_d)s_{kd} + (k_f - \hat{k}_f)s_{kf}\\ &\quad -k_d\left[(\dot{\phi}_d - \dot{\phi}) - (\dot{q}_r - \dot{q})\right]^2\\ &\quad -k_f\left[(\phi_d - \phi) - (q_r - q)\right]\left[(\dot{\phi}_d - \dot{\phi}) - (\dot{q}_r - \dot{q})\right]\end{aligned} \tag{6.82}$$

holds.

Let the non-negative accompanying function for the motor - flexible transmission assembly in the joint torque control be

$$\nu_{\mathcal{T}} = \nu + \frac{1}{2c_{\mathcal{T}}} \varrho \lambda_{\mathcal{T}} (\tilde{\tau}_{td} - \tilde{\tau}_t)^2 + \frac{1}{2} k_f \left[(\phi_d - \phi) - (q_r - q) \right]^2$$
$$+ \frac{1}{2} (k_d - \hat{k}_d)^2 / \rho_{kd} + \frac{1}{2} (k_f - \hat{k}_f)^2 / \rho_{kf} \tag{6.83}$$

with ν being defined by (6.39).

Equations (6.1)–(6.12), (6.14)–(6.29), and (6.35)–(6.38) result in (6.43) from Lemma 6.1. Equations (6.57) and (6.58) result in (6.64) and (6.65) from Lemma 2.9.

Finally, substituting (6.80) and (6.82) into (6.43) and using (6.64) and (6.65) yields

$$\begin{aligned} \dot{\nu}_{\mathcal{T}} \leqslant & - \left({}^{\mathbf{T}}V_r - {}^{\mathbf{T}}V \right)^T \mathbf{K_T} \left({}^{\mathbf{T}}V_r - {}^{\mathbf{T}}V \right) \\ & - \left({}^{\mathbf{A}}V_r - {}^{\mathbf{A}}V \right)^T \mathbf{K_A} \left({}^{\mathbf{A}}V_r - {}^{\mathbf{A}}V \right) \\ & - \left({}^{\mathbf{B}}V_r - {}^{\mathbf{B}}V \right)^T \mathbf{K_B} \left({}^{\mathbf{B}}V_r - {}^{\mathbf{B}}V \right) \\ & - \varrho \lambda_{\mathcal{T}} (\tilde{\tau}_{td} - \tilde{\tau}_t)^2 \\ & - \varrho k_{vq} (\dot{q}_r - \dot{q})^2 - \varrho k_{v\phi} (\dot{\phi}_r - \dot{\phi})^2 \\ & - k_d \left[(\dot{\phi}_d - \dot{\phi}) - (\dot{q}_r - \dot{q}) \right]^2 \\ & + p_{\mathbf{B}} - p_{\mathbf{T}}. \end{aligned} \tag{6.84}$$

Consider the fact that this assembly has one *driving cutting point* associated with frame $\{\mathbf{T}\}$ and one *driven cutting point* associated with frame $\{\mathbf{B}\}$. Using (6.39)–(6.42), (6.83), (6.84), and Definition 2.17 completes the proof. ■

When every subsystem of the rest of the robot qualifies to be *virtually stable* in the sense of Definition 2.17, it follows form Theorem 2.1 that

$${}^{\mathbf{T}}V_r - {}^{\mathbf{T}}V \in L_2 \bigcap L_\infty \tag{6.85}$$
$${}^{\mathbf{A}}V_r - {}^{\mathbf{A}}V \in L_2 \bigcap L_\infty \tag{6.86}$$
$${}^{\mathbf{B}}V_r - {}^{\mathbf{B}}V \in L_2 \bigcap L_\infty \tag{6.87}$$
$$\tilde{\tau}_{td} - \tilde{\tau}_t \in L_2 \bigcap L_\infty \tag{6.88}$$
$$(\phi_d - q_r) - (\phi - q) \in L_\infty \tag{6.89}$$

hold. It further follows from (6.1), (6.2), (6.14), (6.15), and (6.75) that

$$\dot{q}_r - \dot{q} \in L_2 \bigcap L_\infty \tag{6.90}$$
$$\dot{\phi}_r - \dot{\phi} \in L_2 \bigcap L_\infty \tag{6.91}$$
$$\dot{\phi}_d - \dot{\phi} \in L_2 \bigcap L_\infty \tag{6.92}$$

hold.

The asymptotic stability can be achieved once the time derivatives of all L_2 functions are bounded, in view of Lemma 2.8. The detailed procedure is given below.

Given $q_r \in L_\infty$, $\dot{q}_r \in L_\infty$, $\ddot{q}_r \in L_\infty$, and $\tau_{td} \in L_\infty$, it leads to $\phi_d \in L_\infty$ and $\dot{\phi}_d \in L_\infty$ from (6.74) and Lemma 2.5 with $\underline{k}_d > 0$ and $\underline{k}_f > 0$. Then, it follows from (6.89), (6.90), and (6.92) that $\phi - q \in L_\infty$, $\dot{q} \in L_\infty$, and $\dot{\phi} \in L_\infty$ hold, which results in $\tau_t \in L_\infty$ from (6.53) and $\dot{\hat{\tau}}_{td} - \dot{\hat{\tau}}_t \in L_\infty$ from (6.77).

When $\tau_t \in L_\infty$ holds for all the joints, $\ddot{q} \in L_\infty$ together with $\frac{d}{dt}({}^{\mathbf{T}}V) \in L_\infty$ and $\frac{d}{dt}({}^{\mathbf{B}}V) \in L_\infty$ can be ensured, in view of the forward dynamics developed in the previous chapters.

Subtracting (6.10) from (6.29) and using (6.12) and (6.80) yields

$$\mathbf{z}_\tau^T \left({}^{\mathbf{A}}F_{\phi r} - {}^{\mathbf{A}}F_\phi\right) = -(\tau_{td} - \tau_t) - k_{v\phi}\left(\dot{\phi}_r - \dot{\phi}\right) + \mathbf{Y}_m(\dot{\phi})(\boldsymbol{\theta}_m - \hat{\boldsymbol{\theta}}_m) \quad (6.93)$$

which leads to $\mathbf{z}_\tau^T \left({}^{\mathbf{A}}F_{\phi r} - {}^{\mathbf{A}}F_\phi\right) = \mathbf{z}_\tau^T \left({}^{\mathbf{A}}F_r^* - {}^{\mathbf{A}}F^*\right) \in L_\infty$ from (6.7) and (6.26). Furthermore, subtracting (6.4) from (6.17) and using (2.78), (6.1), and (6.14) yields

$$\begin{aligned} {}^{\mathbf{A}}F_r^* - {}^{\mathbf{A}}F^* &= \mathbf{M}_{\mathbf{A}}\left[\frac{d}{dt}({}^{\mathbf{A}}V_r) - \frac{d}{dt}({}^{\mathbf{A}}V)\right] + \mathbf{C}_{\mathbf{A}}({}^{\mathbf{A}}\omega)\left({}^{\mathbf{A}}V_r - {}^{\mathbf{A}}V\right) \\ &\quad -\mathbf{Y}_{\mathbf{A}}\left(\boldsymbol{\theta}_{\mathbf{A}} - \hat{\boldsymbol{\theta}}_{\mathbf{A}}\right) + \mathbf{K}_{\mathbf{A}}\left({}^{\mathbf{A}}V_r - {}^{\mathbf{A}}V\right) \\ &= \mathbf{M}_{\mathbf{A}}{}^{\mathbf{B}}\mathbf{U}_{\mathbf{A}}^T\left[\frac{d}{dt}({}^{\mathbf{B}}V_r) - \frac{d}{dt}({}^{\mathbf{B}}V)\right] + \varrho\mathbf{M}_{\mathbf{A}}\mathbf{z}_\tau\left(\ddot{\phi}_r - \ddot{\phi}\right) \\ &\quad +\boldsymbol{\epsilon}_h \end{aligned} \quad (6.94)$$

with $\boldsymbol{\epsilon}_h$ being a bounded vector.

When $\ddot{q}_r \in L_\infty$ holds for all joints, it ensures $\frac{d}{dt}({}^{\mathbf{B}}V_r) \in L_\infty$. Premultiplying (6.94) by $\mathbf{z}_\tau^T$ and using $\frac{d}{dt}({}^{\mathbf{B}}V_r) - \frac{d}{dt}({}^{\mathbf{B}}V) \in L_\infty$ yields $\ddot{\phi}_r - \ddot{\phi} \in L_\infty$, since $\varrho\mathbf{z}_\tau^T\mathbf{M}_{\mathbf{A}}\mathbf{z}_\tau > 0$ holds. Finally, the boundedness of $\frac{d}{dt}({}^{\mathbf{B}}V_r) - \frac{d}{dt}({}^{\mathbf{B}}V)$ and $\ddot{\phi}_r - \ddot{\phi}$ ensures $\frac{d}{dt}({}^{\mathbf{A}}V_r) - \frac{d}{dt}({}^{\mathbf{A}}V) \in L_\infty$ and $\ddot{\phi}_d - \ddot{\phi} \in L_\infty$ from (6.1), (6.14), and (6.75).

6.10 Case Study - Harmonic Drives

Some recent experimental results on adaptive control of harmonic drives can be found in [221, 224, 225]. Four types of harmonic drives from HD Systems Inc. (Harmonic Drive LLC since 2006) have been tested. Superior performances on both joint torque control and joint position control with an order-of-magnitude improvement over a PID control have been demonstrated. More importantly, when a joint is commanded to move at an ultra-low speed of 0.001 (rad/s), the effectiveness of the adaptive friction compensation has been witnessed by obtaining the position tracking errors with encoder-resolution accuracy.

6.11 Concluding Remarks

A typical motor/transmission assembly has been treated in this chapter in detail, aimed at conducting rigorous control design and analysis without using Assumptions 2 and 3 in Chapter 4. The motor/transmission assembly is virtually decomposed into three rigid bodies so that the joint dynamics are rigorously replaced by the dynamics of the three rigid bodies: one for the base, one for the output shaft of the joint, and the other for the motor rotor. Consequently, it is valid to state that the control of complex robots can be performed by solely using rigid body dynamics.

Both rigid and flexible transmissions have been handled within the same framework by properly imposing a kinematic or dynamic relationship on the positions and velocities of the joint and the motor rotor. The *virtual stability* of the motor/transmission assembly by using either a rigid transmission or a flexible transmission is ensured through theorems.

When every subsystem of a complex robot qualifies to be *virtually stable* in the sense of Definition 2.17, the L_2 and L_∞ stability of the entire robot can be guaranteed. The boundedness of the time derivatives of all L_2 signals further ensures the asymptotic stability of the robot. Some successful applications of the adaptive control of harmonic drives have been discussed.

7

Control of Hydraulic Robots

7.1 Introduction

Hydraulic actuation has been widely used in industrial applications, attributed to its characteristic of having a large power to mass ratio. Although this feature dramatically increases the operational capability of a hydraulically driven robot, the inherent nonlinear dynamics associated with hydraulic cylinders substantially challenge the controller design. It is obvious that the specific difficulties existed in the control of hydraulically driven robots arise from the specific dynamics of the hydraulic cylinders.

Although most industrial hydraulic robots are still using the proportional and integral (PI) control in velocity command modes, advanced control designs aimed at substantially improving motion (position or velocity) tracking performance have indeed begun to be reported in literature, ranging from nonlinear feedback control [72], model-based adaptive motion control [17, 151, 194], to input–output linearized robust control design [130].

The *virtual decomposition control* (VDC) approach allows a hydraulic cylinder/motor or a hydraulic assembly to be controlled independently from the rest of the robot provided that the *virtual stability* defined in Chapter 2 is ensured. The *virtual stability* of every subsystem eventually leads to the L_2 and L_∞ stability of the entire hydraulic robot. Consider the fact that the control of robot links and joints has been well handled in Chapter 5. Therefore, in this chapter, special attentions will be paid to the dynamics and control design of a typical hydraulic actuator assembly.

7.2 Prior Work

Researches on advanced control of hydraulic robots have been extensively reported. In most approaches developed by far, the hydraulic nonlinear dynamics are directly incorporated into the robot models to perform control design [53, 72, 127, 165, 229]. Adaptive control that takes the dynamic parameter

W.-H. Zhu: Virtual Decomposition Control, STAR 60, pp. 167–192.
springerlink.com

uncertainties into account has also been developed in [17, 116, 151, 164, 194] with remarkably achieved control performances [17, 194]. Owing to the specific dynamics of hydraulic cylinders, the derived control laws are usually much more complex than those derived for electrically driven robots in both the free motion [130] and the contact motion [131]. Thus, the need for using the VDC approach (aimed at isolating the hydraulic dynamics from the robot dynamics in control design) is justified.

7.3 A Hydraulic Actuator Assembly with Virtual Decomposition

A typical hydraulic actuator assembly is illustrated in Figure 7.1, in which a mechanism converting the linear motion of the hydraulic cylinder to a rotational motion is used. Being virtually decomposed from a hydraulic manipulator at the two *cutting points* located at its both ends, this hydraulic actuator assembly consists of two rigid links connected by a hydraulic cylinder. Therefore, this assembly has four rigid bodies (two representing the links and the other two representing the hydraulic cylinder), three unactuated rotational joints, and one actuated linear joint.

With additional four subsidiary *cutting points*, this assembly can be viewed as two *open chains* connected with two zero-mass *objects* placed at both ends,[1] see Figures 7.2 and 7.3.

As illustrated in Figures 7.1 or 7.2, nine body-fixed frames are assigned to this assembly. Frames $\{\mathbf{T}\}$, $\{\mathbf{T}_1\}$, and $\{\mathbf{T}_2\}$ are coinciding with each other, fixed to the first link, and located at a *cutting point* that is called the *driving cutting point* of this assembly. Frames $\{\mathbf{B}\}$, $\{\mathbf{B}_1\}$, and $\{\mathbf{B}_2\}$ are coinciding with each other, fixed to the second link, and located at another *cutting point* that is called the *driven cutting point* of this assembly. Frame $\{\mathbf{B}_{11}\}$ is fixed to the first link and also located at the joint connected to the second link. Frame $\{\mathbf{B}_{21}\}$ is fixed to the hydraulic cylinder base, and frame $\{\mathbf{B}_{22}\}$ is fixed to the piston. The x–y planes of all nine frames coincide with the linkage plane, and the x-axis of each frame aligns with the respective link axis.

In view of Figures 7.2 and 7.3, it follows that

- The first zero-mass *object* has one *driving cutting point* associated with frame $\{\mathbf{T}\}$ and two *driven cutting points* associated with frames $\{\mathbf{T}_1\}$ and $\{\mathbf{T}_2\}$, respectively.
- The jth *open chain* has one *driving cutting point* associated with frame $\{\mathbf{T}_j\}$ and one *driven cutting point* associated with frame $\{\mathbf{B}_j\}$ for all $j \in \{1, 2\}$.
- The second zero-mass *object* has two *driving cutting points* associated with frames $\{\mathbf{B}_1\}$ and $\{\mathbf{B}_2\}$, respectively, and one *driven cutting point* associated with frame $\{\mathbf{B}\}$.

To make the presentation straightforward, the following assumption is used.

[1] No motion control specification is assigned to the two zero-mass *objects*.

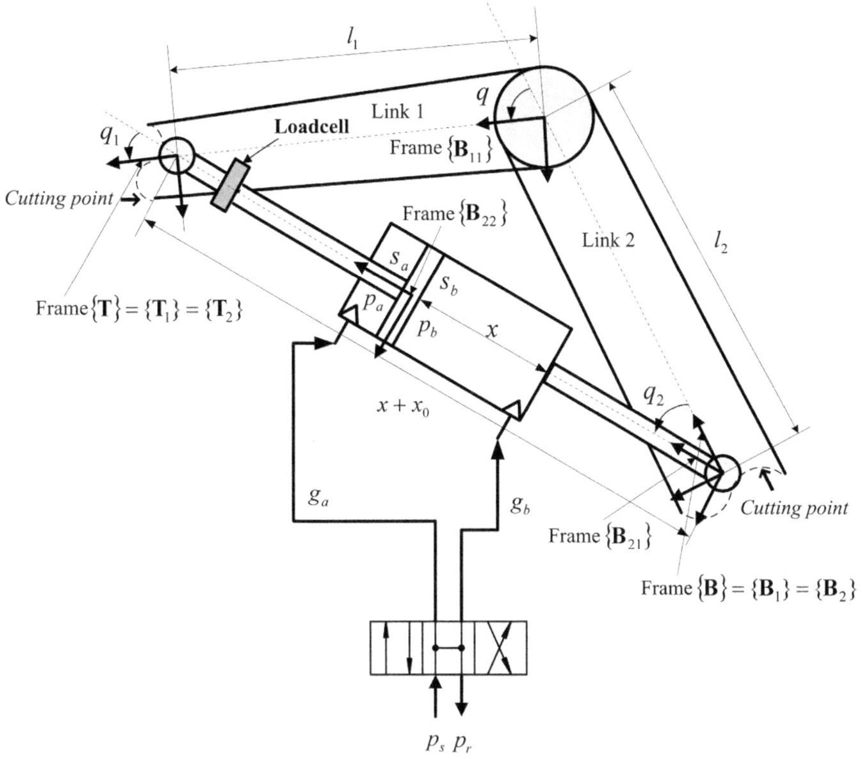

Fig. 7.1. A hydraulic actuator assembly.

Assumption 4. *The friction torques of the three unactuated rotational joints are zero.*

Remark 7.1. In hydraulic manipulators, the friction in motion is usually dominated by the piston friction between the piston seal and the cylinder. Therefore, the bearing frictions can be negligible. However, if a precise modeling is necessary, the bearing frictions can be easily incorporated by following the same treatments presented in the last three chapters.

7.4 Kinematics and Dynamics

Both kinematics and dynamics of the hydraulic actuator assembly are to be presented in this section.

7.4.1 Kinematics

This assembly consists of four rigid bodies (two representing the links and the other two representing the hydraulic cylinder) connected by three unactuated

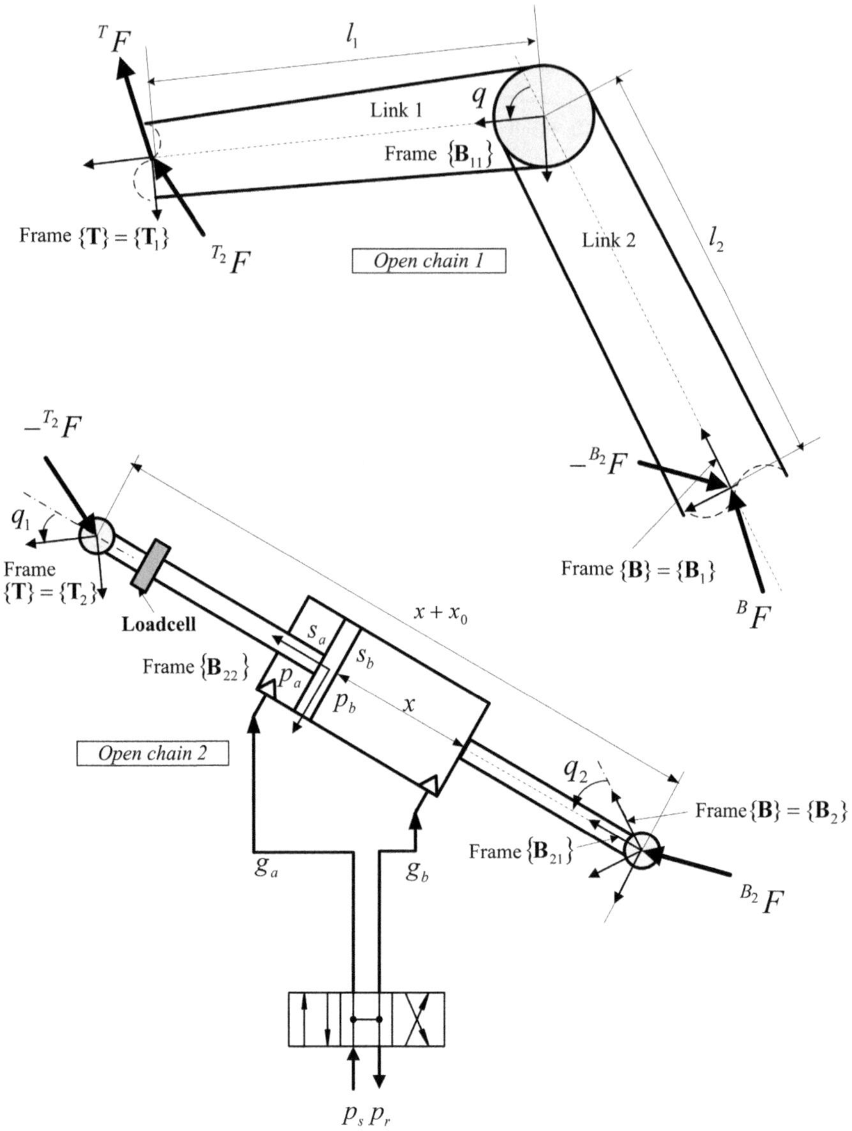

Fig. 7.2. Two *open chains* after the virtual decomposition.

rotational joints and one actuated linear joint. This configuration makes the assembly with one degree of freedom (DOF) in relative motion. Therefore, the angles of the three unactuated joint are univalent to the linear position of the cylinder piston. More specifically, it follows from Figure 7.1 that

$$(x + x_0)^2 = l_1^2 + l_2^2 + 2l_1 l_2 \cos(q) \tag{7.1}$$

$$l_2^2 = (x + x_0)^2 + l_1^2 - 2(x + x_0) l_1 \cos(q_1) \tag{7.2}$$

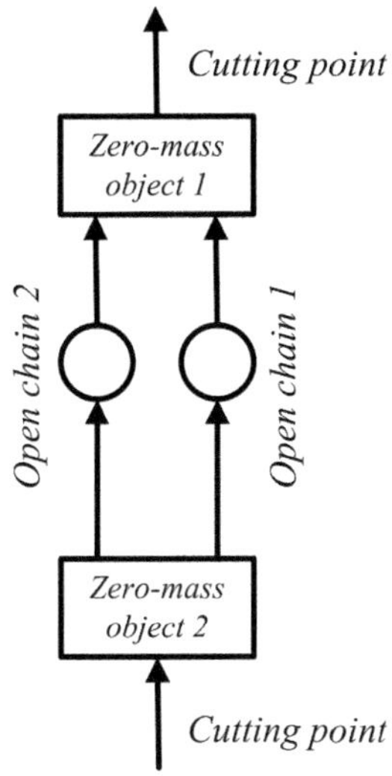

Fig. 7.3. Graphic representation of the system.

$$l_1^2 = (x + x_0)^2 + l_2^2 - 2(x + x_0)l_2 \cos(q_2) \tag{7.3}$$

hold, where l_1 and l_2 denote the lengths of the two links and x_0 denotes the initial length of the cylinder. For a given x, all q, q_1, and q_2 can be computed; or for a given q, all x, q_1, and q_2 can be computed.

Remark 7.2. Equations (7.1)–(7.3) imply that there exists only one degree of freedom in relative motion for the assembly, represented by either x or q, in addition to the degrees of freedom of motion associated with frame $\{\mathbf{B}\}$.

Taking the time derivatives of (7.1)–(7.3) yields

$$\dot{q} = -\frac{x + x_0}{l_1 l_2 \sin(q)}\dot{x} \tag{7.4}$$

$$\dot{q}_1 = -\frac{(x + x_0) - l_1 \cos(q_1)}{(x + x_0)l_1 \sin(q_1)}\dot{x} \tag{7.5}$$

$$\dot{q}_2 = -\frac{(x + x_0) - l_2 \cos(q_2)}{(x + x_0)l_2 \sin(q_2)}\dot{x}. \tag{7.6}$$

The conditions to avoid the singularity are $q_1 > 0$ and $q_2 > 0$ such that $q = q_1 + q_2 > 0$ holds. These conditions can be satisfied by properly planning the motion trajectory. Again, for a given $\dot{x}$, all $\dot{q}$, $\dot{q}_1$, and $\dot{q}_2$ can be computed; or for a given $\dot{q}$, all $\dot{x}$, $\dot{q}_1$, and $\dot{q}_2$ can be computed.

In view of (2.63) and Figure 7.1, the relationships among the linear/angular velocity vectors of the nine frames can be written as

$$^{\mathbf{T}}V = {}^{\mathbf{T}_1}V = {}^{\mathbf{T}_2}V \tag{7.7}$$

$$^{\mathbf{T}_1}V = {}^{\mathbf{B}_{11}}\mathbf{U}_{\mathbf{T}_1}^T \, {}^{\mathbf{B}_{11}}V \tag{7.8}$$

$$ {}^{\mathbf{B}_{11}}V = {}^{\mathbf{B}_1}\mathbf{U}_{\mathbf{B}_{11}}^T\ {}^{\mathbf{B}_1}V + \mathbf{z}_\tau \dot{q} \tag{7.9} $$
$$ {}^{\mathbf{T}_2}V = {}^{\mathbf{B}_{22}}\mathbf{U}_{\mathbf{T}_2}^T\ {}^{\mathbf{B}_{22}}V + \mathbf{z}_\tau \dot{q}_1 \tag{7.10} $$
$$ {}^{\mathbf{B}_{22}}V = {}^{\mathbf{B}_{21}}\mathbf{U}_{\mathbf{B}_{22}}^T\ {}^{\mathbf{B}_{21}}V + \mathbf{x}_f \dot{x} \tag{7.11} $$
$$ {}^{\mathbf{B}_{21}}V = {}^{\mathbf{B}_2}\mathbf{U}_{\mathbf{B}_{21}}^T\ {}^{\mathbf{B}_2}V + \mathbf{z}_\tau \dot{q}_2 \tag{7.12} $$
$$ {}^{\mathbf{B}}V = {}^{\mathbf{B}_1}V = {}^{\mathbf{B}_2}V \tag{7.13} $$

with

$$ \mathbf{z}_\tau = [0,0,0,0,0,1]^T \in \mathbb{R}^6 \tag{7.14} $$
$$ \mathbf{x}_f = [1,0,0,0,0,0]^T \in \mathbb{R}^6. \tag{7.15} $$

7.4.2 Dynamics

Recall (2.74) to express the dynamics of the four rigid bodies as

$$ \mathbf{M}_{\mathbf{B}_1}\frac{d}{dt}({}^{\mathbf{B}_1}V) + \mathbf{C}_{\mathbf{B}_1}({}^{\mathbf{B}_1}\omega){}^{\mathbf{B}_1}V + \mathbf{G}_{\mathbf{B}_1} = {}^{\mathbf{B}_1}F^* \tag{7.16} $$
$$ \mathbf{M}_{\mathbf{B}_{11}}\frac{d}{dt}({}^{\mathbf{B}_{11}}V) + \mathbf{C}_{\mathbf{B}_{11}}({}^{\mathbf{B}_{11}}\omega){}^{\mathbf{B}_{11}}V + \mathbf{G}_{\mathbf{B}_{11}} = {}^{\mathbf{B}_{11}}F^* \tag{7.17} $$
$$ \mathbf{M}_{\mathbf{B}_{21}}\frac{d}{dt}({}^{\mathbf{B}_{21}}V) + \mathbf{C}_{\mathbf{B}_{21}}({}^{\mathbf{B}_{21}}\omega){}^{\mathbf{B}_{21}}V + \mathbf{G}_{\mathbf{B}_{21}} = {}^{\mathbf{B}_{21}}F^* \tag{7.18} $$
$$ \mathbf{M}_{\mathbf{B}_{22}}\frac{d}{dt}({}^{\mathbf{B}_{22}}V) + \mathbf{C}_{\mathbf{B}_{22}}({}^{\mathbf{B}_{22}}\omega){}^{\mathbf{B}_{22}}V + \mathbf{G}_{\mathbf{B}_{22}} = {}^{\mathbf{B}_{22}}F^* \tag{7.19} $$

in which frames $\{\mathbf{B}_1\}$, $\{\mathbf{B}_{11}\}$, $\{\mathbf{B}_{21}\}$, and $\{\mathbf{B}_{22}\}$ are substituted for frame $\{\mathbf{A}\}$, respectively.

The force resultant equation of the first zero-mass *object* is

$$ {}^{\mathbf{T}}F = {}^{\mathbf{T}_1}F + {}^{\mathbf{T}_2}F. \tag{7.20} $$

Let ${}^{\mathbf{T}}\boldsymbol{\eta} \in \mathbb{R}^6$ be the internal force vector between the two *open chains* with its reference direction pointing from the cylinder to the first link, expressed in Frame $\{\mathbf{T}\}$. It follows from (7.20) that

$$ {}^{\mathbf{T}_1}F = \alpha_1 {}^{\mathbf{T}}F + {}^{\mathbf{T}}\boldsymbol{\eta} \tag{7.21} $$
$$ {}^{\mathbf{T}_2}F = \alpha_2 {}^{\mathbf{T}}F - {}^{\mathbf{T}}\boldsymbol{\eta} \tag{7.22} $$

hold with $\alpha_1 + \alpha_2 = 1$.

Then, the force resultant equations of the two rigid bodies affiliated with the first *open chain* can be written as

$$ {}^{\mathbf{B}_{11}}F^* = {}^{\mathbf{B}_{11}}F - {}^{\mathbf{B}_{11}}\mathbf{U}_{\mathbf{T}_1}{}^{\mathbf{T}_1}F \tag{7.23} $$
$$ {}^{\mathbf{B}_1}F^* = {}^{\mathbf{B}_1}F - {}^{\mathbf{B}_1}\mathbf{U}_{\mathbf{B}_{11}}{}^{\mathbf{B}_{11}}F. \tag{7.24} $$

Furthermore, the force resultant equations of the two rigid bodies affiliated with the second *open chain* can be written as

$$ {}^{\mathbf{B}_{22}}F^* = {}^{\mathbf{B}_{22}}F - {}^{\mathbf{B}_{22}}\mathbf{U}_{\mathbf{T}_2}{}^{\mathbf{T}_2}F \tag{7.25} $$

$$ \begin{aligned} {}^{\mathbf{B}_{21}}F^* &= {}^{\mathbf{B}_{21}}F - {}^{\mathbf{B}_{21}}\mathbf{U}_{\mathbf{B}_{22}}{}^{\mathbf{B}_{22}}F \\ &= {}^{\mathbf{B}_{21}}\mathbf{U}_{\mathbf{B}_2}{}^{\mathbf{B}_2}F - {}^{\mathbf{B}_{21}}\mathbf{U}_{\mathbf{B}_{22}}{}^{\mathbf{B}_{22}}F. \end{aligned} \tag{7.26} $$

Finally, the force resultant equation of the second zero-mass *object* is

$$ {}^{\mathbf{B}}F = {}^{\mathbf{B}_1}F + {}^{\mathbf{B}_2}F. \tag{7.27} $$

The actuation force of the cylinder is expressed by

$$ f_c = \mathbf{x}_f^{T\,\mathbf{B}_{22}}F. \tag{7.28} $$

In view of Assumption 4, the torque constraints at the three unactuated joints can be expressed by

$$ \mathbf{z}_\tau^{T\,\mathbf{T}_2}F = 0 \tag{7.29} $$

$$ \mathbf{z}_\tau^{T\,\mathbf{B}_{11}}F = 0 \tag{7.30} $$

$$ \mathbf{z}_\tau^{T\,\mathbf{B}_{21}}F = 0. \tag{7.31} $$

Remark 7.3. After the load distribution factors α_1 and α_2 being specified, the three meaningful elements (the two force elements in x and y and the moment element in z) in ${}^{\mathbf{T}}\boldsymbol{\eta} \in \mathbb{R}^6$ are determined by satisfying the constraints (7.29)–(7.31). Specifically, the moment element in z is determined by (7.29), and the two force elements in x and y are determined by (7.30) and (7.31) through (7.23)–(7.26). Finally, f_c is determined from (7.22), (7.25), and (7.28).

7.5 Control Equations

This section presents control equations for the hydraulic actuator assembly.

7.5.1 Required Velocities

To validate (2.86), the relationships among the linear/angular velocity vectors described by (7.4)–(7.13) are also applied to the required linear/angular velocity vectors as well. It follows that

$$ \dot{q}_r = -\frac{x + x_0}{l_1 l_2 \sin(q)}\dot{x}_r \tag{7.32} $$

$$ \dot{q}_{1r} = -\frac{(x + x_0) - l_1\cos(q_1)}{(x + x_0) l_1 \sin(q_1)}\dot{x}_r \tag{7.33} $$

$$ \dot{q}_{2r} = -\frac{(x + x_0) - l_2\cos(q_2)}{(x + x_0) l_2 \sin(q_2)}\dot{x}_r \tag{7.34} $$

$$ {}^{\mathbf{T}}V_r = {}^{\mathbf{T}_1}V_r = {}^{\mathbf{T}_2}V_r \tag{7.35} $$

$$ {}^{\mathbf{T}_1}V_r = {}^{\mathbf{B}_{11}}\mathbf{U}_{\mathbf{T}_1}^T\,{}^{\mathbf{B}_{11}}V_r \tag{7.36} $$

$$ {}^{\mathbf{B}_{11}}V_r = {}^{\mathbf{B}_1}\mathbf{U}_{\mathbf{B}_{11}}^T \, {}^{\mathbf{B}_1}V_r + \mathbf{z}_\tau \dot{q}_r \tag{7.37} $$
$$ {}^{\mathbf{T}_2}V_r = {}^{\mathbf{B}_{22}}\mathbf{U}_{\mathbf{T}_2}^T \, {}^{\mathbf{B}_{22}}V_r + \mathbf{z}_\tau \dot{q}_{1r} \tag{7.38} $$
$$ {}^{\mathbf{B}_{22}}V_r = {}^{\mathbf{B}_{21}}\mathbf{U}_{\mathbf{B}_{22}}^T \, {}^{\mathbf{B}_{21}}V_r + \mathbf{x}_f \dot{x}_r \tag{7.39} $$
$$ {}^{\mathbf{B}_{21}}V_r = {}^{\mathbf{B}_2}\mathbf{U}_{\mathbf{B}_{21}}^T \, {}^{\mathbf{B}_2}V_r + \mathbf{z}_\tau \dot{q}_{2r} \tag{7.40} $$
$$ {}^{\mathbf{B}}V_r = {}^{\mathbf{B}_1}V_r = {}^{\mathbf{B}_2}V_r \tag{7.41} $$

hold.

Remark 7.4. In view of Remark 7.2, there exists only one independent design variable in (7.32)–(7.34) to characterize the only single DOF in relative motion of the assembly, represented by either $\dot{x}_r$ or $\dot{q}_r$.

Remark 7.5. The required joint velocity $\dot{q}_r \in \mathbb{R}$ and the required linear/angular velocity vector ${}^{\mathbf{B}}V_r \in \mathbb{R}^6$ can be obtained from the required independent velocity vector as described by (4.13) or (4.17)–(4.18) in Chapter 4 or by (5.14) or (5.15)–(5.16) in Chapter 5. Once $\dot{q}_r \in \mathbb{R}$ and ${}^{\mathbf{B}}V_r \in \mathbb{R}^6$ are specified, all the required linear/angular velocity vectors of this assembly can be computed from (7.32)–(7.41).

7.5.2 Required Net Force/Moment Vectors with Parameter Adaptation

Similar to (4.32), the required net force/moment vectors of the four rigid bodies are computed as

$$ {}^{\mathbf{B}_1}F_r^* = \mathbf{Y}_{\mathbf{B}_1}\hat{\boldsymbol{\theta}}_{\mathbf{B}_1} + \mathbf{K}_{\mathbf{B}_1}\left({}^{\mathbf{B}_1}V_r - {}^{\mathbf{B}_1}V\right) \tag{7.42} $$
$$ {}^{\mathbf{B}_{11}}F_r^* = \mathbf{Y}_{\mathbf{B}_{11}}\hat{\boldsymbol{\theta}}_{\mathbf{B}_{11}} + \mathbf{K}_{\mathbf{B}_{11}}\left({}^{\mathbf{B}_{11}}V_r - {}^{\mathbf{B}_{11}}V\right) \tag{7.43} $$
$$ {}^{\mathbf{B}_{21}}F_r^* = \mathbf{Y}_{\mathbf{B}_{21}}\hat{\boldsymbol{\theta}}_{\mathbf{B}_{21}} + \mathbf{K}_{\mathbf{B}_{21}}\left({}^{\mathbf{B}_{21}}V_r - {}^{\mathbf{B}_{21}}V\right) \tag{7.44} $$
$$ {}^{\mathbf{B}_{22}}F_r^* = \mathbf{Y}_{\mathbf{B}_{22}}\hat{\boldsymbol{\theta}}_{\mathbf{B}_{22}} + \mathbf{K}_{\mathbf{B}_{22}}\left({}^{\mathbf{B}_{22}}V_r - {}^{\mathbf{B}_{22}}V\right) \tag{7.45} $$

with appropriate frame substitutions, where

$\mathbf{K}_{\mathbf{B}_1} \in \mathbb{R}^{6\times 6}$	Positive-definite gain matrix.
$\mathbf{K}_{\mathbf{B}_{11}} \in \mathbb{R}^{6\times 6}$	Positive-definite gain matrix.
$\mathbf{K}_{\mathbf{B}_{21}} \in \mathbb{R}^{6\times 6}$	Positive-definite gain matrix.
$\mathbf{K}_{\mathbf{B}_{22}} \in \mathbb{R}^{6\times 6}$	Positive-definite gain matrix.
$\hat{\boldsymbol{\theta}}_{\mathbf{B}_1} \in \mathbb{R}^{13}$	Parameter estimate of $\boldsymbol{\theta}_{\mathbf{B}_1} \in \mathbb{R}^{13}$.
$\hat{\boldsymbol{\theta}}_{\mathbf{B}_{11}} \in \mathbb{R}^{13}$	Parameter estimate of $\boldsymbol{\theta}_{\mathbf{B}_{11}} \in \mathbb{R}^{13}$.
$\hat{\boldsymbol{\theta}}_{\mathbf{B}_{21}} \in \mathbb{R}^{13}$	Parameter estimate of $\boldsymbol{\theta}_{\mathbf{B}_{21}} \in \mathbb{R}^{13}$.
$\hat{\boldsymbol{\theta}}_{\mathbf{B}_{22}} \in \mathbb{R}^{13}$	Parameter estimate of $\boldsymbol{\theta}_{\mathbf{B}_{22}} \in \mathbb{R}^{13}$.
$\mathbf{Y}_{\mathbf{B}_1}\boldsymbol{\theta}_{\mathbf{B}_1} \in \mathbb{R}^6$	Model-based feedforward term defined by (2.78) and expressed in Appendix A with frame $\{\mathbf{B}_1\}$ being substituted for frame $\{\mathbf{A}\}$.

$\mathbf{Y}_{\mathbf{B}_{11}}\boldsymbol{\theta}_{\mathbf{B}_{11}} \in \mathbb{R}^6$ Model-based feedforward term defined by (2.78) and expressed in Appendix A with frame $\{\mathbf{B}_{11}\}$ being substituted for frame $\{\mathbf{A}\}$.

$\mathbf{Y}_{\mathbf{B}_{21}}\boldsymbol{\theta}_{\mathbf{B}_{21}} \in \mathbb{R}^6$ Model-based feedforward term defined by (2.78) and expressed in Appendix A with frame $\{\mathbf{B}_{21}\}$ being substituted for frame $\{\mathbf{A}\}$.

$\mathbf{Y}_{\mathbf{B}_{22}}\boldsymbol{\theta}_{\mathbf{B}_{22}} \in \mathbb{R}^6$ Model-based feedforward term defined by (2.78) and expressed in Appendix A with frame $\{\mathbf{B}_{22}\}$ being substituted for frame $\{\mathbf{A}\}$.

With respect to (7.42)–(7.45), define

$$\mathbf{s}_{\mathbf{B}_1} = \mathbf{Y}_{\mathbf{B}_1}^T \left({}^{\mathbf{B}_1}V_r - {}^{\mathbf{B}_1}V\right) \tag{7.46}$$

$$\mathbf{s}_{\mathbf{B}_{11}} = \mathbf{Y}_{\mathbf{B}_{11}}^T \left({}^{\mathbf{B}_{11}}V_r - {}^{\mathbf{B}_{11}}V\right) \tag{7.47}$$

$$\mathbf{s}_{\mathbf{B}_{21}} = \mathbf{Y}_{\mathbf{B}_{21}}^T \left({}^{\mathbf{B}_{21}}V_r - {}^{\mathbf{B}_{21}}V\right) \tag{7.48}$$

$$\mathbf{s}_{\mathbf{B}_{22}} = \mathbf{Y}_{\mathbf{B}_{22}}^T \left({}^{\mathbf{B}_{22}}V_r - {}^{\mathbf{B}_{22}}V\right). \tag{7.49}$$

The $\mathcal{P}$ function defined by (2.79) is used to update each of the 13 parameters of every rigid body as

$$\hat{\theta}_{\mathbf{B}_1\gamma} = \mathcal{P}\left(s_{\mathbf{B}_1\gamma}, \rho_{\mathbf{B}_1\gamma}, \underline{\theta}_{\mathbf{B}_1\gamma}, \overline{\theta}_{\mathbf{B}_1\gamma}, t\right) \tag{7.50}$$

$$\hat{\theta}_{\mathbf{B}_{11}\gamma} = \mathcal{P}\left(s_{\mathbf{B}_{11}\gamma}, \rho_{\mathbf{B}_{11}\gamma}, \underline{\theta}_{\mathbf{B}_{11}\gamma}, \overline{\theta}_{\mathbf{B}_{11}\gamma}, t\right) \tag{7.51}$$

$$\hat{\theta}_{\mathbf{B}_{21}\gamma} = \mathcal{P}\left(s_{\mathbf{B}_{21}\gamma}, \rho_{\mathbf{B}_{21}\gamma}, \underline{\theta}_{\mathbf{B}_{21}\gamma}, \overline{\theta}_{\mathbf{B}_{21}\gamma}, t\right) \tag{7.52}$$

$$\hat{\theta}_{\mathbf{B}_{22}\gamma} = \mathcal{P}\left(s_{\mathbf{B}_{22}\gamma}, \rho_{\mathbf{B}_{22}\gamma}, \underline{\theta}_{\mathbf{B}_{22}\gamma}, \overline{\theta}_{\mathbf{B}_{22}\gamma}, t\right) \tag{7.53}$$

for all $\gamma \in \{1, 13\}$, where

$\hat{\theta}_{\mathbf{B}_1\gamma}$ γth element of $\hat{\boldsymbol{\theta}}_{\mathbf{B}_1}$.
$\hat{\theta}_{\mathbf{B}_{11}\gamma}$ γth element of $\hat{\boldsymbol{\theta}}_{\mathbf{B}_{11}}$.
$\hat{\theta}_{\mathbf{B}_{21}\gamma}$ γth element of $\hat{\boldsymbol{\theta}}_{\mathbf{B}_{21}}$.
$\hat{\theta}_{\mathbf{B}_{22}\gamma}$ γth element of $\hat{\boldsymbol{\theta}}_{\mathbf{B}_{22}}$.
$s_{\mathbf{B}_1\gamma}$ γth element of $\mathbf{s}_{\mathbf{B}_1}$.
$s_{\mathbf{B}_{11}\gamma}$ γth element of $\mathbf{s}_{\mathbf{B}_{11}}$.
$s_{\mathbf{B}_{21}\gamma}$ γth element of $\mathbf{s}_{\mathbf{B}_{21}}$.
$s_{\mathbf{B}_{22}\gamma}$ γth element of $\mathbf{s}_{\mathbf{B}_{22}}$.
$\rho_{\mathbf{B}_1\gamma}$ Parameter update gain.
$\rho_{\mathbf{B}_{11}\gamma}$ Parameter update gain.
$\rho_{\mathbf{B}_{21}\gamma}$ Parameter update gain.
$\rho_{\mathbf{B}_{22}\gamma}$ Parameter update gain.
$\underline{\theta}_{\mathbf{B}_1\gamma}$ Lower bound of $\theta_{\mathbf{B}_1\gamma}$.
$\underline{\theta}_{\mathbf{B}_{11}\gamma}$ Lower bound of $\theta_{\mathbf{B}_{11}\gamma}$.
$\underline{\theta}_{\mathbf{B}_{21}\gamma}$ Lower bound of $\theta_{\mathbf{B}_{21}\gamma}$.
$\underline{\theta}_{\mathbf{B}_{22}\gamma}$ Lower bound of $\theta_{\mathbf{B}_{22}\gamma}$.
$\overline{\theta}_{\mathbf{B}_1\gamma}$ Upper bound of $\theta_{\mathbf{B}_1\gamma}$.
$\overline{\theta}_{\mathbf{B}_{11}\gamma}$ Upper bound of $\theta_{\mathbf{B}_{11}\gamma}$.

$\overline{\theta}_{\mathbf{B}_{21}\gamma}$	Upper bound of $\theta_{\mathbf{B}_{21}\gamma}$.
$\overline{\theta}_{\mathbf{B}_{22}\gamma}$	Upper bound of $\theta_{\mathbf{B}_{22}\gamma}$.
$\theta_{\mathbf{B}_{1}\gamma}$	γth element of $\boldsymbol{\theta}_{\mathbf{B}_1}$.
$\theta_{\mathbf{B}_{11}\gamma}$	γth element of $\boldsymbol{\theta}_{\mathbf{B}_{11}}$.
$\theta_{\mathbf{B}_{21}\gamma}$	γth element of $\boldsymbol{\theta}_{\mathbf{B}_{21}}$.
$\theta_{\mathbf{B}_{22}\gamma}$	γth element of $\boldsymbol{\theta}_{\mathbf{B}_{22}}$.

7.5.3 Required Force/Moment Vector Transformations

Consider (7.21) and (7.22). Given a required force/moment vector in frame $\{\mathbf{T}\}$ (the *driving cutting point* of the hydraulic assembly), denoted as ${}^{\mathbf{T}}F_r \in \mathbb{R}^6$, the required force/moment vectors at the two *driven cutting points* of the first zero-mass *object* are computed as

$$ {}^{\mathbf{T}_1}F_r = \alpha_1 {}^{\mathbf{T}}F_r + {}^{\mathbf{T}}\boldsymbol{\eta}_r \tag{7.54} $$

$$ {}^{\mathbf{T}_2}F = \alpha_2 {}^{\mathbf{T}}F_r - {}^{\mathbf{T}}\boldsymbol{\eta}_r \tag{7.55} $$

where ${}^{\mathbf{T}}\boldsymbol{\eta}_r \in \mathbb{R}^6$ denotes the required internal force/moment vector to be specified later. Then, the required force/moment vectors of the first *open chain* are computed as

$$ {}^{\mathbf{B}_{11}}F_r = {}^{\mathbf{B}_{11}}F_r{}^* + {}^{\mathbf{B}_{11}}\mathbf{U}_{\mathbf{T}_1}\, {}^{\mathbf{T}_1}F_r \tag{7.56} $$

$$ {}^{\mathbf{B}_{1}}F_r = {}^{\mathbf{B}_{1}}F_r{}^* + {}^{\mathbf{B}_{1}}\mathbf{U}_{\mathbf{B}_{11}}\, {}^{\mathbf{B}_{11}}F_r \tag{7.57} $$

and the required force/moment vectors of the second *open chain* are computed as

$$ {}^{\mathbf{B}_{22}}F_r = {}^{\mathbf{B}_{22}}F_r{}^* + {}^{\mathbf{B}_{22}}\mathbf{U}_{\mathbf{T}_2}\, {}^{\mathbf{T}_2}F_r \tag{7.58} $$

$$ {}^{\mathbf{B}_{2}}F_r = {}^{\mathbf{B}_{2}}\mathbf{U}_{\mathbf{B}_{21}}\, {}^{\mathbf{B}_{21}}F_r{}^* + {}^{\mathbf{B}_{2}}\mathbf{U}_{\mathbf{B}_{22}}\, {}^{\mathbf{B}_{22}}F_r. \tag{7.59} $$

Finally, the required force/moment vector in frame $\{\mathbf{B}\}$ (the *driven cutting point* of this hydraulic assembly) is computed as

$$ {}^{\mathbf{B}}F_r = {}^{\mathbf{B}_1}F_r + {}^{\mathbf{B}_2}F_r. \tag{7.60} $$

Remark 7.6. Equations (7.54)–(7.60) indicate that the required internal force vector ${}^{\mathbf{T}}\boldsymbol{\eta}_r \in \mathbb{R}^6$ exists only along the two *open chains* and is cancelled in forming ${}^{\mathbf{B}}F_r$. Therefore, ${}^{\mathbf{B}}F_r$ can be obtained from ${}^{\mathbf{T}}F_r$, ${}^{\mathbf{B}}F_r^*$, ${}^{\mathbf{B}_{11}}F_r^*$, ${}^{\mathbf{B}_{22}}F_r^*$, and ${}^{\mathbf{B}_{21}}F_r^*$, without involving ${}^{\mathbf{T}}\boldsymbol{\eta}_r$.

According to (7.28), the required actuation force along the cylinder axis is designed as

$$ f_{cr} = \mathbf{x}_f^T\, {}^{\mathbf{B}_{22}}F_r \tag{7.61} $$

and furthermore, according to (7.29)–(7.31), three constraints are imposed on the required torques of the three unactuated joints, leading to

$$ \mathbf{z}_\tau^T\, {}^{\mathbf{T}_2}F_r = 0 \tag{7.62} $$

$$ \mathbf{z}_\tau^T\, {}^{\mathbf{B}_{11}}F_r = 0 \tag{7.63} $$

$$ \mathbf{z}_\tau^T\, {}^{\mathbf{B}_{21}}F_r = \mathbf{z}_\tau^T\, {}^{\mathbf{B}_{21}}\mathbf{U}_{\mathbf{B}_2}\, {}^{\mathbf{B}_2}F_r = 0. \tag{7.64} $$

Remark 7.7. Three meaningful elements in ${}^{\mathbf{T}}\boldsymbol{\eta}_r \in \mathbb{R}^6$ are determined by satisfying the three constraints (7.62)–(7.64). The three elements are the two required forces in x and y and the required moment in z. Specifically, the required moment in z is determined by (7.62), and the two required forces in x and y are determined by (7.63) and (7.64) through (7.54)–(7.59). Once the required forces in x and y are specified, f_{cr} can be computed from (7.55), (7.58), and (7.61).

The overall computation procedure for this hydraulic actuator assembly is described as follows:

Step 1: For a given $\dot{x}_r$ or $\dot{q}_r$, compute all required velocities in terms of (7.32)–(7.41).
Step 2: Compute ${}^{\mathbf{B}_1}F_r^*$, ${}^{\mathbf{B}_{11}}F_r^*$, ${}^{\mathbf{B}_{21}}F_r^*$, and ${}^{\mathbf{B}_{22}}F_r^*$ from (7.42)–(7.45). Update unknown parameters by using (7.46)–(7.53).
Step 3: For a given ${}^{\mathbf{T}}F_r$, compute ${}^{\mathbf{B}}F_r$ in terms of (7.54)–(7.60). Note that ${}^{\mathbf{T}}\boldsymbol{\eta}_r$ is cancelled out in forming ${}^{\mathbf{B}}F_r$.
Step 4: Use (7.62)–(7.64) to specify the three meaningful elements in ${}^{\mathbf{T}}\boldsymbol{\eta}_r \in \mathbb{R}^6$ (two required forces in x and y and one required moment in z).
Step 5: Compute f_{cr} from (7.61).

7.6 Virtual Stability

The *virtual stability* with respect to the two zero-mass *objects* and to the two *open chains*, in the sense of Definition 2.17, will be presented in this section.

Theorem 7.1. *The first zero-mass object described by (7.7) and (7.20)–(7.22), combined with its control equations (7.35), (7.54), and (7.55), is virtually stable in the sense of Definition 2.17.*

Proof: It follows from (7.20) and the summation of (7.54) and (7.55) that

$$ {}^{\mathbf{T}}F_r - {}^{\mathbf{T}}F = ({}^{\mathbf{T}_1}F_r - {}^{\mathbf{T}_1}F) + ({}^{\mathbf{T}_2}F_r - {}^{\mathbf{T}_2}F) \tag{7.65} $$

holds. Let the non-negative accompanying function be *zero*. Premultiplying (7.65) by $({}^{\mathbf{T}}V_r - {}^{\mathbf{T}}V)^T$ and using (2.85), (7.7), and (7.35) yields

$$ 0 = p_{\mathbf{T}_1} + p_{\mathbf{T}_2} - p_{\mathbf{T}} \tag{7.66} $$

which proves the theorem[2] in view of Definition 2.17. ∎

Theorem 7.2. *The first open chain described by (7.8), (7.9), (7.16), (7.17), (7.23), (7.24), and (7.30), combined with its respective control equations (7.36), (7.37), (7.42), (7.43), (7.56), (7.57), and (7.63) and with the parameter adaptation (7.46), (7.47), (7.50), and (7.51), is virtually stable with its affiliated vectors ${}^{\mathbf{B}_1}V_r - {}^{\mathbf{B}_1}V$ and ${}^{\mathbf{B}_{11}}V_r - {}^{\mathbf{B}_{11}}V$ being virtual functions in both L_2 and L_∞, in the sense of Definition 2.17.*

[2] Note the fact that the first zero-mass *object* has one *driving cutting point* associated with frame $\{\mathbf{T}\}$ and two *driven cutting points* associated with frames $\{\mathbf{T}_1\}$ and $\{\mathbf{T}_2\}$.

Proof: Select the non-negative accompanying function of the first *open chain* as

$$\nu = \nu_{\mathbf{B}_1} + \nu_{\mathbf{B}_{11}} \tag{7.67}$$

where

$$\nu_{\mathbf{B}_1} = \frac{1}{2}\left({}^{\mathbf{B}_1}V_r - {}^{\mathbf{B}_1}V\right)^T \mathbf{M}_{\mathbf{B}_1}\left({}^{\mathbf{B}_1}V_r - {}^{\mathbf{B}_1}V\right) + \frac{1}{2}\sum_{\gamma=1}^{13}\left(\theta_{\mathbf{B}_1\gamma} - \hat{\theta}_{\mathbf{B}_1\gamma}\right)^2 / \rho_{\mathbf{B}_1\gamma} \tag{7.68}$$

$$\nu_{\mathbf{B}_{11}} = \frac{1}{2}\left({}^{\mathbf{B}_{11}}V_r - {}^{\mathbf{B}_{11}}V\right)^T \mathbf{M}_{\mathbf{B}_{11}}\left({}^{\mathbf{B}_{11}}V_r - {}^{\mathbf{B}_{11}}V\right) + \frac{1}{2}\sum_{\gamma=1}^{13}\left(\theta_{\mathbf{B}_{11}\gamma} - \hat{\theta}_{\mathbf{B}_{11}\gamma}\right)^2 / \rho_{\mathbf{B}_{11}\gamma} \tag{7.69}$$

are the two non-negative accompanying functions assigned to the two rigid links affiliated with the first *open chain*.

With appropriate frame substitutions, it follows from (7.16), (7.17), (7.42), (7.43), (7.46), (7.47), (7.50), (7.51), and Lemma 4.1 that

$$\dot{\nu}_{\mathbf{B}_1} \leqslant -\left({}^{\mathbf{B}_1}V_r - {}^{\mathbf{B}_1}V\right)^T \mathbf{K}_{\mathbf{B}_1}\left({}^{\mathbf{B}_1}V_r - {}^{\mathbf{B}_1}V\right) + \left({}^{\mathbf{B}_1}V_r - {}^{\mathbf{B}_1}V\right)^T \left({}^{\mathbf{B}_1}F_r^* - {}^{\mathbf{B}_1}F^*\right) \tag{7.70}$$

$$\dot{\nu}_{\mathbf{B}_{11}} \leqslant -\left({}^{\mathbf{B}_{11}}V_r - {}^{\mathbf{B}_{11}}V\right)^T \mathbf{K}_{\mathbf{B}_{11}}\left({}^{\mathbf{B}_{11}}V_r - {}^{\mathbf{B}_{11}}V\right) + \left({}^{\mathbf{B}_{11}}V_r - {}^{\mathbf{B}_{11}}V\right)^T \left({}^{\mathbf{B}_{11}}F_r^* - {}^{\mathbf{B}_{11}}F^*\right) \tag{7.71}$$

hold.

In view of (2.85), (7.8), (7.9), (7.23), (7.24), (7.30), (7.36), (7.37), (7.56), (7.57), and (7.63), it results in

$$\begin{aligned}
&\left({}^{\mathbf{B}_1}V_r - {}^{\mathbf{B}_1}V\right)^T \left({}^{\mathbf{B}_1}F_r^* - {}^{\mathbf{B}_1}F^*\right) \\
&= \left({}^{\mathbf{B}_1}V_r - {}^{\mathbf{B}_1}V\right)^T \left[\left({}^{\mathbf{B}_1}F_r - {}^{\mathbf{B}_1}F\right) - {}^{\mathbf{B}_1}\mathbf{U}_{\mathbf{B}_{11}}\left({}^{\mathbf{B}_{11}}F_r - {}^{\mathbf{B}_{11}}F\right)\right] \\
&= p_{\mathbf{B}_1} - \left[{}^{\mathbf{B}_{11}}\mathbf{U}_{\mathbf{B}_1}^T\left({}^{\mathbf{B}_{11}}V_r - {}^{\mathbf{B}_{11}}V\right) - {}^{\mathbf{B}_{11}}\mathbf{U}_{\mathbf{B}_1}^T \mathbf{z}_\tau\left(\dot{q}_r - \dot{q}\right)\right]^T \\
&\qquad \times {}^{\mathbf{B}_1}\mathbf{U}_{\mathbf{B}_{11}}\left({}^{\mathbf{B}_{11}}F_r - {}^{\mathbf{B}_{11}}F\right) \\
&= p_{\mathbf{B}_1} - p_{\mathbf{B}_{11}} + \left(\dot{q}_r - \dot{q}\right)\mathbf{z}_\tau^T\left({}^{\mathbf{B}_{11}}F_r - {}^{\mathbf{B}_{11}}F\right) \\
&= p_{\mathbf{B}_1} - p_{\mathbf{B}_{11}}
\end{aligned} \tag{7.72}$$

$$\begin{aligned}
&\left({}^{\mathbf{B}_{11}}V_r - {}^{\mathbf{B}_{11}}V\right)^T \left({}^{\mathbf{B}_{11}}F_r^* - {}^{\mathbf{B}_{11}}F^*\right) \\
&= \left({}^{\mathbf{B}_{11}}V_r - {}^{\mathbf{B}_{11}}V\right)^T \left[\left({}^{\mathbf{B}_{11}}F_r - {}^{\mathbf{B}_{11}}F\right) - {}^{\mathbf{B}_{11}}\mathbf{U}_{\mathbf{T}_1}\left({}^{\mathbf{T}_1}F_r - {}^{\mathbf{T}_1}F\right)\right] \\
&= p_{\mathbf{B}_{11}} - \left[{}^{\mathbf{B}_{11}}\mathbf{U}_{\mathbf{T}_1}^T\left({}^{\mathbf{B}_{11}}V_r - {}^{\mathbf{B}_{11}}V\right)\right]^T \left({}^{\mathbf{T}_1}F_r - {}^{\mathbf{T}_1}F\right) \\
&= p_{\mathbf{B}_{11}} - p_{\mathbf{T}_1}.
\end{aligned} \tag{7.73}$$

Substituting (7.72) and (7.73) into (7.70) and (7.71) yields

$$\begin{aligned}\dot{\nu} &= \dot{\nu}_{\mathbf{B}_1} + \dot{\nu}_{\mathbf{B}_{11}} \\ &\leqslant -\left({}^{\mathbf{B}_1}V_r - {}^{\mathbf{B}_1}V\right)^T \mathbf{K}_{\mathbf{B}_1}\left({}^{\mathbf{B}_1}V_r - {}^{\mathbf{B}_1}V\right) \\ &\quad -\left({}^{\mathbf{B}_{11}}V_r - {}^{\mathbf{B}_{11}}V\right)^T \mathbf{K}_{\mathbf{B}_{11}}\left({}^{\mathbf{B}_{11}}V_r - {}^{\mathbf{B}_{11}}V\right) \\ &\quad + p_{\mathbf{B}_1} - p_{\mathbf{T}_1}.\end{aligned} \tag{7.74}$$

Consider the fact that the first *open chain* has one *driving cutting point* associated with frame $\{\mathbf{T}_1\}$ and one *driven cutting point* associated with frame $\{\mathbf{B}_1\}$. Using (7.67)–(7.69), (7.74), and Definition 2.17 completes the proof. ■

Lemma 7.1. *Consider the second open chain described by (7.10)–(7.12), (7.18), (7.19), (7.25), (7.26), (7.28), (7.29), and (7.31) and combined with its respective control equations (7.38)–(7.40), (7.44), (7.45), (7.58), (7.59), (7.61), (7.62), and (7.64) and with the parameter adaptation (7.48), (7.49), (7.52), and (7.53). Let*

$$\nu_2 = \nu_{\mathbf{B}_{21}} + \nu_{\mathbf{B}_{22}} \tag{7.75}$$

be the non-negative accompanying functions assigned to the second open chain, where

$$\begin{aligned}\nu_{\mathbf{B}_{21}} &= \frac{1}{2}\left({}^{\mathbf{B}_{21}}V_r - {}^{\mathbf{B}_{21}}V\right)^T \mathbf{M}_{\mathbf{B}_{21}}\left({}^{\mathbf{B}_{21}}V_r - {}^{\mathbf{B}_{21}}V\right) \\ &\quad + \frac{1}{2}\sum_{\gamma=1}^{13}\left(\theta_{\mathbf{B}_{21}\gamma} - \hat{\theta}_{\mathbf{B}_{21}\gamma}\right)^2 / \rho_{\mathbf{B}_{21}\gamma}\end{aligned} \tag{7.76}$$

$$\begin{aligned}\nu_{\mathbf{B}_{22}} &= \frac{1}{2}\left({}^{\mathbf{B}_{22}}V_r - {}^{\mathbf{B}_{22}}V\right)^T \mathbf{M}_{\mathbf{B}_{22}}\left({}^{\mathbf{B}_{22}}V_r - {}^{\mathbf{B}_{22}}V\right) \\ &\quad + \frac{1}{2}\sum_{\gamma=1}^{13}\left(\theta_{\mathbf{B}_{22}\gamma} - \hat{\theta}_{\mathbf{B}_{22}\gamma}\right)^2 / \rho_{\mathbf{B}_{22}\gamma}\end{aligned} \tag{7.77}$$

are the two non-negative accompanying functions assigned to the two rigid bodies (piston rod and cylinder base) affiliated with the second open chain. Then, the time derivative of (7.75) can be expressed by

$$\begin{aligned}\dot{\nu}_2 &= \dot{\nu}_{\mathbf{B}_{21}} + \dot{\nu}_{\mathbf{B}_{22}} \\ &\leqslant -\left({}^{\mathbf{B}_{21}}V_r - {}^{\mathbf{B}_{21}}V\right)^T \mathbf{K}_{\mathbf{B}_{21}}\left({}^{\mathbf{B}_{21}}V_r - {}^{\mathbf{B}_{21}}V\right) \\ &\quad -\left({}^{\mathbf{B}_{22}}V_r - {}^{\mathbf{B}_{22}}V\right)^T \mathbf{K}_{\mathbf{B}_{22}}\left({}^{\mathbf{B}_{22}}V_r - {}^{\mathbf{B}_{22}}V\right) \\ &\quad + p_{\mathbf{B}_2} - p_{\mathbf{T}_2} \\ &\quad + (\dot{x}_r - \dot{x})(f_{cr} - f_c)\end{aligned} \tag{7.78}$$

where $p_{\mathbf{B}_2}$ and $p_{\mathbf{T}_2}$ denote the two virtual power flows by Definition 2.16 at the two cutting points of the second open chain.

Proof: It follows from (7.18), (7.19), (7.44), (7.45), (7.48), (7.49), (7.52), (7.53), and Lemma 4.1 that

$$\dot{\nu}_{\mathbf{B}_{21}} \leqslant -\left({}^{\mathbf{B}_{21}}V_r - {}^{\mathbf{B}_{21}}V\right)^T \mathbf{K}_{\mathbf{B}_{21}} \left({}^{\mathbf{B}_{21}}V_r - {}^{\mathbf{B}_{21}}V\right) + \left({}^{\mathbf{B}_{21}}V_r - {}^{\mathbf{B}_{21}}V\right)^T \left({}^{\mathbf{B}_{21}}F_r^* - {}^{\mathbf{B}_{21}}F^*\right) \tag{7.79}$$

$$\dot{\nu}_{\mathbf{B}_{22}} \leqslant -\left({}^{\mathbf{B}_{22}}V_r - {}^{\mathbf{B}_{22}}V\right)^T \mathbf{K}_{\mathbf{B}_{22}} \left({}^{\mathbf{B}_{22}}V_r - {}^{\mathbf{B}_{22}}V\right) + \left({}^{\mathbf{B}_{22}}V_r - {}^{\mathbf{B}_{22}}V\right)^T \left({}^{\mathbf{B}_{22}}F_r^* - {}^{\mathbf{B}_{22}}F^*\right) \tag{7.80}$$

hold.

In view of (2.85), (7.10)–(7.12), (7.25), (7.26), (7.28), (7.29), (7.31), (7.38)–(7.40), (7.58), (7.59), (7.61), (7.62), and (7.64), it results in

$$\begin{aligned}
&\left({}^{\mathbf{B}_{21}}V_r - {}^{\mathbf{B}_{21}}V\right)^T \left({}^{\mathbf{B}_{21}}F_r^* - {}^{\mathbf{B}_{21}}F^*\right) \\
&= \left({}^{\mathbf{B}_{21}}V_r - {}^{\mathbf{B}_{21}}V\right)^T \\
&\qquad \times \left[{}^{\mathbf{B}_{21}}\mathbf{U}_{\mathbf{B}_2}\left({}^{\mathbf{B}_2}F_r - {}^{\mathbf{B}_2}F\right) - {}^{\mathbf{B}_{21}}\mathbf{U}_{\mathbf{B}_{22}}\left({}^{\mathbf{B}_{22}}F_r - {}^{\mathbf{B}_{22}}F\right)\right] \\
&= \left[{}^{\mathbf{B}_2}\mathbf{U}_{\mathbf{B}_{21}}^T\left({}^{\mathbf{B}_2}V_r - {}^{\mathbf{B}_2}V\right) + \mathbf{z}_\tau\left(\dot{q}_{2r} - \dot{q}_2\right)\right]^T \\
&\qquad \times {}^{\mathbf{B}_{21}}\mathbf{U}_{\mathbf{B}_2}\left({}^{\mathbf{B}_2}F_r - {}^{\mathbf{B}_2}F\right) \\
&\quad - \left[{}^{\mathbf{B}_{22}}\mathbf{U}_{\mathbf{B}_{21}}^T\left({}^{\mathbf{B}_{22}}V_r - {}^{\mathbf{B}_{22}}V\right) - {}^{\mathbf{B}_{22}}\mathbf{U}_{\mathbf{B}_{21}}^T\mathbf{x}_f\left(\dot{x}_r - \dot{x}\right)\right]^T \\
&\qquad \times {}^{\mathbf{B}_{21}}\mathbf{U}_{\mathbf{B}_{22}}\left({}^{\mathbf{B}_{22}}F_r - {}^{\mathbf{B}_{22}}F\right) \\
&= p_{\mathbf{B}_2} + \left(\dot{q}_{2r} - \dot{q}_2\right)\mathbf{z}_\tau^T\left({}^{\mathbf{B}_{21}}F_r - {}^{\mathbf{B}_{21}}F\right) \\
&\quad - p_{\mathbf{B}_{22}} + \left(\dot{x}_r - \dot{x}\right)\mathbf{x}_f^T\left({}^{\mathbf{B}_{22}}F_r - {}^{\mathbf{B}_{22}}F\right) \\
&= p_{\mathbf{B}_2} - p_{\mathbf{B}_{22}} + \left(\dot{x}_r - \dot{x}\right)\left(f_{cr} - f_c\right)
\end{aligned} \tag{7.81}$$

$$\begin{aligned}
&\left({}^{\mathbf{B}_{22}}V_r - {}^{\mathbf{B}_{22}}V\right)^T \left({}^{\mathbf{B}_{22}}F_r^* - {}^{\mathbf{B}_{22}}F^*\right) \\
&= \left({}^{\mathbf{B}_{22}}V_r - {}^{\mathbf{B}_{22}}V\right)^T \left[\left({}^{\mathbf{B}_{22}}F_r - {}^{\mathbf{B}_{22}}F\right) - {}^{\mathbf{B}_{22}}\mathbf{U}_{\mathbf{T}_2}\left({}^{\mathbf{T}_2}F_r - {}^{\mathbf{T}_2}F\right)\right] \\
&= p_{\mathbf{B}_{22}} - \left[{}^{\mathbf{T}_2}\mathbf{U}_{\mathbf{B}_{22}}^T\left({}^{\mathbf{T}_2}V_r - {}^{\mathbf{T}_2}V\right) - {}^{\mathbf{T}_2}\mathbf{U}_{\mathbf{B}_{22}}^T\mathbf{z}_\tau\left(\dot{q}_{1r} - \dot{q}_1\right)\right]^T \\
&\qquad \times {}^{\mathbf{B}_{22}}\mathbf{U}_{\mathbf{T}_2}\left({}^{\mathbf{T}_2}F_r - {}^{\mathbf{T}_2}F\right) \\
&= p_{\mathbf{B}_{22}} - p_{\mathbf{T}_2} + \left(\dot{q}_{1r} - \dot{q}_1\right)\mathbf{z}_\tau^T\left({}^{\mathbf{T}_2}F_r - {}^{\mathbf{T}_2}F\right) \\
&= p_{\mathbf{B}_{22}} - p_{\mathbf{T}_2}.
\end{aligned} \tag{7.82}$$

Substituting (7.81) and (7.82) into (7.79) and (7.80) yields (7.78). ■

Theorem 7.3. *The second zero-mass object described by (7.13) and (7.27), combined with the control equations (7.41) and (7.60), is virtually stable in the sense of Definition 2.17.*

Proof: It follows from (7.27) and (7.60) that

$${}^{\mathbf{B}}F_r - {}^{\mathbf{B}}F = \left({}^{\mathbf{B}_1}F_r - {}^{\mathbf{B}_1}F\right) + \left({}^{\mathbf{B}_2}F_r - {}^{\mathbf{B}_2}F\right) \tag{7.83}$$

holds.

Let the non-negative accompanying function be *zero*. Premultiplying (7.83) by $({}^{\mathbf{T}}V_r - {}^{\mathbf{T}}V)^T$ and using (2.85), (7.13), and (7.41) yields

$$0 = p_{\mathbf{B}} - p_{\mathbf{B}_1} - p_{\mathbf{B}_2} \tag{7.84}$$

which proves the theorem[3] in view of Definition 2.17. ■

Remark 7.8. Theorems 7.1–7.3 ensure the *virtual stability* of both the two zero-mass *objects* and the first *open chain*. However, the appearance of $(\dot{x}_r - \dot{x})(f_{cr} - f_c)$ in the right hand side of (7.78) prevents the *virtual stability* of the second *open chain* (representing the cylinder) from being held. This term will be addressed in the next section.

7.7 Hydraulic Cylinder Dynamics and Control

This section presents the specific dynamics and control design associated with the hydraulic cylinder, aimed at addressing the term $(\dot{x}_r - \dot{x})(f_{cr} - f_c)$ appeared in the right hand side of (7.78).

7.7.1 Friction Model

In a hydraulic cylinder, the piston friction makes a large difference between the cylinder output force and the chamber pressure-induced force (a product of the chamber pressure and the piston area). Let $f_f \in \mathbb{R}$ be the piston friction force and let f_p be the chamber pressure force. It follows that

$$f_p = f_c + f_f \tag{7.85}$$

holds, where f_c defined by (7.28) is the cylinder output force. The friction force f_f can be expressed by a linear parametrization form as

$$f_f = \mathbf{Y}_f \boldsymbol{\theta}_f \tag{7.86}$$

with the regressor $\mathbf{Y}_f$ being differentiable. In [222], a typical friction model was proposed to ensure the cylinder output force control.

7.7.2 Hydraulic Fluid Dynamics

When a high-bandwidth servo valve is used for a hydraulic cylinder, it is reasonable to ignore the servo valve's dynamics so that the valve position is considered to be proportional to its control voltage within a frequency range of interest [156, 194, 222].

[3] Note the fact that the second zero-mass *object* has two *driving cutting points* associated with frames $\{\mathbf{B}_1\}$ and $\{\mathbf{B}_2\}$ and one *driven cutting point* associated with frame $\{\mathbf{B}\}$.

Based on the Bernoulli's static flow equation [120, page 41], the difference of the two flow speed squares across an orifice is proportional to the difference of pressures. As a result, the rate of flow passing through an orifice, denoted as g, is proportional to the product of the valve control voltage and the square root of the pressure drop across the orifice, that is

$$g = c\sqrt{\Delta p}u \tag{7.87}$$

where $c > 0$ is a constant, $\Delta p > 0$ denotes the pressure drop across the valve orifice, and u is the valve control voltage.

In terms of the definition of the fluid bulk modulus, denoted as β, the dynamic equation for the fluid compressibility inside a chamber can be written as [151, 156]

$$\dot{p} = \frac{\beta}{v_c}(g - \dot{v}_c) \tag{7.88}$$

where p denotes the chamber pressure, v_c denotes the chamber volume, and g denotes the rate of flow entering the chamber.

Define a selective function

$$\varepsilon(x) \overset{def}{=} \begin{cases} 1 & \text{if } x > 0 \\ 0 & \text{if } x \leqslant 0 \end{cases} \tag{7.89}$$

a sign function

$$\text{sign}(x) \overset{def}{=} \begin{cases} 1 & \text{if } x > 0 \\ 0 & \text{if } x = 0 \\ -1 & \text{if } x < 0 \end{cases} \tag{7.90}$$

and a pressure-drop related function

$$\upsilon(x) \overset{def}{=} \sqrt{|x|}\text{sign}(x). \tag{7.91}$$

Remark 7.9. Note that function $\upsilon(x)$ is monotonically increasing.

With respect to a typical cylinder as illustrated in Figure 7.1, let g_a and g_b be the rates of flows entering the left and right chambers, respectively, and let p_a and p_b be the pressures inside the left and right chambers, respectively. It follows from (7.87) that

$$g_a = c_{p1}\upsilon(p_s - p_a)u\varepsilon(u) + c_{n1}\upsilon(p_a - p_r)u\varepsilon(-u) \tag{7.92}$$

$$g_b = -c_{n2}\upsilon(p_b - p_r)u\varepsilon(u) - c_{p2}\upsilon(p_s - p_b)u\varepsilon(-u) \tag{7.93}$$

hold, where $c_{p1} > 0$, $c_{n1} > 0$, $c_{p2} > 0$, and $c_{n2} > 0$ are four constants[4] and $p_s > 0$ and $p_r > 0$ denote the supply and return pressures with $p_s \gg p_r$.

According to (7.88), the pressure dynamic equations of the two chambers can be written as

[4] For an ideal valve, the four parameters are equal to each other. Otherwise, they keep different values for an asymmetric valve.

$$\dot{p}_a = \frac{\beta}{s_a(l_o - x)}(g_a + s_a\dot{x}) \tag{7.94}$$

$$\dot{p}_b = \frac{\beta}{s_b x}(g_b - s_b\dot{x}) \tag{7.95}$$

where $s_a > 0$ and $s_b > 0$ denote the piston areas at both chambers with $s_a < s_b$, x is the cylinder displacement, and l_o is the cylinder effective length.

The pressures of the two chambers can be converted into the net pressure force as

$$f_p = s_b p_b - s_a p_a. \tag{7.96}$$

Premultiplying s_a and s_b to (7.94) and (7.95), respectively, and using (7.92), (7.93), and (7.96) yields

$$\dot{f}_p = \beta\left[u_f - \left(\frac{s_a}{l_o - x} + \frac{s_b}{x}\right)\dot{x}\right] \tag{7.97}$$

where

$$\begin{aligned} u_f &= \frac{g_b}{x} - \frac{g_a}{l_o - x} \\ &= -\left(\frac{c_{p1}\upsilon(p_s - p_a)}{l_o - x} + \frac{c_{n2}\upsilon(p_b - p_r)}{x}\right)u\varepsilon(u) \\ &\quad -\left(\frac{c_{n1}\upsilon(p_a - p_r)}{l_o - x} + \frac{c_{p2}\upsilon(p_s - p_b)}{x}\right)u\varepsilon(-u) \\ &\stackrel{def}{=} -\mathbf{Y}_\upsilon(u)\boldsymbol{\theta}_\upsilon \end{aligned} \tag{7.98}$$

with

$$\mathbf{Y}_\upsilon(u) = \begin{bmatrix} \frac{\upsilon(p_s - p_a)}{l_o - x}u\varepsilon(u) \\ \frac{\upsilon(p_a - p_r)}{l_o - x}u\varepsilon(-u) \\ \frac{\upsilon(p_s - p_b)}{x}u\varepsilon(-u) \\ \frac{\upsilon(p_b - p_r)}{x}u\varepsilon(u) \end{bmatrix}^T \in \mathbb{R}^{1\times 4} \tag{7.99}$$

$$\boldsymbol{\theta}_\upsilon = \begin{bmatrix} c_{p1} & c_{n1} & c_{p2} & c_{n2} \end{bmatrix}^T \in \mathbb{R}^4. \tag{7.100}$$

The following assumption makes sure that the piston never reaches its two ends.

Assumption 5. *The following relationship holds*

$$0 < x < l_o. \tag{7.101}$$

In view of (7.98) and Assumption 5, the univalence between u and u_f exists, provided that

$$\frac{c_{p1}\upsilon(p_s - p_a)}{l_o - x} + \frac{c_{n2}\upsilon(p_b - p_r)}{x} > 0 \tag{7.102}$$

$$\frac{c_{n1}\upsilon(p_a - p_r)}{l_o - x} + \frac{c_{p2}\upsilon(p_s - p_b)}{x} > 0 \tag{7.103}$$

hold. This statement implies that for a given u_f one can find a unique valve control voltage u as

$$
\begin{aligned}
u = & -\frac{1}{\frac{c_{p1}v(p_s-p_a)}{l_o-x}+\frac{c_{n2}v(p_b-p_r)}{x}}u_f\varepsilon(-u_f) \\
& -\frac{1}{\frac{c_{n1}v(p_a-p_r)}{l_o-x}+\frac{c_{p2}v(p_s-p_b)}{x}}u_f\varepsilon(u_f)
\end{aligned}
\tag{7.104}
$$

when both (7.102) and (7.103) are satisfied.

7.7.3 Cylinder Control Equations

In view of the friction model defined by (7.85) and (7.86) and the fluid dynamics defined by (7.97), the control equations are designed as

$$
f_{pr} = f_{cr} + \mathbf{Y}_f\hat{\boldsymbol{\theta}}_f \tag{7.105}
$$

$$
\begin{aligned}
u_{fd} &= \widehat{\left(\frac{1}{\beta}\right)}\dot{f}_{pr} + \left(\frac{\hat{s}_a}{l_o - x} + \frac{\hat{s}_b}{x}\right)\dot{x} + k_{fp}(f_{pr} - f_p) + k_x(\dot{x}_r - \dot{x}) \\
&= \mathbf{Y}_c\hat{\boldsymbol{\theta}}_c + k_{fp}(f_{pr} - f_p) + k_x(\dot{x}_r - \dot{x})
\end{aligned}
\tag{7.106}
$$

$$
\begin{aligned}
u = & -\frac{1}{\frac{\hat{c}_{p1}v(p_s-p_a)}{l_o-x}+\frac{\hat{c}_{n2}v(p_b-p_r)}{x}}u_{fd}\varepsilon(-u_{fd}) \\
& -\frac{1}{\frac{\hat{c}_{n1}v(p_a-p_r)}{l_o-x}+\frac{\hat{c}_{p2}v(p_s-p_b)}{x}}u_{fd}\varepsilon(u_{fd})
\end{aligned}
\tag{7.107}
$$

with

$$
\mathbf{Y}_c = \begin{bmatrix} \dot{f}_{pr} & \frac{\dot{x}}{l_o-x} & \frac{\dot{x}}{x} \end{bmatrix} \in \mathbb{R}^{1\times 3} \tag{7.108}
$$

$$
\boldsymbol{\theta}_c = \begin{bmatrix} \frac{1}{\beta} & s_a & s_b \end{bmatrix}^T \in \mathbb{R}^3 \tag{7.109}
$$

where f_{cr} is obtained from (7.61), $\hat{\boldsymbol{\theta}}_f$ and $\hat{\boldsymbol{\theta}}_c$ denote the estimates of $\boldsymbol{\theta}_f$ and $\boldsymbol{\theta}_c$, respectively, $k_{fp} > 0$ and $k_x > 0$ are two feedback gains, and $\dot{x}_r$ is determined from (7.32)–(7.34).

In line with (7.102) and (7.103), the following conditions

$$
\frac{\hat{c}_{p1}v(p_s - p_a)}{l_o - x} + \frac{\hat{c}_{n2}v(p_b - p_r)}{x} > 0 \tag{7.110}
$$

$$
\frac{\hat{c}_{n1}v(p_a - p_r)}{l_o - x} + \frac{\hat{c}_{p2}v(p_s - p_b)}{x} > 0 \tag{7.111}
$$

need to be satisfied to make (7.107) executable. Then, (7.107) can be inversely written as

$$
u_{fd} = -\mathbf{Y}_v(u)\hat{\boldsymbol{\theta}}_v \tag{7.112}
$$

in view of (7.98).

The three estimated parameter vectors, denoted as $\hat{\boldsymbol{\theta}}_f$, $\hat{\boldsymbol{\theta}}_c$, and $\hat{\boldsymbol{\theta}}_v$, need to be updated. Define

$$\mathbf{s}_f = (\dot{x}_r - \dot{x})\mathbf{Y}_f^T \tag{7.113}$$
$$\mathbf{s}_c = (f_{pr} - f_p)\mathbf{Y}_c^T \tag{7.114}$$
$$\mathbf{s}_v = (f_{pr} - f_p)\mathbf{Y}_v^T. \tag{7.115}$$

The γth elements of $\hat{\boldsymbol{\theta}}_f$, $\hat{\boldsymbol{\theta}}_c$, and $\hat{\boldsymbol{\theta}}_v$ are updated by using the $\mathcal{P}$ function defined by (2.79) as

$$\hat{\theta}_{f\gamma} = \mathcal{P}\left(s_{f\gamma}, \rho_{f\gamma}, \underline{\theta}_{f\gamma}, \overline{\theta}_{f\gamma}, t\right), \quad \forall\gamma \tag{7.116}$$
$$\hat{\theta}_{c\gamma} = \mathcal{P}\left(s_{c\gamma}, \rho_{c\gamma}, \underline{\theta}_{c\gamma}, \overline{\theta}_{c\gamma}, t\right), \quad \gamma = 1, 2, 3 \tag{7.117}$$
$$\hat{\theta}_{v\gamma} = \mathcal{P}\left(s_{v\gamma}, \rho_{v\gamma}, \underline{\theta}_{v\gamma}, \overline{\theta}_{v\gamma}, t\right), \quad \gamma = 1, 2, 3, 4 \tag{7.118}$$

where

$\hat{\theta}_{f\gamma}$	γth element of $\hat{\boldsymbol{\theta}}_f$.
$\hat{\theta}_{c\gamma}$	γth element of $\hat{\boldsymbol{\theta}}_c$.
$\hat{\theta}_{v\gamma}$	γth element of $\hat{\boldsymbol{\theta}}_v$.
$s_{f\gamma}$	γth element of $\mathbf{s}_f$.
$s_{c\gamma}$	γth element of $\mathbf{s}_c$.
$s_{v\gamma}$	γth element of $\mathbf{s}_v$.
$\rho_{f\gamma}$	Parameter update gain.
$\rho_{c\gamma}$	Parameter update gain.
$\rho_{v\gamma}$	Parameter update gain.
$\underline{\theta}_{f\gamma}$	Lower bound of $\theta_{f\gamma}$.
$\underline{\theta}_{c\gamma}$	Lower bound of $\theta_{c\gamma}$.
$\underline{\theta}_{v\gamma}$	Lower bound of $\theta_{v\gamma}$.
$\overline{\theta}_{f\gamma}$	Upper bound of $\theta_{f\gamma}$.
$\overline{\theta}_{c\gamma}$	Upper bound of $\theta_{c\gamma}$.
$\overline{\theta}_{v\gamma}$	Upper bound of $\theta_{v\gamma}$.
$\theta_{f\gamma}$	γth element of $\boldsymbol{\theta}_f$ defined in (7.86).
$\theta_{c\gamma}$	γth element of $\boldsymbol{\theta}_c$ defined by (7.109).
$\theta_{v\gamma}$	γth element of $\boldsymbol{\theta}_v$ defined by (7.100).

7.7.4 Non-negative Accompanying Function for Fluid Dynamics

The following lemma gives a non-negative accompanying function and its time derivative, with respect to the fluid dynamics and the respective control equations.

Lemma 7.2. *Consider the hydraulic cylinder dynamics described by (7.85), (7.86), (7.97), and (7.98) and combined with the control equations (7.105), (7.106), and (7.108)–(7.118). The time derivative of*

$$
\begin{aligned}
\nu_c = &\frac{1}{2\beta}(f_{pr} - f_p)^2 + \frac{k_x}{2}\sum_{\gamma}(\theta_{f\gamma} - \hat{\theta}_{f\gamma})^2/\rho_{f\gamma} \\
&+\frac{1}{2}\sum_{\gamma=1}^{3}(\theta_{c\gamma} - \hat{\theta}_{c\gamma})^2/\rho_{c\gamma} + \frac{1}{2}\sum_{\gamma=1}^{4}(\theta_{v\gamma} - \hat{\theta}_{v\gamma})^2/\rho_{v\gamma}
\end{aligned} \tag{7.119}
$$

is

$$
\dot{\nu}_c \leqslant -k_{fp}(f_{pr} - f_p)^2 - k_x(f_{cr} - f_c)(\dot{x}_r - \dot{x}). \tag{7.120}
$$

Proof: It follows from (7.97), (7.106), (7.108), and (7.109) that

$$
\begin{aligned}
u_{fd} - u_f = &\frac{1}{\beta}(\dot{f}_{pr} - \dot{f}_p) - \mathbf{Y}_c(\boldsymbol{\theta}_c - \hat{\boldsymbol{\theta}}_c) \\
&+k_{fp}(f_{pr} - f_p) + k_x(\dot{x}_r - \dot{x})
\end{aligned} \tag{7.121}
$$

holds. Differentiating (7.119) with respect to time and using (7.85), (7.86), (7.98), (7.105), (7.110)–(7.118), (7.121), and Lemma 2.9 yields

$$
\begin{aligned}
\dot{\nu}_c = &(f_{pr} - f_p)\frac{1}{\beta}(\dot{f}_{pr} - \dot{f}_p) - \sum_{\gamma} k_x(\theta_{f\gamma} - \hat{\theta}_{f\gamma})\frac{\dot{\hat{\theta}}_{f\gamma}}{\rho_{f\gamma}} \\
&-\sum_{\gamma=1}^{3}(\theta_{c\gamma} - \hat{\theta}_{c\gamma})\frac{\dot{\hat{\theta}}_{c\gamma}}{\rho_{c\gamma}} - \sum_{\gamma=1}^{4}(\theta_{v\gamma} - \hat{\theta}_{v\gamma})\frac{\dot{\hat{\theta}}_{v\gamma}}{\rho_{v\gamma}} \\
= &-k_{fp}(f_{pr} - f_p)^2 + (f_{pr} - f_p)\mathbf{Y}_c(\boldsymbol{\theta}_c - \hat{\boldsymbol{\theta}}_c) \\
&+(f_{pr} - f_p)(u_{fd} - u_f) - k_x(f_{pr} - f_p)(\dot{x}_r - \dot{x}) \\
&-\sum_{\gamma} k_x(\theta_{f\gamma} - \hat{\theta}_{f\gamma})\frac{\dot{\hat{\theta}}_{f\gamma}}{\rho_{f\gamma}} - \sum_{\gamma=1}^{3}(\theta_{c\gamma} - \hat{\theta}_{c\gamma})\frac{\dot{\hat{\theta}}_{c\gamma}}{\rho_{c\gamma}} \\
&-\sum_{\gamma=1}^{4}(\theta_{v\gamma} - \hat{\theta}_{v\gamma})\frac{\dot{\hat{\theta}}_{v\gamma}}{\rho_{v\gamma}} \\
= &-k_{fp}(f_{pr} - f_p)^2 - k_x(f_{cr} - f_c)(\dot{x}_r - \dot{x}) \\
&+k_x(\dot{x}_r - \dot{x})\mathbf{Y}_f(\boldsymbol{\theta}_f - \hat{\boldsymbol{\theta}}_f) - \sum_{\gamma} k_x(\theta_{f\gamma} - \hat{\theta}_{f\gamma})\frac{\dot{\hat{\theta}}_{f\gamma}}{\rho_{f\gamma}} \\
&+(f_{pr} - f_p)\mathbf{Y}_c(\boldsymbol{\theta}_c - \hat{\boldsymbol{\theta}}_c) - \sum_{\gamma=1}^{3}(\theta_{c\gamma} - \hat{\theta}_{c\gamma})\frac{\dot{\hat{\theta}}_{c\gamma}}{\rho_{c\gamma}} \\
&+(f_{pr} - f_p)\mathbf{Y}_v(\boldsymbol{\theta}_v - \hat{\boldsymbol{\theta}}_v) - \sum_{\gamma=1}^{4}(\theta_{v\gamma} - \hat{\theta}_{v\gamma})\frac{\dot{\hat{\theta}}_{v\gamma}}{\rho_{v\gamma}} \\
\leqslant &-k_{fp}(f_{pr} - f_p)^2 - k_x(f_{cr} - f_c)(\dot{x}_r - \dot{x}).
\end{aligned} \tag{7.122}
$$

■

7.8 Virtual Stability of the Hydraulic Actuator Assembly

The following theorem ensures the *virtual stability* (in the sense of Definition 2.17) of the hydraulic cylinder (the second *open chain*) comprised of two rigid bodies driven by the hydraulic fluid.

Theorem 7.4. *The second open chain described by (7.10)–(7.12), (7.18), (7.19), (7.25), (7.26), (7.28), (7.29), (7.31), (7.85), (7.86), (7.97), and (7.98), combined with the control equations (7.38)–(7.40), (7.44), (7.45), (7.58), (7.59), (7.61), (7.62), (7.64), (7.105), (7.106), and (7.108)–(7.112) and with the parameter adaptation (7.48), (7.49), (7.52), (7.53), and (7.113)–(7.118), is virtually stable with its affiliated vectors and variable* ${}^{\mathbf{B}_{21}}V_r - {}^{\mathbf{B}_{21}}V$, ${}^{\mathbf{B}_{22}}V_r - {}^{\mathbf{B}_{22}}V$, *and* $f_{pr} - f_p$ *being virtual functions in both* L_2 *and* L_∞, *in the sense of Definition 2.17.*

Proof: The proof follows directly from Lemmas 7.1 and 7.2. Define the non-negative accompanying function of the second *open chain* as

$$\nu = \nu_2 + \frac{\nu_c}{k_x} \tag{7.123}$$

where ν_2 and ν_c are defined by (7.75) and (7.119), respectively. It follows from (7.78) and (7.120) that

$$\begin{aligned}\dot{\nu} &= \dot{\nu}_2 + \frac{\dot{\nu}_c}{k_x}\\ &\leqslant -\left({}^{\mathbf{B}_{21}}V_r - {}^{\mathbf{B}_{21}}V\right)^T \mathbf{K}_{\mathbf{B}_{21}} \left({}^{\mathbf{B}_{21}}V_r - {}^{\mathbf{B}_{21}}V\right)\\ &\quad -\left({}^{\mathbf{B}_{22}}V_r - {}^{\mathbf{B}_{22}}V\right)^T \mathbf{K}_{\mathbf{B}_{22}} \left({}^{\mathbf{B}_{22}}V_r - {}^{\mathbf{B}_{22}}V\right)\\ &\quad -\frac{k_{fp}}{k_x}(f_{pr} - f_p)^2 + p_{\mathbf{B}_2} - p_{\mathbf{T}_2}\end{aligned} \tag{7.124}$$

holds.

Consider the fact that the second *open chain* has one *driving cutting point* associated with frame $\{\mathbf{T}_2\}$ and one *driven cutting point* associated with frame $\{\mathbf{B}_2\}$. Using (7.75), (7.119), (7.123), (7.124), and Definition 2.17 completes the proof. ∎

Theorems 7.1–7.4 ensure that the two zero-mass *objects* and the two *open chains* combined with their respective control equations are *virtually stable* in the sense of Definition 2.17. Therefore, the complete hydraulic actuation assembly is *virtually stable*, in view of Lemma 2.11.

When every subsystem of the rest of the robot qualifies to be *virtually stable* in the sense of Definition 2.17, it follows form Theorem 2.1 that

$$f_{pr} - f_p \in L_2 \bigcap L_\infty \tag{7.125}$$

$${}^{\mathbf{B}_1}V_r - {}^{\mathbf{B}_1}V \in L_2 \bigcap L_\infty \tag{7.126}$$

$${}^{\mathbf{B}_{11}}V_r - {}^{\mathbf{B}_{11}}V \in L_2 \bigcap L_\infty \tag{7.127}$$

$$^{\mathbf{B}_{21}}V_r - {}^{\mathbf{B}_{21}}V \in L_2 \bigcap L_\infty \tag{7.128}$$

$$^{\mathbf{B}_{22}}V_r - {}^{\mathbf{B}_{22}}V \in L_2 \bigcap L_\infty \tag{7.129}$$

are guaranteed.

Expressions (7.126) and (7.127) imply

$$\dot{q}_r - \dot{q} \in L_2 \bigcap L_\infty \tag{7.130}$$

in view of (7.9) and (7.37).

Given a bounded $\dot{q}_r$, the boundedness of $\dot{q}$ is ensured from (7.130). It in turn guarantees the boundedness of $\mathbf{Y}_f$ in (7.86). Then, when $\ddot{q}_r \in L_\infty$ holds for all the joints of the entire robot, it follows from Remark 7.7 that f_{cr} in (7.61) is bounded, which implies $f_{pr} \in L_\infty$ from (7.105). Thus, $f_p \in L_\infty$ is ensured from (7.125) such that

$$|f_p| \leqslant \gamma_p \tag{7.131}$$

holds with γ_p being a positive constant.

The boundedness of f_p and $\dot{q}$ implies the boundedness of f_c from (7.85). When all the actuation forces and torques of a robotic system are bounded, it yields a bounded $\ddot{q}$ from (4.136). Eventually, the asymptotic convergence of all L_2 signals with bounded time derivatives can be ensured from Lemma 2.8.

7.9 Internal Dynamics Stability

In the last section, the boundedness of the pressure induced force f_p is ensured by (7.131). In this section, it will be shown that the pressures inside the two chambers, denoted as p_a and p_b, are bounded as well, leading to the satisfaction of (7.102) and (7.103).

Define

$$f_{ps} = s_a(l_o - x)p_a + s_b x p_b. \tag{7.132}$$

It follows from (7.96) and (7.132) that

$$p_a = \frac{f_{ps} - x f_p}{s_a l_o} \tag{7.133}$$

$$p_b = \frac{f_{ps} + (l_o - x) f_p}{s_b l_o} \tag{7.134}$$

hold.

Since f_p is bounded, the boundedness of p_a and p_b is entirely dependent on the boundedness of f_{ps}, in view of Assumption 5.

Differentiating (7.132) with respect to time yields

$$\dot{f}_{ps} = s_a(l_o - x)\dot{p}_a + s_b x \dot{p}_b + (p_b s_b - p_a s_a)\,\dot{x}. \tag{7.135}$$

Substituting (7.94)-(7.96) into (7.135) yields

$$\dot{f}_{ps} = \beta(g_a + g_b) + [\beta(s_a - s_b) + f_p]\,\dot{x}. \tag{7.136}$$

Furthermore, substituting (7.92) and (7.93) into (7.136) and using (7.133) and (7.134) yields

$$\begin{aligned}
\dot{f}_{ps} &= \beta u\left[c_{p1}\upsilon(p_s - p_a) - c_{n2}\upsilon(p_b - p_r)\right]\varepsilon(u) \\
&\quad + \beta u\left[c_{n1}\upsilon(p_a - p_r) - c_{p2}\upsilon(p_s - p_b)\right]\varepsilon(-u) \\
&\quad + \left[\beta(s_a - s_b) + f_p\right]\dot{x} \\
&= \beta u\left[c_{p1}\upsilon\left(p_s - \frac{f_{ps} - xf_p}{s_a l_o}\right) - c_{n2}\upsilon\left(\frac{f_{ps} + (l_o - x)f_p}{s_b l_o} - p_r\right)\right]\varepsilon(u) \\
&\quad + \beta u\left[c_{n1}\upsilon\left(\frac{f_{ps} - xf_p}{s_a l_o} - p_r\right) - c_{p2}\upsilon\left(p_s - \frac{f_{ps} + (l_o - x)f_p}{s_b l_o}\right)\right]\varepsilon(-u) \\
&\quad + \left[\beta(s_a - s_b) + f_p\right]\dot{x}.
\end{aligned} \tag{7.137}$$

By defining

$$\delta_c = \frac{\max\{c_{p1}, c_{n1}, c_{p2}, c_{n2}\}}{\min\{c_{p1}, c_{n1}, c_{p2}, c_{n2}\}} \tag{7.138}$$

the following three lemmas ensure the boundedness of p_a and p_b.

Lemma 7.3. *When $\dot{x} = 0$ and*

$$p_s > 2\delta_c^2 \frac{2\gamma_p + s_b p_r}{s_a} \tag{7.139}$$

hold, it follows from (7.137) that

$$\dot{f}_{ps} > 0 \quad \text{if } f_{ps} \leqslant p_r s_b l_o + \gamma_p l_o \text{ and } u \neq 0 \tag{7.140}$$

$$\dot{f}_{ps} < 0 \quad \text{if } f_{ps} \geqslant p_s s_b l_o - \gamma_p l_o \text{ and } u \neq 0 \tag{7.141}$$

$$\dot{f}_{ps} = 0 \quad \text{if } u = 0 \tag{7.142}$$

are valid. ■

The proof is given in Appendix B.4.

Lemma 7.4. *When $\dot{x} \neq 0$ and*

$$p_s > 3\delta_c^2 \frac{2\gamma_p + s_b p_r}{s_a} + 2\left\{\frac{\left[\beta(s_b - s_a) + \gamma_p\right](1 + \alpha_s)}{|k_u(t)|\beta\min\{c_{p1}, c_{n1}, c_{p2}, c_{n2}\}}\right\}^2 \tag{7.143}$$

hold together with

$$\alpha_s > 0 \tag{7.144}$$

$$k_u(t) \stackrel{def}{=} \frac{\frac{\hat{s}_a}{l_o - x} + \frac{\hat{s}_b}{x}}{\zeta_u} \tag{7.145}$$

$$\begin{aligned}
\zeta_u \stackrel{def}{=} &\left[\hat{c}_{p1}\frac{\upsilon(p_s - p_a)}{l_o - x} + \hat{c}_{n2}\frac{\upsilon(p_b - p_r)}{x}\right]\varepsilon(u) \\
&+ \left[\hat{c}_{n1}\frac{\upsilon(p_a - p_r)}{l_o - x} + \hat{c}_{p2}\frac{\upsilon(p_s - p_b)}{x}\right]\varepsilon(-u)
\end{aligned} \tag{7.146}$$

it follows from (7.137) that

$$\dot{f}_{ps} > 0 \quad \text{if } f_{ps} \leqslant p_r s_b l_o + \gamma_p l_o \text{ and } |u| \geqslant \frac{1}{\alpha_s}|u_s| \tag{7.147}$$

$$\dot{f}_{ps} < 0 \quad \text{if } f_{ps} \geqslant p_s s_b l_o - \gamma_p l_o \text{ and } |u| \geqslant \frac{1}{\alpha_s}|u_s| \tag{7.148}$$

are valid with

$$u_s(t) \overset{def}{=} \frac{\widehat{\left(\frac{1}{\beta}\right)}\dot{f}_{pr} + k_{fp}(f_{pr} - f_p) + k_x(\dot{x}_r - \dot{x})}{\zeta_u}. \tag{7.149}$$

■

The proof is given in Appendix B.5.

Lemma 7.5. *The inequalities*

$$p_r < p_a < \frac{s_b}{s_a} p_s \tag{7.150}$$

$$p_r < p_b < p_s \tag{7.151}$$

hold under the inequality

$$p_r s_b l_o + \gamma_p l_o \leqslant f_{ps} \leqslant p_s s_b l_o - \gamma_p l_o. \tag{7.152}$$

Proof: Given $f_{ps} \geqslant p_r s_b l_o + \gamma_p l_o$, it follows from (7.131), (7.133), (7.134), and Assumption 5 that

$$\begin{aligned} p_a &= \frac{f_{ps} - f_p x}{s_a l_o} \\ &\geqslant \frac{p_r s_b l_o + \gamma_p l_o - f_p x}{s_a l_o} > \frac{p_r s_b l_o}{s_a l_o} > p_r \end{aligned} \tag{7.153}$$

$$\begin{aligned} p_b &= \frac{f_{ps} + f_p(l_o - x)}{s_b l_o} \\ &\geqslant \frac{p_r s_b l_o + \gamma_p l_o + f_p(l_o - x)}{s_b l_o} > \frac{p_r s_b l_o}{s_b l_o} = p_r \end{aligned} \tag{7.154}$$

hold.

Given $f_{ps} \leqslant p_s s_b l_o - \gamma_p l_o$, it follows from (7.131), (7.133), (7.134), and Assumption 5 that

$$\begin{aligned} p_a &= \frac{f_{ps} - f_p x}{s_a l_o} \\ &\leqslant \frac{p_s s_b l_o - \gamma_p l_o - f_p x}{s_a l_o} < \frac{s_b}{s_a} p_s \end{aligned} \tag{7.155}$$

$$\begin{aligned} p_b &= \frac{f_{ps} + f_p(l_o - x)}{s_b l_o} \\ &\leqslant \frac{p_s s_b l_o - \gamma_p l_o + f_p(l_o - x)}{s_b l_o} < \frac{p_s s_b l_o}{s_b l_o} = p_s \end{aligned} \tag{7.156}$$

hold. ■

Definition 7.1. *A compact interval, denoted as* $[a, b]$ *with* $a < b$*, is said to be attractive for continuous variable* $x(t) \in \mathbb{R}$*, if*

$$\dot{x}(t) \geqslant 0 \ \text{ for } x(t) \leqslant a \tag{7.157}$$

$$\dot{x}(t) \leqslant 0 \ \text{ for } x(t) \geqslant b. \tag{7.158}$$

Based on Definition 7.1, the following theorem ensures the internal stability of the hydraulic cylinder, characterized by the boundedness of both p_a and p_b.

Theorem 7.5. *With the validity of (7.131), the interval* $[(p_r s_b l_o + \gamma_p l_o), (p_s s_b l_o - \gamma_p l_o)]$ *is attractive for* f_{ps} *defined by (7.132), provided that (7.139) holds when* $\dot{x} = 0$ *or (7.143) and* $|u| \geqslant \frac{1}{\alpha_s}|u_s|$ *(with* u_s *being defined by (7.149)) hold when* $\dot{x} \neq 0$*. The boundedness of the pressures in the sense of (7.150) and (7.151) is guaranteed when* f_{ps} *moves into its attractive interval.*

Proof: In view of Definition 7.1 and Lemmas 7.3 and 7.4, the attractive interval for f_{ps} is $[(p_r s_b l_o + \gamma_p l_o), (p_s s_b l_o - \gamma_p l_o)]$. Then, the boundedness of p_a and p_b is guaranteed by Lemma 7.5. ■

Remark 7.10. In normal circumstances, the velocity related term dominates the valve control voltage such that $u_{fd} \approx \left(\frac{\hat{s}_a}{l_o - x} + \frac{\hat{s}_b}{x}\right)\dot{x}$ generally holds in (7.106), leading to a very small u_s in (7.149). An example in [222] gives $u_s \approx 0.001$ (V).

Finally, the satisfaction of the conditions described by (7.102) and (7.103) (which ensure the univalence between u and u_f) and by (7.110) and (7.111) (which make the valve control signal available) needs to be examined. Note that inequalities (7.150) and (7.151) ensure (7.103) and (7.111). The satisfaction of (7.102) and (7.110) is based on the following sufficient condition

$$0 < x < \frac{x_b}{1 + x_b} l_o \tag{7.159}$$

with

$$x_b \overset{def}{=} \frac{\underline{c}_{n2}}{\overline{c}_{p1}} \sqrt{\frac{s_a}{s_b - s_a} \frac{p_s - \frac{\gamma_p}{s_b} - p_r}{p_s}}$$

where $\underline{c}_{n2}$ denotes the lower bound of $\hat{c}_{n2}$ and $\overline{c}_{p1}$ denotes the upper bound of $\hat{c}_{p1}$.

7.10 Output Force Control

With respect to an application of the space system emulation [219], the dynamic equivalence between electrical actuations and hydraulic actuations within a frequency range of interest is required. A possible solution is to conduct the output force control of hydraulic cylinders by placing loadcell sensors at the output shafts [222]. Note the fact that for each hydraulic cylinder, the piston friction that determines the difference between the output force and the pressure-induced force can be significantly large (a few thousands of Newton) and somewhat uncertain. This fact gives rise to difficulties when conducting output force control by solely using the chamber pressure measurements [2, 104]. The experimental results in [222] suggested that a well-performed pressure-induced force control does not necessarily ensure a satisfactory output force control, whereas a satisfactory output force control has to be based on a well-performed pressure-induced force control. The experimental results in [222] also demonstrated that the online parameter adaptation with high gains plays a vital role to ensure an accurate output force tracking result that, otherwise, cannot be achieved by using a fixed-parameter-based friction compensation method.

7.11 Concluding Remarks

In this chapter, the control of hydraulic actuators with parameter uncertainties has been presented, within the framework of the VDC. Contents in this chapter can be generally divided into two parts. In the first part, the kinematics, dynamics, and control of a conventional hydraulic actuator assembly are presented and the *virtual stability* of each subsystem combined with its respective control equations is ensured. In the second part, the hydraulic fluid dynamics are addressed and the fluid control equations are presented. As a result, the *virtual stability* of the addressed hydraulic actuator assembly is guaranteed, and so is the boundedness of the chamber pressures under certain mild conditions.

The concept of the *virtual stability* allows the dynamics and control problem of a hydraulic actuator assembly to be handled separately from the dynamics and control problem of the rest of the hydraulic robot. When every subsystem of the robot is *virtually stable*, the L_2 and L_∞ stability of the entire robot can be guaranteed, as stated by Theorem 2.1 in Chapter 2.

8

Control of Coordinated Multiple Robot Manipulators

8.1 Introduction

The use of coordinated multiple robot manipulators to hold a common object is motivated by applications where the object to be transported is either too large or too heavy and, therefore, beyond the capacity of a single robot. When multiple robot manipulators holding a common object altogether, the number of the degrees of freedom (DOFs) of motion is often less than the number of total actuators. This over-actuation gives rise to the creation of the internal forces. Therefore, not only the motion but also the internal forces resulting from the over-actuation need to be controlled.

By using the *virtual decomposition control* (VDC) approach, a coordinated multiple-manipulator system can be virtually decomposed into the held object and several individual manipulators. The concept of the *virtual stability* allows the control problem of the original system to be converted to the motion and force control problem of the held object and the motion control problem of each individual manipulator. Consider the fact that the dynamics and control issues on manipulators with different structures and with different types of actuators have been precisely studied in Chapters 5–7. The attention in this chapter, therefore, will be focused on the held object with an emphasis on its motion control and internal force control.

8.2 Prior Work

The use of multiple robot manipulators to perform coordinated tasks in transportations and assembly lines can be traced back to the mid 1980s. Since then, there has been a tremendous number of publications in this particular area. Some pioneer work includes the master-slave control [9, 111, 208], nonlinear feedback linearization control [169], hybrid position/force control [40, 68, 175], and payload focused adaptive control [180]. During 1990s, the research on the control of coordinated multiple robot manipulators reached its mature stage by

W.-H. Zhu: Virtual Decomposition Control, STAR 60, pp. 193–219.
springerlink.com

producing a variety of approaches ranging from nonlinear feedback linearization control [93, 169, 186, 188, 204], internal force based optimization [26, 73, 74], variable structure control [193], hybrid control [201], decentralized control [75], impedance control [14, 91, 144], to adaptive control [74, 80].

Specifically, the so-called leader/follower type of control approaches dealing with two industrial robots moving an object and performing assembly tasks was studied in [111, 167]. In these approaches, the desired motion of the leader robot is first planned, and then the follower robot is planned according to the constraints between them. The concept of the nonlinear feedback was applied to the multiple-manipulator systems aimed at converting their nonlinear models to linear models [169, 201, 204]. In [73], the control problems of the coordinated multiple redundant robots were addressed. Both local and global optimizations with respect to the joint redundancy were developed and the robust analysis was carried out. In [93, 177], the joint control space is partitioned into a motion subspace and an internal force subspace, and then a decoupled control law was designed so that the motion control and the internal force control can be performed independently. The concept of the "virtual stick" was introduced in [40, 175], and the relationship between the internal force and a small deformation of the object was established. This technique led to a symmetric hybrid position/force control scheme in which the control directions can be selected either as position control or as force control. A hierarchical control strategy with modular structure was proposed in [75]. This approach is able to control both the motion and the internal force simultaneously. In [186], a unified perspective approach was presented with respect to the motion/force control problem and to the stability analysis, which however is only limited to the set-point control. An important statement that the number of lost DOFs in a closed chain system is just equal to the number of dimensions of the internal force space was made in [96]. In [180], the relationship between the load distribution and the internal force was explored. A simple form of the pseudo-inverse matrix was also given. The concept of the *manipulability* was proposed in [125] for the multi-finger systems. Nonlinear programming was utilized to obtain the optimal contact forces. The friction constraints at the finger tips were considered and the so-called "friction cone" was introduced to enhance the "contact stability." An interesting joint-space PD control scheme was proposed in [20] based on the so-called "kinetostatic filtering." Meanwhile, readers who are interested in the control of under-actuated coordinated multiple robot manipulators are suggested to refer to [106].

Many industrial applications usually involve tasks in which the manipulated object is in contact with the environment [68, 191, 193, 205]. Typical applications include contour following, grinding, scribing, writing, and deburring. In these cases, the motion of the object is constrained so that both the motion and the constraint force need to be controlled. In [144], an object-oriented impedance control scheme was proposed for a cooperative dual robot manipulator system. The grasped object was isolated from the two manipulators which serve as the "actuators" so that the impedance control design can be performed with respect to the grasped object. Meanwhile, two invariant hybrid position/force control designs were developed in [1, 47].

There are two main characteristics associated with the use of the coordinated manipulator systems. The first characteristic is the trade-off between the payload capability and the operational space. When multiple robot manipulators hold a common object, the overall payload capability is increased, while the effective operational space with which all manipulators are singularity-free is reduced [30]. In other words, the payload capability is characterized by the "sets of union," while the operational space is characterized by the "sets of intersection." The second characteristic is the over-actuation, that is, the number of total actuators is larger than the number of the DOFs of motion. This characteristic results in the so-called internal forces which are defined as the forces that do not affect the object motion [186]. This fact allows the motion control to be conducted independently from the internal force control. Therefore, the motion/internal force control problem of the coordinated multiple robot manipulators is similar to the motion/constraint force control problem of the single-arm constrained robot manipulators, but the dimensions. Unlike the constraint forces that have to be controlled to desired values for operational purposes, the internal forces can be designed to satisfy a variety of control objectives ranging from load distributions [125, 180] to control torque optimizations [73].

8.3 Virtual Decomposition

A typical coordinated multiple manipulator system comprised of $n > 1$ manipulators rigidly holding a common object, as illustrated in Figure 8.1, is being studied in this chapter.

To make the *virtual decomposition control* applicable, n *cutting points* are placed between the n end-effectors of the manipulators and the held object. Thus, the original system is virtually decomposed into one held object and n individual manipulators, which results in one *object* and n *open chains* as expressed by a simple oriented graph in Figure 8.2.

Since the motion control problem of each manipulator has been well studied in Chapters 5–7, this chapter will particularly focus on the motion and internal/constraint force control problem of the held object.

There are totally $n + 2$ frames attached to the held object, see Figure 8.1. Frame $\{\mathbf{O}\}$ is used for the motion and force control specifications. Frame $\{\mathbf{S}\}$ is located at a possible contact point with the environment. Frame $\{\mathbf{T}_j\}$ is located at the jth *cutting point* for all $j \in \{1, n\}$.

Note the fact that the held object is an *object* without *driving cutting points*. Therefore, the *object* (held object) has n *driven cutting points* associated with frame $\{\mathbf{T}_j\}$ for all $j \in \{1, n\}$.

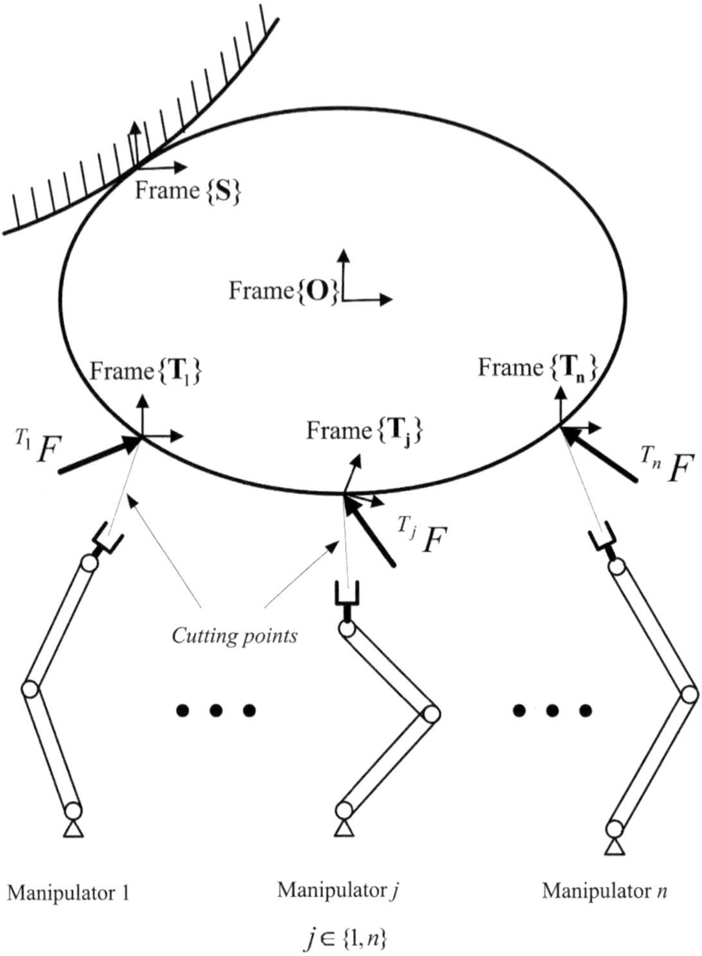

Fig. 8.1. Coordinated multiple manipulator system.

8.4 Kinematics and Dynamics of the Held Object in Free Motion

With the same structure as that of Chapter 5, the held object in the free motion is studied first, followed by the held object in the constrained motion as will be presented in Section 8.9.

8.4.1 Kinematics of the Held Object

Since the held object is in the free motion, the body frame referenced linear/angular velocity vector of frame $\{\mathbf{O}\}$ can be expressed by

$$^{\mathbf{O}}V = \begin{bmatrix} ^{\mathbf{O}}\mathbf{R}_{\mathbf{I}}\dot{\mathbf{x}} \\ ^{\mathbf{O}}\mathbf{R}_{\mathbf{I}}\omega \end{bmatrix} \in \mathbb{R}^6 \tag{8.1}$$

where $\mathbf{x} \in \mathbb{R}^3$ denotes the position vector of the origin of frame $\{\mathbf{O}\}$ expressed in the inertial frame and $\boldsymbol{\omega} \in \mathbb{R}^3$ denotes the angular velocity vector of frame $\{\mathbf{O}\}$ expressed in the inertial frame.

Note the fact that frames $\{\mathbf{O}\}$ and $\{\mathbf{T}_j\}$ for all $j \in \{1, n\}$ are attached to the same rigid body. It follows that

$$\mathcal{V}_T \overset{def}{=} \begin{bmatrix} {}^{\mathbf{T}_1}V \\ \vdots \\ {}^{\mathbf{T}_j}V \\ \vdots \\ {}^{\mathbf{T}_n}V \end{bmatrix} = {}^{\mathbf{O}}\mathcal{U}_{\mathbf{T}}^T \mathcal{H}^T {}^{\mathbf{O}}V \tag{8.2}$$

holds with

$$\begin{aligned} {}^{\mathbf{O}}\mathcal{U}_{\mathbf{T}} &= \operatorname{diag}({}^{\mathbf{O}}\mathbf{U}_{\mathbf{T}_1}, \cdots, {}^{\mathbf{O}}\mathbf{U}_{\mathbf{T}_j}, \cdots, {}^{\mathbf{O}}\mathbf{U}_{\mathbf{T}_n}) \in \mathbb{R}^{6n \times 6n} && (8.3) \\ \mathcal{H} &\overset{def}{=} \underbrace{[\mathbf{I}_6, \cdots, \mathbf{I}_6, \cdots, \mathbf{I}_6]}_{n} \in \mathbb{R}^{6 \times 6n} && (8.4) \end{aligned}$$

being constant. Thus, it follows from (8.2) that

$$\frac{d}{dt}\left({}^{\mathbf{T}_j}V\right) = {}^{\mathbf{O}}\mathbf{U}_{\mathbf{T}_j}^T \frac{d}{dt}\left({}^{\mathbf{O}}V\right) \tag{8.5}$$

holds for all $j \in \{1, n\}$.

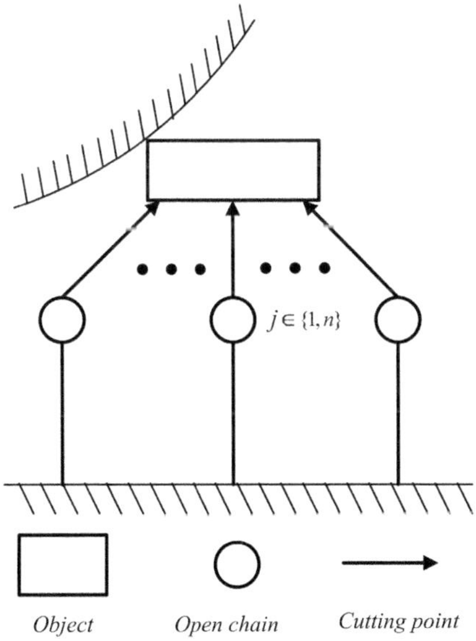

Fig. 8.2. Graphic expression of Figure 8.1.

8.4.2 Dynamics of the Held Object

The held object is a special *object* defined in Chapter 4 with

$$\mathcal{B}_i = \emptyset \tag{8.6}$$

$$\mathcal{T}_i = \{\mathbf{T}_1, \cdots, \mathbf{T}_n\}. \tag{8.7}$$

Recall (2.74) to obtain the dynamics of the held object as

$$\mathbf{M_O}\frac{d}{dt}({}^{\mathbf{O}}V) + \mathbf{C_O}({}^{\mathbf{O}}\omega){}^{\mathbf{O}}V + \mathbf{G_O} = {}^{\mathbf{O}}F^* \tag{8.8}$$

where the two matrices $\mathbf{M_O} \in \mathbb{R}^{6\times 6}$ and $\mathbf{C_O} \in \mathbb{R}^{6\times 6}$ and the one vector $\mathbf{G_O} \in \mathbb{R}^6$ are defined by (2.75)–(2.77) with frame $\{\mathbf{O}\}$ being substituted for frame $\{\mathbf{A}\}$.

The net force/moment vector ${}^{\mathbf{O}}F^* \in \mathbb{R}^6$ is governed by

$${}^{\mathbf{O}}F^* = \sum_{j=1}^{n} {}^{\mathbf{O}}\mathbf{U}_{\mathbf{T}_j}\, {}^{\mathbf{T}_j}F \tag{8.9}$$

where ${}^{\mathbf{O}}\mathbf{U}_{\mathbf{T}_j} \in \mathbb{R}^{6\times 6}$ appeared in (8.3) denotes a force/moment transformation matrix and ${}^{\mathbf{T}_j}F \in \mathbb{R}^6$ (defined by (2.62) with frame $\{\mathbf{T}_j\}$ being substituted for frame $\{\mathbf{A}\}$) denotes the force/moment vector in frame $\{\mathbf{T}_j\}$ at the jth *cutting point* for all $j \in \{1, n\}$.

By introducing an internal force vector $\boldsymbol{\eta} \in \mathbb{R}^{6(n-1)}$, the force/moment vectors at the n *cutting points* can be resolved from (8.9) as

$$\mathcal{F}_T \overset{def}{=} \begin{bmatrix} {}^{\mathbf{T}_1}F \\ \vdots \\ {}^{\mathbf{T}_j}F \\ \vdots \\ {}^{\mathbf{T}_n}F \end{bmatrix} = {}^{\mathbf{T}}\mathcal{U}_{\mathbf{O}} \begin{bmatrix} \boldsymbol{\Omega}_m & \boldsymbol{\Omega}_f \end{bmatrix} \begin{bmatrix} {}^{\mathbf{O}}F^* \\ \boldsymbol{\eta} \end{bmatrix} \tag{8.10}$$

where ${}^{\mathbf{T}}\mathcal{U}_{\mathbf{O}} = {}^{\mathbf{O}}\mathcal{U}_{\mathbf{T}}^{-1}$ is defined by (8.3) and matrices $\boldsymbol{\Omega}_m \in \mathbb{R}^{6n\times 6}$ and $\boldsymbol{\Omega}_f \in \mathbb{R}^{6n\times(6n-6)}$ are governed by

$$\mathcal{H}\boldsymbol{\Omega}_m = \mathbf{I}_6 \tag{8.11}$$

$$\mathcal{H}\boldsymbol{\Omega}_f = \mathbf{0} \tag{8.12}$$

with $\mathcal{H} \in \mathbb{R}^{6\times 6n}$ being defined by (8.4).

Remark 8.1. Matrix $\boldsymbol{\Omega}_m$ determines the load distribution strategy, and matrix $\boldsymbol{\Omega}_f$ determines the physical meaning of the independent internal force coordinates.

A general expression of $\boldsymbol{\Omega}_m \in \mathbb{R}^{6n\times 6}$ is

$$\boldsymbol{\Omega}_m = [\gamma_1(t)\mathbf{I}_6, \cdots, \gamma_j(t)\mathbf{I}_6, \cdots, \gamma_n(t)\mathbf{I}_6]^T \in \mathbb{R}^{6n\times 6} \tag{8.13}$$

subject to

$$\sum_{j=1}^{n} \gamma_j(t) = 1. \tag{8.14}$$

The scalar $\gamma_j(t)$ determines the portion of the payload (the held object) to be shared by the jth manipulators. Equal load distribution in [30] leads to

$$\gamma_j(t) = \frac{1}{n}, \quad \forall j \in \{1, n\}. \tag{8.15}$$

A possible way to define the internal forces is to let the internal forces characterize the squeeze force/moment vectors among the n end-effectors. As in [223], matrix $\mathbf{\Omega}_f$ is specified as

$$\mathbf{\Omega}_f = \begin{bmatrix} \mathbf{I}_6 & 0 & \cdots & \cdots & 0 \\ -\mathbf{I}_6 & \mathbf{I}_6 & \ddots & & \vdots \\ 0 & -\mathbf{I}_6 & \ddots & \ddots & \vdots \\ \vdots & \ddots & \ddots & \ddots & 0 \\ \vdots & & \ddots & \ddots & \mathbf{I}_6 \\ 0 & \cdots & \cdots & 0 & -\mathbf{I}_6 \end{bmatrix} \in \mathbb{R}^{6n \times (6n-6)}$$

with $\boldsymbol{\eta} \in \mathbb{R}^{6(n-1)}$ being defined by

$$\boldsymbol{\eta} = [\boldsymbol{\eta}_{1,2}^T, \boldsymbol{\eta}_{2,3}^T, \cdots, \boldsymbol{\eta}_{n-1,n}^T]^T \in \mathbb{R}^{6(n-1)}$$

where $\boldsymbol{\eta}_{j,j+1} \in \mathbb{R}^6$, $j \in \{1, n-1\}$, denotes the internal force/moment vector exerted from the $(j+1)$th end-effector to the jth end-effector, expressed in frame $\{\mathbf{O}\}$.

Note that matrix $\begin{bmatrix} \mathbf{\Omega}_m & \mathbf{\Omega}_f \end{bmatrix} \in \mathbb{R}^{6n \times 6n}$ is of full rank. There must exist a matrix $\mathbf{\Phi} \in \mathbb{R}^{6(n-1) \times 6n}$ such that

$$\begin{bmatrix} \mathcal{H} \\ \mathbf{\Phi} \end{bmatrix} = \begin{bmatrix} \mathbf{\Omega}_m & \mathbf{\Omega}_f \end{bmatrix}^{-1} \tag{8.16}$$

holds. Thus, the independent internal force coordinates can be obtained from (8.10) by using the force/moment measurements at the n end-effectors as

$$\boldsymbol{\eta} = \mathbf{\Phi}\, {}^{\mathbf{O}}\mathcal{U}_{\mathbf{T}} \mathcal{F}_T. \tag{8.17}$$

8.5 Forward Dynamics

In this section, equations that relate the control forces/torques of the n manipulators to the independent acceleration coordinates of motion and to the internal forces are to be developed.

Assume that each manipulator is a six-joint manipulator as studied in Chapter 5. The dynamics of the jth manipulator can be written as

$$\mathcal{M}_j \frac{d}{dt}\left({}^{\mathbf{T}_j}V\right) + \mathcal{C}_j {}^{\mathbf{T}_j}V + \mathcal{G}_j = \mathbf{J}_j^{-T}\boldsymbol{\tau}_j - {}^{\mathbf{T}_j}F \tag{8.18}$$

for all $j \in \{1, n\}$. This equation is obtained from (5.79) in Subsection 5.8 by substituting $\mathcal{M}_j$, $\mathcal{C}_j$, $\mathcal{G}_j$, ${}^{\mathbf{T}_j}V$, $\mathbf{J}_j^{-T}$, and $\boldsymbol{\tau}_j$ for $\mathcal{M}$, $\mathcal{C}$, $\mathcal{G}$, $\mathcal{V}$, $\mathbf{J}_q^{-T}$, and $\boldsymbol{\tau}$, respectively, and by adding the end-effector exerting force/moment vector ${}^{\mathbf{T}_j}F$. In (8.18), $\mathbf{J}_j \in \mathbb{R}^{6\times 6}$ is a Jacobian matrix that maps the joint velocity coordinates of the jth manipulator to ${}^{\mathbf{T}_j}V \in \mathbb{R}^6$. The two matrices $\mathcal{M}_j$ and $\mathcal{C}_j$ and the vector $\mathcal{G}_j$ have the same expressions as $\mathcal{M}$, $\mathcal{C}$, and $\mathcal{G}$ in (5.80)–(5.87) subject to $\mathbf{M_O} = \mathbf{0}$ in (5.84), $\mathbf{C_O} = \mathbf{0}$ in (5.85), and $\mathbf{G_O} = \mathbf{0}$ in (5.87).

In view of (8.2), (8.5), (8.8), (8.9), and (8.18), the forward dynamics of the entire system can be written as

$$\mathcal{M}_m \frac{d}{dt}\left({}^{\mathbf{O}}V\right) + \mathcal{C}_m {}^{\mathbf{O}}V + \mathcal{G}_m = \sum_{j=1}^{n} {}^{\mathbf{O}}\mathbf{U}_{\mathbf{T}_j}\mathbf{J}_j^{-T}\boldsymbol{\tau}_j \tag{8.19}$$

where

$$\mathcal{M}_m = \mathbf{M_O} + \sum_{j=1}^{n} {}^{\mathbf{O}}\mathbf{U}_{\mathbf{T}_j}\mathcal{M}_j {}^{\mathbf{O}}\mathbf{U}_{\mathbf{T}_j}^T \tag{8.20}$$

$$\mathcal{C}_m = \mathbf{C_O} + \sum_{j=1}^{n} {}^{\mathbf{O}}\mathbf{U}_{\mathbf{T}_j}\mathcal{C}_j {}^{\mathbf{O}}\mathbf{U}_{\mathbf{T}_j}^T \tag{8.21}$$

$$\mathcal{G}_m = \mathbf{G_O} + \sum_{j=1}^{n} {}^{\mathbf{O}}\mathbf{U}_{\mathbf{T}_j}\mathcal{G}_j. \tag{8.22}$$

Equation (8.19) can be re-expressed as

$$\mathcal{M}_o \frac{d}{dt}\left({}^{\mathbf{O}}V\right) + \varepsilon_o = \mathcal{J}_o\boldsymbol{\tau} \tag{8.23}$$

with

$$\mathcal{M}_o = \mathcal{M}_m \tag{8.24}$$

$$\mathcal{J}_o = \left[{}^{\mathbf{O}}\mathbf{U}_{\mathbf{T}_1}\mathbf{J}_1^{-T}, \cdots, {}^{\mathbf{O}}\mathbf{U}_{\mathbf{T}_j}\mathbf{J}_j^{-T}, \cdots, {}^{\mathbf{O}}\mathbf{U}_{\mathbf{T}_n}\mathbf{J}_n^{-T}\right] \tag{8.25}$$

$$\boldsymbol{\tau} = \left[\boldsymbol{\tau}_1^T, \cdots, \boldsymbol{\tau}_j^T, \cdots, \boldsymbol{\tau}_n^T\right]^T \tag{8.26}$$

$$\varepsilon_o = \mathcal{C}_m {}^{\mathbf{O}}V + \mathcal{G}_m. \tag{8.27}$$

Furthermore, in view of (8.3), (8.5), (8.10), (8.17), and (8.18), the internal force vector can be written as

$$\begin{aligned}\boldsymbol{\eta} = {} & \boldsymbol{\Phi}\mathrm{col}_{j=1}^n\left({}^{\mathbf{O}}\mathbf{U}_{\mathbf{T}_j}\mathbf{J}_j^{-T}\boldsymbol{\tau}_j\right) - \boldsymbol{\Phi}\mathrm{col}_{j=1}^n\left({}^{\mathbf{O}}\mathbf{U}_{\mathbf{T}_j}\mathcal{M}_j {}^{\mathbf{O}}\mathbf{U}_{\mathbf{T}_j}^T\right)\frac{d}{dt}\left({}^{\mathbf{O}}V\right) \\ & - \boldsymbol{\Phi}\mathrm{col}_{j=1}^n\left({}^{\mathbf{O}}\mathbf{U}_{\mathbf{T}_j}\mathcal{C}_j {}^{\mathbf{O}}\mathbf{U}_{\mathbf{T}_j}^T\right){}^{\mathbf{O}}V - \boldsymbol{\Phi}\mathrm{col}_{j=1}^n\left({}^{\mathbf{O}}\mathbf{U}_{\mathbf{T}_j}\mathcal{G}_j\right).\end{aligned} \tag{8.28}$$

Equation (8.28) can be re-expressed as

$$\boldsymbol{\eta} + \mathcal{M}_\eta \frac{d}{dt}\left({}^{\mathbf{O}}V\right) + \boldsymbol{\varepsilon}_\eta = \mathcal{J}_\eta \boldsymbol{\tau} \tag{8.29}$$

where

$$\mathcal{M}_\eta = \boldsymbol{\Phi}\mathrm{col}_{j=1}^{n}\left({}^{\mathbf{O}}\mathbf{U}_{\mathbf{T}_j}\mathcal{M}_j{}^{\mathbf{O}}\mathbf{U}_{\mathbf{T}_j}^T\right) \tag{8.30}$$

$$\mathcal{J}_\eta = \boldsymbol{\Phi}\mathrm{diag}\left({}^{\mathbf{O}}\mathbf{U}_{\mathbf{T}_1}\mathbf{J}_1^{-T},\cdots,{}^{\mathbf{O}}\mathbf{U}_{\mathbf{T}_j}\mathbf{J}_j^{-T},\cdots,{}^{\mathbf{O}}\mathbf{U}_{\mathbf{T}_n}\mathbf{J}_n^{-T}\right) \tag{8.31}$$

$$\boldsymbol{\varepsilon}_\eta = \boldsymbol{\Phi}\mathrm{col}_{j=1}^{n}\left({}^{\mathbf{O}}\mathbf{U}_{\mathbf{T}_j}\mathcal{C}_j{}^{\mathbf{O}}\mathbf{U}_{\mathbf{T}_j}^T\right){}^{\mathbf{O}}V + \boldsymbol{\Phi}\mathrm{col}_{j=1}^{n}\left({}^{\mathbf{O}}\mathbf{U}_{\mathbf{T}_j}\mathcal{G}_j\right). \tag{8.32}$$

Equation (8.19) or (8.23) defines the forward dynamics that yield $\frac{d}{dt}\left({}^{\mathbf{O}}V\right) \in \mathbb{R}^6$ from $\boldsymbol{\tau} \in \mathbb{R}^{6n}$. Meanwhile, (8.28) or (8.29) yields the internal force vector $\boldsymbol{\eta} \in \mathbb{R}^{6(n-1)}$ from both $\frac{d}{dt}\left({}^{\mathbf{O}}V\right) \in \mathbb{R}^6$ and $\boldsymbol{\tau} \in \mathbb{R}^{6n}$.

8.6 Motion and Internal Force Control in Free Motion

After giving the motion and internal force control specifications, this section presents the design of the required velocity vectors into which both the position and orientation errors and the filtered internal force errors are incorporated to accomplish both the position and orientation control and the filtered internal force control based on the velocity convergences. Then, the control equations for the held object are presented. This section ends with a discussion on the internal force optimization.

8.6.1 Motion and Internal Force Control Specifications

The following assumption is applied throughout this chapter.

Assumption 6. *Each manipulator allows its end-effector to possess six degrees of freedom of motion without encountering kinematics singularity. The intersection of the singularity-free operation spaces of the n end-effectors, denoted as $\mathcal{W}$, is not empty.*

Based on Assumption 6, the held object can be freely controlled in $\mathcal{W}$. Define a frame $\{\mathbf{D}\}$, whose origin and orientation represent the desired position and orientation of frame $\{\mathbf{O}\}$. The motion control objective is to make frame $\{\mathbf{O}\}$ track $\{\mathbf{D}\}$.

Meanwhile, let $\boldsymbol{\eta}_d \in \mathbb{R}^{6n-6}$ be the desired internal force coordinate vector. The internal force control objective is

$$\boldsymbol{\eta}_d - \boldsymbol{\eta} \rightarrow \mathbf{0}. \tag{8.33}$$

8.6.2 Required Velocity Vectors

Similar to (5.13), the required velocity vector of frame $\{\mathbf{O}\}$ is designed as

$$^{\mathbf{O}}V_r = \mathcal{V}_{pd} + \mathcal{E} \tag{8.34}$$

with

$$\mathcal{V}_{pd} = \begin{bmatrix} ^{\mathbf{O}}\mathbf{R}_{\mathbf{I}}\dot{\mathbf{x}}_d \\ ^{\mathbf{O}}\mathbf{R}_{\mathbf{I}}\boldsymbol{\omega}_d \end{bmatrix} \in \mathbb{R}^6 \tag{8.35}$$

$$\mathcal{E} = \begin{bmatrix} k_x {}^{\mathbf{O}}\mathbf{R}_{\mathbf{I}} (\mathbf{x}_d - \mathbf{x}) \\ k_\omega {}^{\mathbf{O}}\boldsymbol{\mu} \end{bmatrix} \tag{8.36}$$

where $k_x > 0$ and $k_\omega > 0$ are two positive constants, $\mathbf{x}_d \in \mathbb{R}^3$ denotes the desired position (the origin of frame $\{\mathbf{D}\}$) vector, $\boldsymbol{\omega}_d \in \mathbb{R}^3$ denotes the desired angular velocity vector, and $^{\mathbf{O}}\boldsymbol{\mu} \in SO^3$ denotes the vector part of a quaternion representing the rotation from frame $\{\mathbf{O}\}$ to frame $\{\mathbf{D}\}$. It can be verified that (4.176) and (4.177) are satisfied by equalizing $\mathcal{V}_p = {}^{\mathbf{O}}V$ and using Lemmas 4.4 and 4.5.

According to (8.2), the required linear/angular velocity vectors of the n end-effectors are designed as

$$\mathcal{V}_{Tr} \overset{def}{=} \begin{bmatrix} ^{\mathbf{T}_1}V_r \\ \vdots \\ ^{\mathbf{T}_j}V_r \\ \vdots \\ ^{\mathbf{T}_n}V_r \end{bmatrix} = {}^{\mathbf{O}}\mathcal{U}_{\mathbf{T}}^T \left[\mathcal{H}^T {}^{\mathbf{O}}V_r + \boldsymbol{\Phi}^T \mathcal{A}_{f\eta} \left(\tilde{\boldsymbol{\eta}}_d - \tilde{\boldsymbol{\eta}} \right) \right] \tag{8.37}$$

where $\mathcal{H} \in \mathbb{R}^{6\times 6n}$ and $\boldsymbol{\Phi} \in \mathbb{R}^{6(n-1)\times 6n}$ are defined in (8.16), $\mathcal{A}_{f\eta} \in \mathbb{R}^{6(n-1)\times 6(n-1)}$ is a diagonal positive-definite matrix, and $\tilde{\boldsymbol{\eta}}_d$ and $\tilde{\boldsymbol{\eta}}$ are subject to

$$\dot{\tilde{\boldsymbol{\eta}}}_d = \mathcal{C}_{f\eta} \left(\boldsymbol{\eta}_d - \tilde{\boldsymbol{\eta}}_d \right) \tag{8.38}$$

$$\dot{\tilde{\boldsymbol{\eta}}} = \mathcal{C}_{f\eta} \left(\boldsymbol{\eta} - \tilde{\boldsymbol{\eta}} \right) \tag{8.39}$$

with $\mathcal{C}_{f\eta} \in \mathbb{R}^{6(n-1)\times 6(n-1)}$ being a diagonal positive-definite matrix.

Remark 8.2. Note that the second term in the right hand side of (8.37) contains the filtered internal force error vector that is added to the complementary motion configuration space spanned by $\boldsymbol{\Phi}^T$. This term is used to achieve the filtered internal force control characterized by $\tilde{\boldsymbol{\eta}}_d - \tilde{\boldsymbol{\eta}} \to \mathbf{0}$.

Remark 8.3. Equations (8.38) and (8.39) define two low-pass filters to be used to filter out the desired and actual internal forces, respectively. The cut-off frequency of the two filters is determined by matrix $\mathcal{C}_{f\eta}$.

8.6.3 Control of the Held Object

Recall (4.32) by omitting the subscript i. The required net force/moment vector of the held object is written as

$$ {}^{\mathbf{O}}F_r^* = \mathbf{Y}_{\mathbf{O}}\hat{\boldsymbol{\theta}}_{\mathbf{O}} + \mathbf{K}_{\mathbf{O}}\left({}^{\mathbf{O}}V_r - {}^{\mathbf{O}}V\right) \tag{8.40} $$

where $\mathbf{K}_{\mathbf{O}} \in \mathbb{R}^{6\times 6}$ is a positive-definite gain matrix and $\hat{\boldsymbol{\theta}}_{\mathbf{O}} \in \mathbb{R}^{13}$ denotes the estimate of $\boldsymbol{\theta}_{\mathbf{O}} \in \mathbb{R}^{13}$ with $\mathbf{Y}_{\mathbf{O}}\boldsymbol{\theta}_{\mathbf{O}} \in \mathbb{R}^6$ being defined by (2.78) and further expressed in Appendix A by substituting frame $\{\mathbf{O}\}$ for frame $\{\mathbf{A}\}$.

The parameter estimate vector $\hat{\boldsymbol{\theta}}_{\mathbf{O}} \in \mathbb{R}^{13}$ needs to be updated. Define

$$ \mathbf{s}_{\mathbf{O}} = \mathbf{Y}_{\mathbf{O}}^T\left({}^{\mathbf{O}}V_r - {}^{\mathbf{O}}V\right). \tag{8.41} $$

Each of the 13 elements of $\hat{\boldsymbol{\theta}}_{\mathbf{O}} \in \mathbb{R}^{13}$ is updated by using the $\mathcal{P}$ function defined by (2.79) as

$$ \hat{\theta}_{\mathbf{O}\gamma} = \mathcal{P}\left(s_{\mathbf{O}\gamma}, \rho_{\mathbf{O}\gamma}, \underline{\theta}_{\mathbf{O}\gamma}, \overline{\theta}_{\mathbf{O}\gamma}, t\right) \tag{8.42} $$

for all $\gamma \in \{1, 13\}$, where

$s_{\mathbf{O}\gamma}$	γth element of $\mathbf{s}_{\mathbf{O}}$.
$\rho_{\mathbf{O}\gamma}$	Parameter update gain.
$\underline{\theta}_{\mathbf{O}\gamma}$	Lower bound of $\theta_{\mathbf{O}\gamma}$.
$\overline{\theta}_{\mathbf{O}\gamma}$	Upper bound of $\theta_{\mathbf{O}\gamma}$.
$\theta_{\mathbf{O}\gamma}$	γth element of $\boldsymbol{\theta}_{\mathbf{O}}$.

8.6.4 Required Force/Moment Vectors

After ${}^{\mathbf{O}}F_r^* \in \mathbb{R}^6$ being obtained from (8.40) and $\boldsymbol{\eta}_d \in \mathbb{R}^{6(n-1)}$ being obtained from the internal force control specification, the required force/moment vectors at the n end-effectors are designed from (8.10) as

$$ \mathcal{F}_{Tr} \overset{def}{=} \begin{bmatrix} {}^{\mathbf{T}_1}F_r \\ \vdots \\ {}^{\mathbf{T}_j}F_r \\ \vdots \\ {}^{\mathbf{T}_n}F_r \end{bmatrix} = {}^{\mathbf{T}}\mathcal{U}_{\mathbf{O}}\left[\boldsymbol{\Omega}_m \;\; \boldsymbol{\Omega}_f\right]\begin{bmatrix} {}^{\mathbf{O}}F_r^* \\ \boldsymbol{\eta}_d \end{bmatrix} \tag{8.43} $$

with ${}^{\mathbf{T}}\mathcal{U}_{\mathbf{O}} = {}^{\mathbf{O}}\mathcal{U}_{\mathbf{T}}^{-1} \in \mathbb{R}^{6n\times 6n}$ being defined by (8.3).

8.6.5 Internal Force Optimization

The desired internal force coordinate vector $\boldsymbol{\eta}_d \in \mathbb{R}^{6(n-1)}$ can be specified in different ways. Most commonly, it can be used to minimize the contact forces/moments at the n end-effectors, that is, find $\boldsymbol{\eta}_d$ to minimize $\mathcal{F}_T^T\mathbf{W}_\eta\mathcal{F}_T$

under (8.33), where $\mathbf{W}_\eta$ is a weighting matrix and $\mathcal{F}_T$ is being defined by (8.10). For this purpose, re-write (8.10) as

$$\mathcal{F}_T = \mathcal{A}_\eta \boldsymbol{\eta} + \mathcal{B}_\eta \tag{8.44}$$

with

$$\mathcal{A}_\eta = {}^{\mathbf{T}}\mathcal{U}_{\mathbf{O}} \boldsymbol{\Omega}_f \tag{8.45}$$
$$\mathcal{B}_\eta = {}^{\mathbf{T}}\mathcal{U}_{\mathbf{O}} \boldsymbol{\Omega}_m {}^{\mathbf{O}}F^*. \tag{8.46}$$

The least-square solution for $\boldsymbol{\eta}$ in the sense of

$$\min \left(\mathcal{F}_T^T \mathbf{W}_\eta \mathcal{F}_T \right) \tag{8.47}$$

is

$$\boldsymbol{\eta} = - \left(\mathcal{A}_\eta^T \mathbf{W}_\eta \mathcal{A}_\eta \right)^{-1} \mathcal{A}_\eta^T \mathbf{W}_\eta \mathcal{B}_\eta. \tag{8.48}$$

In line with (8.48), the desired internal force coordinate vector $\boldsymbol{\eta}_d$ is designed as

$$\boldsymbol{\eta}_d = - \left(\mathcal{A}_\eta^T \mathbf{W}_\eta \mathcal{A}_\eta \right)^{-1} \mathcal{A}_\eta^T \mathbf{W}_\eta \mathcal{B}_{\eta r} \tag{8.49}$$

with

$$\mathcal{B}_{\eta r} \stackrel{def}{=} {}^{\mathbf{T}}\mathcal{U}_{\mathbf{O}} \boldsymbol{\Omega}_m {}^{\mathbf{O}}F_r^* \tag{8.50}$$

where ${}^{\mathbf{O}}F_r^* \in \mathbb{R}^6$ is obtained from (8.40).

Remark 8.4. Note that the *zero internal load* in [180] does not correspond to $\boldsymbol{\eta} = \mathbf{0}$. Instead, it corresponds to a set of special internal force values obtained through a similar optimization process aimed at minimizing the internal load.

8.7 Virtual Stability

In this section, it will be shown that the held object combined with its respective control equations qualifies to be *virtually stable* in the sense of Definition 2.17.

Theorem 8.1. *The held object described by (8.2), (8.8), (8.10), and (8.16), combined with its respective control equations (8.37)–(8.40) and (8.43) and with the parameter adaptation (8.41) and (8.42), is virtually stable with its affiliated vectors* ${}^{\mathbf{O}}V_r - {}^{\mathbf{O}}V$ *and* $\tilde{\boldsymbol{\eta}}_d - \tilde{\boldsymbol{\eta}}$ *being virtual functions in both* L_2 *and* L_∞*, in the sense of Definition 2.17.*

Proof: Assign the held object the following non-negative accompanying function

$$\nu = \nu_{\mathbf{O}} + \nu_{\boldsymbol{\eta}} \tag{8.51}$$

with

$$\nu_{\mathrm{O}} = \frac{1}{2}\left({}^{\mathrm{O}}V_r - {}^{\mathrm{O}}V\right)^T \mathbf{M}_{\mathrm{O}} \left({}^{\mathrm{O}}V_r - {}^{\mathrm{O}}V\right)$$

$$+\frac{1}{2}\sum_{\gamma=1}^{13}\left(\theta_{\mathrm{O}\gamma} - \hat{\theta}_{\mathrm{O}\gamma}\right)^2 / \rho_{\mathrm{O}\gamma} \tag{8.52}$$

$$\nu_{\boldsymbol{\eta}} = \frac{1}{2}\left(\tilde{\boldsymbol{\eta}}_d - \tilde{\boldsymbol{\eta}}\right)^T \mathcal{A}_{f\eta}\mathcal{C}_{f\eta}^{-1}\left(\tilde{\boldsymbol{\eta}}_d - \tilde{\boldsymbol{\eta}}\right). \tag{8.53}$$

It follows from (8.8), (8.40)–(8.42), and Lemma 4.1 that

$$\dot{\nu}_{\mathrm{O}} \leqslant -\left({}^{\mathrm{O}}V_r - {}^{\mathrm{O}}V\right)^T \mathbf{K}_{\mathrm{O}} \left({}^{\mathrm{O}}V_r - {}^{\mathrm{O}}V\right)$$

$$+\left({}^{\mathrm{O}}V_r - {}^{\mathrm{O}}V\right)^T \left({}^{\mathrm{O}}F_r^* - {}^{\mathrm{O}}F^*\right) \tag{8.54}$$

holds.

In view of (8.2), (8.10), (8.16), (8.37)–(8.39), and (8.43), premultiplying $(\mathcal{F}_{Tr} - \mathcal{F}_T)$ by $(\mathcal{V}_{Tr} - \mathcal{V}_T)^T$ yields

$$\begin{aligned}
&(\mathcal{V}_{Tr} - \mathcal{V}_T)^T (\mathcal{F}_{Tr} - \mathcal{F}_T) \\
&= \left({}^{\mathrm{O}}V_r - {}^{\mathrm{O}}V\right)^T \left({}^{\mathrm{O}}F_r^* - {}^{\mathrm{O}}F^*\right) + \left(\tilde{\boldsymbol{\eta}}_d - \tilde{\boldsymbol{\eta}}\right)^T \mathcal{A}_{f\eta}\left(\boldsymbol{\eta}_d - \boldsymbol{\eta}\right) \\
&= \left({}^{\mathrm{O}}V_r - {}^{\mathrm{O}}V\right)^T \left({}^{\mathrm{O}}F_r^* - {}^{\mathrm{O}}F^*\right) + \left(\tilde{\boldsymbol{\eta}}_d - \tilde{\boldsymbol{\eta}}\right)^T \mathcal{A}_{f\eta}\mathcal{C}_{f\eta}^{-1}\left(\dot{\tilde{\boldsymbol{\eta}}}_d - \dot{\tilde{\boldsymbol{\eta}}}\right) \\
&\quad + \left(\tilde{\boldsymbol{\eta}}_d - \tilde{\boldsymbol{\eta}}\right)^T \mathcal{A}_{f\eta}\left(\tilde{\boldsymbol{\eta}}_d - \tilde{\boldsymbol{\eta}}\right).
\end{aligned} \tag{8.55}$$

On the other hand, the definitions of $(\mathcal{F}_{Tr} - \mathcal{F}_T)$ and $(\mathcal{V}_{Tr} - \mathcal{V}_T)$ give

$$(\mathcal{V}_{Tr} - \mathcal{V}_T)^T (\mathcal{F}_{Tr} - \mathcal{F}_T) = \sum_{j=1}^{n} p_{\mathbf{T}_j} \tag{8.56}$$

in view of Definition 2.16.

Substituting (8.54)–(8.56) into the time derivative of (8.51) yields

$$\dot{\nu} \leqslant -\left({}^{\mathrm{O}}V_r - {}^{\mathrm{O}}V\right)^T \mathbf{K}_{\mathrm{O}} \left({}^{\mathrm{O}}V_r - {}^{\mathrm{O}}V\right)$$

$$-\left(\tilde{\boldsymbol{\eta}}_d - \tilde{\boldsymbol{\eta}}\right)^T \mathcal{A}_{f\eta}\left(\tilde{\boldsymbol{\eta}}_d - \tilde{\boldsymbol{\eta}}\right) + \sum_{j=1}^{n} p_{\mathbf{T}_j}. \tag{8.57}$$

Finally, consider the fact that the held object has n *driven cutting points* associated with frame $\{\mathbf{T}_j\}$ for all $i \in \{1, n\}$. Using (8.51)–(8.53), (8.57), and Definition 2.17 completes the proof. ∎

8.8 Stability of the Entire System

In this section, both the L_2/L_∞ stability and the asymptotic stability of the complete coordinated multiple-manipulator system are presented.

8.8.1 L_2 and L_∞ Stability

The jth manipulator is treated as a base-fixed *open chain* with only one *driving cutting point* at frame $\{\mathbf{T}_j\}$. With the same treatments as that in Chapters 5–7, the jth manipulator can be qualified to be *virtually stable* in the sense of Definition 2.17 for all $j \in \{1, n\}$. Therefore, it follows from Theorems 8.1 and 2.1 that

$$ {}^{\mathbf{O}}V_r - {}^{\mathbf{O}}V \in L_2 \bigcap L_\infty \tag{8.58} $$

$$ \tilde{\boldsymbol{\eta}}_d - \tilde{\boldsymbol{\eta}} \in L_2 \bigcap L_\infty \tag{8.59} $$

hold.

In view of (8.1), (8.34)–(8.36), and the satisfaction of conditions (4.176) and (4.177), it follows from (8.58) and Theorem 4.6 that

$$ \dot{\mathbf{x}}_d - \dot{\mathbf{x}} \in L_2 \bigcap L_\infty \tag{8.60} $$

$$ \mathbf{x}_d - \mathbf{x} \in L_2 \bigcap L_\infty \tag{8.61} $$

and

$$ \boldsymbol{\omega}_d - \boldsymbol{\omega} \in L_2 \bigcap L_\infty \tag{8.62} $$

$$ {}^{\mathbf{O}}\boldsymbol{\mu} \in L_2 \bigcap L_\infty \tag{8.63} $$

hold.

8.8.2 Algebraic Loop Issue

Differentiating (8.37) with respect to time and using (8.39) yields

$$ \begin{aligned} \dot{\mathcal{V}}_{Tr} &= \begin{bmatrix} \frac{d}{dt}\left({}^{\mathbf{T}_1}V_r\right) \\ \vdots \\ \frac{d}{dt}\left({}^{\mathbf{T}_j}V_r\right) \\ \vdots \\ \frac{d}{dt}\left({}^{\mathbf{T}_n}V_r\right) \end{bmatrix} \\ &= {}^{\mathbf{O}}\mathcal{U}_{\mathbf{T}}^T \left[\mathcal{H}^T \frac{d}{dt}\left({}^{\mathbf{O}}V_r\right) + \boldsymbol{\Phi}^T \mathcal{A}_{f\eta} \left(\dot{\tilde{\boldsymbol{\eta}}}_d - \dot{\tilde{\boldsymbol{\eta}}} \right) \right] \\ &= -{}^{\mathbf{O}}\mathcal{U}_{\mathbf{T}}^T \boldsymbol{\Phi}^T \mathcal{A}_{f\eta} \mathcal{C}_{f\eta} \boldsymbol{\eta} + \zeta_{\eta a} \end{aligned} \tag{8.64} $$

with

$$ \zeta_{\eta a} = {}^{\mathbf{O}}\mathcal{U}_{\mathbf{T}}^T \left[\mathcal{H}^T \frac{d}{dt}\left({}^{\mathbf{O}}V_r\right) + \boldsymbol{\Phi}^T \mathcal{A}_{f\eta} \left(\dot{\tilde{\boldsymbol{\eta}}}_d + \mathcal{C}_{f\eta} \tilde{\boldsymbol{\eta}} \right) \right] . \tag{8.65} $$

Assume that the jth manipulator is a six-joint manipulator as studied in Chapter 5. It follows from Lemma 5.1 that the desired joint torque vector can be expressed by

$$ {}^{\mathbf{O}}\mathbf{U}_{\mathbf{T}_j}\mathbf{J}_j^{-T}\boldsymbol{\tau}_{jd} = \left({}^{\mathbf{O}}\mathbf{U}_{\mathbf{T}_j}\hat{\mathcal{M}}_j{}^{\mathbf{O}}\mathbf{U}_{\mathbf{T}_j}^T\right){}^{\mathbf{T}_j}\mathbf{U}_{\mathbf{O}}^T\frac{d}{dt}({}^{\mathbf{T}_j}V_r) + \mathbf{b}_{\tau j} \tag{8.66} $$

where ${}^{\mathbf{O}}\mathbf{U}_{\mathbf{T}_j}\mathbf{J}_j^{-T}$, $\boldsymbol{\tau}_{jd}$, ${}^{\mathbf{O}}\mathbf{U}_{\mathbf{T}_j}\hat{\mathcal{M}}_j{}^{\mathbf{O}}\mathbf{U}_{\mathbf{T}_j}^T$, ${}^{\mathbf{T}_j}V$, and $\mathbf{b}_{\tau j}$ correspond respectively to $\mathbf{J}_q^{-T}$, $\boldsymbol{\tau}_d$, $\hat{\mathcal{M}}$, ${}^{\mathbf{C}}V$, and $\mathbf{b}_\tau$ in (5.88).

In view of Section 5.7, all three joint motor control modes lead to the fact that the joint control torques can be expressed as an affine function of the joint desired control torques as

$$ \boldsymbol{\tau}_j = \mathcal{K}_{\tau j}\boldsymbol{\tau}_{jd} + \boldsymbol{\zeta}_{\tau j} \tag{8.67} $$

where matrix $\mathcal{K}_{\tau j}$ and vector $\boldsymbol{\zeta}_{\tau j}$ hold different values for different joint motor control modes.[1]

Re-write (8.23) and (8.29) in a form as

$$ \begin{bmatrix} \mathcal{M}_o & \mathbf{0} \\ \mathcal{M}_\eta & \mathbf{I}_{6(n-1)} \end{bmatrix} \begin{bmatrix} \frac{d}{dt}({}^{\mathbf{O}}V) \\ \boldsymbol{\eta} \end{bmatrix} = \begin{bmatrix} \mathcal{J}_o \\ \mathcal{J}_\eta \end{bmatrix}\boldsymbol{\tau} - \begin{bmatrix} \varepsilon_o \\ \varepsilon_\eta \end{bmatrix}. \tag{8.68} $$

Thus, in view of (8.64) and (8.66)–(8.68), an algebraic loop about $\boldsymbol{\eta}$ through $\boldsymbol{\tau}$ and $\dot{\mathcal{V}}_{Tr}$ is formed in the sense of Definition 2.18. The gain of the algebraic loop, as defined in (2.109), can be expressed by

$$ \mathbf{K}(t) = -\begin{bmatrix} \mathbf{0} & \mathbf{I}_{6n-6} \end{bmatrix} \begin{bmatrix} \mathcal{M}_o & \mathbf{0} \\ \mathcal{M}_\eta & \mathbf{I}_{6(n-1)} \end{bmatrix}^{-1} \begin{bmatrix} \mathcal{H} \\ \mathbf{\Phi} \end{bmatrix} \mathcal{J}_v^{-T}\mathcal{K}_\tau\mathcal{J}_v^T\hat{\mathcal{M}}_\tau\mathbf{\Phi}^T\mathcal{A}_{f\eta}\mathcal{C}_{f\eta} \tag{8.69} $$

with

$$ \mathcal{J}_v = \mathrm{diag}\left({}^{\mathbf{T}_1}\mathbf{U}_{\mathbf{O}}^T\mathbf{J}_1, \cdots, {}^{\mathbf{T}_j}\mathbf{U}_{\mathbf{O}}^T\mathbf{J}_j, \cdots, {}^{\mathbf{T}_n}\mathbf{U}_{\mathbf{O}}^T\mathbf{J}_n\right) \tag{8.70} $$
$$ \mathcal{K}_\tau = \mathrm{diag}(\mathcal{K}_{\tau 1}, \cdots, \mathcal{K}_{\tau j}, \cdots, \mathcal{K}_{\tau n}) \tag{8.71} $$
$$ \hat{\mathcal{M}}_\tau = \mathrm{diag}({}^{\mathbf{O}}\mathbf{U}_{\mathbf{T}_1}\hat{\mathcal{M}}_1{}^{\mathbf{O}}\mathbf{U}_{\mathbf{T}_1}^T, \cdots, {}^{\mathbf{O}}\mathbf{U}_{\mathbf{T}_j}\hat{\mathcal{M}}_j{}^{\mathbf{O}}\mathbf{U}_{\mathbf{T}_j}^T, \cdots, {}^{\mathbf{O}}\mathbf{U}_{\mathbf{T}_n}\hat{\mathcal{M}}_n{}^{\mathbf{O}}\mathbf{U}_{\mathbf{T}_n}^T) \tag{8.72} $$

where $\begin{bmatrix} \mathcal{J}_o \\ \mathcal{J}_\eta \end{bmatrix} = \begin{bmatrix} \mathcal{H} \\ \mathbf{\Phi} \end{bmatrix} \mathcal{J}_v^{-T}$ is used.

To make (2.115) valid, the norm of $\mathcal{A}_{f\eta}\mathcal{C}_{f\eta}$ must be limited by

$$ \|\mathcal{A}_{f\eta}\mathcal{C}_{f\eta}\| \leqslant \gamma_\eta \tag{8.73} $$

with γ_η being a positive constant that is inversely proportional to $\|\hat{\mathcal{M}}_\tau\|$.

[1] When the *motor torque control* mode is used for all the joints of the jth manipulator, it follows that $\mathcal{K}_{\tau j} = \mathbf{I}$ and $\boldsymbol{\zeta}_{\tau j} = \mathbf{0}$ hold. When the *motor current control* mode is used for all the joints of the jth manipulator, $\boldsymbol{\zeta}_{\tau j} = \mathbf{0}$ holds and the expression of $\mathcal{K}_{\tau j}$ can be obtained from (5.59), (5.60), and (5.63). When the *motor voltage control* mode is used for all the joints of the jth manipulator, the expressions of $\mathcal{K}_{\tau j}$ and $\boldsymbol{\zeta}_{\tau j}$ can be obtained from (5.59), (5.60), and (5.155) subject to (5.74).

8.8.3 Asymptotic Stability

With a stable algebraic loop about $\boldsymbol{\eta}$ and with bounded reference signals, the boundedness of $\boldsymbol{\tau}$ can be ensured. This result further ensures the boundedness of $\frac{d}{dt}\left({}^{\mathbf{O}}V\right) \in \mathbb{R}^6$ and $\boldsymbol{\eta} \in \mathbb{R}^{6n-6}$ from (8.23) and (8.29). It then follows from (8.58), (8.59), and Lemma 2.8 that

$$
{}^{\mathbf{O}}V_r - {}^{\mathbf{O}}V \to \mathbf{0} \tag{8.74}
$$

$$
\tilde{\boldsymbol{\eta}}_d - \tilde{\boldsymbol{\eta}} \to \mathbf{0} \tag{8.75}
$$

hold, leading to

$$
\dot{\mathbf{x}}_d - \dot{\mathbf{x}} \to \mathbf{0} \tag{8.76}
$$

$$
\mathbf{x}_d - \mathbf{x} \to \mathbf{0} \tag{8.77}
$$

$$
\boldsymbol{\omega}_d - \boldsymbol{\omega} \to \mathbf{0} \tag{8.78}
$$

$$
{}^{\mathbf{O}}\boldsymbol{\mu} \to \mathbf{0} \tag{8.79}
$$

from (8.34)–(8.36).

8.9 Coordinated Multiple-Arm Systems in Constrained Motion

When the held object is in contact with the environment, the number of the degrees of freedom of motion is reduced to m $(0 < m < 6)$, resulting in a configuration space having $n_f = 6 - m$ dimensions for the constraint forces/moments. Most treatments developed in Chapter 5 for the hybrid motion/force control can be applied straightforwardly.

Let $\{\mathbf{S}\}$ be a frame located at the contact point between the held object and the environment, and let $\{\mathbf{S}_b\}$ be a frame having the same origin as frame $\{\mathbf{S}\}$. In frame $\{\mathbf{S}_b\}$, some of its axes describe the motion and others the constraints. Let $\mathbf{T}_s \in \mathbb{R}^{6\times(6-n_f)}$ and $\mathbf{T}_f \in \mathbb{R}^{6\times n_f}$ be two selection mapping matrices of full column-rank. Each column contains a single *one* and five *zeros*, subject to $\mathbf{T}_s^T\mathbf{T}_s = \mathbf{I}_{6-n_f}$, $\mathbf{T}_f^T\mathbf{T}_f = \mathbf{I}_{n_f}$, and $\mathbf{T}_s^T\mathbf{T}_f = \mathbf{0}$.

8.9.1 Kinematics in Constrained Motion

Let the independent velocity coordinate vector of the held object be

$$
\mathcal{V} = \chi \in \mathbb{R}^{6-n_f}. \tag{8.80}
$$

The linear/angular velocity vector of frame $\{\mathbf{S}\}$ is expressed by

$$
{}^{\mathbf{S}}V = \mathrm{diag}({}^{\mathbf{S}}\mathbf{R}_{\mathbf{S}_b}, {}^{\mathbf{S}}\mathbf{R}_{\mathbf{S}_b})\mathbf{T}_s\chi. \tag{8.81}
$$

Furthermore, the linear/angular velocity vector of frame $\{\mathbf{O}\}$ can be written as

$$
\begin{aligned}
{}^{\mathbf{O}}V &= {}^{\mathbf{S}}\mathbf{U}_{\mathbf{O}}^T\,{}^{\mathbf{S}}V \\
&= {}^{\mathbf{S}}\mathbf{U}_{\mathbf{O}}^T\mathrm{diag}({}^{\mathbf{S}}\mathbf{R}_{\mathbf{S}_b}, {}^{\mathbf{S}}\mathbf{R}_{\mathbf{S}_b})\mathbf{T}_s\chi \\
&= {}^{\mathbf{S}}\mathbf{U}_{\mathbf{O}}^T\mathrm{diag}({}^{\mathbf{S}}\mathbf{R}_{\mathbf{S}_b}, {}^{\mathbf{S}}\mathbf{R}_{\mathbf{S}_b})\mathbf{T}_s\mathcal{V}.
\end{aligned} \tag{8.82}
$$

8.9.2 Dynamics in Constrained Motion

With a similar approach leading to (5.112), the contact force/moment vector in frame $\{\mathbf{S}\}$ can thus be written as

$$^{\mathbf{S}}F = \mathrm{diag}(^{\mathbf{S}}\mathbf{R}_{\mathbf{S}_b}, {}^{\mathbf{S}}\mathbf{R}_{\mathbf{S}_b})\,(\mathbf{T}_s\boldsymbol{\psi} + \mathbf{T}_f\boldsymbol{\varphi}) \tag{8.83}$$

where $\boldsymbol{\varphi} \in \mathbb{R}^{n_f}$ denotes the independent constraint force/moment coordinate vector at the contact point and

$$\boldsymbol{\psi} = \mathbf{Y}_s\boldsymbol{\theta}_s \tag{8.84}$$

denotes the force coordinate vector in the motion configuration space with $\boldsymbol{\theta}_s$ being a parameter vector.

With $^{\mathbf{O}}F^*$ being expressed by (8.8), the force resultant equation of the held object changes from (8.9) to

$$^{\mathbf{O}}F^* = \sum_{j=1}^{n} {}^{\mathbf{O}}\mathbf{U}_{\mathbf{T}_j}\,{}^{\mathbf{T}_j}F - {}^{\mathbf{O}}\mathbf{U}_{\mathbf{S}}\,{}^{\mathbf{S}}F \tag{8.85}$$

in which a constraint force term is added. Accordingly, (8.10) should be rewritten as

$$\mathcal{F}_T \overset{def}{=} \begin{bmatrix} ^{\mathbf{T}_1}F \\ \vdots \\ ^{\mathbf{T}_j}F \\ \vdots \\ ^{\mathbf{T}_n}F \end{bmatrix} = {}^{\mathbf{T}}\mathcal{U}_{\mathbf{O}}\begin{bmatrix} \boldsymbol{\Omega}_m & \boldsymbol{\Omega}_f \end{bmatrix}\begin{bmatrix} ^{\mathbf{O}}F^* + {}^{\mathbf{O}}\mathbf{U}_{\mathbf{S}}\,{}^{\mathbf{S}}F \\ \boldsymbol{\eta} \end{bmatrix}. \tag{8.86}$$

8.9.3 Forward Dynamics in Constrained Motion

Noting the difference between (8.9) and (8.85) yields a contact force term $^{\mathbf{O}}\mathbf{U}_{\mathbf{S}}\,{}^{\mathbf{S}}F$ being added to the left hand side of (8.19) as

$$\mathcal{M}_m\frac{d}{dt}\left(^{\mathbf{O}}V\right) + \mathcal{C}_m{}^{\mathbf{O}}V + \mathcal{G}_m + {}^{\mathbf{O}}\mathbf{U}_{\mathbf{S}}\,{}^{\mathbf{S}}F = \sum_{j=1}^{n} {}^{\mathbf{O}}\mathbf{U}_{\mathbf{T}_j}\mathbf{J}_j^{-T}\boldsymbol{\tau}_j. \tag{8.87}$$

In view of (8.83), premultiplying (8.87) by $\mathbf{T}_s^T\mathrm{diag}(^{\mathbf{S}_b}\mathbf{R}_{\mathbf{S}}, {}^{\mathbf{S}_b}\mathbf{R}_{\mathbf{S}})^{\mathbf{S}}\mathbf{U}_{\mathbf{O}}$ and using (8.82), $\mathbf{T}_s^T\mathbf{T}_s = \mathbf{I}_{6-n_f}$, and $\mathbf{T}_s^T\mathbf{T}_f = \mathbf{0}$ yields

$$\mathcal{M}_s\dot{\boldsymbol{\chi}} + \boldsymbol{\varepsilon}_s + \boldsymbol{\psi} = \mathcal{J}_s\boldsymbol{\tau} \tag{8.88}$$

where

$$\mathcal{M}_s = \mathbf{T}_s^T\mathcal{M}_{ss}\mathbf{T}_s \tag{8.89}$$

$$\boldsymbol{\varepsilon}_s = \mathbf{T}_s^T\boldsymbol{\varepsilon}_{ss} \tag{8.90}$$

$$\mathcal{J}_s = \mathbf{T}_s^T\mathcal{J}_{ss} \tag{8.91}$$

with

$$\mathcal{M}_{ss} = \mathrm{diag}\left({}^{\mathbf{S}_b}\mathbf{R}_{\mathbf{S}}, {}^{\mathbf{S}_b}\mathbf{R}_{\mathbf{S}}\right) {}^{\mathbf{S}}\mathbf{U}_{\mathbf{O}} \mathcal{M}_m {}^{\mathbf{S}}\mathbf{U}_{\mathbf{O}}^T \mathrm{diag}\left({}^{\mathbf{S}}\mathbf{R}_{\mathbf{S}_b}, {}^{\mathbf{S}}\mathbf{R}_{\mathbf{S}_b}\right) \tag{8.92}$$

$$\begin{aligned} \varepsilon_{ss} = {} & \mathrm{diag}\left({}^{\mathbf{S}_b}\mathbf{R}_{\mathbf{S}}, {}^{\mathbf{S}_b}\mathbf{R}_{\mathbf{S}}\right) {}^{\mathbf{S}}\mathbf{U}_{\mathbf{O}} \mathcal{M}_m \\ & \times \frac{d}{dt}\left[{}^{\mathbf{S}}\mathbf{U}_{\mathbf{O}}^T \mathrm{diag}\left({}^{\mathbf{S}}\mathbf{R}_{\mathbf{S}_b}, {}^{\mathbf{S}}\mathbf{R}_{\mathbf{S}_b}\right)\right] \mathbf{T}_s \boldsymbol{\chi} \\ & + \mathrm{diag}\left({}^{\mathbf{S}_b}\mathbf{R}_{\mathbf{S}}, {}^{\mathbf{S}_b}\mathbf{R}_{\mathbf{S}}\right) {}^{\mathbf{S}}\mathbf{U}_{\mathbf{O}} \left(\mathcal{C}_m {}^{\mathbf{O}}\mathcal{V} + \mathcal{G}_m\right) \end{aligned} \tag{8.93}$$

$$\begin{aligned} \mathcal{J}_{ss} = {} & \mathrm{diag}\left({}^{\mathbf{S}_b}\mathbf{R}_{\mathbf{S}}, {}^{\mathbf{S}_b}\mathbf{R}_{\mathbf{S}}\right) {}^{\mathbf{S}}\mathbf{U}_{\mathbf{O}} \\ & \times \left[{}^{\mathbf{O}}\mathbf{U}_{\mathbf{T}_1} \mathbf{J}_1^{-T}, \cdots, {}^{\mathbf{O}}\mathbf{U}_{\mathbf{T}_j} \mathbf{J}_j^{-T}, \cdots, {}^{\mathbf{O}}\mathbf{U}_{\mathbf{T}_n} \mathbf{J}_n^{-T}\right]. \end{aligned} \tag{8.94}$$

Note that $\boldsymbol{\tau}$ is defined by (8.26), $\mathcal{M}_m$, $\mathcal{C}_m$, and $\mathcal{G}_m$ are defined by (8.20)–(8.22), and $\boldsymbol{\psi}$ is defined by (8.84).

The independent constraint force coordinate vector $\boldsymbol{\varphi} \in \mathbb{R}^{n_f}$ can be obtained by premultiplying (8.87) by $\mathbf{T}_f^T \mathrm{diag}({}^{\mathbf{S}_b}\mathbf{R}_{\mathbf{S}}, {}^{\mathbf{S}_b}\mathbf{R}_{\mathbf{S}}) {}^{\mathbf{S}}\mathbf{U}_{\mathbf{O}}$ and by using (8.82), (8.83), $\mathbf{T}_f^T \mathbf{T}_f = \mathbf{I}_{n_f}$, and $\mathbf{T}_s^T \mathbf{T}_f = \mathbf{0}$. It follows that

$$\boldsymbol{\varphi} + \mathcal{M}_f \dot{\boldsymbol{\chi}} + \varepsilon_f = \mathcal{J}_f \boldsymbol{\tau} \tag{8.95}$$

holds, where

$$\mathcal{M}_f = \mathbf{T}_f^T \mathcal{M}_{ss} \mathbf{T}_s \tag{8.96}$$

$$\varepsilon_f = \mathbf{T}_f^T \varepsilon_{ss} \tag{8.97}$$

$$\mathcal{J}_f = \mathbf{T}_f^T \mathcal{J}_{ss}. \tag{8.98}$$

Note that $\mathcal{M}_{ss}$, ε_{ss}, and $\mathcal{J}_{ss}$ are defined by (8.92)–(8.94).

In summary, the motion dynamics are presented by (8.88), the constraint force dynamics are presented by (8.95), and the internal force dynamics are presented by (8.29).

8.9.4 Motion Control and Constraint/Internal Force Control

Let

$$\mathcal{V}_r = \boldsymbol{\chi}_r \in \mathbb{R}^{6-n_f} \tag{8.99}$$

be the required (design) independent velocity vector in the motion configuration space,[2] and let $\boldsymbol{\varphi}_d \in \mathbb{R}^{n_f}$ be the desired (design) constraint force vector.

Similar to (5.115) and (5.116), the required linear/angular velocity vector of frame $\{\mathbf{O}\}$ can be written as

$$\begin{aligned} {}^{\mathbf{O}}V_r &= {}^{\mathbf{S}}\mathbf{U}_{\mathbf{O}}^T {}^{\mathbf{S}}V_r \\ &= {}^{\mathbf{S}}\mathbf{U}_{\mathbf{O}}^T \mathrm{diag}\left({}^{\mathbf{S}}\mathbf{R}_{\mathbf{S}_b}, {}^{\mathbf{S}}\mathbf{R}_{\mathbf{S}_b}\right) \left[\mathbf{T}_s \boldsymbol{\chi}_r + \mathbf{T}_f \mathbf{A}_f \left(\tilde{\varphi}_d - \tilde{\varphi}\right)\right] \end{aligned} \tag{8.100}$$

---

[2] The specific design of $\chi_r \in \mathbb{R}^{6-n_f}$ for the purpose of position control will be given by (8.137).

where $\mathbf{A}_f \in \mathbb{R}^{n_f \times n_f}$ is a diagonal positive-definite matrix. The other two vectors, denoted as $\tilde{\varphi}_d \in \mathbb{R}^{n_f}$ and $\tilde{\varphi} \in \mathbb{R}^{n_f}$, are obtained by

$$\dot{\tilde{\varphi}}_d = -\mathbf{C}_f \tilde{\varphi}_d + \mathbf{C}_f \varphi_d \tag{8.101}$$

$$\dot{\tilde{\varphi}} = -\mathbf{C}_f \tilde{\varphi} + \mathbf{C}_f \varphi \tag{8.102}$$

from (5.117) and (5.118).

Remark 8.5. The second term in the right hand side of (8.100) is added to achieve the filtered constraint force control characterized by $\tilde{\varphi}_d - \tilde{\varphi} \rightarrow \mathbf{0}$.

After ${}^{\mathbf{O}}V_r$ being obtained, $\mathcal{V}_{Tr}$ can be computed from (8.37)–(8.39). It follows that

$$\mathcal{V}_{Tr} \stackrel{def}{=} \begin{bmatrix} {}^{\mathbf{T}_1}V_r \\ \vdots \\ {}^{\mathbf{T}_j}V_r \\ \vdots \\ {}^{\mathbf{T}_n}V_r \end{bmatrix}$$

$$= {}^{\mathbf{O}}\mathcal{U}_{\mathbf{T}}^T \left[\mathcal{H}^T {}^{\mathbf{O}}V_r + \mathbf{\Phi}^T \mathcal{A}_{f\eta} \left(\tilde{\boldsymbol{\eta}}_d - \tilde{\boldsymbol{\eta}} \right) \right] \tag{8.103}$$

holds, subject to

$$\dot{\tilde{\boldsymbol{\eta}}}_d = \mathcal{C}_{f\eta} \left(\boldsymbol{\eta}_d - \tilde{\boldsymbol{\eta}}_d \right) \tag{8.104}$$

$$\dot{\tilde{\boldsymbol{\eta}}} = \mathcal{C}_{f\eta} \left(\boldsymbol{\eta} - \tilde{\boldsymbol{\eta}} \right) \tag{8.105}$$

where $\mathcal{H} \in \mathbb{R}^{6 \times 6n}$ and $\mathbf{\Phi} \in \mathbb{R}^{6(n-1) \times 6n}$ are defined in (8.16) and $\mathcal{A}_{f\eta} \in \mathbb{R}^{6(n-1) \times 6(n-1)}$ and $\mathcal{C}_{f\eta} \in \mathbb{R}^{6(n-1) \times 6(n-1)}$ are two diagonal positive-definite matrices defined by (8.37)–(8.39).

Remark 8.6. After the required independent velocity vector $\chi_r \in \mathbb{R}^{6-n_f}$ being obtained, a term $\tilde{\varphi}_d - \tilde{\varphi} \in \mathbb{R}^{n_f}$ is added to form ${}^{\mathbf{O}}V_r \in \mathbb{R}^6$ in (8.100), aimed at achieving the filtered constraint force control. Then, a term $\tilde{\boldsymbol{\eta}}_d - \tilde{\boldsymbol{\eta}} \in \mathbb{R}^{6(n-1)}$ is further added to form $\mathcal{V}_{Tr} \in \mathbb{R}^{6n}$ in (8.103), aimed at achieving the filtered internal force control.

In line with (8.83), the required constraint force/moment vector in frame $\{\mathbf{S}\}$, denoted as ${}^{\mathbf{S}}F_r \in \mathbb{R}^6$, is designed as

$${}^{\mathbf{S}}F_r = \mathrm{diag}\left({}^{\mathbf{S}}\mathbf{R}_{\mathbf{S}_b}, {}^{\mathbf{S}}\mathbf{R}_{\mathbf{S}_b} \right) \left(\mathbf{T}_s \hat{\psi} + \mathbf{T}_f \varphi_d \right) \tag{8.106}$$

with

$$\hat{\psi} = \mathbf{Y}_s \hat{\boldsymbol{\theta}}_s \tag{8.107}$$

as being defined by (5.120). Similarly, by defining

$$\mathbf{s}_s = \mathbf{Y}_s^T \left(\chi_r - \chi \right) \tag{8.108}$$

the γth element of $\boldsymbol{\theta}_s$ is updated by using the $\mathcal{P}$ function defined by (2.79) as

$$\hat{\theta}_{s\gamma} = \mathcal{P}\left(s_{s\gamma}, \rho_{s\gamma}, \underline{\theta}_{s\gamma}, \overline{\theta}_{s\gamma}, t\right) \tag{8.109}$$

where $\hat{\theta}_{s\gamma}$ is the γth element of $\hat{\boldsymbol{\theta}}_s$, $s_{s\gamma}$ is the γth element of $\mathbf{s}_s$, $\rho_{s\gamma} > 0$ is an update gain, and $\underline{\theta}_{s\gamma}$ and $\overline{\theta}_{s\gamma}$ denote the lower and upper bounds of $\theta_{s\gamma}$—the γth element of $\boldsymbol{\theta}_s$.

Finally, after ${}^{\mathbf{S}}F_r \in \mathbb{R}^6$ being obtained from (8.106) and $\boldsymbol{\eta}_d \in \mathbb{R}^{6(n-1)}$ being obtained from the internal force control specification, the required forces/moments at the n end-effectors are computed as

$$\mathcal{F}_{Tr} \stackrel{def}{=} \begin{bmatrix} {}^{\mathbf{T}_1}F_r \\ \vdots \\ {}^{\mathbf{T}_j}F_r \\ \vdots \\ {}^{\mathbf{T}_n}F_r \end{bmatrix} = {}^{\mathbf{T}}\mathcal{U}_{\mathbf{O}} \begin{bmatrix} \boldsymbol{\Omega}_m & \boldsymbol{\Omega}_f \end{bmatrix} \begin{bmatrix} {}^{\mathbf{O}}F_r^* + {}^{\mathbf{O}}\mathbf{U}_{\mathbf{S}}{}^{\mathbf{S}}F_r \\ \boldsymbol{\eta}_d \end{bmatrix} \tag{8.110}$$

with ${}^{\mathbf{T}}\mathcal{U}_{\mathbf{O}} = {}^{\mathbf{O}}\mathcal{U}_{\mathbf{T}}^{-1} \in \mathbb{R}^{6n \times 6n}$ being defined by (8.3) and ${}^{\mathbf{O}}F_r^*$ being defined by (8.40).

8.9.5 Virtual Stability

In this subsection, it will be shown that the constrained held object combined with its respective control equations qualifies to be *virtually stable* in the sense of Definition 2.17.

Theorem 8.2. *The held object in contact with the environment described by (8.2), (8.8), and (8.81)–(8.86), combined with the control equations (8.40), (8.100)–(8.107), and (8.110) and with the parameter adaptation (8.41), (8.42), (8.108), and (8.109), is virtually stable with its affiliated vectors* ${}^{\mathbf{O}}V_r - {}^{\mathbf{O}}V$, $\tilde{\varphi}_d - \tilde{\varphi}$, *and* $\tilde{\boldsymbol{\eta}}_d - \tilde{\boldsymbol{\eta}}$ *being virtual functions in both* L_2 *and* L_∞, *in the sense of Definition 2.17.*

Proof: Assign the following non-negative accompanying function to the held object

$$\nu = \nu_{\mathbf{O}} + \nu_s + \nu_\eta \tag{8.111}$$

with

$$\begin{aligned} \nu_{\mathbf{O}} &= \frac{1}{2}\left({}^{\mathbf{O}}V_r - {}^{\mathbf{O}}V\right)^T \mathbf{M}_{\mathbf{O}} \left({}^{\mathbf{O}}V_r - {}^{\mathbf{O}}V\right) \\ &\quad + \frac{1}{2}\sum_{\gamma=1}^{13}\left(\theta_{\mathbf{O}\gamma} - \hat{\theta}_{\mathbf{O}\gamma}\right)^2 / \rho_{\mathbf{O}\gamma} \end{aligned} \tag{8.112}$$

$$\begin{aligned} \nu_s &= \frac{1}{2}\sum_{\gamma}\left(\boldsymbol{\theta}_{s\gamma} - \hat{\boldsymbol{\theta}}_{s\gamma}\right)^2 / \rho_{s\gamma} \\ &\quad + \frac{1}{2}\left(\tilde{\varphi}_d - \tilde{\varphi}\right)^T \mathbf{A}_f \mathbf{C}_f^{-1}\left(\tilde{\varphi}_d - \tilde{\varphi}\right) \end{aligned} \tag{8.113}$$

$$\nu_\eta = \frac{1}{2}\left(\tilde{\boldsymbol{\eta}}_d - \tilde{\boldsymbol{\eta}}\right)^T \mathcal{A}_{f\eta}\mathcal{C}_{f\eta}^{-1}\left(\tilde{\boldsymbol{\eta}}_d - \tilde{\boldsymbol{\eta}}\right). \tag{8.114}$$

It follows from (8.8), (8.40), (8.41), (8.42), and Lemma 4.1 that

$$\dot{\nu}_{\mathbf{O}} \leqslant -\left({}^{\mathbf{O}}V_r - {}^{\mathbf{O}}V\right)^T \mathbf{K}_{\mathbf{O}} \left({}^{\mathbf{O}}V_r - {}^{\mathbf{O}}V\right) + \left({}^{\mathbf{O}}V_r - {}^{\mathbf{O}}V\right)^T \left({}^{\mathbf{O}}F_r^* - {}^{\mathbf{O}}F^*\right) \tag{8.115}$$

holds.

Then, premultiplying $(\mathcal{F}_{Tr} - \mathcal{F}_T)$ by $(\mathcal{V}_{Tr} - \mathcal{V}_T)^T$ and using (8.2), (8.16), (8.81)–(8.86), (8.100)–(8.108), and (8.110) yields

$$\begin{aligned}
&(\mathcal{V}_{Tr} - \mathcal{V}_T)^T (\mathcal{F}_{Tr} - \mathcal{F}_T) \\
&\quad = \left({}^{\mathbf{O}}V_r - {}^{\mathbf{O}}V\right)^T \left({}^{\mathbf{O}}F_r^* - {}^{\mathbf{O}}F^*\right) + \left({}^{\mathbf{O}}V_r - {}^{\mathbf{O}}V\right)^T {}^{\mathbf{O}}\mathbf{U}_{\mathbf{S}} \left({}^{\mathbf{S}}F_r - {}^{\mathbf{S}}F\right) \\
&\qquad + (\tilde{\boldsymbol{\eta}}_d - \tilde{\boldsymbol{\eta}})^T \mathcal{A}_{f\eta} (\boldsymbol{\eta}_d - \boldsymbol{\eta}) \\
&\quad = \left({}^{\mathbf{O}}V_r - {}^{\mathbf{O}}V\right)^T \left({}^{\mathbf{O}}F_r^* - {}^{\mathbf{O}}F^*\right) + \left({}^{\mathbf{S}}V_r - {}^{\mathbf{S}}V\right)^T \left({}^{\mathbf{S}}F_r - {}^{\mathbf{S}}F\right) \\
&\qquad + (\tilde{\boldsymbol{\eta}}_d - \tilde{\boldsymbol{\eta}})^T \mathcal{A}_{f\eta} (\boldsymbol{\eta}_d - \boldsymbol{\eta}) \\
&\quad = \left({}^{\mathbf{O}}V_r - {}^{\mathbf{O}}V\right)^T \left({}^{\mathbf{O}}F_r^* - {}^{\mathbf{O}}F^*\right) - (\boldsymbol{\chi}_r - \boldsymbol{\chi})^T \mathbf{Y}_s(\boldsymbol{\theta}_s - \hat{\boldsymbol{\theta}}_s) \\
&\qquad + (\tilde{\boldsymbol{\varphi}}_d - \tilde{\boldsymbol{\varphi}})^T \mathbf{A}_f(\boldsymbol{\varphi}_d - \boldsymbol{\varphi}) + (\tilde{\boldsymbol{\eta}}_d - \tilde{\boldsymbol{\eta}})^T \mathcal{A}_{f\eta} (\boldsymbol{\eta}_d - \boldsymbol{\eta}) \\
&\quad = \left({}^{\mathbf{O}}V_r - {}^{\mathbf{O}}V\right)^T \left({}^{\mathbf{O}}F_r^* - {}^{\mathbf{O}}F^*\right) - \mathbf{s}_s^T(\boldsymbol{\theta}_s - \hat{\boldsymbol{\theta}}_s) \\
&\qquad + (\tilde{\boldsymbol{\varphi}}_d - \tilde{\boldsymbol{\varphi}})^T \mathbf{A}_f \mathbf{C}_f^{-1} \left(\dot{\tilde{\boldsymbol{\varphi}}}_d - \dot{\tilde{\boldsymbol{\varphi}}}\right) + (\tilde{\boldsymbol{\varphi}}_d - \tilde{\boldsymbol{\varphi}})^T \mathbf{A}_f (\tilde{\boldsymbol{\varphi}}_d - \tilde{\boldsymbol{\varphi}}) \\
&\qquad + (\tilde{\boldsymbol{\eta}}_d - \tilde{\boldsymbol{\eta}})^T \mathcal{A}_{f\eta} \mathcal{C}_{f\eta}^{-1} \left(\dot{\tilde{\boldsymbol{\eta}}}_d - \dot{\tilde{\boldsymbol{\eta}}}\right) + (\tilde{\boldsymbol{\eta}}_d - \tilde{\boldsymbol{\eta}})^T \mathcal{A}_{f\eta} (\tilde{\boldsymbol{\eta}}_d - \tilde{\boldsymbol{\eta}}) .
\end{aligned} \tag{8.116}$$

Furthermore, substituting (8.115) and (8.116) into the time derivative of (8.111) and using (8.56), (8.109), and Lemma 2.9 yields

$$\begin{aligned}
\dot{\nu} &= \dot{\nu}_{\mathbf{O}} - \sum_{\gamma} \left(\theta_{s\gamma} - \hat{\theta}_{s\gamma}\right) \frac{\dot{\hat{\theta}}_{s\gamma}}{\rho_{s\gamma}} \\
&\quad + (\tilde{\boldsymbol{\varphi}}_d - \tilde{\boldsymbol{\varphi}})^T \mathbf{A}_f \mathbf{C}_f^{-1} \left(\dot{\tilde{\boldsymbol{\varphi}}}_d - \dot{\tilde{\boldsymbol{\varphi}}}\right) + (\tilde{\boldsymbol{\eta}}_d - \tilde{\boldsymbol{\eta}})^T \mathcal{A}_{f\eta} \mathcal{C}_{f\eta}^{-1} \left(\dot{\tilde{\boldsymbol{\eta}}}_d - \dot{\tilde{\boldsymbol{\eta}}}\right) \\
&\leqslant -\left({}^{\mathbf{O}}V_r - {}^{\mathbf{O}}V\right)^T \mathbf{K}_{\mathbf{O}} \left({}^{\mathbf{O}}V_r - {}^{\mathbf{O}}V\right) + \sum_{\gamma} \left(\theta_{s\gamma} - \hat{\theta}_{s\gamma}\right) \left[s_{s\gamma} - \frac{\dot{\hat{\theta}}_{s\gamma}}{\rho_{s\gamma}} \right] \\
&\quad - (\tilde{\boldsymbol{\varphi}}_d - \tilde{\boldsymbol{\varphi}})^T \mathbf{A}_f (\tilde{\boldsymbol{\varphi}}_d - \tilde{\boldsymbol{\varphi}}) - (\tilde{\boldsymbol{\eta}}_d - \tilde{\boldsymbol{\eta}})^T \mathcal{A}_{f\eta} (\tilde{\boldsymbol{\eta}}_d - \tilde{\boldsymbol{\eta}}) \\
&\quad + (\mathcal{V}_{Tr} - \mathcal{V}_T)^T (\mathcal{F}_{Tr} - \mathcal{F}_T) \\
&\leqslant -\left({}^{\mathbf{O}}V_r - {}^{\mathbf{O}}V\right)^T \mathbf{K}_{\mathbf{O}} \left({}^{\mathbf{O}}V_r - {}^{\mathbf{O}}V\right) \\
&\quad - (\tilde{\boldsymbol{\varphi}}_d - \tilde{\boldsymbol{\varphi}})^T \mathbf{A}_f (\tilde{\boldsymbol{\varphi}}_d - \tilde{\boldsymbol{\varphi}}) \\
&\quad - (\tilde{\boldsymbol{\eta}}_d - \tilde{\boldsymbol{\eta}})^T \mathcal{A}_{f\eta} (\tilde{\boldsymbol{\eta}}_d - \tilde{\boldsymbol{\eta}}) \\
&\quad + \sum_{j=1}^{n} p_{\mathbf{T}_j} .
\end{aligned} \tag{8.117}$$

Finally, consider the fact that the held object has n *driven cutting points* associated with frame $\{\mathbf{T}_j\}$ for all $i \in \{1, n\}$. Using (8.111)–(8.114), (8.117), and Definition 2.17 completes the proof. ■

8.9.6 L_2 and L_∞ Stability

When the jth manipulator combined with its control equations qualifies to be *virtually stable* in the sense of Definition 2.17 for all $j \in \{1, n\}$, it follows from Theorems 8.2 and 2.1 that

$$ {}^{\mathbf{O}}V_r - {}^{\mathbf{O}}V \in L_2 \bigcap L_\infty \tag{8.118} $$

$$ \tilde{\varphi}_d - \tilde{\varphi} \in L_2 \bigcap L_\infty \tag{8.119} $$

$$ \tilde{\eta}_d - \tilde{\eta} \in L_2 \bigcap L_\infty \tag{8.120} $$

hold. Expression (8.118) further implies

$$ \chi_r - \chi \in L_2 \bigcap L_\infty \tag{8.121} $$

in view of (8.82) and (8.100).

8.9.7 Algebraic Loop Issue

In this subsection, it will be shown that an algebraic loop about both the constraint forces and the internal forces is formed by using the control equations outlined in this section. Stability conditions are then given.

Substituting (8.100) into (8.103), differentiating it with respect to time, and using (8.102) and (8.105) yields

$$ \begin{aligned} \dot{\mathcal{V}}_{Tr} &= \begin{bmatrix} \frac{d}{dt}\left({}^{\mathbf{T}_1}V_r\right) \\ \vdots \\ \frac{d}{dt}\left({}^{\mathbf{T}_j}V_r\right) \\ \vdots \\ \frac{d}{dt}\left({}^{\mathbf{T}_n}V_r\right) \end{bmatrix} \\ &= {}^{\mathbf{O}}\mathcal{U}_{\mathbf{T}}^T \left[\mathcal{H}^T \frac{d}{dt}\left({}^{\mathbf{O}}V_r\right) + \mathbf{\Phi}^T \mathcal{A}_{f\eta} \left(\dot{\tilde{\eta}}_d - \dot{\tilde{\eta}}\right) \right] \\ &= {}^{\mathbf{O}}\mathcal{U}_{\mathbf{T}}^T \left(-\mathcal{H}^T \mathbf{J}_\Psi \mathbf{A}_f \mathbf{C}_f \varphi - \mathbf{\Phi}^T \mathcal{A}_{f\eta} \mathcal{C}_{f\eta} \eta + \zeta_a \right) \end{aligned} \tag{8.122} $$

where

$$ \begin{aligned} \mathbf{J}_\Psi &= {}^{\mathbf{S}}\mathbf{U}_{\mathbf{O}}^T \mathrm{diag}\left({}^{\mathbf{S}}\mathbf{R}_{\mathbf{S}_b}, {}^{\mathbf{S}}\mathbf{R}_{\mathbf{S}_b}\right) \mathbf{T}_f \\ \zeta_a &= \mathcal{H}^{T\,\mathbf{S}}\mathbf{U}_{\mathbf{O}}^T \mathrm{diag}\left({}^{\mathbf{S}}\mathbf{R}_{\mathbf{S}_b}, {}^{\mathbf{S}}\mathbf{R}_{\mathbf{S}_b}\right) \\ &\quad \times \left[\mathbf{T}_s \dot{\chi}_r + \mathbf{T}_f \mathbf{A}_f \left(\dot{\tilde{\varphi}}_d + \mathbf{C}_f \tilde{\varphi} \right) \right] \end{aligned} \tag{8.123} $$

$$+\mathcal{H}^T \frac{d}{dt}\left[{}^{\mathbf{S}}\mathbf{U}_{\mathbf{O}}^T \mathrm{diag}\left({}^{\mathbf{S}}\mathbf{R}_{\mathbf{S}_b}, {}^{\mathbf{S}}\mathbf{R}_{\mathbf{S}_b}\right)\right]$$
$$\times \left[\mathbf{T}_s \boldsymbol{\chi}_r + \mathbf{T}_f \mathbf{A}_f \left(\tilde{\boldsymbol{\varphi}}_d - \tilde{\boldsymbol{\varphi}}\right)\right]$$
$$+\boldsymbol{\Phi}^T \mathcal{A}_{f\eta}\left(\dot{\tilde{\boldsymbol{\eta}}}_d + \mathcal{C}_{f\eta}\tilde{\boldsymbol{\eta}}\right). \quad (8.124)$$

Recall (8.66) and (8.67) as

$$ {}^{\mathbf{O}}\mathbf{U}_{\mathbf{T}_j}\mathbf{J}_j^{-T}\boldsymbol{\tau}_{jd} = \left({}^{\mathbf{O}}\mathbf{U}_{\mathbf{T}_j}\hat{\mathcal{M}}_j {}^{\mathbf{O}}\mathbf{U}_{\mathbf{T}_j}^T\right){}^{\mathbf{T}_j}\mathbf{U}_{\mathbf{O}}^T \frac{d}{dt}({}^{\mathbf{T}_j}V_r) + \mathbf{b}_{\tau j} \quad (8.125)$$
$$\boldsymbol{\tau}_j = \mathcal{K}_{\tau j}\boldsymbol{\tau}_{jd} + \boldsymbol{\zeta}_{\tau j}. \quad (8.126)$$

Then, in view of (8.82), express (8.29), (8.88), and (8.95) in a compact form as

$$\begin{bmatrix} \mathcal{M}_s & \mathbf{0} & \mathbf{0} \\ \mathcal{M}_f & \mathbf{I}_{n_f} & \mathbf{0} \\ \mathcal{M}_\eta \mathbf{J}_\Omega & \mathbf{0} & \mathbf{I}_{6(n-1)} \end{bmatrix} \begin{bmatrix} \dot{\boldsymbol{\chi}} \\ \boldsymbol{\varphi} \\ \boldsymbol{\eta} \end{bmatrix} = \begin{bmatrix} \mathcal{J}_s \\ \mathcal{J}_f \\ \mathcal{J}_\eta \end{bmatrix} \boldsymbol{\tau} - \begin{bmatrix} \boldsymbol{\varepsilon}_s + \boldsymbol{\psi} \\ \boldsymbol{\varepsilon}_f \\ \boldsymbol{\varepsilon}_{f\eta} \end{bmatrix} \quad (8.127)$$

with

$$\mathbf{J}_\Omega = {}^{\mathbf{S}}\mathbf{U}_{\mathbf{O}}^T \mathrm{diag}\left({}^{\mathbf{S}}\mathbf{R}_{\mathbf{S}_b}, {}^{\mathbf{S}}\mathbf{R}_{\mathbf{S}_b}\right)\mathbf{T}_s \quad (8.128)$$
$$\boldsymbol{\varepsilon}_{f\eta} = \boldsymbol{\varepsilon}_\eta + \mathcal{M}_\eta \frac{d}{dt}\left[{}^{\mathbf{S}}\mathbf{U}_{\mathbf{O}}^T \mathrm{diag}\left({}^{\mathbf{S}}\mathbf{R}_{\mathbf{S}_b}, {}^{\mathbf{S}}\mathbf{R}_{\mathbf{S}_b}\right)\right]\mathbf{T}_s \boldsymbol{\chi} \quad (8.129)$$

where $\mathcal{M}_s$ is defined by (8.89) and (8.92), $\mathcal{M}_f$ is defined by (8.96), $\mathcal{M}_\eta$ is defined by (8.30), $\mathcal{J}_s$ is defined by (8.91) and (8.94), $\mathcal{J}_f$ is defined by (8.98), $\mathcal{J}_\eta$ is defined by (8.31), $\boldsymbol{\varepsilon}_s$ is defined by (8.90) and (8.93), $\boldsymbol{\varepsilon}_f$ is defined by (8.97), and $\boldsymbol{\varepsilon}_\eta$ is defined by (8.32).

Substituting (8.122), (8.125), and (8.126) into (8.127) yields an algebraic loop about $[\boldsymbol{\varphi}^T, \boldsymbol{\eta}^T]^T$ through $\boldsymbol{\tau}$ and $\dot{\mathcal{V}}_{Tr}$, in the sense of Definition 2.18. The gain of the algebraic loop, as defined in (2.109), can be expressed as

$$\mathbf{K}(t) = -\begin{bmatrix} \mathbf{0} & \mathbf{I}_{n_f} & \mathbf{0} \\ \mathbf{0} & \mathbf{0} & \mathbf{I}_{6(n-1)} \end{bmatrix} \begin{bmatrix} \mathcal{M}_s & \mathbf{0} & \mathbf{0} \\ \mathcal{M}_f & \mathbf{I}_{n_f} & \mathbf{0} \\ \mathcal{M}_\eta \mathbf{J}_\Omega & \mathbf{0} & \mathbf{I}_{6(n-1)} \end{bmatrix}^{-1}$$
$$\times \begin{bmatrix} \mathbf{J}_\Omega^T \mathcal{H} \\ \mathbf{J}_\Psi^T \mathcal{H} \\ \boldsymbol{\Phi} \end{bmatrix} \mathcal{J}_v^{-T}\mathcal{K}_\tau \mathcal{J}_v^T \hat{\mathcal{M}}_\tau \left[\mathcal{H}^T \mathbf{J}_\Psi \mathbf{A}_f \mathbf{C}_f \quad \boldsymbol{\Phi}^T \mathcal{A}_{f\eta}\mathcal{C}_{f\eta}\right] \quad (8.130)$$

with $\mathcal{J}_v$ being defined by (8.70), $\mathcal{K}_\tau$ being defined by (8.71), $\hat{\mathcal{M}}_\tau$ being defined by (8.72), $\mathbf{J}_\Psi$ being defined by (8.123), and $\mathbf{J}_\Omega$ being defined by (8.128). In (8.130), $\begin{bmatrix} \mathcal{J}_s \\ \mathcal{J}_f \\ \mathcal{J}_\eta \end{bmatrix} = \begin{bmatrix} \mathbf{J}_\Omega^T \mathcal{H} \\ \mathbf{J}_\Psi^T \mathcal{H} \\ \boldsymbol{\Phi} \end{bmatrix} \mathcal{J}_v^{-T}$ is used.

To make (2.115) valid, the norms of both $\mathbf{A}_f\mathbf{C}_f$ and $\mathcal{A}_{f\eta}\mathcal{C}_{f\eta}$ must be limited by

$$\|\mathbf{A}_f\mathbf{C}_f\| \leqslant \gamma_\varphi \quad (8.131)$$
$$\|\mathcal{A}_{f\eta}\mathcal{C}_{f\eta}\| \leqslant \gamma_\eta \quad (8.132)$$

with γ_φ and γ_η being two positive constants inversely proportional to $\|\hat{\mathcal{M}}_\tau\|$.

8.9.8 Asymptotic Stability

With a stable algebraic loop and with bounded reference signals, the boundedness of $\boldsymbol{\tau} \in \mathbb{R}^{6n}$ can be ensured from (8.122), (8.125), and (8.126). This result further ensures the boundedness of $\dot{\boldsymbol{\chi}} \in \mathbb{R}^{6-n_f}$, $\boldsymbol{\varphi} \in \mathbb{R}^{n_f}$, and $\boldsymbol{\eta} \in \mathbb{R}^{6n-6}$ from (8.127). Consequently, it follows from (8.118)–(8.121) and Lemma 2.8 that

$$^{\mathbf{O}}V_r - {}^{\mathbf{O}}V \to \mathbf{0} \tag{8.133}$$

$$\tilde{\boldsymbol{\varphi}}_d - \tilde{\boldsymbol{\varphi}} \to \mathbf{0} \tag{8.134}$$

$$\tilde{\boldsymbol{\eta}}_d - \tilde{\boldsymbol{\eta}} \to \mathbf{0} \tag{8.135}$$

$$\boldsymbol{\chi}_r - \boldsymbol{\chi} \to \mathbf{0} \tag{8.136}$$

hold.

8.9.9 Position Control

Expressions (8.121) and (8.136) ensure both the L_2/L_∞ stability and the asymptotic stability of the velocity control. It has been stated in Chapter 4 that position control can be achieved by re-designing the required independent velocity coordinate vector as

$$\boldsymbol{\chi}_r = \boldsymbol{\chi}_d + \boldsymbol{\Lambda}\mathbf{e} \tag{8.137}$$

where $\boldsymbol{\chi}_d \in \mathbb{R}^{6-n_f}$ denotes the desired independent velocity vector, $\boldsymbol{\Lambda} \in \mathbb{R}^{(6-n_f)\times(6-n_f)}$ is a diagonal positive-definite matrix, and $\mathbf{e} \in \mathbb{R}^{6-n_f}$ characterizes the position errors, subject to

$$\int_0^t (\boldsymbol{\chi}_d - \boldsymbol{\chi})^T \boldsymbol{\Lambda}\mathbf{e} d\tau \geqslant -\gamma_{fm} \tag{8.138}$$

$$\boldsymbol{\chi}_r - \boldsymbol{\chi} \in L_\infty \Rightarrow \begin{cases} \boldsymbol{\chi}_d - \boldsymbol{\chi} \in L_\infty \\ \mathbf{e} \in L_\infty \end{cases} \tag{8.139}$$

$$\boldsymbol{\chi}_d - \boldsymbol{\chi} \in L_\infty \Rightarrow \dot{\mathbf{e}} \in L_\infty \tag{8.140}$$

with γ_{fm} being a positive constant.

It then follows from (8.121), (8.138), and (8.139) that

$$\boldsymbol{\chi}_d - \boldsymbol{\chi} \in L_2 \bigcap L_\infty \tag{8.141}$$

$$\mathbf{e} \in L_2 \bigcap L_\infty \tag{8.142}$$

hold, leading to

$$\boldsymbol{\chi}_d - \boldsymbol{\chi} \to \mathbf{0} \tag{8.143}$$

$$\mathbf{e} \to \mathbf{0} \tag{8.144}$$

from (8.136), (8.140), and Lemma 2.8.

8.10 Case Study - KUMA 361 and KUMA 160

In practical applications, (8.100) should be modified to

$$\begin{aligned} {}^{\mathbf{O}}V_r &= {}^{\mathbf{S}}\mathbf{U}_{\mathbf{O}}^{T}\,{}^{\mathbf{S}}V_r \\ &= {}^{\mathbf{S}}\mathbf{U}_{\mathbf{O}}^{T}\mathrm{diag}\left({}^{\mathbf{S}}\mathbf{R}_{\mathbf{S}_b}, {}^{\mathbf{S}}\mathbf{R}_{\mathbf{S}_b}\right) \\ &\quad \times \left[\mathbf{T}_s\boldsymbol{\chi}_r\left(\boldsymbol{\chi}_r - \mathbf{A}_m\tilde{\boldsymbol{\psi}}_z\right) + \mathbf{T}_f\mathbf{A}_f\left(\tilde{\varphi}_d - \tilde{\varphi}\right)\right] \end{aligned} \tag{8.145}$$

where a new term $\mathbf{A}_m\tilde{\boldsymbol{\psi}}_z$ is added to the motion configuration space with $\mathbf{A}_m \in \mathbb{R}^{(6-n_f)\times(6-n_f)}$ being positive-definite and $\tilde{\boldsymbol{\psi}}_z = [\tilde{\psi}_{z1}, \cdots, \tilde{\psi}_{z(6-n_f)}]^T \in \mathbb{R}^{6-n_f}$ being defined by (5.182) and (5.183).

The introduction of the term $\mathbf{A}_m\tilde{\boldsymbol{\psi}}_z$ in (8.145) gives the coordinated multiple-robot system the capability of handing unexpected collisions with unknown obstacles and the capability of working with unknown contact geometry. In normal circumstances, the thresholds in the dead-zone filters are designed in such a way that $\tilde{\boldsymbol{\psi}}_z = \mathbf{0}$ holds. When a collision with an unexpected obstacle happens, the collision forces in the motion configuration space become large enough, resulting in a nonzero $\tilde{\boldsymbol{\psi}}_z$ being added to the required velocity vector through (8.145) aimed at reducing the collision forces. Therefore, the developed controller behaves as hybrid control for known contact geometry and also behaves as impedance control for unknown contact geometry.

Two interesting experiments of using two industrial robots KUKA 160 and KUMA 361 performing coordinated tasks have been reported in [215, 218].

In [218], two six-joint robot manipulators have their end-effectors fixed together, leading to $m = 6$ and $n = 2$ such that

$$\begin{aligned} \mathcal{H} &= \begin{bmatrix} \mathbf{I}_6 & \mathbf{I}_6 \end{bmatrix} \in \mathbb{R}^{6\times 12} \\ \boldsymbol{\Psi} &= \begin{bmatrix} \mathbf{0} & \mathbf{I}_6 \end{bmatrix} \in \mathbb{R}^{6\times 12} \\ \boldsymbol{\Omega}_m &= \begin{bmatrix} \mathbf{I}_6 \\ \mathbf{0} \end{bmatrix} \in \mathbb{R}^{12\times 6} \\ \boldsymbol{\Omega}_f &= \begin{bmatrix} -\mathbf{I}_6 \\ \mathbf{I}_6 \end{bmatrix} \in \mathbb{R}^{12\times 6} \end{aligned}$$

hold.

In [215], the two robots are rigidly holding a raw egg, as illustrated in Figure 8.3. The egg is assumed to be an ideal ball with $r > 0$ being its radius. The two end-effectors are two vertical plates holding the egg by applying a squeeze force. Having a two-point contact with the two robots, the egg is subject to a three dimensional position control and a two-dimensional orientation control excluding the rotation along the line connecting the two contact points, since a zero moment can be applied along this connection line. Assign a body frame $\{\mathbf{O}\}$ to the center of the egg with the z-axis aligning with the line connecting the two contact points.

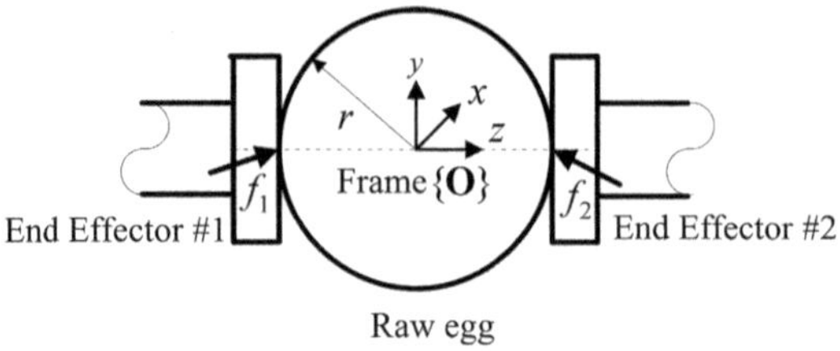

Fig. 8.3. Two robots hold an egg.

Let

$$\mathbf{f}_o^* = [f_x^*, f_y^*, f_z^*]^T \in \mathbb{R}^3$$
$$\mathbf{m}_o^* = [m_x^*, m_y^*]^T \in \mathbb{R}^2$$

be the net forces and moments applied to the center of the egg, and let

$$\eta \in \mathbb{R}$$

be the squeeze force applied to the egg. Then, let

$$\mathbf{v}_o = [v_x, v_y, v_z]^T \in \mathbb{R}^3$$
$$\boldsymbol{\omega}_o = [\omega_x, \omega_y]^T \in \mathbb{R}^2$$

be the linear and angular velocities at the center of the egg. Furthermore, let $\mathbf{f}_j \in \mathbb{R}^3$ be the forces applied from the jth end-effector to the egg, and let $\mathbf{v}_j \in \mathbb{R}^3$ be the linear velocities of the jth end-effector for all $j \in \{1, 2\}$. The force transformation from $[\mathbf{f}_1^T, \mathbf{f}_2^T]^T \in \mathbb{R}^6$ to $[\mathbf{f}_o^{*T}, \mathbf{m}_o^{*T}, \eta]^T \in \mathbb{R}^6$ can be expressed by

$$\begin{bmatrix} \mathbf{f}_o^* \\ \mathbf{m}_o^* \\ \eta \end{bmatrix} = \begin{bmatrix} \mathbf{I}_3 & \mathbf{I}_3 \\ -\mathbf{A} & \mathbf{A} \\ -\frac{1}{2}\mathbf{B} & \frac{1}{2}\mathbf{B} \end{bmatrix} \begin{bmatrix} \mathbf{f}_1 \\ \mathbf{f}_2 \end{bmatrix} \tag{8.146}$$

with

$$\mathbf{A} = \begin{bmatrix} 0 & -r & 0 \\ r & 0 & 0 \end{bmatrix} \in \mathbb{R}^{2\times 3}$$
$$\mathbf{B} = \begin{bmatrix} 0 & 0 & 1 \end{bmatrix} \in \mathbb{R}^{1\times 3}.$$

Inversely, (8.146) can be expressed as

$$\begin{bmatrix} \mathbf{f}_1 \\ \mathbf{f}_2 \end{bmatrix} = \begin{bmatrix} \frac{1}{2}\mathbf{I}_3 & -\frac{1}{2}\mathbf{A}_p^T & -\mathbf{B}^T \\ \frac{1}{2}\mathbf{I}_3 & \frac{1}{2}\mathbf{A}_p^T & \mathbf{B}^T \end{bmatrix} \begin{bmatrix} \mathbf{f}_o^* \\ \mathbf{m}_o^* \\ \eta \end{bmatrix} \tag{8.147}$$

with

$$\mathbf{A}_p = \begin{bmatrix} 0 & -\frac{1}{r} & 0 \\ \frac{1}{r} & 0 & 0 \end{bmatrix} \in \mathbb{R}^{2\times 3}.$$

Finally, applying the duality between velocity and force transformations yields

$$\begin{bmatrix} \mathbf{v}_o \\ \boldsymbol{\omega}_o \end{bmatrix} = \begin{bmatrix} \frac{1}{2}\mathbf{I}_3 & \frac{1}{2}\mathbf{I}_3 \\ -\frac{1}{2}\mathbf{A}_p & \frac{1}{2}\mathbf{A}_p \end{bmatrix} \begin{bmatrix} \mathbf{v}_1 \\ \mathbf{v}_2 \end{bmatrix} \tag{8.148}$$

and inversely

$$\begin{bmatrix} \mathbf{v}_1 \\ \mathbf{v}_2 \end{bmatrix} = \begin{bmatrix} \mathbf{I}_3 & -\mathbf{A}^T \\ \mathbf{I}_3 & \mathbf{A}^T \end{bmatrix} \begin{bmatrix} \mathbf{v}_o \\ \boldsymbol{\omega}_o \end{bmatrix}. \tag{8.149}$$

The remaining designs are straightforward, starting with the orientation control of the two end-effectors.

8.11 Concluding Remarks

In this chapter, the *virtual decomposition control* (VDC) approach has been applied to the systems of coordinated multiple robot manipulators holding a common object. The original system is virtually decomposed into the held object and several manipulators. The advantage of using the VDC approach has been demonstrated by allowing the control design to be specially focused on the held object to make it *virtually stable*, whereas the *virtual stability* of each manipulator interpreted as a base-fixed *open chain* is well ensured in Chapters 4–7.

Without having *driving cutting points*, the held object is a special *object* with possibly available force measurements at its *driven cutting points* where force sensors can be mounted on all the end-effectors of the n manipulators. Consequently, the control equations of the held object are slightly different from what have been designed for the standard *objects* described in Chapter 4. The difference is the introduction of a filtered internal force term to the required linear/angular velocity vectors at the n end-effectors to ensure that not only the motion control, but also the filtered internal force control can be achieved.

When the held object is in contact with an environment, the introduction of a filtered constraint force term to the required velocity vector of the held object makes sure that the motion control, the filtered internal force control, and the filtered constraint force control can all be achieved. This result suggests that the internal force and the constraint force can be handled with a similar manner [227], since they share the same properties in the sense that they do not affect the motion of the systems.

However, the introduction of the filtered constraint and internal force errors to the required velocity vectors forms an algebraic loop about the constraint and internal forces through the feedback control in the sense of Definition 2.18. The gains associated with the filtered constraint and internal force terms must be limited to ensure the stability of the algebraic loop.

An interesting experiment performed by two industrial robots rigidly holding a raw egg has been reported, after the presentation of a coordinated task performed by the two robots subject to a rigid contact.

9

Control of Space Robots

9.1 Introduction

Space robot, referring to a robot manipulator or a group of robot manipulators mounted on a platform in a space orbit, possesses unique challenges to control and motion planning due to the nature of its underactuation leading to the existence of the nonholonomic constraints governed by the momentum conservation law. In general, there exist six nonholonomic constraints on each orbital space robot. These six nonholonomic constraints comprise three constraints for the linear motion and the other three constraints for the rotational motion. If thrusters are enabled in all three directions, the number of nonholonomic constraints reduces to three. In the control and motion planning, the three- or six-dimensional nonholonomic constraints that make the kinematics no longer independent of the dynamics need to be properly handled.

The nonholonomic constraints can be handled by releasing a certain number of the degrees of freedom (DOFs) of motion,[1] that is, the number of the DOFs that can be freely planned and controlled is less than the number of the total DOFs of an orbital space robot. The difference in number is exactly determined by the number of the nonholonomic constraints. For instance, an orbital space robot without using thrusters has six less DOFs that can be freely planned and controlled. The linear motion of the platform and the rotational motion of the three reaction wheels are usually to be released to satisfy the six nonholonomic constraints governed by the momentum conservation law. In case thrusters are used to propel the platform in all three directions, the number of nonholonomic constraints is reduced to three. Consequently, the rotational motion of the three reaction wheels needs to be released to satisfy the three nonholonomic constraints.

Without the need of using a force/moment sensor to measure the dynamic interactions between the robot manipulator and the platform, the *virtual decomposition control* (VDC) approach is to be used in this chapter to handle the adaptive

[1] The released variables become dependent variables that cannot be subject to independent position and orientation control.

W.-H. Zhu: Virtual Decomposition Control, STAR 60, pp. 221–250.
springerlink.com

control problem of orbital space robots. As a unique feature of the VDC approach, the nonholonomic constraints are applied to the design (required accelerations) variables as opposed to the actual system variables. By zeroing the required (design) forces/moments from the environment to the platform, a zero *virtual power flow* to the system is ensured, which in turn guarantees the convergence of the actual forces/moments and velocities to their desired/required values.

It has to be noted that the applicability of the VDC approach to an orbital space robot depends on the solvability of the equations that zero the required exerting forces/moments to the platform. Practical conditions that ensure the solvability are to be given.

9.2 Prior Work

Witnessed by Canadarm 2 and Dextre operated in the International Space Station (ISS), space robots have become important tools in space operations performed in a weightless environment. In general, space robots can be classified into three categories [51, 108]: (a) free-floating, where neither the position nor the orientation of the spacecraft platform is controlled for the purpose of energy saving, (b) free-flying, where only the orientation of the spacecraft platform is controlled through reaction wheels for the purposes of keeping a communication link with the ground and generating electrical power from the solar panels, (c) full-controlling, where both the position and the orientation of the spacecraft platform are controlled through jet thrusts and reaction wheels. In any one of the three cases in which only one manipulator is mounted on the platform, the number of the total actuators is always less than the number of the DOFs of the entire system. This underactuation gives rise to the nonholonomic constraints being imposed on the space robot [126].

Aimed at establishing the similarity between ground-based manipulators and space robots, the research in [135] suggested that nearly any control algorithm used for ground-based manipulators can be employed in the control of space robots provided that the dynamic singularity defined in [136] is avoided. In [52], the problem of motion decoupling between the end-effector and the spacecraft platform was addressed by using the resolved acceleration control based on a recursive calculation of kinematics and dynamics. Derived from the momentum conservation principle, the resolved rate control of space robots was proposed in [176] based on the concept of Generalized Jacobian Matrix that incorporates the dynamics into the kinematics. In [133], a coordinated control scheme, in which both the platform (satellite) and the manipulator have their own controllers, was proposed. A feedforward signal is sent to the platform controller to compensate for the momentum resulting from the motion of the manipulator. This technology led to a successful satellite capturing in 1999 with the Japanese test satellite ETS-VII [76].

Viewing a space robot as an underactuated system, a general approach for acquiring both forward and inverse kinematics and dynamics was developed in [77]. Again, recursive computations were used.

Researches concerning the control and simulations of multiple manipulators being mounted on a common platform were also conducted [121, 124, 197, 199]. In these approaches, dynamic interactions between the platform and the multiple manipulators were considered and recursive computations were used.

Aimed at automatically accommodating the dynamic parameter uncertainties, adaptive control of space robots was investigated by using the Lagrangian models [183, 190]. Addressing the implementation difficulties, the VDC of space robots was presented in [216].

Readers who are interested in the coordinated control of two orbital space robots capturing a common object are suggested to explore two early references [50, 209].

9.3 Space Robot with Virtual Decomposition

A typical space robotic system is illustrated in Figure 9.1. Without loss of generality, a robot manipulator holding an object (space payload) is mounded on a spacecraft platform equipped with three reaction wheels that are used to keep the orientation of the spacecraft platform when the robot arm is in motion. This robot manipulator has six single-DOF joints allowing its two ends to have a relative motion with *six* DOFs without encountering kinematics singularity. By inserting a massless six-DOF virtual manipulator[2] between the spacecraft platform and the ground, this system becomes a complex robotic system as defined in Definition 4.1 without involving any closed kinematic loop. After being placed six *cutting points*, the complete system is virtually decomposed into two *objects* and five *open chains*, as illustrated in Figure 9.2.

The two *objects* represent the held object (payload) and the spacecraft platform, respectively. The five *open chains* represent the six-joint manipulator, the three reaction wheels, and the massless virtual manipulator, respectively.

Several coordinate frames are defined. Frame $\{\mathbf{O}\}$ is attached to the payload and frame $\{\mathbf{B}\}$ is attached to the (spacecraft) platform. Then, frame $\{\mathbf{C}\}$ is placed at the *cutting point* between the held object and the manipulator. Frame $\{\mathbf{A}_m\}$ is placed at the *cutting point* that separates the manipulator from the (spacecraft) platform with its z axis aligning with the first joint axis of the manipulator. Furthermore, frame $\{\mathbf{A}_j\}$ is attached to the (spacecraft) platform with its z axis aligning with the jth reaction wheel axis for all $j \in \{1, 3\}$.

For each *open chain* representing a reaction wheel, there exists a subsidiary *cutting point* that virtually decomposes the reaction wheel from its joint. Let frame $\{\mathbf{W}_j\}$ be attached to the jth reaction wheel with the z axis aligning with the jth reaction wheel axis for all $j \in \{1, 3\}$.

With the same treatment as that in Chapter 5, the manipulator is further virtually decomposed into six links and six joints by placing eleven subsidiary

[2] It could be a four-joint robot. The first three joints are prismatic joints with single-DOF and are arranged orthogonally to provide three position coordinates. The last joint is a three-DOF spherical joint to provide rotations in 3D.

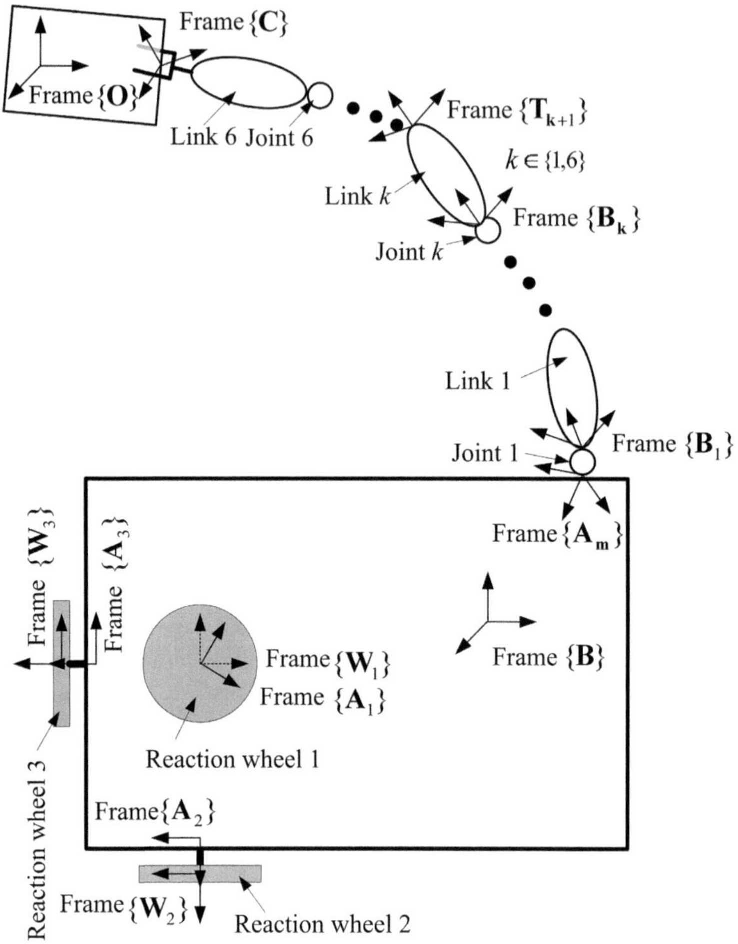

Fig. 9.1. A robot manipulator mounted on a spacecraft with reaction wheels.

cutting points. The six links and joints of the manipulator are numbered sequentially from the (spacecraft) platform toward the payload with joint k connecting link k and link $k-1$ for all[3] $k \in \{2, 6\}$. There are additional eleven frames assigned to the manipulator. Frame $\{\mathbf{B}_k\}$ is attached to the kth link of the manipulator with its z axis aligning with the kth joint axis for all $k \in \{1, 6\}$. Frame $\{\mathbf{T}_k\}$ is attached to the $(k-1)$th link of the manipulator with its z axis aligning with the kth joint axis for all $k \in \{2, 6\}$.

In view of Definition 2.13, the *cutting point* associated with frame $\{\mathbf{C}\}$ is called the *driven cutting point* of the held object and is simultaneously called the *driving cutting point* of link 6 of the manipulator. The *cutting point* associated with frame $\{\mathbf{B}_k\}$ is called the *driven cutting point* of link k of the manipulator and

[3] Joint 1 connects the first link of the manipulator with the spacecraft platform.

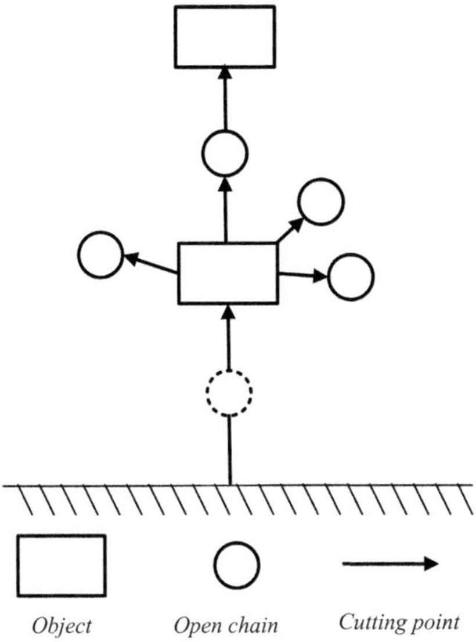

Fig. 9.2. Graphic expression of Figure 9.1.

is simultaneously called the *driving cutting point* of joint k of the manipulator for all $k \in \{1, 6\}$. The *cutting point* associated with frame $\{\mathbf{T}_k\}$ is called the *driven cutting point* of joint k of the manipulator and is simultaneously called the *driving cutting point* of link $k-1$ of the manipulator for all $k \in \{2, 6\}$. The *cutting point* associated with frame $\{\mathbf{A}_m\}$ is called the *driven cutting point* of the first joint of the manipulator and is simultaneously called the *driving cutting point* of the (spacecraft) platform. The *cutting point* associated with frame $\{\mathbf{W}_j\}$ is called the *driven cutting point* of the jth reaction wheel and is simultaneously called the *driving cutting point* of the jth reaction wheel joint for all $j \in \{1, 3\}$. The *cutting point* associated with frame $\{\mathbf{A}_j\}$ is called the *driven cutting point* of the jth reaction wheel joint for all $j \in \{1, 3\}$ and is simultaneously called the *driving cutting point* of the (spacecraft) platform. It follows that

- The held object has only one *driven cutting point* associated with frame $\{\mathbf{C}\}$.
- The 6th link of the manipulator has one *driving cutting point* associated with frame $\{\mathbf{C}\}$ and one *driven cutting point* associated with frame $\{\mathbf{B}_6\}$.
- The kth link of the manipulator has one *driving cutting point* associated with frame $\{\mathbf{T}_{k+1}\}$ and one *driven cutting point* associated with frame $\{\mathbf{B}_k\}$ for all $k \in \{1, 5\}$.
- The kth joint of the manipulator has one *driving cutting point* associated with frame $\{\mathbf{B}_k\}$ and one *driven cutting point* associated with frame $\{\mathbf{T}_k\}$ for all $k \in \{2, 6\}$.
- The first joint of the manipulator has one *driving cutting point* associated with frame $\{\mathbf{B}_1\}$ and one *driven cutting point* associated with frame $\{\mathbf{A}_m\}$.
- The jth reaction wheel has one *driven cutting point* associated with frame $\{\mathbf{W}_j\}$ for all $j \in \{1, 3\}$.

- The jth reaction wheel joint has one *driving cutting point* associated with frame $\{\mathbf{W}_j\}$ and one *driven cutting point* associated with frame $\{\mathbf{A}_j\}$ for all $j \in \{1, 3\}$.
- The (spacecraft) platform has four *driving cutting points* associated with frames $\{\mathbf{A}_m\}$ and $\{\mathbf{A}_j\}$ for all $j \in \{1, 3\}$ and one *driven cutting point* associated with frame $\{\mathbf{B}\}$.
- The massless virtual manipulator has one *driving cutting point* associated with frame $\{\mathbf{B}\}$.

9.4 Kinematics

This section comprises two subsections specifying the independent velocity coordinates and the velocity mapping matrices.

9.4.1 Independent Velocities

Consider the fact that this system has fifteen DOFs of motion: six for the (spacecraft) platform, six for the manipulator, and three for the reaction wheels. The independent velocity coordinate vector $\mathcal{V} \in \mathbb{R}^{15}$ can be expressed as either

$$\mathcal{V} = \begin{bmatrix} \dot{\mathbf{q}} \\ \dot{\mathbf{q}}_w \\ {}^{\mathbf{B}}V \end{bmatrix} \in \mathbb{R}^{15}$$

or

$$\mathcal{V} = \begin{bmatrix} {}^{\mathbf{O}}V \\ \dot{\mathbf{q}}_w \\ {}^{\mathbf{B}}V \end{bmatrix} \in \mathbb{R}^{15}$$

where $\mathbf{q} = [q_1, \cdots, q_6]^T \in \mathbb{R}^6$ denotes the joint position vector of the manipulator, ${}^{\mathbf{O}}V \in \mathbb{R}^6$ denotes the linear/angular velocity vector of frame $\{\mathbf{O}\}$, ${}^{\mathbf{B}}V \in \mathbb{R}^6$ denotes the linear/angular velocity vector of frame $\{\mathbf{B}\}$, and $\mathbf{q}_w = [q_{w1}, q_{w2}, q_{w3}]^T \in \mathbb{R}^3$ denotes the joint coordinate vector of the three reaction wheels.

With the Jacobian matrix $\mathbf{J}_q \in \mathbb{R}^{6\times 6}$ being defined by (5.3) as

$$\mathbf{J}_q = \begin{bmatrix} {}^{\mathbf{B}_1}\mathbf{U}_{\mathbf{O}}^T\mathbf{z} & {}^{\mathbf{B}_2}\mathbf{U}_{\mathbf{O}}^T\mathbf{z} & \cdots & {}^{\mathbf{B}_6}\mathbf{U}_{\mathbf{O}}^T\mathbf{z} \end{bmatrix} \in \mathbb{R}^{6\times 6} \tag{9.1}$$

where $\mathbf{z} = [0, 0, 1, 0, 0, 0]^T \in \mathbb{R}^6$ holds for a prismatic joint and $\mathbf{z} = [0, 0, 0, 0, 0, 1]^T \in \mathbb{R}^6$ holds for a revolute joint, the relationship among ${}^{\mathbf{B}}V \in \mathbb{R}^6$, ${}^{\mathbf{O}}V \in \mathbb{R}^6$, and $\dot{\mathbf{q}} \in \mathbb{R}^6$ is governed by

$${}^{\mathbf{O}}V = \mathbf{J}_q\dot{\mathbf{q}} + {}^{\mathbf{B}}\mathbf{U}_{\mathbf{O}}^T\,{}^{\mathbf{B}}V \tag{9.2}$$

or equivalently

$$\dot{\mathbf{q}} = \mathbf{J}_q^{-1}\,{}^{\mathbf{O}}V - \mathbf{J}_q^{-1}\,{}^{\mathbf{B}}\mathbf{U}_{\mathbf{O}}^T\,{}^{\mathbf{B}}V. \tag{9.3}$$

Consequently, it follows that

$$\begin{bmatrix} {}^{\mathbf{O}}V \\ \dot{\mathbf{q}}_w \\ {}^{\mathbf{B}}V \end{bmatrix} = \begin{bmatrix} \mathbf{J}_q & \mathbf{0} & {}^{\mathbf{B}}\mathbf{U}_{\mathbf{O}}^T \\ \mathbf{0} & \mathbf{I}_3 & \mathbf{0} \\ \mathbf{0} & \mathbf{0} & \mathbf{I}_6 \end{bmatrix} \begin{bmatrix} \dot{\mathbf{q}} \\ \dot{\mathbf{q}}_w \\ {}^{\mathbf{B}}V \end{bmatrix} \tag{9.4}$$

and

$$\begin{bmatrix} \dot{\mathbf{q}} \\ \dot{\mathbf{q}}_w \\ {}^{\mathbf{B}}V \end{bmatrix} = \begin{bmatrix} \mathbf{J}_q^{-1} & \mathbf{0} & -\mathbf{J}_q^{-1}{}^{\mathbf{B}}\mathbf{U}_{\mathbf{O}}^T \\ \mathbf{0} & \mathbf{I}_3 & \mathbf{0} \\ \mathbf{0} & \mathbf{0} & \mathbf{I}_6 \end{bmatrix} \begin{bmatrix} {}^{\mathbf{O}}V \\ \dot{\mathbf{q}}_w \\ {}^{\mathbf{B}}V \end{bmatrix} \tag{9.5}$$

hold.

9.4.2 Velocity Mappings

Let

$$\mathcal{V} = \mathcal{V}_a \overset{def}{=} \begin{bmatrix} \dot{\mathbf{q}} \\ \dot{\mathbf{q}}_w \\ {}^{\mathbf{B}}V \end{bmatrix} \tag{9.6}$$

be the independent velocity coordinate vector. The extended velocity vector containing linear/angular velocity vectors of all rigid bodies[4] in the system, defined by

$$\mathcal{V}_b = [{}^{\mathbf{O}}V^T, {}^{\mathbf{B}_1}V^T, \cdots, {}^{\mathbf{B}_6}V^T, {}^{\mathbf{W}_1}V^T, \cdots, {}^{\mathbf{W}_3}V^T, {}^{\mathbf{B}}V^T]^T \in \mathbb{R}^{66} \tag{9.7}$$

can be obtained from the independent velocity coordinate vector as

$$\mathcal{V}_b = \mathcal{J}_b \mathcal{V} = \mathcal{J}_{ba} \mathcal{J}_a \mathcal{V} = \mathcal{J}_{ba} \mathcal{V}_a \tag{9.8}$$

where

$$\mathcal{J}_a = \mathbf{I}_{15} \tag{9.9}$$

$$\mathcal{J}_b = \mathcal{J}_{ba} = \begin{bmatrix} \mathbf{J}_q & \mathbf{0} & {}^{\mathbf{B}}\mathbf{U}_{\mathbf{O}}^T \\ \begin{bmatrix} \mathbf{z} & \mathbf{0} & \cdots & \mathbf{0} \\ {}^{\mathbf{B}_1}\mathbf{U}_{\mathbf{B}_2}^T\mathbf{z} & \mathbf{z} & & \mathbf{0} \\ \vdots & \cdots & \ddots & \vdots \\ {}^{\mathbf{B}_1}\mathbf{U}_{\mathbf{B}_6}^T\mathbf{z} & {}^{\mathbf{B}_2}\mathbf{U}_{\mathbf{B}_6}^T\mathbf{z} & \cdots & \mathbf{z} \end{bmatrix} & \mathbf{0} & \begin{matrix} {}^{\mathbf{B}}\mathbf{U}_{\mathbf{B}_1}^T \\ {}^{\mathbf{B}}\mathbf{U}_{\mathbf{B}_2}^T \\ \vdots \\ {}^{\mathbf{B}}\mathbf{U}_{\mathbf{B}_6}^T \end{matrix} \\ \mathbf{0} & \begin{bmatrix} \mathbf{z}_w & \mathbf{0} & \mathbf{0} \\ \mathbf{0} & \mathbf{z}_w & \mathbf{0} \\ \mathbf{0} & \mathbf{0} & \mathbf{z}_w \end{bmatrix} & \begin{matrix} {}^{\mathbf{B}}\mathbf{U}_{\mathbf{W}_1}^T \\ {}^{\mathbf{B}}\mathbf{U}_{\mathbf{W}_2}^T \\ {}^{\mathbf{B}}\mathbf{U}_{\mathbf{W}_3}^T \end{matrix} \\ \mathbf{0} & \mathbf{0} & \mathbf{I}_6 \end{bmatrix} \in \mathbb{R}^{66 \times 15} \tag{9.10}$$

with $\mathbf{z}_w = [0, 0, 0, 0, 0, 1]^T$.

[4] There are totally eleven rigid bodies: the (spacecraft) platform, the three reaction wheels, the six rigid links of the manipulator, and the payload.

Alternatively, the overall extended velocity vector can be defined by

$$\mathcal{V}_e = [{}^{\mathbf{O}}V^T, \dot{\mathbf{q}}^T, {}^{\mathbf{B}_1}V^T, \cdots, {}^{\mathbf{B}_6}V^T, \dot{\mathbf{q}}_w^T, {}^{\mathbf{W}_1}V^T, \cdots, {}^{\mathbf{W}_3}V^T, {}^{\mathbf{B}}V^T]^T \in \mathbb{R}^{75}$$

such that

$$\mathcal{V}_e = \mathcal{J}_e \mathcal{V} \tag{9.11}$$

holds, where $\mathcal{J}_e \in \mathbb{R}^{75}$ can be apparently constructed form $\mathcal{J}_b \in \mathbb{R}^{66\times 15}$ defined by (9.10).

9.5 Motion Control Specifications

This section presents the position control specifications in both the joint space and the Cartesian space. Then, the required independent velocity coordinate vector is defined, leading to the computations of all required linear/angular velocity vectors of the eleven rigid bodies.

9.5.1 Position Control

Among the fifteen DOFs of motion, six of them will be released to satisfy the nonholonomic constraints imposed by the momentum conservation law on the system. Thus, only nine DOFs of motion are subject to the position/orientation control through specifying $\mathcal{V}_{pr} \in \mathbb{R}^9$ (a sub-vector of $\mathcal{V}_r \in \mathbb{R}^{15}$). There are two options to choose the nine variables for the position control.

- Joint space control: the six joint positions of the manipulator and the orientation vector of the (spacecraft) platform.
- Cartesian space control: the position/orientation vector of the held object and the orientation vector of the (spacecraft) platform.

Joint Space Control

Let $\mathbf{q}_d = [q_{1d}, \cdots, q_{6d}]^T \in \mathbb{R}^6$ be the desired joint positions of the manipulator, and let $\{\mathbf{B}_d\}$ be a frame representing the desired orientation of the (spacecraft) platform. When the joint positions of the manipulator and the orientation of the (spacecraft) platform are chosen for the position control, the required joint velocities are designed as

$$\dot{\mathbf{q}}_r = \dot{\mathbf{q}}_d + \Lambda(\mathbf{q}_d - \mathbf{q}) \tag{9.12}$$

where $\Lambda = \mathrm{diag}(\lambda_1, \cdots, \lambda_6) \in \mathbb{R}^{6\times 6}$ is a diagonal positive-definite matrix. Meanwhile, the required angular velocity vector of the (spacecraft) platform is designed as

$${}^{\mathbf{B}}\boldsymbol{\omega}_r = {}^{\mathbf{B}}\mathbf{R}_{\mathbf{I}}\boldsymbol{\omega}_{\mathbf{B}_d} + k_{\mathbf{B}}{}^{\mathbf{B}}\boldsymbol{\mu} \tag{9.13}$$

where $\boldsymbol{\omega}_{\mathbf{B}_d} \in \mathbb{R}^3$ denotes the angular velocity vector of frame $\{\mathbf{B}_d\}$ expressed in the inertial frame, $k_{\mathbf{B}} > 0$ is a constant, and ${}^{\mathbf{B}}\boldsymbol{\mu} \in \mathbb{R}^3$ denotes the vector part of the quaternion representing the rotation from frame $\{\mathbf{B}\}$ to frame $\{\mathbf{B}_d\}$, expressed in frame $\{\mathbf{B}\}$. Then, it follows that

$$\mathcal{V}_{pr} = \begin{bmatrix} \dot{\mathbf{q}}_r \\ {}^{\mathbf{B}}\boldsymbol{\omega}_r \end{bmatrix} \tag{9.14}$$

holds.

Cartesian Space Control

Let $\{\mathbf{O}_d\}$ be the reference frame of frame $\{\mathbf{O}\}$. When the position/orientation vector of the held object and the orientation vector of the (spacecraft) platform are chosen for the position control, the required linear/angular velocity vector of frame $\{\mathbf{O}\}$ is designed as

$$^{\mathbf{O}}V_r = \begin{bmatrix} {}^{\mathbf{O}}\mathbf{R}_{\mathbf{I}}\dot{\mathbf{x}}_{\mathbf{O}_d} + k_{\mathbf{O}x}{}^{\mathbf{O}}\mathbf{R}_{\mathbf{I}}\left(\mathbf{x}_{\mathbf{O}_d} - \mathbf{x}_{\mathbf{O}}\right) \\ {}^{\mathbf{O}}\mathbf{R}_{\mathbf{I}}\boldsymbol{\omega}_{\mathbf{O}_d} + k_{\mathbf{O}\omega}{}^{\mathbf{O}}\boldsymbol{\mu} \end{bmatrix} \tag{9.15}$$

where $k_{\mathbf{O}x} > 0$ and $k_{\mathbf{O}\omega} > 0$ are two positive gains, $\mathbf{x}_{\mathbf{O}_d} \in \mathbb{R}^3$ and $\mathbf{x}_{\mathbf{O}} \in \mathbb{R}^3$ denote the origin position vectors of frames $\{\mathbf{O}_d\}$ and $\{\mathbf{O}\}$, respectively, expressed in the inertial frame, $\boldsymbol{\omega}_{\mathbf{O}_d} \in \mathbb{R}^3$ denotes the angular velocity vector of frame $\{\mathbf{O}_d\}$ expressed in the inertial frame, and ${}^{\mathbf{O}}\boldsymbol{\mu} \in SO^3$ denotes the vector part of the quaternion representing the rotation from frame $\{\mathbf{O}\}$ to frame $\{\mathbf{O}_d\}$.

The required angular velocity vector of the (spacecraft) platform, denoted as ${}^{\mathbf{B}}\boldsymbol{\omega}_r$, is computed from (9.13).

9.5.2 Required Independent Velocities

When $\dot{\mathbf{q}}_r \in \mathbb{R}^6$ and ${}^{\mathbf{B}}\boldsymbol{\omega}_r \in \mathbb{R}^3$ are chosen for the position control, the required independent velocity coordinate vector is formed as

$$\mathcal{V}_r = \begin{bmatrix} \dot{\mathbf{q}}_r \\ \dot{\mathbf{q}}_{wr} \\ {}^{\mathbf{B}}V_r \end{bmatrix} \tag{9.16}$$

with

$$^{\mathbf{B}}V_r = \begin{bmatrix} {}^{\mathbf{B}}\mathbf{v}_r \\ {}^{\mathbf{B}}\boldsymbol{\omega}_r \end{bmatrix}.$$

In (9.16), ${}^{\mathbf{B}}\mathbf{v}_r \in \mathbb{R}^3$ (the required linear velocity vector of frame $\{\mathbf{B}\}$) and $\dot{\mathbf{q}}_{wr} \in \mathbb{R}^3$ are to be released to satisfy the six nonholonomic constraints imposed by the momentum conservation law.

When ${}^{\mathbf{O}}V_r \in \mathbb{R}^6$ and ${}^{\mathbf{B}}\boldsymbol{\omega}_r \in \mathbb{R}^3$ are chosen for the position control, the required independent velocity coordinate vector is formed as

$$\mathcal{V}_r = \begin{bmatrix} {}^{\mathbf{O}}V_r \\ \dot{\mathbf{q}}_{wr} \\ {}^{\mathbf{B}}V_r \end{bmatrix} \tag{9.17}$$

such that

$$\begin{bmatrix} \dot{\mathbf{q}}_r \\ \dot{\mathbf{q}}_{wr} \\ {}^{\mathbf{B}}V_r \end{bmatrix} = \begin{bmatrix} \mathbf{J}_q^{-1} & \mathbf{0} & -\mathbf{J}_q^{-1\mathbf{B}}\mathbf{U}_{\mathbf{O}}^T \\ \mathbf{0} & \mathbf{I}_3 & \mathbf{0} \\ \mathbf{0} & \mathbf{0} & \mathbf{I}_6 \end{bmatrix} \begin{bmatrix} {}^{\mathbf{O}}V_r \\ \dot{\mathbf{q}}_{wr} \\ {}^{\mathbf{B}}V_r \end{bmatrix} \tag{9.18}$$

holds from (9.5). Again, ${}^{\mathbf{B}}\mathbf{v}_r \in \mathbb{R}^3$ (the required linear velocity vector of frame $\{\mathbf{B}\}$) and $\dot{\mathbf{q}}_{wr} \in \mathbb{R}^3$ are to be released to satisfy the six nonholonomic constraints imposed by the momentum conservation law.

9.5.3 Required Velocity Mappings

Given

$$\begin{bmatrix} \dot{\mathbf{q}}_r \\ \dot{\mathbf{q}}_{wr} \\ {}^{\mathbf{B}}V_r \end{bmatrix} \in \mathbb{R}^{15}$$

the required extended velocity vector of all rigid bodies, defined by

$$\mathcal{V}_{br} = [{}^{\mathbf{O}}V_r^T, {}^{\mathbf{B}_1}V_r^T, \cdots, {}^{\mathbf{B}_6}V_r^T, {}^{\mathbf{W}_1}V_r^T, \cdots, {}^{\mathbf{W}_3}V_r^T, {}^{\mathbf{B}}V_r^T]^T \in \mathbb{R}^{66} \tag{9.19}$$

can be obtained by

$$\mathcal{V}_{br} = \mathcal{J}_b \begin{bmatrix} \dot{\mathbf{q}}_r \\ \dot{\mathbf{q}}_{wr} \\ {}^{\mathbf{B}}V_r \end{bmatrix} \tag{9.20}$$

with $\mathcal{J}_b \in \mathbb{R}^{66\times 15}$ being defined by (9.10).

Accordingly, the time-derivative of $\mathcal{V}_{br}$ is obtained by differentiating (9.20).

9.6 Dynamics and Control of the Held Object

This section deals with the dynamics and control issue and the *virtual stability* of the held object (payload).

9.6.1 Kinematics and Dynamics

In view of (2.63), the velocity transformation equation of the held object can be written as

$${}^{\mathbf{O}}V = {}^{\mathbf{C}}\mathbf{U}_{\mathbf{O}}^{T\,\mathbf{C}}V. \tag{9.21}$$

Recall (2.74) and substitute frame $\{\mathbf{O}\}$ for frame $\{\mathbf{A}\}$ leading to the dynamics of the held object as

$$\mathbf{M_O}\frac{d}{dt}({}^{\mathbf{O}}V) + \mathbf{C_O}({}^{\mathbf{O}}\omega){}^{\mathbf{O}}V + \mathbf{G_O} = {}^{\mathbf{O}}F^* \tag{9.22}$$

where ${}^{\mathbf{O}}F^* \in \mathbb{R}^6$ denotes the net force/moment vector of the held object, which is governed by

$${}^{\mathbf{O}}F^* = {}^{\mathbf{O}}\mathbf{U_C}{}^{\mathbf{C}}F. \tag{9.23}$$

9.6.2 Control Equations

The transformation equation for the required linear/angular velocity vectors is written as

$${}^{\mathbf{C}}V_r = {}^{\mathbf{O}}\mathbf{U}_{\mathbf{C}}^T{}^{\mathbf{O}}V_r. \tag{9.24}$$

Accordingly, recall (4.32) (by omitting the subscript i) to compute the required net force/moment vector of the held object as

$${}^{\mathbf{O}}F_r^* = \mathbf{Y_O}\hat{\boldsymbol{\theta}}_{\mathbf{O}} + \mathbf{K_O}\left({}^{\mathbf{O}}V_r - {}^{\mathbf{O}}V\right) \tag{9.25}$$

where $\mathbf{K_O} \in \mathbb{R}^{6\times6}$ is a symmetric positive-definite matrix and $\hat{\boldsymbol{\theta}}_{\mathbf{O}} \in \mathbb{R}^{13}$ denotes the estimate of $\boldsymbol{\theta}_{\mathbf{O}} \in \mathbb{R}^{13}$ with $\mathbf{Y_O}\boldsymbol{\theta}_{\mathbf{O}}$ being defined by (2.78) and expressed in Appendix A by substituting frame $\{\mathbf{O}\}$ for frame $\{\mathbf{A}\}$.

The parameter estimates need to be updated. Define

$$\mathbf{s_O} = \mathbf{Y}_{\mathbf{O}}^T\left({}^{\mathbf{O}}V_r - {}^{\mathbf{O}}V\right). \tag{9.26}$$

The $\mathcal{P}$ function defined by (2.79) is used to update each of the 13 parameter estimates of the held object as

$$\hat{\theta}_{\mathbf{O}\gamma} = \mathcal{P}\left(s_{\mathbf{O}\gamma}, \rho_{\mathbf{O}\gamma}, \underline{\theta}_{\mathbf{O}\gamma}, \overline{\theta}_{\mathbf{O}\gamma}, t\right) \tag{9.27}$$

for all $\gamma \in \{1, 13\}$, where

$\hat{\theta}_{\mathbf{O}\gamma}$	γth element of $\hat{\boldsymbol{\theta}}_{\mathbf{O}}$.
$s_{\mathbf{O}\gamma}$	γth elements of $\mathbf{s_O}$.
$\rho_{\mathbf{O}\gamma}$	Parameter update gain.
$\underline{\theta}_{\mathbf{O}\gamma}$	Lower bound of $\theta_{\mathbf{O}\gamma}$.
$\overline{\theta}_{\mathbf{O}\gamma}$	Upper bound of $\theta_{\mathbf{O}\gamma}$.
$\theta_{\mathbf{O}\gamma}$	γth element of $\boldsymbol{\theta}_{\mathbf{O}}$.

After ${}^{\mathbf{O}}F_r^* \in \mathbb{R}^6$ being obtained from (9.25), the required force/moment vector in frame $\{\mathbf{C}\}$ is computed as

$${}^{\mathbf{C}}F_r = {}^{\mathbf{C}}\mathbf{U_O}{}^{\mathbf{O}}F_r^*. \tag{9.28}$$

9.6.3 Virtual Stability

The *virtual stability* of the held object, subject to the respective control equations, is presented in this subsection.

Theorem 9.1. *The held object described by (9.21)–(9.23), combined with its respective control equations (9.24), (9.25), and (9.28) and with the parameter adaptation (9.26) and (9.27), is virtually stable with its affiliated vector* ${}^{\mathbf{O}}V_r - {}^{\mathbf{O}}V$ *being a virtual function in both* L_2 *and* L_∞*, in the sense of Definition 2.17.*

Proof: The non-negative accompanying function of the held object is chosen as

$$\begin{aligned}\nu_{\mathbf{O}} = &\frac{1}{2}\left({}^{\mathbf{O}}V_r - {}^{\mathbf{O}}V\right)^T \mathbf{M}_{\mathbf{O}} \left({}^{\mathbf{O}}V_r - {}^{\mathbf{O}}V\right) \\ &+\frac{1}{2}\sum_{\gamma=1}^{13}\left(\theta_{\mathbf{O}\gamma} - \hat{\theta}_{\mathbf{O}\gamma}\right)^2 / \rho_{\mathbf{O}\gamma}. \end{aligned} \tag{9.29}$$

It follows from (9.22), (9.25)–(9.27), and Lemma 4.1 that

$$\begin{aligned}\dot{\nu}_{\mathbf{O}} \leqslant &-\left({}^{\mathbf{O}}V_r - {}^{\mathbf{O}}V\right)^T \mathbf{K}_{\mathbf{O}} \left({}^{\mathbf{O}}V_r - {}^{\mathbf{O}}V\right) \\ &+\left({}^{\mathbf{O}}V_r - {}^{\mathbf{O}}V\right)^T \left({}^{\mathbf{O}}F_r^* - {}^{\mathbf{O}}F^*\right) \end{aligned} \tag{9.30}$$

holds.

In view of (2.85), (9.21), (9.23), (9.24), and (9.28), it yields

$$\begin{aligned}&\left({}^{\mathbf{O}}V_r - {}^{\mathbf{O}}V\right)^T \left({}^{\mathbf{O}}F_r^* - {}^{\mathbf{O}}F^*\right) \\ &= \left({}^{\mathbf{O}}V_r - {}^{\mathbf{O}}V\right)^T {}^{\mathbf{O}}\mathbf{U}_{\mathbf{C}} \left({}^{\mathbf{C}}F_r - {}^{\mathbf{C}}F\right) \\ &= p_{\mathbf{C}}. \end{aligned} \tag{9.31}$$

Consider the fact that the held object has only one *driven cutting point* associated with frame $\{\mathbf{C}\}$. Substituting (9.31) into (9.30) and using (9.29) and Definition 2.17 completes the proof. ■

9.7 Dynamics and Control of the Manipulator

The manipulator consists of six links and six joints. It will be treated with an approach similar to that in Chapter 5.

9.7.1 Kinematics and Dynamics

In view of (2.63) and (9.8), the velocities of the links and joints of the manipulator are governed by

$$ {}^{\mathbf{B}_6}V = {}^{\mathbf{C}}\mathbf{U}_{\mathbf{B}_6}^T \, {}^{\mathbf{C}}V \tag{9.32} $$

$$ {}^{\mathbf{B}_k}V = {}^{\mathbf{T}_{k+1}}\mathbf{U}_{\mathbf{B}_k}^T \, {}^{\mathbf{T}_{k+1}}V, \; \forall k \in \{1,5\} \tag{9.33} $$

$$ {}^{\mathbf{B}_k}V = \mathbf{z}\dot{q}_k + {}^{\mathbf{T}_k}\mathbf{U}_{\mathbf{B}_k}^T \, {}^{\mathbf{T}_k}V, \; \forall k \in \{2,6\} \tag{9.34} $$

$$ {}^{\mathbf{B}_1}V = \mathbf{z}\dot{q}_1 + {}^{\mathbf{A}_m}\mathbf{U}_{\mathbf{B}_1}^T \, {}^{\mathbf{A}_m}V. \tag{9.35} $$

Recall (2.74). The dynamics of the six links are written as

$$ \mathbf{M}_{\mathbf{B}_k}\frac{d}{dt}({}^{\mathbf{B}_k}V) + \mathbf{C}_{\mathbf{B}_k}({}^{\mathbf{B}_k}\omega){}^{\mathbf{B}_k}V + \mathbf{G}_{\mathbf{B}_k} = {}^{\mathbf{B}_k}F^* \tag{9.36} $$

by substituting frame $\{\mathbf{B}_k\}$ for frame $\{\mathbf{A}\}$ for all $k \in \{1,6\}$.

Recall (4.74). The dynamics of the six joints are written as

$$ J_{mk}\ddot{q}_k + \xi_k(q_k, \dot{q}_k) = \tau_k - \mathbf{z}^T{}^{\mathbf{B}_k}F \tag{9.37} $$

for all $k \in \{1,6\}$, where $J_{mk} \in \mathbb{R}$ is the equivalent mass or moment of inertia, $\xi_k(q_k, \dot{q}_k) \in \mathbb{R}$ is a term representing the friction force/torque, $\tau_k \in \mathbb{R}$ is the control force/torque, and $\mathbf{z}^T{}^{\mathbf{B}_k}F$ is the projection of the force/moment vector in frame $\{\mathbf{B}_k\}$ onto the kth joint axis.

Furthermore, the force resultant equations of the six links are written as

$$ {}^{\mathbf{B}_6}F^* = {}^{\mathbf{B}_6}F - {}^{\mathbf{B}_6}\mathbf{U}_{\mathbf{C}} \, {}^{\mathbf{C}}F \tag{9.38} $$

$$ {}^{\mathbf{B}_k}F^* = {}^{\mathbf{B}_k}F - {}^{\mathbf{B}_k}\mathbf{U}_{\mathbf{T}_{j+1}} \, {}^{\mathbf{T}_{j+1}}F, \; \forall k \in \{1,5\} \tag{9.39} $$

with

$$ {}^{\mathbf{T}_k}F = {}^{\mathbf{T}_k}\mathbf{U}_{\mathbf{B}_k} \, {}^{\mathbf{B}_k}F, \; \forall k \in \{2,6\} \tag{9.40} $$

$$ {}^{\mathbf{A}_m}F = {}^{\mathbf{A}_m}\mathbf{U}_{\mathbf{B}_1} \, {}^{\mathbf{B}_1}F. \tag{9.41} $$

9.7.2 Control Equations

In view of (9.20), the required velocities of the links and joints of the manipulator can be respectively written as

$$ {}^{\mathbf{B}_6}V_r = {}^{\mathbf{C}}\mathbf{U}_{\mathbf{B}_6}^T \, {}^{\mathbf{C}}V_r \tag{9.42} $$

$$ {}^{\mathbf{B}_k}V_r = {}^{\mathbf{T}_{k+1}}\mathbf{U}_{\mathbf{B}_k}^T \, {}^{\mathbf{T}_{k+1}}V_r, \; \forall k \in \{1,5\} \tag{9.43} $$

$$ {}^{\mathbf{B}_k}V_r = \mathbf{z}\dot{q}_{kr} + {}^{\mathbf{T}_k}\mathbf{U}_{\mathbf{B}_k}^T \, {}^{\mathbf{T}_k}V_r, \; \forall k \in \{2,6\} \tag{9.44} $$

$$ {}^{\mathbf{B}_1}V_r = \mathbf{z}\dot{q}_{1r} + {}^{\mathbf{A}_m}\mathbf{U}_{\mathbf{B}_1}^T \, {}^{\mathbf{A}_m}V_r \tag{9.45} $$

where $\dot{q}_{kr}$ is the kth element of $\dot{\mathbf{q}}_r \in \mathbb{R}^6$ defined by (9.12) or (9.16).

Recall (4.32) with frame $\{\mathbf{B}_k\}$ being substituted for frame $\{\mathbf{O}_i\}$. The required net force/moment vectors of the six links are computed as

$$ {}^{\mathbf{B}_k}F_r^* = \mathbf{Y}_{\mathbf{B}_k}\hat{\boldsymbol{\theta}}_{\mathbf{B}_k} + \mathbf{K}_{\mathbf{B}_k}\left({}^{\mathbf{B}_k}V_r - {}^{\mathbf{B}_k}V\right), \; \forall k \in \{1,6\} \tag{9.46} $$

where $\mathbf{K}_{\mathbf{B}_k} \in \mathbb{R}^{6\times 6}$ is a symmetric positive-definite matrix and $\hat{\boldsymbol{\theta}}_{\mathbf{B}_k} \in \mathbb{R}^{13}$ denotes the estimate of $\boldsymbol{\theta}_{\mathbf{B}_k} \in \mathbb{R}^{13}$ with $\mathbf{Y}_{\mathbf{B}_k}\boldsymbol{\theta}_{\mathbf{B}_k}$ being defined by (2.78) and expressed in Appendix A by substituting frame $\{\mathbf{B}_k\}$ for frame $\{\mathbf{A}\}$ for all $k \in \{1, 6\}$.

The parameter estimates need to be updated. Define

$$\mathbf{s}_{\mathbf{B}_k} = \mathbf{Y}_{\mathbf{B}_k}^T \left({}^{\mathbf{B}_k}V_r - {}^{\mathbf{B}_k}V\right), \ \forall k \in \{1, 6\}. \tag{9.47}$$

The $\mathcal{P}$ function defined by (2.79) is used to update each of the 13 parameters of link k as

$$\hat{\theta}_{\mathbf{B}_k\gamma} = \mathcal{P}\left(s_{\mathbf{B}_k\gamma}, \rho_{\mathbf{B}_k\gamma}, \underline{\theta}_{\mathbf{B}_k\gamma}, \overline{\theta}_{\mathbf{B}_k\gamma}, t\right) \tag{9.48}$$

for all $k \in \{1, 6\}$ and $\gamma \in \{1, 13\}$, where

$\hat{\theta}_{\mathbf{B}_k\gamma}$	γth element of $\hat{\boldsymbol{\theta}}_{\mathbf{B}_k}$.
$s_{\mathbf{B}_k\gamma}$	γth element of $\mathbf{s}_{\mathbf{B}_k}$.
$\rho_{\mathbf{B}_k\gamma}$	Parameter update gain.
$\underline{\theta}_{\mathbf{B}_k\gamma}$	Lower bound of $\theta_{\mathbf{B}_k\gamma}$.
$\overline{\theta}_{\mathbf{B}_k\gamma}$	Upper bound of $\theta_{\mathbf{B}_k\gamma}$.
$\theta_{\mathbf{B}_k\gamma}$	γth element of $\boldsymbol{\theta}_{\mathbf{B}_k}$.

Furthermore, the required force/moment vectors at the six joints are computed as

$$\begin{aligned}
{}^{\mathbf{B}_6}F_r &= {}^{\mathbf{B}_6}F_r^* + {}^{\mathbf{B}_6}\mathbf{U}_{\mathbf{C}}\,{}^{\mathbf{C}}F_r && (9.49)\\
{}^{\mathbf{T}_k}F_r &= {}^{\mathbf{T}_k}\mathbf{U}_{\mathbf{B}_k}\,{}^{\mathbf{B}_k}F_r, \ k = 6, \cdots, 2 && (9.50)\\
{}^{\mathbf{B}_k}F_r &= {}^{\mathbf{B}_k}F_r^* + {}^{\mathbf{B}_k}\mathbf{U}_{\mathbf{T}_{k+1}}\,{}^{\mathbf{T}_{k+1}}F_r, \ k = 5, \cdots, 1 && (9.51)\\
{}^{\mathbf{A}_m}F_r &= {}^{\mathbf{A}_m}\mathbf{U}_{\mathbf{B}_1}\,{}^{\mathbf{B}_1}F_r. && (9.52)
\end{aligned}$$

Finally, recall (4.77) and (4.78). The desired joint force/torque for the kth joint is designed as

$$\begin{aligned}
\tau_{kd}^* &= \mathbf{Y}_{\tau k}\hat{\boldsymbol{\theta}}_{\tau k} + k_{\tau k}(\dot{q}_{kr} - \dot{q}_k) && (9.53)\\
\tau_{kd} &= \tau_{kd}^* + \mathbf{z}^T\,{}^{\mathbf{B}_k}F_r \\
&= \mathbf{Y}_{\tau k}\hat{\boldsymbol{\theta}}_{\tau k} + \mathbf{z}^T\,{}^{\mathbf{B}_k}F_r + k_{\tau k}(\dot{q}_{kr} - \dot{q}_k) && (9.54)
\end{aligned}$$

with $k_{\tau k} > 0$, subject to

$$\mathbf{Y}_{\tau k}\boldsymbol{\theta}_{\tau k} = J_{mk}\ddot{q}_{kr} + \xi_k(q_k, \dot{q}_k) \tag{9.55}$$

where $\dot{q}_{kr}$ is the kth element of $\dot{\mathbf{q}}_r \in \mathbb{R}^6$ for all $k \in \{1, 6\}$.

Similar to (5.52), define

$$\mathbf{s}_{\tau k} = \mathbf{Y}_{\tau k}^T\left(\dot{q}_{kr} - \dot{q}_k\right). \tag{9.56}$$

The estimate of $\boldsymbol{\theta}_{\tau k}$ is updated by using the $\mathcal{P}$ function defined by (2.79) as

$$\hat{\theta}_{\tau k\gamma} = \mathcal{P}\left(s_{\tau k\gamma}, \rho_{\tau k\gamma}, \underline{\theta}_{\tau k\gamma}, \overline{\theta}_{\tau k\gamma}, t\right) \tag{9.57}$$

where

$\theta_{\mathcal{T}k\gamma}$	γth element of $\boldsymbol{\theta}_{\mathcal{T}k}$.
$\hat{\theta}_{\mathcal{T}k\gamma}$	Estimate of $\theta_{\mathcal{T}k\gamma}$.
$s_{\mathcal{T}k\gamma}$	γth element of $\mathbf{s}_{\mathcal{T}k}$.
$\rho_{\mathcal{T}k\gamma}$	Parameter update gain.
$\underline{\theta}_{\mathcal{T}k\gamma}$	Lower bound of $\theta_{\mathcal{T}k\gamma}$.
$\overline{\theta}_{\mathcal{T}k\gamma}$	Upper bound of $\theta_{\mathcal{T}k\gamma}$.

9.7.3 Virtual Stability

The following theorems indicate that the manipulator combined with its respective control equations developed in this section is *virtually stable* in the sense of Definition 2.17.

Theorem 9.2. *The kth link, $k \in \{1, 6\}$, described by (9.32), (9.33), (9.36), (9.38), and (9.39), combined with the control equations (9.42), (9.43), (9.46), (9.49), and (9.51) and with the parameter adaptation (9.47) and (9.48), is virtually stable with its affiliated vector ${}^{\mathbf{B}_k}V_r - {}^{\mathbf{B}_k}V$ being a virtual function in both L_2 and L_∞, in the sense of Definition 2.17.*

Proof: Let the non-negative accompanying function assigned to the kth link be

$$\nu_{\mathbf{B}_k} = \frac{1}{2}\left({}^{\mathbf{B}_k}V_r - {}^{\mathbf{B}_k}V\right)^T \mathbf{M}_{\mathbf{B}_k}\left({}^{\mathbf{B}_k}V_r - {}^{\mathbf{B}_k}V\right) + \frac{1}{2}\sum_{\gamma=1}^{13}\left(\theta_{\mathbf{B}_k\gamma} - \hat{\theta}_{\mathbf{B}_k\gamma}\right)^2 / \rho_{\mathbf{B}_k\gamma} \tag{9.58}$$

for all $k \in \{1, 6\}$.

By substituting frame $\{\mathbf{B}_k\}$ for frame $\{\mathbf{O}_i\}$, it follows from (9.36), (9.46)–(9.48), and Lemma 4.1 that

$$\dot{\nu}_{\mathbf{B}_k} \leqslant -\left({}^{\mathbf{B}_k}V_r - {}^{\mathbf{B}_k}V\right)^T \mathbf{K}_{\mathbf{B}_k}\left({}^{\mathbf{B}_k}V_r - {}^{\mathbf{B}_k}V\right) + \left({}^{\mathbf{B}_k}V_r - {}^{\mathbf{B}_k}V\right)^T\left({}^{\mathbf{B}_k}F_r^* - {}^{\mathbf{B}_k}F^*\right) \tag{9.59}$$

holds for all $k \in \{1, 6\}$.

In view of (2.85), (9.32), (9.33), (9.38), (9.39), (9.42), (9.43), (9.49), and (9.51), it then yields

$$\begin{aligned}
&\left({}^{\mathbf{B}_6}V_r - {}^{\mathbf{B}_6}V\right)^T\left({}^{\mathbf{B}_6}F_r^* - {}^{\mathbf{B}_6}F^*\right)\\
&- \left({}^{\mathbf{B}_6}V_r - {}^{\mathbf{B}_6}V\right)^T\left[\left({}^{\mathbf{B}_6}F_r - {}^{\mathbf{B}_6}F\right) - {}^{\mathbf{B}_6}\mathbf{U}_{\mathbf{C}}\left({}^{\mathbf{C}}F_r - {}^{\mathbf{C}}F\right)\right]\\
&= p_{\mathbf{B}_6} - \left[{}^{\mathbf{B}_6}\mathbf{U}_{\mathbf{C}}^T\left({}^{\mathbf{B}_6}V_r - {}^{\mathbf{B}_6}V\right)\right]^T\left({}^{\mathbf{C}}F_r - {}^{\mathbf{C}}F\right)\\
&= p_{\mathbf{B}_6} - p_{\mathbf{C}}
\end{aligned} \tag{9.60}$$

$$\begin{aligned}
&\left({}^{\mathbf{B}_k}V_r - {}^{\mathbf{B}_k}V\right)^T\left({}^{\mathbf{B}_k}F_r^* - {}^{\mathbf{B}_k}F^*\right)\\
&= \left({}^{\mathbf{B}_k}V_r - {}^{\mathbf{B}_k}V\right)^T\left[\left({}^{\mathbf{B}_k}F_r - {}^{\mathbf{B}_k}F\right) - {}^{\mathbf{B}_k}\mathbf{U}_{\mathbf{T}_{k+1}}\left({}^{\mathbf{T}_{k+1}}F_r - {}^{\mathbf{T}_{k+1}}F\right)\right]\\
&= p_{\mathbf{B}_k} - \left[{}^{\mathbf{B}_k}\mathbf{U}_{\mathbf{T}_{k+1}}^T\left({}^{\mathbf{B}_k}V_r - {}^{\mathbf{B}_k}V\right)\right]^T\left({}^{\mathbf{T}_{k+1}}F_r - {}^{\mathbf{T}_{k+1}}F\right)\\
&= p_{\mathbf{B}_k} - p_{\mathbf{T}_{k+1}}, \ \forall\, k \in \{1, 5\}.
\end{aligned} \tag{9.61}$$

Consider the fact that the kth link of the manipulator has one *driving cutting point* associated with frame $\{\mathbf{T}_{k+1}\}$ for all $k \in \{1,5\}$ or associated with frame $\{\mathbf{C}\}$ for $k = 6$ and one *driven cutting point* associated with frame $\{\mathbf{B}_k\}$ for all $k \in \{1,6\}$. Substituting (9.60) and (9.61) into (9.59) and using (9.58) and Definition 2.17 completes the proof. ∎

Theorem 9.3. *The kth joint of the manipulator, $k \in \{1,6\}$, described by (9.34), (9.35), (9.37), (9.40), and (9.41), combined with its control equations (9.44), (9.45), (9.50), (9.52)–(9.55), and $\tau_k = \tau_{kd}$ and with the parameter adaptation (9.56) and (9.57), is virtually stable with its affiliated variable $\dot{q}_{kr} - \dot{q}_k$ being a virtual function in both L_2 and L_∞, in the sense of Definition 2.17.*

Proof: The proof is similar to the process leading to Theorem 4.3. Subtracting (9.37) from (9.54) and using (9.53) and (9.55) yields

$$\begin{aligned} J_{mk}(\ddot{q}_{kr} - \ddot{q}_k) &= \tau_{kd} - \tau_k + \mathbf{Y}_{\tau k}(\boldsymbol{\theta}_{\tau k} - \hat{\boldsymbol{\theta}}_{\tau k}) \\ &\quad -\mathbf{z}^T({}^{\mathbf{B}_k}F_r - {}^{\mathbf{B}_k}F) - k_{\tau k}(\dot{q}_{kr} - \dot{q}_k). \end{aligned} \tag{9.62}$$

Let

$$\nu_{\tau k} = \frac{1}{2} J_{mk}(\dot{q}_{kr} - \dot{q}_k)^2 + \frac{1}{2}\sum_{\gamma}\left(\theta_{\tau k\gamma} - \hat{\theta}_{\tau k\gamma}\right)^2 / \rho_{\tau k\gamma} \tag{9.63}$$

be the non-negative accompanying function assigned to joint k. It follows from (9.56), (9.57), (9.62), and Lemma 2.9 that

$$\begin{aligned} \dot{\nu}_{\tau k} &= -k_{\tau k}(\dot{q}_{kr} - \dot{q}_k)^2 + (\dot{q}_{kr} - \dot{q}_k)(\tau_{kd} - \tau_k) \\ &\quad -(\dot{q}_{kr} - \dot{q}_k)\mathbf{z}^T({}^{\mathbf{B}_k}F_r - {}^{\mathbf{B}_k}F) \\ &\quad +(\boldsymbol{\theta}_{\tau k} - \hat{\boldsymbol{\theta}}_{\tau k})^T \mathbf{Y}_{\tau k}^T(\dot{q}_{kr} - \dot{q}_k) - \sum_{\gamma}\left(\theta_{\tau k\gamma} - \hat{\theta}_{\tau k\gamma}\right)\dot{\hat{\theta}}_{\tau k\gamma}/\rho_{\tau k\gamma} \\ &= -k_{\tau k}(\dot{q}_{kr} - \dot{q}_k)^2 + (\dot{q}_{kr} - \dot{q}_k)(\tau_{kd} - \tau_k) \\ &\quad -(\dot{q}_{kr} - \dot{q}_k)\mathbf{z}^T({}^{\mathbf{B}_k}F_r - {}^{\mathbf{B}_k}F) \\ &\quad +\sum_{\gamma}\left(\theta_{\tau k\gamma} - \hat{\theta}_{\tau k\gamma}\right)\left(s_{\tau k\gamma} - \dot{\hat{\theta}}_{\tau k\gamma}/\rho_{\tau k\gamma}\right) \\ &\leqslant -k_{\tau k}(\dot{q}_{kr} - \dot{q}_k)^2 + (\dot{q}_{kr} - \dot{q}_k)(\tau_{kd} - \tau_k) \\ &\quad -(\dot{q}_{kr} - \dot{q}_k)\mathbf{z}^T({}^{\mathbf{B}_k}F_r - {}^{\mathbf{B}_k}F) \end{aligned} \tag{9.64}$$

holds.

Furthermore, in view of (2.85), (9.34), (9.35), (9.40), (9.41), (9.44), (9.45), (9.50), and (9.52), the last term in the right hand side of (9.64) can be written as

$$\begin{aligned} &-(\dot{q}_{kr} - \dot{q}_k)\mathbf{z}^T({}^{\mathbf{B}_k}F_r - {}^{\mathbf{B}_k}F) \\ &= -\left[({}^{\mathbf{B}_k}V_r - {}^{\mathbf{B}_k}V) - {}^{\mathbf{T}_k}\mathbf{U}_{\mathbf{B}_k}^T({}^{\mathbf{T}_k}V_r - {}^{\mathbf{T}_k}V)\right]^T({}^{\mathbf{B}_k}F_r - {}^{\mathbf{B}_k}F) \\ &= -p_{\mathbf{B}_k} + p_{\mathbf{T}_k} \end{aligned} \tag{9.65}$$

for all $k \in \{2, 6\}$ and

$$\begin{aligned}
&-(\dot{q}_{1r} - \dot{q}_1)\mathbf{z}^T({}^{\mathbf{B}_1}F_r - {}^{\mathbf{B}_1}F) \\
&= -\left[({}^{\mathbf{B}_1}V_r - {}^{\mathbf{B}_1}V) - {}^{\mathbf{A}_m}\mathbf{U}_{\mathbf{B}_1}^T\left({}^{\mathbf{A}_m}V_r - {}^{\mathbf{A}_m}V\right)\right]^T({}^{\mathbf{B}_1}F_r - {}^{\mathbf{B}_1}F) \\
&= -p_{\mathbf{B}_1} + p_{\mathbf{A}_m}.
\end{aligned} \tag{9.66}$$

Consider the fact that the kth joint of the manipulator has one *driving cutting point* associated with frame $\{\mathbf{B}_k\}$ for all $k \in \{1, 6\}$ and one *driven cutting point* associated with frame $\{\mathbf{A}_m\}$ for $k = 1$ or associated with frame $\{\mathbf{T}_k\}$ for all $k \in \{2, 6\}$. With $\tau_k = \tau_{kd}$, substituting (9.65) and (9.66) into (9.64) and using (9.63) and Definition 2.17 completes the proof. ■

Theorem 9.4. *The manipulator comprised of six links and six joints as described by (9.32)–(9.41), combined with the control equations (9.42)–(9.46), (9.49)–(9.55), and $\tau_k = \tau_{kd}$ for all $k \in \{1, 6\}$ and with the parameter adaptation (9.47), (9.48), (9.56), and (9.57), is virtually stable with its affiliated vector ${}^{\mathbf{B}_k}V_r - {}^{\mathbf{B}_k}V$ and variable $\dot{q}_{kr} - \dot{q}_k$ being virtual functions in both L_2 and L_∞ for all $k \in \{1, 6\}$, in the sense of Definition 2.17.*

Proof: Theorem 9.2 ensures that each link with its control equations is *virtually stable*, and Theorem 9.3 ensures that each joint with its control equations is *virtually stable*. In view of Lemma 2.11, the fact that the *driven cutting point* of link 6 connects the *driving cutting point* of joint 6 at frame $\{\mathbf{B}_6\}$ guarantees that the new subsystem comprised of both link 6 and joint 6 is *virtually stable*. Continuing this process leads to the proof of the theorem. ■

9.8 Dynamics and Control of the Reaction Wheels

Each reaction wheel consists of a rigid body (the reaction wheel) and a joint. Therefore, it can be treated as a special *open chain* with only one joint.

9.8.1 Kinematics and Dynamics

In view of (9.8), the velocity relationship of the jth reaction wheel is governed by

$${}^{\mathbf{W}_j}V = \mathbf{z}_w\dot{q}_{wj} + {}^{\mathbf{A}_j}\mathbf{U}_{\mathbf{W}_j}^T\,{}^{\mathbf{A}_j}V \tag{9.67}$$

for all $j \in \{1, 3\}$, where $\mathbf{z}_w = [0, 0, 0, 0, 0, 1]^T$.

Recall (2.74). The dynamics of the jth reaction wheel are written as

$$\mathbf{M}_{\mathbf{W}_j}\frac{d}{dt}({}^{\mathbf{W}_j}V) + \mathbf{C}_{\mathbf{W}_j}({}^{\mathbf{W}_j}\omega){}^{\mathbf{W}_j}V + \mathbf{G}_{\mathbf{W}_j} = {}^{\mathbf{W}_j}F^* \tag{9.68}$$

by substituting frame $\{\mathbf{W}_j\}$ for frame $\{\mathbf{A}\}$ for all $j \in \{1, 3\}$.

With a similar approach leading to (9.37), the dynamics of the jth joint are written as

$$J_{wj}\ddot{q}_{wj} + \xi_{wj}(q_{wj}, \dot{q}_{wj}) = \tau_{wj} - \mathbf{z}_w^T {}^{\mathbf{W}_j}F \tag{9.69}$$

where $J_{wj} \in \mathbb{R}$ is the equivalent moment of inertia, $\xi_{wj}(q_{wj}, \dot{q}_{wj}) \in \mathbb{R}$ is a term representing the friction torque, $\tau_{wj} \in \mathbb{R}$ is the control torque, and $\mathbf{z}^T {}^{\mathbf{W}_j}F$ is the projection of the force/moment vector in frame $\{\mathbf{W}_j\}$ onto the jth joint axis.

Again, the force resultant equations of the jth reaction wheel are

$$ {}^{\mathbf{W}_j}F^* = {}^{\mathbf{W}_j}F \tag{9.70}$$

$$ {}^{\mathbf{W}_j}F = {}^{\mathbf{W}_j}\mathbf{U}_{\mathbf{A}_j} {}^{\mathbf{A}_j}F \tag{9.71}$$

for all $j \in \{1, 3\}$.

9.8.2 Control Equations

Both reaction wheel control equations and joint control equations are presented in this subsection.

Reaction Wheel Control

In view of (9.20), the required velocities of the jth reaction wheel can be written as

$$ {}^{\mathbf{W}_j}V_r = \mathbf{z}_w \dot{q}_{wjr} + {}^{\mathbf{A}_j}\mathbf{U}_{\mathbf{W}_j}^T {}^{\mathbf{A}_j}V_r \tag{9.72}$$

for all $j \in \{1, 3\}$.

Recall (4.32). The required net force/moment vector of the jth reaction wheel is computed as

$$ {}^{\mathbf{W}_j}F_r^* = \mathbf{Y}_{\mathbf{W}_j}\hat{\boldsymbol{\theta}}_{\mathbf{W}_j} + \mathbf{K}_{\mathbf{W}_j}\left({}^{\mathbf{W}_j}V_r - {}^{\mathbf{W}_j}V\right) \tag{9.73}$$

where $\mathbf{K}_{\mathbf{W}_j} \in \mathbb{R}^{6\times 6}$ is a symmetric positive-definite matrix and $\hat{\boldsymbol{\theta}}_{\mathbf{W}_j} \in \mathbb{R}^{13}$ denotes the estimate of $\boldsymbol{\theta}_{\mathbf{W}_j} \in \mathbb{R}^{13}$ with $\mathbf{Y}_{\mathbf{W}_j}\boldsymbol{\theta}_{\mathbf{W}_j}$ being defined by (2.78) and expressed in Appendix A by substituting frame $\{\mathbf{W}_j\}$ for frame $\{\mathbf{A}\}$ for all $j \in \{1, 3\}$.

The parameter estimate vector $\hat{\boldsymbol{\theta}}_{\mathbf{W}_j} \in \mathbb{R}^{13}$ needs to be updated. Define

$$\mathbf{s}_{\mathbf{W}_j} = \mathbf{Y}_{\mathbf{W}_j}^T\left({}^{\mathbf{W}_j}V_r - {}^{\mathbf{W}_j}V\right), \ \forall j \in \{1, 3\}. \tag{9.74}$$

The $\mathcal{P}$ function defined by (2.79) is used to update each of the 13 parameters of the jth reaction wheel as

$$\hat{\theta}_{\mathbf{W}_j\gamma} = \mathcal{P}\left(s_{\mathbf{W}_j\gamma}, \rho_{\mathbf{W}_j\gamma}, \underline{\theta}_{\mathbf{W}_j\gamma}, \overline{\theta}_{\mathbf{W}_j\gamma}, t\right) \tag{9.75}$$

for all $j \in \{1, 3\}$ and $\gamma \in \{1, 13\}$, where

$\hat{\theta}_{\mathbf{W}_j\gamma}$	Estimate of $\theta_{\mathbf{W}_j\gamma}$.
$s_{\mathbf{W}_j\gamma}$	γth element of $\mathbf{s}_{\mathbf{W}_j}$.
$\rho_{\mathbf{W}_j\gamma}$	Parameter update gain.
$\underline{\theta}_{\mathbf{W}_j\gamma}$	Lower bound of $\theta_{\mathbf{W}_j\gamma}$.
$\overline{\theta}_{\mathbf{W}_j\gamma}$	Upper bound of $\theta_{\mathbf{W}_j\gamma}$.
$\theta_{\mathbf{W}_j\gamma}$	γth element of $\boldsymbol{\theta}_{\mathbf{W}_j}$.

Then, the required force/moment vectors of the jth reaction wheel are computed as

$$
\begin{aligned}
{}^{\mathbf{W}_j}F_r &= {}^{\mathbf{W}_j}F_r^* &(9.76)\\
{}^{\mathbf{A}_j}F_r &= {}^{\mathbf{A}_j}\mathbf{U}_{\mathbf{W}_j}\,{}^{\mathbf{W}_j}F_r &(9.77)
\end{aligned}
$$

for all $j \in \{1, 3\}$.

Joint Control

Recall (4.77) and (4.78). The desired joint torque for the jth reaction wheel joint is computed as

$$
\begin{aligned}
\tau_{wjd}^* &= \mathbf{Y}_{w\tau j}\hat{\boldsymbol{\theta}}_{w\tau j} + k_{w\tau j}(\dot{q}_{wjr} - \dot{q}_{wj}) &(9.78)\\
\tau_{wjd} &= \tau_{wjd}^* + \mathbf{z}_w^T\,{}^{\mathbf{W}_j}F_r\\
&= \mathbf{Y}_{w\tau j}\hat{\boldsymbol{\theta}}_{w\tau j} + \mathbf{z}_w^T\,{}^{\mathbf{W}_j}F_r + k_{w\tau j}(\dot{q}_{wjr} - \dot{q}_{wj}) &(9.79)
\end{aligned}
$$

with $k_{w\tau j} > 0$, subject to

$$
\mathbf{Y}_{w\tau j}\boldsymbol{\theta}_{w\tau c} = J_{wj}\ddot{q}_{wjr} + \xi_{wj}(q_{wj}, \dot{q}_{wj}) \tag{9.80}
$$

for all $j \in \{1, 3\}$, where $\dot{q}_{wjr}$ is the jth element of $\dot{\mathbf{q}}_{wr} \in \mathbb{R}^3$.

Define

$$
\mathbf{s}_{w\tau j} = \mathbf{Y}_{w\tau j}^T\left(\dot{q}_{wjr} - \dot{q}_{wj}\right). \tag{9.81}
$$

The estimate of $\boldsymbol{\theta}_{w\tau j}$ is updated by using the $\mathcal{P}$ function defined by (2.79) as

$$
\hat{\theta}_{w\tau j\gamma} = \mathcal{P}\left(s_{w\tau j\gamma}, \rho_{w\tau j\gamma}, \underline{\theta}_{w\tau j\gamma}, \overline{\theta}_{w\tau j\gamma}, t\right) \tag{9.82}
$$

where

$\theta_{w\tau j\gamma}$	γth element of $\boldsymbol{\theta}_{w\tau j}$.
$\hat{\theta}_{w\tau j\gamma}$	Estimate of $\theta_{w\tau j\gamma}$.
$s_{w\tau j\gamma}$	γth element of $\mathbf{s}_{w\tau j}$.
$\rho_{w\tau j\gamma}$	Parameter update gain.
$\underline{\theta}_{w\tau j\gamma}$	Lower bound of $\theta_{w\tau j\gamma}$.
$\overline{\theta}_{w\tau j\gamma}$	Upper bound of $\theta_{w\tau j\gamma}$.

9.8.3 Virtual Stability

The *virtual stability* of the jth reaction wheel (with its joint), subject to the control equations, is presented in this subsection.

Theorem 9.5. *The jth reaction wheels with its joint, $j \in \{1, 3\}$, described by (9.67)–(9.71), combined with the respective control equations (9.72), (9.73), (9.76)–(9.80), and $\tau_{wj} = \tau_{wjd}$ and with the parameter adaptation (9.74), (9.75), (9.81), and (9.82), is virtually stable with its affiliated vector ${}^{\mathbf{W}_j}V_r - {}^{\mathbf{W}_j}V$ and variable $\dot{q}_{wjr} - \dot{q}_{wj}$ being virtual functions in both L_2 and L_∞, in the sense of Definition 2.17.* ■

The proof procedure follows straightforwardly to that for Theorem 9.4, but with one rigid body (reaction wheel) and one joint.

9.9 Dynamics and Control of the Spacecraft Platform

9.9.1 Kinematics and Dynamics

The velocity relationship of the (spacecraft) platform can be expressed by

$$ {}^{\mathbf{B}}V = {}^{\mathbf{A}_m}\mathbf{U}_{\mathbf{B}}^{T}\,{}^{\mathbf{A}_m}V = {}^{\mathbf{A}_j}\mathbf{U}_{\mathbf{B}}^{T}\,{}^{\mathbf{A}_j}V, \ \forall j \in \{1, 3\}. \tag{9.83} $$

Accordingly, the dynamics are written as

$$ \mathbf{M}_{\mathbf{B}}\frac{d}{dt}({}^{\mathbf{B}}V) + \mathbf{C}_{\mathbf{B}}({}^{\mathbf{B}}\omega){}^{\mathbf{B}}V + \mathbf{G}_{\mathbf{B}} = {}^{\mathbf{B}}F^* \tag{9.84} $$

which is obtained from (2.74) by substituting frame $\{\mathbf{B}\}$ for frame $\{\mathbf{A}\}$, with the net force/moment vector in frame $\{\mathbf{B}\}$ being expressed by

$$ {}^{\mathbf{B}}F^* = {}^{\mathbf{B}}F - {}^{\mathbf{B}}\mathbf{U}_{\mathbf{A}_m}\,{}^{\mathbf{A}_m}F - \sum_{j=1}^{3} {}^{\mathbf{B}}\mathbf{U}_{\mathbf{A}_j}\,{}^{\mathbf{A}_j}F \tag{9.85} $$

where ${}^{\mathbf{B}}F \in \mathbb{R}^6$ denotes the force/moment vector applied from the environment to the (spacecraft) platform. Therefore, ${}^{\mathbf{B}}F = \mathbf{0}$ holds indefinitely (see next section).

9.9.2 Control Equations

The relationship (9.83) also applies to the required velocities as

$$ {}^{\mathbf{B}}V_r = {}^{\mathbf{A}_m}\mathbf{U}_{\mathbf{B}}^{T}\,{}^{\mathbf{A}_m}V_r = {}^{\mathbf{A}_j}\mathbf{U}_{\mathbf{B}}^{T}\,{}^{\mathbf{A}_j}V_r, \ \forall j \in \{1, 3\}. \tag{9.86} $$

Accordingly, recall (4.32) with frame $\{\mathbf{B}\}$ being substituted for frame $\{\mathbf{O}_i\}$ to compute the required net force/moment vector of the (spacecraft) platform as

$$ {}^{\mathbf{B}}F_r^* = \mathbf{Y_B}\hat{\boldsymbol{\theta}}_\mathbf{B} + \mathbf{K_B}\left({}^{\mathbf{B}}V_r - {}^{\mathbf{B}}V\right) \tag{9.87} $$

where $\mathbf{K_B} \in \mathbb{R}^{6\times 6}$ is a symmetric positive-definite matrix and $\hat{\boldsymbol{\theta}}_\mathbf{B} \in \mathbb{R}^{13}$ denotes the estimate of $\boldsymbol{\theta}_\mathbf{B} \in \mathbb{R}^{13}$ with $\mathbf{Y_B}\boldsymbol{\theta}_\mathbf{B}$ being defined by (2.78) and expressed in Appendix A by substituting frame $\{\mathbf{B}\}$ for frame $\{\mathbf{A}\}$.

Define

$$ \mathbf{s_B} = \mathbf{Y_B}^T\left({}^{\mathbf{B}}V_r - {}^{\mathbf{B}}V\right). \tag{9.88} $$

The $\mathcal{P}$ function defined by (2.79) is used to update each of the 13 parameters of the (spacecraft) platform as

$$ \hat{\theta}_{\mathbf{B}\gamma} = \mathcal{P}\left(s_{\mathbf{B}\gamma}, \rho_{\mathbf{B}\gamma}, \underline{\theta}_{\mathbf{B}\gamma}, \overline{\theta}_{\mathbf{B}\gamma}, t\right) \tag{9.89} $$

for all $\gamma \in \{1, 13\}$, where

$\hat{\theta}_{\mathbf{B}\gamma}$	Estimate of $\theta_{\mathbf{B}\gamma}$.
$s_{\mathbf{B}\gamma}$	γth elements of $\mathbf{s_B}$.
$\rho_{\mathbf{B}\gamma}$	Parameter update gain.
$\underline{\theta}_{\mathbf{B}\gamma}$	Lower bound of $\theta_{\mathbf{B}\gamma}$.
$\overline{\theta}_{\mathbf{B}\gamma}$	Upper bound of $\theta_{\mathbf{B}\gamma}$.
$\theta_{\mathbf{B}\gamma}$	γth element of $\boldsymbol{\theta}_\mathbf{B}$.

Finally, the required force/moment vector from the environment to the (spacecraft) platform is computed by

$$ {}^{\mathbf{B}}F_r = {}^{\mathbf{B}}F_r^* + {}^{\mathbf{B}}\mathbf{U}_{\mathbf{A}_m}\,{}^{\mathbf{A}_m}F_r + \sum_{j=1}^{3} {}^{\mathbf{B}}\mathbf{U}_{\mathbf{A}_j}\,{}^{\mathbf{A}_j}F_r \tag{9.90} $$

where ${}^{\mathbf{B}}F_r^* \in \mathbb{R}^6$ is obtained from (9.87), ${}^{\mathbf{A}_m}F_r \in \mathbb{R}^6$ is obtained from (9.49)–(9.52), and ${}^{\mathbf{A}_j}F_r \in \mathbb{R}^6$ is obtained from (9.76) and (9.77).

9.9.3 Virtual Stability

The *virtual stability* of the (spacecraft) platform, subject to its respective control equations, is presented in this subsection.

Theorem 9.6. *The (spacecraft) platform described by (9.83)–(9.85), combined with its respective control equations (9.86), (9.87), and (9.90) and with the parameter adaptation (9.88) and (9.89), is virtually stable with its affiliated vector ${}^{\mathbf{B}}V_r - {}^{\mathbf{B}}V$ being a virtual function in both L_2 and L_∞, in the sense of Definition 2.17.*

Proof: The non-negative accompanying function assigned to the (spacecraft) platform is chosen as

$$\nu_{\mathbf{B}} = \frac{1}{2}\left({}^{\mathbf{B}}V_r - {}^{\mathbf{B}}V\right)^T \mathbf{M}_{\mathbf{B}} \left({}^{\mathbf{B}}V_r - {}^{\mathbf{B}}V\right)$$
$$+\frac{1}{2}\sum_{\gamma=1}^{13}\left(\theta_{\mathbf{B}\gamma} - \hat{\theta}_{\mathbf{B}\gamma}\right)^2 / \rho_{\mathbf{B}\gamma}. \tag{9.91}$$

It follows from (9.84), (9.87)–(9.89), and Lemma 4.1 that

$$\dot{\nu}_{\mathbf{B}} \leqslant -\left({}^{\mathbf{B}}V_r - {}^{\mathbf{B}}V\right)^T \mathbf{K}_{\mathbf{B}} \left({}^{\mathbf{B}}V_r - {}^{\mathbf{B}}V\right)$$
$$+\left({}^{\mathbf{B}}V_r - {}^{\mathbf{B}}V\right)^T \left({}^{\mathbf{B}}F_r^* - {}^{\mathbf{B}}F^*\right) \tag{9.92}$$

holds.

In view of (2.85), (9.83), (9.85), (9.86), and (9.90), it yields

$$\left({}^{\mathbf{B}}V_r - {}^{\mathbf{B}}V\right)^T \left({}^{\mathbf{B}}F_r^* - {}^{\mathbf{B}}F^*\right)$$
$$= \left({}^{\mathbf{B}}V_r - {}^{\mathbf{B}}V\right)^T \Big[\left({}^{\mathbf{B}}F_r - {}^{\mathbf{B}}F\right) - {}^{\mathbf{B}}\mathbf{U}_{\mathbf{A}_m}\left({}^{\mathbf{A}_m}F_r - {}^{\mathbf{A}_m}F\right)$$
$$-\sum_{j=1}^{3} {}^{\mathbf{B}}\mathbf{U}_{\mathbf{A}_j}\left({}^{\mathbf{A}_j}F_r - {}^{\mathbf{A}_j}F\right)\Big]$$
$$= p_{\mathbf{B}} - p_{\mathbf{A}_m} - \sum_{j=1}^{3} p_{\mathbf{A}_j}. \tag{9.93}$$

Consider the fact that the (spacecraft) platform has four *driving cutting points* associated with frames $\{\mathbf{A}_m\}$ and $\{\mathbf{A}_j\}$ for all $j \in \{1, 3\}$ and one *driven cutting point* associated with frame $\{\mathbf{B}\}$. Substituting (9.93) into (9.92) and using (9.91) and Definition 2.17 completes the proof. ■

9.10 Dynamics and Control of the Virtual Manipulator

The massless virtual manipulator ensures

$${}^{\mathbf{B}}F = \mathbf{0} \tag{9.94}$$

which characterizes the nonholonomic constraints of the system imposed by the momentum conservation law. Therefore, the solution for ${}^{\mathbf{B}}F_r$ with respect to (9.94) needs to be

$${}^{\mathbf{B}}F_r = -\mathbf{K}_s({}^{\mathbf{B}}V_r - {}^{\mathbf{B}}V) \tag{9.95}$$

where $\mathbf{K}_s \in \mathbb{R}^{6\times 6}$ is a semi-positive-definite matrix.

Remark 9.1. Equation (9.95) includes the case of ${}^{\mathbf{B}}F_r = \mathbf{0}$, since $\mathbf{K}_s = \mathbf{0}$ is permitted.

Theorem 9.7. *The massless virtual manipulator characterized by (9.94), combined with its respective control equation (9.95), is virtually stable in the sense of Definition 2.17.*

Proof: Let the non-negative accompanying function for the massless virtual manipulator be *zero*. It follows from (2.85), (9.94), and (9.95) that

$$\begin{aligned} 0 &\leqslant ({}^{\mathbf{B}}V_r - {}^{\mathbf{B}}V)^T \mathbf{K}_s ({}^{\mathbf{B}}V_r - {}^{\mathbf{B}}V) \\ &= -({}^{\mathbf{B}}V_r - {}^{\mathbf{B}}V)^T ({}^{\mathbf{B}}F_r - {}^{\mathbf{B}}F) \\ &= -p_{\mathbf{B}} \end{aligned} \tag{9.96}$$

holds.

Noting the fact that the virtual manipulator has only one *driving cutting point* associated with frame $\{\mathbf{B}\}$ completes the proof. ■

9.11 Overall System Stability

This section presents the stability of the overall system.

9.11.1 L_2 and L_∞ Stability

Theorem 9.8. *Consider the space robotic system comprised of a (spacecraft) platform, three reaction wheels, and a manipulator holding a rigid object. All virtual functions in L_p are functions in L_p for $p = 2, \infty$.*

Proof: The proof is directly based on Theorem 2.1, since every subsystem is *virtually stable* in the sense of Definition 2.17. ■

Consequently, it follows from Theorems 9.1–9.8 that

$${}^{\mathbf{O}}V_r - {}^{\mathbf{O}}V \in L_2 \bigcap L_\infty \tag{9.97}$$

$${}^{\mathbf{B}_k}V_r - {}^{\mathbf{B}_k}V \in L_2 \bigcap L_\infty, \ \forall k \in \{1, 6\} \tag{9.98}$$

$$\dot{q}_{kr} - \dot{q}_k \in L_2 \bigcap L_\infty, \ \forall k \in \{1, 6\} \tag{9.99}$$

$${}^{\mathbf{W}_j}V_r - {}^{\mathbf{W}_j}V \in L_2 \bigcap L_\infty, \ \forall j \in \{1, 3\} \tag{9.100}$$

$$\dot{q}_{wjr} - \dot{q}_{wj} \in L_2 \bigcap L_\infty, \ \forall j \in \{1, 3\} \tag{9.101}$$

$${}^{\mathbf{B}}V_r - {}^{\mathbf{B}}V \in L_2 \bigcap L_\infty \tag{9.102}$$

hold. It further leads to

$$\mathcal{V}_r - \mathcal{V} \in L_2 \bigcap L_\infty. \tag{9.103}$$

When the joint position control is used, it follows from (9.12), (9.13), (9.99), (9.102), Lemmas 4.3 and 4.5, and Theorem 4.6 that

$$\dot{q}_{kd} - \dot{q}_k \in L_2 \bigcap L_\infty, \ \forall k \in \{1, 6\} \tag{9.104}$$

$$q_{kd} - q_k \subset L_2 \bigcap L_\infty, \ \forall k \in \{1, 6\} \tag{9.105}$$

$$\boldsymbol{\omega}_{\mathbf{B}_d} - \boldsymbol{\omega}_{\mathbf{B}} \in L_2 \bigcap L_\infty \tag{9.106}$$

$${}^{\mathbf{B}}\boldsymbol{\mu} \in L_2 \bigcap L_\infty \tag{9.107}$$

hold.

Alternatively, when the Cartesian position control is used, it follows from (9.13), (9.15), (9.97), (9.102), Lemmas 4.4 and 4.5, and Theorem 4.6 that

$$\dot{\mathbf{x}}_{\mathbf{O}_d} - \dot{\mathbf{x}}_{\mathbf{O}} \in L_2 \bigcap L_\infty \tag{9.108}$$

$$\boldsymbol{\omega}_{\mathbf{O}_d} - \boldsymbol{\omega}_{\mathbf{O}} \in L_2 \bigcap L_\infty \tag{9.109}$$

$$\mathbf{x}_{\mathbf{O}_d} - \mathbf{x}_{\mathbf{O}} \in L_2 \bigcap L_\infty \tag{9.110}$$

$${}^{\mathbf{O}}\boldsymbol{\mu} \in L_2 \bigcap L_\infty \tag{9.111}$$

$$\boldsymbol{\omega}_{\mathbf{B}_d} - \boldsymbol{\omega}_{\mathbf{B}} \in L_2 \bigcap L_\infty \tag{9.112}$$

$${}^{\mathbf{B}}\boldsymbol{\mu} \in L_2 \bigcap L_\infty \tag{9.113}$$

hold.

9.11.2 Forward Dynamics

The forward dynamics of the space robot, which will be used to ensure the asymptotic stability of the entire system, are to be presented in this subsection.

Recall the procedure in Section 4.9 leading to (4.136). The forward dynamics of the space robot can be expressed by

$$\mathcal{M}\dot{\mathcal{V}} + \mathcal{C}\mathcal{V} + \mathcal{G} = \begin{bmatrix} \boldsymbol{\tau} \\ \boldsymbol{\tau}_w \\ \mathbf{0} \end{bmatrix} \tag{9.114}$$

where $\mathcal{V} \in \mathbb{R}^{15}$ is defined by (9.6), $\boldsymbol{\tau} = [\tau_1, \cdots, \tau_k, \cdots, \tau_6]^T \in \mathbb{R}^6$ denotes the joint control force/torque vector of the manipulator, $\boldsymbol{\tau}_w = [\tau_{w1}, \tau_{w2}, \tau_{w3}]^T \in \mathbb{R}^3$ denotes the control torques of the reaction wheels, and

$$\begin{aligned} \mathcal{M} &= \mathcal{J}_b^T \mathcal{M}_b \mathcal{J}_b + \mathcal{M}_a \\ \mathcal{C} &= \mathcal{J}_b^T \mathcal{M}_b \dot{\mathcal{J}}_b + \mathcal{J}_b^T \mathcal{C}_b \mathcal{J}_b \\ \mathcal{G} &= \mathcal{J}_b^T \mathcal{G}_b + \mathcal{G}_a \end{aligned}$$

with $\mathcal{J}_b$ being defined by (9.10) and

$$\begin{aligned} \mathcal{M}_a &= \mathrm{diag}(J_{m1}, \cdots, J_{m6}, J_{w1}, J_{w2}, J_{w3}, \mathbf{0}_{6\times 6}) \\ \mathcal{M}_b &= \mathrm{diag}(\mathbf{M}_{\mathbf{O}}, \mathbf{M}_{\mathbf{B}_1}, \cdots, \mathbf{M}_{\mathbf{B}_6}, \mathbf{M}_{\mathbf{W}_1}, \cdots, \mathbf{M}_{\mathbf{W}_3}, \mathbf{M}_{\mathbf{B}}) \\ \mathcal{C}_b &= \mathrm{diag}(\mathbf{C}_{\mathbf{O}}, \mathbf{C}_{\mathbf{B}_1}, \cdots, \mathbf{C}_{\mathbf{B}_6}, \mathbf{C}_{\mathbf{W}_1}, \cdots, \mathbf{C}_{\mathbf{W}_3}, \mathbf{C}_{\mathbf{B}}) \\ \mathcal{G}_a &= [\xi_1, \cdots, \xi_6, \xi_{w1}, \cdots, \xi_{w3}, 0, \cdots, 0]^T \\ \mathcal{G}_b &= [\mathbf{G}_{\mathbf{O}}^T, \mathbf{G}_{\mathbf{B}_1}^T, \cdots, \mathbf{G}_{\mathbf{B}_6}^T, \mathbf{G}_{\mathbf{W}_1}^T, \cdots, \mathbf{G}_{\mathbf{W}_3}^T, \mathbf{G}_{\mathbf{B}}^T]^T. \end{aligned}$$

9.11.3 Asymptotic Stability

When $\mathcal{V}_r$ and $\dot{\mathcal{V}}_r$ are bounded, the boundedness of ${}^{\mathbf{O}}F_r^* \in \mathbb{R}^6$, ${}^{\mathbf{B}_k}F_r^* \in \mathbb{R}^6$ for all $k \in \{1, 6\}$, and ${}^{\mathbf{W}_j}F_r^* \in \mathbb{R}^6$ for all $j \in \{1, 3\}$ can be ensured. This result ensures the boundedness of ${}^{\mathbf{B}_k}F_r$ for all $k \in \{1,6\}$ and ${}^{\mathbf{W}_j}F_r$ for all $j \in \{1,3\}$, which further ensures the boundedness of $\tau_{kd} = \tau_k$ for all $k \in \{1,6\}$ and $\tau_{wjd} = \tau_{wj}$ for all $j \in \{1,3\}$. Thus, the boundedness of $\dot{\mathcal{V}}$ is guaranteed from (9.114). Finally, it follows from (9.103) and Lemma 2.8 that

$$\mathcal{V}_r - \mathcal{V} \to \mathbf{0} \tag{9.115}$$

holds.

When the joint position control is used, it follows from (9.104)–(9.107) and (9.115) that

$$\dot{q}_{kd} - \dot{q}_k \to 0, \ \forall k \in \{1,6\} \tag{9.116}$$

$$q_{kd} - q_k \to 0, \ \forall k \in \{1,6\} \tag{9.117}$$

$$\boldsymbol{\omega}_{\mathbf{B}_d} - \boldsymbol{\omega}_{\mathbf{B}} \to \mathbf{0} \tag{9.118}$$

$${}^{\mathbf{B}}\boldsymbol{\mu} \to \mathbf{0} \tag{9.119}$$

hold.

When the Cartesian position control is used, it follows from (9.108)–(9.113) and (9.115) that

$$\dot{\mathbf{x}}_{\mathbf{O}_d} - \dot{\mathbf{x}}_{\mathbf{O}} \to \mathbf{0} \tag{9.120}$$

$$\boldsymbol{\omega}_{\mathbf{O}_d} - \boldsymbol{\omega}_{\mathbf{O}} \to \mathbf{0} \tag{9.121}$$

$$\mathbf{x}_{\mathbf{O}_d} - \mathbf{x}_{\mathbf{O}} \to \mathbf{0} \tag{9.122}$$

$${}^{\mathbf{O}}\boldsymbol{\mu} \to \mathbf{0} \tag{9.123}$$

$$\boldsymbol{\omega}_{\mathbf{B}_d} - \boldsymbol{\omega}_{\mathbf{B}} \to \mathbf{0} \tag{9.124}$$

$${}^{\mathbf{B}}\boldsymbol{\mu} \to \mathbf{0} \tag{9.125}$$

hold.

9.12 Nonholonomic Constraints

The nonholonomic constraints imposed on the space robot are characterized by ${}^{\mathbf{B}}F = \mathbf{0}$ in (9.94). Unlike many other approaches, the nonholonomic constraints in this chapter are equivalently converted to the six constraints imposed on the required force/moment vector in frame $\{\mathbf{B}\}$ by equalizing (9.90) and (9.95).

Remark 9.2. The advantage of representing the nonholonomic constraints with respect to the required force/moment vector rather than to the physical force/moment vector is that the required force/moment vector has numerical values stored in computer memory. They are always available for computations without the need of having additional force sensors. As a result, the adaptive control algorithm developed in this chapter does not require a force/torque sensor to be placed between the manipulator and the (spacecraft) platform.

In view of (9.28), (9.49)–(9.52), (9.76), and (9.77), express (9.90) as

$$^{\mathbf{B}}F_r = \mathcal{U}_{\mathbf{B}}\mathcal{F}_r^* \tag{9.126}$$

with

$$\mathcal{U}_{\mathbf{B}} = \left[{}^{\mathbf{B}}\mathbf{U}_{\mathbf{O}}, {}^{\mathbf{B}}\mathbf{U}_{\mathbf{B}_1}, \cdots, {}^{\mathbf{B}}\mathbf{U}_{\mathbf{B}_6}, {}^{\mathbf{B}}\mathbf{U}_{\mathbf{W}_1}, \cdots, {}^{\mathbf{B}}\mathbf{U}_{\mathbf{W}_3}, \mathbf{I}_6\right] \in \mathbb{R}^{6\times 66} \tag{9.127}$$

$$\mathcal{F}_r^* = \left[{}^{\mathbf{O}}F_r^{*T}, {}^{\mathbf{B}_1}F_r^{*T}, \cdots, {}^{\mathbf{B}_6}F_r^{*T}, {}^{\mathbf{W}_1}F_r^{*T}, \cdots, {}^{\mathbf{W}_3}F_r^{*T}, {}^{\mathbf{B}}F_r^{*T}\right]^T \in \mathbb{R}^{66}. \tag{9.128}$$

Then, rewrite (9.128) as

$$\mathcal{F}_r^* = \hat{\mathcal{M}}\mathcal{J}_b\dot{\mathcal{V}}_r + \mathcal{B}(t) \tag{9.129}$$

where $\hat{\mathcal{M}}$ denotes the estimate of

$$\mathcal{M} = \mathrm{diag}\left(\mathbf{M}_{\mathbf{O}}, \mathbf{M}_{\mathbf{B}_1}, \cdots, \mathbf{M}_{\mathbf{B}_6}, \mathbf{M}_{\mathbf{W}_1}, \cdots, \mathbf{M}_{\mathbf{W}_3}, \mathbf{M}_{\mathbf{B}}\right) \tag{9.130}$$

by using estimated dynamic parameters. These block inertial matrices are defined by (2.75) with appropriate frame substitutions. In (9.129), $\mathcal{J}_b$ is defined by (9.10), and $\mathcal{B}(t)$ is a remaining vector.

Substituting (9.129) into (9.126) and using (9.95) yields

$$\mathcal{U}_{\mathbf{B}}\hat{\mathcal{M}}\mathcal{J}_b\dot{\mathcal{V}}_r + \mathcal{U}_{\mathbf{B}}\mathcal{B}(t) + \mathbf{K}_s({}^{\mathbf{B}}V_r - {}^{\mathbf{B}}V) = \mathbf{0}. \tag{9.131}$$

Thus, the nonholonomic constraints of the space robot are completely represented by (9.131) which implies that six elements in

$$\dot{\mathcal{V}}_r = \frac{d}{dt}\begin{bmatrix} \dot{\mathbf{q}}_r \\ \dot{\mathbf{q}}_{wr} \\ {}^{\mathbf{B}}V_r \end{bmatrix} \in \mathbb{R}^{15}$$

must be released to satisfy (9.131). Let $\mathcal{R} \in \mathbb{R}^{15\times 15}$ be a re-ordering matrix such that

$$\mathcal{R}\dot{\mathcal{V}}_r = \left[\dot{\mathcal{V}}_{pr}^T, \dot{\mathcal{V}}_{ur}^T\right]^T \tag{9.132}$$

$$\mathcal{U}_{\mathbf{B}}\hat{\mathcal{M}}\mathcal{J}_b\mathcal{R}^T = \left[\mathcal{M}_p \quad \mathcal{M}_u\right] \tag{9.133}$$

hold with $\mathcal{M}_u \in \mathbb{R}^{6\times 6}$ being invertible, where

$$\dot{\mathcal{V}}_{pr} = \begin{bmatrix} \dot{\mathbf{q}}_r \\ {}^{\mathbf{B}}\boldsymbol{\omega}_r \end{bmatrix} \in \mathbb{R}^9 \tag{9.134}$$

$$\dot{\mathcal{V}}_{ur} = \begin{bmatrix} \dot{\mathbf{q}}_{wr} \\ {}^{\mathbf{B}}\mathbf{v}_r \end{bmatrix} \in \mathbb{R}^6 \tag{9.135}$$

denote the independent and dependent acceleration coordinate vectors, respectively.

Substituting (9.132) and (9.133) into (9.131) yields

$$\mathcal{M}_p \dot{\mathcal{V}}_{pr} + \mathcal{M}_u \dot{\mathcal{V}}_{ur} + \mathcal{U}_{\mathbf{B}} \mathcal{B}(t) + \mathbf{K}_s({}^{\mathbf{B}}V_r - {}^{\mathbf{B}}V) = \mathbf{0} \tag{9.136}$$

which further leads to

$$\dot{\mathcal{V}}_{ur} = -\mathcal{M}_u^{-1} \mathcal{M}_p \dot{\mathcal{V}}_{pr} - \mathcal{M}_u^{-1} \mathcal{U}_{\mathbf{B}} \mathcal{B}(t) - \mathcal{M}_u^{-1} \mathbf{K}_s({}^{\mathbf{B}}V_r - {}^{\mathbf{B}}V). \tag{9.137}$$

Thus, the dependence of $\dot{\mathcal{V}}_{ur} \in \mathbb{R}^6$ on $\dot{\mathcal{V}}_{pr} \in \mathbb{R}^9$ is determined. This statement implies that only the time derivative of the required independent velocity coordinate vector $\dot{\mathcal{V}}_{pr} \in \mathbb{R}^9$ can be freely designed for the position control as outlined in Section 4.11.

The choice of the re-ordering matrix $\mathcal{R}$ is to make $\mathcal{M}_u \in \mathbb{R}^{6\times 6}$ be of full-rank. Two possible cases are discussed below.

The first case is to release the required linear and angular accelerations of the (spacecraft) platform. The re-ordering matrix $\mathcal{R}$ is chosen as

$$\mathcal{R} = \mathbf{I}_{15}. \tag{9.138}$$

It follows from (9.10), (9.132), and (9.133) that

$$\mathcal{M}_u = \mathcal{U}_{\mathbf{B}} \hat{\mathcal{M}} \mathcal{U}_{\mathbf{B}}^T \tag{9.139}$$

holds. Therefore, as long as the lower and upper bounds of the dynamic parameters are properly chosen, matrix $\hat{\mathcal{M}}$ is positive-definition, which ensures $\mathcal{M}_u$ to be positive-definite as well. The physical interpretation of choosing $\mathcal{R}$ by (9.138) is that the required linear and angular accelerations of the (spacecraft) platform are released so that the position and orientation of the (spacecraft) platform are not controllable.

The second case is to release the required linear accelerations of the (spacecraft) platform and the required accelerations of the three reaction wheels so that the orientation of the (spacecraft) platform can be controlled. The re-ordering matrix $\mathcal{R}$ is chosen as

$$\mathcal{R} = \begin{bmatrix} \mathbf{I}_6 & \mathbf{0}_{6\times 6} & \mathbf{0}_{6\times 3} \\ \mathbf{0}_{3\times 6} & \mathbf{0}_{3\times 6} & \mathbf{I}_3 \\ \mathbf{0}_{6\times 6} & \mathbf{I}_6 & \mathbf{0}_{6\times 3} \end{bmatrix}. \tag{9.140}$$

It follows from (9.10), (9.127), (9.130), and (9.133) that

$$\mathcal{M}_u = \begin{bmatrix} \mathbf{m}_1 & \mathbf{m}_2 & \mathbf{m}_3 & \mathbf{M}_f \end{bmatrix} \in \mathbb{R}^{6\times 6} \tag{9.141}$$

holds with

$$\mathbf{m}_j = {}^{\mathbf{B}}\mathbf{U}_{\mathbf{W}_j} \hat{\mathbf{M}}_{\mathbf{W}_j} \mathbf{z}_w \in \mathbb{R}^6 \text{ for } j = 1, 2, 3 \tag{9.142}$$

$$\mathbf{M}_f = \left({}^{\mathbf{B}}\mathbf{U}_{\mathbf{O}} \hat{\mathbf{M}}_{\mathbf{O}} {}^{\mathbf{B}}\mathbf{U}_{\mathbf{O}}^T + \sum_{k=1}^{6} {}^{\mathbf{B}}\mathbf{U}_{\mathbf{B}_k} \hat{\mathbf{M}}_{\mathbf{B}_k} {}^{\mathbf{B}}\mathbf{U}_{\mathbf{B}_k}^T \right.$$
$$\left. + \sum_{j=1}^{3} {}^{\mathbf{B}}\mathbf{U}_{\mathbf{W}_j} \hat{\mathbf{M}}_{\mathbf{W}_j} {}^{\mathbf{B}}\mathbf{U}_{\mathbf{W}_j}^T + \hat{\mathbf{M}}_{\mathbf{B}} \right) \begin{bmatrix} \mathbf{I}_3 \\ \mathbf{0}_{3\times 3} \end{bmatrix} \in \mathbb{R}^{6\times 3}. \tag{9.143}$$

Re-examining the structure of the force/moment transformation matrix defined by (2.65) and of the body-frame referenced inertia matrix defined by (2.75) reveals that the upper-left 3×3 sub-matrix of ${}^{\mathbf{B}}\mathbf{U}_{\mathbf{O}}\hat{\mathbf{M}}_{\mathbf{O}}{}^{\mathbf{B}}\mathbf{U}_{\mathbf{O}}^{T}$ can be written as $\hat{m}_{\mathbf{O}}\mathbf{I}_3$, where $\hat{m}_{\mathbf{O}}$ denotes the estimated mass of the held object. Consequently, the upper 3×3 sub-matrix of $\mathbf{M}_f \in \mathbb{R}^{6\times 3}$ can be written as

$$[\mathbf{I}_3 \quad \mathbf{0}_{3\times 3}]\mathbf{M}_f = \left(\hat{m}_{\mathbf{O}} + \sum_{k=1}^{6} \hat{m}_{\mathbf{B}_k} + \sum_{j=1}^{3} \hat{m}_{\mathbf{W}_j} + \hat{m}_{\mathbf{B}}\right)\mathbf{I}_3 \tag{9.144}$$

where $\hat{m}_{\mathbf{B}_k}$ denotes the estimated mass of the kth robot link, $\hat{m}_{\mathbf{W}_j}$ denotes the estimated mass of the jth reaction wheel, and $\hat{m}_{\mathbf{B}}$ denotes the estimated mass of the (spacecraft) platform, respectively.

On the other hand, consider the fact that each reaction wheel is symmetric about its principal axis which is in line with the z axis of the corresponding frame. It turns out that the elements in the sixth column of $\hat{\mathbf{M}}_{\mathbf{W}_j}$ are all zeros except the last one. Thus, $\mathbf{m}_j \in \mathbb{R}^6$ in (9.142) can be re-written as

$$\mathbf{m}_j = \begin{bmatrix} \mathbf{0}_{3\times 1} \\ {}^{\mathbf{B}}\mathbf{z}_{\mathbf{W}_j}\hat{I}_{zzj} \end{bmatrix} \tag{9.145}$$

for all $j \in \{1,3\}$, where $\hat{I}_{zzj} \in \mathbb{R}$ denotes the (6,6)th element of $\hat{\mathbf{M}}_{\mathbf{W}_j}$ and ${}^{\mathbf{B}}\mathbf{z}_{\mathbf{W}_j} \in \mathbb{R}^3$ denote the z axis of frame $\{\mathbf{W}_j\}$ expressed in frame $\{\mathbf{B}\}$ for $j = 1,2,3$.

Substituting (9.144) and (9.145) into (9.141) yields

$$\mathcal{M}_u = \begin{bmatrix} \mathbf{0}_{3\times 3} & \hat{m}\mathbf{I}_3 \\ \left[{}^{\mathbf{B}}\mathbf{z}_{\mathbf{W}_1}\hat{I}_{zz1} \quad {}^{\mathbf{B}}\mathbf{z}_{\mathbf{W}_2}\hat{I}_{zz2} \quad {}^{\mathbf{B}}\mathbf{z}_{\mathbf{W}_3}\hat{I}_{zz3}\right] & \mathbf{M}_{f2} \end{bmatrix} \in \mathbb{R}^{6\times 6} \tag{9.146}$$

with

$$\hat{m} = \hat{m}_{\mathbf{O}} + \sum_{k=1}^{6} \hat{m}_{\mathbf{B}_k} + \sum_{j=1}^{3} \hat{m}_{\mathbf{W}_j} + \hat{m}_{\mathbf{B}} \tag{9.147}$$

being the estimated mass of the entire space robot and $\mathbf{M}_{f2} \in \mathbb{R}^{3\times 3}$ being a matrix.

The geometry of the three reaction wheels is arranged such that

$$\mathbf{M}_m \stackrel{def}{=} \left[{}^{\mathbf{B}}\mathbf{z}_{\mathbf{W}_1}, {}^{\mathbf{B}}\mathbf{z}_{\mathbf{W}_2}, {}^{\mathbf{B}}\mathbf{z}_{\mathbf{W}_3}\right] \in \mathbb{R}^{3\times 3} \tag{9.148}$$

is of full-rank. Thus, $\mathcal{M}_u$ given by (9.146) is of full rank, as long as the lower bounds of the masses and the moments of inertia of the three reaction wheels are larger than zero, or mathematically

$$rank\,(\mathbf{M}_m) = 3 \tag{9.149}$$

$$\underline{m}_{\mathbf{O}} + \sum_{k=1}^{6} \underline{m}_{\mathbf{B}_k} + \sum_{j=1}^{3} \underline{m}_{\mathbf{W}_j} + \underline{m}_{\mathbf{B}} > 0 \tag{9.150}$$

$$\underline{I}_{zzj} > 0, \ \forall j \in \{1,3\}. \tag{9.151}$$

Remark 9.3. The invertibility of matrix $\mathcal{M}_u$ defined in (9.133) is the key to make the *virtual decomposition control* algorithm executable such that the L_2 and L_∞ stability in the sense of (9.103), as well as the asymptotic stability in the sense of (9.115), can be ensured. It is worth noting that matrix $\mathcal{M}_u$ in this chapter has the same meaning as matrix $\mathcal{A}_f$ in [216]. The difference is that the conditions to guarantee the invertibility of $\mathcal{M}_u$, as described by (9.149)–(9.151), are much simple, explicit, and easy to satisfy with mathematical certainty.

9.13 Control Algorithm

The overall control algorithm is summarized into seven steps as follows:

Step 1: Determine to use either the joint position control or the Cartesian position control.

- When the joint position control is chosen, compute $\dot{\mathbf{q}}_r \in \mathbb{R}^6$ and ${}^{\mathbf{B}}\boldsymbol{\omega}_r \in \mathbb{R}^3$ from (9.12) and (9.13), respectively.
- When the Cartesian position control is chosen, compute ${}^{\mathbf{O}}V_r \in \mathbb{R}^6$ and ${}^{\mathbf{B}}\boldsymbol{\omega}_r \in \mathbb{R}^3$ from (9.15) and (9.13), respectively.

Step 2: Construct

$$\begin{bmatrix} \dot{\mathbf{q}}_r \\ \dot{\mathbf{q}}_{wr} \\ {}^{\mathbf{B}}V_r \end{bmatrix} \in \mathbb{R}^{15}$$

as an affine function of

$$\begin{bmatrix} \dot{\mathbf{q}}_{wr} \\ {}^{\mathbf{B}}\mathbf{v}_r \end{bmatrix} \in \mathbb{R}^6.$$

Then, compute $\mathcal{V}_{br} \in \mathbb{R}^{66}$ as an affine function of

$$\begin{bmatrix} \dot{\mathbf{q}}_{wr} \\ {}^{\mathbf{B}}\mathbf{v}_r \end{bmatrix} \in \mathbb{R}^6$$

from (9.20). Furthermore, express $\dot{\mathcal{V}}_{br} \in \mathbb{R}^{66}$ as an affine function of

$$\begin{bmatrix} \ddot{\mathbf{q}}_{wr} \\ \frac{d}{dt}\left({}^{\mathbf{B}}\mathbf{v}_r\right) \end{bmatrix} \in \mathbb{R}^6.$$

Step 3: Compute ${}^{\mathbf{O}}F_r^* \in \mathbb{R}^6$, ${}^{\mathbf{B}_k}F_r^* \in \mathbb{R}^6$ for all $k \in \{1, 6\}$, ${}^{\mathbf{W}_j}F_r^* \in \mathbb{R}^6$ for all $j \in \{1, 3\}$, and ${}^{\mathbf{B}}F_r^* \in \mathbb{R}^6$ as affine functions of

$$\begin{bmatrix} \ddot{\mathbf{q}}_{wr} \\ \frac{d}{dt}\left({}^{\mathbf{B}}\mathbf{v}_r\right) \end{bmatrix} \in \mathbb{R}^6$$

by using (9.25), (9.46), (9.73), and (9.87).

Step 4: Solve

$$\dot{\mathcal{V}}_{ur} = \begin{bmatrix} \ddot{\mathbf{q}}_{wr} \\ \frac{d}{dt}\left({}^{\mathbf{B}}\mathbf{v}_r\right) \end{bmatrix} \in \mathbb{R}^6$$

from (9.137) under the conditions (9.149)–(9.151). Substituting the solution into step 3 obtains the numerical values of ${}^{\mathbf{O}}F_r^* \in \mathbb{R}^6$, ${}^{\mathbf{B}_k}F_r^* \in \mathbb{R}^6$ for all $k \in \{1, 6\}$, ${}^{\mathbf{W}_j}F_r^* \in \mathbb{R}^6$ for all $j \in \{1, 3\}$, and ${}^{\mathbf{B}}F_r^* \in \mathbb{R}^6$.

Step 5: Compute ${}^{\mathbf{C}}F_r \in \mathbb{R}^6$, ${}^{\mathbf{B}_k}F_r \in \mathbb{R}^6$ for all $k \in \{1, 6\}$, ${}^{\mathbf{A}_m}F_r \in \mathbb{R}^6$, ${}^{\mathbf{W}_j}F_r \in \mathbb{R}^6$ and ${}^{\mathbf{A}_j}F_r \in \mathbb{R}^6$ for all $j \in \{1, 3\}$, and ${}^{\mathbf{B}}F_r \in \mathbb{R}^6$ by using (9.28), (9.49)–(9.52), (9.76), (9.77), and (9.90).

Step 6: Perform parameter adaptation for all eleven rigid bodies representing the held object, the six manipulator links, the three reaction wheels, and the (spacecraft) platform by using (9.26), (9.27), (9.47), (9.48), (9.74), (9.75), (9.88), and (9.89).

Step 7: Compute the joint control equations and perform corresponding parameter adaptation for all nine joints: six for the manipulator and three for the reaction wheels by using (9.53)–(9.57) and (9.78)–(9.82).

9.14 Concluding Remarks

In this chapter, the VDC approach has been applied to orbital space robots. The zero force/moment applied to the floating (spacecraft) platform results in the so-called nonholonomic constraints governed by the momentum conservation law. With the VDC approach, the six nonholonomic constraints are, in fact, applied to the required (design) force/moment vector of the (spacecraft) platform rather than to the actual force/moment vector for having computational advantages without needing additional force/moment measurements. Consequently, the solvability of the six constraint equations becomes the key to the successful application of the VDC approach. The solvability has been rendered to three easy-to-satisfy conditions that require the lower bounds of the mass estimates of all rigid bodies and the lower bounds of the moment of inertia estimates of all reaction wheels to be positive.

10

Control of Humanoid Robots

10.1 Introduction

Since "Asimo" was introduced by Honda in 2000, humanoid robots have drawn increasing attentions of both academics and general public. To make a humanoid robot function as expected, control plays a vital role. Unfortunately, due to the difficulties in implementing a full-dynamics-based real-time control for a humanoid robot with hyper degrees of freedom, most state-of-the-art control approaches use ZMP (zero moment point) technique in which the dynamic interactions among different parts of a robot are treated as disturbances and only the gravity compensation with a rough dynamic force prediction is integrated into the decentralized joint PID controller. The lack of a precise modeling, without doubting, limits the control performances of biped systems and consequently limits the speeds in dynamic walking.

In this chapter, the application of the *virtual decomposition control* (VDC) to humanoid robots will be discussed. The VDC approach fully explores the dynamic potentials of humanoid robots by incorporating a subsystem dynamics-based adaptive control that is mathematically equivalent to any full-dynamics-based control. After the dynamics based control issue is completely addressed, the walking gait planning and control problem becomes a pure kinematics issue involving merely algebraic computations.

10.2 Prior Work

Control of bipeds has been extensively studied in the past two decades. The walking gait of a biped consists of repeated and alternative phases of single support (only one leg touches the ground) and double support (both legs touch the ground). The central concept in a single-support phase is the so-called *zero moment point* (ZMP) [179]. In a stable walking, the ZMP of a biped has to be located at any point within the contact surface of the stance foot or exactly located at the ankle if no foot is used. The constraint of locating the ZMP at

W.-H. Zhu: Virtual Decomposition Control, STAR 60, pp. 251–301.
springerlink.com

the ankle makes a biped an underactuated system in the single-support phase. A solution of using a time-scaling control to handle this underactuated system was proposed in [29]. The same problem can also be treated with zero dynamics [187].

Although being simple, a biped without feet walking in the sagittal plane captures the main characteristics of a realistic biped [28]. With respect to this particular type of system, a control method that guarantees the asymptotic stability of a cyclic walking motion was proposed in [62]. In this approach, the knees are supposed to be fixed during a single-support phase, and a stroke of the swing leg on the ground is considered as part of the walking cycle. The velocities of the system after the swing leg impact are determined by the approach developed in [207].

Readers who are interested in passive dynamic walking should refer to [118], for instance. Three-dimensional passive dynamic walking was reported in [37, 97] as well.

Without doubting, the remarkable Honda Asimo robot made a break-through in the practical development of three-dimensional bipeds. Readers are suggested to refer to [69] for its control approach.

10.3 Biped Walking

A biped walking motion possesses the following three characteristics [179]:

(i) There exists underactuation at ankles.
(ii) The motion is periodic.
(iii) The walking motion involves an alternation between a single-support phase and a double-support phase.

In biped locomotion, the overall ground reaction forces are equivalent to a single force applied to the robot at the so-called ZMP.[1] The concept of ZMP plays a crucial role in biped walking. In the phase of a single-support, the ZMP has to be located at the stance foot (or at the ankle when no foot is used). The location of ZMP itself imposes a dynamic constraint on the biped. For instance, when no foot is used the torque at the ankle joint of the stance leg must be zero. This scenario is equivalent to the use of an unactuated ankle joint. When feet are used, the allowable torque at the ankle joint of the stance leg is also restricted to keep the ZMP within the foot contact area. Thus, the maximum torque will be limited by the product of the reaction force and the foot length.

Periodic motion is certainly a basic characteristic to sustain a walking pattern. It requires that the system states at the end of a cycle are the same as the system states at the beginning of the cycle. A double-support phase usually is an intermediate state that allows both stance leg and swing leg to be alternated.

The unique difficulty in the trajectory planning of a biped is to find a cyclic motion that keeps the ZMP at the ankle/foot of the stance leg. In other words,

[1] The sum of all active force induced moments with respect to this point is equal to zero.

keeping the ZMP at the ankle/foot of the stance leg imposes a constraint on the dynamic equation at the acceleration level, whereas its integral (velocity) and double integral (position) must fulfill the requirements of a cyclic motion.

Remark 10.1. The feet have generally three functions: 1) it allows the ZMP to be allocated at any point of the contact surface between the stance foot and the ground, 2) a foot can absorb somehow the impact energy of the swing leg when it contacts with the ground, and 3) a foot can propel somehow before the swing leg leaves the ground. All these functions are supportive. They do not affect the main characteristics of a biped.

10.4 A Five-Bar Biped in Sagittal Plane

10.4.1 System Description

Without loss of generality, a five-bar biped moving in the sagittal plane by ignoring its lateral motion is studied in this section. This is a very abstractive and widely used representation of a biped. Regardless of its simplicity, a biped in the sagittal plane without feet captures the main characteristics of a realistic biped in three-dimensional motion.

Figure 10.1 illustrates a typical five-bar biped in a double-support phase in which both legs are used to support the biped. This system consists of two legs and one trunk. Each leg has two bars simulating the shank and the thigh, respectively. The trunk is used to represent the body. There are totally four electrical motors in the system: two at the two knees, and the other two at the torso between the trunk and the two thighs.

Figure 10.2 illustrates the five-bar biped in a single-support phase in which only one leg is used to support the biped and the other leg behaves as a swing leg. In a biped locomotion, both double-support and single-support phases occur alternatively.

The following two assumptions apply to the five-bar biped:

Assumption 7. *There is no slippage between the stance leg(s) and the ground.*

Assumption 8. *Zero torque can be transferred between the stance leg(s) and the ground.*

10.4.2 Double Support Phases

In a double-support phase as illustrated in Figure 10.1, there exist three degrees of freedom (DOFs) of motion: the x and y positions of the torso constituting two DOFs and the trunk angle constituting the third degree of freedom. Thus, the system becomes an over-actuated system with four motors controlling three DOFs of motion, leading to the creation of a one-dimensional controllable internal force.

Based on Assumption 7, the five-bar biped in the double-support phase can be viewed as two legs holding the trunk through the torso joint. When each ankle

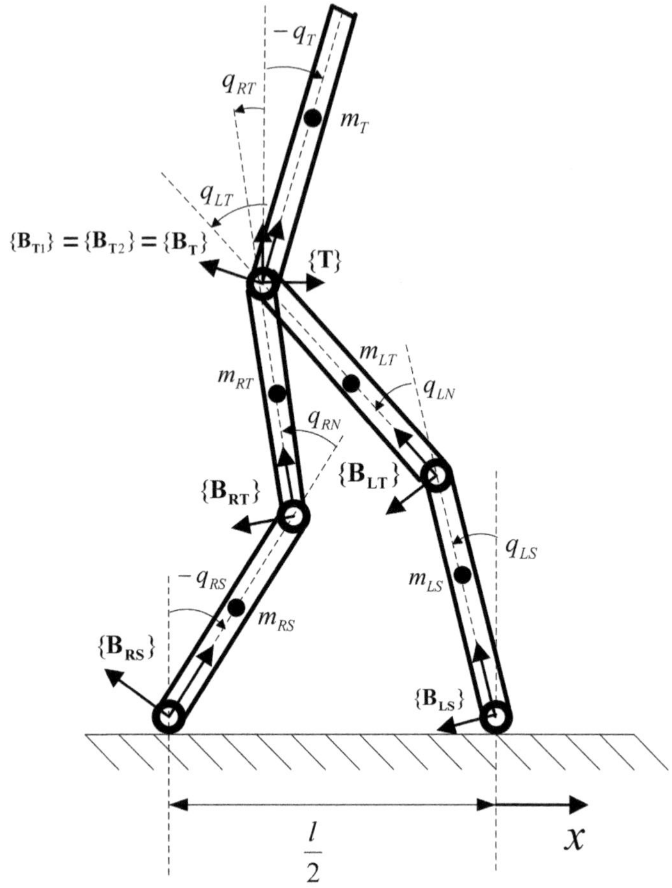

Fig. 10.1. Five-bar biped in double-support.

joint is treated as a revolution joint, the system is equivalent to two three-joint planar manipulators holding a common object (trunk), resulting in three DOFs of motion and three dimensions of the internal force. Among the six joints, four of them are actuated and the other two joints at the ankles are unactuated. Thus, two of the three internal force dimensions have to be released to satisfy the torque constraints at the two ankle joints, leaving one dimension for the controllable internal force.

10.4.3 Single Support Phases

In a single-support phase as illustrated in Figure 10.2, there exist five DOFs of motion: the x and y positions of the torso constituting two DOFs, the x and y positions of the swing leg tip constituting another two DOFs, and the trunk angle constituting the fifth degree of freedom. By locating four electrical motors to the system: two for the knees and two between the trunk and the two

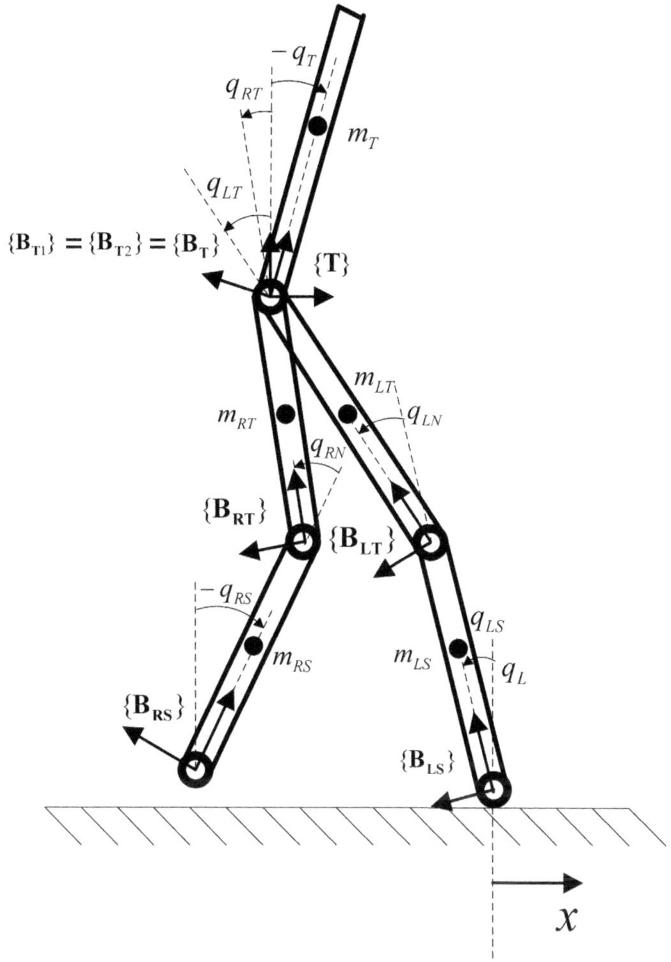

Fig. 10.2. Five-bar biped in single-support.

thighs, it creates a single underactuated DOF of motion, because only four out of five DOFs can be directly controlled. This one-dimensional underactuation is characterized by the zero torque at the ankle joint of the stance leg through placing the ZMP exactly at the ankle joint, based on Assumption 8.

Remark 10.2. The five-bar biped is characterized as an under-actuated system with five DOFs of motion subject to four control dimensions in a single-support phase. This system is also characterized as an over-actuated system with three DOFs of motion subject to four control dimensions in a double-support phase. Therefore, the five-bar biped locomotion involves alternations of under-actuation in a single-support phase and over-actuation in a double-support phase.

Because the five-bar biped in a double-support phase is an over-actuated system, the techniques developed in Chapter 8 for controlling the coordinated multiple manipulators can be directly applied. However, the five-bar biped in a single-support phase needs special treatments as will be seen below.

To concentrate on special features of a biped, the following assumption is made throughout this chapter.

Assumption 9. *All actuated joints are direct-drive joints.*

Remark 10.3. If conventional joints with transmissions are used, the results developed in Chapters 5 and 6 can be applied straightforwardly.

10.5 Kinematics of the Five-Bar Biped

This section presents the kinematics of the five-bar biped, including the definition of coordinate frames, the torso position and its relation to the center of mass (CoM), and the velocity mappings.

10.5.1 Coordinate Frames

The inertial frame $\{\mathbf{I}\}$ is mainly used to describe the biped motion. The x-axis and y-axis of frame $\{\mathbf{I}\}$ are placed horizontally and vertically. Frame $\{\mathbf{T}\}$ is located at the torso, having the same orientation as the inertia frame.

As illustrated in Figure 10.1, frames $\{\mathbf{B}_T\}$, $\{\mathbf{B}_{T1}\}$, and $\{\mathbf{B}_{T2}\}$ are fixed to the trunk, having their x axes aligned with the trunk and their origins located at the torso. Frames $\{\mathbf{B}_{LT}\}$ and $\{\mathbf{B}_{RT}\}$ are fixed to the left and right thighs, respectively, having their x axes aligned with the thigh axes and their origins located at the corresponding knees. Frames $\{\mathbf{B}_{LS}\}$ and $\{\mathbf{B}_{RS}\}$ are fixed to the left and right shanks, respectively, having their x axes aligned with the shank axes and their origins located at the corresponding ankles.

The same as in Chapter 3, the z axis of each frame is perpendicular to the motion (sagittal) plane to describe one-dimensional rotations and torques. Consequently, all rotation matrices are 2×2 matrices and all force/moment transformation matrices (such as ${}^{\mathbf{A}}\mathbf{U}_{\mathbf{B}}$ in (2.65)) are 3×3 matrices.

10.5.2 Torso Position

Let l_T and l_S be the lengths of thighs and shanks, respectively. Let $l > 0$ be the step length. Let $[x_t, y_t]^T$ denote the coordinates of the torso in the inertial frame $\{\mathbf{I}\}$. The following relationships

$$x_t = l_T \cos\left(q_{LT} + \frac{\pi}{2}\right) + l_S \cos\left(q_{LS} + \frac{\pi}{2}\right) \tag{10.1}$$

$$y_t = l_T \sin\left(q_{LT} + \frac{\pi}{2}\right) + l_S \sin\left(q_{LS} + \frac{\pi}{2}\right) \tag{10.2}$$

hold when the left ankle touches the ground at the origin $[0,0]^T$ or the following relationships

$$x_t = l_T \cos\left(q_{RT} + \frac{\pi}{2}\right) + l_S \cos\left(q_{RS} + \frac{\pi}{2}\right) - \frac{l}{2} \tag{10.3}$$

$$y_t = l_T \sin\left(q_{RT} + \frac{\pi}{2}\right) + l_S \sin\left(q_{RS} + \frac{\pi}{2}\right) \tag{10.4}$$

hold when the right ankle touches the ground at $\left[-\frac{l}{2}, 0\right]^T$, with

$$q_{LT} = q_{LS} + q_{LN} \tag{10.5}$$

$$q_{RT} = q_{RS} + q_{RN}. \tag{10.6}$$

The time derivatives of (10.1)–(10.4) are

$$\begin{bmatrix} \dot{x}_t \\ \dot{y}_t \end{bmatrix} = \mathbf{J}_L \begin{bmatrix} \dot{q}_{LT} \\ \dot{q}_{LS} \end{bmatrix} \tag{10.7}$$

with

$$\mathbf{J}_L = \begin{bmatrix} -l_T \sin\left(q_{LT} + \frac{\pi}{2}\right) & -l_S \sin\left(q_{LS} + \frac{\pi}{2}\right) \\ l_T \cos\left(q_{LT} + \frac{\pi}{2}\right) & l_S \cos\left(q_{LS} + \frac{\pi}{2}\right) \end{bmatrix} \in \mathbb{R}^{2\times 2}$$

when the left ankle touches the ground or

$$\begin{bmatrix} \dot{x}_t \\ \dot{y}_t \end{bmatrix} = \mathbf{J}_R \begin{bmatrix} \dot{q}_{RT} \\ \dot{q}_{RS} \end{bmatrix} \tag{10.8}$$

with

$$\mathbf{J}_R = \begin{bmatrix} -l_T \sin\left(q_{RT} + \frac{\pi}{2}\right) & -l_S \sin\left(q_{RS} + \frac{\pi}{2}\right) \\ l_T \cos\left(q_{RT} + \frac{\pi}{2}\right) & l_S \cos\left(q_{RS} + \frac{\pi}{2}\right) \end{bmatrix} \in \mathbb{R}^{2\times 2}$$

when the right ankle touches the ground.

Consequently, the body-frame referenced linear/angular velocity vector of frame $\{\mathbf{B}_T\}$ in the two-dimensional planar motion is expressed by

$$^{\mathbf{B}_T}V = \begin{bmatrix} \begin{bmatrix} \cos\left(q_T + \frac{\pi}{2}\right) & \sin\left(q_T + \frac{\pi}{2}\right) \\ -\sin\left(q_T + \frac{\pi}{2}\right) & \cos\left(q_T + \frac{\pi}{2}\right) \end{bmatrix} \begin{bmatrix} \dot{x}_t \\ \dot{y}_t \end{bmatrix} \\ \dot{q}_T \end{bmatrix} \in \mathbb{R}^3. \tag{10.9}$$

10.5.3 Relationship between Torso Position and CoM

The following assumption gives the locations of the centers of mass of the five bars.

Assumption 10. *The centers of mass of the truck, the left and right thighs, and the left and right shanks lie in the extensions of the x axes of frames $\{\mathbf{B}_T\}$, $\{\mathbf{B}_{LT}\}$, $\{\mathbf{B}_{RT}\}$, $\{\mathbf{B}_{LS}\}$, and $\{\mathbf{B}_{RS}\}$, respectively.*

Some symbols are defined:

x_c — x coordinate of the center of mass (CoM) of the biped expressed in the inertial frame $\{\mathbf{I}\}$.
y_c — y coordinate of the center of mass (CoM) of the biped expressed in the inertial frame $\{\mathbf{I}\}$.
x_t — x coordinate of the torso expressed in the inertial frame $\{\mathbf{I}\}$.
y_t — y coordinate of the torso expressed in the inertial frame $\{\mathbf{I}\}$.
d_T — Distance from the torso to the center of mass of the trunk.
d_{LT} — Distance from the left knee to the center of mass of the left thigh.
d_{RT} — Distance from the right knee to the center of mass of the right thigh.
d_{LS} — Distance from the left ankle to the center of mass of the left shank.
d_{RS} — Distance from the right ankle to the center of mass of the right shank.
m_T — Mass of the trunk.
m_{LT} — Mass of the left thigh.
m_{RT} — Mass of the right thigh.
m_{LS} — Mass of the left shank.
m_{RS} — Mass of the right shank.

Assume that the left ankle is located at $[0,0]^T$ in the inertial frame $\{\mathbf{I}\}$. Let the total mass of the system be

$$m_c = m_T + m_{LT} + m_{RT} + m_{LS} + m_{RS}. \tag{10.10}$$

Expressing the CoM (center of mass) with respect to the torso yields

$$\begin{aligned}
&(m_T + m_{LT} + m_{LS} + m_{RT} + m_{RS})(x_c - x_t) \\
&\quad = m_T d_T \cos\left(q_T + \frac{\pi}{2}\right) + m_{LT}(l_T - d_{LT})\cos\left(q_{LT} - \frac{\pi}{2}\right) \\
&\quad\quad + m_{LS}\left[l_T \cos\left(q_{LT} - \frac{\pi}{2}\right) + (l_S - d_{LS})\cos\left(q_{LS} - \frac{\pi}{2}\right)\right] \\
&\quad\quad + m_{RT}(l_T - d_{RT})\cos\left(q_{RT} - \frac{\pi}{2}\right) \\
&\quad\quad + m_{RS}\left[l_T \cos\left(q_{RT} - \frac{\pi}{2}\right) + (l_S - d_{RS})\cos\left(q_{RS} - \frac{\pi}{2}\right)\right]
\end{aligned} \tag{10.11}$$

$$\begin{aligned}
&(m_T + m_{LT} + m_{LS} + m_{RT} + m_{RS})(y_c - y_t) \\
&\quad = m_T d_T \sin\left(q_T + \frac{\pi}{2}\right) + m_{LT}(l_T - d_{LT})\sin\left(q_{LT} - \frac{\pi}{2}\right) \\
&\quad\quad + m_{LS}\left[l_T \sin\left(q_{LT} - \frac{\pi}{2}\right) + (l_S - d_{LS})\sin\left(q_{LS} - \frac{\pi}{2}\right)\right] \\
&\quad\quad + m_{RT}(l_T - d_{RT})\sin\left(q_{RT} - \frac{\pi}{2}\right) \\
&\quad\quad + m_{RS}\left[l_T \sin\left(q_{RT} - \frac{\pi}{2}\right) + (l_S - d_{RS})\sin\left(q_{RS} - \frac{\pi}{2}\right)\right].
\end{aligned} \tag{10.12}$$

The time derivatives are

$$(m_T + m_{LT} + m_{LS} + m_{RT} + m_{RS}) \begin{bmatrix} \dot{x}_c - \dot{x}_t \\ \dot{y}_c - \dot{y}_t \end{bmatrix}$$
$$= \begin{bmatrix} -m_T d_T \sin\left(q_T + \frac{\pi}{2}\right) \\ m_T d_T \cos\left(q_T + \frac{\pi}{2}\right) \end{bmatrix} \dot{q}_T$$
$$+ \begin{bmatrix} a_{11} & a_{12} \\ a_{21} & a_{22} \end{bmatrix} \begin{bmatrix} \dot{q}_{LT} \\ \dot{q}_{LS} \end{bmatrix} + \begin{bmatrix} b_{11} & b_{12} \\ b_{21} & b_{22} \end{bmatrix} \begin{bmatrix} \dot{q}_{RT} \\ \dot{q}_{RS} \end{bmatrix} \tag{10.13}$$

where

$$\begin{aligned}
a_{11} &= -[m_{LT}(l_T - d_{LT}) + m_{LS} l_T] \sin\left(q_{LT} - \frac{\pi}{2}\right) \\
a_{12} &= -m_{LS}(l_S - d_{LS}) \sin\left(q_{LS} - \frac{\pi}{2}\right) \\
a_{21} &= [m_{LT}(l_T - d_{LT}) + m_{LS} l_T] \cos\left(q_{LT} - \frac{\pi}{2}\right) \\
a_{22} &= m_{LS}(l_S - d_{LS}) \cos\left(q_{LS} - \frac{\pi}{2}\right) \\
b_{11} &= -[m_{RT}(l_T - d_{RT}) + m_{RS} l_T] \sin\left(q_{RT} - \frac{\pi}{2}\right) \\
b_{12} &= -m_{RS}(l_S - d_{RS}) \sin\left(q_{RS} - \frac{\pi}{2}\right) \\
b_{21} &= [m_{RT}(l_T - d_{RT}) + m_{RS} l_T] \cos\left(q_{RT} - \frac{\pi}{2}\right) \\
b_{22} &= m_{RS}(l_S - d_{RS}) \cos\left(q_{RS} - \frac{\pi}{2}\right).
\end{aligned}$$

Equation (10.13) defines the relationship between the velocities of the CoM $[\dot{x}_c, \dot{y}_c]^T$ and the velocities of the torso $[\dot{x}_t, \dot{y}_t]^T$ for given joint velocities of the biped.

10.5.4 Mapping Matrices for Velocities and Forces

In the two-dimensional planar motion, both linear/angular velocity vectors and force/moment vectors are 3×1 vectors having the same data structure as (3.14) and (3.15) with appropriate frame substitutions (for frame $\{\mathbf{B}_1\}$). In view of Figure 10.1, the force/moment transformation matrices in the two-dimensional planar motion can be expressed by

$${}^{\mathbf{B}_T}\mathbf{U}_\mathbf{T} = \begin{bmatrix} {}^{\mathbf{B}_T}\mathbf{R}_\mathbf{T} & \begin{bmatrix} 0 \\ 0 \end{bmatrix} \\ \begin{bmatrix} 0 & 0 \end{bmatrix} & 1 \end{bmatrix} \in \mathbb{R}^{3\times 3} \tag{10.14}$$

$${}^{\mathbf{B}_{LS}}\mathbf{U}_{\mathbf{B}_{LT}} = \begin{bmatrix} {}^{\mathbf{B}_{LS}}\mathbf{R}_{\mathbf{B}_{LT}} & \begin{bmatrix} 0 \\ 0 \end{bmatrix} \\ \begin{bmatrix} l_S \sin(q_{LN}) & l_S \cos(q_{LN}) \end{bmatrix} & 1 \end{bmatrix} \in \mathbb{R}^{3\times 3} \tag{10.15}$$

$${}^{\mathbf{B}_{LT}}\mathbf{U}_\mathbf{T} = \begin{bmatrix} {}^{\mathbf{B}_{LT}}\mathbf{R}_\mathbf{T} & \begin{bmatrix} 0 \\ 0 \end{bmatrix} \\ \begin{bmatrix} -l_T \cos(q_{LT}) & -l_T \sin(q_{LT}) \end{bmatrix} & 1 \end{bmatrix} \in \mathbb{R}^{3\times 3} \tag{10.16}$$

$$
{}^{\mathbf{B}_{LS}}\mathbf{U}_{\mathbf{T}} = \begin{bmatrix} {}^{\mathbf{B}_{LS}}\mathbf{R}_{\mathbf{T}} & \begin{bmatrix} 0 \\ 0 \end{bmatrix} \\ \begin{bmatrix} -l_{L1}(q_{LT}, q_{LS}) & -l_{L2}(q_{LT}, q_{LS}) \end{bmatrix} & 1 \end{bmatrix} \in \mathbb{R}^{3\times 3} \tag{10.17}
$$

$$
{}^{\mathbf{B}_{RS}}\mathbf{U}_{\mathbf{B}_{RT}} = \begin{bmatrix} {}^{\mathbf{B}_{RS}}\mathbf{R}_{\mathbf{B}_{RT}} & \begin{bmatrix} 0 \\ 0 \end{bmatrix} \\ \begin{bmatrix} l_S \sin(q_{RN}) & l_S \cos(q_{RN}) \end{bmatrix} & 1 \end{bmatrix} \in \mathbb{R}^{3\times 3} \tag{10.18}
$$

$$
{}^{\mathbf{B}_{RT}}\mathbf{U}_{\mathbf{T}} = \begin{bmatrix} {}^{\mathbf{B}_{RT}}\mathbf{R}_{\mathbf{T}} & \begin{bmatrix} 0 \\ 0 \end{bmatrix} \\ \begin{bmatrix} -l_T \cos(q_{RT}) & -l_T \sin(q_{RT}) \end{bmatrix} & 1 \end{bmatrix} \in \mathbb{R}^{3\times 3} \tag{10.19}
$$

$$
{}^{\mathbf{B}_{RS}}\mathbf{U}_{\mathbf{T}} = \begin{bmatrix} {}^{\mathbf{B}_{RS}}\mathbf{R}_{\mathbf{T}} & \begin{bmatrix} 0 \\ 0 \end{bmatrix} \\ \begin{bmatrix} -l_{R1}(q_{RT}, q_{RS}) & -l_{R2}(q_{RT}, q_{RS}) \end{bmatrix} & 1 \end{bmatrix} \in \mathbb{R}^{3\times 3} \tag{10.20}
$$

with

$$
\begin{aligned}
{}^{\mathbf{B}_{T}}\mathbf{R}_{\mathbf{T}} &= \begin{bmatrix} \cos\left(q_T + \frac{\pi}{2}\right) & \sin\left(q_T + \frac{\pi}{2}\right) \\ -\sin\left(q_T + \frac{\pi}{2}\right) & \cos\left(q_T + \frac{\pi}{2}\right) \end{bmatrix} \in \mathbb{R}^{2\times 2} \\
{}^{\mathbf{B}_{LS}}\mathbf{R}_{\mathbf{B}_{LT}} &= \begin{bmatrix} \cos(q_{LN}) & -\sin(q_{LN}) \\ \sin(q_{LN}) & \cos(q_{LN}) \end{bmatrix} \in \mathbb{R}^{2\times 2} \\
{}^{\mathbf{B}_{LT}}\mathbf{R}_{\mathbf{T}} &= \begin{bmatrix} \cos\left(q_{LT} + \frac{\pi}{2}\right) & \sin\left(q_{LT} + \frac{\pi}{2}\right) \\ -\sin\left(q_{LT} + \frac{\pi}{2}\right) & \cos\left(q_{LT} + \frac{\pi}{2}\right) \end{bmatrix} \in \mathbb{R}^{2\times 2} \\
{}^{\mathbf{B}_{LS}}\mathbf{R}_{\mathbf{T}} &= \begin{bmatrix} \cos\left(q_{LS} + \frac{\pi}{2}\right) & \sin\left(q_{LS} + \frac{\pi}{2}\right) \\ -\sin\left(q_{LS} + \frac{\pi}{2}\right) & \cos\left(q_{LS} + \frac{\pi}{2}\right) \end{bmatrix} \in \mathbb{R}^{2\times 2} \\
{}^{\mathbf{B}_{RS}}\mathbf{R}_{\mathbf{B}_{RT}} &= \begin{bmatrix} \cos(q_{RN}) & -\sin(q_{RN}) \\ \sin(q_{RN}) & \cos(q_{RN}) \end{bmatrix} \in \mathbb{R}^{2\times 2} \\
{}^{\mathbf{B}_{RT}}\mathbf{R}_{\mathbf{T}} &= \begin{bmatrix} \cos\left(q_{RT} + \frac{\pi}{2}\right) & \sin\left(q_{RT} + \frac{\pi}{2}\right) \\ -\sin\left(q_{RT} + \frac{\pi}{2}\right) & \cos\left(q_{RT} + \frac{\pi}{2}\right) \end{bmatrix} \in \mathbb{R}^{2\times 2} \\
{}^{\mathbf{B}_{RS}}\mathbf{R}_{\mathbf{T}} &= \begin{bmatrix} \cos\left(q_{RS} + \frac{\pi}{2}\right) & \sin\left(q_{RS} + \frac{\pi}{2}\right) \\ -\sin\left(q_{RS} + \frac{\pi}{2}\right) & \cos\left(q_{RS} + \frac{\pi}{2}\right) \end{bmatrix} \in \mathbb{R}^{2\times 2} \\
l_{L1}(q_{LT}, q_{LS}) &= l_T \cos(q_{LT}) + l_S \cos(q_{LS}) \\
l_{L2}(q_{LT}, q_{LS}) &= l_T \sin(q_{LT}) + l_S \sin(q_{LS}) \\
l_{R1}(q_{RT}, q_{RS}) &= l_T \cos(q_{RT}) + l_S \cos(q_{RS}) \\
l_{R2}(q_{RT}, q_{RS}) &= l_T \sin(q_{RT}) + l_S \sin(q_{RS}).
\end{aligned}
$$

These matrices will be frequently used in the following sections to handle the velocity and force transformations among different frames.

10.6 Dynamics and Control of the Five-Bar Biped in Double-Support Phase

With respect to the five-bar biped in a double-support phase as illustrated in Figure 10.1, the system has three DOFs of motion and six joints comprised of four actuated joints and two unactuated joints. Under Assumption 7, this system is equivalent to two three-joint planar manipulators (each represents a leg) holding a common object (the trunk). Thus, it has three DOFs of motion and three dimensions of internal forces. Two out of the three dimensions of the internal forces are to be determined by satisfying the torque constraints at the two unactuated ankle joints. The remaining one dimensional internal force is controllable.

Special force/moment vectors in the two-dimensional planar motion are defined as follows:

${}^{\mathbf{B}_T}F^* \in \mathbb{R}^3$	Net force/moment vector of the trunk expressed in frame $\{\mathbf{B}_T\}$.
${}^{\mathbf{B}_{LT}}F^* \in \mathbb{R}^3$	Net force/moment vector of the left thigh expressed in frame $\{\mathbf{B}_{LT}\}$.
${}^{\mathbf{B}_{LS}}F^* \in \mathbb{R}^3$	Net force/moment vector of the left shank expressed in frame $\{\mathbf{B}_{LS}\}$.
${}^{\mathbf{B}_{RT}}F^* \in \mathbb{R}^3$	Net force/moment vector of the right thigh expressed in frame $\{\mathbf{B}_{RT}\}$.
${}^{\mathbf{B}_{RS}}F^* \in \mathbb{R}^3$	Net force/moment vector of the right shank expressed in frame $\{\mathbf{B}_{RS}\}$.
${}^{\mathbf{B}_{LS}}F \in \mathbb{R}^3$	Ground force/moment vector applied to the left leg expressed in frame $\{\mathbf{B}_{LS}\}$.
${}^{\mathbf{B}_{RS}}F \in \mathbb{R}^3$	Ground force/moment vector applied to the right leg expressed in frame $\{\mathbf{B}_{RS}\}$.
${}^{\mathbf{B}_{LT}}F \in \mathbb{R}^3$	Force/moment vector applied from the left shank to the left thigh expressed in frame $\{\mathbf{B}_{LT}\}$.
${}^{\mathbf{B}_{RT}}F \in \mathbb{R}^3$	Force/moment vector applied from the right shank to the right thigh expressed in frame $\{\mathbf{B}_{RT}\}$.
${}^{\mathbf{B}_{T1}}F \in \mathbb{R}^3$	Force/moment vector applied from the left thigh to the trunk expressed in frame $\{\mathbf{B}_{T1}\} = \{\mathbf{B}_T\}$.
${}^{\mathbf{B}_{T2}}F \in \mathbb{R}^3$	Force/moment vector applied from the right thigh to the trunk expressed in frame $\{\mathbf{B}_{T2}\} = \{\mathbf{B}_T\}$.

10.6.1 Virtual Decomposition

The biped in a double-support phase can be virtually decomposed into the trunk and the two legs, as illustrated in Figure 10.3, with its graphic expression being illustrated in Figure 10.4.

There are totally two *cutting points* being placed in the system. The two *cutting points* associated with frames $\{\mathbf{B}_{T1}\}$ and $\{\mathbf{B}_{T2}\}$ are the *driven cutting points* of the trunk and also the *driving cutting points* of the two legs.

As will be shown below, the VDC allows the dynamics and control problem of a biped to be converted to the dynamics and control problem of each subsystem, namely the trunk, the left leg, and the right leg.

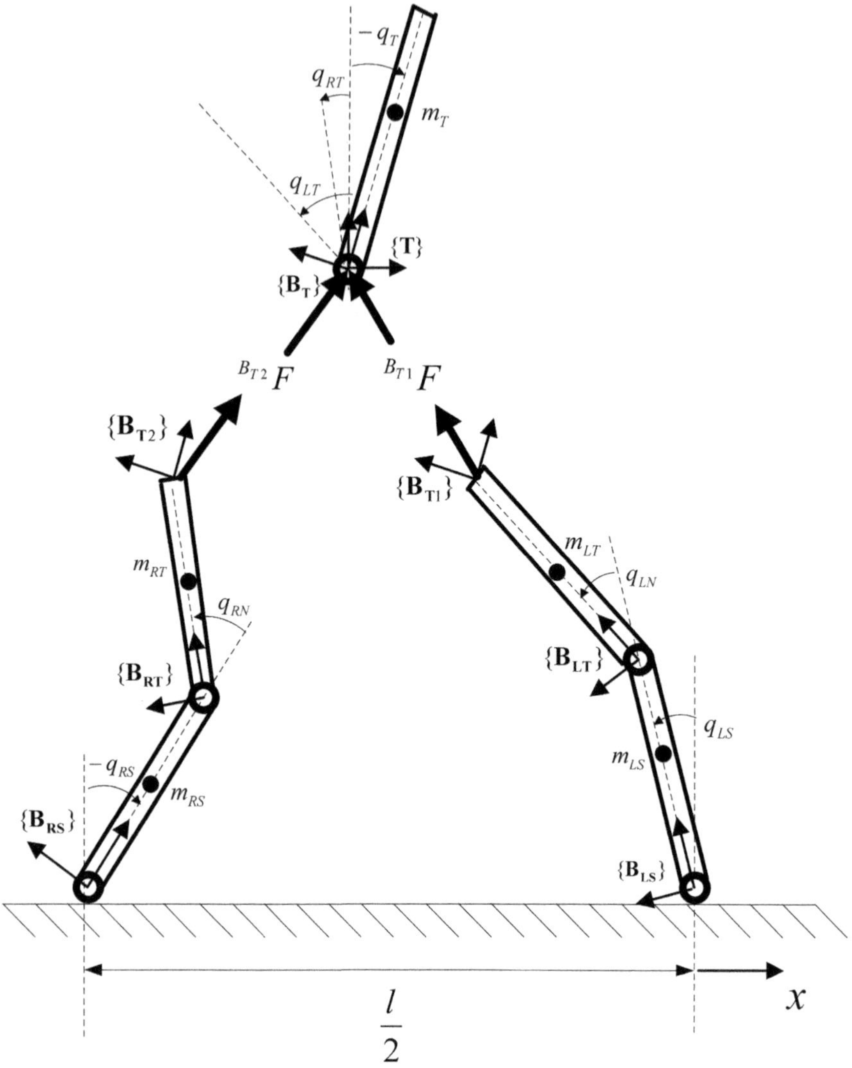

Fig. 10.3. Five-bar biped in double-support after virtual decomposition.

10.6.2 Control Specifications

In a double-support phase, let $[q_T, q_{LT}, q_{LS}]^T \in \mathbb{R}^3$ be the independent motion coordinates. Given a desired trajectory vector $[q_{Td}, q_{LTd}, q_{LSd}]^T \in \mathbb{R}^3$ and its time derivative $[\dot{q}_{Td}, \dot{q}_{LTd}, \dot{q}_{LSd}]^T \in \mathbb{R}^3$, the required velocity vector is designed as

$$\begin{bmatrix} \dot{q}_{Tr} \\ \dot{q}_{LTr} \\ \dot{q}_{LSr} \end{bmatrix} = \begin{bmatrix} \dot{q}_{Td} \\ \dot{q}_{LTd} \\ \dot{q}_{LSd} \end{bmatrix} + \begin{bmatrix} \lambda_1(q_{Td} - q_T) \\ \lambda_2(q_{LTd} - q_{LT}) \\ \lambda_3(q_{LSd} - q_{LS}) \end{bmatrix} \tag{10.21}$$

with $\lambda_i > 0$ being a position control gain for all $i \in \{1, 3\}$.

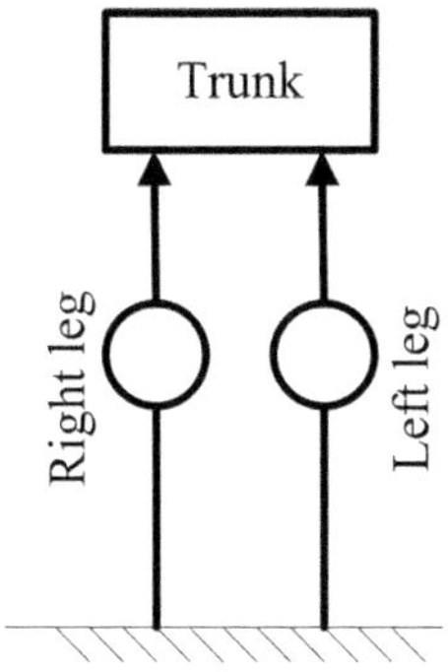

Fig. 10.4. Graphic expression of Figure 10.3.

In view of (10.7) and (10.8), the required velocities of the right leg can be obtained from the required velocities of the left leg as

$$\begin{bmatrix} \dot{q}_{RTr} \\ \dot{q}_{RSr} \end{bmatrix} = \mathbf{J}_R^{-1}\mathbf{J}_L \begin{bmatrix} \dot{q}_{LTr} \\ \dot{q}_{LSr} \end{bmatrix} \tag{10.22}$$

by noting the fact that both legs are touching the ground.

Remark 10.4. Having $[q_T, q_{LT}, q_{LS}]^T \in \mathbb{R}^3$ (the trunk and the left leg) as the independent coordinates is only one of the options. The other two options are to select either $[q_T, q_{RT}, q_{RS}]^T \in \mathbb{R}^3$ (the trunk and the right leg) or $[q_T, x_t, y_t]^T \in \mathbb{R}^3$ (the trunk and the torso) as the independent coordinates.

Accordingly, the required linear/angular velocity vector of the trunk expressed in frame $\{\mathbf{B}_T\}$ can be obtained from (10.7) and (10.9) as

$$\begin{aligned} {}^{\mathbf{B}_T}V_r &= \begin{bmatrix} \begin{bmatrix} \cos\left(q_T + \frac{\pi}{2}\right) & \sin\left(q_T + \frac{\pi}{2}\right) \\ -\sin\left(q_T + \frac{\pi}{2}\right) & \cos\left(q_T + \frac{\pi}{2}\right) \end{bmatrix} \begin{bmatrix} \dot{x}_{tr} \\ \dot{y}_{tr} \end{bmatrix} \\ \dot{q}_{Tr} \end{bmatrix} \\ &= \begin{bmatrix} \begin{bmatrix} \cos\left(q_T + \frac{\pi}{2}\right) & \sin\left(q_T + \frac{\pi}{2}\right) \\ -\sin\left(q_T + \frac{\pi}{2}\right) & \cos\left(q_T + \frac{\pi}{2}\right) \end{bmatrix} \mathbf{J}_L \begin{bmatrix} \dot{q}_{LTr} \\ \dot{q}_{LSr} \end{bmatrix} \\ \dot{q}_{Tr} \end{bmatrix} \in \mathbb{R}^3. \end{aligned} \tag{10.23}$$

10.6.3 Dynamics and Control of the Trunk

This subsection presents the dynamics and control equations of the trunk to ensure its *virtual stability*.

Dynamics

Recall (2.74) to obtain the dynamics of the trunk with frame $\{\mathbf{B}_T\}$ being substituted for frame $\{\mathbf{A}\}$. After projecting the dynamics into the sagittal plane with an approach similar to that in Section 3.4, it yields

$$\mathbf{M}_{\mathbf{B}_T}\frac{d}{dt}({}^{\mathbf{B}_T}V) + \mathbf{C}_{\mathbf{B}_T}({}^{\mathbf{B}_T}\omega){}^{\mathbf{B}_T}V + \mathbf{G}_{\mathbf{B}_T} = {}^{\mathbf{B}_T}F^* \tag{10.24}$$

where ${}^{\mathbf{B}_T}V \in \mathbb{R}^3$ is defined by (10.9) and

$$\mathbf{M}_{\mathbf{B}_T} = \begin{bmatrix} m_T & 0 & 0 \\ 0 & m_T & m_T d_T \\ 0 & m_T d_T & I_{oT} + m_T d_T^2 \end{bmatrix} \tag{10.25}$$

$$\mathbf{C}_{\mathbf{B}_T} = \begin{bmatrix} 0 & -m_T & -m_T d_T \\ m_T & 0 & 0 \\ m_T d_T & 0 & 0 \end{bmatrix} \dot{q}_T \tag{10.26}$$

$$\mathbf{G}_{\mathbf{B}_T} = \begin{bmatrix} m_T \sin\left(q_T + \frac{\pi}{2}\right) g \\ m_T \cos\left(q_T + \frac{\pi}{2}\right) g \\ m_T d_T \cos\left(q_T + \frac{\pi}{2}\right) g \end{bmatrix} \tag{10.27}$$

in which I_{oT} is the moment of inertia of the trunk around its center of mass and $g \approx 9.8$ is the standard gravity.

In (10.24), the net force/moment vector of the trunk can be expressed by

$${}^{\mathbf{B}_T}F^* = {}^{\mathbf{B}_{T1}}F + {}^{\mathbf{B}_{T2}}F. \tag{10.28}$$

Control Equations

With a similar procedure leading to (3.36), the required net force/moment vector of the trunk in the sagittal plane can be written as

$${}^{\mathbf{B}_T}F_r^* = \mathbf{Y}_{\mathbf{B}_T}\hat{\boldsymbol{\theta}}_{\mathbf{B}_T} + \mathbf{K}_{\mathbf{B}_T}\left({}^{\mathbf{B}_T}V_r - {}^{\mathbf{B}_T}V\right) \tag{10.29}$$

where $\mathbf{K}_{\mathbf{B}_T} \in \mathbb{R}^{3\times 3}$ is a symmetric positive-definite matrix

$$\boldsymbol{\theta}_{\mathbf{B}_T} = [m_T, m_T d_T, I_{oT} + m_T d_T^2]^T \in \mathbb{R}^3 \tag{10.30}$$

is a 3×1 parameter vector, and $\mathbf{Y}_{\mathbf{B}_T}$ is a 3×3 regressor matrix defined by

$$\mathbf{Y}_{\mathbf{B}_T} = \begin{bmatrix} \mathbf{y}_{\mathbf{B}_T 1} & \mathbf{y}_{\mathbf{B}_T 2} & \mathbf{y}_{\mathbf{B}_T 3} \end{bmatrix} \tag{10.31}$$

with

$$\mathbf{y}_{\mathbf{B}_T 1} = \begin{bmatrix} \frac{d}{dt}(v_{\mathbf{B}_T r})(1) - \dot{q}_T v_{\mathbf{B}_T r}(2) + \sin\left(q_T + \frac{\pi}{2}\right) g \\ \frac{d}{dt}(v_{\mathbf{B}_T r})(2) + \dot{q}_T v_{\mathbf{B}_T r}(1) + \cos\left(q_T + \frac{\pi}{2}\right) g \\ 0 \end{bmatrix}$$

$$\mathbf{y}_{\mathbf{B}_T 2} = \begin{bmatrix} -\dot{q}_T v_{\mathbf{B}_T r}(3) \\ \frac{d}{dt}(v_{\mathbf{B}_T r})(3) \\ \frac{d}{dt}(v_{\mathbf{B}_T r})(2) + \dot{q}_T v_{\mathbf{B}_T r}(1) + \cos\left(q_T + \frac{\pi}{2}\right) g \end{bmatrix}$$

$$\mathbf{y}_{\mathbf{B}_T 3} = \begin{bmatrix} 0 \\ 0 \\ \frac{d}{dt}(v_{\mathbf{B}_T r})(3) \end{bmatrix}$$

in which $\frac{d}{dt}(v_{\mathbf{B}_T r})(j) \in \mathbb{R}$ denotes the jth element of $\frac{d}{dt}\left({}^{\mathbf{B}_T}V_r\right) \in \mathbb{R}^3$ and $v_{\mathbf{B}_T r}(j) \in \mathbb{R}$ denotes the jth element of ${}^{\mathbf{B}_T}V_r \in \mathbb{R}^3$ for all $j \in \{1, 3\}$, such that

$$\mathbf{Y}_{\mathbf{B}_T}\boldsymbol{\theta}_{\mathbf{B}_T} = \mathbf{M}_{\mathbf{B}_T}\frac{d}{dt}\left({}^{\mathbf{B}_T}V_r\right) + \mathbf{C}_{\mathbf{B}_T}{}^{\mathbf{B}_T}V_r + \mathbf{G}_{\mathbf{B}_T} \tag{10.32}$$

holds indefinitely.

The estimates of the three parameters defined by (10.30) need to be updated. Define

$$\mathbf{s}_{\mathbf{B}_T} = \mathbf{Y}_{\mathbf{B}_T}^T\left({}^{\mathbf{B}_T}V_r - {}^{\mathbf{B}_T}V\right). \tag{10.33}$$

The $\mathcal{P}$ function defined by (2.79) is used to update each of the parameter estimates of the trunk as

$$\hat{\theta}_{\mathbf{B}_T\gamma} = \mathcal{P}\left(s_{\mathbf{B}_T\gamma}, \rho_{\mathbf{B}_T\gamma}, \underline{\theta}_{\mathbf{B}_T\gamma}, \overline{\theta}_{\mathbf{B}_T\gamma}, t\right) \tag{10.34}$$

for all $\gamma \in \{1, 3\}$, where

$\theta_{\mathbf{B}_T\gamma}$	γth element of $\boldsymbol{\theta}_{\mathbf{B}_T}$.
$\hat{\theta}_{\mathbf{B}_T\gamma}$	Estimate of $\theta_{\mathbf{B}_T\gamma}$.
$s_{\mathbf{B}_T\gamma}$	γth element of $\mathbf{s}_{\mathbf{B}_T}$.
$\rho_{\mathbf{B}_T\gamma}$	Parameter update gain.
$\underline{\theta}_{\mathbf{B}_T\gamma}$	Lower bound of $\theta_{\mathbf{B}_T\gamma}$.
$\overline{\theta}_{\mathbf{B}_T\gamma}$	Upper bound of $\theta_{\mathbf{B}_T\gamma}$.

In line with (10.28), the required net force/moment vector of the trunk can also be expressed by

$${}^{\mathbf{B}_T}F_r^* = {}^{\mathbf{B}_{T1}}F_r + {}^{\mathbf{B}_{T2}}F_r. \tag{10.35}$$

By introducing a three-dimensional required internal force vector $\boldsymbol{\eta}_{br} \in \mathbb{R}^3$, it follows from (10.35) that

$${}^{\mathbf{B}_{T1}}F_r = \alpha{}^{\mathbf{B}_T}F_r^* + \boldsymbol{\eta}_{br} \tag{10.36}$$

$${}^{\mathbf{B}_{T2}}F_r = (1-\alpha){}^{\mathbf{B}_T}F_r^* - \boldsymbol{\eta}_{br} \tag{10.37}$$

hold with $0 \leqslant \alpha \leqslant 1$.

Virtual Stability

In the sense of Definition 2.17, the *virtual stability* of the trunk combined with its respective control equations is given in this subsection.

Lemma 10.1. *Consider the trunk described by (10.24)–(10.27) and combined with its respective control equations (10.29)–(10.32) and with the parameter adaptation (10.33) and (10.34). If a non-negative accompanying function of the trunk is chosen as*

$$
\begin{aligned}
\nu_{\mathbf{B}_T} = & \frac{1}{2}\left({}^{\mathbf{B}_T}V_r - {}^{\mathbf{B}_T}V\right)^T \mathbf{M}_{\mathbf{B}_T}\left({}^{\mathbf{B}_T}V_r - {}^{\mathbf{B}_T}V\right) \\
& + \frac{1}{2}\sum_{\gamma=1}^{3}\left(\theta_{\mathbf{B}_T\gamma} - \hat{\theta}_{\mathbf{B}_T\gamma}\right)^2 / \rho_{\mathbf{B}_T\gamma}
\end{aligned} \tag{10.38}
$$

then it follows that

$$
\begin{aligned}
\dot{\nu}_{\mathbf{B}_T} \leqslant & -\left({}^{\mathbf{B}_T}V_r - {}^{\mathbf{B}_T}V\right)^T \mathbf{K}_{\mathbf{B}_T}\left({}^{\mathbf{B}_T}V_r - {}^{\mathbf{B}_T}V\right) \\
& + \left({}^{\mathbf{B}_T}V_r - {}^{\mathbf{B}_T}V\right)^T \left({}^{\mathbf{B}_T}F_r^* - {}^{\mathbf{B}_T}F^*\right)
\end{aligned} \tag{10.39}
$$

holds.

The proof follows exactly the same procedure as that for Lemma 4.1. ■

Theorem 10.1. *The trunk described by (10.24)–(10.28), combined with its control equations (10.29)–(10.32) and (10.35) and with the parameter adaptation (10.33) and (10.34), is virtually stable with its affiliated vector* ${}^{\mathbf{B}_T}V_r - {}^{\mathbf{B}_T}V$ *being a virtual function in both* L_2 *and* L_∞*, in the sense of Definition 2.17.*

Proof: In view of (2.85), (10.28), (10.35), and $\{\mathbf{B}_T\} = \{\mathbf{B}_{T1}\} = \{\mathbf{B}_{T2}\}$, it follows that

$$
\begin{aligned}
& \left({}^{\mathbf{B}_T}V_r - {}^{\mathbf{B}_T}V\right)^T \left({}^{\mathbf{B}_T}F_r^* - {}^{\mathbf{B}_T}F^*\right) \\
& = \left({}^{\mathbf{B}_T}V_r - {}^{\mathbf{B}_T}V\right)^T \left({}^{\mathbf{B}_{T1}}F_r - {}^{\mathbf{B}_{T1}}F\right) \\
& \quad + \left({}^{\mathbf{B}_T}V_r - {}^{\mathbf{B}_T}V\right)^T \left({}^{\mathbf{B}_{T2}}F_r - {}^{\mathbf{B}_{T2}}F\right) \\
& = \left({}^{\mathbf{B}_{T1}}V_r - {}^{\mathbf{B}_{T1}}V\right)^T \left({}^{\mathbf{B}_{T1}}F_r - {}^{\mathbf{B}_{T1}}F\right) \\
& \quad + \left({}^{\mathbf{B}_{T2}}V_r - {}^{\mathbf{B}_{T2}}V\right)^T \left({}^{\mathbf{B}_{T2}}F_r - {}^{\mathbf{B}_{T2}}F\right) \\
& = p_{\mathbf{B}_{T1}} + p_{\mathbf{B}_{T2}}
\end{aligned} \tag{10.40}
$$

holds.

Consider the fact that the trunk has two *driven cutting points* associated with frames $\{\mathbf{B}_{T1}\}$ and $\{\mathbf{B}_{T2}\}$. Substituting (10.40) into (10.39) and using (10.38) and Definition 2.17 completes the proof. ■

10.6.4 Dynamics and Control of the Left Leg

The left leg consists of two rigid links, namely the thigh and the shank, and three one-dimensional joints located between the trunk and the thigh, between the thigh and the shank, and between the shank and the ground, respectively.

Kinematics and Dynamics

Because the left leg touches the ground, the velocity relationships of the left leg are

$$
\begin{aligned}
{}^{\mathbf{B}_{T1}}V &= {}^{\mathbf{B}_T}V \\
&= \mathbf{z}(\dot{q}_T - \dot{q}_{LT}) + {}^{\mathbf{B}_{LT}}\mathbf{U}_{\mathbf{B}_T}^T\, {}^{\mathbf{B}_{LT}}V && (10.41)\\
{}^{\mathbf{B}_{LT}}V &= \mathbf{z}\dot{q}_{LN} + {}^{\mathbf{B}_{LS}}\mathbf{U}_{\mathbf{B}_{LT}}^T\, {}^{\mathbf{B}_{LS}}V && (10.42)\\
{}^{\mathbf{B}_{LS}}V &= \mathbf{z}\dot{q}_{LS} && (10.43)\\
\dot{q}_{LT} &= \dot{q}_{LN} + \dot{q}_{LS} && (10.44)
\end{aligned}
$$

with $\mathbf{z} = [0, 0, 1]^T$ and ${}^{\mathbf{B}_{LT}}\mathbf{U}_{\mathbf{B}_T} = {}^{\mathbf{B}_{LT}}\mathbf{U}_{\mathbf{T}}{}^{\mathbf{B}_T}\mathbf{U}_{\mathbf{T}}^{-1}$.

With a similar procedure leading to (10.24) by substituting frames $\{\mathbf{B}_{LT}\}$ and $\{\mathbf{B}_{LS}\}$ for frame $\{\mathbf{B}_T\}$, respectively, the dynamics of the two links representing the thigh and the shank are written as

$$
\mathbf{M}_{\mathbf{B}_{LT}} \frac{d}{dt}({}^{\mathbf{B}_{LT}}V) + \mathbf{C}_{\mathbf{B}_{LT}}({}^{\mathbf{B}_{LT}}\omega){}^{\mathbf{B}_{LT}}V + \mathbf{G}_{\mathbf{B}_{LT}} = {}^{\mathbf{B}_{LT}}F^* \qquad (10.45)
$$

$$
\mathbf{M}_{\mathbf{B}_{LS}} \frac{d}{dt}({}^{\mathbf{B}_{LS}}V) + \mathbf{C}_{\mathbf{B}_{LS}}({}^{\mathbf{B}_{LS}}\omega){}^{\mathbf{B}_{LS}}V + \mathbf{G}_{\mathbf{B}_{LS}} = {}^{\mathbf{B}_{LS}}F^*. \qquad (10.46)
$$

In (10.45), $\mathbf{M}_{\mathbf{B}_{LT}} \in \mathbb{R}^{3\times3}$, $\mathbf{C}_{\mathbf{B}_{LT}} \in \mathbb{R}^{3\times3}$, and $\mathbf{G}_{\mathbf{B}_{LT}} \in \mathbb{R}^3$ take the same forms as that in (10.25)-(10.27) with

$$m_{LT},\ d_{LT},\ I_{oLT},\ \dot{q}_{LT},\ \text{and}\ q_{LT}$$

being substituted for

$$m_T,\ d_T,\ I_{oT},\ \dot{q}_T,\ \text{and}\ q_T$$

respectively, where I_{oLT} denotes the moment of inertia of the left thigh around its center of mass. In (10.46), $\mathbf{M}_{\mathbf{B}_{LS}} \in \mathbb{R}^{3\times3}$, $\mathbf{C}_{\mathbf{B}_{LS}} \in \mathbb{R}^{3\times3}$, and $\mathbf{G}_{\mathbf{B}_{LS}} \in \mathbb{R}^3$ take the same forms as that in (10.25)–(10.27) with

$$m_{LS},\ d_{LS},\ I_{oLS},\ \dot{q}_{LS},\ \text{and}\ q_{LS}$$

being substituted for

$$m_T,\ d_T,\ I_{oT},\ \dot{q}_T,\ \text{and}\ q_T$$

respectively, where I_{oLS} denotes the moment of inertia of the left shank around its center of mass.

The two net force/moment vectors in (10.45) and (10.46) are governed by

$$
\begin{aligned}
{}^{\mathbf{B}_{LT}}F^* &= {}^{\mathbf{B}_{LT}}F - {}^{\mathbf{B}_{LT}}\mathbf{U}_{\mathbf{B}_T}{}^{\mathbf{B}_{T1}}F && (10.47)\\
{}^{\mathbf{B}_{LS}}F^* &= {}^{\mathbf{B}_{LS}}F - {}^{\mathbf{B}_{LS}}\mathbf{U}_{\mathbf{B}_{LT}}{}^{\mathbf{B}_{LT}}F && (10.48)
\end{aligned}
$$

where ${}^{\mathbf{B}_{T1}}F \in \mathbb{R}^3$ appears in (10.28).

In view of Assumption 9, the torques of the three joints associated with the left leg can be written as

$$
\begin{aligned}
\tau_{LT} &= \mathbf{z}^{T\,\mathbf{B}_{T1}}F && (10.49)\\
\tau_{LN} &= \mathbf{z}^{T\,\mathbf{B}_{LT}}F && (10.50)\\
\tau_{LS} &= \mathbf{z}^{T\,\mathbf{B}_{LS}}F = 0. && (10.51)
\end{aligned}
$$

Control Equations

According to (10.41)–(10.44), the required velocity vectors are designed as

$$\begin{aligned} {}^{\mathbf{B}_{T1}}V_r &= {}^{\mathbf{B}_T}V_r \\ &= \mathbf{z}(\dot{q}_{Tr} - \dot{q}_{LTr}) + {}^{\mathbf{B}_{LT}}\mathbf{U}_{\mathbf{B}_T}^T \, {}^{\mathbf{B}_{LT}}V_r \end{aligned} \tag{10.52}$$

$${}^{\mathbf{B}_{LT}}V_r = \mathbf{z}\dot{q}_{LNr} + {}^{\mathbf{B}_{LS}}\mathbf{U}_{\mathbf{B}_{LT}}^T \, {}^{\mathbf{B}_{LS}}V_r \tag{10.53}$$

$${}^{\mathbf{B}_{LS}}V_r = \mathbf{z}\dot{q}_{LSr} \tag{10.54}$$

$$\dot{q}_{LNr} = \dot{q}_{LTr} - \dot{q}_{LSr}. \tag{10.55}$$

Recall (10.29) to compute the required net force/moment vectors of the two links with frames $\{\mathbf{B}_{LT}\}$ and $\{\mathbf{B}_{LS}\}$ being substituted for frame $\{\mathbf{B}_T\}$, respectively. It yields

$${}^{\mathbf{B}_{LT}}F_r^* = \mathbf{Y}_{\mathbf{B}_{LT}}\hat{\boldsymbol{\theta}}_{\mathbf{B}_{LT}} + \mathbf{K}_{\mathbf{B}_{LT}}\left({}^{\mathbf{B}_{LT}}V_r - {}^{\mathbf{B}_{LT}}V\right) \tag{10.56}$$

$${}^{\mathbf{B}_{LS}}F_r^* = \mathbf{Y}_{\mathbf{B}_{LS}}\hat{\boldsymbol{\theta}}_{\mathbf{B}_{LS}} + \mathbf{K}_{\mathbf{B}_{LS}}\left({}^{\mathbf{B}_{LS}}V_r - {}^{\mathbf{B}_{LS}}V\right) \tag{10.57}$$

where $\hat{\boldsymbol{\theta}}_{\mathbf{B}_{LT}} \in \mathbb{R}^3$ and $\hat{\boldsymbol{\theta}}_{\mathbf{B}_{LS}} \in \mathbb{R}^3$ denote the estimates of $\boldsymbol{\theta}_{\mathbf{B}_{LT}} \in \mathbb{R}^3$ and $\boldsymbol{\theta}_{\mathbf{B}_{LS}} \in \mathbb{R}^3$, respectively, and $\mathbf{K}_{\mathbf{B}_{LT}}$ and $\mathbf{K}_{\mathbf{B}_{LS}}$ are two symmetric positive-definite matrices. In (10.56), $\boldsymbol{\theta}_{\mathbf{B}_{LT}} \in \mathbb{R}^3$ and $\mathbf{Y}_{\mathbf{B}_{LT}} \in \mathbb{R}^{3\times3}$ take the same forms as that in (10.30) and (10.31) with

$$m_{LT},\ d_{LT},\ I_{oLT},\ \dot{q}_{LT},\ q_{LT},\ \tfrac{d}{dt}({}^{\mathbf{B}_{LT}}V_r),\ \text{and}\ {}^{\mathbf{B}_{LT}}V_r$$

being substituted for

$$m_T,\ d_T,\ I_{oT},\ \dot{q}_T,\ q_T,\ \tfrac{d}{dt}({}^{\mathbf{B}_T}V_r),\ \text{and}\ {}^{\mathbf{B}_T}V_r$$

respectively. In (10.57), $\boldsymbol{\theta}_{\mathbf{B}_{LS}} \in \mathbb{R}^3$ and $\mathbf{Y}_{\mathbf{B}_{LS}} \in \mathbb{R}^{3\times3}$ take the same forms as that in (10.30) and (10.31) with

$$m_{LS},\ d_{LS},\ I_{oLS},\ \dot{q}_{LS},\ q_{LS},\ \tfrac{d}{dt}({}^{\mathbf{B}_{LS}}V_r),\ \text{and}\ {}^{\mathbf{B}_{LS}}V_r$$

being substituted for

$$m_T,\ d_T,\ I_{oT},\ \dot{q}_T,\ q_T,\ \tfrac{d}{dt}({}^{\mathbf{B}_T}V_r),\ \text{and}\ {}^{\mathbf{B}_T}V_r$$

respectively.

The parameter estimates need to be updated. Define

$$\mathbf{s}_{\mathbf{B}_{LT}} = \mathbf{Y}_{\mathbf{B}_{LT}}^T\left({}^{\mathbf{B}_{LT}}V_r - {}^{\mathbf{B}_{LT}}V\right) \tag{10.58}$$

$$\mathbf{s}_{\mathbf{B}_{LS}} = \mathbf{Y}_{\mathbf{B}_{LS}}^T\left({}^{\mathbf{B}_{LS}}V_r - {}^{\mathbf{B}_{LS}}V\right). \tag{10.59}$$

The $\mathcal{P}$ function defined by (2.79) is used to update each of the parameter estimates of the left thigh and shank as

$$\hat{\theta}_{\mathbf{B}_{LT}\gamma} = \mathcal{P}\left(s_{\mathbf{B}_{LT}\gamma}, \rho_{\mathbf{B}_{LT}\gamma}, \underline{\theta}_{\mathbf{B}_{LT}\gamma}, \overline{\theta}_{\mathbf{B}_{LT}\gamma}, t\right) \tag{10.60}$$

$$\hat{\theta}_{\mathbf{B}_{LS}\gamma} = \mathcal{P}\left(s_{\mathbf{B}_{LS}\gamma}, \rho_{\mathbf{B}_{LS}\gamma}, \underline{\theta}_{\mathbf{B}_{LS}\gamma}, \overline{\theta}_{\mathbf{B}_{LS}\gamma}, t\right) \tag{10.61}$$

for all $\gamma \in \{1, 3\}$, where

$\theta_{\mathbf{B}_{LT}\gamma}$	γth element of $\boldsymbol{\theta}_{\mathbf{B}_{LT}}$.
$\theta_{\mathbf{B}_{LS}\gamma}$	γth element of $\boldsymbol{\theta}_{\mathbf{B}_{LS}}$.
$\hat{\theta}_{\mathbf{B}_{LT}\gamma}$	Estimate of $\theta_{\mathbf{B}_{LT}\gamma}$.
$\hat{\theta}_{\mathbf{B}_{LS}\gamma}$	Estimate of $\theta_{\mathbf{B}_{LS}\gamma}$.
$s_{\mathbf{B}_{LT}\gamma}$	γth element of $\mathbf{s}_{\mathbf{B}_{LT}}$.
$s_{\mathbf{B}_{LS}\gamma}$	γth element of $\mathbf{s}_{\mathbf{B}_{LS}}$.
$\rho_{\mathbf{B}_{LT}\gamma}$	Parameter update gain.
$\rho_{\mathbf{B}_{LS}\gamma}$	Parameter update gain.
$\underline{\theta}_{\mathbf{B}_{LT}\gamma}$	Lower bound of $\theta_{\mathbf{B}_{LT}\gamma}$.
$\overline{\theta}_{\mathbf{B}_{LT}\gamma}$	Upper bound of $\theta_{\mathbf{B}_{LT}\gamma}$.
$\underline{\theta}_{\mathbf{B}_{LS}\gamma}$	Lower bound of $\theta_{\mathbf{B}_{LS}\gamma}$.
$\overline{\theta}_{\mathbf{B}_{LS}\gamma}$	Upper bound of $\theta_{\mathbf{B}_{LS}\gamma}$.

At the end, the required force/moment vectors of the left leg are computed as

$$^{\mathbf{B}_{LT}}F_r = {}^{\mathbf{B}_{LT}}F_r^* + {}^{\mathbf{B}_{LT}}\mathbf{U}_{\mathbf{B}_T}\, {}^{\mathbf{B}_{T1}}F_r \tag{10.62}$$
$$^{\mathbf{B}_{LS}}F_r = {}^{\mathbf{B}_{LS}}F_r^* + {}^{\mathbf{B}_{LS}}\mathbf{U}_{\mathbf{B}_{LT}}\, {}^{\mathbf{B}_{LT}}F_r \tag{10.63}$$

with $^{\mathbf{B}_{T1}}F_r \in \mathbb{R}^3$ being obtained by (10.36), and the joint control equations are

$$\tau_{LT} = \mathbf{z}^{T\,\mathbf{B}_{T1}}F_r + k_{LT}\left[(\dot{q}_{Tr} - \dot{q}_{LTr}) - (\dot{q}_T - \dot{q}_{LT})\right] \tag{10.64}$$
$$\tau_{LN} = \mathbf{z}^{T\,\mathbf{B}_{LT}}F_r + k_{LN}\left(\dot{q}_{LNr} - \dot{q}_{LN}\right) \tag{10.65}$$
$$\begin{aligned}\tau_{LS} &= \mathbf{z}^{T\,\mathbf{B}_{LS}}F_r + k_{LS}\left(\dot{q}_{LSr} - \dot{q}_{LS}\right)\\ &= 0\end{aligned} \tag{10.66}$$

where $k_{LT} > 0$, $k_{LN} > 0$, and $k_{LS} > 0$ are three velocity feedback gains.

Virtual Stability

It will be shown that the left leg combined with its respective control equations qualifies to be *virtually stable* in the sense of Definition 2.17.

Theorem 10.2. *The left leg described by (10.41)–(10.51), combined with its control equations (10.52)–(10.57) and (10.62)–(10.66) and with the parameter adaptation (10.58)–(10.61), is virtually stable with its affiliated vectors* $^{\mathbf{B}_{LT}}V_r - {}^{\mathbf{B}_{LT}}V$ *and* $^{\mathbf{B}_{LS}}V_r - {}^{\mathbf{B}_{LS}}V$ *being virtual functions in both* L_2 *and* L_∞*, in the sense of Definition 2.17.*

Proof: Let the non-negative accompanying function assigned to the left leg be

$$\nu_L = \nu_{\mathbf{B}_{LT}} + \nu_{\mathbf{B}_{LS}} \tag{10.67}$$

with

$$\nu_{\mathbf{B}_{LT}} = \frac{1}{2}\left({}^{\mathbf{B}_{LT}}V_r - {}^{\mathbf{B}_{LT}}V\right)^T \mathbf{M}_{\mathbf{B}_{LT}} \left({}^{\mathbf{B}_{LT}}V_r - {}^{\mathbf{B}_{LT}}V\right) + \frac{1}{2}\sum_{\gamma=1}^{3}\left(\theta_{\mathbf{B}_{LT\gamma}} - \hat{\theta}_{\mathbf{B}_{LT\gamma}}\right)^2 / \rho_{\mathbf{B}_{LT\gamma}} \tag{10.68}$$

$$\nu_{\mathbf{B}_{LS}} = \frac{1}{2}\left({}^{\mathbf{B}_{LS}}V_r - {}^{\mathbf{B}_{LS}}V\right)^T \mathbf{M}_{\mathbf{B}_{LS}} \left({}^{\mathbf{B}_{LS}}V_r - {}^{\mathbf{B}_{LS}}V\right) + \frac{1}{2}\sum_{\gamma=1}^{3}\left(\theta_{\mathbf{B}_{LS\gamma}} - \hat{\theta}_{\mathbf{B}_{LS\gamma}}\right)^2 / \rho_{\mathbf{B}_{LS\gamma}}. \tag{10.69}$$

With frames $\{\mathbf{B}_{LT}\}$ and $\{\mathbf{B}_{LS}\}$ being substituted for frame $\{\mathbf{B}_T\}$, it follows from (10.45), (10.46), (10.56)–(10.61), and Lemma 10.1 that

$$\dot{\nu}_{\mathbf{B}_{LT}} \leqslant -\left({}^{\mathbf{B}_{LT}}V_r - {}^{\mathbf{B}_{LT}}V\right)^T \mathbf{K}_{\mathbf{B}_{LT}} \left({}^{\mathbf{B}_{LT}}V_r - {}^{\mathbf{B}_{LT}}V\right) + \left({}^{\mathbf{B}_{LT}}V_r - {}^{\mathbf{B}_{LT}}V\right)^T \left({}^{\mathbf{B}_{LT}}F_r^* - {}^{\mathbf{B}_{LT}}F^*\right) \tag{10.70}$$

$$\dot{\nu}_{\mathbf{B}_{LS}} \leqslant -\left({}^{\mathbf{B}_{LS}}V_r - {}^{\mathbf{B}_{LS}}V\right)^T \mathbf{K}_{\mathbf{B}_{LS}} \left({}^{\mathbf{B}_{LS}}V_r - {}^{\mathbf{B}_{LS}}V\right) + \left({}^{\mathbf{B}_{LS}}V_r - {}^{\mathbf{B}_{LS}}V\right)^T \left({}^{\mathbf{B}_{LS}}F_r^* - {}^{\mathbf{B}_{LS}}F^*\right) \tag{10.71}$$

hold.

In view of (2.85), (10.41)–(10.44), (10.47)–(10.55), and (10.62)–(10.66), it yields

$$\left({}^{\mathbf{B}_{LT}}V_r - {}^{\mathbf{B}_{LT}}V\right)^T \left({}^{\mathbf{B}_{LT}}F_r^* - {}^{\mathbf{B}_{LT}}F^*\right) = p_{\mathbf{B}_{LT}} - p_{\mathbf{B}_{T1}} - k_{LT}\left((\dot{q}_{Tr} - \dot{q}_T) - (\dot{q}_{LTr} - \dot{q}_{LT})\right)^2 \tag{10.72}$$

$$\left({}^{\mathbf{B}_{LS}}V_r - {}^{\mathbf{B}_{LS}}V\right)^T \left({}^{\mathbf{B}_{LS}}F_r^* - {}^{\mathbf{B}_{LS}}F^*\right) = -p_{\mathbf{B}_{LT}} - k_{LS}\left(\dot{q}_{LSr} - \dot{q}_{LS}\right)^2 - k_{LN}\left(\dot{q}_{LNr} - \dot{q}_{LN}\right)^2. \tag{10.73}$$

The derivation of (10.72) and (10.73) is given in Appendix B.6.

Substituting (10.72) and (10.73) into (10.70) and (10.71) yields

$$\begin{aligned}\dot{\nu}_L &= \dot{\nu}_{\mathbf{B}_{LT}} + \dot{\nu}_{\mathbf{B}_{LS}} \\ &\leqslant -\left({}^{\mathbf{B}_{LT}}V_r - {}^{\mathbf{B}_{LT}}V\right)^T \mathbf{K}_{\mathbf{B}_{LT}} \left({}^{\mathbf{B}_{LT}}V_r - {}^{\mathbf{B}_{LT}}V\right) \\ &\quad -\left({}^{\mathbf{B}_{LS}}V_r - {}^{\mathbf{B}_{LS}}V\right)^T \mathbf{K}_{\mathbf{B}_{LS}} \left({}^{\mathbf{B}_{LS}}V_r - {}^{\mathbf{B}_{LS}}V\right) \\ &\quad -k_{LT}\left((\dot{q}_{Tr} - \dot{q}_T) - (\dot{q}_{LTr} - \dot{q}_{LT})\right)^2 \\ &\quad -k_{LS}\left(\dot{q}_{LSr} - \dot{q}_{LS}\right)^2 - k_{LN}\left(\dot{q}_{LNr} - \dot{q}_{LN}\right)^2 \\ &\quad -p_{\mathbf{B}_{T1}}.\end{aligned} \tag{10.74}$$

Consider the fact that the left leg has one *driving cutting point* associated with frame $\{\mathbf{B}_{T1}\}$. It completes the proof, in view of (10.67)–(10.69), (10.74), and Definition 2.17. ■

10.6.5 Dynamics and Control of the Right Leg

The dynamics and control problem of the right leg follows the same procedure as that of the left leg, since both legs touch the ground in a double-support phase.

The right leg consists of two rigid links, namely the thigh and the shank, and three one-dimensional joints located between the trunk and the thigh, between the thigh and the shank, and between the shank and the ground, respectively.

Kinematics and Dynamics

When the right leg touches the ground, the velocity relationships can be written as

$$\begin{aligned} {}^{\mathbf{B}_{T2}}V &= {}^{\mathbf{B}_T}V \\ &= \mathbf{z}(\dot{q}_T - \dot{q}_{RT}) + {}^{\mathbf{B}_{RT}}\mathbf{U}_{\mathbf{B}_T}^T \, {}^{\mathbf{B}_{RT}}V \qquad (10.75) \\ {}^{\mathbf{B}_{RT}}V &= \mathbf{z}\dot{q}_{RN} + {}^{\mathbf{B}_{RS}}\mathbf{U}_{\mathbf{B}_{RT}}^T \, {}^{\mathbf{B}_{RS}}V \qquad (10.76) \\ {}^{\mathbf{B}_{RS}}V &= \mathbf{z}\dot{q}_{RS} \qquad (10.77) \\ \dot{q}_{RT} &= \dot{q}_{RN} + \dot{q}_{RS} \qquad (10.78) \end{aligned}$$

with $\mathbf{z} = [0, 0, 1]^T$ and ${}^{\mathbf{B}_{RT}}\mathbf{U}_{\mathbf{B}_T} = {}^{\mathbf{B}_{RT}}\mathbf{U}_{\mathbf{T}}{}^{\mathbf{B}_T}\mathbf{U}_{\mathbf{T}}^{-1}$.

With a similar procedure leading to (10.24) by substituting frames $\{\mathbf{B}_{RT}\}$ and $\{\mathbf{B}_{RS}\}$ for frame $\{\mathbf{B}_T\}$, respectively, the dynamics of the two links representing the thigh and the shank are written as

$$\mathbf{M}_{\mathbf{B}_{RT}} \frac{d}{dt}({}^{\mathbf{B}_{RT}}V) + \mathbf{C}_{\mathbf{B}_{RT}}({}^{\mathbf{B}_{RT}}\omega){}^{\mathbf{B}_{RT}}V + \mathbf{G}_{\mathbf{B}_{RT}} = {}^{\mathbf{B}_{RT}}F^* \qquad (10.79)$$

$$\mathbf{M}_{\mathbf{B}_{RS}} \frac{d}{dt}({}^{\mathbf{B}_{RS}}V) + \mathbf{C}_{\mathbf{B}_{RS}}({}^{\mathbf{B}_{RS}}\omega){}^{\mathbf{B}_{RS}}V + \mathbf{G}_{\mathbf{B}_{RS}} = {}^{\mathbf{B}_{RS}}F^*. \qquad (10.80)$$

In (10.79), $\mathbf{M}_{\mathbf{B}_{RT}} \in \mathbb{R}^{3\times 3}$, $\mathbf{C}_{\mathbf{B}_{RT}} \in \mathbb{R}^{3\times 3}$, and $\mathbf{G}_{\mathbf{B}_{RT}} \in \mathbb{R}^3$ take the same forms as that in (10.25)–(10.27) with

$$m_{RT},\ d_{RT},\ I_{oRT},\ \dot{q}_{RT},\ \text{and } q_{RT}$$

being substituted for

$$m_T,\ d_T,\ I_{oT},\ \dot{q}_T,\ \text{and } q_T$$

respectively, where I_{oRT} denotes the moment of inertia of the right thigh around its center of mass. In (10.80), $\mathbf{M}_{\mathbf{B}_{RS}} \in \mathbb{R}^{3\times 3}$, $\mathbf{C}_{\mathbf{B}_{RS}} \in \mathbb{R}^{3\times 3}$, and $\mathbf{G}_{\mathbf{B}_{RS}} \in \mathbb{R}^3$ take the same forms as that in (10.25)–(10.27) with

$$m_{RS},\ d_{RS},\ I_{oRS},\ \dot{q}_{RS},\ \text{and } q_{RS}$$

being substituted for

$$m_T,\ d_T,\ I_{oT},\ \dot{q}_T,\ \text{and } q_T$$

respectively, where I_{oRS} denotes the moment of inertia of the right shank around its center of mass.

The two net force/moment vectors are governed by the following equations

$$ {}^{\mathbf{B}_{RT}}F^* = {}^{\mathbf{B}_{RT}}F - {}^{\mathbf{B}_{RT}}\mathbf{U}_{\mathbf{B}_T} {}^{\mathbf{B}_{T2}}F \tag{10.81} $$

$$ {}^{\mathbf{B}_{RS}}F^* = {}^{\mathbf{B}_{RS}}F - {}^{\mathbf{B}_{RS}}\mathbf{U}_{\mathbf{B}_{RT}} {}^{\mathbf{B}_{RT}}F \tag{10.82} $$

where ${}^{\mathbf{B}_{T2}}F \in \mathbb{R}^3$ appears in (10.28).

In view of Assumption 9, the torques of the three joints of the right leg can be written as

$$ \tau_{RT} = \mathbf{z}^T {}^{\mathbf{B}_{T2}}F \tag{10.83} $$

$$ \tau_{RN} = \mathbf{z}^T {}^{\mathbf{B}_{RT}}F \tag{10.84} $$

$$ \tau_{RS} = \mathbf{z}^T {}^{\mathbf{B}_{RS}}F = 0. \tag{10.85} $$

Control Equations

According to (10.75)–(10.78), the required velocity vectors are designed as

$$ \begin{aligned} {}^{\mathbf{B}_{T2}}V_r &= {}^{\mathbf{B}_T}V_r \\ &= \mathbf{z}(\dot{q}_{Tr} - \dot{q}_{RTr}) + {}^{\mathbf{B}_{RT}}\mathbf{U}_{\mathbf{B}_T}^T {}^{\mathbf{B}_{RT}}V_r \end{aligned} \tag{10.86} $$

$$ {}^{\mathbf{B}_{RT}}V_r = \mathbf{z}\dot{q}_{RNr} + {}^{\mathbf{B}_{RS}}\mathbf{U}_{\mathbf{B}_{RT}}^T {}^{\mathbf{B}_{RS}}V_r \tag{10.87} $$

$$ {}^{\mathbf{B}_{RS}}V_r = \mathbf{z}\dot{q}_{RSr} \tag{10.88} $$

$$ \dot{q}_{RNr} = \dot{q}_{RTr} - \dot{q}_{RSr}. \tag{10.89} $$

Recall (10.29) to compute the required net force/moment vectors of the two links with frames $\{\mathbf{B}_{RT}\}$ and $\{\mathbf{B}_{RS}\}$ being substituted for frame $\{\mathbf{B}_T\}$, respectively. It yields

$$ {}^{\mathbf{B}_{RT}}F_r^* = \mathbf{Y}_{\mathbf{B}_{RT}}\hat{\boldsymbol{\theta}}_{\mathbf{B}_{RT}} + \mathbf{K}_{\mathbf{B}_{RT}}\left({}^{\mathbf{B}_{RT}}V_r - {}^{\mathbf{B}_{RT}}V\right) \tag{10.90} $$

$$ {}^{\mathbf{B}_{RS}}F_r^* = \mathbf{Y}_{\mathbf{B}_{RS}}\hat{\boldsymbol{\theta}}_{\mathbf{B}_{RS}} + \mathbf{K}_{\mathbf{B}_{RS}}\left({}^{\mathbf{B}_{RS}}V_r - {}^{\mathbf{B}_{RS}}V\right) \tag{10.91} $$

where $\hat{\boldsymbol{\theta}}_{\mathbf{B}_{RT}} \in \mathbb{R}^3$ and $\hat{\boldsymbol{\theta}}_{\mathbf{B}_{RS}} \in \mathbb{R}^3$ denote the estimates of $\boldsymbol{\theta}_{\mathbf{B}_{RT}} \in \mathbb{R}^3$ and $\boldsymbol{\theta}_{\mathbf{B}_{RS}} \in \mathbb{R}^3$, respectively, and $\mathbf{K}_{\mathbf{B}_{RT}}$ and $\mathbf{K}_{\mathbf{B}_{RS}}$ are two symmetric positive-definite matrices. In (10.90), $\boldsymbol{\theta}_{\mathbf{B}_{RT}} \in \mathbb{R}^3$ and $\mathbf{Y}_{\mathbf{B}_{RT}} \in \mathbb{R}^{3\times 3}$ take the same forms as that in (10.30) and (10.31) with

$$ m_{RT},\ d_{RT},\ I_{oRT},\ \dot{q}_{RT},\ q_{RT},\ \tfrac{d}{dt}({}^{\mathbf{B}_{RT}}V_r),\ \text{and}\ {}^{\mathbf{B}_{RT}}V_r $$

being substituted for

$$ m_T,\ d_T,\ I_{oT},\ \dot{q}_T,\ q_T,\ \tfrac{d}{dt}({}^{\mathbf{B}_T}V_r),\ \text{and}\ {}^{\mathbf{B}_T}V_r $$

respectively. In (10.91), $\boldsymbol{\theta}_{\mathbf{B}_{RS}} \in \mathbb{R}^3$ and $\mathbf{Y}_{\mathbf{B}_{RS}} \in \mathbb{R}^{3\times 3}$ take the same forms as that in (10.30) and (10.31) with

$$ m_{RS},\ d_{RS},\ I_{oRS},\ \dot{q}_{RS},\ q_{RS},\ \tfrac{d}{dt}({}^{\mathbf{B}_{RS}}V_r),\ \text{and}\ {}^{\mathbf{B}_{RS}}V_r $$

being substituted for

$$m_T,\ d_T,\ I_{oT},\ \dot{q}_T,\ q_T,\ \tfrac{d}{dt}({}^{\mathbf{B}_T}V_r),\ \text{and}\ {}^{\mathbf{B}_T}V_r$$

respectively.

The parameter estimates need to be updated. Define

$$\mathbf{s}_{\mathbf{B}_{RT}} = \mathbf{Y}^T_{\mathbf{B}_{RT}} \left({}^{\mathbf{B}_{RT}}V_r - {}^{\mathbf{B}_{RT}}V\right) \tag{10.92}$$

$$\mathbf{s}_{\mathbf{B}_{RS}} = \mathbf{Y}^T_{\mathbf{B}_{RS}} \left({}^{\mathbf{B}_{RS}}V_r - {}^{\mathbf{B}_{RS}}V\right). \tag{10.93}$$

The $\mathcal{P}$ function defined by (2.79) is used to update each of the parameter estimates of the thigh and shank as

$$\hat{\theta}_{\mathbf{B}_{RT}\gamma} = \mathcal{P}\left(s_{\mathbf{B}_{RT}\gamma}, \rho_{\mathbf{B}_{RT}\gamma}, \underline{\theta}_{\mathbf{B}_{RT}\gamma}, \overline{\theta}_{\mathbf{B}_{RT}\gamma}, t\right) \tag{10.94}$$

$$\hat{\theta}_{\mathbf{B}_{RS}\gamma} = \mathcal{P}\left(s_{\mathbf{B}_{RS}\gamma}, \rho_{\mathbf{B}_{RS}\gamma}, \underline{\theta}_{\mathbf{B}_{RS}\gamma}, \overline{\theta}_{\mathbf{B}_{RS}\gamma}, t\right) \tag{10.95}$$

for all $\gamma \in \{1, 3\}$, where

$\theta_{\mathbf{B}_{RT}\gamma}$	γth element of $\boldsymbol{\theta}_{\mathbf{B}_{RT}}$.
$\theta_{\mathbf{B}_{RS}\gamma}$	γth element of $\boldsymbol{\theta}_{\mathbf{B}_{RS}}$.
$\hat{\theta}_{\mathbf{B}_{RT}\gamma}$	Estimate of $\theta_{\mathbf{B}_{RT}\gamma}$.
$\hat{\theta}_{\mathbf{B}_{RS}\gamma}$	Estimate of $\theta_{\mathbf{B}_{RS}\gamma}$.
$s_{\mathbf{B}_{RT}\gamma}$	γth element of $\mathbf{s}_{\mathbf{B}_{RT}}$.
$s_{\mathbf{B}_{RS}\gamma}$	γth element of $\mathbf{s}_{\mathbf{B}_{RS}}$.
$\rho_{\mathbf{B}_{RT}\gamma}$	Parameter update gain.
$\rho_{\mathbf{B}_{RS}\gamma}$	Parameter update gain.
$\underline{\theta}_{\mathbf{B}_{RT}\gamma}$	Lower bound of $\theta_{\mathbf{B}_{RT}\gamma}$.
$\overline{\theta}_{\mathbf{B}_{RT}\gamma}$	Upper bound of $\theta_{\mathbf{B}_{RT}\gamma}$.
$\underline{\theta}_{\mathbf{B}_{RS}\gamma}$	Lower bound of $\theta_{\mathbf{B}_{RS}\gamma}$.
$\overline{\theta}_{\mathbf{B}_{RS}\gamma}$	Upper bound of $\theta_{\mathbf{B}_{RS}\gamma}$.

Finally, the required force/moment vectors of the right leg are computed as

$${}^{\mathbf{B}_{RT}}F_r = {}^{\mathbf{B}_{RT}}F_r^* + {}^{\mathbf{B}_{RT}}\mathbf{U}_{\mathbf{B}_T}\, {}^{\mathbf{B}_{T2}}F_r \tag{10.96}$$

$${}^{\mathbf{B}_{RS}}F_r = {}^{\mathbf{B}_{RS}}F_r^* + {}^{\mathbf{B}_{RS}}\mathbf{U}_{\mathbf{B}_{RT}}\, {}^{\mathbf{B}_{RT}}F_r \tag{10.97}$$

with ${}^{\mathbf{B}_{T2}}F_r \in \mathbb{R}^3$ being obtained by (10.37), and the joint control equations are

$$\tau_{RT} = \mathbf{z}^T\, {}^{\mathbf{B}_{T2}}F_r + k_{RT}\left[(\dot{q}_{Tr} - \dot{q}_{RTr}) - (\dot{q}_T - \dot{q}_{RT})\right] \tag{10.98}$$

$$\tau_{RN} = \mathbf{z}^T\, {}^{\mathbf{B}_{RT}}F_r + k_{RN}\left(\dot{q}_{RNr} - \dot{q}_{RN}\right) \tag{10.99}$$

$$\begin{aligned}\tau_{RS} &= \mathbf{z}^T\, {}^{\mathbf{B}_{RS}}F_r + k_{RS}\left(\dot{q}_{RSr} - \dot{q}_{RS}\right)\\ &= 0\end{aligned} \tag{10.100}$$

where $k_{RT} > 0$, $k_{RN} > 0$, and $k_{RS} > 0$ are three velocity feedback gains.

Virtual Stability

The *virtual stability*, in the sense of Definition 2.17, of the right leg combined with its respective control equations is presented in the following theorem.

Theorem 10.3. *The right leg described by (10.75)–(10.85), combined with its control equations (10.86)–(10.91) and (10.96)–(10.100) and with the parameter adaptation (10.92)–(10.95), is virtually stable with its affiliated vectors* ${}^{\mathbf{B}_{RT}}V_r - {}^{\mathbf{B}_{RT}}V$ *and* ${}^{\mathbf{B}_{RS}}V_r - {}^{\mathbf{B}_{RS}}V$ *being virtual functions in both* L_2 *and* L_∞*, in the sense of Definition 2.17.*

Proof: Let the non-negative accompanying function assigned to the right leg be

$$\nu_R = \nu_{\mathbf{B}_{RT}} + \nu_{\mathbf{B}_{RS}} \tag{10.101}$$

with

$$\begin{aligned} \nu_{\mathbf{B}_{RT}} &= \frac{1}{2}\left({}^{\mathbf{B}_{RT}}V_r - {}^{\mathbf{B}_{RT}}V\right)^T \mathbf{M}_{\mathbf{B}_{RT}} \left({}^{\mathbf{B}_{RT}}V_r - {}^{\mathbf{B}_{RT}}V\right) \\ &\quad + \frac{1}{2}\sum_{\gamma=1}^{3}\left(\theta_{\mathbf{B}_{RT\gamma}} - \hat{\theta}_{\mathbf{B}_{RT\gamma}}\right)^2 / \rho_{\mathbf{B}_{RT\gamma}} \end{aligned} \tag{10.102}$$

$$\begin{aligned} \nu_{\mathbf{B}_{RS}} &= \frac{1}{2}\left({}^{\mathbf{B}_{RS}}V_r - {}^{\mathbf{B}_{RS}}V\right)^T \mathbf{M}_{\mathbf{B}_{RS}} \left({}^{\mathbf{B}_{RS}}V_r - {}^{\mathbf{B}_{RS}}V\right) \\ &\quad + \frac{1}{2}\sum_{\gamma=1}^{3}\left(\theta_{\mathbf{B}_{RS\gamma}} - \hat{\theta}_{\mathbf{B}_{RS\gamma}}\right)^2 / \rho_{\mathbf{B}_{RS\gamma}}. \end{aligned} \tag{10.103}$$

With frames $\{\mathbf{B}_{RT}\}$ and $\{\mathbf{B}_{RS}\}$ being substituted for frame $\{\mathbf{B}_T\}$, it follows from (10.79), (10.80), (10.90)–(10.95), and Lemma 10.1 that

$$\begin{aligned} \dot{\nu}_{\mathbf{B}_{RT}} &\leqslant -\left({}^{\mathbf{B}_{RT}}V_r - {}^{\mathbf{B}_{RT}}V\right)^T \mathbf{K}_{\mathbf{B}_{RT}} \left({}^{\mathbf{B}_{RT}}V_r - {}^{\mathbf{B}_{RT}}V\right) \\ &\quad + \left({}^{\mathbf{B}_{RT}}V_r - {}^{\mathbf{B}_{RT}}V\right)^T \left({}^{\mathbf{B}_{RT}}F_r^* - {}^{\mathbf{B}_{RT}}F^*\right) \end{aligned} \tag{10.104}$$

$$\begin{aligned} \dot{\nu}_{\mathbf{B}_{RS}} &\leqslant -\left({}^{\mathbf{B}_{RS}}V_r - {}^{\mathbf{B}_{RS}}V\right)^T \mathbf{K}_{\mathbf{B}_{RS}} \left({}^{\mathbf{B}_{RS}}V_r - {}^{\mathbf{B}_{RS}}V\right) \\ &\quad + \left({}^{\mathbf{B}_{RS}}V_r - {}^{\mathbf{B}_{RS}}V\right)^T \left({}^{\mathbf{B}_{RS}}F_r^* - {}^{\mathbf{B}_{RS}}F^*\right) \end{aligned} \tag{10.105}$$

hold.

In view of (2.85), (10.75)–(10.78), (10.81)–(10.89), and (10.96)–(10.100), it yields

$$\begin{aligned} &\left({}^{\mathbf{B}_{RT}}V_r - {}^{\mathbf{B}_{RT}}V\right)^T \left({}^{\mathbf{B}_{RT}}F_r^* - {}^{\mathbf{B}_{RT}}F^*\right) \\ &\quad = p_{\mathbf{B}_{RT}} - p_{\mathbf{B}_{T2}} - k_{RT}\left((\dot{q}_{Tr} - \dot{q}_T) - (\dot{q}_{RTr} - \dot{q}_{RT})\right)^2 \end{aligned} \tag{10.106}$$

$$\begin{aligned} &\left({}^{\mathbf{B}_{RS}}V_r - {}^{\mathbf{B}_{RS}}V\right)^T \left({}^{\mathbf{B}_{RS}}F_r^* - {}^{\mathbf{B}_{RS}}F^*\right) \\ &\quad = -p_{\mathbf{B}_{RT}} - k_{RS}\left(\dot{q}_{RSr} - \dot{q}_{RS}\right)^2 - k_{RN}\left(\dot{q}_{RNr} - \dot{q}_{RN}\right)^2. \end{aligned} \tag{10.107}$$

The derivation of (10.106) and (10.107) is given in Appendix B.7.

Substituting (10.106) and (10.107) into (10.104) and (10.105) yields

$$\begin{aligned}\dot{\nu}_R &= \dot{\nu}_{\mathbf{B}_{RT}} + \dot{\nu}_{\mathbf{B}_{RS}} \\ &\leqslant -\left({}^{\mathbf{B}_{RT}}V_r - {}^{\mathbf{B}_{RT}}V\right)^T \mathbf{K}_{\mathbf{B}_{RT}} \left({}^{\mathbf{B}_{RT}}V_r - {}^{\mathbf{B}_{RT}}V\right) \\ &\quad -\left({}^{\mathbf{B}_{RS}}V_r - {}^{\mathbf{B}_{RS}}V\right)^T \mathbf{K}_{\mathbf{B}_{RS}} \left({}^{\mathbf{B}_{RS}}V_r - {}^{\mathbf{B}_{RS}}V\right) \\ &\quad -k_{RT}\left((\dot{q}_{Tr} - \dot{q}_T) - (\dot{q}_{RTr} - \dot{q}_{RT})\right)^2 \\ &\quad -k_{RS}\left(\dot{q}_{RSr} - \dot{q}_{RS}\right)^2 - k_{RN}\left(\dot{q}_{RNr} - \dot{q}_{RN}\right)^2 \\ &\quad -p_{\mathbf{B}_{T2}}. \end{aligned} \tag{10.108}$$

Consider the fact that the right leg has one *driving cutting point* associated with frame $\{\mathbf{B}_{T2}\}$. It completes the proof, in view of (10.101)–(10.103), (10.108), and Definition 2.17. ■

10.6.6 L_2 and L_∞ Stability

It follows from Theorems 10.1–10.3 that all three subsystems of the five-bar biped in the sagittal plane are *virtually stable* in the sense of Definition 2.17. Therefore, it follows from Theorem 2.1 that

$${}^{\mathbf{B}_T}V_r - {}^{\mathbf{B}_T}V \in L_2 \bigcap L_\infty \tag{10.109}$$

$${}^{\mathbf{B}_{LT}}V_r - {}^{\mathbf{B}_{LT}}V \in L_2 \bigcap L_\infty \tag{10.110}$$

$${}^{\mathbf{B}_{LS}}V_r - {}^{\mathbf{B}_{LS}}V \in L_2 \bigcap L_\infty \tag{10.111}$$

$${}^{\mathbf{B}_{RT}}V_r - {}^{\mathbf{B}_{RT}}V \in L_2 \bigcap L_\infty \tag{10.112}$$

$${}^{\mathbf{B}_{RS}}V_r - {}^{\mathbf{B}_{RS}}V \in L_2 \bigcap L_\infty \tag{10.113}$$

hold.

Expressions (10.111) and (10.113) imply

$$\dot{q}_{LSr} - \dot{q}_{LS} \in L_2 \bigcap L_\infty \tag{10.114}$$

$$\dot{q}_{RSr} - \dot{q}_{RS} \in L_2 \bigcap L_\infty \tag{10.115}$$

from (10.43), (10.54), (10.77), and (10.88). Meanwhile, (10.110)–(10.113) imply

$$\dot{q}_{LNr} - \dot{q}_{LN} \in L_2 \bigcap L_\infty \tag{10.116}$$

$$\dot{q}_{RNr} - \dot{q}_{RN} \in L_2 \bigcap L_\infty \tag{10.117}$$

from (10.42), (10.53), (10.76), and (10.87). Then, it follows from (10.44), (10.55), (10.78), (10.89), and (10.114)–(10.117) that

$$\dot{q}_{LTr} - \dot{q}_{LT} \in L_2 \bigcap L_\infty \tag{10.118}$$

$$\dot{q}_{RTr} - \dot{q}_{RT} \in L_2 \bigcap L_\infty \tag{10.119}$$

hold. Furthermore, (10.109) and (10.110) imply

$$(\dot{q}_{Tr} - \dot{q}_T) - (\dot{q}_{LTr} - \dot{q}_{LT}) \in L_2 \bigcap L_\infty \tag{10.120}$$

from (10.41) and (10.52). Adding (10.118) to (10.120) yields

$$\dot{q}_{Tr} - \dot{q}_T \in L_2 \bigcap L_\infty. \tag{10.121}$$

Finally, it follows from (10.21), (10.114), (10.118), (10.121), and Lemma 2.4 that

$$\dot{q}_{Td} - \dot{q}_T \in L_2 \bigcap L_\infty \tag{10.122}$$

$$\dot{q}_{LTd} - \dot{q}_{LT} \in L_2 \bigcap L_\infty \tag{10.123}$$

$$\dot{q}_{LSd} - \dot{q}_{LS} \in L_2 \bigcap L_\infty \tag{10.124}$$

and

$$q_{Td} - q_T \in L_2 \bigcap L_\infty \tag{10.125}$$

$$q_{LTd} - q_{LT} \in L_2 \bigcap L_\infty \tag{10.126}$$

$$q_{LSd} - q_{LS} \in L_2 \bigcap L_\infty \tag{10.127}$$

hold.

10.6.7 Forward Dynamics

The forward dynamics determine the independent acceleration (or time-derivative of velocity) coordinates, given the control torques at the torso and the knees.

Let the independent velocity coordinate vector be the linear/angular velocity vector of the trunk, that is

$$\mathcal{V} = {}^{\mathbf{B}_T}V. \tag{10.128}$$

Following the procedure leading to (4.136) in Section 4.9 gives the forward dynamics as

$$\mathcal{M}_D \dot{\mathcal{V}} + \mathcal{C}_D \mathcal{V} + \mathcal{G}_D = \mathbf{J}_D^T \boldsymbol{\tau} \tag{10.129}$$

where

$$\boldsymbol{\tau} = [\tau_{LT}, \tau_{LN}, \tau_{RT}, \tau_{RN}]^T \in \mathbb{R}^4 \tag{10.130}$$

$$\mathcal{M}_D = \mathcal{J}_D^T \mathcal{M}_b \mathcal{J}_D \tag{10.131}$$

$$\mathcal{C}_D = \mathcal{J}_D^T \mathcal{M}_b \dot{\mathcal{J}}_D + \mathcal{J}_D^T \mathcal{C}_b \mathcal{J}_D \tag{10.132}$$

$$\mathcal{G}_D = \mathcal{J}_D^T \mathcal{G}_b \tag{10.133}$$

with

$$\mathcal{M}_b = \text{diag}\left(\mathbf{M}_{\mathbf{B}_T}, \mathbf{M}_{\mathbf{B}_{LT}}, \mathbf{M}_{\mathbf{B}_{LS}}, \mathbf{M}_{\mathbf{B}_{RT}}, \mathbf{M}_{\mathbf{B}_{RS}}\right) \tag{10.134}$$

$$\mathcal{C}_b = \text{diag}\left(\mathbf{C}_{\mathbf{B}_T}, \mathbf{C}_{\mathbf{B}_{LT}}, \mathbf{C}_{\mathbf{B}_{LS}}, \mathbf{C}_{\mathbf{B}_{RT}}, \mathbf{C}_{\mathbf{B}_{RS}}\right) \tag{10.135}$$

$$\mathcal{G}_b = [\mathbf{G}_{\mathbf{B}_T}^T, \mathbf{G}_{\mathbf{B}_{LT}}^T, \mathbf{G}_{\mathbf{B}_{LS}}^T, \mathbf{G}_{\mathbf{B}_{RT}}^T, \mathbf{G}_{\mathbf{B}_{RS}}^T]^T. \tag{10.136}$$

Note that $\mathbf{J}_D \in \mathbb{R}^{4\times 3}$ in (10.129) and $\mathcal{J}_D \in \mathbb{R}^{15\times 3}$ in (10.131)–(10.133) map ${}^{\mathbf{B}_T}V \in \mathbb{R}^3$ to $[(\dot{q}_T - \dot{q}_{LT}), \dot{q}_{LN}, (\dot{q}_T - \dot{q}_{RT}), \dot{q}_{RN}]^T \in \mathbb{R}^4$ and to $[{}^{\mathbf{B}_T}V^T, {}^{\mathbf{B}_{LT}}V^T, {}^{\mathbf{B}_{LS}}V^T, {}^{\mathbf{B}_{RT}}V^T, {}^{\mathbf{B}_{RS}}V^T]^T \in \mathbb{R}^{15}$, respectively. The two matrices are to be obtained below.

It follows from (10.42), (10.43), (10.76), and (10.77) that

$$\begin{bmatrix} {}^{\mathbf{B}_{LT}}V \\ {}^{\mathbf{B}_{LS}}V \\ {}^{\mathbf{B}_{RT}}V \\ {}^{\mathbf{B}_{RS}}V \end{bmatrix} = \begin{bmatrix} \begin{bmatrix} \mathbf{z} & {}^{\mathbf{B}_{LS}}\mathbf{U}_{\mathbf{B}_{LT}}^T \mathbf{z} \\ \mathbf{0}_{3\times 1} & \mathbf{z} \end{bmatrix} & \mathbf{0}_{6\times 2} \\ \mathbf{0}_{6\times 2} & \begin{bmatrix} \mathbf{z} & {}^{\mathbf{B}_{RS}}\mathbf{U}_{\mathbf{B}_{RT}}^T \mathbf{z} \\ \mathbf{0}_{3\times 1} & \mathbf{z} \end{bmatrix} \end{bmatrix} \begin{bmatrix} \dot{q}_{LN} \\ \dot{q}_{LS} \\ \dot{q}_{RN} \\ \dot{q}_{RS} \end{bmatrix} \tag{10.137}$$

holds. Then, it follows from (10.44) and (10.78) that

$$\begin{bmatrix} \dot{q}_{LN} \\ \dot{q}_{LS} \\ \dot{q}_{RN} \\ \dot{q}_{RS} \end{bmatrix} = \begin{bmatrix} 1 & -1 & 0 & 0 \\ 0 & 1 & 0 & 0 \\ 0 & 0 & 1 & -1 \\ 0 & 0 & 0 & 1 \end{bmatrix} \begin{bmatrix} \dot{q}_{LT} \\ \dot{q}_{LS} \\ \dot{q}_{RT} \\ \dot{q}_{RS} \end{bmatrix} \tag{10.138}$$

$$\begin{bmatrix} -\dot{q}_{LT} \\ \dot{q}_{LN} \\ -\dot{q}_{RT} \\ \dot{q}_{RN} \end{bmatrix} = \begin{bmatrix} -1 & 0 & 0 & 0 \\ 1 & -1 & 0 & 0 \\ 0 & 0 & -1 & 0 \\ 0 & 0 & 1 & -1 \end{bmatrix} \begin{bmatrix} \dot{q}_{LT} \\ \dot{q}_{LS} \\ \dot{q}_{RT} \\ \dot{q}_{RS} \end{bmatrix} \tag{10.139}$$

hold. Furthermore, it follows from (10.7)–(10.9) and (10.14) that

$$\begin{bmatrix} \dot{q}_{LT} \\ \dot{q}_{LS} \\ \dot{q}_{RT} \\ \dot{q}_{RS} \end{bmatrix} = \begin{bmatrix} \mathbf{J}_L^{-1} & \mathbf{0}_{2\times 1} \\ \mathbf{J}_R^{-1} & \mathbf{0}_{2\times 1} \end{bmatrix} \begin{bmatrix} {}^{\mathbf{B}_T}\mathbf{R}_{\mathbf{T}}^T & \mathbf{0}_{2\times 1} \\ \mathbf{0}_{1\times 2} & 1 \end{bmatrix} {}^{\mathbf{B}_T}V \tag{10.140}$$

holds. Finally, using (10.137)–(10.140) yields

$$\mathbf{J}_D = \begin{bmatrix} -1 & 0 & 0 & 0 \\ 1 & -1 & 0 & 0 \\ 0 & 0 & -1 & 0 \\ 0 & 0 & 1 & -1 \end{bmatrix} \begin{bmatrix} \mathbf{J}_L^{-1} & \mathbf{0}_{2\times 1} \\ \mathbf{J}_R^{-1} & \mathbf{0}_{2\times 1} \end{bmatrix} \begin{bmatrix} {}^{\mathbf{B}_T}\mathbf{R}_{\mathbf{T}}^T & \mathbf{0}_{2\times 1} \\ \mathbf{0}_{1\times 2} & 1 \end{bmatrix} + \begin{bmatrix} 0 & 0 & 1 \\ 0 & 0 & 0 \\ 0 & 0 & 1 \\ 0 & 0 & 0 \end{bmatrix} \tag{10.141}$$

$$\mathcal{J}_D = \begin{bmatrix} \mathbf{I}_3 \\ \mathbf{J}_{bb} \begin{bmatrix} 1 & -1 & 0 & 0 \\ 0 & 1 & 0 & 0 \\ 0 & 0 & 1 & -1 \\ 0 & 0 & 0 & 1 \end{bmatrix} \begin{bmatrix} \mathbf{J}_L^{-1} & \mathbf{0}_{2\times 1} \\ \mathbf{J}_R^{-1} & \mathbf{0}_{2\times 1} \end{bmatrix} \begin{bmatrix} {}^{\mathbf{B}_T}\mathbf{R}_{\mathbf{T}}^T & \mathbf{0}_{2\times 1} \\ \mathbf{0}_{1\times 2} & 1 \end{bmatrix} \end{bmatrix} \tag{10.142}$$

with

$$\mathbf{J}_{bb} = \begin{bmatrix} \begin{bmatrix} \mathbf{z} & {}^{\mathbf{B}_{LS}}\mathbf{U}_{\mathbf{B}_{LT}}^T \mathbf{z} \\ \mathbf{0}_{3\times 1} & \mathbf{z} \end{bmatrix} & \mathbf{0}_{6\times 2} \\ \mathbf{0}_{6\times 2} & \begin{bmatrix} \mathbf{z} & {}^{\mathbf{B}_{RS}}\mathbf{U}_{\mathbf{B}_{RT}}^T \mathbf{z} \\ \mathbf{0}_{3\times 1} & \mathbf{z} \end{bmatrix} \end{bmatrix}.$$

10.6.8 Asymptotic Stability

Given bounded signals $[\ddot{q}_{Td}, \ddot{q}_{LTd}, \ddot{q}_{LSd}]^T \in \mathbb{R}^3 \bigcap L_\infty$ and $\boldsymbol{\eta}_{br} \in \mathbb{R}^3 \bigcap L_\infty$, the control equations comprised of (10.29), (10.35)–(10.37), (10.56), (10.57), (10.62)–(10.66), (10.90), (10.91), and (10.96)–(10.100), together with the expressions (10.122)–(10.127) and the time derivatives of (10.21), (10.22), (10.52)–(10.55), and (10.86)–(10.89), guarantee $\boldsymbol{\tau} = [\tau_{LT}, \tau_{LN}, \tau_{RT}, \tau_{RN}]^T \in L_\infty$. In view of (10.129), the boundedness of $\boldsymbol{\tau} \in \mathbb{R}^4 \bigcap L_\infty$ ensures the boundedness of $\frac{d}{dt}\left({}^{\mathbf{B}_T}V\right)$. Consequently, it follows from (10.109)–(10.113) and Lemma 2.8 that

$$
\begin{aligned}
{}^{\mathbf{B}_T}V_r - {}^{\mathbf{B}_T}V &\to \mathbf{0} && (10.143)\\
{}^{\mathbf{B}_{LT}}V_r - {}^{\mathbf{B}_{LT}}V &\to \mathbf{0} && (10.144)\\
{}^{\mathbf{B}_{LS}}V_r - {}^{\mathbf{B}_{LS}}V &\to \mathbf{0} && (10.145)\\
{}^{\mathbf{B}_{RT}}V_r - {}^{\mathbf{B}_{RT}}V &\to \mathbf{0} && (10.146)\\
{}^{\mathbf{B}_{RS}}V_r - {}^{\mathbf{B}_{RS}}V &\to \mathbf{0} && (10.147)
\end{aligned}
$$

hold, leading to

$$
\begin{aligned}
\dot{q}_{Td} - \dot{q}_T &\to 0 && (10.148)\\
\dot{q}_{LTd} - \dot{q}_{LT} &\to 0 && (10.149)\\
\dot{q}_{LSd} - \dot{q}_{LS} &\to 0 && (10.150)
\end{aligned}
$$

and

$$
\begin{aligned}
q_{Td} - q_T &\to 0 && (10.151)\\
q_{LTd} - q_{LT} &\to 0 && (10.152)\\
q_{LSd} - q_{LS} &\to 0. && (10.153)
\end{aligned}
$$

10.6.9 Torque Constraints and Optimization

The torque constraints expressed by (10.66) and (10.100) determine two components in $\boldsymbol{\eta}_{br} \in \mathbb{R}^3$ introduced by (10.36) and (10.37). The remaining component can be used for torque optimization. In this subsection, both torque constraint and optimization issues are handled by using a unified procedure as to be seen below.

Express τ_{LS} and τ_{RS} as affine functions of $\boldsymbol{\eta}_{br} \in \mathbb{R}^3$. It follows from (10.66) and (10.100) that

$$
\begin{aligned}
\tau_{LS} &= \mathbf{a}_L^T \boldsymbol{\eta}_{br} + b_L = 0 && (10.154)\\
\tau_{RS} &= \mathbf{a}_R^T \boldsymbol{\eta}_{br} + b_R = 0 && (10.155)
\end{aligned}
$$

hold with $\mathbf{a}_L^T = \mathbf{z}^{T\mathbf{B}_{LS}}\mathbf{U}_{\mathbf{B}_{T1}} \in \mathbb{R}^{1\times 3}$ and $\mathbf{a}_R^T = -\mathbf{z}^{T\mathbf{B}_{RS}}\mathbf{U}_{\mathbf{B}_{T2}} \in \mathbb{R}^{1\times 3}$. Then, express $\boldsymbol{\tau} \in \mathbb{R}^4$ defined by (10.130) as

$$\boldsymbol{\tau} = \mathbf{A}_\tau \boldsymbol{\eta}_{br} + \mathbf{b}_\tau. \tag{10.156}$$

Furthermore, define a performance cost function as

$$\psi = \frac{1}{2}\boldsymbol{\tau}^T \mathbf{W}_\tau \boldsymbol{\tau} \tag{10.157}$$

where $\mathbf{W}_\tau \in \mathbb{R}^{4\times 4}$ is a symmetric weighting matrix. Substituting (10.154)–(10.156) into (10.157) yields

$$\begin{aligned}\psi = &\frac{1}{2}(\mathbf{A}_\tau\boldsymbol{\eta}_{br} + \mathbf{b}_\tau)^T \mathbf{W}_\tau (\mathbf{A}_\tau\boldsymbol{\eta}_{br} + \mathbf{b}_\tau) \\ &+\lambda_L(\mathbf{a}_L^T\boldsymbol{\eta}_{br} + b_L) + \lambda_R(\mathbf{a}_R^T\boldsymbol{\eta}_{br} + b_R)\end{aligned} \tag{10.158}$$

where λ_L and λ_R are two multipliers. Take

$$\frac{\partial\psi}{\partial\boldsymbol{\eta}_{br}} = \mathbf{0}$$

leading to

$$\boldsymbol{\eta}_{br} = -\left(\mathbf{A}_\tau^T\mathbf{W}_\tau\mathbf{A}_\tau\right)^{-1}\mathbf{A}_\tau^T\mathbf{W}_\tau\mathbf{b}_\tau - \left(\mathbf{A}_\tau^T\mathbf{W}_\tau\mathbf{A}_\tau\right)^{-1}(\mathbf{a}_L\lambda_L + \mathbf{a}_R\lambda_R). \tag{10.159}$$

Substituting (10.159) into (10.154) and (10.155) yields

$$\begin{aligned}&\begin{bmatrix} \mathbf{a}_L^T\left(\mathbf{A}_\tau^T\mathbf{W}_\tau\mathbf{A}_\tau\right)^{-1}\mathbf{a}_L & \mathbf{a}_L^T\left(\mathbf{A}_\tau^T\mathbf{W}_\tau\mathbf{A}_\tau\right)^{-1}\mathbf{a}_R \\ \mathbf{a}_R^T\left(\mathbf{A}_\tau^T\mathbf{W}_\tau\mathbf{A}_\tau\right)^{-1}\mathbf{a}_L & \mathbf{a}_R^T\left(\mathbf{A}_\tau^T\mathbf{W}_\tau\mathbf{A}_\tau\right)^{-1}\mathbf{a}_R \end{bmatrix}\begin{bmatrix}\lambda_L \\ \lambda_R\end{bmatrix} \\ &= \begin{bmatrix} b_L - \mathbf{a}_L^T\left(\mathbf{A}_\tau^T\mathbf{W}_\tau\mathbf{A}_\tau\right)^{-1}\mathbf{A}_\tau^T\mathbf{W}_\tau\mathbf{b}_\tau \\ b_R - \mathbf{a}_R^T\left(\mathbf{A}_\tau^T\mathbf{W}_\tau\mathbf{A}_\tau\right)^{-1}\mathbf{A}_\tau^T\mathbf{W}_\tau\mathbf{b}_\tau \end{bmatrix}.\end{aligned} \tag{10.160}$$

Note that matrix

$$\begin{bmatrix} \mathbf{a}_L^T\left(\mathbf{A}_\tau^T\mathbf{W}_\tau\mathbf{A}_\tau\right)^{-1}\mathbf{a}_L & \mathbf{a}_L^T\left(\mathbf{A}_\tau^T\mathbf{W}_\tau\mathbf{A}_\tau\right)^{-1}\mathbf{a}_R \\ \mathbf{a}_R^T\left(\mathbf{A}_\tau^T\mathbf{W}_\tau\mathbf{A}_\tau\right)^{-1}\mathbf{a}_L & \mathbf{a}_R^T\left(\mathbf{A}_\tau^T\mathbf{W}_\tau\mathbf{A}_\tau\right)^{-1}\mathbf{a}_R \end{bmatrix} \in \mathbb{R}^{2\times 2}$$

is of full-rank.[2] Therefore, substituting the solution of $[\lambda_L, \lambda_R]^T$ from (10.160) into (10.159) gives

$$\begin{aligned}\boldsymbol{\eta}_{br} = &-\left(\mathbf{A}_\tau^T\mathbf{W}_\tau\mathbf{A}_\tau\right)^{-1}\mathbf{A}_\tau^T\mathbf{W}_\tau\mathbf{b}_\tau \\ &-\left(\mathbf{A}_\tau^T\mathbf{W}_\tau\mathbf{A}_\tau\right)^{-1}\begin{bmatrix}\mathbf{a}_L & \mathbf{a}_R\end{bmatrix} \\ &\times\begin{bmatrix} \mathbf{a}_L^T\left(\mathbf{A}_\tau^T\mathbf{W}_\tau\mathbf{A}_\tau\right)^{-1}\mathbf{a}_L & \mathbf{a}_L^T\left(\mathbf{A}_\tau^T\mathbf{W}_\tau\mathbf{A}_\tau\right)^{-1}\mathbf{a}_R \\ \mathbf{a}_R^T\left(\mathbf{A}_\tau^T\mathbf{W}_\tau\mathbf{A}_\tau\right)^{-1}\mathbf{a}_L & \mathbf{a}_R^T\left(\mathbf{A}_\tau^T\mathbf{W}_\tau\mathbf{A}_\tau\right)^{-1}\mathbf{a}_R \end{bmatrix}^{-1} \\ &\times\begin{bmatrix} b_L - \mathbf{a}_L^T\left(\mathbf{A}_\tau^T\mathbf{W}_\tau\mathbf{A}_\tau\right)^{-1}\mathbf{A}_\tau^T\mathbf{W}_\tau\mathbf{b}_\tau \\ b_R - \mathbf{a}_R^T\left(\mathbf{A}_\tau^T\mathbf{W}_\tau\mathbf{A}_\tau\right)^{-1}\mathbf{A}_\tau^T\mathbf{W}_\tau\mathbf{b}_\tau \end{bmatrix}\end{aligned} \tag{10.161}$$

[2] This property is equivalent to $\begin{bmatrix}\mathbf{a}_L & \mathbf{a}_R\end{bmatrix} \in \mathbb{R}^{3\times 2}$ being of full column rank and is further equivalent to $\begin{bmatrix}\mathbf{z} & -{}^{\mathbf{B}_{RS}}\mathbf{U}_{\mathbf{B}_{LS}}^T\mathbf{z}\end{bmatrix} \in \mathbb{R}^{3\times 2}$ being of full column rank, which is ensured as long as the two legs are not standing on a single point.

which not only satisfies (10.66) and (10.100), but also minimizes ψ defined by (10.158).

10.7 Dynamics and Control of the Five-Bar Biped in Single-Support Phase

In the last section, the dynamics and control problem of a five-bar biped in a double-support phase was addressed. In this section, attention will be focused on the dynamics and control problem of the same biped in a single-support phase. Without loss of generality, it is assumed that the left leg is the stance leg and the right leg is the swing leg, see Figure 10.2. To make the VDC applicable, a virtual leg is connected between the swing (right) leg and the ground.

10.7.1 Model of Quasi-Inverted Pendulum

As being analyzed above, a five-bar biped in a single-support phase has five DOFs of motion with four control dimensions. This fact results in a one-dimensional underactuation that imposes a constraint on the system by locating the ZMP at the ankle joint of the stance leg.

To demonstrate the main characteristics of a biped in a single-support phase, a model of quasi-inverted pendulum, as illustrated in Figure 10.5, is studied in this subsection. Although being roughly approximate, the solutions yield a relatively *good guess* about the true results.

The following assumption is used in this subsection.

Assumption 11. *The following two conditions are assumed:*

(i) The CoM (center of mass) keeps a constant height from the ground.
(ii) The moment of inertia of each link is ignored.

Let all four actuated joints be position controlled such that the center of mass (CoM) of the system is kept at a constant height from the ground, as illustrated in Figure 10.5. The horizontal movement of the CoM is, therefore, governed by

$$\ddot{x}(t) = \frac{g}{h}x(t) \tag{10.162}$$

to place the ZMP at point O in Figure 10.5, where $h > 0$ denotes the height of the CoM, $g \approx 9.8$ denotes the standard gravity, and $x(t)$ denotes the horizontal movement of the CoM. The analytical solution of (10.162) is

$$x(t) = c_1 e^{-\sqrt{\frac{g}{h}}t} + c_2 e^{\sqrt{\frac{g}{h}}t} \tag{10.163}$$

with c_1 and c_2 being two constants to be specified by the boundary (initial and end) conditions.

The following lemma gives a condition to ensure a stable walking in the sagittal plane.

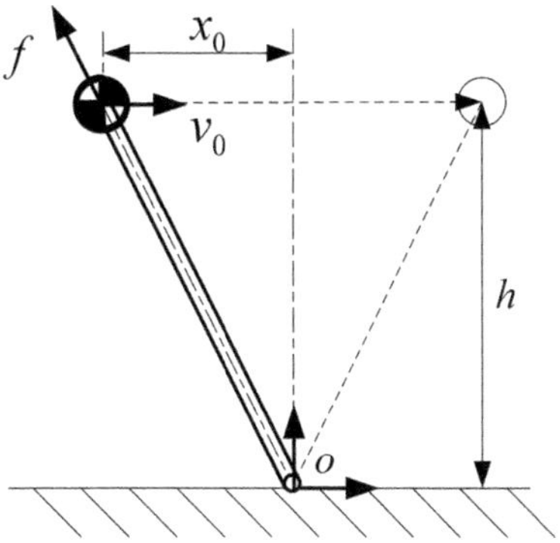

Fig. 10.5. Model of quasi-inverted-pendulum.

Lemma 10.2. *Consider (10.162) with $g > 0$ and $h > 0$ under the initial conditions*

$$x(0) = -x_0 \tag{10.164}$$
$$\dot{x}(0) = v_0 \tag{10.165}$$

with $x_0 > 0$ and $v_0 > 0$. The necessary and sufficient condition to ensure $x(t) < 0$ for $t > 0$ is

$$v_0 < \sqrt{\frac{g}{h}} x_0. \tag{10.166}$$

Proof: The solution of (10.162) with the boundary conditions given by (10.164) and (10.165) is

$$x(t) = c_1 e^{-\sqrt{\frac{g}{h}}t} + c_2 e^{\sqrt{\frac{g}{h}}t} \tag{10.167}$$
$$\dot{x}(t) = \sqrt{\frac{g}{h}} \left(-c_1 e^{-\sqrt{\frac{g}{h}}t} + c_2 e^{\sqrt{\frac{g}{h}}t} \right) \tag{10.168}$$

with

$$c_1 = -\frac{1}{2}\sqrt{\frac{h}{g}} \left(v_0 + \sqrt{\frac{g}{h}} x_0 \right) \tag{10.169}$$
$$c_2 = \frac{1}{2}\sqrt{\frac{h}{g}} \left(v_0 - \sqrt{\frac{g}{h}} x_0 \right). \tag{10.170}$$

Note that (10.166) is equivalent to $c_2 < 0$. Since $c_1 < 0$, it follows from (10.167) that the necessary and sufficient condition to ensure $x(t) < 0$ for $t > 0$ is $c_2 < 0$, which is equivalent to (10.166). ■

In a steady walk, the periodic motion requires that the horizontal speed of the CoM at the beginning of a single-support phase should be equal to the horizontal speed of the CoM at the end of the single-support phase. In addition, the CoM should move a half of the step length. Let $x_0 > 0$ be the initial horizontal distance from the CoM to the stance leg ankle joint, let $v_0 > 0$ be the initial horizontal velocity of the CoM, let $l > 0$ be the step length,[3] and let $t_s > 0$ be the time duration of the single-support phase. Thus, the initial conditions of the CoM

$$\begin{cases} x(0) = -x_0 \\ \dot{x}(0) = v_0 \end{cases} \tag{10.171}$$

and the ending conditions of the CoM

$$\begin{cases} x(t_s) = \frac{l}{2} - x_0 \\ \dot{x}(t_s) = v_0 \end{cases} \tag{10.172}$$

are applied. The solutions of (10.167)–(10.170) subject to (10.172)[4] are

$$x_0 = \frac{l}{4} \tag{10.173}$$

$$e^{\sqrt{\frac{g}{h}}t_s} = \frac{v_0 + \sqrt{\frac{g}{h}}x_0}{v_0 - \sqrt{\frac{g}{h}}x_0}. \tag{10.174}$$

Remark 10.5. Equations (10.173) and (10.174) impose two constraints on four parameters x_0, v_0, l, and t_s. Therefore, only two of the four parameters are independent. Let l and t_s be the two free design parameters. Given l and t_s, the remaining two parameters x_0, and v_0 can be determined from (10.173) and (10.174).

Remark 10.6. Equation (10.163) demonstrates that the complete single-support phase can be divided into two time sections. The first time section is from the beginning to the time instance when the CoM is right above the support point (point O in Figure 10.5). During this time section, the horizontal speed is decreasing. The second time section is from the time instance when the CoM is right above the support point to the end. During this time section, the horizontal speed is increasing. Therefore, it can be concluded that the complete movement of a single-support phase comprises a deceleration section and an acceleration section.

Remark 10.7. In view of Lemma 10.2, the necessary condition to guarantee a steady walking is

$$v_0 - \sqrt{\frac{g}{h}}x_0 > 0. \tag{10.175}$$

[3] The step length is defined as the distance covered by the swing leg.
[4] The solutions of (10.167)–(10.170) satisfy (10.171).

Remark 10.7 implies that the initial horizontal velocity has to be large enough to make the CoM pass over the support point.

Remark 10.8. Equation (10.173) is valid for a steady walk in ideal situations (without friction and air resistance). This equation allows the deceleration section and the acceleration section to be symmetric to ensure the same horizontal speed at the beginning and at the end of a single-support phase. When acceleration is desired, the acceleration section has to be made longer than the deceleration section, leading to $x_0 < \frac{l}{4}$. When deceleration is desired, the deceleration section has to be made longer than the acceleration section, leading to $x_0 > \frac{l}{4}$. It makes

$$\begin{cases} x_0 < l/4 & \text{acceleration} \\ x_0 = l/4 & \text{steady walk} \\ x_0 > l/4 & \text{deceleration.} \end{cases} \tag{10.176}$$

10.7.2 Time Scaling

As indicated in Remark 10.5, parameters x_0 and v_0 can be computed from (10.173) and (10.174) for given step length l and step time t_s. However, this calculation is only valid under the two specific conditions in Assumption 11 leading to (10.162). In a general case, when the two conditions are not satisfied, the solutions (10.173) and (10.174) are only approximate rather than exact. Thus, the actual step time will be different from the given value.

The main challenge to the control of a five-bar biped in a single-support phase is to properly handle the one-dimensional underactuation at the ankle joint of the stance leg. This underactuation is caused by using four actuators to control five DOFs of motion. In the previous chapters, such as in Chapters 4 and 9, the force/moment constraints resulting from the underactuation are satisfied by releasing an equal number of accelerations and internal forces. Unfortunately, this approach cannot be applied here, due to the requirement on the periodic motion in a steady walk.

The solution is to use the time scaling method.[5] Instead of releasing the acceleration of one of the five independent variables characterizing the five DOFs of motion, the time scaling factor for all of the five independent variables is released to satisfy the torque constraint at the ankle joint of the stance leg.

Definition 10.1. *Let $t \geqslant 0$ be the actual time sequence defined on a compact interval $[0, \infty)$ and let $t_p \geqslant 0$ be the planned time sequence defined on a compact interval $[0, \infty)$ such that the function $t = t(t_p)$ or $t_p = t_p(t)$ is monotonically increasing and*

$$\frac{dt_p}{dt} \overset{def}{=} s(t) \tag{10.177}$$

exists.

[5] A similar approach can be found in [29].

Consider a real function $x(t)$. It follows from (10.177) that

$$\begin{aligned}\frac{dx}{dt} &= \frac{dx}{dt_p}\frac{dt_p}{dt} \\ &= \frac{dx}{dt_p}s(t) \qquad (10.178)\\ \frac{d^2x}{dt^2} &= \frac{d}{dt}\left(\frac{dx}{dt_p}s(t)\right) \\ &= \frac{d^2x}{dt_p dt}s(t) + \frac{dx}{dt_p}\frac{ds}{dt} \\ &= \frac{d^2x}{dt_p^2}s(t)^2 + \frac{dx}{dt_p}\frac{ds}{dt} \qquad (10.179)\end{aligned}$$

hold.

Given a real function x with its planned first- and second-order time derivatives, denoted as dx/dt_p and d^2x/dt_p^2, the actual first- and second-order time derivatives can be obtained from (10.178) and (10.179).

Meanwhile, given a time scaling function $s(t)$, the planned time sequence can be obtained from (10.177) as

$$t_p(t) = \int_0^t dt_p(\tau) = \int_0^t s(\tau)d\tau \qquad (10.180)$$

with $t_p(0) = 0$.

10.7.3 Motion Planning in a Single-Support Phase

A single-support phase starts from and ends with double-support phases. Therefore, the initial and final configurations of a single-support phase are determined by the double-support configurations. In view of Figure 10.1, let the biped be in a double-support phase with its two legs being distanced by a half of the step length $l/2$. Adjust the trunk angle q_T and the horizontal position of the torso x_t such that the horizontal position of the CoM, denoted as x_c, is exactly placed in the middle of the two leg ankles. This process involves the use of (10.1)–(10.4), (10.11), and (10.12). Record joint angles as q^*_{LT}, q^*_{LS}, q^*_{RT}, q^*_{RS}, and compute the vertical positions of both the torso and the CoM, denoted as y_t and y_c.

Then, the quasi-inverted pendulum model under Assumptions 11 presented in Subsection 10.7.1 is used to determine the walking parameters. Given a step length $l > 0$ and a step time $t_s > 0$, the initial horizontal position x_0 and the initial horizontal velocity v_0 are determined from (10.173) and (10.174), respectively, subject to (10.175). Meanwhile, $h = y_c$ is used and the two parameters c_1 and c_2 are computed from (10.169) and (10.170).

Furthermore, for a given $\dot{q}_T$, substituting (10.7) and (10.8) into (10.13) yields the relationship between $[\dot{x}_t, \dot{y}_t]^T$ and $[\dot{x}_c, \dot{y}_c]^T$, which determines $[\dot{x}_t, \dot{y}_t]^T$ from known $\dot{x}_c = v_0$ and $\dot{y}_c = 0$. Substituting $[\dot{x}_t, \dot{y}_t]^T$ into (10.7) and (10.8) yields $\dot{q}^*_{LT}$, $\dot{q}^*_{LS}$, $\dot{q}^*_{RT}$, and $\dot{q}^*_{RS}$.

Let the actual step time[6] be $t_f > 0$. The trajectory planning is to find the desired joint angles of the two legs, denoted as q_{LTd}, q_{LSd}, q_{RTd}, and q_{RSd}, with respect to the planned time sequence $t_p \in [0, t_s]$ which can be stored in computer memories. The cyclic motion requires

$$q_{LTd}\big|_{t_p=0} = q^*_{LT} \tag{10.181}$$

$$q_{LSd}\big|_{t_p=0} = q^*_{LS} \tag{10.182}$$

$$q_{RTd}\big|_{t_p=0} = q^*_{RT} \tag{10.183}$$

$$q_{RSd}\big|_{t_p=0} = q^*_{RS} \tag{10.184}$$

at the beginning and

$$q_{LTd}\big|_{t_p=t_s} = q^*_{RT} \tag{10.185}$$

$$q_{LSd}\big|_{t_p=t_s} = q^*_{RS} \tag{10.186}$$

$$q_{RTd}\big|_{t_p=t_s} = q^*_{LT} \tag{10.187}$$

$$q_{RSd}\big|_{t_p=t_s} = q^*_{LS} \tag{10.188}$$

at the end with their time derivatives being

$$\left.\frac{dq_{LTd}}{dt_p}\right|_{t_p=0} = \dot{q}^*_{LT} \tag{10.189}$$

$$\left.\frac{dq_{LSd}}{dt_p}\right|_{t_p=0} = \dot{q}^*_{LS} \tag{10.190}$$

$$\left.\frac{dq_{RTd}}{dt_p}\right|_{t_p=0} = \dot{q}^*_{RT} \tag{10.191}$$

$$\left.\frac{dq_{RSd}}{dt_p}\right|_{t_p=0} = \dot{q}^*_{RS} \tag{10.192}$$

at the beginning and

$$\left.\frac{dq_{LTd}}{dt_p}\right|_{t_p=t_s} = \dot{q}^*_{RT} \tag{10.193}$$

$$\left.\frac{dq_{LSd}}{dt_p}\right|_{t_p=t_s} = \dot{q}^*_{RS} \tag{10.194}$$

$$\left.\frac{dq_{RTd}}{dt_p}\right|_{t_p=t_s} = \dot{q}^*_{LT} \tag{10.195}$$

$$\left.\frac{dq_{RSd}}{dt_p}\right|_{t_p=t_s} = \dot{q}^*_{LS} \tag{10.196}$$

at the end.

[6] The actual (unknown) step time $t_f > 0$ could be different from the given step time $t_s > 0$. Note that t reaches t_f at the exact time instant when t_p reaches t_s.

Subject to the boundary conditions specified by (10.181)–(10.196), the trajectory planning needs to make sure that the swing (right) leg does not touch the ground, that is

$$[l_T \cos(q_{LTd}(t_p)) + l_S \cos(q_{LSd}(t_p))] > [l_T \cos(q_{RTd}(t_p)) + l_S \cos(q_{RSd}(t_p))] \tag{10.197}$$

holds for all $t_p \in (0, t_s)$. A possible solution toward generating optimal trajectories has been reported in [27].

After the desired joint velocities with respect to the planned time sequence $t_p \in [0, t_s]$ being obtained, the desired joint velocities and accelerations with respect to the actual time sequence $t \in [0, t_f]$ can be obtained from (10.178) and (10.179) as

$$\dot{q}_{Td}(t) = \frac{dq_{Td}}{dt_p} s(t) \tag{10.198}$$

$$\dot{q}_{LTd}(t) = \frac{dq_{LTd}}{dt_p} s(t) \tag{10.199}$$

$$\dot{q}_{LSd}(t) = \frac{dq_{LSd}}{dt_p} s(t) \tag{10.200}$$

$$\dot{q}_{RTd}(t) = \frac{dq_{RTd}}{dt_p} s(t) \tag{10.201}$$

$$\dot{q}_{RSd}(t) = \frac{dq_{RSd}}{dt_p} s(t) \tag{10.202}$$

and

$$\ddot{q}_{Td}(t) = \frac{d^2 q_{Td}}{dt_p^2} s(t)^2 + \frac{dq_{Td}}{dt_p} \dot{s}(t) \tag{10.203}$$

$$\ddot{q}_{LTd}(t) = \frac{d^2 q_{LTd}}{dt_p^2} s(t)^2 + \frac{dq_{LTd}}{dt_p} \dot{s}(t) \tag{10.204}$$

$$\ddot{q}_{LSd}(t) = \frac{d^2 q_{LSd}}{dt_p^2} s(t)^2 + \frac{dq_{LSd}}{dt_p} \dot{s}(t) \tag{10.205}$$

$$\ddot{q}_{RTd}(t) = \frac{d^2 q_{RTd}}{dt_p^2} s(t)^2 + \frac{dq_{RTd}}{dt_p} \dot{s}(t) \tag{10.206}$$

$$\ddot{q}_{RSd}(t) = \frac{d^2 q_{RSd}}{dt_p^2} s(t)^2 + \frac{dq_{RSd}}{dt_p} \dot{s}(t) \tag{10.207}$$

where $s(t) > 0$ represents the time scaling factor defined by (10.177) with $s(0) = 1$.

Remark 10.9. In executing (10.198)-(10.207), the planned time sequence $t_p \in [0, t_s]$ is determined by using $s(t)$ for all $t \in [0, t_f]$, according to (10.180).

Remark 10.10. Note that the dynamic parameter uncertainty is the main factor leading to $s(t_f) \neq 1$.

Remark 10.11. When all joint tracking errors tend to zero, the biped keeps its walking pattern regardless of what the time scaling factor $s(t)$ is. Also, the velocities of the swing leg tip are always zero at the end of a single-support phase characterized by $t = t_f$ or $t_p = t_s$. However, the value of $s(t_f)$ indicates whether the biped is in acceleration with $s(t_f) > 1$ or in deceleration with $s(t_f) < 1$.

Finally, the required independent velocity coordinates are designed as

$$\begin{bmatrix} \dot{q}_{Tr} \\ \dot{q}_{LTr} \\ \dot{q}_{LSr} \\ \dot{q}_{RTr} \\ \dot{q}_{RSr} \end{bmatrix} = \begin{bmatrix} \dot{q}_{Td} \\ \dot{q}_{LTd} \\ \dot{q}_{LSd} \\ \dot{q}_{RTd} \\ \dot{q}_{RSd} \end{bmatrix} + \begin{bmatrix} \lambda_1(q_{Td} - q_T) \\ \lambda_2(q_{LTd} - q_{LT}) \\ \lambda_3(q_{LSd} - q_{LS}) \\ \lambda_4(q_{RTd} - q_{RT}) \\ \lambda_5(q_{RSd} - q_{RS}) \end{bmatrix} \tag{10.208}$$

with $\lambda_i > 0$ being a position control gain for all $i \in \{1, 5\}$.

10.7.4 Virtual Decomposition

The biped in a single-support phase can be virtually decomposed into the trunk, the two physical legs, and a virtual leg, as illustrated in Figure 10.6, with its graphic expression in Figure 10.7.

There are totally three *cutting points* being placed in the system. Two of the *cutting points* are placed between the trunk and the two physical legs. The two *cutting points* are interpreted as the *driven cutting points* of the trunk and also the *driving cutting points* of the two physical legs. The remaining one *cutting point* is placed between the swing (right) leg and the virtual leg. This *cutting point* is interpreted as the *driven cutting point* of the swing leg and also the *driving cutting point* of the virtual leg.

10.7.5 Dynamics and Control

Let the left leg be the stance leg. It follows from Figure 10.7 that the control equations for both the trunk and the left leg keep unchanged such that Theorems 10.1 and 10.2 are still valid. However, the right leg becomes a swing leg. In this subsection, both the right (swing) leg and the virtual leg will be specifically addressed.

Kinematics and Dynamics of the Swing Leg

The velocity relationships for the swing (right) leg are described by (10.75)–(10.78) excluding (10.77).

Remark 10.12. As a swing leg, the velocities are determined from the torso toward the ankle. Therefore, (10.77) is no longer valid.

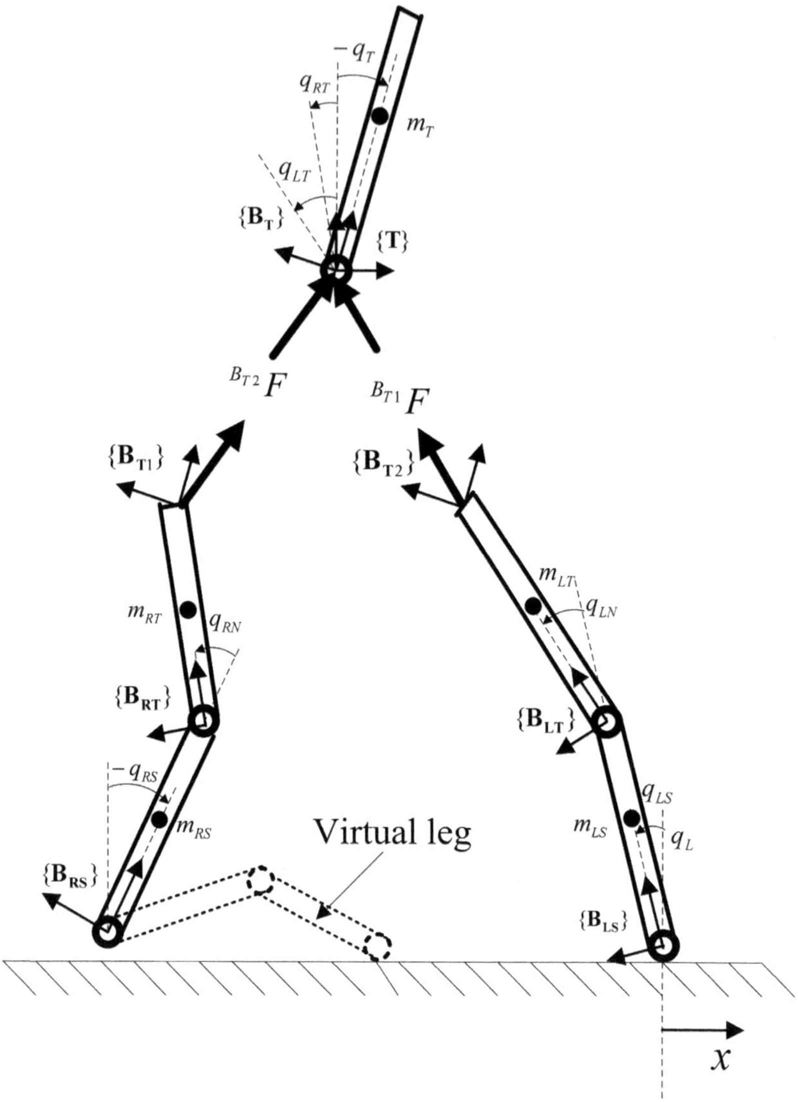

Fig. 10.6. Five-bar biped in single-support after virtual decomposition.

The dynamics of the two links described by (10.79) and (10.80) remain valid, so do the two force/moment resultant equations (10.81) and (10.82). In addition, the following force constraint equation

$$^{\mathbf{B}_{RS}}F = \mathbf{0} \tag{10.209}$$

is applied with the joint torque equations (10.83)–(10.85) being unchanged.

Remark 10.13. Equation (10.209) characterizes the swing status, because no force is applied to the (ankle joint) tip of the swing leg.

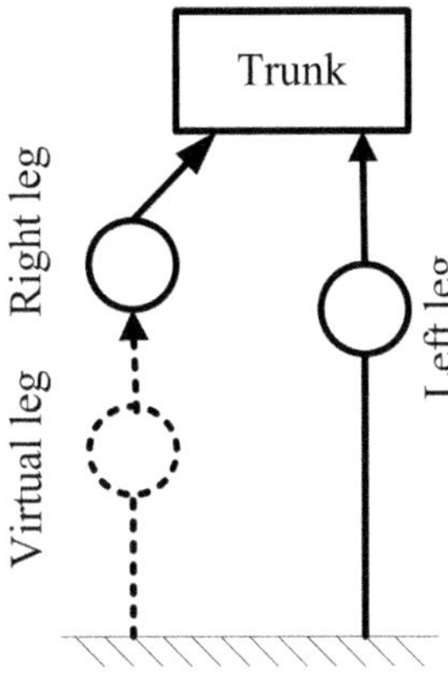

Fig. 10.7. Graphic expression of Figure 10.6.

Control of the Swing Leg

The required velocity vectors are governed by (10.86)–(10.89) excluding (10.88).

Remark 10.14. The computations of the required velocity vectors in a single-support phase are executed from the stance leg all the way to the swing leg by incorporating (10.52)–(10.55) and (10.86)–(10.89) excluding (10.88).

In the meantime, keep the required net force/moment vector equations (10.90) and (10.91) and the corresponding parameter adaptation (10.92)–(10.95) unchanged.

The required force/moment vectors are computed from (10.96) and (10.97), subject to

$$^{\mathbf{B}_{RS}}F_r = -\mathbf{K}_S \left(^{\mathbf{B}_{RS}}V_r - {}^{\mathbf{B}_{RS}}V \right) \tag{10.210}$$

where $\mathbf{K}_S \in \mathbb{R}^{3\times 3}$ is a semi-positive-definite matrix.

Remark 10.15. Equation (10.210) determines $\eta_{br} \in \mathbb{R}^3$ introduced by (10.37).

Finally, the same joint control equations (10.98) and (10.99) are used.

Virtual Stability of the Swing Leg

The *virtual stability* of the swing leg combined with its respective control equations is given by the following theorem.

Theorem 10.4. *The right leg described by (10.75), (10.76), and (10.78)–(10.84), combined with its respective control equations (10.86), (10.87), (10.89)–(10.91), and (10.96)–(10.99) and with the parameter adaptation (10.92)–(10.95), is virtually stable with its affiliated vectors* $^{\mathbf{B}_{RT}}V_r - {}^{\mathbf{B}_{RT}}V$ *and* $^{\mathbf{B}_{RS}}V_r - {}^{\mathbf{B}_{RS}}V$ *being virtual functions in both* L_2 *and* L_∞*, in the sense of Definition 2.17.*

Proof: Let the non-negative accompanying function be the same as ν_R in (10.101). It follows that (10.104) and (10.105) hold.

In view of (10.75), (10.76), (10.78), (10.81)–(10.84), (10.86), (10.87), (10.89), and (10.96)–(10.99), the validity of (10.106) still holds. However, (10.107) has to be modified to

$$\begin{aligned} &\left({}^{\mathbf{B}_{RS}}V_r - {}^{\mathbf{B}_{RS}}V\right)^T \left({}^{\mathbf{B}_{RS}}F_r^* - {}^{\mathbf{B}_{RS}}F^*\right) \\ &= p_{\mathbf{B}_{RS}} - p_{\mathbf{B}_{RT}} - k_{RN}\left(\dot{q}_{RNr} - \dot{q}_{RN}\right)^2 \end{aligned} \quad (10.211)$$

where a term $p_{\mathbf{B}_{RS}}$, representing the *virtual power flow* at the (ankle joint) tip of the swing leg, is added. The derivation of (10.211) is given in Appendix B.8.

Substituting (10.106) and (10.211) into (10.104) and (10.105) results in the addition of the *virtual power flow* term $p_{\mathbf{B}_{RS}}$ to the right hand side of (10.108). This treatment completes the proof, in view of Definition 2.17, by noting the fact that the right (swing) leg has one *driving cutting point* associated with frame $\{\mathbf{B}_{T2}\}$ and one *driven cutting point* associated with frame $\{\mathbf{B}_{RS}\}$. ■

Virtual Stability of the Virtual Leg

The *virtual stability* of the virtual leg combined with its respective control equations is given by the following theorem.

Theorem 10.5. *The virtual leg under zero control is virtually stable in the sense of Definition 2.17.*

Proof: Let the non-negative accompanying function assigned to the virtual leg be *zero*. It follows from (2.85), (10.209), and (10.210) that

$$\begin{aligned} 0 &\leqslant \left({}^{\mathbf{B}_{RS}}V_r - {}^{\mathbf{B}_{RS}}V\right)^T \mathbf{K}_S \left({}^{\mathbf{B}_{RS}}V_r - {}^{\mathbf{B}_{RS}}V\right) \\ &= -\left({}^{\mathbf{B}_{RS}}V_r - {}^{\mathbf{B}_{RS}}V\right)^T \left({}^{\mathbf{B}_{RS}}F_r - {}^{\mathbf{B}_{RS}}F\right) \\ &= -p_{\mathbf{B}_{RS}} \end{aligned} \quad (10.212)$$

holds.

Consider the fact that the virtual leg has only one *driving cutting point* associated with frame $\{\mathbf{B}_{RS}\}$. It completes the proof from Definition 2.17. ■

10.7.6 L_2 and L_∞ Stability

Theorems 10.1, 10.2, 10.4, and 10.5 suggest that all four subsystems of the biped are *virtually stable* in the sense of Definition 2.17. Therefore, it follows from Theorem 2.1 that

$$^{\mathbf{B}_T}V_r - {}^{\mathbf{B}_T}V \in L_2 \bigcap L_\infty \quad (10.213)$$

$$^{\mathbf{B}_{LT}}V_r - {}^{\mathbf{B}_{LT}}V \in L_2 \bigcap L_\infty \quad (10.214)$$

$$^{\mathbf{B}_{LS}}V_r - {}^{\mathbf{B}_{LS}}V \in L_2 \bigcap L_\infty \quad (10.215)$$

$$^{\mathbf{B}_{RT}}V_r - {}^{\mathbf{B}_{RT}}V \in L_2 \bigcap L_\infty \quad (10.216)$$

$$^{\mathbf{B}_{RS}}V_r - {}^{\mathbf{B}_{RS}}V \in L_2 \bigcap L_\infty \quad (10.217)$$

hold.

Note that (10.215) implies

$$\dot{q}_{LSr} - \dot{q}_{LS} \in L_2 \bigcap L_\infty \tag{10.218}$$

from (10.43) and (10.54). Also, (10.214) and (10.215) imply

$$\dot{q}_{LNr} - \dot{q}_{LN} \in L_2 \bigcap L_\infty \tag{10.219}$$

from (10.42) and (10.53). Thus, it follows from (10.44), (10.55), (10.218), and (10.219) that

$$\dot{q}_{LTr} - \dot{q}_{LT} \in L_2 \bigcap L_\infty \tag{10.220}$$

holds.

On the other hand, (10.213), (10.214), and (10.220) imply

$$\dot{q}_{Tr} - \dot{q}_{T} \in L_2 \bigcap L_\infty \tag{10.221}$$

from (10.41) and (10.52). Then, using (10.213), (10.216), and (10.221) yields

$$\dot{q}_{RTr} - \dot{q}_{RT} \in L_2 \bigcap L_\infty \tag{10.222}$$

from (10.75) and (10.86). Furthermore, using (10.216) and (10.217) results in

$$\dot{q}_{RNr} - \dot{q}_{RN} \in L_2 \bigcap L_\infty \tag{10.223}$$

from (10.76) and (10.87). Finally, it follows from (10.78), (10.89), (10.222), and (10.223) that

$$\dot{q}_{RSr} - \dot{q}_{RS} \in L_2 \bigcap L_\infty \tag{10.224}$$

holds.

In view of the required velocity design in (10.208), it follows from (10.218)–(10.224) and Lemma 2.4 that

$$\dot{q}_{Td} - \dot{q}_{T} \in L_2 \bigcap L_\infty \tag{10.225}$$

$$\dot{q}_{LTd} - \dot{q}_{LT} \in L_2 \bigcap L_\infty \tag{10.226}$$

$$\dot{q}_{LSd} - \dot{q}_{LS} \in L_2 \bigcap L_\infty \tag{10.227}$$

$$\dot{q}_{RTd} - \dot{q}_{RT} \in L_2 \bigcap L_\infty \tag{10.228}$$

$$\dot{q}_{RSd} - \dot{q}_{RS} \in L_2 \bigcap L_\infty \tag{10.229}$$

and

$$q_{Td} - q_T \in L_2 \bigcap L_\infty \tag{10.230}$$

$$q_{LTd} - q_{LT} \in L_2 \bigcap L_\infty \tag{10.231}$$

$$q_{LSd} - q_{LS} \in L_2 \bigcap L_\infty \tag{10.232}$$

$$q_{RTd} - q_{RT} \in L_2 \bigcap L_\infty \tag{10.233}$$

$$q_{RSd} - q_{RS} \in L_2 \bigcap L_\infty \tag{10.234}$$

hold.

10.7.7 Forward Dynamics

In a single-support phase, there are five DOFs of motion driven by four actuators. Therefore, this is an underactuated system.

Let the independent velocity vector be

$$\mathcal{V} = \begin{bmatrix} \dot{q}_T \\ \dot{q}_{LT} \\ \dot{q}_{LS} \\ \dot{q}_{RT} \\ \dot{q}_{RS} \end{bmatrix} \in \mathbb{R}^5. \tag{10.235}$$

Similarly, following the procedure leading to (4.136) in Section 4.9 gives the forward dynamics as

$$\mathcal{M}_S \dot{\mathcal{V}} + \mathcal{C}_S \mathcal{V} + \mathcal{G}_S = \mathbf{J}_S^T \boldsymbol{\tau} \tag{10.236}$$

where

$$\boldsymbol{\tau} = [\tau_{LT}, \tau_{LN}, \tau_{RT}, \tau_{RN}]^T \in \mathbb{R}^4 \tag{10.237}$$

$$\mathcal{M}_S = \mathcal{J}_S^T \mathcal{M}_b \mathcal{J}_S \tag{10.238}$$

$$\mathcal{C}_S = \mathcal{J}_S^T \mathcal{M}_b \dot{\mathcal{J}}_S + \mathcal{J}_S^T \mathcal{C}_b \mathcal{J}_S \tag{10.239}$$

$$\mathcal{G}_S = \mathcal{J}_S^T \mathcal{G}_b \tag{10.240}$$

with

$$\mathcal{M}_b = \mathrm{diag}\left(\mathbf{M}_{\mathbf{B}_T}, \mathbf{M}_{\mathbf{B}_{LT}}, \mathbf{M}_{\mathbf{B}_{LR}}, \mathbf{M}_{\mathbf{B}_{RT}}, \mathbf{M}_{\mathbf{B}_{RS}}\right) \tag{10.241}$$

$$\mathcal{C}_b = \mathrm{diag}\left(\mathbf{C}_{\mathbf{B}_T}, \mathbf{C}_{\mathbf{B}_{LT}}, \mathbf{C}_{\mathbf{B}_{LR}}, \mathbf{C}_{\mathbf{B}_{RT}}, \mathbf{C}_{\mathbf{B}_{RS}}\right) \tag{10.242}$$

$$\mathcal{G}_b = [\mathbf{G}_{\mathbf{B}_T}^T, \mathbf{G}_{\mathbf{B}_{LT}}^T, \mathbf{G}_{\mathbf{B}_{LR}}^T, \mathbf{G}_{\mathbf{B}_{RT}}^T, \mathbf{G}_{\mathbf{B}_{RS}}^T]^T. \tag{10.243}$$

Matrices $\mathbf{J}_S \in \mathbb{R}^{4\times5}$ in (10.236) and $\mathcal{J}_S \in \mathbb{R}^{15\times5}$ in (10.238)-(10.240) map $\mathcal{V} \in \mathbb{R}^5$ to $[\dot{q}_T - \dot{q}_{LT}, \dot{q}_{LN}, \dot{q}_T - \dot{q}_{RT}, \dot{q}_{RN}]^T \in \mathbb{R}^4$ and to $[{}^{\mathbf{B}_T}V^T, {}^{\mathbf{B}_{LT}}V^T, {}^{\mathbf{B}_{LS}}V^T, {}^{\mathbf{B}_{RT}}V^T, {}^{\mathbf{B}_{RS}}V^T]^T \in \mathbb{R}^{15}$, respectively. They are obtained from (10.41)-(10.43), (10.75), and (10.76) as

$$\mathbf{J}_S = \begin{bmatrix} 1 & -1 & 0 & 0 & 0 \\ 0 & 1 & -1 & 0 & 0 \\ 1 & 0 & 0 & -1 & 0 \\ 0 & 0 & 0 & 1 & -1 \end{bmatrix} \tag{10.244}$$

$$\mathcal{J}_S = \begin{bmatrix} \mathbf{z} & {}^{\mathbf{B}_{LT}}\mathbf{U}^T_{\mathbf{B}_T}\mathbf{z} - \mathbf{z} & {}^{\mathbf{B}_{LS}}\mathbf{U}^T_{\mathbf{B}_T}\mathbf{z} - {}^{\mathbf{B}_{LT}}\mathbf{U}^T_{\mathbf{B}_T}\mathbf{z} & \mathbf{0}_{3\times1} & \mathbf{0}_{3\times1} \\ \mathbf{0}_{3\times1} & \mathbf{z} & {}^{\mathbf{B}_{LS}}\mathbf{U}^T_{\mathbf{B}_{LT}}\mathbf{z} - \mathbf{z} & \mathbf{0}_{3\times1} & \mathbf{0}_{3\times1} \\ \mathbf{0}_{3\times1} & \mathbf{0}_{3\times1} & \mathbf{z} & \mathbf{0}_{3\times1} & \mathbf{0}_{3\times1} \\ \mathbf{0}_{3\times1} & {}^{\mathbf{B}_{LT}}\mathbf{U}^T_{\mathbf{B}_{RT}}\mathbf{z} - {}^{\mathbf{B}_T}\mathbf{U}^T_{\mathbf{B}_{RT}}\mathbf{z} & {}^{\mathbf{B}_{LS}}\mathbf{U}^T_{\mathbf{B}_{RT}}\mathbf{z} - {}^{\mathbf{B}_{LT}}\mathbf{U}^T_{\mathbf{B}_{RT}}\mathbf{z} & {}^{\mathbf{B}_T}\mathbf{U}^T_{\mathbf{B}_{RT}}\mathbf{z} & \mathbf{0}_{3\times1} \\ \mathbf{0}_{3\times1} & {}^{\mathbf{B}_{LT}}\mathbf{U}^T_{\mathbf{B}_{RS}}\mathbf{z} - {}^{\mathbf{B}_T}\mathbf{U}^T_{\mathbf{B}_{RS}}\mathbf{z} & {}^{\mathbf{B}_{LS}}\mathbf{U}^T_{\mathbf{B}_{RS}}\mathbf{z} - {}^{\mathbf{B}_{LT}}\mathbf{U}^T_{\mathbf{B}_{RS}}\mathbf{z} & {}^{\mathbf{B}_T}\mathbf{U}^T_{\mathbf{B}_{RS}}\mathbf{z} - {}^{\mathbf{B}_{RT}}\mathbf{U}^T_{\mathbf{B}_{RS}}\mathbf{z} & {}^{\mathbf{B}_{RT}}\mathbf{U}^T_{\mathbf{B}_{RS}}\mathbf{z} \end{bmatrix} \tag{10.245}$$

with $\mathbf{z} = [0, 0, 1]^T$.

10.7.8 Asymptotic Stability

Given a bounded $[\ddot{q}_{Td}, \ddot{q}_{LTd}, \ddot{q}_{LSd}, \ddot{q}_{RTd}, \ddot{q}_{RSd}]^T \in \mathbb{R}^5$, the control equations comprised of (10.29), (10.35)–(10.37), (10.56), (10.57), (10.62)–(10.65), (10.90), (10.91), (10.96)–(10.99), together with the time derivatives of (10.52)–(10.55), (10.86)–(10.89), and (10.208), guarantee the boundedness of $\boldsymbol{\tau} \in \mathbb{R}^4$. In view of (10.236), the boundedness of $\boldsymbol{\tau} \in \mathbb{R}^4$ ensures the boundedness of $\dot{\mathcal{V}} \in \mathbb{R}^5$. It, therefore, follows from (10.213)–(10.217) and Lemma 2.8 that

$$^{\mathbf{B}_T}V_r - {}^{\mathbf{B}_T}V \to \mathbf{0} \tag{10.246}$$

$$^{\mathbf{B}_{LT}}V_r - {}^{\mathbf{B}_{LT}}V \to \mathbf{0} \tag{10.247}$$

$$^{\mathbf{B}_{LS}}V_r - {}^{\mathbf{B}_{LS}}V \to \mathbf{0} \tag{10.248}$$

$$^{\mathbf{B}_{RT}}V_r - {}^{\mathbf{B}_{RT}}V \to \mathbf{0} \tag{10.249}$$

$$^{\mathbf{B}_{RS}}V_r - {}^{\mathbf{B}_{RS}}V \to \mathbf{0} \tag{10.250}$$

hold. Consequently, it yields

$$\dot{q}_{LSr} - \dot{q}_{LS} \to 0 \tag{10.251}$$

$$\dot{q}_{LNr} - \dot{q}_{LN} \to 0 \tag{10.252}$$

$$\dot{q}_{LTr} - \dot{q}_{LT} \to 0 \tag{10.253}$$

$$\dot{q}_{Tr} - \dot{q}_{T} \to 0 \tag{10.254}$$

$$\dot{q}_{RTr} - \dot{q}_{RT} \to 0 \tag{10.255}$$

$$\dot{q}_{RNr} - \dot{q}_{RN} \to 0 \tag{10.256}$$

$$\dot{q}_{RSr} - \dot{q}_{RS} \to 0 \tag{10.257}$$

which further lead to

$$\dot{q}_{Td} - \dot{q}_T \to 0 \tag{10.258}$$
$$\dot{q}_{LTd} - \dot{q}_{LT} \to 0 \tag{10.259}$$
$$\dot{q}_{LSd} - \dot{q}_{LS} \to 0 \tag{10.260}$$
$$\dot{q}_{RTd} - \dot{q}_{RT} \to 0 \tag{10.261}$$
$$\dot{q}_{RSd} - \dot{q}_{RS} \to 0 \tag{10.262}$$

and

$$q_{Td} - q_T \to 0 \tag{10.263}$$
$$q_{LTd} - q_{LT} \to 0 \tag{10.264}$$
$$q_{LSd} - q_{LS} \to 0 \tag{10.265}$$
$$q_{RTd} - q_{RT} \to 0 \tag{10.266}$$
$$q_{RSd} - q_{RS} \to 0 \tag{10.267}$$

from Lemma 2.4.

10.7.9 Torque Constraint and Solution of Time Scaling

Equations (10.64)–(10.66), (10.98), and (10.99) determine τ_{LT}, τ_{LN}, τ_{RT}, and τ_{RN} and also specify

$$\tau_{LS} = 0. \tag{10.268}$$

This constraint on the torque of the ankle joint of the stance leg is to be satisfied by releasing the time derivative of the time scaling factor $\dot{s}(t)$.

From (10.203) to (10.207), it can be found that $\ddot{q}_{Td}(t)$, $\ddot{q}_{LTd}(t)$, $\ddot{q}_{LSd}(t)$, $\ddot{q}_{RTd}(t)$, and $\ddot{q}_{RSd}(t)$ are affine functions of $\dot{s}(t)$, so are $\ddot{q}_{Tr}(t)$, $\ddot{q}_{LTr}(t)$, $\ddot{q}_{LSr}(t)$, $\ddot{q}_{RTr}(t)$, and $\ddot{q}_{RSr}(t)$ from (10.208). Consequently, vectors $\frac{d}{dt}\left({}^{\mathbf{B}_T}V_r\right)$, $\frac{d}{dt}\left({}^{\mathbf{B}_{LT}}V_r\right)$, $\frac{d}{dt}\left({}^{\mathbf{B}_{LS}}V_r\right)$, $\frac{d}{dt}\left({}^{\mathbf{B}_{RT}}V_r\right)$, and $\frac{d}{dt}\left({}^{\mathbf{B}_{RS}}V_r\right)$ are affine functions of $\dot{s}(t)$, in view of (10.52)–(10.55), (10.86), (10.87), and (10.89). Furthermore, ${}^{\mathbf{B}_T}F_r^*$, ${}^{\mathbf{B}_{LT}}F_r^*$, ${}^{\mathbf{B}_{LS}}F_r^*$, ${}^{\mathbf{B}_{RT}}F_r^*$, and ${}^{\mathbf{B}_{RS}}F_r^*$, defined by (10.29), (10.56), (10.57), (10.90), and (10.91), are affine functions of $\dot{s}(t)$ as well, so are ${}^{\mathbf{B}_T}F_r$, ${}^{\mathbf{B}_{T1}}F_r$, ${}^{\mathbf{B}_{T2}}F_r$, ${}^{\mathbf{B}_{RT}}F_r$, ${}^{\mathbf{B}_{RS}}F_r$, ${}^{\mathbf{B}_{LT}}F_r$, and ${}^{\mathbf{B}_{LS}}F_r$, in view of (10.36), (10.37), (10.62), (10.63), (10.96), and (10.97), after determining $\boldsymbol{\eta}_{br} \in \mathbb{R}^3$ by satisfying (10.210). Finally, it follows from (10.51) that τ_{LS} is an affine function of $\dot{s}(t)$ and can be expressed by

$$\tau_{LS} = \varrho_a(t)\dot{s}(t) + \varrho_b(t) \tag{10.269}$$

with

$$\varrho_a(t) \neq 0 \tag{10.270}$$

for all $t \in [0, \infty)$. Substituting (10.268) into (10.269) determines $\dot{s}(t)$ for $t \geqslant 0$. With $s(0) = 1$, the solution of $s(t)$ for $t > 0$ is therefore obtained through an integral computation.

The necessary condition to satisfy (10.268) by releasing $\dot{s}(t)$ is given by (10.270). Fortunately, the problem's nature guarantees that this necessary condition can always be satisfied in normal circumstances. An explanation is given below.

Let $\mathbf{q}_b = [q_T, q_{LT}, q_{LS}, q_{RT}, q_{RS}]^T \in \mathbb{R}^5$ be the joint position vector, and let

$$L(t) = \mathbf{c}(\mathbf{q}_b)^T \dot{\mathbf{q}}_b = \mathbf{c}(\mathbf{q}_b)^T \frac{d\mathbf{q}_b}{dt_p} s(t)$$

be the angular momentum of the biped with respect to the ankle joint of the stance leg, where $\mathbf{c}(\mathbf{q}_b) \in \mathbb{R}^5$ is a coefficient vector. Thus, the time derivative of $L(t)$ can be written as

$$\dot{L}(t) = \left[\mathbf{c}(\mathbf{q}_b)^T \frac{d\mathbf{q}_b}{dt_p}\right] \dot{s}(t) + \frac{d}{dt}\left[\mathbf{c}(\mathbf{q}_b)^T \frac{d\mathbf{q}_b}{dt_p}\right] s(t). \tag{10.271}$$

On the other hand, note that $\dot{L}(t)$ equals the total external torques applied to the biped, that is

$$\dot{L}(t) = \tau_{LS} + \tau_g(\mathbf{q}_b) \tag{10.272}$$

where $\tau_g(\mathbf{q}_b) \in \mathbb{R}$ denotes the gravitational moment applied to the biped, which is a function of the system configuration $\mathbf{q}_b$. Obviously, $\mathbf{c}(\mathbf{q}_b)^T \frac{d\mathbf{q}_b}{dt_p}$ represents the angular momentum with respect to the planned time sequence for all $t_p \in [0, t_s]$. This term is never being zero to keep a consistent walking. Thus, substituting (10.268) and (10.272) into (10.271) yields

$$\varrho_a(t) = \mathbf{c}(\mathbf{q}_b)^T \frac{d\mathbf{q}_b}{dt_p} \neq 0$$

which ensures (10.268).

Remark 10.16. Equation (10.180) relates the planned time sequence $t_p \in [0, t_s]$ to the actual time sequence $t \in [0, t_f]$. The actual step time (of a single-support phase) t_f is determined at the time t when the planned time sequence t_p reaches the planned step time t_s. In view of (10.181)–(10.188) and (10.264)–(10.267), it is clear that $q_{LT}(t)$ and $q_{LS}(t)$ approach q_{RT}^* and q_{RS}^* when $q_{RT}(t)$ and $q_{RS}(t)$ approach q_{LT}^* and q_{LS}^*. Alternatively, $q_{LT}(t)$ and $q_{LS}(t)$ approach q_{LT}^* and q_{LS}^* when $q_{RT}(t)$ and $q_{RS}(t)$ approach q_{RT}^* and q_{RS}^*. A repeated walking pattern is, therefore, formed.

Remark 10.17. There is no guarantee to achieve $s(t_f) = 1$ at the end of a step. Instead, $s(t_f) > 1$ indicates that the biped is accelerating and $s(t_f) < 1$ indicates that the biped is decelerating. A walking speed correction using (10.176) is expected. The correction is to adjust the horizontal position of the swing leg at landing.

Remark 10.18. Despite $s(t_f) \neq 1$ might hold, the velocities of the swing leg tip at landing are still zero provided that position control errors are negligible. This is because $s(t_f)$ acts as a multiplier to the planned velocities. As a result, all velocities remain continuous during walking phase switches, and no impact of the swing leg on the ground is expected. This is one of the characteristics of this approach, when compared to other approaches [62, 207].

The computational algorithm for a single support phase is summarized as follows:

Step 1: A single-support phase is initialized from a double-support phase at $t = 0$ when the swing leg tip (at the ankle joint) is going to leave the ground. Given a step length $l > 0$, make the initial horizontal positions of the stance leg tip, the swing leg tip, and the CoM at 0, $-l/2$, and $-x_0 = -l/4$, respectively. Then, for a given q_T, find q_{LT}^*, q_{LS}^*, q_{RT}^*, and q_{RS}^* from (10.1)–(10.4), (10.11), and (10.12). Furthermore, given a planned step time $t_s > 0$ and the height of the CoM $y_c = h > 0$, compute the initial horizontal speed of the CoM, denoted as $\dot{x}_c(0) = v_0$, from (10.174) to satisfy (10.175). Finally, compute $\dot{q}_{LT}^*$, $\dot{q}_{LS}^*$, $\dot{q}_{RT}^*$, and $\dot{q}_{RS}^*$ from (10.7), (10.8), and (10.13) and from given $\dot{y}_c = 0$ and $\dot{q}_T$.

Step 2: Plan the trajectory in the joint space for all $t_p \in [0, t_s]$, where $t_s > 0$ denotes the planned step time, subject to (10.181)–(10.196) and (10.197). Equations (10.181)–(10.196) make a cyclic motion possible and (10.197) makes sure that the swing leg does not touch the ground before a leg switch is expected.

Step 3: Express $\dot{q}_{Td}(t)$, $\dot{q}_{LTd}(t)$, $\dot{q}_{LSd}(t)$, $\dot{q}_{RTd}(t)$, and $\dot{q}_{RSd}(t)$ in terms of (10.198)–(10.202) and express $\ddot{q}_{Td}(t)$, $\ddot{q}_{LTd}(t)$, $\ddot{q}_{LSd}(t)$, $\ddot{q}_{RTd}(t)$, and $\ddot{q}_{RSd}(t)$ in terms of (10.203)–(10.207). Then, compute $\dot{q}_{Tr}(t)$, $\dot{q}_{LTr}(t)$, $\dot{q}_{LSr}(t)$, $\dot{q}_{RTr}(t)$, and $\dot{q}_{RSr}(t)$ in terms of (10.208). Furthermore, compute ${}^{\mathbf{B}_T}V_r$, ${}^{\mathbf{B}_{LT}}V_r$, ${}^{\mathbf{B}_{LS}}V_r$, ${}^{\mathbf{B}_{RT}}V_r$, and ${}^{\mathbf{B}_{RS}}V_r$, and their time derivatives from (10.52)–(10.55), (10.86), (10.87), and (10.89).

Step 4: Compute the net force/moment vectors of the five links ${}^{\mathbf{B}_T}F_r^*$, ${}^{\mathbf{B}_{LT}}F_r^*$, ${}^{\mathbf{B}_{LS}}F_r^*$, ${}^{\mathbf{B}_{RT}}F_r^*$, and ${}^{\mathbf{B}_{RS}}F_r^*$ from (10.29), (10.56), (10.57), (10.90), and (10.91), respectively. Then, compute the corresponding parameter adaptation for $\hat{\boldsymbol{\theta}}_{\mathbf{B}_T}$, $\hat{\boldsymbol{\theta}}_{\mathbf{B}_{LT}}$, $\hat{\boldsymbol{\theta}}_{\mathbf{B}_{LS}}$, $\hat{\boldsymbol{\theta}}_{\mathbf{B}_{RT}}$, and $\hat{\boldsymbol{\theta}}_{\mathbf{B}_{RS}}$ from (10.33), (10.34), (10.58)–(10.61), and (10.92)–(10.95).

Step 5: Compute ${}^{\mathbf{B}_{T1}}F_r$, ${}^{\mathbf{B}_{T2}}F_r$, ${}^{\mathbf{B}_{LT}}F_r$, ${}^{\mathbf{B}_{LS}}F_r$, ${}^{\mathbf{B}_{RT}}F_r$, and ${}^{\mathbf{B}_{RS}}F_r$ from (10.36), (10.37), (10.62), (10.63), (10.96), and (10.97), respectively, and determine $\eta_{br} \in \mathbb{R}^3$ from (10.210).

Step 6: Determine the time derivative of the temporal scaling factor $\dot{s}(t)$ by solving (10.66) or equivalently solving (10.268) and (10.269). Then, obtain $s(t)$ through an integral computation. Substitute $\dot{s}(t)$ and $s(t)$ into $\dot{q}_{Td}(t)$, $\dot{q}_{LTd}(t)$, $\dot{q}_{LSd}(t)$, $\dot{q}_{RTd}(t)$, $\dot{q}_{RSd}(t)$, $\ddot{q}_{Td}(t)$, $\ddot{q}_{LTd}(t)$, $\ddot{q}_{LSd}(t)$, $\ddot{q}_{RTd}(t)$, and $\ddot{q}_{RSd}(t)$ in step 3 and re-compute steps 3–5 to obtain all variables in numerical values.

Step 7: Finally, compute τ_{LT}, τ_{LN}, τ_{RT}, and τ_{RN} in terms of (10.64), (10.65), (10.98), and (10.99), respectively. Return to step 3 for the next sampling period.

Step 8: At $t_p = t_s$ or $t = t_f$, the swing leg touches the ground with a zero velocity at the tip (ankle joint). If $s(t_f) = 1$ holds, continue the walking motion; otherwise, a walking speed adjustment is needed. This can be done by changing the initial horizontal position of the CoM of the next walking cycle, according to (10.176). Technically, this is done by either changing the trunk angle q_T or adjusting the relative horizontal position between the torso and the stance leg for the next walking cycle. When a walking speed adjustment occurs, the computation in step 1 needs to be repeated to obtain new values of q^*_{LT}, q^*_{LS}, q^*_{RT}, and q^*_{RS}, as well as $\dot{q}^*_{LT}$, $\dot{q}^*_{LS}$, $\dot{q}^*_{RT}$, and $\dot{q}^*_{RS}$.

10.8 Bipeds in 3D

In a three-dimensional (3D) motion, each leg has at least six DOFs of motion comprised of two DOFs of actuated motion at the torso joint, one DOF of actuated motion at the knee joint, and three DOFs of unactuated motion at the ankle joint.

The inertial frame is established in such a way that the x-axis points to the walking direction, the y-axis points laterally, and the z-axis points upward (opposite to the gravity).

10.8.1 Estimation of CoM

Being crucial for the lateral walking stability, the estimated position of the CoM is being obtained in this subsection.

For a given configuration, let the 3D net force/moment vector of each rigid body, defined by (2.74), take only the gravitational term, that is

$$ {}^{\mathbf{A}}F^* = \begin{bmatrix} \hat{m}_{\mathbf{A}} {}^{\mathbf{A}}\mathbf{R}_{\mathbf{I}}\mathbf{g} \\ \hat{m}_{\mathbf{A}}({}^{\mathbf{A}}\mathbf{r}_{\mathbf{AB}}\times){}^{\mathbf{A}}\mathbf{R}_{\mathbf{I}}\mathbf{g} \end{bmatrix} \in \mathbb{R}^6 \tag{10.273} $$

with

$$ \mathbf{g} = \begin{bmatrix} 0 \\ 0 \\ 9.8 \end{bmatrix} \in \mathbb{R}^3 \tag{10.274} $$

where $\hat{m}_{\mathbf{A}}$ denotes the estimated mass of the rigid body (link) to which frame $\{\mathbf{A}\}$ is attached for all $\mathbf{A} \in \{\mathbf{B}_T, \mathbf{B}_{LT}, \mathbf{B}_{LS}, \mathbf{B}_{RT}, \mathbf{B}_{RS}\}$ and ${}^{\mathbf{A}}\mathbf{r}_{\mathbf{AB}} \in \mathbb{R}^3$ denotes a vector from the origin of frame $\{\mathbf{A}\}$ to the center of mass of the link, expressed in frame $\{\mathbf{A}\}$.

Then, recall (10.28), (10.47), (10.48), (10.81), and (10.82) to compute ${}^{\mathbf{B}_{LS}}F \in \mathbb{R}^6$ in 3D subject to ${}^{\mathbf{B}_{RS}}F = \mathbf{0}$, leading to

$$ {}^{\mathbf{I}}\mathbf{R}_{\mathbf{B}_{LS}} \begin{bmatrix} \mathbf{0}_{3\times 3} & \mathbf{I}_3 \end{bmatrix} {}^{\mathbf{B}_{LS}}F \stackrel{def}{=} \begin{bmatrix} m_x \\ m_y \\ m_z \end{bmatrix} = \begin{bmatrix} 0 & -r_z & r_y \\ r_z & 0 & -r_x \\ -r_y & r_x & 0 \end{bmatrix} \hat{m}_c \mathbf{g} \tag{10.275} $$

where $\hat{m}_c$ denotes the estimate of the total mass of the system defined by (10.10); m_x, m_y, and m_z represent the three moments of ${}^{\mathbf{B}_{LS}}F \in \mathbb{R}^6$, expressed in the inertial frame; and r_x, r_y, and r_z are the coordinates of the CoM with respect to the tip (ankle joint) of the left leg, expressed in the inertial frame. Since $\hat{m}_c\mathbf{g} = [0, 0, 9.8\hat{m}_c]^T \in \mathbb{R}^3$ is known, it yields r_y and r_x from m_x and m_y. Then, change $\mathbf{g}$ to

$$\mathbf{g}_x = \begin{bmatrix} 9.8 \\ 0 \\ 0 \end{bmatrix} \in \mathbb{R}^3. \tag{10.276}$$

Repeating the above process leads to

$$m_y = 9.8 r_z \hat{m}_c \tag{10.277}$$

which yields r_z from the newly obtained m_y.

10.8.2 Swing Leg Planning with Lateral Walking Stability

With the model of the quasi-inverted pendulum representation under Assumption 11, the rear and top views of the CoM trajectory in a typical single-support phase are shown in Figure 10.8. The motion is basically a superposition of both motions in the sagittal plane and in the lateral plane. As described early in the five-bar system, the CoM trajectory in the sagittal plane has to pass the stance leg to realize a locomotion. In contrast, a stable walking requires that the CoM trajectory in the lateral plane should never go over the stance leg. Let the initial conditions of the CoM in the sagittal plane be expressed by

$$x_c(0) = -x_0 \tag{10.278}$$

$$\dot{x}_c(0) = v_{x0} \tag{10.279}$$

with $x_0 > 0$ and $v_{x0} > 0$ and let the initial conditions of the CoM in the lateral plane be expressed by

$$y_c(0) = -y_0 \tag{10.280}$$

$$\dot{y}_c(0) = v_{y0} \tag{10.281}$$

with $y_0 > 0$ and $v_{y0} > 0$. It follows from Lemma 10.2 that the following inequalities

$$v_{x0} > \sqrt{\frac{g}{h}} x_0 \tag{10.282}$$

$$v_{y0} < \sqrt{\frac{g}{h}} y_0 \tag{10.283}$$

must be held to ensure a stable walk.

Remark 10.19. Inequality (10.282) ensures that the CoM trajectory in the sagittal plane passes the stance leg to realize a locomotion, and (10.283) ensures that the CoM trajectory in the lateral plane does not go over the stance leg to keep the walking stability.

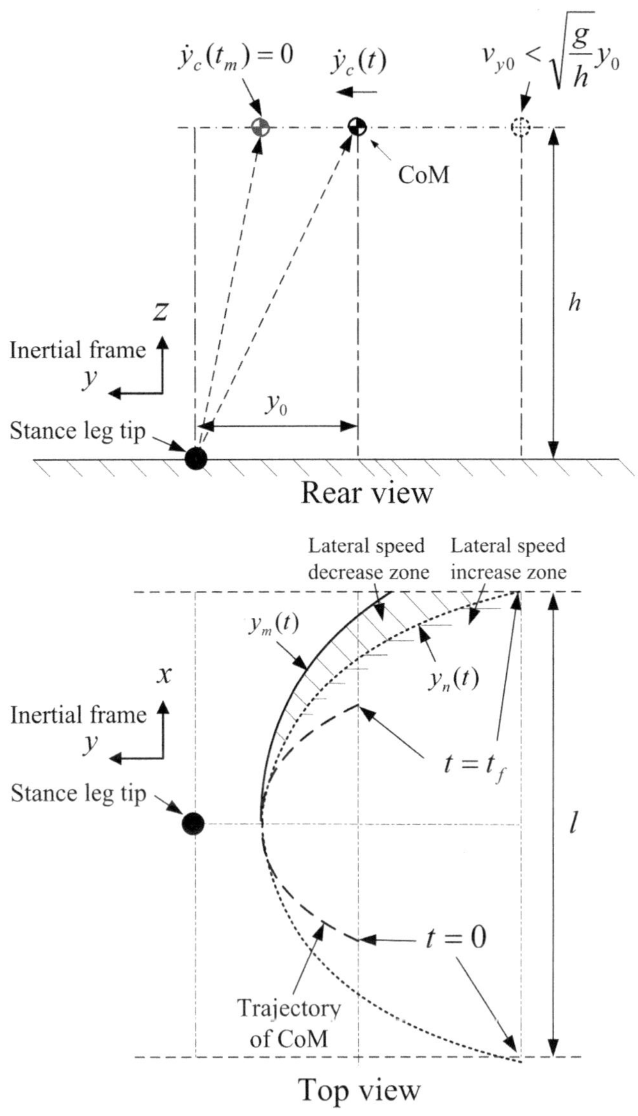

Fig. 10.8. Model of quasi-inverted-pendulum for lateral walking.

The CoM trajectory in both the sagittal and the lateral planes is described as

$$x_c(t) = c_{x1} e^{-\sqrt{\frac{g}{h}} t} + c_{x2} e^{\sqrt{\frac{g}{h}} t} \tag{10.284}$$

$$y_c(t) = c_{y1} e^{-\sqrt{\frac{g}{h}} t} + c_{y2} e^{\sqrt{\frac{g}{h}} t} \tag{10.285}$$

with

$$c_{x1} = -\frac{1}{2}\sqrt{\frac{h}{g}}\left(v_{x0} + \sqrt{\frac{g}{h}}x_0\right) \tag{10.286}$$

$$c_{x2} = \frac{1}{2}\sqrt{\frac{h}{g}}\left(v_{x0} - \sqrt{\frac{g}{h}}x_0\right) \tag{10.287}$$

$$c_{y1} = -\frac{1}{2}\sqrt{\frac{h}{g}}\left(v_{y0} + \sqrt{\frac{g}{h}}y_0\right) \tag{10.288}$$

$$c_{y2} = \frac{1}{2}\sqrt{\frac{h}{g}}\left(v_{y0} - \sqrt{\frac{g}{h}}y_0\right). \tag{10.289}$$

Substituting (10.282) and (10.283) into (10.286)–(10.289) yields

$$c_{x1} < 0 \tag{10.290}$$

$$c_{x2} > 0 \tag{10.291}$$

$$c_{y1} < 0 \tag{10.292}$$

$$c_{y2} < 0. \tag{10.293}$$

The trajectory of the swing leg needs to be planned in a way that once it becomes the stance leg, the lateral stability still maintains. Owing to the time-scaling factor, the actual step time may be slightly different from the planned step time. As a result, the desired trajectory of the swing leg in the lateral motion is set as a function of the CoM such that the lateral stability always maintains regardless of the exact time the leg switch occurs. The lateral marginal trajectory of the swing leg tip is defined by

$$\begin{aligned} y_m(t) &\stackrel{def}{=} y_c(t) + \sqrt{\frac{h}{g}}\dot{y}_c(t) \\ &= 2c_{y2}e^{\sqrt{\frac{g}{h}}t} \end{aligned} \tag{10.294}$$

for $t \geqslant t_m$, where t_m is the time instant when $\dot{y}_c(t)|_{t=t_m} = 0$ holds. A steady walking motion yields $t_m \approx \frac{t_f}{2}$.

Remark 10.20. The marginal lateral trajectory for the swing leg defined by (10.294) serves as the boundary for the swing leg to achieve a stable walk. If the swing leg does land on its marginal lateral trajectory, then the y coordinate of the CoM will eventually converge to the y coordinate of the landed swing leg that becomes the stance leg after a leg switch, which makes the biped marginally stable.

In view of (10.283), the nominal trajectory of the swing leg tip for the lateral motion is, therefore, defined by

$$y_n(t) \stackrel{def}{=} y_c(t) + \frac{y_0}{v_{y0}}\dot{y}_c(t). \tag{10.295}$$

Remark 10.21. Equation (10.295) yields $\frac{y_n(t)-y_c(t)}{\dot{y}_c(t)} = \frac{y_0}{v_{y0}}$, which implies that the ratio of the lateral distance between the stance leg and the CoM to the lateral velocity at the moment of leg switch is constant. Consequently, the step time determined by

$$t_s = \sqrt{\frac{h}{g}} ln \left(\frac{y_0 + \sqrt{\frac{h}{g}} v_{y0}}{y_0 - \sqrt{\frac{h}{g}} v_{y0}} \right) = \sqrt{\frac{h}{g}} ln \left(\frac{\frac{y_0}{v_{y0}} + \sqrt{\frac{h}{g}}}{\frac{y_0}{v_{y0}} - \sqrt{\frac{h}{g}}} \right) \tag{10.296}$$

is constant.

The maximum lateral speed can be adjusted by deviating the lateral trajectory of the swing leg tip from its nominal trajectory. If it is leaning toward the marginal trajectory, the maximum lateral speed is decreasing; if it is leaning away from the marginal trajectory, the maximum lateral speed is increasing.

10.8.3 Rotation Stability

When arms are used, a zero moment around the vertical axis of the stance leg can be achieved by properly controlling the movements of the arms as opposite to the movement of the swing leg.

10.9 Concluding Remarks

One important aspect in biped walking is the dynamics and control. Owing to the complexity of biped dynamics, the full-dynamics-based control is still way beyond reality. In this chapter, the VDC approach has been applied to biped robots for the first time. The VDC approach allows the use of the dynamics of subsystems (links or legs) to achieve full-dynamics-based control performances. Although the detailed analysis was focused on a two-dimensional biped, this approach can be straightforwardly applied to any three-dimensional (3D) biped robot.

In this chapter, special attentions have been paid to the trajectory planning that takes into account the underactuated motion associated with the ankle joint of the stance leg. The trajectory planning in a single-support phase starts from and ends with double-support phases. This planning technique allows the tip (ankle joint) of the swing leg to start from and end with a zero velocity, which prevents a stroke from happening under the asymptotic stability.

A quasi-inverted pendulum model developed in this chapter has played a central role for the motion planning. Conditions for keeping a continuous locomotion in the sagittal plane and for keeping a stable motion in the lateral plane have been given.

11

Control of Force-Reflected Bilateral Teleoperation

11.1 Introduction

A teleoperated system can extend human's reach to a remote site which is either hazard or impossible for humans to present. Examples include operations in nuclear facilities, in deep sea or deep space. Teleoperation systems can also enhance a human's capability to handle both the macro and the micro worlds by scaling up or scaling down operation tasks for easy and accurate operations. Examples include hydraulic excavation and eye/brain surgeries.

A typical teleoperation system is illustrated in Figure 11.1. It consists of a master manipulator, a slave manipulator, the human operator, the operated environment, and a communication network that connects the master and the slave. The goal of the control design is to let the slave faithfully track the motion of the master, while letting the human operator feel precisely the contact force at the remote site [217]. The forward motion channel plus the feedback force channel forms a force-reflected bilateral teleoperation system.

In this chapter, the *virtual decomposition control* (VDC) approach is applied to bilateral force-reflected teleoperation systems [217]. The master and slave robots are handled separately. Unlike the conventional two-port model-based approaches, this VDC approach uses the adaptive control framework and, therefore, possesses

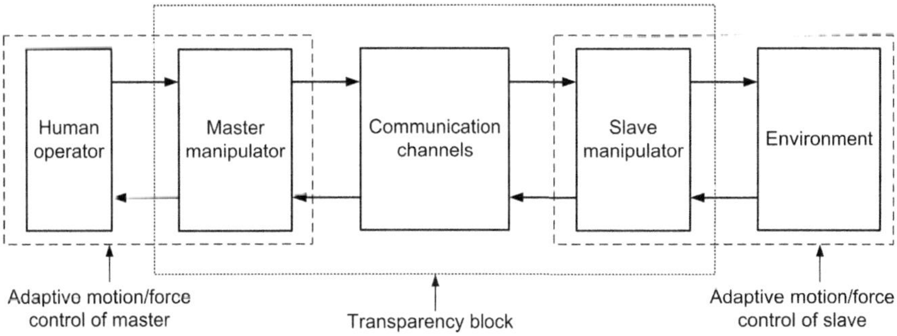

Fig. 11.1. A typical force-reflected bilateral teleoperation system.

W.-H. Zhu: Virtual Decomposition Control, STAR 60, pp. 303–324.
springerlink.com

the following advantages: (1) It allows arbitrary motion/force scaling in both position and rate modes. (2) It takes the full non-linear dynamics of the master and slave robots into account, rather than using the linearized models. (3) It is able to handle the parametric uncertainties of the master and slave robots by applying independent adaptive control. (4) It is stability guaranteed against any time-delay. (5) It allows rigid contact control between the slave robot and the environment. (6) The entire teleoperation system behaves as a mass and a damper specified by the control and scaling parameters. It means that the operator feels like he/she is operating a rigid tool with mass and damping to interact with the environment directly. (7) Most importantly, the asymptotic (position/velocity) motion tracking between the master and slave robots is guaranteed in both the free motion and the contact motion with an environment.

11.2 Prior Work

Teleoperation control has been extensively studied, motivated by a large variety of applications [148] ranging from nuclear operations and space exploration [12, 65, 70], forestry-related tasks [58], to medical applications [56].

Conventionally, a two-port model from circuit theory was used to conduct the system analysis for teleoperation systems. In [64], ideal transfer functions were given with respect to a two-port model. These ideal transfer functions were obtained by using an impractical cancellation of the complete mechanism impedances. This approach addressed more issues of control architecture than design and no stability or performance measures were given. An approach to control a single-master and multi-slave manipulator system by using a *virtual internal model* was proposed in [94]. The stability issue for bilateral teleoperation subject to time delays was first studied in [5]. In this approach, scattering operators were used to design the controller in such a way that the stability is guaranteed with strictly passive operators and environments. Although the asymptotic stability is ensured in [6], no motion/force tracking control of the master and slave robots has been properly achieved.

Without requiring the passivity for both the operator and the environment, an H_∞ controller with μ-synthesis was developed in [101]. This approach guarantees the stability for a pre-specified time-delay margin. Other approaches that used H_∞ optimization result in high-dimensional matrices with large computational burdens [83, 192].

The conflicting issue between stability and transparency was revealed in [98], and a unified four-channel control structure based on the complete cancellation of the master and slave robot dynamics was proposed. The perfect transparency with ideal kinesthetic feedback was reported in [198] provided that the master and slave manipulator dynamics are known and their accelerations can be measured.

In [36], an impedance-shaping control scheme was proposed to deal with different impedances and different geometric scales between the task and the human operator, and the robust stability criteria were discussed. To avert the stability issue caused by using force feedback in time-delayed teleoperation systems, visual force feedback with telemonitoring technique was proposed in [99]. This approach explored a trade-off between the robustness and the performance to achieve an optimization goal.

Applying the adaptive motion/force control approach that was originally developed for the constrained robot manipulators (as presented in Chapter 5) to force-reflected bilateral teleoperation was initialized in [217]. The master and slave robots are handled separately, subject to independent adaptive motion/force control. The states of a robot (master/slave) are sent to the other robot (slave/master) as motion/force tracking commands rather than as control actions. This approach ensures the asymptotic motion tracking between the master and slave robots in both free motion and contact tasks. Furthermore, the combination of the master and slave robots is equivalent to a pure mass plus a damper. The technical details are to be outlined in this chapter.

11.3 Dynamics and Control of the Slave Robot with Environment

The dynamics and control of the slave robot incorporating with an environment are presented in this section.

11.3.1 The Environment

Without loss of generality, assume that the environment is unilateral and is flexible in an $n_f \geqslant 0$ dimensional space and rigid in an $n_c \geqslant 0$ dimensional space with

$$n_f + n_c = 6.$$

In general, every environment possesses a certain amount of flexibility. To determine an environment as flexible or rigid is completely dependent on its mechanical impedance. If an environment possesses a comparable or lower mechanical impedance than that of a controlled robot, it is considered as a flexible environment. Soft springs and soft tissues are flexible environments. If an environment possesses a much higher mechanical impedance than that of a controlled robot, it is considered as a rigid environment. Table-mounted aluminium and steel may be considered as rigid environments in most cases.

Two binary selective factors σ_f and σ_c are defined by

$$\sigma_f = \begin{cases} 0 & \text{free motion} \\ 1 & \text{contact with flexible constraints} \end{cases}$$

$$\sigma_c = \begin{cases} 0 & \text{free motion} \\ 1 & \text{contact with rigid constraints.} \end{cases}$$

Let $\mathbf{f}_s = [\mathbf{f}_{sf}^T, \mathbf{f}_{sc}^T]^T \in \mathbb{R}^6$ be a vector of the independent coordinates representing the contact forces exerted from the slave robot toward the flexible/rigid environment, and let $\mathbf{v}_s = [\mathbf{v}_{sf}^T, \mathbf{v}_{sc}^T]^T \in \mathbb{R}^6$ be a vector of the independent coordinates representing the velocities of the slave robot at the contact point with the flexible/rigid environment, where $\mathbf{f}_{sf} \in \mathbb{R}^{n_f}$ and $\mathbf{v}_{sf} \in \mathbb{R}^{n_f}$ are associated with the flexible constraints and $\mathbf{f}_{sc} \in \mathbb{R}^{n_c}$ and $\mathbf{v}_{sc} \in \mathbb{R}^{n_c}$ are associated with the rigid constraints. The rigid contact dynamics are governed by

$$\mathbf{v}_{sc} = (1 - \sigma_c)\, \mathbf{v}_{sc} \tag{11.1}$$

$$\mathbf{f}_{sc} = \sigma_c\, \mathbf{f}_{sc} \tag{11.2}$$

whereas the flexible contact dynamics are governed by

$$\sigma_f [\mathbf{M}_f \ddot{\mathbf{x}}_{sf} + \mathbf{D}_f \dot{\mathbf{x}}_{sf} + \delta_z \mathbf{K}_f \mathbf{x}_{sf} + (1 - \delta_z)\varphi_f] = \mathbf{f}_{sf} \tag{11.3}$$

$$\dot{\mathbf{x}}_{sf} = \sigma_f \mathbf{v}_{sf} \tag{11.4}$$

where

$$\mathbf{v}_{sf} = \dot{\mathbf{x}}_{sf}$$

is used and

$$\delta_z \overset{def}{=} \begin{cases} 1 & \text{position–force control mode} \\ 0 & \text{rate–force control mode} \end{cases}$$

denotes a binary selective factor specifying either the position-force control mode or the rate-force control mode.[1] In (11.3), $\varphi_f \in \mathbb{R}^{n_f} \bigcap L_\infty$ is a bounded vector with a known structure and unknown parameters, $\mathbf{M}_f$, $\mathbf{D}_f$, and $\mathbf{K}_f$ are $n_f \times n_f$ time-invariant and symmetric positive-definite matrices associated with the inertia, damping, and stiffness of the flexible environment, respectively, and $\mathbf{x}_{sf}$ denotes the deformation of the flexible environment, that is, the difference between the current position/orientation and the position/orientation at the initial contact.

Remark 11.1. In (11.3), no stiffness term is assumed when the rate-force control mode is used. This is true for some applications such as soil excavation where the position-dependent resistant force in the soil sliding direction can be bounded. For other applications, if a stiffness term is necessarily involved with the rate-force control mode and the slave's working range is limited, then this stiffness term is bounded and can be included in φ_f.

11.3.2 The Slave Robot

Assume that the slave robot is a robot manipulator with six single-degree-of-freedom (single-DOF) joints as studied in Chapter 5. This robot manipulator

[1] In the position-force mode, the positions of the slave robot are proportional to the positions of the master robot, and the contact forces between the slave robot and the environment are fed back to the master robot. In the rate-force mode, the velocities of the slave robot are proportional to the positions of the master robot, and the contact forces between the slave robot and the environment are fed back to the master robot.

allows its end-effector to possess *six* degrees of freedom (DOFs) of motion without encountering kinematics singularity.

A 6×6 reordering matrix $[\mathbf{S}_f \ \ \mathbf{S}_c]$ is formed such that

$$\begin{bmatrix}\mathbf{S}_f & \mathbf{S}_c\end{bmatrix}^{-1} = \begin{bmatrix}\mathbf{S}_f^T \\ \mathbf{S}_c^T\end{bmatrix}$$

holds, where $\mathbf{S}_f \in \mathbb{R}^{6\times n_f}$ and $\mathbf{S}_c \in \mathbb{R}^{6\times n_c}$ contain unit, orthogonal, and constant vectors that span the flexible contact subspace and the rigid contact subspace, respectively.

Let frame $\{\mathbf{S}_s\}$ be located at the contact point with the environment and simultaneously fixed to the tool (payload) of the slave robot. It follows that

$$^{\mathbf{S}_s}F = \mathbf{T}_s^{-T}\begin{bmatrix}\mathbf{S}_f & \mathbf{S}_c\end{bmatrix}\begin{bmatrix}\sigma_f \mathbf{f}_{sf} \\ \sigma_c \mathbf{f}_{sc}\end{bmatrix} \tag{11.5}$$

$$^{\mathbf{S}_s}V = \mathbf{T}_s\begin{bmatrix}\mathbf{S}_f & \mathbf{S}_c\end{bmatrix}\begin{bmatrix}\mathbf{v}_{sf} \\ (1-\sigma_c)\mathbf{v}_{sc}\end{bmatrix} \tag{11.6}$$

hold, where $\mathbf{T}_s \in \mathbb{R}^{6\times 6}$ is an invertible transformation matrix that relates the units of $\mathbf{f}_s$ and $\mathbf{v}_s$ to the SI units of $^{\mathbf{S}_s}F$ and $^{\mathbf{S}_s}V$.

The remaining process for modeling the slave robot follows exactly the same procedure developed in Chapter 5. Noticeably, let frame $\{\mathbf{O}_s\}$ be attached to the tool. Let frame $\{\mathbf{C}_s\}$ be located at the *cutting point* between the tool and the slave manipulator and be fixed to both of them. It yields (5.19), (5.111), and (5.114) with frames $\{\mathbf{O}_s\}$, $\{\mathbf{S}_s\}$, and $\{\mathbf{C}_s\}$ being substituted for frames $\{\mathbf{O}\}$, $\{\mathbf{S}\}$, and $\{\mathbf{C}\}$, respectively, that is

$$^{\mathbf{O}_s}F^* = {}^{\mathbf{O}_s}\mathbf{U}_{\mathbf{C}_s}\,{}^{\mathbf{C}_s}F - {}^{\mathbf{O}_s}\mathbf{U}_{\mathbf{S}_s}\,{}^{\mathbf{S}_s}F \tag{11.7}$$

$$^{\mathbf{O}_s}V = {}^{\mathbf{C}_s}\mathbf{U}_{\mathbf{O}_s}^T\,{}^{\mathbf{C}_s}V = {}^{\mathbf{S}_s}\mathbf{U}_{\mathbf{O}_s}^T\,{}^{\mathbf{S}_s}V \tag{11.8}$$

where $^{\mathbf{O}_s}F^* \in \mathbb{R}^6$ is defined by (5.20) with frame $\{\mathbf{O}_s\}$ being substituted for frame $\{\mathbf{O}\}$.

Using the same procedure leading to (5.79) with $\mathcal{V} = {}^{\mathbf{O}_s}V \in \mathbb{R}^6$ and incorporating (11.7) yields the dynamics of the slave manipulator as

$$\mathcal{M}_s\frac{d}{dt}({}^{\mathbf{O}_s}V) + \mathcal{C}_s\,{}^{\mathbf{O}_s}V + \mathcal{G}_s = \mathbf{J}_{qs}^{-T}\boldsymbol{\tau}_s - {}^{\mathbf{O}_s}\mathbf{U}_{\mathbf{S}_s}\,{}^{\mathbf{S}_s}F \tag{11.9}$$

where $\mathcal{M}_s$, $\mathcal{C}_s$, $\mathcal{G}_s$, and $\mathbf{J}_{qs}$ have the same structures as $\mathcal{M}$, $\mathcal{C}$, $\mathcal{G}$, and $\mathbf{J}_q$ in (5.79) with appropriate frame substitutions. The last term in the right hand side of (11.9) is added to represent the contact forces exerted from the slave robot to the environment. Substituting (11.5), (11.6), and (11.8) into (11.9) yields

$$(\mathcal{T}_s^T\mathcal{M}_s\mathcal{T}_s)\dot{\mathbf{v}}_s + (\mathcal{T}_s^T\mathcal{C}_s\mathcal{T}_s + \mathcal{T}_s^T\mathcal{M}_s\dot{\mathcal{T}}_s)\mathbf{v}_s + \mathcal{T}_s^T\mathcal{G}_s = \mathcal{T}_s^T\mathbf{J}_{qs}^{-T}\boldsymbol{\tau}_s - \begin{bmatrix}\sigma_f\mathbf{f}_{sf} \\ \sigma_c\mathbf{f}_{sc}\end{bmatrix} \tag{11.10}$$

with

$$\begin{aligned}\mathbf{v}_s &= \begin{bmatrix}\mathbf{v}_{sf}^T, & (1-\sigma_c)\mathbf{v}_{sc}^T\end{bmatrix}^T \\ \mathcal{T}_s &= {}^{\mathbf{S}_s}\mathbf{U}_{\mathbf{O}_s}^T\mathbf{T}_s\begin{bmatrix}\mathbf{S}_f & \mathbf{S}_c\end{bmatrix}.\end{aligned} \tag{11.11}$$

11.3.3 Control and Stability

Let $\mathbf{f}_{sd} = [\mathbf{f}_{sfd}^T, \mathbf{f}_{scd}^T]^T \in \mathbb{R}^6$ be the desired vector of $\mathbf{f}_s$ with $\mathbf{f}_{sfd} \in \mathbb{R}^{n_f}$ and $\mathbf{f}_{scd} \in \mathbb{R}^{n_c}$ being two sub-vectors, and let $\mathbf{v}_{sd} = [\mathbf{v}_{sfd}^T, \mathbf{v}_{scd}^T]^T \in \mathbb{R}^6$ be the desired vector of $\mathbf{v}_s$ with $\mathbf{v}_{sfd} \in \mathbb{R}^{n_f}$ and $\mathbf{v}_{scd} \in \mathbb{R}^{n_c}$ being two sub-vectors.

Then, define a required vector $\mathbf{v}_{sr} = [\mathbf{v}_{sfr}^T, \mathbf{v}_{scr}^T]^T \in \mathbb{R}^6$ with $\mathbf{v}_{sfr} \in \mathbb{R}^{n_f}$ and $\mathbf{v}_{scr} \in \mathbb{R}^{n_c}$ being its two sub-vectors such that

$$\mathbf{v}_{sr} = \mathbf{v}_{sd} - \mathbf{A}\tilde{\mathbf{f}}_s \tag{11.12}$$

holds, where $\mathbf{A} = \mathrm{diag}\,(\mathbf{A}_f, \mathbf{A}_c) \in \mathbb{R}^{6\times 6}$ is a diagonal positive-definite matrix with $\mathbf{A}_f \in \mathbb{R}^{n_f\times n_f}$ and $\mathbf{A}_c \in \mathbb{R}^{n_c\times n_c}$ being its two sub-matrices and $\tilde{\mathbf{f}}_s = [\tilde{\mathbf{f}}_{sf}^T, \tilde{\mathbf{f}}_{sc}^T]^T \in \mathbb{R}^6$ (with $\tilde{\mathbf{f}}_{sf} \in \mathbb{R}^{n_f}$ and $\tilde{\mathbf{f}}_{sc} \in \mathbb{R}^{n_c}$ being its two sub-vectors) denotes a filtered contact force vector governed by

$$\dot{\tilde{\mathbf{f}}}_s + \mathbf{C}\tilde{\mathbf{f}}_s = \mathbf{C}\mathbf{f}_s \tag{11.13}$$

where $\mathbf{C} = \mathrm{diag}\,(\mathbf{C}_f, \mathbf{C}_c) \in \mathbb{R}^{6\times 6}$ is a diagonal positive-definite matrix with $\mathbf{C}_f \in \mathbb{R}^{n_f\times n_f}$ and $\mathbf{C}_c \in \mathbb{R}^{n_c\times n_c}$ being its two sub-matrices.

Remark 11.2. Incorporating a filtered contact force term into the required velocity vector to handle the rigid contact control of robot manipulators was originally reported in [213, 215, 218] and systematically presented in Chapter 5. In this chapter, it will be shown that incorporating a filtered contact force term into the required velocity vector of each (master/slave) robot guarantees the asymptotic motion tracking between the master and slave robots even in the contact directions. This result has not been reported before.

Based on the dynamics of the flexible environment described by (11.3), the term $\mathbf{f}_{sfd} \in \mathbb{R}^{n_f}$ in $\mathbf{f}_{sd} \in \mathbb{R}^6$ is designed as

$$\mathbf{f}_{sfd} = \mathbf{Y}_{sf}\hat{\boldsymbol{\theta}}_{sf} \tag{11.14}$$

where $\hat{\boldsymbol{\theta}}_{sf}$ denotes the estimate of $\boldsymbol{\theta}_{sf}$, subject to

$$\mathbf{Y}_{sf}\boldsymbol{\theta}_{sf} = \sigma_f[\mathbf{M}_f\dot{\mathbf{v}}_{sfr} + \mathbf{D}_f\dot{\mathbf{x}}_{sf} + \delta_z\mathbf{K}_f\mathbf{x}_{sf} + (1-\delta_z)\varphi_f]. \tag{11.15}$$

The parameter estimates need to be updated. Define

$$\mathbf{s}_{sf} = \sigma_f\mathbf{Y}_{sf}^T\left(\mathbf{v}_{sfr} - \mathbf{v}_{sf}\right). \tag{11.16}$$

The $\mathcal{P}$ function defined by (2.79) is used to update each of the elements of $\hat{\boldsymbol{\theta}}_{sf}$ as

$$\hat{\theta}_{sf\gamma} = \mathcal{P}\left(s_{sf\gamma}, \rho_{sf\gamma}, \underline{\theta}_{sf\gamma}, \overline{\theta}_{sf\gamma}, t\right) \tag{11.17}$$

where $\hat{\theta}_{sf\gamma}$ denotes the γth element of $\hat{\boldsymbol{\theta}}_{sf}$ and $s_{sf\gamma}$ denotes the γth element of $\mathbf{s}_{sf}$ defined by (11.16), $\rho_{sf\gamma} > 0$ is an update gain, and $\underline{\theta}_{sf\gamma}$ and $\overline{\theta}_{sf\gamma}$ denote the lower and upper bounds of $\theta_{sf\gamma}$—the γth element of $\boldsymbol{\theta}_{sf}$.

Meanwhile, the desired contact force vector associated with the rigid environment is designed to be a function of the corresponding desired velocity vector, that is

$$\mathbf{f}_{scd} = \sigma_c \mathbf{A}_c^{-1}(\mathbf{C}_c^{-1}\dot{\mathbf{v}}_{scd} + \mathbf{v}_{scd}). \tag{11.18}$$

After $\mathbf{v}_{sr} \in \mathbb{R}^6$ being obtained from (11.12) and $\mathbf{f}_{sd} \in \mathbb{R}^6$ being obtained from (11.14) and (11.18), the required force/moment vector in frame $\{\mathbf{S}_s\}$ and the required linear/angular velocity vector of frame $\{\mathbf{S}_s\}$ are computed as

$$^{\mathbf{S}_s}F_r = \mathbf{T}_s^{-T}\begin{bmatrix}\mathbf{S}_f & \mathbf{S}_c\end{bmatrix}\begin{bmatrix}\sigma_f \mathbf{f}_{sfd} \\ \sigma_c \mathbf{f}_{scd}\end{bmatrix} \tag{11.19}$$

$$^{\mathbf{S}_s}V_r = \mathbf{T}_s\begin{bmatrix}\mathbf{S}_f & \mathbf{S}_c\end{bmatrix}\begin{bmatrix}\mathbf{v}_{sfr} \\ \mathbf{v}_{scr}\end{bmatrix}. \tag{11.20}$$

Furthermore, $^{\mathbf{S}_s}F_r$ and $^{\mathbf{S}_s}V_r$ are related to the required net force/moment vector in frame $\{\mathbf{O}_s\}$ and to the required linear/angular velocity vector of frame $\{\mathbf{O}_s\}$ by

$$^{\mathbf{O}_s}F_r^* = {}^{\mathbf{O}_s}\mathbf{U}_{\mathbf{C}_s}\, {}^{\mathbf{C}_s}F_r - {}^{\mathbf{O}_s}\mathbf{U}_{\mathbf{S}_s}\, {}^{\mathbf{S}_s}F_r \tag{11.21}$$

$$^{\mathbf{O}_s}V_r = {}^{\mathbf{C}_s}\mathbf{U}_{\mathbf{O}_s}^T\, {}^{\mathbf{C}_s}V_r = {}^{\mathbf{S}_s}\mathbf{U}_{\mathbf{O}_s}^T\, {}^{\mathbf{S}_s}V_r \tag{11.22}$$

where $^{\mathbf{O}_s}F_r^*$ is defined by (5.23) with frame $\{\mathbf{O}_s\}$ being substituted for frame $\{\mathbf{O}\}$.

Lemma 11.1. *Consider the flexible/rigid environment described by (11.1)–(11.6) and combined with their respective control equations (11.12)–(11.20). It follows that*

$$\int_0^\infty p_{\mathbf{S}_s}(t)dt \geqslant -\gamma_s \tag{11.23}$$

holds with

$$p_{\mathbf{S}_s} \overset{def}{=} ({}^{\mathbf{S}_s}V_r - {}^{\mathbf{S}_s}V)^T({}^{\mathbf{S}_s}F_r - {}^{\mathbf{S}_s}F) \tag{11.24}$$

and $0 \leqslant \gamma_s < \infty$. ■

The proof is given in Appendix B.9.

Equations (11.7), (11.8), (11.21), and (11.22) validate

$$\left({}^{\mathbf{O}_s}V_r - {}^{\mathbf{O}_s}V\right)^T \left({}^{\mathbf{O}_s}F_r^* - {}^{\mathbf{O}_s}F^*\right) = p_{\mathbf{C}_s} - p_{\mathbf{S}_s}. \tag{11.25}$$

Equation (11.25) incorporating Lemma 11.1 allows Theorem 5.6 to be used with appropriate frame substitutions to ensure the *virtual stability* of the (payload) tool of the slave robot (combined with its respective control equations), in the sense of Definition 2.17. The control of the rest of the slave robot (links and joints) can follow exactly what has been developed in Chapter 5 to make sure that every subsystem of the slave robot is *virtually stable* in the sense of Definition 2.17. Consequently, it follows from Theorem 2.1 that

$$^{\mathbf{O}_s}V_r - {}^{\mathbf{O}_s}V \in L_2 \bigcap L_\infty \tag{11.26}$$

holds. This result further leads to

$$(\mathbf{v}_{sd} - \mathbf{v}_s - \mathbf{A}\tilde{\mathbf{f}}_s) \in L_2 \bigcap L_\infty \tag{11.27}$$

from (11.1), (11.6), (11.8), (11.12), (11.20), and (11.22).

Remark 11.3. Like the results in Chapter 5, (11.27) is valid only for constant σ_f and σ_c (either *one* or *zero*).

11.4 Dynamics and Control of the Master Robot with Human Operator

The dynamics and control of the master robot incorporating with the human operator are presented in this section.

11.4.1 The Human Operator

Based on the research in [38, 84], the following two-order LTI model is used for the human operator

$$\mathbf{M}_h\ddot{\mathbf{x}}_h + \mathbf{D}_h\dot{\mathbf{x}}_h + \mathbf{K}_h\mathbf{x}_h = \mathbf{f}_m - \mathbf{f}_h^* \tag{11.28}$$

where $\mathbf{M}_h$, $\mathbf{D}_h$, and $\mathbf{K}_h$ are three 6×6 time-invariant and symmetric positive-definite matrices associated with the inertia, damping, and stiffness of the human operator, $\mathbf{f}_m \in \mathbb{R}^6$ is the reaction force/moment vector from the master robot toward the human hand, and $\mathbf{f}_h^* \in \mathbb{R}^6$ denotes the exogenous force/moment vector generated by the operator, subject to

$$\|\mathbf{f}_h^*\|_\infty \leqslant \alpha_h < +\infty \tag{11.29}$$

with α_h being a positive constant. In (11.28), $\mathbf{x}_h = [\mathbf{p}_h^T,\ \boldsymbol{\mu}_h^T]^T \in \mathbb{R}^6$ denotes the position/orientation vector of the human operator, $\mathbf{p}_h \in \mathbb{R}^3$ is the linear position vector, and $\boldsymbol{\mu}_h \in \mathbb{R}^3$ is the vector part of a quaternion $[\xi_h,\ \boldsymbol{\mu}_h^T]^T \in \mathbb{R}^4$ representing the rotation from a given reference orientation to the current operator hand orientation.

Remark 11.4. The dynamics of a real human operator are very complicated and usually are posture-dependent, time-dependent, and subject-dependent. The LTI model (11.28) is only an approximation at certain operation points within certain parameter ranges. Further research will take into account the adaptive properties of the human operator with respect to the object control dynamics.

Remark 11.5. The quaternion expression is used to express the operator hand orientation. Since a rotation of π is not happening in most cases, this quaternion expression is assumed to be singularity-free.

Assume that a frame $\{\mathbf{S}_m\}$ is fixed to the handle of the master robot. The velocity transformation from $\dot{\mathbf{x}}_h$ to ${}^{\mathbf{S}_m}V \in \mathbb{R}^6$ (the linear/angular velocity vector of frame $\{\mathbf{S}_m\}$) can be expressed by

$$ {}^{\mathbf{S}_m}V = \mathbf{T}_h \dot{\mathbf{x}}_h \tag{11.30} $$

with

$$ \mathbf{T}_h = \begin{bmatrix} {}^{\mathbf{S}_m}\mathbf{R}_\mathbf{I} & \mathbf{0} \\ \mathbf{0} & \left[\frac{1}{2}[\xi_h \mathbf{I}_3 - (\boldsymbol{\mu}_h \times)]\right]^{-1} \end{bmatrix} \in \mathbb{R}^6 $$

where ${}^{\mathbf{S}_m}\mathbf{R}_\mathbf{I} \in \mathbb{R}^{3\times 3}$ is a rotation matrix that transforms a vector expressed in the inertial frame to the same vector expressed in frame $\{\mathbf{S}_m\}$. In $\mathbf{T}_h$, matrix $\frac{1}{2}[\xi_h \mathbf{I}_3 - (\boldsymbol{\mu}_h \times)]$ maps ${}^{\mathbf{S}_m}\boldsymbol{\omega} \in \mathbb{R}^3$ to $\dot{\boldsymbol{\mu}}_h \in \mathbb{R}^3$, in view of (2.49). Note that matrix $\mathbf{T}_h \in \mathbb{R}^{6\times 6}$ remains to be full-rank as long as $\xi_h \neq 0$. On the other hand, the duality on velocity and force transformations implies

$$ {}^{\mathbf{S}_m}F = \mathbf{T}_h^{-T} \mathbf{f}_m. \tag{11.31} $$

11.4.2 The Master Robot

Assume that the master robot is a robot manipulator with six single-DOF joints as studied in Chapter 5. This robot manipulator allows its end-effector to possess *six* DOFs of motion without encountering kinematics singularity.

Let frames $\{\mathbf{O}_m\}$ and $\{\mathbf{C}_m\}$ be attached to the handle of the master robot with frame $\{\mathbf{C}_m\}$ being located at a *cutting point* that virtually isolates the handle from the master robot. It follows that

$$ {}^{\mathbf{O}_m}F^* = {}^{\mathbf{O}_m}\mathbf{U}_{\mathbf{C}_m} {}^{\mathbf{C}_m}F - {}^{\mathbf{O}_m}\mathbf{U}_{\mathbf{S}_m} {}^{\mathbf{S}_m}F \tag{11.32} $$

$$ {}^{\mathbf{O}_m}V = {}^{\mathbf{C}_m}\mathbf{U}_{\mathbf{O}_m}^T {}^{\mathbf{C}_m}V = {}^{\mathbf{S}_m}\mathbf{U}_{\mathbf{O}_m}^T {}^{\mathbf{S}_m}V \tag{11.33} $$

hold, where ${}^{\mathbf{O}_m}F^* \in \mathbb{R}^6$ is defined by (5.20) with frame $\{\mathbf{O}_m\}$ being substituted for frame $\{\mathbf{O}\}$.

Meanwhile, by setting

$$ \mathbf{x}_m = \mathbf{x}_h \tag{11.34} $$

$$ \mathbf{v}_m = \dot{\mathbf{x}}_h \tag{11.35} $$

$$ \dot{\mathbf{v}}_m = \ddot{\mathbf{x}}_h \tag{11.36} $$

and following an approach similar to that leads to (11.10), the dynamic equation of the master robot incorporating the human operator can be obtained as

$$ (\mathcal{T}_m^T \mathcal{M}_m \mathcal{T}_m)\dot{\mathbf{v}}_m + (\mathcal{T}_m^T \mathcal{C}_m \mathcal{T}_m + \mathcal{T}_m^T \mathcal{M}_m \dot{\mathcal{T}}_m)\mathbf{v}_m + \mathcal{T}_m^T \mathcal{G}_m = \mathcal{T}_m^T \mathbf{J}_{qm}^{-T} \boldsymbol{\tau}_m - \mathbf{f}_m \tag{11.37} $$

with

$$ \mathcal{T}_m = {}^{\mathbf{S}_m}\mathbf{U}_{\mathbf{O}_m}^T \mathbf{T}_h \tag{11.38} $$

where $\mathcal{M}_m$, $\mathcal{C}_m$, $\mathcal{G}_m$, and $\mathbf{J}_{qm}$ have the same structures as $\mathcal{M}$, $\mathcal{C}$, $\mathcal{G}$, and $\mathbf{J}_q$ in (5.79) with appropriate frame substitutions and $\boldsymbol{\tau}_m \in \mathbb{R}^6$ is the joint control force/torque vector.

11.4.3 Control and Stability

Let $\mathbf{v}_{md} \in \mathbb{R}^6$ be the desired vector of $\mathbf{v}_m \in \mathbb{R}^6$, and let $\mathbf{v}_{mr} \in \mathbb{R}^6$ be the required vector of $\mathbf{v}_m \in \mathbb{R}^6$, subject to

$$\mathbf{v}_{mr} = \mathbf{v}_{md} - \mathbf{A}\tilde{\mathbf{f}}_m \tag{11.39}$$

where $\mathbf{A} \in \mathbb{R}^{6\times 6}$ is the same diagonal positive-definite matrix as defined in (11.12) and $\tilde{\mathbf{f}}_m \in \mathbb{R}^6$ denotes a filtered contact force vector governed by

$$\dot{\tilde{\mathbf{f}}}_m + \mathbf{C}\tilde{\mathbf{f}}_m = \mathbf{C}\mathbf{f}_m \tag{11.40}$$

with $\mathbf{C} \in \mathbb{R}^{6\times 6}$ being the same diagonal positive-definite matrix as defined in (11.13).

The desired contact force vector from the master robot handle to the operator is designed as

$$\mathbf{f}_{md} = \mathbf{Y}_m\hat{\boldsymbol{\theta}}_m + \alpha_h \text{sign}(\mathbf{v}_{mr} - \mathbf{v}_m) \tag{11.41}$$

subject to

$$\mathbf{Y}_m\boldsymbol{\theta}_m \stackrel{def}{=} \mathbf{M}_h\dot{\mathbf{v}}_{mr} + \mathbf{D}_h\dot{\mathbf{x}}_h + \mathbf{K}_h\mathbf{x}_h. \tag{11.42}$$

Remark 11.6. Equation (11.41) comprises two parts. The first part, denoted as $\mathbf{Y}_m\hat{\boldsymbol{\theta}}_m$, represents the estimated dynamics of the operator hand. The second part is a velocity feedback term.

The parameter estimates need to be updated. Define

$$\mathbf{s}_m = \mathbf{Y}_m^T(\mathbf{v}_{mr} - \mathbf{v}_m). \tag{11.43}$$

The $\mathcal{P}$ function defined by (2.79) is used to update each of the elements of $\hat{\boldsymbol{\theta}}_m$ as

$$\hat{\theta}_{m\gamma} = \mathcal{P}\left(s_{m\gamma}, \rho_{m\gamma}, \underline{\theta}_{m\gamma}, \overline{\theta}_{m\gamma}, t\right) \tag{11.44}$$

where $\hat{\theta}_{m\gamma}$ denotes the γth element of $\hat{\boldsymbol{\theta}}_m$ and $s_{m\gamma}$ denotes the γth element of $\mathbf{s}_m$ defined by (11.43), $\rho_{m\gamma} > 0$ is an update gain, and $\underline{\theta}_{m\gamma}$ and $\overline{\theta}_{m\gamma}$ denote the lower and upper bounds of $\theta_{m\gamma}$—the γth element of $\boldsymbol{\theta}_m$.

After $\mathbf{v}_{mr}$ and $\mathbf{f}_{md}$ being obtained from (11.39) and (11.41), respectively, vectors ${}^{\mathbf{S}_m}V_r$ and ${}^{\mathbf{S}_m}F_r$ are computed as

$$ {}^{\mathbf{S}_m}V_r = \mathbf{T}_h\mathbf{v}_{mr} \tag{11.45}$$

$$ {}^{\mathbf{S}_m}F_r = \mathbf{T}_h^{-T}\mathbf{f}_{md}. \tag{11.46}$$

Both ${}^{\mathbf{S}_m}F_r$ and ${}^{\mathbf{S}_m}V_r$ are related to the required net force/moment vector in frame $\{\mathbf{O}_m\}$ and to the required linear/angular velocity vector of frame $\{\mathbf{O}_m\}$ by

$$ {}^{\mathbf{O}_m}F_r^* = {}^{\mathbf{O}_m}\mathbf{U}_{\mathbf{C}_m}\,{}^{\mathbf{C}_m}F_r - {}^{\mathbf{O}_m}\mathbf{U}_{\mathbf{S}_m}\,{}^{\mathbf{S}_m}F_r \tag{11.47}$$

$$ {}^{\mathbf{O}_m}V_r = {}^{\mathbf{C}_m}\mathbf{U}_{\mathbf{O}_m}^T\,{}^{\mathbf{C}_m}V_r = {}^{\mathbf{S}_m}\mathbf{U}_{\mathbf{O}_m}^T\,{}^{\mathbf{S}_m}V_r \tag{11.48}$$

where ${}^{\mathbf{O}_m}F_r^*$ is defined by (5.23) with frame $\{\mathbf{O}_m\}$ being substituted for frame $\{\mathbf{O}\}$.

Lemma 11.2. *Consider the human operator described by (11.28)–(11.31) and (11.34)–(11.36) and combined with its respective control equations (11.41)–(11.46). It follows that*

$$\int_0^\infty p_{\mathbf{S}_m}(t)dt \geqslant -\gamma_m \tag{11.49}$$

holds with

$$p_{\mathbf{S}_m} \overset{def}{=} ({}^{\mathbf{S}_m}V_r - {}^{\mathbf{S}_m}V)^T({}^{\mathbf{S}_m}F_r - {}^{\mathbf{S}_m}F) \tag{11.50}$$

and $0 \leqslant \gamma_m < \infty$. ■

The proof is given in Appendix B.10.

Equations (11.32), (11.33), (11.47), and (11.48) validate

$$\left({}^{\mathbf{O}_m}V_r - {}^{\mathbf{O}_m}V\right)^T \left({}^{\mathbf{O}_m}F_r^* - {}^{\mathbf{O}_m}F^*\right) = p_{\mathbf{C}_m} - p_{\mathbf{S}_m}. \tag{11.51}$$

Equation (11.51) incorporating Lemma 11.2 allows Theorem 5.6 to be used with appropriate frame substitutions to ensure the *virtual stability* of the handle (the held object) of the master robot (combined with its respective control equations), in the sense of Definition 2.17. The control of the rest of the master robot (links and joints) follows exactly what has been developed in Chapter 5 to make sure that every subsystem of the master robot is *virtually stable* in the sense of Definition 2.17. It follows from Theorem 2.1 that

$${}^{\mathbf{O}_m}V_r - {}^{\mathbf{O}_m}V \in L_2 \bigcap L_\infty \tag{11.52}$$

holds. This result further leads to

$$(\mathbf{v}_{md} - \mathbf{v}_m - \mathbf{A}\tilde{\mathbf{f}}_m) \in L_2 \bigcap L_\infty \tag{11.53}$$

from (11.30), (11.33), (11.35), (11.39), (11.45), and (11.48).

Finally, re-write (11.27) and (11.53) as

$$\boldsymbol{\rho}_s \overset{def}{=} \mathbf{v}_{sd} - \mathbf{v}_s - \mathbf{A}\tilde{\mathbf{f}}_s \quad \in \quad L_2 \bigcap L_\infty \tag{11.54}$$

$$\boldsymbol{\rho}_m \overset{def}{=} \mathbf{v}_{md} - \mathbf{v}_m - \mathbf{A}\tilde{\mathbf{f}}_m \quad \in \quad L_2 \bigcap L_\infty. \tag{11.55}$$

These two expressions form the foundation for the force-reflected bilateral teleoperation design to develop in the next section.

11.5 Adaptive Bilateral Teleoperation Control

After the adaptive motion and force control for both the master and slave robots being presented independently, this section presents the bilateral teleoperation design with respect to the communication block aimed at creating the desired/command velocity signals for both the master and slave robots.

11.5.1 Teleoperation Control Design

In a bilateral motion/force teleoperation system, let $\boldsymbol{\kappa}_p \in \mathbb{R}^{6\times 6}$ and $\boldsymbol{\kappa}_f \in \mathbb{R}^{6\times 6}$ be two diagonal matrices for the motion scaling and the force scaling, respectively, from the master robot toward the slave robot. Based on (11.54) and (11.55), the two design vectors $\mathbf{v}_{sd} \in \mathbb{R}^6$ and $\mathbf{v}_{md} \in \mathbb{R}^6$ are specified as

$$\mathbf{v}_{sd} = \boldsymbol{\kappa}_p \left(\tilde{\mathbf{v}}_m + \boldsymbol{\Lambda}\tilde{\mathbf{p}}_m\right) - \delta_z \boldsymbol{\Lambda}\mathbf{p}_s - \mathbf{A}\boldsymbol{\kappa}_f \tilde{\mathbf{f}}_m \tag{11.56}$$

$$\mathbf{v}_{md} = \boldsymbol{\kappa}_p^{-1} \left\{\tilde{\mathbf{v}}_s + \delta_z \boldsymbol{\Lambda}\tilde{\mathbf{p}}_s - \boldsymbol{\kappa}_p \boldsymbol{\Lambda}\mathbf{p}_m - \mathbf{A}\left[\tilde{\mathbf{f}}_s + (\boldsymbol{\kappa}_f - \boldsymbol{\kappa}_p)\,\tilde{\mathbf{f}}_m\right]\right\} \tag{11.57}$$

where δ_z is a binary selective factor between the position-force control mode ($\delta_z = 1$) and the rate-force control mode ($\delta_z = 0$), $\boldsymbol{\Lambda} \in \mathbb{R}^{6\times 6}$ is a diagonal positive-definite matrix, and $\mathbf{p}_m$ and $\mathbf{p}_s$ denote the position/orientation vectors of the master and slave robots, respectively, subject to $\dot{\mathbf{p}}_m = \mathbf{v}_m$ and $\dot{\mathbf{p}}_s = \mathbf{v}_s$. Let $\mathbf{z} \in \{\mathbf{v}_m,\ \mathbf{v}_s,\ \mathbf{p}_m,\ \mathbf{p}_s\}$ represent any one of the four vectors, and $\tilde{\mathbf{z}}$ is obtained by using a first order filter

$$\dot{\tilde{\mathbf{z}}} + \mathbf{C}\tilde{\mathbf{z}} = \mathbf{C}\mathbf{z} \tag{11.58}$$

where $\mathbf{C} \in \mathbb{R}^6$ denotes the same diagonal positive-definite matrix as defined in (11.13).

Remark 11.7. The teleoperation design equations (11.56) and (11.57) make $\dot{\mathbf{v}}_{sd}$ and $\dot{\mathbf{v}}_{md}$ the functions of $\mathbf{v}_m$, $\mathbf{v}_s$, $\mathbf{f}_m$, and $\mathbf{f}_s$, in terms of (11.13), (11.40), and (11.58). Therefore, no acceleration measurement is required.

Remark 11.8. Note that $\mathbf{p}_s$ may involve Euler parameters (quaternion). In order to avoid the unique singularity point corresponding to a rotation of π, the rate-force control mode should be employed for large rotations in the free motion.

Remark 11.9. The teleoperation controller design is simpler than these in many other approaches. The only design variables include the two velocity commanding vectors $\mathbf{v}_{md}$ and $\mathbf{v}_{sd}$, the two scaling parameter matrices $\boldsymbol{\kappa}_p$ and $\boldsymbol{\kappa}_f$, and the three control parameter matrices $\mathbf{A}$, $\mathbf{C}$, and $\boldsymbol{\Lambda}$.

Substitute (11.56) and (11.57) into (11.54) and (11.55) by using $\dot{\mathbf{p}}_m = \mathbf{v}_m$ and $\dot{\mathbf{p}}_s = \mathbf{v}_s$. Then, doing summation and subtraction yields

$$\begin{aligned}\boldsymbol{\rho}_s - \boldsymbol{\kappa}_p\boldsymbol{\rho}_m &= \boldsymbol{\kappa}_p\tilde{\mathbf{v}}_m - \tilde{\mathbf{v}}_s + \boldsymbol{\Lambda}\left[\boldsymbol{\kappa}_p\tilde{\mathbf{p}}_m - \delta_z\tilde{\mathbf{p}}_s\right] \\ &\quad + \boldsymbol{\kappa}_p\mathbf{v}_m - \mathbf{v}_s + \boldsymbol{\Lambda}\left[\boldsymbol{\kappa}_p\mathbf{p}_m - \delta_z\mathbf{p}_s\right]\end{aligned} \tag{11.59}$$

$$\begin{aligned}\boldsymbol{\rho}_s + \boldsymbol{\kappa}_p\boldsymbol{\rho}_m &= \boldsymbol{\kappa}_p(\tilde{\mathbf{v}}_m - \mathbf{v}_m) + \tilde{\mathbf{v}}_s - \mathbf{v}_s \\ &\quad + \boldsymbol{\Lambda}\boldsymbol{\kappa}_p(\tilde{\mathbf{p}}_m - \mathbf{p}_m) + \delta_z\boldsymbol{\Lambda}(\tilde{\mathbf{p}}_s - \mathbf{p}_s) - 2\mathbf{A}(\tilde{\mathbf{f}}_s + \boldsymbol{\kappa}_f\tilde{\mathbf{f}}_m) \\ &= -\mathbf{C}^{-1}\left(\boldsymbol{\kappa}_p\dot{\tilde{\mathbf{v}}}_m + \dot{\tilde{\mathbf{v}}}_s + \boldsymbol{\Lambda}\boldsymbol{\kappa}_p\tilde{\mathbf{v}}_m + \delta_z\boldsymbol{\Lambda}\tilde{\mathbf{v}}_s\right) \\ &\quad - 2\mathbf{A}(\tilde{\mathbf{f}}_s + \boldsymbol{\kappa}_f\tilde{\mathbf{f}}_m).\end{aligned} \tag{11.60}$$

Remark 11.10. Equations (11.59) and (11.60) are the two key equations in this chapter. Equation (11.59) will be used to ensure the asymptotic motion tracking between the master and slave robots. Equation (11.60) will be used to characterize the transparency of the entire teleoperation system.

11.5.2 L_2 and L_∞ Stability of Master/Slave Motion Tracking

Equation (11.59) can be re-written as

$$\tilde{\mathbf{e}} + \mathbf{e} = \boldsymbol{\rho}_s - \boldsymbol{\kappa}_p \boldsymbol{\rho}_m \tag{11.61}$$

with

$$\mathbf{e} \stackrel{def}{=} \boldsymbol{\kappa}_p \mathbf{v}_m - \mathbf{v}_s + \boldsymbol{\Lambda}\left(\boldsymbol{\kappa}_p \mathbf{p}_m - \delta_z \mathbf{p}_s\right) \tag{11.62}$$

subject to

$$\dot{\tilde{\mathbf{e}}} + \mathbf{C}\tilde{\mathbf{e}} = \mathbf{C}\mathbf{e} \tag{11.63}$$

where $\mathbf{C} \in \mathbb{R}^6$ denotes the same diagonal positive-definite matrix as defined in (11.13).

Substituting (11.63) into (11.61) and using Lemma 2.6 yields $\dot{\tilde{\mathbf{e}}} \in L_2 \bigcap L_\infty$, $\tilde{\mathbf{e}} \in L_2 \bigcap L_\infty$, and further

$$\mathbf{e} \in L_2 \bigcap L_\infty. \tag{11.64}$$

When the position-force control mode ($\delta_z = 1$) is used, it follows from (11.62), (11.64), and Lemma 2.6 that

$$\boldsymbol{\rho}_v \stackrel{def}{=} \boldsymbol{\kappa}_p \mathbf{v}_m - \mathbf{v}_s \in L_2 \bigcap L_\infty \tag{11.65}$$

$$\boldsymbol{\rho}_p \stackrel{def}{=} \boldsymbol{\kappa}_p \mathbf{p}_m - \mathbf{p}_s \in L_2 \bigcap L_\infty \tag{11.66}$$

hold. These two equations guarantee that the velocity and position errors in both the free motion and the contact motion with a flexible/rigid environment are L_2 and L_∞ signals.

When the rate-force control ($\delta_z = 0$) mode is used, it follows from (11.62) that

$$\boldsymbol{\rho}_r \stackrel{def}{=} \boldsymbol{\kappa}_p(\mathbf{v}_m + \boldsymbol{\Lambda}\mathbf{p}_m) - \mathbf{v}_s \in L_2 \bigcap L_\infty \tag{11.67}$$

holds. This equation guarantees that the error vector between the slave velocity $\mathbf{v}_s$ and a combination of the master velocity and position (denoted as $\boldsymbol{\kappa}_p(\mathbf{v}_m + \boldsymbol{\Lambda}\mathbf{p}_m)$) belongs to L_2 and L_∞.

Expressions (11.65)–(11.67) give the motion control results which are valid in both the free motion and the contact motion with a flexible/rigid environment.

Substituting (11.65)–(11.67) into (11.60) yields

$$-\frac{1}{2}\boldsymbol{\rho} = \mathbf{A}^{-1}\mathbf{C}^{-1}(s\mathbf{I} + \boldsymbol{\Lambda})\boldsymbol{\kappa}_p \tilde{\mathbf{v}}_m + (\tilde{\mathbf{f}}_s + \boldsymbol{\kappa}_f \tilde{\mathbf{f}}_m) \tag{11.68}$$

with

$$\boldsymbol{\rho} \stackrel{def}{=} \mathbf{A}^{-1}\left[-\delta_z \mathbf{C}^{-1}(s\mathbf{I} + \boldsymbol{\Lambda})\tilde{\boldsymbol{\rho}}_v - (1 - \delta_z)\mathbf{C}^{-1} s \tilde{\boldsymbol{\rho}}_r + (\boldsymbol{\rho}_s + \boldsymbol{\kappa}_p \boldsymbol{\rho}_m)\right] \tag{11.69}$$

where s denotes the Laplace operator and $\tilde{\boldsymbol{\rho}}_v$ and $\tilde{\boldsymbol{\rho}}_r$ are governed by

$$\dot{\tilde{\boldsymbol{\rho}}}_v + \mathbf{C}\tilde{\boldsymbol{\rho}}_v = \mathbf{C}\boldsymbol{\rho}_v$$
$$\dot{\tilde{\boldsymbol{\rho}}}_r + \mathbf{C}\tilde{\boldsymbol{\rho}}_r = \mathbf{C}\boldsymbol{\rho}_r$$

with $\boldsymbol{\rho}_v$ and $\boldsymbol{\rho}_r$ being defined by (11.65) and (11.67), respectively, and $\mathbf{C} \in \mathbb{R}^6$ being defined in (11.13). It further follows from (11.54), (11.55), (11.65), (11.67), and Lemma 2.6 that

$$\boldsymbol{\rho} \in L_2 \bigcap L_\infty \tag{11.70}$$

holds.

Remark 11.11. Equation (11.68) will be used below to characterize the transparency of the entire teleoperation system, after being derived from (11.60).

11.5.3 Stability of the Entire Teleoperation System

Passivity theory is commonly used to ensure the overall stability of a teleoperation system. The concept is to make sure that the complete teleoperation system is energy dissipating. In this subsection, a different approach is used. It will be shown that the master and slave velocities are bounded for bounded exogenous forces/moments $\mathbf{f}_h^* \in \mathbb{R}^6$. Four cases are studied as follows.

Case 1: When the slave robot is in the free motion with $\sigma_f = 0$ and $\sigma_c = 0$, it yields $\mathbf{f}_s = \mathbf{0}$. It follows from (11.28), (11.34)–(11.36), and (11.68) that

$$\boldsymbol{\rho}_a = \mathbf{M}_a \dot{\tilde{\mathbf{v}}}_m + \mathbf{D}_a \tilde{\mathbf{v}}_m + \mathbf{K}_a \tilde{\mathbf{p}}_m \tag{11.71}$$

holds, where

$$\boldsymbol{\rho}_a \stackrel{def}{=} -\frac{1}{2}\boldsymbol{\rho} - \boldsymbol{\kappa}_f \tilde{\mathbf{f}}_h^*$$
$$\mathbf{M}_a \stackrel{def}{=} \mathbf{A}^{-1}\mathbf{C}^{-1}\boldsymbol{\kappa}_p + \boldsymbol{\kappa}_f \mathbf{M}_h$$
$$\mathbf{D}_a \stackrel{def}{=} \mathbf{A}^{-1}\mathbf{C}^{-1}\boldsymbol{\Lambda}\boldsymbol{\kappa}_p + \boldsymbol{\kappa}_f \mathbf{D}_h$$
$$\mathbf{K}_a \stackrel{def}{=} \boldsymbol{\kappa}_f \mathbf{K}_h$$

with $\frac{d}{dt}\tilde{\mathbf{f}}_h^* + \mathbf{C}\tilde{\mathbf{f}}_h^* = \mathbf{C}\mathbf{f}_h^*$. In term of Lemma 2.7, it follows from $\boldsymbol{\rho}_a \in L_\infty$ that $\tilde{\mathbf{p}}_m \in L_\infty$, $\tilde{\mathbf{v}}_m \in L_\infty$, and $\dot{\tilde{\mathbf{v}}}_m \in L_\infty$ hold, leading to

$$\mathbf{v}_m \in L_\infty \tag{11.72}$$
$$\mathbf{p}_m \in L_\infty \tag{11.73}$$
$$\tilde{\mathbf{f}}_m \in L_\infty \tag{11.74}$$
$$\mathbf{v}_s \in L_\infty. \tag{11.75}$$

Case 2: When the slave robot is in contact with the flexible environment with $\sigma_f = 1$ and $\sigma_c = 0$, it follows from (11.65) and (11.67) that

$$\mathbf{v}_s = \boldsymbol{\kappa}_p \mathbf{v}_m - \delta_z \boldsymbol{\rho}_v + (1 - \delta_z)(\boldsymbol{\Lambda}\boldsymbol{\kappa}_p \mathbf{p}_m - \boldsymbol{\rho}_r) \tag{11.76}$$

holds. Then, it follows from (11.3), (11.65)–(11.68), and (11.76) that

$$\boldsymbol{\rho}_b = \mathbf{M}_b \dot{\tilde{\mathbf{v}}}_m + \mathbf{D}_b \tilde{\mathbf{v}}_m + \mathbf{K}_b \tilde{\mathbf{p}}_m \tag{11.77}$$

holds, where

$$\begin{aligned}
\boldsymbol{\rho}_b &\stackrel{def}{=} \boldsymbol{\rho}_a + \mathrm{diag}\,(\mathbf{M}_f, \mathbf{0})\,[\delta_z \dot{\tilde{\boldsymbol{\rho}}}_v + (1-\delta_z)\,\dot{\tilde{\boldsymbol{\rho}}}_r] \\
&\quad + \mathrm{diag}\,(\mathbf{D}_f, \mathbf{0})\,[\delta_z \tilde{\boldsymbol{\rho}}_v + (1-\delta_z)\tilde{\boldsymbol{\rho}}_r] \\
&\quad + \mathrm{diag}\,(\mathbf{K}_f, \mathbf{0})\,\delta_z \tilde{\boldsymbol{\rho}}_p - (1-\delta_z)[\tilde{\boldsymbol{\varphi}}_f^T, \mathbf{0}^T]^T \\
\mathbf{M}_b &\stackrel{def}{=} \mathbf{M}_a + \mathrm{diag}\,(\mathbf{M}_f, \mathbf{0})\,\boldsymbol{\kappa}_p \\
\mathbf{D}_b &\stackrel{def}{=} \mathbf{D}_a + (1-\delta_z)\mathrm{diag}\,(\mathbf{M}_f, \mathbf{0})\,\boldsymbol{\Lambda}\boldsymbol{\kappa}_p + \mathrm{diag}\,(\mathbf{D}_f, \mathbf{0})\,\boldsymbol{\kappa}_p \\
\mathbf{K}_b &\stackrel{def}{=} \mathbf{K}_a + (1-\delta_z)\mathrm{diag}\,(\mathbf{D}_f, \mathbf{0})\,\boldsymbol{\Lambda}\boldsymbol{\kappa}_p + \delta_z \mathrm{diag}\,(\mathbf{K}_f, \mathbf{0})\,\boldsymbol{\kappa}_p
\end{aligned}$$

with

$$\dot{\tilde{\mathbf{z}}}_\rho + \mathbf{C}\tilde{\mathbf{z}}_\rho = \mathbf{C}\mathbf{z}_\rho$$

for $\mathbf{z}_\rho \in \{\boldsymbol{\rho}_v,\ \boldsymbol{\rho}_r,\ \boldsymbol{\rho}_p,\ [\boldsymbol{\varphi}_f^T, \mathbf{0}^T]^T\}$ and $\mathbf{C} \in \mathbb{R}^6$ being defined in (11.13).

Note that $\boldsymbol{\rho}_b \in L_\infty$ holds. It can be checked that (11.72)-(11.75) are still valid. Furthermore, it follows from (11.68) and (11.70) that

$$\tilde{\mathbf{f}}_s \in L_\infty \tag{11.78}$$

holds.

Case 3: When the slave robot is in contact with the rigid environment with $\sigma_f = 0$ and $\sigma_c = 1$, it yields $\mathbf{v}_{sc} = [\mathbf{0}_{n_c \times n_f},\ \mathbf{I}_{n_c}]\mathbf{v}_s = \mathbf{0}$ and $[\mathbf{0}_{n_c \times n_f},\ \mathbf{I}_{n_c}]\mathbf{p}_s \in L_\infty$. In view of (11.65)–(11.67), it follows that $[\mathbf{0}_{n_c \times n_f},\ \mathbf{I}_{n_c}]\mathbf{v}_m \in L_\infty$ and $[\mathbf{0}_{n_c \times n_f},\ \mathbf{I}_{n_c}]\mathbf{p}_m \in L_\infty$ hold. By following the process of case 1 and reducing the number of dimensions from 6 to n_f, $[\mathbf{I}_{n_f},\ \mathbf{0}_{n_f \times n_c}]\mathbf{v}_m \in L_\infty$ and $[\mathbf{I}_{n_f},\ \mathbf{0}_{n_f \times n_c}]\mathbf{p}_m \in L_\infty$ can be obtained. This yields $\tilde{\mathbf{f}}_m \in L_\infty$ from (11.28) and (11.36) and $\mathbf{v}_s \in L_\infty$ from (11.65) and (11.67). Furthermore, it follows from (11.68) that $\tilde{\mathbf{f}}_s \in L_\infty$ holds. Thus, (11.72)–(11.75) and (11.78) hold.

Case 4: When the slave robot is in contact with the flexible and rigid environment with $\sigma_f = 1$ and $\sigma_c = 1$, results (11.72)–(11.75) and (11.78) are still valid by following the processes in cases 2 and 3.

Remark 11.12. Expressions (11.72)–(11.75) and (11.78) hold for constant σ_f and σ_c.

11.5.4 Algebraic Loop Issue

When the joint torque control is used, it follows from Lemma 5.3 that the joint control vector of the master robot $\boldsymbol{\tau}_m \in \mathbb{R}^6$ is an affine function of $\frac{d}{dt}({}^{\mathbf{O}_m}V_r) \in \mathbb{R}^6$, that is

$$\mathbf{J}_{qm}^{-T}\boldsymbol{\tau}_m = \hat{\mathcal{M}}_m \frac{d}{dt}({}^{\mathbf{O}_m}V_r) + {}^{\mathbf{O}_m}\mathbf{U}_{\mathbf{S}_m}\,{}^{\mathbf{S}_m}F_r + \mathbf{b}_{\tau m} \tag{11.79}$$

where $\hat{\mathcal{M}}_m \in \mathbb{R}^{6\times 6}$ denotes the estimated inertial matrix of the master robot (see (5.90) for its data structure with appropriate frame/subscript substitutions) and

$\mathbf{b}_{\tau m} \in \mathbb{R}^6$ is a state-dependent vector. In view of (11.1)–(11.3), (11.13), (11.28), (11.36), (11.38)–(11.42), (11.45), (11.46), (11.48), and (11.57), premultiplying (11.79) by $\mathcal{T}_m^T$ yields

$$\begin{aligned}
\mathcal{T}_m^T \mathbf{J}_{qm}^{-T} \boldsymbol{\tau}_m &= (\mathcal{T}_m^T \hat{\mathcal{M}}_m \mathcal{T}_m)\dot{\mathbf{v}}_{mr} + \mathcal{T}_m^T {}^{\mathbf{O}_m}\mathbf{U}_{\mathbf{S}_m} {}^{\mathbf{S}_m}F_r + \mathbf{b}_{m1} \\
&= (\mathcal{T}_m^T \hat{\mathcal{M}}_m \mathcal{T}_m)\dot{\mathbf{v}}_{mr} + \mathbf{f}_{md} + \mathbf{b}_{m2} \\
&= (\mathcal{T}_m^T \hat{\mathcal{M}}_m \mathcal{T}_m + \hat{\mathbf{M}}_h)(\dot{\mathbf{v}}_{md} - \mathbf{A}\dot{\tilde{\mathbf{f}}}_m) + \mathbf{b}_{m3} \\
&= -(\mathcal{T}_m^T \hat{\mathcal{M}}_m \mathcal{T}_m + \hat{\mathbf{M}}_h)\boldsymbol{\kappa}_p^{-1}\mathbf{A}(\boldsymbol{\kappa}_f \dot{\tilde{\mathbf{f}}}_m + \dot{\tilde{\mathbf{f}}}_s) + \mathbf{b}_{m4} \\
&= -(\mathcal{T}_m^T \hat{\mathcal{M}}_m \mathcal{T}_m + \hat{\mathbf{M}}_h)\boldsymbol{\kappa}_p^{-1}\mathbf{A}\mathbf{C}(\boldsymbol{\kappa}_f \mathbf{f}_m + \mathbf{f}_s) + \mathbf{b}_{m5} \\
&= -\mathbf{A}_m^* \mathbf{A}\mathbf{C}\dot{\mathbf{v}}_m - \mathbf{A}_{ms}^* \mathbf{A}\mathbf{C}\begin{bmatrix} \dot{\mathbf{v}}_{sf} \\ (1-\sigma_c)\dot{\mathbf{v}}_{sc} + \sigma_c \mathbf{f}_{sc} \end{bmatrix} + \mathbf{b}_{m6}
\end{aligned} \tag{11.80}$$

with

$$\mathbf{A}_m^* \stackrel{def}{=} (\mathcal{T}_m^T \hat{\mathcal{M}}_m \mathcal{T}_m + \hat{\mathbf{M}}_h)\boldsymbol{\kappa}_p^{-1}\boldsymbol{\kappa}_f \mathbf{M}_h \in \mathbb{R}^{6\times 6}$$
$$\mathbf{A}_{ms}^* \stackrel{def}{=} (\mathcal{T}_m^T \hat{\mathcal{M}}_m \mathcal{T}_m + \hat{\mathbf{M}}_h)\boldsymbol{\kappa}_p^{-1}\begin{bmatrix} \sigma_f \mathbf{M}_f & \mathbf{0} \\ \mathbf{0} & \sigma_c \mathbf{I}_{n_c} \end{bmatrix} \in \mathbb{R}^{6\times 6}$$

where $\hat{\mathbf{M}}_h \in \mathbb{R}^{6\times 6}$ denotes $\mathbf{M}_h \in \mathbb{R}^{6\times 6}$ using estimated parameters and $\mathbf{b}_{m1} \in \mathbb{R}^6$, $\mathbf{b}_{m2} \in \mathbb{R}^6$, $\mathbf{b}_{m3} \in \mathbb{R}^6$, $\mathbf{b}_{m4} \in \mathbb{R}^6$, $\mathbf{b}_{m5} \in \mathbb{R}^6$, and $\mathbf{b}_{m6} \in \mathbb{R}^6$ are six state-dependent vectors.

With the same procedure, the joint control vector of the slave robot $\boldsymbol{\tau}_s \in \mathbb{R}^6$ can be expressed as an affine function of $\frac{d}{dt}({}^{\mathbf{O}_s}V_r) \in \mathbb{R}^6$, that is

$$\mathbf{J}_{qs}^{-T}\boldsymbol{\tau}_s = \hat{\mathcal{M}}_s \frac{d}{dt}({}^{\mathbf{O}_s}V_r) + {}^{\mathbf{O}_s}\mathbf{U}_{\mathbf{S}_s} {}^{\mathbf{S}_s}F_r + \mathbf{b}_{\tau s} \tag{11.81}$$

where $\hat{\mathcal{M}}_s \in \mathbb{R}^{6\times 6}$ denotes the estimated inertial matrix of the slave robot (see (5.90) for its data structure with appropriate frame/subscript substitutions) and $\mathbf{b}_{\tau s} \in \mathbb{R}^6$ is a state-dependent vector. In view of (11.1)–(11.3), (11.11)–(11.15), (11.18)–(11.20), (11.22), (11.28), (11.40), and (11.56), premultiplying (11.81) by $\mathcal{T}_s^T$ yields

$$\begin{aligned}
\mathcal{T}_s^T \mathbf{J}_{qs}^{-T}\boldsymbol{\tau}_s &= (\mathcal{T}_s^T \hat{\mathcal{M}}_s \mathcal{T}_s)\dot{\mathbf{v}}_{sr} + \mathcal{T}_s^T {}^{\mathbf{O}_s}\mathbf{U}_{\mathbf{S}_s} {}^{\mathbf{S}_s}F_r + \mathbf{b}_{s1} \\
&= (\mathcal{T}_s^T \hat{\mathcal{M}}_s \mathcal{T}_s)\dot{\mathbf{v}}_{sr} + \begin{bmatrix} \sigma_f \mathbf{f}_{sfd} \\ \sigma_c \mathbf{f}_{scd} \end{bmatrix} + \mathbf{b}_{s2} \\
&= \left[\mathcal{T}_s^T \hat{\mathcal{M}}_s \mathcal{T}_s + \mathrm{diag}(\sigma_f \hat{\mathbf{M}}_f, \mathbf{0})\right]\dot{\mathbf{v}}_{sr} + [\mathbf{0}^T,\ \sigma_c \mathbf{f}_{scd}^T]^T + \mathbf{b}_{s3} \\
&= \left[\mathcal{T}_s^T \hat{\mathcal{M}}_s \mathcal{T}_s + \mathrm{diag}(\sigma_f \hat{\mathbf{M}}_f, \mathbf{0})\right](\dot{\mathbf{v}}_{sd} - \mathbf{A}\dot{\tilde{\mathbf{f}}}_s) \\
&\quad + [\mathbf{0}^T,\ \sigma_c \mathbf{A}_c^{-1}\mathbf{C}_c^{-1}\dot{\mathbf{v}}_{scd}^T]^T + \mathbf{b}_{s4} \\
&= \left[\mathcal{T}_s^T \hat{\mathcal{M}}_s \mathcal{T}_s + \mathrm{diag}(\sigma_f \hat{\mathbf{M}}_f, \sigma_c \mathbf{A}_c^{-1}\mathbf{C}_c^{-1})\right]\dot{\mathbf{v}}_{sd}
\end{aligned}$$

$$
\begin{aligned}
& - \left[\mathcal{T}_s^T \hat{\mathcal{M}}_s \mathcal{T}_s + \mathrm{diag}(\sigma_f \hat{\mathbf{M}}_f, \mathbf{0})\right] \mathbf{A}\dot{\hat{\mathbf{f}}}_s + \mathbf{b}_{s4} \\
= & - \left[\mathcal{T}_s^T \hat{\mathcal{M}}_s \mathcal{T}_s + \mathrm{diag}(\sigma_f \hat{\mathbf{M}}_f, \sigma_c \mathbf{A}_c^{-1}\mathbf{C}_c^{-1})\right] \mathbf{A}\boldsymbol{\kappa}_f \dot{\hat{\mathbf{f}}}_m \\
& - \left[\mathcal{T}_s^T \hat{\mathcal{M}}_s \mathcal{T}_s + \mathrm{diag}(\sigma_f \hat{\mathbf{M}}_f, \mathbf{0})\right] \mathbf{A}\dot{\hat{\mathbf{f}}}_s + \mathbf{b}_{s5} \\
= & - \left[\mathcal{T}_s^T \hat{\mathcal{M}}_s \mathcal{T}_s + \mathrm{diag}(\sigma_f \hat{\mathbf{M}}_f, \sigma_c \mathbf{A}_c^{-1}\mathbf{C}_c^{-1})\right] \boldsymbol{\kappa}_f \mathbf{A}\mathbf{C}\mathbf{f}_m \\
& - \left[\mathcal{T}_s^T \hat{\mathcal{M}}_s \mathcal{T}_s + \mathrm{diag}(\sigma_f \hat{\mathbf{M}}_f, \mathbf{0})\right] \mathbf{A}\mathbf{C}\mathbf{f}_s + \mathbf{b}_{s6} \\
= & -\mathbf{A}_{sm}^* \mathbf{A}\mathbf{C}\dot{\mathbf{v}}_m - \mathbf{A}_s^* \mathbf{A}\mathbf{C} \begin{bmatrix} \dot{\mathbf{v}}_{sf} \\ (1-\sigma_c)\dot{\mathbf{v}}_{sc} + \sigma_c \mathbf{f}_{sc} \end{bmatrix} + \mathbf{b}_{s7}
\end{aligned}
\tag{11.82}
$$

with

$$
\begin{aligned}
\mathbf{A}_{sm}^* &\stackrel{def}{=} \left[\mathcal{T}_s^T \hat{\mathcal{M}}_s \mathcal{T}_s + \mathrm{diag}(\sigma_f \hat{\mathbf{M}}_f, \sigma_c \mathbf{A}_c^{-1}\mathbf{C}_c^{-1})\right] \boldsymbol{\kappa}_f \mathbf{M}_h \in \mathbb{R}^{6\times 6} \\
\mathbf{A}_s^* &\stackrel{def}{=} \left[\mathcal{T}_s^T \hat{\mathcal{M}}_s \mathcal{T}_s + \mathrm{diag}(\sigma_f \hat{\mathbf{M}}_f, \mathbf{0})\right] \begin{bmatrix} \sigma_f \mathbf{M}_f & \mathbf{0} \\ \mathbf{0} & \sigma_c \mathbf{I}_{n_c} \end{bmatrix} \in \mathbb{R}^{6\times 6}
\end{aligned}
$$

where $\hat{\mathbf{M}}_f \in \mathbb{R}^{n_f \times n_f}$ denotes $\mathbf{M}_f \in \mathbb{R}^{n_f \times n_f}$ using estimated parameters and $\mathbf{b}_{s1} \in \mathbb{R}^6$, $\mathbf{b}_{s2} \in \mathbb{R}^6$, $\mathbf{b}_{s3} \in \mathbb{R}^6$, $\mathbf{b}_{s4} \in \mathbb{R}^6$, $\mathbf{b}_{s5} \in \mathbb{R}^6$, $\mathbf{b}_{s6} \in \mathbb{R}^6$, and $\mathbf{b}_{s7} \in \mathbb{R}^6$ are seven state-dependent vectors.

After $\mathbf{f}_m \in \mathbb{R}^6$ being moved from the right-hand side to the left-hand side of (11.37) and $\mathbf{f}_s \in \mathbb{R}^6$ being moved from the right-hand side to the left-hand side of (11.10), substituting (11.80) and (11.82) into (11.37) and (11.10), respectively, and using (11.1)–(11.3), (11.28), and (11.36) yields

$$
\begin{aligned}
& \begin{bmatrix} \mathbf{A}_m & \mathbf{0} \\ \mathbf{0} & \mathbf{A}_s \end{bmatrix} \begin{bmatrix} \dot{\mathbf{v}}_m \\ \begin{bmatrix} \dot{\mathbf{v}}_{sf} \\ (1-\sigma_c)\dot{\mathbf{v}}_{sc} + \sigma_c \mathbf{f}_{sc} \end{bmatrix} \end{bmatrix} \\
& = - \begin{bmatrix} \mathbf{A}_m^* & \mathbf{A}_{ms}^* \\ \mathbf{A}_{sm}^* & \mathbf{A}_s^* \end{bmatrix} \mathrm{diag}(\mathbf{A}\mathbf{C}, \mathbf{A}\mathbf{C}) \begin{bmatrix} \dot{\mathbf{v}}_m \\ \begin{bmatrix} \dot{\mathbf{v}}_{sf} \\ (1-\sigma_c)\dot{\mathbf{v}}_{sc} + \sigma_c \mathbf{f}_{sc} \end{bmatrix} \end{bmatrix} \\
& \quad + \begin{bmatrix} \mathbf{b}_m \\ \mathbf{b}_s \end{bmatrix}
\end{aligned}
\tag{11.83}
$$

where

$$
\begin{aligned}
\mathbf{A}_m &\stackrel{def}{=} (\mathcal{T}_m^T \mathcal{M}_m \mathcal{T}_m + \mathbf{M}_h) \in \mathbb{R}^{6\times 6} \\
\mathbf{A}_s &\stackrel{def}{=} (\mathcal{T}_s^T \mathcal{M}_s \mathcal{T}_s) \begin{bmatrix} \mathbf{I}_{n_f} & \mathbf{0} \\ \mathbf{0} & (1-\sigma_c)\mathbf{I}_{n_c} \end{bmatrix} + \begin{bmatrix} \sigma_f \mathbf{M}_f & \\ & \sigma_c \mathbf{I}_{n_c} \end{bmatrix} \in \mathbb{R}^{6\times 6}
\end{aligned}
$$

and $\mathbf{b}_m \in \mathbb{R}^6$ and $\mathbf{b}_s \in \mathbb{R}^6$ are two state-dependent vectors.

Thus, an algebraic loop about

$$
\begin{bmatrix} \dot{\mathbf{v}}_m \\ \begin{bmatrix} \dot{\mathbf{v}}_{sf} \\ (1-\sigma_c)\dot{\mathbf{v}}_{sc} + \sigma_c \mathbf{f}_{sc} \end{bmatrix} \end{bmatrix}
$$

is formed in the sense of Definition 2.18 with

$$\mathbf{K}(t) = -\begin{bmatrix} \mathbf{A}_m^{-1} & \mathbf{0} \\ \mathbf{0} & \mathbf{A}_s^{-1} \end{bmatrix} \begin{bmatrix} \mathbf{A}_m^* & \mathbf{A}_{ms}^* \\ \mathbf{A}_{sm}^* & \mathbf{A}_s^* \end{bmatrix} \mathrm{diag}(\mathbf{AC}, \mathbf{AC}). \tag{11.84}$$

To make (2.115) valid, the norm of $\mathbf{AC}$ must be limited by

$$\|\mathbf{AC}\| \leqslant \sigma_t \tag{11.85}$$

with $\sigma_t > 0$ being a positive constant determined by the maximum singular value of $\begin{bmatrix} \mathbf{A}_m^{-1} & \mathbf{0} \\ \mathbf{0} & \mathbf{A}_s^{-1} \end{bmatrix} \begin{bmatrix} \mathbf{A}_m^* & \mathbf{A}_{ms}^* \\ \mathbf{A}_{sm}^* & \mathbf{A}_s^* \end{bmatrix}$ in (11.84).

11.5.5 Asymptotic Stability

When the stability of the algebraic loop is guaranteed by (11.85), it follows that $\dot{\mathbf{v}}_m \in L_\infty$, $\dot{\mathbf{v}}_s \in L_\infty$, and $\mathbf{f}_{sc} \in L_\infty$ hold, leading to $\mathbf{f}_m \in L_\infty$ and $\mathbf{f}_s \in L_\infty$. Consequently, the boundedness of $\dot{\mathbf{v}}_{sd}$ and $\dot{\mathbf{v}}_{md}$ can be ensured from (11.13), (11.40), and (11.56)–(11.58), leading to $\dot{\boldsymbol{\rho}}_s \in L_\infty$ and $\dot{\boldsymbol{\rho}}_m \in L_\infty$ from (11.54) and (11.55). This result implies that $\boldsymbol{\rho}_s$ and $\boldsymbol{\rho}_m$ defined by (11.54) and (11.55) are uniformly continuous functions in both L_2 and L_∞, leading to

$$\boldsymbol{\rho}_s \to \mathbf{0} \tag{11.86}$$

$$\boldsymbol{\rho}_m \to \mathbf{0} \tag{11.87}$$

from Lemma 2.8 for constant σ_f and σ_c. Consequently, it follows that

$$\boldsymbol{\rho}_v = \boldsymbol{\kappa}_p \mathbf{v}_m - \mathbf{v}_s \to \mathbf{0} \tag{11.88}$$

$$\boldsymbol{\rho}_p = \boldsymbol{\kappa}_p \mathbf{p}_m - \mathbf{p}_s \to \mathbf{0} \tag{11.89}$$

hold in the position-force control mode and

$$\boldsymbol{\rho}_r = \boldsymbol{\kappa}_p(\mathbf{v}_m + \boldsymbol{\Lambda}\mathbf{p}_m) - \mathbf{v}_s \to \mathbf{0} \tag{11.90}$$

holds in the rate-force control mode.

Remark 11.13. The developed bilateral teleoperation control ensures the asymptotic position control in the sense of (11.89) and the asymptotic rate control in the sense of (11.90) for both the free motion and the contact motion with a flexible/rigid environment. This result distinguishes this developed teleoperation control approach from other approaches in which the asymptotic position tracking control usually can not be ensured in the contact directions.

11.5.6 Transparency

To study the transparency of the complete teleoperation system, re-write (11.68) as

$$-\tilde{\mathbf{f}}_m = \boldsymbol{\kappa}_f^{-1}\tilde{\mathbf{f}}_s + \boldsymbol{\kappa}_f^{-1}\boldsymbol{\kappa}_p \mathbf{A}^{-1}\mathbf{C}^{-1}(s\mathbf{I} + \boldsymbol{\Lambda})\tilde{\mathbf{v}}_m + \frac{1}{2}\boldsymbol{\kappa}_f^{-1}\boldsymbol{\rho}. \tag{11.91}$$

Remark 11.14. Equation (11.91) summarizes the transparency properties of the complete teleoperation system. Within a certain frequency range bounded by $\underline{\sigma}(\mathbf{C})/2\pi$, it results in $\tilde{\mathbf{f}}_m \approx \mathbf{f}_m$, $\tilde{\mathbf{f}}_s \approx \mathbf{f}_s$, and $\tilde{\mathbf{v}}_m \approx \mathbf{v}_m$. Meanwhile, $\boldsymbol{\rho} \to \mathbf{0}$ holds, in view of (11.69), (11.86)–(11.88), and (11.90). Note that the term in the left hand side denotes the force vector exerted from the operator toward the master robot. The first term in the right hand side represents the contribution of the operator force to the task execution. The second term indicates that the teleoperation system behaves as a free-floating mass $\boldsymbol{\kappa}_f^{-1}\boldsymbol{\kappa}_p\mathbf{A}^{-1}\mathbf{C}^{-1}$ plus a linear damper $\boldsymbol{\kappa}_f^{-1}\boldsymbol{\kappa}_p\mathbf{A}^{-1}\mathbf{C}^{-1}\boldsymbol{\Lambda}$. Contributing to the transparency error, the equivalent mass and damper are completely specified by the control and scaling parameters rather than by the system parameters. In order to reduce the transparency error, large $\mathbf{A}$ and $\mathbf{C}$ are desirable, subject to (11.85) to maintain the stability of the algebraic loop specified by (11.83). In fact, the choice of $\mathbf{A}$ and $\mathbf{C}$ specifies the trade-off between the stability and the transparency. Given $\mathbf{A}$ and $\mathbf{C}$, the transparency error can be further reduced by using small motion scaling factor $\boldsymbol{\kappa}_p$ and large force scaling factor $\boldsymbol{\kappa}_f$, subject to task specifications.

11.6 Time Delays

When time delays present in the communication channels between the master robot and the slave robot, (11.56) and (11.57) have to be modified to

$$\mathbf{v}_{sd} = e^{-sT}\boldsymbol{\kappa}_p(\tilde{\mathbf{v}}_m + \boldsymbol{\Lambda}\tilde{\mathbf{p}}_m) - \delta_z\boldsymbol{\Lambda}\mathbf{p}_s - e^{-sT}\mathbf{A}\boldsymbol{\kappa}_f\tilde{\mathbf{f}}_m \tag{11.92}$$

$$\mathbf{v}_{md} = \boldsymbol{\kappa}_p^{-1}\left\{e^{-sT}(\tilde{\mathbf{v}}_s + \delta_z\boldsymbol{\Lambda}\tilde{\mathbf{p}}_s) - \boldsymbol{\kappa}_p\boldsymbol{\Lambda}\mathbf{p}_m - \mathbf{A}[e^{-sT}\tilde{\mathbf{f}}_s + (\boldsymbol{\kappa}_f - \boldsymbol{\kappa}_p)\tilde{\mathbf{f}}_m]\right\} \tag{11.93}$$

where T denotes the one-way time delay. Substituting (11.92) and (11.93) into (11.54) and (11.55) yields

$$\mathbf{v}_s + \delta_z\boldsymbol{\Lambda}\mathbf{p}_s + \mathbf{A}\tilde{\mathbf{f}}_s = e^{-sT}[\boldsymbol{\kappa}_p(\tilde{\mathbf{v}}_m + \boldsymbol{\Lambda}\tilde{\mathbf{p}}_m) - \mathbf{A}\boldsymbol{\kappa}_f\tilde{\mathbf{f}}_m] - \boldsymbol{\rho}_s \tag{11.94}$$

$$\boldsymbol{\kappa}_p(\mathbf{v}_m + \boldsymbol{\Lambda}\mathbf{p}_m) + \mathbf{A}\boldsymbol{\kappa}_f\tilde{\mathbf{f}}_m = e^{-sT}(\tilde{\mathbf{v}}_s + \delta_z\boldsymbol{\Lambda}\tilde{\mathbf{p}}_s - \mathbf{A}\tilde{\mathbf{f}}_s) - \boldsymbol{\kappa}_p\boldsymbol{\rho}_m. \tag{11.95}$$

Assume that

$$\mathbf{f}_s(s) = \mathbf{Z}_e(s)\mathbf{v}_s(s) \tag{11.96}$$

$$\mathbf{f}_m(s) = \mathbf{Z}_h(s)\mathbf{v}_m(s) + \mathbf{f}_h^*(s) \tag{11.97}$$

hold, that is, no rigid contact is considered. It follows from (11.94) and (11.95) that

$$\mathbf{G}_{1s}(s)\tilde{\mathbf{p}}_s(s) = \mathbf{G}_{2s}(s)e^{-sT}\boldsymbol{\kappa}_p\tilde{\mathbf{p}}_m(s) - \boldsymbol{\rho}_{s1}(s) \tag{11.98}$$

$$\mathbf{G}_{1m}(s)\boldsymbol{\kappa}_p\tilde{\mathbf{p}}_m(s) = \mathbf{G}_{2m}(s)e^{-sT}\tilde{\mathbf{p}}_s(s) - \boldsymbol{\kappa}_p\boldsymbol{\rho}_{m1}(s) \tag{11.99}$$

hold, where

$$\boldsymbol{\rho}_{s1}(s) \stackrel{def}{=} \boldsymbol{\rho}_s(s) + e^{-sT}\mathbf{A}\boldsymbol{\kappa}_f\tilde{\mathbf{f}}_h^*(s) \tag{11.100}$$

$$\boldsymbol{\rho}_{m1}(s) \stackrel{def}{=} \boldsymbol{\rho}_m(s) + \mathbf{A}\boldsymbol{\kappa}_f\boldsymbol{\kappa}_p^{-1}\tilde{\mathbf{f}}_h^*(s) \tag{11.101}$$

$$\mathbf{G}_{1s}(s) \stackrel{def}{=} (s\mathbf{I} + \delta_z\boldsymbol{\Lambda})(\mathbf{C}^{-1}s + \mathbf{I}_6) + s\mathbf{A}\mathbf{Z}_e(s) \tag{11.102}$$

$$\mathbf{G}_{2s}(s) \stackrel{def}{=} (s\mathbf{I} + \boldsymbol{\Lambda})\mathbf{I}_6 - s\mathbf{A}\boldsymbol{\kappa}_f\boldsymbol{\kappa}_p^{-1}\mathbf{Z}_h(s) \tag{11.103}$$

$$\mathbf{G}_{1m}(s) \stackrel{def}{=} (s\mathbf{I} + \boldsymbol{\Lambda})(\mathbf{C}^{-1}s + \mathbf{I}_6) + s\mathbf{A}\boldsymbol{\kappa}_f\boldsymbol{\kappa}_p^{-1}\mathbf{Z}_h(s) \tag{11.104}$$

$$\mathbf{G}_{2m}(s) \stackrel{def}{=} (s\mathbf{I} + \delta_z\boldsymbol{\Lambda}) - s\mathbf{A}\mathbf{Z}_e(s). \tag{11.105}$$

Combining (11.98) and (11.99) yields

$$\tilde{\mathbf{p}}_s(s) = e^{-2sT}\mathbf{G}_{1s}(s)^{-1}\mathbf{G}_{2s}(s)\mathbf{G}_{1m}(s)^{-1}\mathbf{G}_{2m}(s)\tilde{\mathbf{p}}_s(s) - \boldsymbol{\rho}_{s2}(s) \tag{11.106}$$

$$\boldsymbol{\kappa}_p\tilde{\mathbf{p}}_m(s) = e^{-2sT}\mathbf{G}_{1m}(s)^{-1}\mathbf{G}_{2m}(s)\mathbf{G}_{1s}(s)^{-1}\mathbf{G}_{2s}(s)\boldsymbol{\kappa}_p\tilde{\mathbf{p}}_m(s) - \boldsymbol{\rho}_{m2}(s) \tag{11.107}$$

with

$$\boldsymbol{\rho}_{s2}(s) = \mathbf{G}_{1s}(s)^{-1}\boldsymbol{\rho}_{s1}(s) + e^{-sT}\mathbf{G}_{1s}(s)^{-1}\mathbf{G}_{2s}(s)\mathbf{G}_{1m}(s)^{-1}\boldsymbol{\kappa}_p\boldsymbol{\rho}_{m1}(s)$$
$$\boldsymbol{\rho}_{m2}(s) = \mathbf{G}_{1m}(s)^{-1}\boldsymbol{\kappa}_p\boldsymbol{\rho}_{m1}(s) + e^{-sT}\mathbf{G}_{1m}(s)^{-1}\mathbf{G}_{2m}(s)\mathbf{G}_{1s}(s)^{-1}\boldsymbol{\rho}_{s1}(s).$$

The sufficient condition to ensure the stability subject to an arbitrary time delay $T > 0$ is

$$\begin{cases} \|\mathbf{G}_{1s}(j\omega)^{-1}\mathbf{G}_{2s}(j\omega)\mathbf{G}_{1m}(j\omega)^{-1}\mathbf{G}_{2m}(j\omega)\|_\infty < 1 \\ \|\mathbf{G}_{1m}(j\omega)^{-1}\mathbf{G}_{2m}(j\omega)\mathbf{G}_{1s}(j\omega)^{-1}\mathbf{G}_{2s}(j\omega)\|_\infty < 1. \end{cases} \tag{11.108}$$

Two examples are given below to demonstrate how (11.108) can be satisfied.

The first example concerns both the master and slave robots in the free motion with $\mathbf{f}_h^* = \mathbf{f}_m = \mathbf{f}_s = \mathbf{0}$ and, therefore, $\mathbf{Z}_h = \mathbf{Z}_e = \mathbf{0}$ holds. It follows from (11.102)–(11.105) and (11.108) that

$$\begin{aligned} &\|\mathbf{G}_{1s}(j\omega)^{-1}\mathbf{G}_{2s}(j\omega)\mathbf{G}_{1m}(j\omega)^{-1}\mathbf{G}_{2m}(j\omega)\|_\infty \\ &= \|\mathbf{G}_{1m}(j\omega)^{-1}\mathbf{G}_{2m}(j\omega)\mathbf{G}_{1s}(j\omega)^{-1}\mathbf{G}_{2s}(j\omega)\|_\infty \\ &= \|(j\omega\mathbf{I}_6 + \mathbf{C})^{-1}\mathbf{C}(j\omega\mathbf{I}_6 + \mathbf{C})^{-1}\mathbf{C}\|_\infty \\ &< 1 \end{aligned} \tag{11.109}$$

holds with $\mathbf{C}$ being the same diagonal positive-definite matrix as defined in (11.13).

The second example is a one-dimensional case with $\mathbf{A} = a$, $\mathbf{C} = c$, $\boldsymbol{\Lambda} = \lambda$, $\boldsymbol{\kappa}_p = \kappa_p$, $\boldsymbol{\kappa}_f = \kappa_f$, and

$$\mathbf{Z}_e(s) = z_e(s) = m_e s + b_e + \frac{k_e}{s} \tag{11.110}$$

$$\mathbf{Z}_h(s) = z_h(s) = m_h s + b_h + \frac{k_h}{s}. \tag{11.111}$$

It follows that

$$
\begin{aligned}
&\|\mathbf{G}_{1s}(j\omega)^{-1}\mathbf{G}_{2s}(j\omega)\mathbf{G}_{1m}(j\omega)^{-1}\mathbf{G}_{2m}(j\omega)\|_\infty \\
&\quad = \|\mathbf{G}_{1m}(j\omega)^{-1}\mathbf{G}_{2m}(j\omega)\mathbf{G}_{1s}(j\omega)^{-1}\mathbf{G}_{2s}(j\omega)\|_\infty \\
&\quad = \max_\omega \left| \frac{(j\omega+\delta_z\lambda) - j\omega a z_e(j\omega)}{(j\omega+\delta_z\lambda)(\frac{j\omega}{c}+1) + j\omega a z_e(j\omega)} \right. \\
&\qquad\qquad \left. \frac{(j\omega+\lambda) - j\omega a \frac{\kappa_f}{\kappa_p} z_h(j\omega)}{(j\omega+\lambda)(\frac{j\omega}{c}+1) + j\omega a \frac{\kappa_f}{\kappa_p} z_h(j\omega)} \right|
\end{aligned}
\tag{11.112}
$$

holds. The sufficient conditions to ensure (11.108) become

$$
\begin{aligned}
k_e &< 2cb_e \\
m_e &< \frac{1}{2c}b_e + \frac{\lambda}{4ac^2} \\
k_h &< 2cb_h \\
m_h &< \max\left(\left(\frac{1}{2c}b_h + \frac{\lambda\kappa_p}{4\kappa_f ac^2}\right), \left(\frac{k_h}{c\lambda} + \sqrt{\left(\frac{k_h}{c\lambda}\right)^2 + \frac{k_h}{ac^2\lambda}\frac{\kappa_p}{\kappa_f}}\right)\right).
\end{aligned}
$$

These conditions can be satisfied by appropriately choosing control parameters a, c, and λ, provided that the damping coefficients b_e and b_h are positive. Noticeably, a large c satisfies the constraints on k_e and k_h, while a small a satisfies the constraints on m_e and m_h.

11.7 Case Study - A One Degree of Freedom System

A case study with a block diagram illustrating the developed bilateral teleoperation control solution with respect to a single-DOF system was given in [217]. The experimental setup includes two motors. A direct-drive DC motor acts as the one-joint master robot and another DC motor equipped with a harmonic drive acts as the one-joint slave robot. Two aluminium bars are connected to the two driving axes. Two force sensors are mounted at the tips of the two aluminium bars to measure the forces between the human operator and the master robot and between the slave robot and the environment. A third aluminium bar is rigidly fixed to the base, allowing a rigid (metal-to-metal) contact with the slave robot.

Despite the high friction uncertainties associated with the harmonic drive, superior position and velocity tracking results in both the free motion and the rigid contact motion have been experimentally demonstrated. The force tracking was excellent for the rigid contact control without showing bouncing behaviours. Furthermore, the experimental teleoperation system kept its stability for any time delay.

11.8 Concluding Remarks

The fundamental requirement of a force-reflected bilateral teleoperation system is to let the slave robot track the motion of the master robot and simultaneously to let the human operator holding the master robot feel the environment contact force at a remote site. In this chapter, a novel solution to this problem has been suggested by using the adaptive motion and force control approach developed in Chapter 5. The master robot and the slave robot are subject to their own independent adaptive motion/force control. A properly designed communication network sends a combination of the motion and force information of one robot to the other as commands.

The developed approach allows both position-force and rate-force control modes to be implemented with arbitrary scaling. Unlike many other linear-model-based teleoperation control approaches, the developed approach bases its control directly on the nonlinear dynamic models of both the (master and slave) robots. The dynamic parameter uncertainties are accommodated by using parameter adaptation. This feature allows a teleoperation system to have superior position tracking performance between the two (master and slave) robots even in the contact directions with a rigid environment.

The entire teleoperation system behaves as a mass plus a damper specified by the scaling and control parameters. The equivalent mass and damping determine the accuracy of transparency. However, the achievable transparency is limited by the stability condition so that a trade-off between the transparency and the stability must be made.

The fact that each (master or slave) robot is subject to its own adaptive control allows each robot to have the tendency of tracking to *what* the command is regardless of *when* it is given. As a result, it makes the entire teleoperation system robust against time delays.

12

Control of Modular Robot Manipulators

12.1 Introduction

Today's robot manipulators are generally called integrated manipulators. Not only their mechanical structure, sensors, and actuators, but also their electronics and computing systems are specifically designed, fabricated, and integrated before being delivered to users. While being operationally effective and precise in well-structured assembly lines, these manipulators, on the other hand, possess weak adaptability, flexibility, and versatility against task variations in unstructured environments. The fact of having a fixed structure limits the capability of these integrated manipulators to perform a variety of tasks that require different robot structures with different degrees of freedom of motion. Besides being unable to have their structures changed on-site applications, the integrated robot manipulators are generally stand-along devices with separate control cabinets. This characteristic makes them difficult to be incorporated into mobile platforms.

The integrated robot manipulators are also vulnerable to component/part failures. When a component, such as a sensor or an actuator, goes out of order, the solution gives rise to either repairing the component or replacing the entire manipulator. Repairing might be a difficult task in a hostile environment where human access is highly restricted.

A possible solution is to use modular robot manipulators. A modular robot manipulator simply consists of robot modules, rather than an integrated robot manipulator itself. These modules incorporate their own actuators, sensors, power electronics, and computing systems into their mechanical assemblies to create all-in-one devices which can be appropriately connected to form any desirable robot manipulator or robot leg. Besides its standard mechanical interface, each module might have a standard electrical and electronics interface comprised of two connectors—the power connector and the data connector.

Modular robot manipulators have at least four features over conventional integrated robot manipulators:

1) The integration of electronics into mechanical assemblies makes the modular robot manipulators particularly suitable for mobile applications. Without

W.-H. Zhu: Virtual Decomposition Control, STAR 60, pp. 325–336.
springerlink.com

requiring additional space for its electronics and controller, a modular robot manipulator can be conveniently mounted on any mobile platform as long as the power and data connectors are accessible.
2) Modular robot manipulators allow on-site changes of robot structure and even changes of the degrees of freedom of motion by adding/removing modules to/from a robot.
3) It makes easy repair possible, since a defected module can be easily replaced.
4) Mass production of modules will eventually reduce the cost of using modular robot manipulators.

Besides for ground applications, above features make the modular robot manipulators particularly suitable for space applications. In space, due to the limited accessibility and the huge cost for human presence, the use of robots becomes more feasible. In the future, it is likely that the robotic technologies will be used not only to acquire information but also to manipulate the space/planetary environment. Thus, the adaptability, flexibility, reliability, and re-formability will be the fundamental requirements. In future space robotic missions, a mission manager might elect to send robot modules instead of the integrated robots into the space and to assemble these modules in the space to form any type of robots needed for operations, with minimum human interventions.

With its concept being initialized two decades ago, the currently existing modular robotics focuses mainly on mechanical functions, such as the reconfigurability and re-formability. The control and communication issues remain to be challenging. This is due to the fact that robot dynamics are centralized in nature, whereas modular robots only permit decentralized control restricted by their modular electro-mechanical structures. This natural restriction gives each module very limited capability of performing local communications to its neighbors and usually allows a very simple local controller (such as a PID controller) to be implemented. While being less sensitive to certain applications such as the snake robot climbing, these issues are substantially important when precise position control and multi-module coordinated control are concerned.

Aimed at giving modular robot manipulators the comparable control performances as these dynamically controlled integrated robot manipulators without jeopardizing their fundamental feasibilities such as the low cost, high flexibility, and easy use and expansion, the *virtual decomposition control* (VDC) approach is applied to modular robot manipulators in this chapter. As an adaptive control approach capable of "thinking globally and acting locally," the VDC approach possesses an undeniable theoretical advantage of applying the full-dynamics-based control to modular robot manipulators by solely using the dynamics of individual modules [211, 225]. When these modules are integrated to form a robot manipulator, the master computer only needs to compute the robot kinematics. All module dynamics are completely handled by these embedded module controllers to guarantee the stability and convergence of the entire modular robot manipulator.

12.2 Prior Work

Modular or re-configurable robots have been studied and developed for more than 15 years [25, 35, 55, 107, 137, 195]. A modular robot manipulator simply consists of robot modules, rather than an integrated robot manipulator itself. The concept of cellular robotics was first introduced in [55]. Then, a re-configurable modular manipulator system (RMMS) with modular links and joints was developed in [137], in which the kinematic configuration optimization was performed analytically for robots with two degrees of freedom and numerically for robots with multiple degrees of freedom. In the meantime, a (kinematic) configuration optimization procedure based on an assembly incidence matrix representation was presented in [25], and a genetic algorithm leading to a task-based design was proposed in [35]. Afterwards, a re-configurable modular robot PolyBot targeting space applications was developed in [195]. Other applications include a modular joint design for a humanoid robot [107], a decentralized kinematics control approach [173], and a decentralized dynamics based control of modular robot manipulators by using joint torque sensing [103], in which the dynamics of adjacent links are fully represented by the joint torque measurements. Being able to ensure the asymptotic stability in the free motion, this approach results in an algebraic loop of gain *one* about the joint torques when a rigid contact occurs.

In the area of *self-reconfigurable robots*, researches have been mainly focusing on robot structural configuration changes for executing relevant tasks [18, 33, 34, 123, 143, 146]. A recent survey on modular and self-reconfigurable robots can be found in [196].

12.3 A Typical Module

A modular robot manipulator comprised of $n > 0$ modules connected serially is studied in this chapter.

A typical module is illustrated in Figure 12.1. It comprises two rigid links connected by a joint. At each of the two ends, a standard mechanical/electrical/electronic interface is placed. This interface mechanically connects a module with another module and provides high-speed communication and electrical power connection to and from the adjacent module. The mechanical connection uses screw–bolt type to maintain the connection rigidity. A high-speed communication network goes through a serial port, and another connector provides the power connection. At each module, a micro-computer board which performs communication and control, a DC (direct current) power distributor which provides the power with different voltages, and a power amplifier which drives a brushless DC motor are embedded on one of the two links. At each joint, a motor-transmission assembly with an encoder attached to the motor side is situated. After all the modules are connected, the overall electrical/electronic interface of the entire robot is a DC power connector and a high-speed data-bus connector.

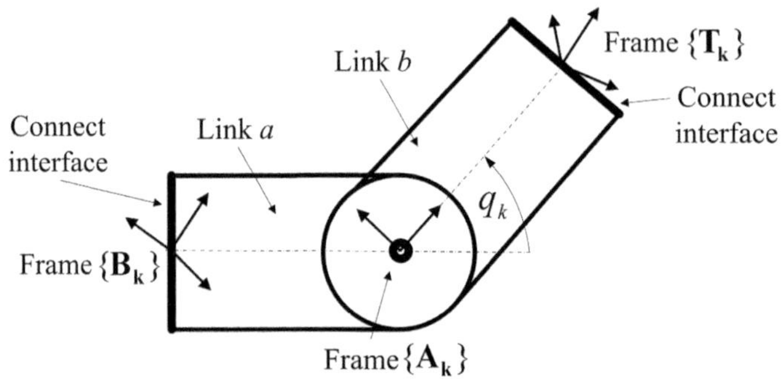

Fig. 12.1. A module assembly.

This feature substantially enhances the mobility of the robot, since no control cabinet is required.

Three coordinate frames are attached to the kth module for all $k \in \{1, n\}$. Frames $\{\mathbf{B}_k\}$ and $\{\mathbf{T}_k\}$ are attached to the two connection interfaces of the kth module. The two connection interfaces also serve as the two *cutting points*. The *cutting point* associated with frame $\{\mathbf{B}_k\}$ is called the *driven cutting point* of the kth module and the *cutting point* associated with frame $\{\mathbf{T}_k\}$ is called the *driving cutting point* of the kth module. Meanwhile, frame $\{\mathbf{A}_k\}$ is attached to the same link to which frame $\{\mathbf{T}_k\}$ is attached, with its z axis aligning with the joint axis.

The complete modular robot manipulator is constructed in such a way that the interface associated with frame $\{\mathbf{B}_1\}$ is attached to the base, the interface associated with frame $\{\mathbf{T}_1\}$ is connected to the interface associated with frame $\{\mathbf{B}_2\}$, the interface associated with frame $\{\mathbf{T}_{k-1}\}$ is connected to the interface associated with frame $\{\mathbf{B}_k\}$, and eventually the interface associated with frame $\{\mathbf{T}_{n-1}\}$ is connected to the interface associated with frame $\{\mathbf{B}_n\}$.

12.4 Virtual Decomposition Control

In this section, the VDC approach is to be applied.

12.4.1 Kinematics and Dynamics

In this subsection, the kinematics and dynamics of the kth module, for all $k \in \{1, n\}$, are presented as the basis for the upcoming dynamics based control to be developed in the next subsection.

In view of (2.63), the velocity transformation from frame $\{\mathbf{B}_k\}$ to frame $\{\mathbf{T}_k\}$ can be written as

$$\begin{aligned} {}^{\mathbf{T}_k}V &= {}^{\mathbf{A}_k}\mathbf{U}_{\mathbf{T}_k}^T \, {}^{\mathbf{A}_k}V \\ &= {}^{\mathbf{A}_k}\mathbf{U}_{\mathbf{T}_k}^T \left(\mathbf{z}\dot{q}_k + {}^{\mathbf{B}_k}\mathbf{U}_{\mathbf{A}_k}^T \, {}^{\mathbf{B}_k}V\right) \\ &= {}^{\mathbf{A}_k}\mathbf{U}_{\mathbf{T}_k}^T \mathbf{z}\dot{q}_k + {}^{\mathbf{B}_k}\mathbf{U}_{\mathbf{T}_k}^T \, {}^{\mathbf{B}_k}V \end{aligned} \tag{12.1}$$

where $\mathbf{z} = [0, 0, 0, 0, 0, 1]^T \in \mathbb{R}^6$ is a vector and $\dot{q}_k \in \mathbb{R}$ denotes the joint velocity.

The dynamics of the two rigid links can be obtained from (2.74) as

$$\mathbf{M}_{\mathbf{B}_k} \frac{d}{dt}({}^{\mathbf{B}_k}V) + \mathbf{C}_{\mathbf{B}_k}({}^{\mathbf{B}_k}\omega){}^{\mathbf{B}_k}V + \mathbf{G}_{\mathbf{B}_k} = {}^{\mathbf{B}_k}F^* \quad (12.2)$$

$$\mathbf{M}_{\mathbf{T}_k} \frac{d}{dt}({}^{\mathbf{T}_k}V) + \mathbf{C}_{\mathbf{T}_k}({}^{\mathbf{T}_k}\omega){}^{\mathbf{T}_k}V + \mathbf{G}_{\mathbf{T}_k} = {}^{\mathbf{T}_k}F^* \quad (12.3)$$

by substituting frames $\{\mathbf{B}_k\}$ and $\{\mathbf{T}_k\}$ for frame $\{\mathbf{A}\}$, respectively. The net force/moment vectors are subject to

$${}^{\mathbf{T}_k}F^* = {}^{\mathbf{T}_k}\mathbf{U}_{\mathbf{A}_k}{}^{\mathbf{A}_k}F - {}^{\mathbf{T}_k}F \quad (12.4)$$

$${}^{\mathbf{B}_k}F^* = {}^{\mathbf{B}_k}F - {}^{\mathbf{B}_k}\mathbf{U}_{\mathbf{A}_k}{}^{\mathbf{A}_k}F. \quad (12.5)$$

In view of (4.74), the joint dynamics are written as

$$J_{mk}\ddot{q}_k + \xi_k(q_k, \dot{q}_k) = \tau_k - \mathbf{z}^T{}^{\mathbf{A}_k}F \quad (12.6)$$

where $J_{mk} \in \mathbb{R}$ denotes the equivalent moment of inertia, $\xi_k(q_k, \dot{q}_k) \in \mathbb{R}$ is a term representing the friction and gravitation torque, $\tau_k \in \mathbb{R}$ denotes the joint control torque, and $\mathbf{z}^T{}^{\mathbf{A}_k}F$ represents the projection of the force/moment vector onto the kth joint axis.

12.4.2 Control Equations

In this subsection, the model-based control design for the kth module is presented.

Given desired joint position, velocity, and acceleration, denoted as $q_{kd} \in \mathbb{R}$, $\dot{q}_{kd} \in \mathbb{R}$, and $\ddot{q}_{kd} \in \mathbb{R}$, respectively, the required joint velocity and acceleration are designed as

$$\dot{q}_{kr} = \dot{q}_{kd} + \lambda_k(q_{kd} - q_k) \quad (12.7)$$

$$\ddot{q}_{kr} = \ddot{q}_{kd} + \lambda_k(\dot{q}_{kd} - \dot{q}_k) \quad (12.8)$$

with $\lambda_k > 0$ being a control parameter.

The required linear/angular velocity vector of frame $\{\mathbf{T}_k\}$ is computed from the given required linear/angular velocity vector of frame $\{\mathbf{B}_k\}$ as

$$\begin{aligned} {}^{\mathbf{T}_k}V_r &= {}^{\mathbf{A}_k}\mathbf{U}_{\mathbf{T}_k}^T\,{}^{\mathbf{A}_k}V_r \\ &= {}^{\mathbf{A}_k}\mathbf{U}_{\mathbf{T}_k}^T\left(\mathbf{z}\dot{q}_{kr} + {}^{\mathbf{B}_k}\mathbf{U}_{\mathbf{A}_k}^T\,{}^{\mathbf{B}_k}V_r\right) \\ &= {}^{\mathbf{A}_k}\mathbf{U}_{\mathbf{T}_k}^T\mathbf{z}\dot{q}_{kr} + {}^{\mathbf{B}_k}\mathbf{U}_{\mathbf{T}_k}^T\,{}^{\mathbf{B}_k}V_r \end{aligned} \quad (12.9)$$

with $\mathbf{z} = [0, 0, 0, 0, 0, 1]^T \in \mathbb{R}^6$.

Accordingly, recall (4.32) with appropriate frame substitutions to compute the required net force/moment vectors of the two rigid links as

$${}^{\mathbf{B}_k}F_r^* = \mathbf{Y}_{\mathbf{B}_k}\hat{\boldsymbol{\theta}}_{\mathbf{B}_k} + \mathbf{K}_{\mathbf{B}_k}\left({}^{\mathbf{B}_k}V_r - {}^{\mathbf{B}_k}V\right) \quad (12.10)$$

$$ {}^{\mathbf{T}_k}F_r^* = \mathbf{Y}_{\mathbf{T}_k}\hat{\boldsymbol{\theta}}_{\mathbf{T}_k} + \mathbf{K}_{\mathbf{T}_k}\left({}^{\mathbf{T}_k}V_r - {}^{\mathbf{T}_k}V\right) \tag{12.11} $$

where $\mathbf{K}_{\mathbf{B}_k} \in \mathbb{R}^{6\times 6}$ and $\mathbf{K}_{\mathbf{T}_k} \in \mathbb{R}^{6\times 6}$ are two positive-definite matrices and $\hat{\boldsymbol{\theta}}_{\mathbf{B}_k} \in \mathbb{R}^{13}$ and $\hat{\boldsymbol{\theta}}_{\mathbf{T}_k} \in \mathbb{R}^{13}$ denote the estimates of $\boldsymbol{\theta}_{\mathbf{B}_k} \in \mathbb{R}^{13}$ and $\boldsymbol{\theta}_{\mathbf{T}_k} \in \mathbb{R}^{13}$, respectively, with $\mathbf{Y}_{\mathbf{B}_k}\boldsymbol{\theta}_{\mathbf{B}_k}$ and $\mathbf{Y}_{\mathbf{T}_k}\boldsymbol{\theta}_{\mathbf{T}_k}$ being defined by (2.78) and expressed in Appendix A by respectively substituting frames $\{\mathbf{B}_k\}$ and $\{\mathbf{T}_k\}$ for frame $\{\mathbf{A}\}$.

The parameter estimates of the two links need to be updated. Define

$$ \mathbf{s}_{\mathbf{B}_k} = \mathbf{Y}_{\mathbf{B}_k}^T\left({}^{\mathbf{B}_k}V_r - {}^{\mathbf{B}_k}V\right) \tag{12.12} $$

$$ \mathbf{s}_{\mathbf{T}_k} = \mathbf{Y}_{\mathbf{T}_k}^T\left({}^{\mathbf{T}_k}V_r - {}^{\mathbf{B}_T}V\right). \tag{12.13} $$

The $\mathcal{P}$ function defined by (2.79) is used to update each of the 13 parameters of each link as

$$ \hat{\theta}_{\mathbf{B}_k\gamma} = \mathcal{P}\left(s_{\mathbf{B}_k\gamma}, \rho_{\mathbf{B}_k\gamma}, \underline{\theta}_{\mathbf{B}_k\gamma}, \overline{\theta}_{\mathbf{B}_k\gamma}, t\right) \tag{12.14} $$

$$ \hat{\theta}_{\mathbf{T}_k\gamma} = \mathcal{P}\left(s_{\mathbf{T}_k\gamma}, \rho_{\mathbf{T}_k\gamma}, \underline{\theta}_{\mathbf{T}_k\gamma}, \overline{\theta}_{\mathbf{T}_k\gamma}, t\right) \tag{12.15} $$

for all $\gamma \in \{1, 13\}$, where

$\theta_{\mathbf{B}_k\gamma}$	γth element of $\boldsymbol{\theta}_{\mathbf{B}_k}$.
$\theta_{\mathbf{T}_k\gamma}$	γth element of $\boldsymbol{\theta}_{\mathbf{T}_k}$.
$\hat{\theta}_{\mathbf{B}_k\gamma}$	Estimate of $\theta_{\mathbf{B}_k\gamma}$.
$\hat{\theta}_{\mathbf{T}_k\gamma}$	Estimate of $\theta_{\mathbf{T}_k\gamma}$.
$s_{\mathbf{B}_k\gamma}$	γth element of $\mathbf{s}_{\mathbf{B}_k}$.
$s_{\mathbf{T}_k\gamma}$	γth element of $\mathbf{s}_{\mathbf{T}_k}$.
$\rho_{\mathbf{B}_k\gamma}$	Parameter update gain.
$\rho_{\mathbf{T}_k\gamma}$	Parameter update gain.
$\underline{\theta}_{\mathbf{B}_k\gamma}$	Lower bound of $\theta_{\mathbf{B}_k\gamma}$.
$\overline{\theta}_{\mathbf{B}_k\gamma}$	Upper bound of $\theta_{\mathbf{B}_k\gamma}$.
$\underline{\theta}_{\mathbf{T}_k\gamma}$	Lower bound of $\theta_{\mathbf{T}_k\gamma}$.
$\overline{\theta}_{\mathbf{T}_k\gamma}$	Upper bound of $\theta_{\mathbf{T}_k\gamma}$.

Then, the required force/moment vector in frame $\{\mathbf{B}_k\}$ can be computed from the given required force/moment vector in frame $\{\mathbf{T}_k\}$ as

$$ {}^{\mathbf{A}_k}F_r = {}^{\mathbf{A}_k}\mathbf{U}_{\mathbf{T}_k}\left({}^{\mathbf{T}_k}F_r^* + {}^{\mathbf{T}_k}F_r\right) \tag{12.16} $$

$$ {}^{\mathbf{B}_k}F_r = {}^{\mathbf{B}_k}F_r^* + {}^{\mathbf{B}_k}\mathbf{U}_{\mathbf{A}_k}{}^{\mathbf{A}_k}F_r. \tag{12.17} $$

Finally, the joint torque control is designed as

$$ \tau_{kd}^* = \mathbf{Y}_{\tau k}\hat{\boldsymbol{\theta}}_{\tau k} + k_{\tau k}(\dot{q}_{kr} - \dot{q}_k) \tag{12.18} $$

$$ \begin{aligned} \tau_k &= \tau_{kd}^* + \mathbf{z}^T\,{}^{\mathbf{A}_k}F_r \\ &= \mathbf{Y}_{\tau k}\hat{\boldsymbol{\theta}}_{\tau k} + \mathbf{z}^T\,{}^{\mathbf{A}_k}F_r + k_{\tau k}(\dot{q}_{kr} - \dot{q}_k) \end{aligned} \tag{12.19} $$

with $k_{\tau k} > 0$, subject to

$$\mathbf{Y}_{\tau k}\boldsymbol{\theta}_{\tau k} = J_{mk}\ddot{q}_{kr} + \xi_k(q_k, \dot{q}_k) \tag{12.20}$$

where $\dot{q}_{kr}$ is defined by (12.7).

The parameter estimates of the joint also need to be updated. Define

$$\mathbf{s}_{\tau k} = \mathbf{Y}_{\tau k}^T \left(\dot{q}_{kr} - \dot{q}_k\right). \tag{12.21}$$

Each element of $\hat{\boldsymbol{\theta}}_{\tau k}$ is updated by using the $\mathcal{P}$ function defined by (2.79) as

$$\hat{\theta}_{\tau k\gamma} = \mathcal{P}\left(s_{\tau k\gamma}, \rho_{\tau k\gamma}, \underline{\theta}_{\tau k\gamma}, \overline{\theta}_{\tau k\gamma}, t\right) \tag{12.22}$$

where $\hat{\theta}_{\tau k\gamma}$ denotes the estimate of $\theta_{\tau k\gamma}$—the γth element of $\boldsymbol{\theta}_{\tau k}$, $s_{\tau k\gamma}$ denotes the γth element of $\mathbf{s}_{\tau k}$ defined by (12.21), $\rho_{\tau k\gamma} > 0$ denotes an update gain, and $\underline{\theta}_{\tau k\gamma}$ and $\overline{\theta}_{\tau k\gamma}$ denote the lower and upper bounds of the parameter $\theta_{\tau k\gamma}$.

12.4.3 Virtual Stability

The following theorem ensures that the *virtual stability* of the kth module combined with its respective control equations holds in the sense of Definition 2.17.

Theorem 12.1. *The kth module described by (12.1)–(12.6), combined with its respective control equations (12.9)–(12.11) and (12.16)–(12.20) and with the parameter adaptation (12.12)–(12.15), (12.21), and (12.22), is virtually stable with its affiliated variable $\dot{q}_{kr} - \dot{q}_k$ and affiliated vectors ${}^{\mathbf{B}_k}V_r - {}^{\mathbf{B}_k}V$ and ${}^{\mathbf{B}_k}V_r - {}^{\mathbf{B}_k}V$ being virtual functions in both L_2 and L_∞, in the sense of Definition 2.17.*

Proof: The non-negative accompanying function for the kth module is chosen as

$$\nu_k = \nu_{\mathbf{B}_k} + \nu_{\mathbf{T}_k} + \nu_{\tau k} \tag{12.23}$$

with

$$\begin{aligned}\nu_{\mathbf{B}_k} &= \frac{1}{2}\left({}^{\mathbf{B}_k}V_r - {}^{\mathbf{B}_k}V\right)^T \mathbf{M}_{\mathbf{B}_k}\left({}^{\mathbf{B}_k}V_r - {}^{\mathbf{B}_k}V\right) \\ &\quad + \frac{1}{2}\sum_{\gamma=1}^{13}\left(\theta_{\mathbf{B}_k\gamma} - \hat{\theta}_{\mathbf{B}_k\gamma}\right)^2 / \rho_{\mathbf{B}_k\gamma}\end{aligned} \tag{12.24}$$

$$\begin{aligned}\nu_{\mathbf{T}_k} &= \frac{1}{2}\left({}^{\mathbf{T}_k}V_r - {}^{\mathbf{T}_k}V\right)^T \mathbf{M}_{\mathbf{T}_k}\left({}^{\mathbf{T}_k}V_r - {}^{\mathbf{T}_k}V\right) \\ &\quad + \frac{1}{2}\sum_{\gamma=1}^{13}\left(\theta_{\mathbf{T}_k\gamma} - \hat{\theta}_{\mathbf{T}_k\gamma}\right)^2 / \rho_{\mathbf{T}_k\gamma}\end{aligned} \tag{12.25}$$

$$\nu_{\tau k} = \frac{1}{2}J_{mk}(\dot{q}_{kr} - \dot{q}_k)^2 + \frac{1}{2}\sum_{\gamma}\left(\theta_{\tau k\gamma} - \hat{\theta}_{\tau k\gamma}\right)^2 / \rho_{\tau k\gamma}. \tag{12.26}$$

In view of (12.2), (12.3), and (12.10)–(12.15), it follows from Lemma 4.1 that

$$\dot{\nu}_{\mathbf{B}_k} \leqslant -\left({}^{\mathbf{B}_k}V_r - {}^{\mathbf{B}_k}V\right)^T \mathbf{K}_{\mathbf{B}_k} \left({}^{\mathbf{B}_k}V_r - {}^{\mathbf{B}_k}V\right) + \left({}^{\mathbf{B}_k}V_r - {}^{\mathbf{B}_k}V\right)^T \left({}^{\mathbf{B}_k}F_r^* - {}^{\mathbf{B}_k}F^*\right) \tag{12.27}$$

$$\dot{\nu}_{\mathbf{T}_k} \leqslant -\left({}^{\mathbf{T}_k}V_r - {}^{\mathbf{T}_k}V\right)^T \mathbf{K}_{\mathbf{T}_k} \left({}^{\mathbf{T}_k}V_r - {}^{\mathbf{T}_k}V\right) + \left({}^{\mathbf{T}_k}V_r - {}^{\mathbf{T}_k}V\right)^T \left({}^{\mathbf{T}_k}F_r^* - {}^{\mathbf{T}_k}F^*\right) \tag{12.28}$$

hold by respectively substituting frames $\{\mathbf{B}_k\}$ and $\{\mathbf{T}_k\}$ for frame $\{\mathbf{O}_i\}$.

On the other hand, it follows from (12.6), (12.18)–(12.22), and Lemma 2.9 that

$$\begin{aligned}
\dot{\nu}_{\mathcal{T}k} &= (\dot{q}_{kr} - \dot{q}_k) J_{mk} (\ddot{q}_{kr} - \ddot{q}_k) - \sum_{\gamma} \left(\theta_{\mathcal{T}k\gamma} - \hat{\theta}_{\mathcal{T}k\gamma}\right) \dot{\hat{\theta}}_{\mathcal{T}k\gamma} / \rho_{\mathcal{T}k\gamma} \\
&= -k_{\mathcal{T}k} (\dot{q}_{kr} - \dot{q}_k)^2 - (\dot{q}_{kr} - \dot{q}_k) \mathbf{z}^T ({}^{\mathbf{A}_k}F_r - {}^{\mathbf{A}_k}F) \\
&\quad + (\dot{q}_{kr} - \dot{q}_k) \mathbf{Y}_{\mathcal{T}k} (\boldsymbol{\theta}_{\mathcal{T}k} - \hat{\boldsymbol{\theta}}_{\mathcal{T}k}) - \sum_{\gamma} \left(\theta_{\mathcal{T}k\gamma} - \hat{\theta}_{\mathcal{T}k\gamma}\right) \dot{\hat{\theta}}_{\mathcal{T}k\gamma} / \rho_{\mathcal{T}k\gamma} \\
&= -k_{\mathcal{T}k} (\dot{q}_{kr} - \dot{q}_k)^2 - (\dot{q}_{kr} - \dot{q}_k) \mathbf{z}^T ({}^{\mathbf{A}_k}F_r - {}^{\mathbf{A}_k}F) \\
&\quad + \sum_{\gamma} \left(\theta_{\mathcal{T}k\gamma} - \hat{\theta}_{\mathcal{T}k\gamma}\right) \left(s_{\mathcal{T}k\gamma} - \dot{\hat{\theta}}_{\mathcal{T}k\gamma} / \rho_{\mathcal{T}k\gamma}\right) \\
&\leqslant -k_{\mathcal{T}k} (\dot{q}_{kr} - \dot{q}_k)^2 - (\dot{q}_{kr} - \dot{q}_k) \mathbf{z}^T ({}^{\mathbf{A}_k}F_r - {}^{\mathbf{A}_k}F)
\end{aligned} \tag{12.29}$$

holds.

In view of (2.85), (12.1), (12.4), (12.5), (12.9), (12.16), and (12.17), it yields

$$\begin{aligned}
&\left({}^{\mathbf{B}_k}V_r - {}^{\mathbf{B}_k}V\right)^T \left({}^{\mathbf{B}_k}F_r^* - {}^{\mathbf{B}_k}F^*\right) \\
&\quad = \left({}^{\mathbf{B}_k}V_r - {}^{\mathbf{B}_k}V\right)^T \left({}^{\mathbf{B}_k}F_r - {}^{\mathbf{B}_k}F\right) \\
&\qquad - \left({}^{\mathbf{B}_k}V_r - {}^{\mathbf{B}_k}V\right)^T {}^{\mathbf{B}_k}\mathbf{U}_{\mathbf{A}_k} \left({}^{\mathbf{A}_k}F_r - {}^{\mathbf{A}_k}F\right) \\
&\quad = p_{\mathbf{B}_k} - \left({}^{\mathbf{B}_k}V_r - {}^{\mathbf{B}_k}V\right)^T {}^{\mathbf{B}_k}\mathbf{U}_{\mathbf{A}_k} \left({}^{\mathbf{A}_k}F_r - {}^{\mathbf{A}_k}F\right)
\end{aligned} \tag{12.30}$$

$$\begin{aligned}
&\left({}^{\mathbf{T}_k}V_r - {}^{\mathbf{T}_k}V\right)^T \left({}^{\mathbf{T}_k}F_r^* - {}^{\mathbf{T}_k}F^*\right) \\
&\quad = -\left({}^{\mathbf{T}_k}V_r - {}^{\mathbf{T}_k}V\right)^T \left({}^{\mathbf{T}_k}F_r - {}^{\mathbf{T}_k}F\right) \\
&\qquad + \left[{}^{\mathbf{A}_k}\mathbf{U}_{\mathbf{T}_k}^T \mathbf{z} \left(\dot{q}_{kr} - \dot{q}_k\right) + {}^{\mathbf{B}_k}\mathbf{U}_{\mathbf{T}_k}^T \left({}^{\mathbf{B}_k}V_r - {}^{\mathbf{B}_k}V\right)\right]^T \\
&\qquad\quad \times {}^{\mathbf{T}_k}\mathbf{U}_{\mathbf{A}_k} \left({}^{\mathbf{A}_k}F_r - {}^{\mathbf{A}_k}F\right) \\
&\quad = -p_{\mathbf{T}_k} + \left(\dot{q}_{kr} - \dot{q}_k\right) \mathbf{z}^T \left({}^{\mathbf{A}_k}F_r - {}^{\mathbf{A}_k}F\right) \\
&\qquad + \left({}^{\mathbf{B}_k}V_r - {}^{\mathbf{B}_k}V\right)^T {}^{\mathbf{B}_k}\mathbf{U}_{\mathbf{A}_k} \left({}^{\mathbf{A}_k}F_r - {}^{\mathbf{A}_k}F\right).
\end{aligned} \tag{12.31}$$

Substituting (12.27)–(12.31) into the time-derivative of (12.23) yields

$$\begin{aligned}
\dot{\nu}_k &= \dot{\nu}_{\mathbf{B}_k} + \dot{\nu}_{\mathbf{T}_k} + \dot{\nu}_{\mathcal{T}k} \\
&\leqslant -\left({}^{\mathbf{B}_k}V_r - {}^{\mathbf{B}_k}V\right)^T \mathbf{K}_{\mathbf{B}_k} \left({}^{\mathbf{B}_k}V_r - {}^{\mathbf{B}_k}V\right)
\end{aligned}$$

$$
\begin{aligned}
&-\left({}^{\mathbf{T}_k}V_r - {}^{\mathbf{T}_k}V\right)^T \mathbf{K}_{\mathbf{T}_k}\left({}^{\mathbf{T}_k}V_r - {}^{\mathbf{T}_k}V\right) \\
&-k_{\tau k}(\dot{q}_{kr} - \dot{q}_k)^2 \\
&+p_{\mathbf{B}_k} - p_{\mathbf{T}_k} \qquad (12.32)
\end{aligned}
$$

where $p_{\mathbf{B}_k}$ and $p_{\mathbf{T}_k}$ denote the two *virtual power flows* at the two *cutting points* of the kth module.

Consider the fact that the *cutting point* associated with frame $\{\mathbf{B}_k\}$ is called the *driven cutting point* of the kth module and the *cutting point* associated with frame $\{\mathbf{T}_k\}$ is called the *driving cutting point* of the kth module. Using (12.23)–(12.26), (12.32), and Definition 2.17 completes the proof. ■

12.5 Stability of the Entire Modular Robot Manipulator

When every module qualifies to be *virtually stable* in the sense of Definition 2.17 for all $k \in \{1, n\}$, it follows form Theorem 2.1 that

$$ {}^{\mathbf{B}_k}V_r - {}^{\mathbf{B}_k}V \in L_2 \bigcap L_\infty \qquad (12.33) $$

$$ {}^{\mathbf{T}_k}V_r - {}^{\mathbf{T}_k}V \in L_2 \bigcap L_\infty \qquad (12.34) $$

$$ \dot{q}_{kr} - \dot{q}_k \in L_2 \bigcap L_\infty \qquad (12.35) $$

hold for all $k \in \{1, n\}$.

Furthermore, it follows from (12.7), (12.35), and Lemma 2.4 that

$$ \dot{q}_{kd} - \dot{q}_k \in L_2 \bigcap L_\infty \qquad (12.36) $$

$$ q_{kd} - q_k \in L_2 \bigcap L_\infty \qquad (12.37) $$

hold for all $k \in \{1, n\}$.

The forward dynamics of the entire modular robot manipulator can be obtained with a form similar to (5.79). Let $\mathbf{q}_d = [q_{1d}, \cdots, q_{kd}, \cdots, q_{nd}]^T \in \mathbb{R}^n$ be the desired joint position vector. Then, the boundedness of $\mathbf{q}_d \in \mathbb{R}^n$, $\dot{\mathbf{q}}_d \in \mathbb{R}^n$, and $\ddot{\mathbf{q}}_d \in \mathbb{R}^n$ ensures the boundedness of $\boldsymbol{\tau} = [\tau_1, \cdots, \tau_k, \cdots, \tau_n]^T \in \mathbb{R}^n$, in view of (12.7)–(12.22). Furthermore, the boundedness of $\boldsymbol{\tau} \in \mathbb{R}^n$ ensures the boundedness of $\ddot{\mathbf{q}} = [\ddot{q}_1, \cdots, \ddot{q}_k, \cdots, \ddot{q}_n]^T \in \mathbb{R}^n$ from the forward dynamics. Thus, the boundedness of $\mathbf{q}_d \in \mathbb{R}^n$, $\dot{\mathbf{q}}_d \in \mathbb{R}^n$, and $\ddot{\mathbf{q}}_d \in \mathbb{R}^n$ ensures the boundedness of $\ddot{\mathbf{q}} \in \mathbb{R}^n$ and further ensures

$$ {}^{\mathbf{B}_k}V_r - {}^{\mathbf{B}_k}V \to \mathbf{0} \qquad (12.38) $$

$$ {}^{\mathbf{T}_k}V_r - {}^{\mathbf{T}_k}V \to \mathbf{0} \qquad (12.39) $$

$$ \dot{q}_{kr} - \dot{q}_k \to 0 \qquad (12.40) $$

$$ \dot{q}_{kd} - \dot{q}_k \to 0 \qquad (12.41) $$

$$ q_{kd} - q_k \to 0 \qquad (12.42) $$

from Lemma 2.8 for all $k \in \{1, n\}$.

12.6 Communication and Control Implementation

A high-speed communication system is required to implement the VDC approach. The overall diagram is illustrated in Figure 12.2 [226]. Each slave node represents the micro-computer-board of a module. The sole master node is a real-time computer that is connected to the first slave node.

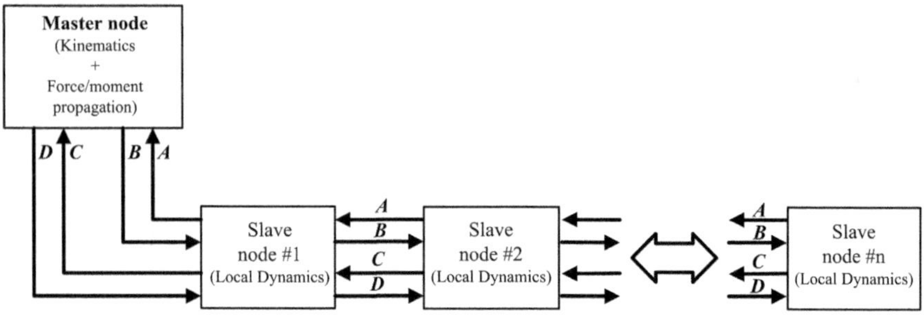

Fig. 12.2. The communication system.

The communication protocol is designed in such a way that the master node is able to broadcast information to all the slave nodes and is also able to extract information from any designated slave node. The data exchange is illustrated in Figure 12.3.

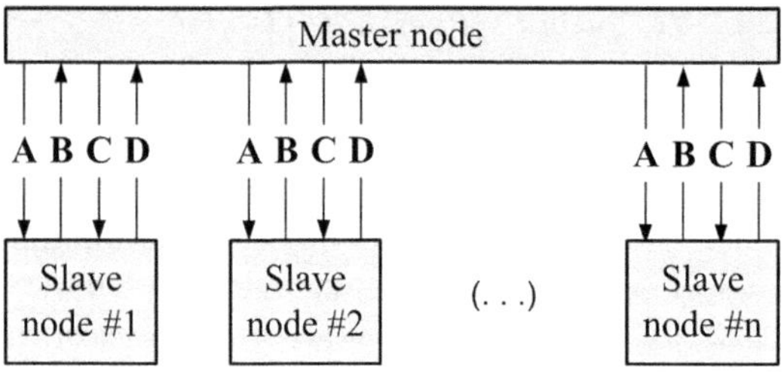

Fig. 12.3. Data exchange in communications.

With respect to this communication protocol, a communication mechanism comprised of two cycles with four communication actions within each sampling period is required to have the VDC implemented. The first communication cycle constitutes *Communications A* and *B*, and the second communication cycle constitutes *Communications C* and *D*.

The first communication cycle starts with *Communication A*. The master node extracts the joint positions and velocities from all the slave nodes. Then, the master node computes the linear/angular velocity vectors and the corresponding required linear/angular velocity vectors of all the slave nodes and broadcasts them back to respective slave nodes through *Communication B*. After receiving its respective linear/angular velocity vector and the required linear/angular velocity vector, each slave node computes its required accelerations and its required net force/moment vectors based on the (local) dynamics of each module.

The second communication cycle starts with *Communication C*. The master node extracts the net force/moment vectors from all the slave nodes. Then, the master node computes the required projected joint torques for all the joints and sends them back to each slave node through *Communication D*. Finally, each slave node computes its control torque based on the dynamics of each individual joint by incorporating the required projected joint torque received from the master node through *Communication D*.

The detailed communication and control procedure is listed below:

Step 1: Perform *Communication A*, through which the master node extracts the joint position $q_k \in \mathbb{R}$ and the joint velocity $\dot{q}_k \in \mathbb{R}$ for all $k \in \{1, n\}$.

Step 2: In the master node, from given $q_{kd} \in \mathbb{R}$, $\dot{q}_{kd} \in \mathbb{R}$, and $\ddot{q}_{kd} \in \mathbb{R}$, compute $\dot{q}_{kr} \in \mathbb{R}$, and $\ddot{q}_{kr} \in \mathbb{R}$ from (12.7) and (12.8) for all $k \in \{1, n\}$. Starting with ${}^{\mathbf{B}_1}V = \mathbf{0}$ and ${}^{\mathbf{B}_1}V_r = \mathbf{0}$, compute ${}^{\mathbf{T}_1}V$ and ${}^{\mathbf{T}_1}V_r$ from (12.1) and (12.9). Then, make ${}^{\mathbf{B}_2}V = {}^{\mathbf{T}_1}V$ and ${}^{\mathbf{B}_2}V_r = {}^{\mathbf{T}_1}V_r$ and compute ${}^{\mathbf{T}_2}V$ and ${}^{\mathbf{T}_2}V_r$ from (12.1) and (12.9). Repeat this process to obtain ${}^{\mathbf{B}_k}V \in \mathbb{R}^6$ and ${}^{\mathbf{B}_k}V_r \in \mathbb{R}^6$ for all $k \in \{1, n\}$.

Step 3: Perform *Communication B*. Transfer ${}^{\mathbf{B}_k}V \in \mathbb{R}^6$, ${}^{\mathbf{B}_k}V_r \in \mathbb{R}^6$, $\dot{q}_{kr} \in \mathbb{R}$, and ${}^{\mathbf{B}_k}\mathbf{R_I g} \in \mathbb{R}^3$ to the kth slave node for all $k \in \{1, n\}$, where $\mathbf{g} = [0, 0, 9.8]^T \in \mathbb{R}^3$ denotes the gravitational vector.

Step 4: In the kth slave node, $k \in \{1, n\}$, compute ${}^{\mathbf{T}_k}V \in \mathbb{R}^6$ and ${}^{\mathbf{T}_k}V_r \in \mathbb{R}^6$ from (12.1) and (12.9). Then, compute $\frac{d}{dt}\left({}^{\mathbf{B}_k}V_r\right) \in \mathbb{R}^6$, $\frac{d}{dt}\left({}^{\mathbf{T}_k}V_r\right) \in \mathbb{R}^6$, and $\ddot{q}_{kr} \in \mathbb{R}$ by using backward differentiation. Furthermore, compute ${}^{\mathbf{B}_k}F_r^* \in \mathbb{R}^6$ and ${}^{\mathbf{T}_k}F_r^* \in \mathbb{R}^6$ from (12.10) and (12.11), and update $\hat{\boldsymbol{\theta}}_{\mathbf{B}_k} \in \mathbb{R}^{13}$ and $\hat{\boldsymbol{\theta}}_{\mathbf{T}_k} \in \mathbb{R}^{13}$ from (12.12)–(12.15).

Step 5: Perform *Communication C*, through which the master node extracts ${}^{\mathbf{B}_k}F_r^* \in \mathbb{R}^6$ and ${}^{\mathbf{T}_k}F_r^* \in \mathbb{R}^6$ from the kth slave nodes for all $k \in \{1, n\}$.

Step 6: In the master node, compute ${}^{\mathbf{A}_k}F_r \in \mathbb{R}^6$ by using (12.16) and (12.17) alternatively for $k = n, n-1, \cdots, 1$ with ${}^{\mathbf{T}_k}F_r = {}^{\mathbf{B}_{k+1}}F_r$ and ${}^{\mathbf{B}_{n+1}}F_r = \mathbf{0}$.

Step 7: Perform *Communication D*, through which the required projected joint torques for all the joints, denoted as $\mathbf{z}^{T\,\mathbf{A}_k}F_r \in \mathbb{R}$ for all $k \in \{1, n\}$, are sent to the respective slave nodes.

Step 8: In the kth slave node, $k \in \{1, n\}$, compute the joint control torque τ_k from (12.18) and (12.19) and update $\hat{\boldsymbol{\theta}}_{\tau k}$ from (12.21) and (12.22).

Remark 12.1. The four communication actions occur in steps 1, 3, 5, and 7. The computation in the master node mainly performs in steps 2 and 6, and the computation in each slave node performs in steps 4 and 8 where parallel computations are permitted.

Remark 12.2. All the module dynamics are handled locally by the respective modules in steps 4 and 8. The master node only handles the kinematics related computation in steps 2 and 6. This is the main feature of using the VDC approach in modular robot applications.

12.7 Communication Data-Bus Rate

The complete communication and control computation has to be completed within a sampling period. The size of the communication package is very critical to meet the timing constraint. Therefore, the information exchange in the communication actions should be kept as minimal as possible. With respect to the particular communication and control implementation outlined in the last section, it can be verified that the numbers of variables need to be transferred between the master node and each slave node in the four communication actions are 2, 16, 12, and 1, respectively. With a 32-bit resolution for the joint position and a 16-bit resolution for all other variables, a 32-byte data length is required. To make the communication package uniform, a 40-byte package can be designed for all communication actions between the master node and each slave node. If the communication time spent on data transfer for a 12-module robot needs to be limited to 200μs, it requires that the data-bus rate should be at least

$$\begin{aligned} &\frac{8 \text{ bits}}{1 \text{ byte}} \times \frac{40 \text{ bytes}}{1 \text{ package}} \times \frac{4 \text{ packages}}{1 \text{ node}} \times 12 \text{ nodes} \times \frac{1}{200\mu s} \\ &= \frac{8 \times 40 \times 4 \times 12}{200 \times 10^{-6}} \approx 76.8 \text{ (MHz)}. \end{aligned}$$

12.8 Concluding Remarks

The lack of dynamics based control performances has been identified as one of the long-time problems in modular robotics. Aimed at effectively addressing this problem, the VDC approach has been applied to modular robot manipulators in this chapter. The VDC approach has an inherent advantage over other approaches on implementing the dynamics based control for modular robots, since it directly makes use of local module dynamics without the need to acquire the dynamics of the complete modular robot. It is theoretically expected that the VDC approach might allow modular robot manipulators to possess the comparable control performances as these dynamically controlled integrated robot manipulators.

13

Control of Flexible Link Robots

13.1 Introduction

Most applications discussed in the previous chapters are based on the rigid link (body) assumption. However, mechanical structures are flexible in nature. Rigidity is only an approximation when the flexibility is ignorable. In robotic applications, lightweight arms are always desirable due to their cost advantages, and long arms are needed for certain applications ranging from the assembly tasks of the International Space Station (ISS) to aircraft cleaning tasks. When either lightweight or long arms are used, the flexibility shows up inevitably regardless of the materials the arm is made of. This fact can be seen from the Euler–Bernoulli equation which implies that the static deformation of a uniform cantilever beam[1] subject to a force at its free end is proportional to the third power of the beam length. By restricting robot control design to rigid models, the operational efficiency can be severely affected due to the extra time needed to damp out vibrations for safe executions of robotic tasks. This situation technically motivates the need of developing appropriate control approaches for flexible link robots.

Unfortunately, the control of flexible link robots has never been easy due to the complicity of the flexible link dynamics. This issue becomes even more difficult when robots with multiple flexible links are addressed. By far, the most successfully control schemes are these either based on simplified models or limited to simple systems, such as a robot with one or two flexible beams in planar motion. The difficult in modeling also drove some researchers to seek model-free intelligent methods with a compromise of the control performances.

In this chapter, the *virtual decomposition control* (VDC) approach is applied for the first time to the tracking control problem of three-dimensional flexible link robots. The VDC approach allows the dynamics and control of each flexible beam to be handled separately from the rest of the system, since only the local subsystem dynamics are required for the control design under the VDC framework. Consider the fact that the kinematics and dynamics of a uniform beam have

[1] In this chapter, the terminologies of "beam" and "link" are equivalent.

W.-H. Zhu: Virtual Decomposition Control, STAR 60, pp. 337–364.
springerlink.com

been well defined from the extended Hamilton's principle. The use of the VDC approach, therefore, highlights the possibility of reviving the full-dynamics-based control for robots with multiple flexible links, without compromising the control performances.

13.2 Prior Work

The control of flexible link robots has been developed for more than two decades [82]. The technical challenges arise from the complexity in deriving the dynamic equations that not only use distributed parameter models, but also take into account the dynamic couplings among the multiple flexible links. The use of distributed parameter models makes the dynamic equations of a multi-body system very complex [114], which, in turn, complicates the developments of control strategies [113]. By far, the most successfully designed controllers are either based on simplified models or limited to a quite small class of systems, see [81]. Examples include robots with one or two flexible beams in planar motion or robots with a specially designed mechanism or structure. Remarkably, a modeling approach with approximations leading to a finite dimensional model was proposed in [184], and the so-called dynamic inversion technique was used in [42]. A strain gauge-based feedback control scheme was reported in [112]. This research based the control on the system model described by the Euler–Bernoulli equation with distributed parameter in nature. An adaptive controller aimed at compensating for an unknown payload mass was developed by incorporating an infinite dimensional model of the flexible link robot [45].

Whereas most contributions focused on planar motion, some specific contributions to the three-dimensional (3D) motion did exist. A 3D lightweight robot with all actuators being located at the base and the arm mass being concentrated at the tip was developed in [157]. The specially designed mechanism makes the motion decoupling possible at the tip. A perturbation method was also used to study the vibration characteristics of an n-degree-of-freedom flexible link manipulator in 3D [86].

Regardless of the numerous developments in the past, the full-dynamics-based tracking control of multiple flexible link robots in 3D, however, has never been precisely developed.

13.3 Flexible Beams in 2D Motion

The VDC approach allows each flexible beam to be treated independently, as long as every flexible link combined with its respective control equations qualifies to be *virtually stable* in the sense of Definition 2.17. Therefore, the focus in this section is on a typical flexible beam with a restricted motion in a two-dimensional plane. The extension to the 3D motion will be addressed in the next section.

The presentation in this section comprises the flexible link dynamics, the respective control, and the *virtual stability*.

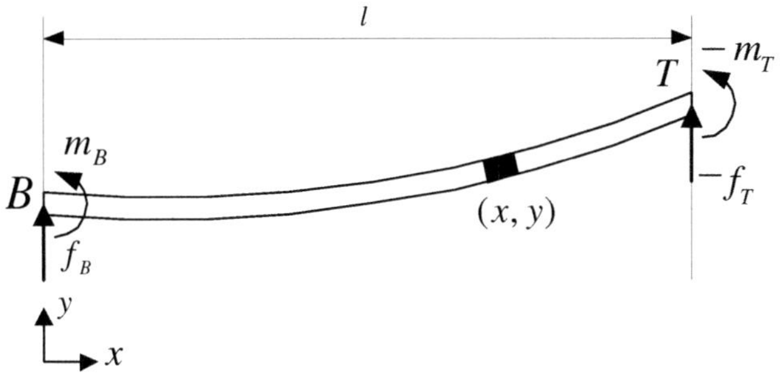

Fig. 13.1. A flexible link virtually decomposed from a planar flexible link robot.

13.3.1 Flexible Link Dynamics

The Flexible Link

An Euler-Bernoulli beam (link) that is virtually decomposed from a two-dimensional (2D) flexible-link robot is illustrated in Figure 13.1. There are two *cutting points* located at both ends. The *cutting point* at point B is interpreted as a *driven cutting point* of the beam, and the *cutting point* at point T is interpreted as a *driving cutting point* of the beam.

This link is subject to the following assumptions:

- The link is slender with uniform geometric and inertial characteristics.
- The link is flexible in the lateral direction and stiff with respect to the axial force and to the axial torsion.
- The link has negligible shear deformation and negligible distributed moment of inertia.
- The link is restricted in a plane without motion in the x axis.
- The axis of the beam aligns with the x axis.
- No gravity is presented.

With respect to this link, two pairs of force and moment applied at both ends are defined by

f_B	Shear force applied to the link at point B.
m_B	Bending moment applied to the link at point B.
f_T	Shear force applied from the link at point T.
m_T	Bending moment applied from the link at point T.

Extended Hamilton's Principle

Denote

$$e_K = \frac{1}{2}\int_0^l \rho \dot{y}(x,t)^2 dx \tag{13.1}$$

as the kinetic energy of the flexible link and

$$e_P = \frac{1}{2}\int_0^l EIy''(x,t)^2 dx \tag{13.2}$$

as the potential energy of the flexible link, where

l	Length of the link.
ρ	Mass per unit length.
E	Young's modulus of elasticity.
I	Cross-sectional area moment of inertia of the link about its neutral axis.
$\dot{y}$	Partial derivative of $y(x,t)$ with respect to time t.
y'	Partial derivative of $y(x,t)$ with respect to spatial variable x.

Meanwhile, let

$$\begin{aligned}\delta\overline{w_E} = & f_B\delta y(0,t) + m_B\delta y'(0,t) \\ & -f_T\delta y(l,t) - m_T\delta y'(l,t)\end{aligned} \tag{13.3}$$

be the external work variation.

The extended Hamilton's principle yields [81, 119]

$$\int_{t_1}^{t_2} (\delta e_K - \delta e_P + \delta\overline{w_E})\, dt = 0 \tag{13.4}$$

for $t_1 < t_2$. The Hamilton's principle states that the integral of the Lagrangian $\mathcal{L} = e_K - e_P$ of an isolated motion system from time t_1 to time t_2 has a stationary value. The addition of the external work to (13.4) extends the Hamilton's principle to non-isolated systems as well. The extended Hamilton's principle described by (13.4) will be used to derive the dynamic equations.

With integration by parts, it follows from (13.1) and (13.2) that

$$\begin{aligned}\int_{t_1}^{t_2} \delta e_K dt &= \int_{t_1}^{t_2}\int_0^l \rho\dot{y}(x,t)\delta\dot{y}(x,t)dxdt \\ &= \int_0^l\int_{t_1}^{t_2} \rho\dot{y}(x,t)\delta\dot{y}(x,t)dtdx \\ &= \int_0^l \rho\dot{y}(x,t)\delta y(x,t)\mid_{t_1}^{t_2} dx \\ &\quad -\int_0^l\int_{t_1}^{t_2} \rho\ddot{y}(x,t)\delta y(x,t)dtdx\end{aligned} \tag{13.5}$$

$$\begin{aligned}\int_{t_1}^{t_2} \delta e_P dt &= \int_{t_1}^{t_2}\int_0^l EIy''(x,t)\delta y''(x,t)dxdt \\ &= \int_{t_1}^{t_2} EIy''(x,t)\delta y'(x,t)\mid_0^l dt\end{aligned}$$

$$
\begin{aligned}
&-\int_{t_1}^{t_2}\int_0^l EIy'''(x,t)\delta y'(x,t)dxdt \\
&=\int_{t_1}^{t_2} EIy''(x,t)\delta y'(x,t)\mid_0^l dt \\
&-\int_{t_1}^{t_2} EIy'''(x,t)\delta y(x,t)\mid_0^l dt \\
&+\int_{t_1}^{t_2}\int_0^l EIy''''(x,t)\delta y(x,t)dxdt
\end{aligned}
\tag{13.6}
$$

hold. Given $\delta y(x,t) = 0$ at $t = t_1$ and $t = t_2$, substituting (13.3), (13.5), and (13.6) into (13.4) yields the flexible link dynamic equation

$$\rho\ddot{y}(x,t) + EIy''''(x,t) = 0 \tag{13.7}$$

for all $x \in [0, l]$, subject to the boundary conditions

$$f_B = EIy'''(0,t) \tag{13.8}$$
$$m_B = -EIy''(0,t) \tag{13.9}$$
$$f_T = EIy'''(l,t) \tag{13.10}$$
$$m_T = -EIy''(l,t). \tag{13.11}$$

13.3.2 Control Equations

In view of (13.7)–(13.11), the model based control is designed as

$$\rho\ddot{y}_r(x,t) + EIy_r''''(x,t) + k_v\left[\dot{y}_r(x,t) - \dot{y}(x,t)\right] = 0 \tag{13.12}$$

for all $x \in [0, l]$ with $k_v \geqslant 0$, subject to the following boundary conditions

$$f_{Br} = EIy_r'''(0,t) \tag{13.13}$$
$$m_{Br} = -EIy_r''(0,t) \tag{13.14}$$
$$f_{Tr} = EIy_r'''(l,t) \tag{13.15}$$
$$m_{Tr} = -EIy_r''(l,t) \tag{13.16}$$

where y_r, f_{Br}, m_{Br}, f_{Tr}, and m_{Tr} represent the required (design) variables of y, f_B, m_B, f_T, and m_T, respectively. All required (design) variables will be defined later.

13.3.3 Virtual Stability

Theorem 13.1. *The uniform flexible beam described by the Euler–Bernoulli equation (13.7) subject to its boundary conditions (13.8)–(13.11), combined with its respective control equations (13.12)–(13.16), is virtually stable in the sense of Definition 2.17.*

Proof: Subtracting (13.7) from (13.12) yields

$$\rho\left[\ddot{y}_r(x,t) - \ddot{y}(x,t)\right] + EI\left[y_r''''(x,t) - y''''(x,t)\right]$$
$$+k_v\left[\dot{y}_r(x,t) - \dot{y}(x,t)\right] = 0. \tag{13.17}$$

As an important step in the VDC approach, the non-negative accompanying function assigned to this flexible link is chosen as

$$\nu = \nu_K + \nu_V \tag{13.18}$$

where

$$\nu_K = \frac{1}{2}\int_0^l \rho\left[\dot{y}_r(x,t) - \dot{y}(x,t)\right]^2 dx \tag{13.19}$$

$$\nu_V = \frac{1}{2}\int_0^l EI\left[y_r''(x,t) - \dot{y}''(x,t)\right]^2 dx. \tag{13.20}$$

With integration by parts, it follows from (13.17) and (13.19) that

$$\begin{aligned}
\dot{\nu}_K &= \int_0^l \rho\left[\dot{y}_r(x,t) - \dot{y}(x,t)\right]\left[\ddot{y}_r(x,t) - \ddot{y}(x,t)\right] dx \\
&= -\int_0^l \left[\dot{y}_r(x,t) - \dot{y}(x,t)\right] EI\left[y_r''''(x,t) - y''''(x,t)\right] dx \\
&\quad -\int_0^l k_v\left[\dot{y}_r(x,t) - \dot{y}(x,t)\right]^2 dx \\
&= -\left[\dot{y}_r(x,t) - \dot{y}(x,t)\right] EI\left[y_r'''(x,t) - y'''(x,t)\right] \Big|_0^l \\
&\quad +\int_0^l \left[\dot{y}_r'(x,t) - \dot{y}'(x,t)\right] EI\left[y_r'''(x,t) - y'''(x,t)\right] dx \\
&\quad -\int_0^l k_v\left[\dot{y}_r(x,t) - \dot{y}(x,t)\right]^2 dx \\
&= -\left[\dot{y}_r(x,t) - \dot{y}(x,t)\right] EI\left[y_r'''(x,t) - y'''(x,t)\right] \Big|_0^l \\
&\quad +\left[\dot{y}_r'(x,t) - \dot{y}'(x,t)\right] EI\left[y_r''(x,t) - y''(x,t)\right] \Big|_0^l \\
&\quad -\int_0^l \left[\dot{y}_r''(x,t) - \dot{y}''(x,t)\right] EI\left[y_r''(x,t) - y''(x,t)\right] dx \\
&\quad -\int_0^l k_v\left[\dot{y}_r(x,t) - \dot{y}(x,t)\right]^2 dx
\end{aligned} \tag{13.21}$$

holds. Substituting (13.21) and the time derivative of (13.20) into the time derivative of (13.18) yields

$$\dot{\nu} = -\int_0^l k_v\left[\dot{y}_r(x,t) - \dot{y}(x,t)\right]^2 dx$$
$$+p_B - p_T \tag{13.22}$$

where

$$\begin{aligned}
p_B &= [\dot{y}_r(0,t) - \dot{y}(0,t)]\, EI\, [y_r'''(0,t) - y'''(0,t)] \\
&\quad - [\dot{y}_r'(0,t) - \dot{y}'(0,t)]\, EI\, [y_r''(0,t) - y''(0,t)] \\
&= [\dot{y}_r(0,t) - \dot{y}(0,t)]\, (f_{Br} - f_B) \\
&\quad + [\dot{y}_r'(0,t) - \dot{y}'(0,t)]\, (m_{Br} - m_B) \qquad (13.23)\\
p_T &= [\dot{y}_r(l,t) - \dot{y}(l,t)]\, EI\, [y_r'''(l,t) - y'''(l,t)] \\
&\quad - [\dot{y}_r'(l,t) - \dot{y}'(l,t)]\, EI\, [y_r''(l,t) - y''(l,t)] \\
&= [\dot{y}_r(l,t) - \dot{y}(l,t)]\, (f_{Tr} - f_T) \\
&\quad + [\dot{y}_r'(l,t) - \dot{y}'(l,t)]\, (m_{Tr} - m_T) \qquad (13.24)
\end{aligned}$$

denote the *virtual power flows* at the two ends of the beam, in view of (13.8)–(13.11) and (13.13)–(13.16).

Consider the fact that the flexible beam has one *driving cutting point* associated with point T and one *driven cutting point* associated with point B. Using (13.18)–(13.20), (13.22), and Definition 2.17 completes the proof. ■

13.3.4 Control Implementations

In this subsection, the implementation of the control equations (13.12)–(13.16) is discussed.

The control algorithm is to find $y_r(0,t)$, $y_r'(0,t)$, f_{Br}, and m_{Br} from given $y_r(l,t)$, $y_r'(l,t)$, f_{Tr}, and m_{Tr}. This process is equivalent to finding $y_r(0,t)$, $y_r'(0,t)$, $y_r''(0,t)$, and $y_r'''(0,t)$ from given $y_r(l,t)$, $y_r'(l,t)$, $y_r''(l,t)$, and $y_r'''(l,t)$ subject to (13.12), since $y_r''(0,t)$ and $y_r'''(0,t)$ yield f_{Br} and m_{Br} from (13.13) and (13.14) and f_{Tr} and m_{Tr} yield $y_r'''(l,t)$ and $y_r''(l,t)$ from (13.15) and (13.16).

In the control implementation, the link is divided into $N > 0$ discrete sections with $x(0) = 0$ and $x(N) = l$.

The backward differentiation is applied to both time and spatial variables. It follows that

$$\dot{y}(x,k) = \frac{y(x,k) - y(x,k-1)}{\Delta T} \qquad (13.25)$$

$$y'(x(j),t) = \frac{y(x(j),t) - y(x(j-1),t)}{\Delta x} \qquad (13.26)$$

are used, where ΔT is the sampling time of the control system with k being a positive integer and $\Delta x = l/N$ is the length of a discrete section with $j \in \{1, N\}$.

The overall computational algorithm is summarized as follows:

Step 1: For a given $\ddot{y}_r(l,k)$, compute

$$y_r''''(l,k) = -\frac{\rho\ddot{y}_r(l,k) + k_v\, [\dot{y}_r(l,k) - \dot{y}(l,k)]}{EI} \qquad (13.27)$$

from (13.12).

Step 2: Given $y_r(l,k)$, $y_r'(l,k)$, $y_r''(l,k)$, $y_r'''(l,k)$, and $y_r''''(l,k)$, compute

$$y_r'''(x(N-1),k) = y_r'''(x(N),k) - \Delta x y_r''''(x(N),k) \tag{13.28}$$
$$y_r''(x(N-1),k) = y_r''(x(N),k) - \Delta x y_r'''(x(N),k) \tag{13.29}$$
$$y_r'(x(N-1),k) = y_r'(x(N),k) - \Delta x y_r''(x(N),k) \tag{13.30}$$
$$y_r(x(N-1),k) = y_r(x(N),k) - \Delta x y_r'(x(N),k). \tag{13.31}$$

Step 3: The required velocity and acceleration at $x(N-1)$ are updated by

$$\dot{y}_r(x(N-1),k) = \frac{y_r(x(N-1),k) - y_r(x(N-1),k-1)}{\Delta T} \tag{13.32}$$

$$\ddot{y}_r(x(N-1),k) = \frac{\dot{y}_r(x(N-1),k) - \dot{y}_r(x(N-1),k-1)}{\Delta T}. \tag{13.33}$$

Step 4: Repeat step 1 to step 3 iteratively from $x(N-1)$ to $x(0)$ to obtain $y_r(0,k)$, $y_r'(0,k)$, $y_r''(0,k)$, and $y_r'''(0,k)$ and further to obtain f_{Br} and m_{Br} from (13.13) and (13.14).

In some applications, the linear position or angle at point B may be subject to certain constraints, such as $y(0,t) = 0$ for a link mounted on a motor rotor or $y'(0,t) = 0$ when only linear motion is permitted. In such a case, the corresponding required variable at point T has to be released by adding a constant. For instance, if $y_r(0,t) = y(0,t) = 0$ is required, then $y_r(l,t)$ has to be released by adding a constant to be specified by satisfying $y_r(0,t) = 0$; or if $y_r'(0,t) = y'(0,t) = 0$ is required, then $y_r'(l,t)$ has to be released by adding a constant to be specified by satisfying $y_r'(0,t) = 0$.

13.3.5 Case Study

A case study with simulation results has been reported in [228]. The simulations used pure link stiffness without imposing additional damping. Consistent responses have been observed with respect to two simulated systems having a significant stiffness difference of more than *four* orders of magnitude. Meanwhile, the simulation results demonstrated very smooth responses without showing oscillation phenomena. This is one of the advantages of using the VDC approach that brings in active damping to the system and guarantees the asymptotic position and velocity convergences.

13.3.6 Adaptive Control in 2D Motion

In (13.12), both dynamic parameters ρ and EI are assumed to be known. When they are not exactly known, the following control equation has to be used instead

$$\hat{\rho}\ddot{y}_r(x,t) + \widehat{EI} y_r''''(x,t) + k_v\left[\dot{y}_r(x,t) - \dot{y}(x,t)\right] = 0 \tag{13.34}$$

for all $x \in [0, l]$ together with the boundary conditions

$$f_{Br} = \widehat{EI} y_r'''(0, t) \tag{13.35}$$

$$m_{Br} = -\widehat{EI} y_r''(0, t) \tag{13.36}$$

$$f_{Tr} = \widehat{EI} y_r'''(l, t) \tag{13.37}$$

$$m_{Tr} = -\widehat{EI} y_r''(l, t) \tag{13.38}$$

where $\hat{\rho}$ and $\widehat{EI}$ denote the estimates of ρ and EI, respectively, and are updated by using the $\mathcal{P}$ function defined by (2.79) as

$$\hat{\rho} = \mathcal{P}(s_\rho, \rho_\rho, \underline{\rho}, \overline{\rho}, t) \tag{13.39}$$

$$\widehat{EI} = \mathcal{P}(s_{EI}, \rho_{EI}, \underline{EI}, \overline{EI}, t) \tag{13.40}$$

with

$$s_\rho = \int_0^l [\dot{y}_r(x, t) - \dot{y}(x, t)] \, \ddot{y}_r(x, t) dx \tag{13.41}$$

$$\begin{aligned} s_{EI} = & \int_0^l [\dot{y}_r(x, t) - \dot{y}(x, t)] \, y_r''''(x, t) dx \\ & - [\dot{y}_r(l, t) - \dot{y}(l, t)] \, y_r'''(l, t) + [\dot{y}_r(0, t) - \dot{y}(0, t)] \, y_r'''(0, t) \\ & + [\dot{y}_r'(l, t) - \dot{y}'(l, t)] \, y_r''(l, t) - [\dot{y}_r'(0, t) - \dot{y}'(0, t)] \, y_r''(0, t). \end{aligned} \tag{13.42}$$

In (13.39) and (13.40), $\rho_\rho > 0$ and $\rho_{EI} > 0$ are two update gains, $\underline{\rho}$ and $\overline{\rho}$ denote the lower and upper bounds of ρ, subject to $\overline{\rho} \geqslant \underline{\rho} > 0$, and $\underline{EI}$ and $\overline{EI}$ denote the lower and upper bounds of EI, subject to $\overline{EI} \geqslant \underline{EI} > 0$.

The following theorem ensures the *virtual stability* of the beam combined with its control equations.

Theorem 13.2. *The uniform flexible beam described by the Euler–Bernoulli equation (13.7) subject to the boundary conditions (13.8)–(13.11), combined with its respective control equations (13.34)–(13.38) and with the parameter adaptation (13.39)-(13.42), is virtually stable in the sense of Definition 2.17.* ■

The proof is given in Appendix B.11.

13.4 Flexible Beams in 3D Motion

In the last section, the motion of a uniform flexible beam is restricted to a two-dimensional plane. These restrictions are removed in this section to allow a free motion of the uniform flexible beam (link) in a three-dimensional geometric space.

The uniform flexible beam (link) is illustrated in Figure 13.2, which is formed by placing two *cutting points* at its both ends. One *cutting point* is called the

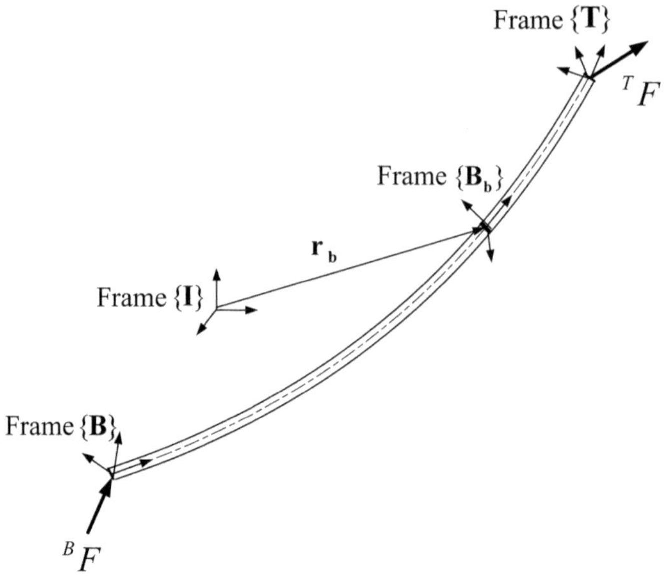

Fig. 13.2. A three-dimensional flexible beam virtually cut from a flexible link robot.

driven cutting point (or initial/base point) to which frame $\{\mathbf{B}\}$ is attached. The other *cutting point* is called the *driving cutting point* (or final/tip point) to which frame $\{\mathbf{T}\}$ is attached. Thus, ${}^{\mathbf{B}}F \in \mathbb{R}^6$ is applied to the beam and ${}^{\mathbf{T}}F \in \mathbb{R}^6$ is applied away from the beam.

A few assumptions are made about this flexible beam.

Assumption 12. *The flexible beam has a uniform mass density along its length l.*

Assumption 13. *The cross-section is either an ellipse or a circle at any distance such that there exist two symmetric and orthogonal axes.*

Assumption 14. *No gravity is considered.*

Remark 13.1. Assumptions 12 and 13 make a practical sense to long robot arms. When a link is long enough, uniform cross-sections with symmetric centers are usually used.

A section of the beam with an infinitesimally small length dx is located at an axial distance of x with $0 \leqslant x \leqslant l$. A body frame $\{\mathbf{B}_b\}$ is attached to the center of this infinitesimally small section in such a way that its x axis aligns with the beam axis, and its y and z axes coincide with the major and minor axes of the cross sectional ellipse (circle). Frame $\{\mathbf{B}_b\}$ coincides with frame $\{\mathbf{B}\}$ at $x = 0$ and coincides with frame $\{\mathbf{T}\}$ at $x = l$.

13.4.1 Extended Hamilton's Principle

The extended Hamilton's principle, as described by (13.4), is again used to derive the dynamic equations of this uniform flexible beam in the 3D motion.

Kinematics

Let $\mathbf{r}_b(x,t) \in \mathbb{R}^3$ represent the coordinates of the origin of frame $\{\mathbf{B}_b\}$, expressed in the inertial frame $\{\mathbf{I}\}$. Meanwhile, let ${}^{\mathbf{B}_b}\mathbf{R}_{\mathbf{I}} \in \mathbb{R}^{3\times 3}$ be the direction cosine matrix that transforms a vector expressed in the inertial frame $\{\mathbf{I}\}$ to the same vector expressed in frame $\{\mathbf{B}_b\}$, and define

$$
{}^{\mathbf{B}_b}\mathbf{r}_b(x,t) = {}^{\mathbf{B}_b}\mathbf{R}_{\mathbf{I}}\mathbf{r}_b(x,t) \stackrel{def}{=} \begin{bmatrix} x_b(x,t) \\ y_b(x,t) \\ z_b(x,t) \end{bmatrix} \in \mathbb{R}^3. \tag{13.43}
$$

Then, let ds be an infinitesimally small area of this infinitesimally small section of the flexible beam, as illustrated in Figure 13.3. The coordinates of this infinitesimally small area referenced and expressed in frame $\{\mathbf{B}_b\}$ can be written as

$$
{}^{\mathbf{B}_b}\mathbf{r}_{sb}(y_s,z_s,x) \stackrel{def}{=} \begin{bmatrix} 0 \\ y_s \\ z_s \end{bmatrix} \in \mathbb{R}^3 \tag{13.44}
$$

such that its coordinates referenced and expressed in the inertial frame $\{\mathbf{I}\}$ are written as

$$
\begin{aligned}
\mathbf{r}_s(y_s,z_s,x,t) &= \mathbf{r}_b(x,t) + \mathbf{r}_{sb}(y_s,z_s,x,t) \\
&= \mathbf{r}_b(x,t) + {}^{\mathbf{I}}\mathbf{R}_{\mathbf{B}_b}\,{}^{\mathbf{B}_b}\mathbf{r}_{sb}(y_s,z_s,x) \in \mathbb{R}^3
\end{aligned} \tag{13.45}
$$

with

$$
{}^{\mathbf{B}_b}\mathbf{r}_{sb}(y_s,z_s,x) = {}^{\mathbf{B}_b}\mathbf{R}_{\mathbf{I}}\mathbf{r}_{sb}(y_s,z_s,x,t).
$$

In (13.44), y_s and z_s denote the y and z coordinates of this infinitesimally small area expressed in frame $\{\mathbf{B}_b\}$. In (13.45), $\mathbf{r}_{sb}(y_s,z_s,x,t) \in \mathbb{R}^3$ denotes the vector pointing from the origin of frame $\{\mathbf{B}_b\}$ to this infinitesimally small area, expressed in the inertial frame $\{\mathbf{I}\}$.

Differentiating $\mathbf{r}_{sb}(y_s,z_s,x,t) \in \mathbb{R}^3$ with respect to time yields

$$
\begin{aligned}
\frac{\partial}{\partial t}\mathbf{r}_{sb}(y_s,z_s,x,t) &= (\boldsymbol{\omega}_b\times)\mathbf{r}_{sb}(y_s,z_s,x,t) \\
&= -(\mathbf{r}_{sb}(y_s,z_s,x,t)\times)\boldsymbol{\omega}_b
\end{aligned} \tag{13.46}
$$

$$
\frac{\partial^2}{\partial t^2}\mathbf{r}_{sb}(y_s,z_s,x,t) = (\boldsymbol{\omega}_b\times)(\boldsymbol{\omega}_b\times)\mathbf{r}_{sb} + (\frac{\partial}{\partial t}\boldsymbol{\omega}_b\times)\mathbf{r}_{sb} \tag{13.47}
$$

since both y_s and z_s are constant, where $\boldsymbol{\omega}_b \in \mathbb{R}^3$ denotes the angular velocity vector of the infinitesimally small section, expressed in the inertial frame $\{\mathbf{I}\}$.

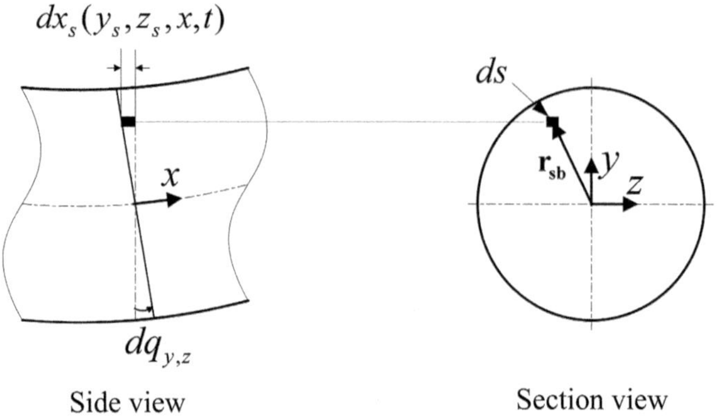

Fig. 13.3. An infinitesimally small section of the flexible beam.

Consequently, the angular velocity vector of the infinitesimally small section, expressed in frame $\{\mathbf{B}_b\}$, is defined by

$$^{\mathbf{B}_b}\boldsymbol{\omega} = {}^{\mathbf{B}_b}\mathbf{R}_{\mathbf{I}}\boldsymbol{\omega}_b \overset{def}{=} \begin{bmatrix} ^{\mathbf{B}_b}\omega_x \\ ^{\mathbf{B}_b}\omega_y \\ ^{\mathbf{B}_b}\omega_z \end{bmatrix} \tag{13.48}$$

and the linear velocity vector of the infinitesimally small section, expressed in frame $\{\mathbf{B}_b\}$, is defined by

$$^{\mathbf{B}_b}\mathbf{R}_{\mathbf{I}}\frac{\partial}{\partial t}\mathbf{r}_b(x,t) = \begin{bmatrix} ^{\mathbf{B}_b}v_{bx} \\ ^{\mathbf{B}_b}v_{by} \\ ^{\mathbf{B}_b}v_{bz} \end{bmatrix} \in \mathbb{R}^3 \tag{13.49}$$

subject to

$$^{\mathbf{B}_b}\omega_y = -\frac{\partial}{\partial x}{}^{\mathbf{B}_b}v_{bz} \tag{13.50}$$

$$^{\mathbf{B}_b}\omega_z = \frac{\partial}{\partial x}{}^{\mathbf{B}_b}v_{by}. \tag{13.51}$$

Kinetic Energy

The kinetic energy of the flexible beam can be expressed as

$$\begin{aligned} e_K &= \int_0^l \int_s \frac{1}{2}\frac{\rho}{S_x}\frac{\partial}{\partial t}\mathbf{r}_s(y_s,z_s,x,t)^T\frac{\partial}{\partial t}\mathbf{r}_s(y_s,z_s,x,t)dsdx \\ &= \int_0^l \int_s \frac{1}{2}\frac{\rho}{S_x}\left[\frac{\partial}{\partial t}\mathbf{r}_b(x,t)^T\frac{\partial}{\partial t}\mathbf{r}_b(x,t)\right. \\ &\qquad \left. + \frac{\partial}{\partial t}\mathbf{r}_{sb}(y_s,z_s,x,t)^T\frac{\partial}{\partial t}\mathbf{r}_{sb}(y_s,z_s,x,t)\right] dsdx \end{aligned}$$

$$= \int_0^l \frac{1}{2}\rho \frac{\partial}{\partial t}\mathbf{r}_b(x,t)^T \frac{\partial}{\partial t}\mathbf{r}_b(x,t)dx$$
$$+ \int_0^l \int_s \frac{1}{2}\frac{\rho}{S_x}\frac{\partial}{\partial t}\mathbf{r}_{sb}(y_s,z_s,x,t)^T \frac{\partial}{\partial t}\mathbf{r}_{sb}(y_s,z_s,x,t)dsdx \qquad (13.52)$$

where ρ denotes the mass per unit length of the beam and S_x is the section area. In deriving (13.52)

$$\int_0^l \int_s \frac{1}{2}\frac{\rho}{S_x}\frac{\partial}{\partial t}\mathbf{r}_b(x,t)^T \frac{\partial}{\partial t}\mathbf{r}_{sb}(y_s,z_s,x,t)dsdx = 0 \qquad (13.53)$$

is used due to the fact that the cross sections are symmetric about their centers.

Integrating the variation of e_K over the time period from t_1 to t_2 yields

$$\begin{aligned}
\int_{t_1}^{t_2} \delta e_K dt &= \int_{t_1}^{t_2}\int_0^l \rho\frac{\partial}{\partial t}\mathbf{r}_b(x,t)^T\delta\frac{\partial}{\partial t}\mathbf{r}_b(x,t)dxdt \\
&\quad + \int_{t_1}^{t_2}\int_0^l\int_s \frac{\rho}{S_x}\frac{\partial}{\partial t}\mathbf{r}_{sb}(y_s,z_s,x,t)^T\delta\frac{\partial}{\partial t}\mathbf{r}_{sb}(y_s,z_s,x,t)dsdxdt \\
&= \int_0^l\int_{t_1}^{t_2} \rho\frac{\partial}{\partial t}\mathbf{r}_b(x,t)^T\delta\frac{\partial}{\partial t}\mathbf{r}_b(x,t)dtdx \\
&\quad + \int_0^l\int_s\int_{t_1}^{t_2} \frac{\rho}{S_x}\frac{\partial}{\partial t}\mathbf{r}_{sb}(y_s,z_s,x,t)^T\delta\frac{\partial}{\partial t}\mathbf{r}_{sb}(y_s,z_s,x,t)dtdsdx \\
&= \int_0^l \rho\frac{\partial}{\partial t}\mathbf{r}_b(x,t)^T\delta\mathbf{r}_b(x,t)\Big|_{t_1}^{t_2} dx \\
&\quad - \int_0^l\int_{t_1}^{t_2} \rho\frac{\partial^2}{\partial t^2}\mathbf{r}_b(x,t)^T\delta\mathbf{r}_b(x,t)dtdx \\
&\quad + \int_0^l\int_s \frac{\rho}{S_x}\frac{\partial}{\partial t}\mathbf{r}_{sb}(y_s,z_s,x,t)^T\delta\mathbf{r}_{sb}(y_s,z_s,x,t)\Big|_{t_1}^{t_2} dsdx \\
&\quad - \int_0^l\int_s\int_{t_1}^{t_2} \frac{\rho}{S_x}\frac{\partial^2}{\partial t^2}\mathbf{r}_{sb}(y_s,z_s,x,t)^T\delta\mathbf{r}_{sb}(y_s,z_s,x,t)dtdsdx \\
&= -\int_0^l\int_{t_1}^{t_2} \rho\frac{\partial^2}{\partial t^2}\mathbf{r}_b(x,t)^T\delta\mathbf{r}_b(x,t)dtdx \\
&\quad - \int_0^l\int_s\int_{t_1}^{t_2} \frac{\rho}{S_x}\frac{\partial^2}{\partial t^2}\mathbf{r}_{sb}(y_s,z_s,x,t)^T\delta\mathbf{r}_{sb}(y_s,z_s,x,t)dtdsdx
\end{aligned} \qquad (13.54)$$

in which $\delta\mathbf{r}_b(x,t) = \mathbf{0}$ and $\delta\mathbf{r}_{sb}(y_s,z_s,x,t) = \mathbf{0}$ at $t = t_1$ and $t = t_2$ are used.

By using (13.46), (13.47), and

$$\delta\mathbf{r}_{sb} - (\delta\mathbf{q}_b\times)\mathbf{r}_{sb} = -(\mathbf{r}_{sb}\times)\delta\mathbf{q}_b \qquad (13.55)$$
$${}^{\mathbf{A}}\mathbf{R}_{\mathbf{I}}(\mathbf{a}\times)\mathbf{b} = ({}^{\mathbf{A}}\mathbf{R}_{\mathbf{I}}\mathbf{a}\times){}^{\mathbf{A}}\mathbf{R}_{\mathbf{I}}\mathbf{b} \qquad (13.56)$$

where $\delta\mathbf{q}_b \in \mathbb{R}^3$ denotes the angular variation vector of frame $\{\mathbf{B}_b\}$ and $\mathbf{a} \in \mathbb{R}^3$ and $\mathbf{b} \in \mathbb{R}^3$ are two (arbitrary) vectors, it follows that

$$\begin{aligned}
&-\int_0^l \int_s \int_{t_1}^{t_2} \frac{\rho}{S_x}\frac{\partial^2}{\partial t^2}\mathbf{r}_{sb}(y_s,z_s,x,t)^T\delta\mathbf{r}_{sb}(y_s,z_s,x,t)dtdsdx \\
&= -\int_0^l \int_{t_1}^{t_2} \int_s \frac{\rho}{S_x}\frac{\partial^2}{\partial t^2}\mathbf{r}_{sb}(y_s,z_s,x,t)^T\delta\mathbf{r}_{sb}(y_s,z_s,x,t)dsdtdx \\
&= -\int_0^l \int_{t_1}^{t_2} \int_s \frac{\rho}{S_x}\left[(\boldsymbol{\omega}_b\times)(\boldsymbol{\omega}_b\times)\mathbf{r}_{sb} + (\frac{\partial}{\partial t}\boldsymbol{\omega}_b\times)\mathbf{r}_{sb}\right]^T (\delta\mathbf{q}_b\times)\mathbf{r}_{sb}dsdtdx \\
&= -\int_0^l \int_{t_1}^{t_2} \int_s \frac{\rho}{S_x}\left[(\boldsymbol{\omega}_b\times)(\mathbf{r}_{sb}\times)\boldsymbol{\omega}_b + (\mathbf{r}_{sb}\times)\frac{\partial}{\partial t}\boldsymbol{\omega}_b\right]^T (\mathbf{r}_{sb}\times)\delta\mathbf{q}_b dsdtdx \\
&= -\int_0^l \int_{t_1}^{t_2} \int_s \frac{\rho}{S_x}\left[(\mathbf{r}_{sb}\times)^T(\boldsymbol{\omega}_b\times)(\mathbf{r}_{sb}\times)\boldsymbol{\omega}_b\right. \\
&\qquad \left. +(\mathbf{r}_{sb}\times)^T(\mathbf{r}_{sb}\times)\frac{\partial}{\partial t}\boldsymbol{\omega}_b\right]^T \delta\mathbf{q}_b dsdtdx \\
&= -\int_0^l \int_{t_1}^{t_2} \left[\int_s \frac{\rho}{S_x}(\mathbf{r}_{sb}\times)^T(\boldsymbol{\omega}_b\times)(\mathbf{r}_{sb}\times)ds\boldsymbol{\omega}_b\right. \\
&\qquad \left. +\int_s \frac{\rho}{S_x}(\mathbf{r}_{sb}\times)^T(\mathbf{r}_{sb}\times)ds\frac{\partial}{\partial t}\boldsymbol{\omega}_b\right]^T \delta\mathbf{q}_b dtdx \\
&= -\int_0^l \int_{t_1}^{t_2} \left[\left(\int_s \frac{\rho}{S_x}({}^{\mathbf{B}_b}\mathbf{r}_{sb}\times)^T({}^{\mathbf{B}_b}\boldsymbol{\omega}\times)({}^{\mathbf{B}_b}\mathbf{r}_{sb}\times)ds\right){}^{\mathbf{B}_b}\boldsymbol{\omega}\right. \\
&\qquad \left. +\left(\int_s \frac{\rho}{S_x}({}^{\mathbf{B}_b}\mathbf{r}_{sb}\times)^T({}^{\mathbf{B}_b}\mathbf{r}_{sb}\times)ds\right){}^{\mathbf{B}_b}\mathbf{R}_{\mathbf{I}}\frac{\partial}{\partial t}\omega_b\right]^T {}^{\mathbf{B}_b}\mathbf{R}_{\mathbf{I}}\delta\mathbf{q}_b dtdx
\end{aligned} \tag{13.57}$$

holds, where ${}^{\mathbf{B}_b}\boldsymbol{\omega} \in \mathbb{R}^3$, as defined by (13.48), denotes the angular velocity vector of the infinitesimally small section, expressed in frame $\{\mathbf{B}_b\}$.

Note that Assumption 13 implies

$$\int_s y_s z_s ds = 0. \tag{13.58}$$

Using (13.44) and (13.58) yields

$$\int_s ({}^{\mathbf{B}_b}\mathbf{r}_{sb}\times)^T({}^{\mathbf{B}_b}\mathbf{r}_{sb}\times)ds = \mathrm{diag}(J_x, I_y, I_z) \tag{13.59}$$

$$\begin{aligned}
&\int_s ({}^{\mathbf{B}_b}\mathbf{r}_{sb}\times)^T({}^{\mathbf{B}_b}\boldsymbol{\omega}\times)({}^{\mathbf{B}_b}\mathbf{r}_{sb}\times)ds \\
&\quad = \begin{bmatrix} 0 & -I_y{}^{\mathbf{B}_b}\omega_z & I_z{}^{\mathbf{B}_b}\omega_y \\ I_y{}^{\mathbf{B}_b}\omega_z & 0 & 0 \\ -I_z{}^{\mathbf{B}_b}\omega_y & 0 & 0 \end{bmatrix}
\end{aligned} \tag{13.60}$$

with

$$I_z = \int_s y_s^2 ds \tag{13.61}$$

$$I_y = \int_s z_s^2 ds \tag{13.62}$$

$$J_x = \int_s (y_s^2 + z_s^2) ds = I_z + I_y. \tag{13.63}$$

Substituting (13.57)–(13.60) into (13.54) yields

$$\begin{aligned}\int_{t_1}^{t_2} \delta e_K dt = &-\int_0^l \int_{t_1}^{t_2} \rho \left({}^{\mathbf{B}_b}\mathbf{R}_\mathbf{I} \frac{\partial^2}{\partial t^2} \mathbf{r}_b(x,t) \right)^T {}^{\mathbf{B}_b}\mathbf{R}_\mathbf{I} \delta \mathbf{r}_b(x,t) dt dx \\ &- \int_0^l \int_{t_1}^{t_2} \left\{ \frac{\rho}{S_x} \mathrm{diag}(J_x, I_y, I_z) {}^{\mathbf{B}_b}\mathbf{R}_\mathbf{I} \frac{\partial}{\partial t} \boldsymbol{\omega}_b \right. \\ &\left. + \frac{\rho}{S_x} \begin{bmatrix} (I_z - I_y) {}^{\mathbf{B}_b}\omega_y {}^{\mathbf{B}_b}\omega_z \\ I_y {}^{\mathbf{B}_b}\omega_x {}^{\mathbf{B}_b}\omega_z \\ -I_z {}^{\mathbf{B}_b}\omega_x {}^{\mathbf{B}_b}\omega_y \end{bmatrix} \right\}^T {}^{\mathbf{B}_b}\mathbf{R}_\mathbf{I} \delta \mathbf{q}_b dt dx. \end{aligned} \tag{13.64}$$

Potential Energy

In view of Figure 13.3, the potential energy of the flexible beam is stored when the beam is in deformation. Let $u_x(x,t)$ be the axial deformation of the flexible beam. Thus, the axial deformation of the infinitesimally small section of length dx can then be written as

$$du_x(x,t) = \frac{\partial u_x(x,t)}{\partial x} dx. \tag{13.65}$$

Meanwhile, the angular deformations around the y and z axes of frame $\{\mathbf{B}_b\}$ can be represented by

$$dq_y(x,t) = -\frac{\partial^2 z_b(x,t)}{\partial x^2} dx \tag{13.66}$$

$$dq_z(x,t) = \frac{\partial^2 y_b(x,t)}{\partial x^2} dx. \tag{13.67}$$

As a result, the deformation along the x axis at the infinitesimally small area ds can be linearly approximated as

$$dx_s(y_s, z_s, x, t) = du_x(x,t) + z_s dq_y(x,t) - y_s dq_z(x,t). \tag{13.68}$$

Let E be the Young's modulus. Thus, the equivalent linear stiffness of the infinitesimally small area ds becomes $E\frac{ds}{dx}$. Consequently, the potential energy stored in this infinitesimally small area ds due to the axial deformation is

$$\frac{1}{2}E\frac{ds}{dx}(dx_s)^2.$$

On the other hand, let G be the shear modulus and $\Theta_x(x,t)$ be the torsional deformation of the beam. Thus, the torsional deformation of the infinitesimally small section of the flexible beam can be expressed as

$$d\Theta_x(x,t) = \frac{\partial\Theta_x(x,t)}{\partial x}dx. \tag{13.69}$$

Then, the shear deformation at the infinitesimally small area ds becomes $d\Theta_x\sqrt{y_s^2+z_s^2}$, leading to a shear force of $dF = Gds\frac{d\Theta_x\sqrt{y_s^2+z_s^2}}{dx}$ and further resulting in a shear torque of $d\tau = Gds\frac{d\Theta_x(y_s^2+z_s^2)}{dx}$. Thus, the torsional stiffness of the infinitesimally small area ds is $G(y_s^2+z_s^2)\frac{ds}{dx}$. Consequently, the potential energy stored in the infinitesimally small area ds due to the shear deformation is

$$\frac{1}{2}G(y_s^2+z_s^2)\frac{ds}{dx}(d\Theta_x)^2.$$

Finally, the overall potential energy of the flexible beam without subject to the gravity can be written as

$$\int_0^l\int_s\frac{1}{2}E\frac{ds}{dx}(dx_s)^2 + \int_0^l\int_s\frac{1}{2}G(y_s^2+z_s^2)\frac{ds}{dx}(d\Theta_x)^2.$$

By using

$$\int_s 2z_s du_x dq_y ds = 0 \tag{13.70}$$

$$\int_s 2y_s du_x dq_z ds = 0 \tag{13.71}$$

$$\int_s 2y_s z_s dq_y dq_z ds = 0 \tag{13.72}$$

from Assumption 13 and using (13.61)–(13.63) and (13.65)–(13.69), the overall potential energy of the flexible beam becomes

$$\begin{aligned}
e_P &= \int_0^l\int_s\frac{1}{2}E\frac{ds}{dx}(dx_s)^2 + \int_0^l\int_s\frac{1}{2}G(y_s^2+z_s^2)\frac{ds}{dx}(d\Theta_x)^2 \\
&= \int_0^l\int_s\frac{1}{2}E\frac{ds}{(dx)^2}\left[du_x^2 + z_s^2(dq_y)^2 + y_s^2(dq_z)^2\right. \\
&\quad \left.+2z_s du_x dq_y - 2y_s du_x dq_z - 2y_s z_s dq_y dq_z\right]dx \\
&\quad + \int_0^l\int_s\frac{1}{2}G(y_s^2+z_s^2)\frac{ds}{(dx)^2}(d\Theta_x)^2 dx \\
&= \int_0^l\int_s\frac{1}{2}E\frac{1}{(dx)^2}\left[du_x^2 + z_s^2(dq_y)^2 + y_s^2(dq_z)^2\right] ds dx
\end{aligned}$$

$$
\begin{aligned}
&+\int_0^l \int_s \frac{1}{2} G(y_s^2+z_s^2) \frac{(d\Theta_x(x,t))^2}{(dx)^2} dsdx \\
&= \int_0^l \frac{1}{2} ES_x \left(\frac{\partial u_x(x,t)}{\partial x}\right)^2 dx + \int_0^l \int_s \frac{1}{2} Ez_s^2 \left(\frac{\partial^2 z_b(x,t)}{\partial x^2}\right)^2 dsdx \\
&+\int_0^l \int_s \frac{1}{2} Ey_s^2 \left(\frac{\partial^2 y_b(x,t)}{\partial x^2}\right)^2 dsdx \\
&+\int_0^l \int_s \frac{1}{2} G(y_s^2+z_s^2) \left(\frac{\partial \Theta_x(x,t)}{\partial x}\right)^2 dsdx \\
&= \int_0^l \frac{1}{2} ES_x \left(\frac{\partial u_x(x,t)}{\partial x}\right)^2 dx + \int_0^l \frac{1}{2} EI_z \left(\frac{\partial^2 y_b(x,t)}{\partial x^2}\right)^2 dx \\
&+\int_0^l \frac{1}{2} EI_y \left(\frac{\partial^2 z_b(x,t)}{\partial x^2}\right)^2 dx + \int_0^l \frac{1}{2} GJ_x \left(\frac{\partial \Theta_x(x,t)}{\partial x}\right)^2 dx.
\end{aligned}
\tag{13.73}
$$

Finally, integrating the variation of e_P over the time period from t_1 to t_2 and using integration by parts yields

$$
\begin{aligned}
\int_{t_1}^{t_2} \delta e_P dt &= \int_{t_1}^{t_2} \int_0^l ES_x \left(\frac{\partial u_x(x,t)}{\partial x}\right) \delta \left(\frac{\partial u_x(x,t)}{\partial x}\right) dxdt \\
&+\int_{t_1}^{t_2} \int_0^l EI_z \left(\frac{\partial^2 y_b(x,t)}{\partial x^2}\right) \delta \left(\frac{\partial^2 y_b(x,t)}{\partial x^2}\right) dxdt \\
&+\int_{t_1}^{t_2} \int_0^l EI_y \left(\frac{\partial^2 z_b(x,t)}{\partial x^2}\right) \delta \left(\frac{\partial^2 z_b(x,t)}{\partial x^2}\right) dxdt \\
&+\int_{t_1}^{t_2} \int_0^l GJ_x \left(\frac{\partial \Theta_x(x,t)}{\partial x}\right) \delta \left(\frac{\partial \Theta_x(x,t)}{\partial x}\right) dxdt \\
&= \int_{t_1}^{t_2} ES_x \left(\frac{\partial u_x(x,t)}{\partial x}\right) \delta u_x(x,t)\Big|_0^l dt \\
&\quad \int_{t_1}^{t_2} \int_0^l ES_x \left(\frac{\partial^2 u_x(x,t)}{\partial x^2}\right) \delta u_x(x,t) dxdt \\
&+\int_{t_1}^{t_2} EI_z \left(\frac{\partial^2 y_b(x,t)}{\partial x^2}\right) \delta \left(\frac{\partial y_b(x,t)}{\partial x}\right)\Big|_0^l dt \\
&-\int_{t_1}^{t_2} \int_0^l EI_z \left(\frac{\partial^3 y_b(x,t)}{\partial x^3}\right) \delta \left(\frac{\partial y_b(x,t)}{\partial x}\right) dxdt \\
&+\int_{t_1}^{t_2} EI_y \left(\frac{\partial^2 z_b(x,t)}{\partial x^2}\right) \delta \left(\frac{\partial z_b(x,t)}{\partial x}\right)\Big|_0^l dt \\
&-\int_{t_1}^{t_2} \int_0^l EI_y \left(\frac{\partial^3 z_b(x,t)}{\partial x^3}\right) \delta \left(\frac{\partial z_b(x,t)}{\partial x}\right) dxdt
\end{aligned}
$$

$$
\begin{aligned}
&+\int_{t_1}^{t_2} GJ_x\left(\frac{\partial \Theta_x(x,t)}{\partial x}\right)\delta\Theta_x(x,t)\Big|_0^l dt \\
&-\int_{t_1}^{t_2}\int_0^l GJ_x\left(\frac{\partial^2 \Theta_x(x,t)}{\partial x^2}\right)\delta\Theta_x(x,t)dxdt \\
=&\int_{t_1}^{t_2} ES_x\left(\frac{\partial u_x(x,t)}{\partial x}\right)\delta u_x(x,t)\Big|_0^l dt \\
&-\int_{t_1}^{t_2}\int_0^l ES_x\left(\frac{\partial^2 u_x(x,t)}{\partial x^2}\right)\delta u_x(x,t)dxdt \\
&+\int_{t_1}^{t_2} EI_z\left(\frac{\partial^2 y_b(x,t)}{\partial x^2}\right)\delta\left(\frac{\partial y_b(x,t)}{\partial x}\right)\Big|_0^l dt \\
&-\int_{t_1}^{t_2}\int_0^l \alpha_z(x,t)EI_z\left(\frac{\partial^3 y_b(x,t)}{\partial x^3}\right)\delta\left(\frac{\partial y_b(x,t)}{\partial x}\right)dxdt \\
&-\int_{t_1}^{t_2}(1-\alpha_z(x,t))EI_z\left(\frac{\partial^3 y_b(x,t)}{\partial x^3}\right)\delta y_b(x,t)\Big|_0^l dt \\
&+\int_{t_1}^{t_2}\int_0^l EI_z\frac{\partial}{\partial x}\left((1-\alpha_z(x,t))\frac{\partial^3 y_b(x,t)}{\partial x^3}\right)\delta y_b(x,t)dxdt \\
&+\int_{t_1}^{t_2} EI_y\left(\frac{\partial^2 z_b(x,t)}{\partial x^2}\right)\delta\left(\frac{\partial z_b(x,t)}{\partial x}\right)\Big|_0^l dt \\
&-\int_{t_1}^{t_2}\int_0^l \alpha_y(x,t)EI_y\left(\frac{\partial^3 z_b(x,t)}{\partial x^3}\right)\delta\left(\frac{\partial z_b(x,t)}{\partial x}\right)dxdt \\
&-\int_{t_1}^{t_2}(1-\alpha_y(x,t))EI_y\left(\frac{\partial^3 z_b(x,t)}{\partial x^3}\right)\delta z_b(x,t)\Big|_0^l dt \\
&+\int_{t_1}^{t_2}\int_0^l EI_y\frac{\partial}{\partial x}\left((1-\alpha_y(x,t))\frac{\partial^3 z_b(x,t)}{\partial x^3}\right)\delta z_b(x,t)dxdt \\
&+\int_{t_1}^{t_2} GJ_x\left(\frac{\partial \Theta_x(x,t)}{\partial x}\right)\delta\Theta_x(x,t)\Big|_0^l dt \\
&-\int_{t_1}^{t_2}\int_0^l GJ_x\left(\frac{\partial^2 \Theta_x(x,t)}{\partial x^2}\right)\delta\Theta_x(x,t)dxdt
\end{aligned}
\tag{13.74}
$$

subject to[2] $|\alpha_z(x,t)| \ll 1$ and $|\alpha_y(x,t)| \ll 1$.

External Work

The external work applied to the beam is given by the rest of the robot through the two *cutting points* located at both ends. The integral of the variation of the external work $\overline{w_E}$ over the time period from t_1 to t_2 can be expressed by

[2] The physical meanings of $\alpha_z(x,t)$ and $\alpha_y(x,t)$ will be discussed later.

$$\begin{aligned}\int_{t_1}^{t_2} \delta\overline{w_E}dt &= \int_{t_1}^{t_2} \mathbf{f}_\mathbf{B}^T\delta\mathbf{r}_\mathbf{B} + \mathbf{m}_\mathbf{B}^T\delta\mathbf{q}_\mathbf{B} - \mathbf{f}_\mathbf{T}^T\delta\mathbf{r}_\mathbf{T} - \mathbf{m}_\mathbf{T}^T\delta\mathbf{q}_\mathbf{T}dt \\ &= \int_{t_1}^{t_2} ({}^\mathbf{B}\mathbf{R}_\mathbf{I}\mathbf{f}_\mathbf{B})^T({}^\mathbf{B}\mathbf{R}_\mathbf{I}\delta\mathbf{r}_\mathbf{B}) + ({}^\mathbf{B}\mathbf{R}_\mathbf{I}\mathbf{m}_\mathbf{B})^T({}^\mathbf{B}\mathbf{R}_\mathbf{I}\delta\mathbf{q}_\mathbf{B}) \\ &\quad -({}^\mathbf{T}\mathbf{R}_\mathbf{I}\mathbf{f}_\mathbf{T})^T({}^\mathbf{T}\mathbf{R}_\mathbf{I}\delta\mathbf{r}_\mathbf{T}) - ({}^\mathbf{T}\mathbf{R}_\mathbf{I}\mathbf{m}_\mathbf{T})^T({}^\mathbf{T}\mathbf{R}_\mathbf{I}\delta\mathbf{q}_\mathbf{T})dt \end{aligned} \tag{13.75}$$

where

$\mathbf{f}_\mathbf{B} \in \mathbb{R}^3$	Force vector applied to frame $\{\mathbf{B}\}$.
$\mathbf{m}_\mathbf{B} \in \mathbb{R}^3$	Moment vector applied to frame $\{\mathbf{B}\}$.
$\mathbf{f}_\mathbf{T} \in \mathbb{R}^3$	Force vector applied away from frame $\{\mathbf{T}\}$.
$\mathbf{m}_\mathbf{T} \in \mathbb{R}^3$	Moment vector applied away from frame $\{\mathbf{T}\}$.
$\mathbf{r}_\mathbf{B} \in \mathbb{R}^3$	Position vector of frame $\{\mathbf{B}\}$.
$\mathbf{q}_\mathbf{B} \in \mathbb{R}^3$	Orientation vector of frame $\{\mathbf{B}\}$.
$\mathbf{r}_\mathbf{T} \in \mathbb{R}^3$	Position vector of frame $\{\mathbf{T}\}$.
$\mathbf{q}_\mathbf{T} \in \mathbb{R}^3$	Orientation vector of frame $\{\mathbf{T}\}$.

These variables are all expressed in the inertial frame $\{\mathbf{I}\}$.

13.4.2 Beam Dynamics

By using

$$\delta x_b(x,t) = \delta u_x(x,t) \tag{13.76}$$

$$^{\mathbf{B}_b}\mathbf{R}_\mathbf{I}\delta\mathbf{r}_b(x,t) = \begin{bmatrix} \delta u_x(x,t) \\ \delta y_b(x,t) \\ \delta z_b(x,t) \end{bmatrix} \tag{13.77}$$

$$^{\mathbf{B}_b}\mathbf{R}_\mathbf{I}\delta\mathbf{q}_b = \begin{bmatrix} \delta\Theta_x(x,t) \\ -\delta\left(\frac{\partial z_b(x,t)}{\partial x}\right) \\ \delta\left(\frac{\partial y_b(x,t)}{\partial x}\right) \end{bmatrix} \tag{13.78}$$

$$^{\mathbf{B}}\mathbf{R}_\mathbf{I}\delta\mathbf{r}_\mathbf{B} = \begin{bmatrix} \delta u_x(x,t)|_{x=0} \\ \delta y_b(x,t)|_{x=0} \\ \delta z_b(x,t)|_{x=0} \end{bmatrix} \tag{13.79}$$

$$^{\mathbf{B}}\mathbf{R}_\mathbf{I}\delta\mathbf{q}_\mathbf{B} = \begin{bmatrix} \delta\Theta_x(x,t)|_{x=0} \\ -\delta\left(\frac{\partial z_b(x,t)}{\partial x}\right)\Big|_{x=0} \\ \delta\left(\frac{\partial y_b(x,t)}{\partial x}\right)\Big|_{x=0} \end{bmatrix} \tag{13.80}$$

$$^{\mathbf{T}}\mathbf{R}_\mathbf{I}\delta\mathbf{r}_\mathbf{T} = \begin{bmatrix} \delta u_x(x,t)|_{x-l} \\ \delta y_b(x,t)|_{x=l} \\ \delta z_b(x,t)|_{x=l} \end{bmatrix} \tag{13.81}$$

$$
{}^{\mathbf{T}}\mathbf{R_I}\delta\mathbf{q_T} = \begin{bmatrix} \delta\Theta_x(x,t)\big|_{x=l} \\ -\delta\left(\frac{\partial z_b(x,t)}{\partial x}\right)\Big|_{x=l} \\ \delta\left(\frac{\partial y_b(x,t)}{\partial x}\right)\Big|_{x=l} \end{bmatrix} \tag{13.82}
$$

substituting (13.64), (13.74), and (13.75) into (13.4) leads to the three-dimensional beam dynamics

$$
\rho\left({}^{\mathbf{B}_b}\mathbf{R_I}\frac{\partial^2}{\partial t^2}\mathbf{r}_b(x,t)\right) + \begin{bmatrix} -ES_x\left(\frac{\partial^2 u_x(x,t)}{\partial x^2}\right) \\ EI_z\frac{\partial}{\partial x}\left((1-\alpha_z(x,t))\frac{\partial^3 y_b(x,t)}{\partial x^3}\right) \\ EI_y\frac{\partial}{\partial x}\left((1-\alpha_y(x,t))\frac{\partial^3 z_b(x,t)}{\partial x^3}\right) \end{bmatrix} = \mathbf{0} \tag{13.83}
$$

$$
\frac{\rho}{S_x}\mathrm{diag}(J_x, I_y, I_z){}^{\mathbf{B}_b}\mathbf{R_I}\frac{\partial}{\partial t}\boldsymbol{\omega}_b + \frac{\rho}{S_x}\begin{bmatrix} (I_z-I_y){}^{\mathbf{B}_b}\omega_y{}^{\mathbf{B}_b}\omega_z \\ I_y{}^{\mathbf{B}_b}\omega_x{}^{\mathbf{B}_b}\omega_z \\ -I_z{}^{\mathbf{B}_b}\omega_x{}^{\mathbf{B}_b}\omega_y \end{bmatrix}
$$

$$
-\begin{bmatrix} GJ_x\left(\frac{\partial^2\Theta_x(x,t)}{\partial x^2}\right) \\ -\alpha_y(x,t)EI_y\left(\frac{\partial^3 z_b(x,t)}{\partial x^3}\right) \\ \alpha_z(x,t)EI_z\left(\frac{\partial^3 y_b(x,t)}{\partial x^3}\right) \end{bmatrix} = \mathbf{0} \tag{13.84}
$$

for all $x \in [0, l]$ with the boundary conditions

$$
{}^{\mathbf{B}}\mathbf{R_I}\mathbf{f_B} = \begin{bmatrix} -ES_x\left(\frac{\partial u_x(x,t)}{\partial x}\right)\Big|_{x=0} \\ (1-\alpha_z(x,t))EI_z\left(\frac{\partial^3 y_b(x,t)}{\partial x^3}\right)\Big|_{x=0} \\ (1-\alpha_y(x,t))EI_y\left(\frac{\partial^3 z_b(x,t)}{\partial x^3}\right)\Big|_{x=0} \end{bmatrix} \tag{13.85}
$$

$$
{}^{\mathbf{B}}\mathbf{R_I}\mathbf{m_B} = \begin{bmatrix} -GJ_x\left(\frac{\partial\Theta_x(x,t)}{\partial x}\right)\Big|_{x=0} \\ EI_y\left(\frac{\partial^2 z_b(x,t)}{\partial x^2}\right)\Big|_{x=0} \\ -EI_z\left(\frac{\partial^2 y_b(x,t)}{\partial x^2}\right)\Big|_{x=0} \end{bmatrix} \tag{13.86}
$$

$$
{}^{\mathbf{T}}\mathbf{R_I}\mathbf{f_T} = \begin{bmatrix} -ES_x\left(\frac{\partial u_x(x,t)}{\partial x}\right)\Big|_{x=l} \\ (1-\alpha_z(x,t))EI_z\left(\frac{\partial^3 y_b(x,t)}{\partial x^3}\right)\Big|_{x=l} \\ (1-\alpha_y(x,t))EI_y\left(\frac{\partial^3 z_b(x,t)}{\partial x^3}\right)\Big|_{x=l} \end{bmatrix} \tag{13.87}
$$

$$
{}^{\mathbf{T}}\mathbf{R_I}\mathbf{m_T} = \begin{bmatrix} -GJ_x\left(\frac{\partial\Theta_x(x,t)}{\partial x}\right)\Big|_{x=l} \\ EI_y\left(\frac{\partial^2 z_b(x,t)}{\partial x^2}\right)\Big|_{x=l} \\ -EI_z\left(\frac{\partial^2 y_b(x,t)}{\partial x^2}\right)\Big|_{x=l} \end{bmatrix}. \tag{13.88}
$$

The derivation is given in Appendix B.12.

Remark 13.2. Equations (13.83) and (13.84) give the dynamics of the linear and angular motions, respectively. The two small parameters $\alpha_z(x,t)$ and $\alpha_y(x,t)$ specify the portions of the shear moments that are used to produce the angular accelerations of the infinitesimally small section located at distance x of the flexible beam. In most text books, such as in [119], $\alpha_z(x,t) \approx 0$ and $\alpha_y(x,t) \approx 0$ are used with a reasonable approximation.

13.4.3 Control Equations

The control equations with respect to the addressed flexible beam in the 3D motion are presented in this subsection.

Let $u_{xr}(x,t) \in \mathbb{R}$, $x_{br}(x,t) \in \mathbb{R}$, $y_{br}(x,t) \in \mathbb{R}$, $z_{br}(x,t) \in \mathbb{R}$, $\alpha_{yr}(x,t) \in \mathbb{R}$, $\alpha_{zr}(x,t) \in \mathbb{R}$, $\Theta_{xr}(x,t) \in \mathbb{R}$, $\mathbf{f}_{\mathbf{B}r} \in \mathbb{R}^3$, $\mathbf{m}_{\mathbf{B}r} \in \mathbb{R}^3$, $\mathbf{f}_{\mathbf{T}r} \in \mathbb{R}^3$, and $\mathbf{m}_{\mathbf{T}r} \in \mathbb{R}^3$ be the required (design) variables/vectors of $u_x(x,t) \in \mathbb{R}$, $x_b(x,t) \in \mathbb{R}$, $y_b(x,t) \in \mathbb{R}$, $z_b(x,t) \in \mathbb{R}$, $\alpha_y(x,t) \in \mathbb{R}$, $\alpha_z(x,t) \in \mathbb{R}$, $\Theta_x(x,t) \in \mathbb{R}$, $\mathbf{f}_{\mathbf{B}} \in \mathbb{R}^3$, $\mathbf{m}_{\mathbf{B}} \in \mathbb{R}^3$, $\mathbf{f}_{\mathbf{T}} \in \mathbb{R}^3$, and $\mathbf{m}_{\mathbf{T}} \in \mathbb{R}^3$, respectively, subject to

$$\frac{\partial x_b(x,t)}{\partial x} = \frac{\partial u_x(x,t)}{\partial x} \tag{13.89}$$

$$\frac{\partial x_{br}(x,t)}{\partial x} = \frac{\partial u_{xr}(x,t)}{\partial x} \tag{13.90}$$

where $x_b(x,t) \in \mathbb{R}$, $y_b(x,t) \in \mathbb{R}$, and $z_b(x,t) \in \mathbb{R}$ are defined in (13.43).

Then, let

$$^{\mathbf{B}_b}\mathbf{R_I}\boldsymbol{\omega}_{br} \overset{def}{=} \begin{bmatrix} ^{\mathbf{B}_b}\omega_{xr} \\ ^{\mathbf{B}_b}\omega_{yr} \\ ^{\mathbf{B}_b}\omega_{zr} \end{bmatrix} \in \mathbb{R}^3 \tag{13.91}$$

be the required (design) vector of

$$^{\mathbf{B}_b}\mathbf{R_I}\boldsymbol{\omega}_b = \begin{bmatrix} ^{\mathbf{B}_b}\omega_x \\ ^{\mathbf{B}_b}\omega_y \\ ^{\mathbf{B}_b}\omega_z \end{bmatrix} \in \mathbb{R}^3$$

defined by (13.48). Meanwhile, let

$$^{\mathbf{B}_b}\mathbf{R_I}\frac{\partial}{\partial t}\mathbf{r}_{br}(x,t) = \begin{bmatrix} ^{\mathbf{B}_b}v_{bxr} \\ ^{\mathbf{B}_b}v_{byr} \\ ^{\mathbf{B}_b}v_{bzr} \end{bmatrix} \in \mathbb{R}^3 \tag{13.92}$$

be the required (design) vector of

$$^{\mathbf{B}_b}\mathbf{R_I}\frac{\partial}{\partial t}\mathbf{r}_b(x,t) = \begin{bmatrix} ^{\mathbf{B}_b}v_{bx} \\ ^{\mathbf{B}_b}v_{by} \\ ^{\mathbf{B}_b}v_{bz} \end{bmatrix} \in \mathbb{R}^3$$

defined by (13.49), subject to

$$ {}^{\mathbf{B}_b}\omega_{yr} = -\frac{\partial}{\partial x} {}^{\mathbf{B}_b}v_{bzr} \tag{13.93} $$

$$ {}^{\mathbf{B}_b}\omega_{zr} = \frac{\partial}{\partial x} {}^{\mathbf{B}_b}v_{byr}. \tag{13.94} $$

Furthermore, the four required variables $x_{br}(x,t) \in \mathbb{R}$, $y_{br}(x,t) \in \mathbb{R}$, $z_{br}(x,t) \in \mathbb{R}$, and $\Theta_{xr}(x,t) \in \mathbb{R}$ are determined by

$$ \begin{bmatrix} {}^{\mathbf{B}_b}v_{bxr} - {}^{\mathbf{B}_b}v_{bx} \\ {}^{\mathbf{B}_b}v_{byr} - {}^{\mathbf{B}_b}v_{by} \\ {}^{\mathbf{B}_b}v_{bzr} - {}^{\mathbf{B}_b}v_{bz} \\ {}^{\mathbf{B}_b}\omega_{xr} - {}^{\mathbf{B}_b}\omega_{x} \end{bmatrix} = \frac{\partial}{\partial t} \begin{bmatrix} x_{br}(x,t) - x_b(x,t) \\ y_{br}(x,t) - y_b(x,t) \\ z_{br}(x,t) - z_b(x,t) \\ \Theta_{xr}(x,t) - \Theta_x(x,t) \end{bmatrix} + \begin{bmatrix} \lambda_x \left(x_{br}(x,t) - x_b(x,t)\right) \\ \lambda_y \left(y_{br}(x,t) - y_b(x,t)\right) \\ \lambda_z \left(z_{br}(x,t) - z_b(x,t)\right) \\ \lambda_\Theta \left(\Theta_{xr}(x,t) - \Theta_x(x,t)\right) \end{bmatrix} \tag{13.95} $$

with $\lambda_x \geqslant 0$, $\lambda_y \geqslant 0$, $\lambda_z \geqslant 0$, and $\lambda_\Theta \geqslant 0$ being four constants.

Remark 13.3. Equation (13.95) makes

$$ {}^{\mathbf{B}_b}\mathbf{R}_\mathbf{I}\mathbf{r}_{br}(x,t) \neq \begin{bmatrix} x_{br}(x,t) \\ y_{br}(x,t) \\ z_{br}(x,t) \end{bmatrix} \in \mathbb{R}^3 $$

$$ {}^{\mathbf{B}_b}\omega_{xr} \neq \frac{\partial}{\partial t}\Theta_{xr}(x,t) \in \mathbb{R} $$

even though

$$ {}^{\mathbf{B}_b}\mathbf{R}_\mathbf{I}\mathbf{r}_{b}(x,t) = \begin{bmatrix} x_{b}(x,t) \\ y_{b}(x,t) \\ z_{b}(x,t) \end{bmatrix} \in \mathbb{R}^3 $$

$$ {}^{\mathbf{B}_b}\omega_{x} = \frac{\partial}{\partial t}\Theta_{x}(x,t) \in \mathbb{R} $$

hold for the physical beam.

The model-based control equations for this three-dimensional flexible beam are therefore designed as

$$ \rho {}^{\mathbf{B}_b}\mathbf{R}_\mathbf{I} \frac{\partial^2}{\partial t^2}\mathbf{r}_{br}(x,t) + \begin{bmatrix} -ES_x \left(\frac{\partial^2 u_{xr}(x,t)}{\partial x^2}\right) \\ EI_z \frac{\partial}{\partial x}\left(\left(1-\alpha_{zr}(x,t)\right)\frac{\partial^3 y_{br}(x,t)}{\partial x^3}\right) \\ EI_y \frac{\partial}{\partial x}\left(\left(1-\alpha_{yr}(x,t)\right)\frac{\partial^3 z_{br}(x,t)}{\partial x^3}\right) \end{bmatrix} + \mathbf{K}_p {}^{\mathbf{B}_b}\mathbf{R}_\mathbf{I}\left(\frac{\partial}{\partial t}\mathbf{r}_{br}(x,t) - \frac{\partial}{\partial t}\mathbf{r}_{b}(x,t)\right) = \mathbf{0} \tag{13.96} $$

$$\frac{\rho}{S_x}\mathrm{diag}(J_x, I_y, I_z)\left[{}^{\mathbf{B}_b}\mathbf{R_I}\frac{\partial}{\partial t}\boldsymbol{\omega}_{br} - {}^{\mathbf{B}_b}\mathbf{R_I}(\boldsymbol{\omega}_b\times)\boldsymbol{\omega}_{br}\right]$$
$$+\frac{\rho}{S_x}\begin{bmatrix} I_z{}^{\mathbf{B}_b}\omega_y{}^{\mathbf{B}_b}\omega_{zr} - I_y{}^{\mathbf{B}_b}\omega_z{}^{\mathbf{B}_b}\omega_{yr} \\ I_y{}^{\mathbf{B}_b}\omega_z{}^{\mathbf{B}_b}\omega_{xr} \\ -I_z{}^{\mathbf{B}_b}\omega_y{}^{\mathbf{B}_b}\omega_{xr} \end{bmatrix}$$
$$-\begin{bmatrix} GJ_x\left(\frac{\partial^2\Theta_{xr}(x,t)}{\partial x^2}\right) \\ -\alpha_{yr}(x,t)EI_y\left(\frac{\partial^3 z_{br}(x,t)}{\partial x^3}\right) \\ \alpha_{zr}(x,t)EI_z\left(\frac{\partial^3 y_{br}(x,t)}{\partial x^3}\right) \end{bmatrix} + \mathbf{K}_\omega{}^{\mathbf{B}_b}\mathbf{R_I}\left(\boldsymbol{\omega}_{br} - \boldsymbol{\omega}_b\right) = \mathbf{0} \tag{13.97}$$

for all $x \in [0, l]$ with $\mathbf{K}_p \in \mathbb{R}^{3\times3}$ and $\mathbf{K}_\omega \in \mathbb{R}^{3\times3}$ being two semi-positive-definite matrices characterizing the feedback control. The corresponding boundary conditions of the required (design) variables are

$$^{\mathbf{B}}\mathbf{R_I}\mathbf{f}_{\mathbf{B}r} = \begin{bmatrix} -ES_x\left(\frac{\partial u_{xr}(x,t)}{\partial x}\right)\Big|_{x=0} \\ (1-\alpha_{zr}(x,t))EI_z\left(\frac{\partial^3 y_{br}(x,t)}{\partial x^3}\right)\Big|_{x=0} \\ (1-\alpha_{yr}(x,t))EI_y\left(\frac{\partial^3 z_{br}(x,t)}{\partial x^3}\right)\Big|_{x=0} \end{bmatrix} \tag{13.98}$$

$$^{\mathbf{B}}\mathbf{R_I}\mathbf{m}_{\mathbf{B}r} = \begin{bmatrix} -GJ_x\left(\frac{\partial\Theta_{xr}(x,t)}{\partial x}\right)\Big|_{x=0} \\ EI_y\left(\frac{\partial^2 z_{br}(x,t)}{\partial x^2}\right)\Big|_{x=0} \\ -EI_z\left(\frac{\partial^2 y_{br}(x,t)}{\partial x^2}\right)\Big|_{x=0} \end{bmatrix} \tag{13.99}$$

$$^{\mathbf{T}}\mathbf{R_I}\mathbf{f}_{\mathbf{T}r} = \begin{bmatrix} -ES_x\left(\frac{\partial u_{xr}(x,t)}{\partial x}\right)\Big|_{x=l} \\ (1-\alpha_{zr}(x,t))EI_z\left(\frac{\partial^3 y_{br}(x,t)}{\partial x^3}\right)\Big|_{x=l} \\ (1-\alpha_{yr}(x,t))EI_y\left(\frac{\partial^3 z_{br}(x,t)}{\partial x^3}\right)\Big|_{x=l} \end{bmatrix} \tag{13.100}$$

$$^{\mathbf{T}}\mathbf{R_I}\mathbf{m}_{\mathbf{T}r} = \begin{bmatrix} -GJ_x\left(\frac{\partial\Theta_{xr}(x,t)}{\partial x}\right)\Big|_{x=l} \\ EI_y\left(\frac{\partial^2 z_{br}(x,t)}{\partial x^2}\right)\Big|_{x=l} \\ -EI_z\left(\frac{\partial^2 y_{br}(x,t)}{\partial x^2}\right)\Big|_{x=l} \end{bmatrix}. \tag{13.101}$$

13.4.4 Virtual Stability

The following theorem ensures that the three-dimensional flexible beam combined with its respective control equations qualifies to be *virtually stable* in the sense of Definition 2.17.

Theorem 13.3. *The three-dimensional uniform flexible beam described by (13.83) and (13.84) subject to (13.48)–(13.51) and the boundary conditions (13.85)–(13.88), combined with its respective control equations (13.89)–(13.101), is virtually stable in the sense of Definition 2.17.* ■

In the proof, a non-negative accompanying function

$$\nu = \nu_a + \nu_a + \nu_c \tag{13.102}$$

with

$$\nu_a = \frac{1}{2}\int_0^l \rho \frac{\partial}{\partial t}\left(\mathbf{r}_{br}(x,t) - \mathbf{r}_b(x,t)\right)^T \frac{\partial}{\partial t}\left(\mathbf{r}_{br}(x,t) - \mathbf{r}_b(x,t)\right) dx \tag{13.103}$$

$$\begin{aligned}\nu_b = &\frac{1}{2}\int_0^l \frac{\rho}{S_x}\left[{}^{\mathbf{B}_b}\mathbf{R}_{\mathbf{I}}\left(\boldsymbol{\omega}_{br} - \boldsymbol{\omega}_b\right)\right]^T \mathrm{diag}(J_x, I_y, I_z) \\ &\times \left[{}^{\mathbf{B}_b}\mathbf{R}_{\mathbf{I}}\left(\boldsymbol{\omega}_{br} - \boldsymbol{\omega}_b\right)\right] dx\end{aligned} \tag{13.104}$$

$$\begin{aligned}\nu_c = &\frac{1}{2}\int_0^l ES_x\left[\frac{\partial}{\partial x}\left(u_{xr}(x,t) - u_x(x,t)\right)\right]^2 dx \\ &+\frac{1}{2}\int_0^l EI_z\left[\frac{\partial^2}{\partial x^2}\left(y_{br}(x,t) - y_b(x,t)\right)\right]^2 dx \\ &+\frac{1}{2}\int_0^l EI_y\left[\frac{\partial^2}{\partial x^2}\left(z_{br}(x,t) - z_b(x,t)\right)\right]^2 dx \\ &+\frac{1}{2}\int_0^l GJ_x\left[\frac{\partial}{\partial x}\left(\Theta_{xr}(x,t) - \Theta_x(x,t)\right)\right]^2 dx\end{aligned} \tag{13.105}$$

is assigned. After mathematical manipulations, it leads to

$$\begin{aligned}\dot{\nu} &= \dot{\nu}_a + \dot{\nu}_b + \dot{\nu}_c \\ &= -\int_0^l \frac{\partial}{\partial t}\left[\mathbf{r}_{br}(x,t) - \mathbf{r}_b(x,t)\right]^T \left({}^{\mathbf{B}_b}\mathbf{R}_{\mathbf{I}}^T \mathbf{K}_p {}^{\mathbf{B}_b}\mathbf{R}_{\mathbf{I}}\right) \\ &\qquad\times \frac{\partial}{\partial t}\left[\mathbf{r}_{br}(x,t) - \mathbf{r}_b(x,t)\right] dx \\ &\quad - \int_0^l \left(\boldsymbol{\omega}_{br} - \boldsymbol{\omega}_b\right)^T \left({}^{\mathbf{B}_b}\mathbf{R}_{\mathbf{I}}^T \mathbf{K}_\omega {}^{\mathbf{B}_b}\mathbf{R}_{\mathbf{I}}\right)\left(\boldsymbol{\omega}_{br} - \boldsymbol{\omega}_b\right) dx \\ &\quad -ES_x\lambda_x\int_0^l \left[\frac{\partial}{\partial x}\left(u_{xr}(x,t) - u_x(x,t)\right)\right]^2 dx \\ &\quad -EI_z\lambda_y\int_0^l \left[\frac{\partial^2}{\partial x^2}\left(y_{br}(x,t) - y_b(x,t)\right)\right]^2 dx \\ &\quad -EI_y\lambda_z\int_0^l \left[\frac{\partial^2}{\partial x^2}\left(z_{br}(x,t) - z_b(x,t)\right)\right]^2 dx \\ &\quad -GJ_x\lambda_\Theta\int_0^l \left[\frac{\partial}{\partial x}\left(\Theta_{xr}(x,t) - \Theta_x(x,t)\right)\right]^2 dx \\ &\quad +p_{\mathbf{B}} - p_{\mathbf{T}}\end{aligned} \tag{13.106}$$

where $p_{\mathbf{B}}$ and $p_{\mathbf{T}}$ denote the *virtual power flows* at the two ends at which the beam is virtually decomposed from the original robotic system. The detail process is given in Appendix B.13.

Consider the fact that the flexible beam has one *driving cutting point* associated with frame $\{\mathbf{T}\}$ and one *driven cutting point* associated with frame $\{\mathbf{B}\}$. Using (13.102)–(13.106) and Definition 2.17 ensures the *virtual stability* of the flexible beam combined with its respective control equations. The stability of the entire flexible link robot is, therefore, governed by Theorem 2.1.

13.4.5 Adaptive Control in 3D Motion

Control Equations

When the dynamic parameters ρ, ES_x, EI_z, EI_y, $\frac{\rho}{S_x}J_x$, $\frac{\rho}{S_x}I_z$, $\frac{\rho}{S_x}I_y$, and GJ_x are unknown, they have to be replaced by their estimates denoted as $\widehat{\rho}$, $\widehat{ES_x}$, $\widehat{EI_z}$, $\widehat{EI_y}$, $\widehat{\frac{\rho}{S_x}J_x}$, $\widehat{\frac{\rho}{S_x}I_z}$, $\widehat{\frac{\rho}{S_x}I_y}$, and $\widehat{GJ_x}$, respectively. In line with these changes, the control equations (13.96) and (13.97) have to be modified to

$$\widehat{\rho}\,{}^{\mathbf{B}_b}\mathbf{R}_{\mathbf{I}}\frac{\partial^2}{\partial t^2}\mathbf{r}_{br}(x,t) + \begin{bmatrix} -\widehat{ES_x}\left(\frac{\partial^2 u_{xr}(x,t)}{\partial x^2}\right) \\ \widehat{EI_z}\frac{\partial}{\partial x}\left((1-\alpha_{zr}(x,t))\frac{\partial^3 y_{br}(x,t)}{\partial x^3}\right) \\ \widehat{EI_y}\frac{\partial}{\partial x}\left((1-\alpha_{yr}(x,t))\frac{\partial^3 z_{br}(x,t)}{\partial x^3}\right)\end{bmatrix}$$
$$+\mathbf{K}_p{}^{\mathbf{B}_b}\mathbf{R}_{\mathbf{I}}\left(\frac{\partial}{\partial t}\mathbf{r}_{br}(x,t) - \frac{\partial}{\partial t}\mathbf{r}_b(x,t)\right) = \mathbf{0} \tag{13.107}$$

$$\mathrm{diag}(\widehat{\frac{\rho}{S_x}J_x}, \widehat{\frac{\rho}{S_x}I_y}, \widehat{\frac{\rho}{S_x}I_z})\left[{}^{\mathbf{B}_b}\mathbf{R}_{\mathbf{I}}\frac{\partial}{\partial t}\boldsymbol{\omega}_{br} - {}^{\mathbf{B}_b}\mathbf{R}_{\mathbf{I}}(\boldsymbol{\omega}_b\times)\boldsymbol{\omega}_{br}\right]$$
$$+\begin{bmatrix} \widehat{\frac{\rho}{S_x}I_z}\,{}^{\mathbf{B}_b}\omega_y\,{}^{\mathbf{B}_b}\omega_{zr} - \widehat{\frac{\rho}{S_x}I_y}\,{}^{\mathbf{B}_b}\omega_z\,{}^{\mathbf{B}_b}\omega_{yr} \\ \widehat{\frac{\rho}{S_x}I_y}\,{}^{\mathbf{B}_b}\omega_z\,{}^{\mathbf{B}_b}\omega_{xr} \\ -\widehat{\frac{\rho}{S_x}I_z}\,{}^{\mathbf{B}_b}\omega_y\,{}^{\mathbf{B}_b}\omega_{xr}\end{bmatrix}$$
$$-\begin{bmatrix} \widehat{GJ_x}\left(\frac{\partial^2\Theta_{xr}(x,t)}{\partial x^2}\right) \\ -\alpha_{yr}(x,t)\widehat{EI_y}\left(\frac{\partial^3 z_{br}(x,t)}{\partial x^3}\right) \\ \alpha_{zr}(x,t)\widehat{EI_z}\left(\frac{\partial^3 y_{br}(x,t)}{\partial x^3}\right)\end{bmatrix} + \mathbf{K}_\omega{}^{\mathbf{B}_b}\mathbf{R}_{\mathbf{I}}\left(\boldsymbol{\omega}_{br} - \boldsymbol{\omega}_b\right) = \mathbf{0} \tag{13.108}$$

for all $x \in [0,l]$ with $\mathbf{K}_p \in \mathbb{R}^{3\times 3}$ and $\mathbf{K}_\omega \in \mathbb{R}^{3\times 3}$ being two semi-positive-definite matrices characterizing the feedback control. Accordingly, the boundary conditions of the required (design) variables change to

$$
{}^{\mathbf{B}}\mathbf{R_I f_{B}}_r = \begin{bmatrix} -\widehat{ES_x}\left(\frac{\partial u_{xr}(x,t)}{\partial x}\right)\Big|_{x=0} \\ (1-\alpha_{zr}(x,t))\widehat{EI_z}\left(\frac{\partial^3 y_{br}(x,t)}{\partial x^3}\right)\Big|_{x=0} \\ (1-\alpha_{yr}(x,t))\widehat{EI_y}\left(\frac{\partial^3 z_{br}(x,t)}{\partial x^3}\right)\Big|_{x=0} \end{bmatrix} \tag{13.109}
$$

$$
{}^{\mathbf{B}}\mathbf{R_I m_{B}}_r = \begin{bmatrix} -\widehat{GJ_x}\left(\frac{\partial \Theta_{xr}(x,t)}{\partial x}\right)\Big|_{x=0} \\ \widehat{EI_y}\left(\frac{\partial^2 z_{br}(x,t)}{\partial x^2}\right)\Big|_{x=0} \\ -\widehat{EI_z}\left(\frac{\partial^2 y_{br}(x,t)}{\partial x^2}\right)\Big|_{x=0} \end{bmatrix} \tag{13.110}
$$

$$
{}^{\mathbf{T}}\mathbf{R_I f_{T}}_r = \begin{bmatrix} -\widehat{ES_x}\left(\frac{\partial u_{xr}(x,t)}{\partial x}\right)\Big|_{x=l} \\ (1-\alpha_{zr}(x,t))\widehat{EI_z}\left(\frac{\partial^3 y_{br}(x,t)}{\partial x^3}\right)\Big|_{x=l} \\ (1-\alpha_{yr}(x,t))\widehat{EI_y}\left(\frac{\partial^3 z_{br}(x,t)}{\partial x^3}\right)\Big|_{x=l} \end{bmatrix} \tag{13.111}
$$

$$
{}^{\mathbf{T}}\mathbf{R_I m_{T}}_r = \begin{bmatrix} -\widehat{GJ_x}\left(\frac{\partial \Theta_{xr}(x,t)}{\partial x}\right)\Big|_{x=l} \\ \widehat{EI_y}\left(\frac{\partial^2 z_{br}(x,t)}{\partial x^2}\right)\Big|_{x=l} \\ -\widehat{EI_z}\left(\frac{\partial^2 y_{br}(x,t)}{\partial x^2}\right)\Big|_{x=l} \end{bmatrix}. \tag{13.112}
$$

The eight parameter estimates are updated by using the $\mathcal{P}$ function defined by (2.79) as

$$\hat{\rho} = \mathcal{P}(s_\rho, \rho_\rho, \underline{\rho}, \overline{\rho}, t) \tag{13.113}$$

$$\widehat{ES_x} = \mathcal{P}(s_{ES_x}, \rho_{ES_x}, \underline{ES_x}, \overline{ES_x}, t) \tag{13.114}$$

$$\widehat{EI_z} = \mathcal{P}(s_{EI_z}, \rho_{EI_z}, \underline{EI_z}, \overline{EI_z}, t) \tag{13.115}$$

$$\widehat{EI_y} = \mathcal{P}(s_{EI_y}, \rho_{EI_y}, \underline{EI_y}, \overline{EI_y}, t) \tag{13.116}$$

$$\widehat{\frac{\rho}{S_x}J_x} = \mathcal{P}(s_{\frac{\rho}{S_x}J_x}, \rho_{\frac{\rho}{S_x}J_x}, \underline{\frac{\rho}{S_x}J_x}, \overline{\frac{\rho}{S_x}J_x}, t) \tag{13.117}$$

$$\widehat{\frac{\rho}{S_x}I_z} = \mathcal{P}(s_{\frac{\rho}{S_x}I_z}, \rho_{\frac{\rho}{S_x}I_z}, \underline{\frac{\rho}{S_x}I_z}, \overline{\frac{\rho}{S_x}I_z}, t) \tag{13.118}$$

$$\widehat{\frac{\rho}{S_x}I_y} = \mathcal{P}(s_{\frac{\rho}{S_x}I_y}, \rho_{\frac{\rho}{S_x}I_y}, \underline{\frac{\rho}{S_x}I_y}, \overline{\frac{\rho}{S_x}I_y}, t) \tag{13.119}$$

$$\widehat{GJ_x} = \mathcal{P}(s_{GJ_x}, \rho_{GJ_x}, \underline{GJ_x}, \overline{GJ_x}, t) \tag{13.120}$$

with

$$
s_\rho = \int_0^l \frac{\partial}{\partial t}\left(\mathbf{r}_{br}(x,t) - \mathbf{r}_b(x,t)\right)^T \frac{\partial^2}{\partial t^2}\mathbf{r}_{br}(x,t)dx \tag{13.121}
$$

$$
s_{ES_x} = \left({}^{\mathbf{B}_b}v_{bxr} - {}^{\mathbf{B}_b}v_{bx}\right)\frac{\partial}{\partial x}u_{xr}(x,t)\Big|_0^l
$$

$$-\int_0^l \left({}^{\mathbf{B}_b}v_{bxr} - {}^{\mathbf{B}_b}v_{bx}\right) \frac{\partial^2}{\partial x^2} u_{xr}(x,t)dx \tag{13.122}$$

$$\begin{aligned}
s_{EI_z} =& -\left({}^{\mathbf{B}_b}v_{byr} - {}^{\mathbf{B}_b}v_{by}\right)(1-\alpha_{zr}(x,t))\frac{\partial^3 y_{br}(x,t)}{\partial x^3}\bigg|_0^l \\
&+ \frac{\partial}{\partial x}\left({}^{\mathbf{B}_b}v_{byr} - {}^{\mathbf{B}_b}v_{by}\right)\frac{\partial^2 y_{br}(x,t)}{\partial x^2}\bigg|_0^l \\
&+ \int_0^l \left({}^{\mathbf{B}_b}v_{byr} - {}^{\mathbf{B}_b}v_{bx}\right)\frac{\partial}{\partial x}\left((1-\alpha_{zr}(x,t))\frac{\partial^3 y_{br}(x,t)}{\partial x^3}\right)dx \\
&- \int_0^l \left({}^{\mathbf{B}_b}\omega_{zr} - {}^{\mathbf{B}_b}\omega_z\right)\alpha_{zr}(x,t)\frac{\partial^3 y_{br}(x,t)}{\partial x^3}dx
\end{aligned} \tag{13.123}$$

$$\begin{aligned}
s_{EI_y} =& -\left({}^{\mathbf{B}_b}v_{bzr} - {}^{\mathbf{B}_b}v_{bz}\right)(1-\alpha_{yr}(x,t))\frac{\partial^3 z_{br}(x,t)}{\partial x^3}\bigg|_0^l \\
&+ \frac{\partial}{\partial x}\left({}^{\mathbf{B}_b}v_{bzr} - {}^{\mathbf{B}_b}v_{bz}\right)\frac{\partial^2 z_{br}(x,t)}{\partial x^2}\bigg|_0^l \\
&+ \int_0^l \left({}^{\mathbf{B}_b}v_{bzr} - {}^{\mathbf{B}_b}v_{bx}\right)\frac{\partial}{\partial x}\left((1-\alpha_{yr}(x,t))\frac{\partial^3 z_{br}(x,t)}{\partial x^3}\right)dx \\
&+ \int_0^l \left({}^{\mathbf{B}_b}\omega_{yr} - {}^{\mathbf{B}_b}\omega_y\right)\alpha_{yr}(x,t)\frac{\partial^3 z_{br}(x,t)}{\partial x^3}dx
\end{aligned} \tag{13.124}$$

$$\begin{aligned}
s_{\frac{\rho}{S_x}J_x} =& \int_0^l \left({}^{\mathbf{B}_b}\omega_{xr} - {}^{\mathbf{B}_b}\omega_x\right)\begin{bmatrix}1 & 0 & 0\end{bmatrix} \\
&\times\left({}^{\mathbf{B}_b}\mathbf{R_I}\frac{\partial}{\partial t}\boldsymbol{\omega}_{br} - {}^{\mathbf{B}_b}\mathbf{R_I}(\boldsymbol{\omega}_b\times)\boldsymbol{\omega}_{br}\right)dx
\end{aligned} \tag{13.125}$$

$$\begin{aligned}
s_{\frac{\rho}{S_x}I_z} =& \int_0^l \left({}^{\mathbf{B}_b}\omega_{zr} - {}^{\mathbf{B}_b}\omega_z\right)\begin{bmatrix}0 & 0 & 1\end{bmatrix} \\
&\times\left({}^{\mathbf{B}_b}\mathbf{R_I}\frac{\partial}{\partial t}\boldsymbol{\omega}_{br} - {}^{\mathbf{B}_b}\mathbf{R_I}(\boldsymbol{\omega}_b\times)\boldsymbol{\omega}_{br}\right)dx \\
&+ \int_0^l \left({}^{\mathbf{B}_b}\omega_{xr} - {}^{\mathbf{B}_b}\omega_x\right){}^{\mathbf{B}_b}\omega_y{}^{\mathbf{B}_b}\omega_{zr}dx \\
&- \int_0^l \left({}^{\mathbf{B}_b}\omega_{zr} - {}^{\mathbf{B}_b}\omega_z\right){}^{\mathbf{B}_b}\omega_y{}^{\mathbf{B}_b}\omega_{xr}dx
\end{aligned} \tag{13.126}$$

$$\begin{aligned}
s_{\frac{\rho}{S_x}I_y} =& \int_0^l \left({}^{\mathbf{B}_b}\omega_{yr} - {}^{\mathbf{B}_b}\omega_y\right)\begin{bmatrix}0 & 1 & 0\end{bmatrix} \\
&\times\left({}^{\mathbf{B}_b}\mathbf{R_I}\frac{\partial}{\partial t}\boldsymbol{\omega}_{br} - {}^{\mathbf{B}_b}\mathbf{R_I}(\boldsymbol{\omega}_b\times)\boldsymbol{\omega}_{br}\right)dx \\
&- \int_0^l \left({}^{\mathbf{B}_b}\omega_{xr} - {}^{\mathbf{B}_b}\omega_x\right){}^{\mathbf{B}_b}\omega_z{}^{\mathbf{B}_b}\omega_{yr}dx
\end{aligned}$$

$$+ \int_0^l \left({}^{\mathbf{B}_b}\omega_{yr} - {}^{\mathbf{B}_b}\omega_y \right) {}^{\mathbf{B}_b}\omega_z {}^{\mathbf{B}_b}\omega_{xr} dx \tag{13.127}$$

$$s_{GJ_x} = \left({}^{\mathbf{B}_b}\omega_{xr} - {}^{\mathbf{B}_b}\omega_x \right) \frac{\partial}{\partial x} \Theta_{xr}(x,t) \Big|_0^l - \int_0^l \left({}^{\mathbf{B}_b}\omega_{xr} - {}^{\mathbf{B}_b}\omega_x \right) \frac{\partial^2}{\partial x^2} \Theta_{xr}(x,t) dx. \tag{13.128}$$

Theorem 13.4. *The three-dimensional uniform flexible beam described by (13.83) and (13.84) subject to (13.48)–(13.51) and the boundary conditions (13.85)–(13.88), combined with its respective control equations (13.89)–(13.95) and (13.107)–(13.112) and with the parameter adaptation (13.113)–(13.128), is virtually stable in the sense of Definition 2.17.* ■

The proof is given in Appendix B.14.

13.5 Concluding Remarks

In this chapter, the VDC approach has been applied to robots with multiple flexible links. The advantage of using the VDC approach is obvious, since the modularity of the VDC approach allows the control design to be focused on each individual flexible link. The dynamics and control issue of a flexible beam in both restricted two-dimensional motion and unrestricted three-dimensional motion has been thoroughly addressed. The *virtual stability* of the flexible beam combined with its respective control equations is ensured, allowing the stability and convergence of the entire robot to be established.

This is the first time for the VDC approach to be successfully applied to systems with distributed parameter models incorporating possible parameter uncertainties.

14

Applications to Electrical Circuits

14.1 Introduction

In the previous chapters, the *virtual decomposition control* (VDC) approach has been extensively applied to a variety of robotic systems. In this chapter, this novel approach will be applied to electrical systems in terms of the duality between mechanical and electrical systems.

14.2 Duality of Mechanical and Electrical Systems

The duality of mechanical and electrical systems is established by corresponding *forces* to *voltages* and *velocities* to *currents*. Thus, a mechanical object is equivalent to an electrical circuit loop. A force balance equation is equivalent to a voltage summation along an electrical circuit loop by corresponding *mass* to *inductance*, *damping* to *resistance*, and *stiffness* to *inverse of capacitance*.[1] The details are illustrated in Figure 14.1.

The duality also implies that a serial connection of electrical components corresponds to a mechanical object subject to multiple forces. Two parallel-connected electrical components correspond to two mechanical objects connected serially. In this sense, a serial electrical connection corresponds to a parallel mechanical connection, whereas a parallel electrical connection corresponds to a serial mechanical connection.

14.3 Virtual Decomposition Control of Four Circuit Loops

In this section, technical details are shown when the VDC approach is applied to an electrical system comprised of four circuit loops.

[1] This mechanical/electrical duality has been widely used in the control of bilateral teleoperation systems, where the passivity concept originally used in electrical systems is successfully used to conduct the stability analysis.

W.-H. Zhu: Virtual Decomposition Control, STAR 60, pp. 365–385.
springerlink.com

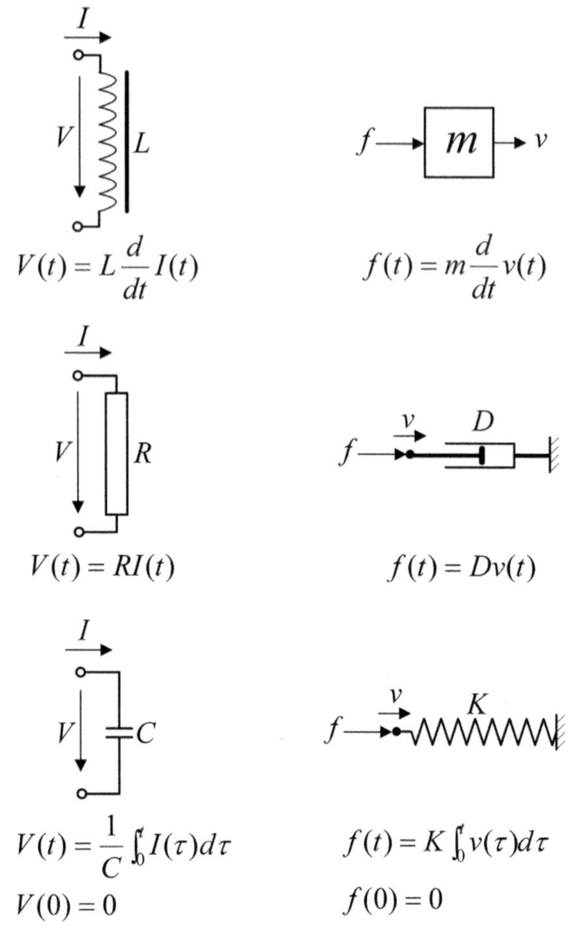

Fig. 14.1. The duality of mechanical and electrical components.

14.3.1 Virtual Decomposition

An electrical circuit comprised of four circuit loops is illustrated in Figure 14.2. The control objective is to control the currents of the four loops by applying control voltages.

The duality between mechanical and electrical systems suggests that a circuit loop can be represented by an *object* and a shared edge can be represented by a one-joint *open chain*. Thus, the original system, as illustrated in Figure 14.2, can be virtually decomposed into four circuit loops and four shared edges as illustrated in Figure 14.3 with its graphic expression in Figure 14.4.

There are totally eight *cutting points* being placed in the system. A *cutting point* represents a port with its reference direction being defined in Figure 14.3 (a pair of dashed arrows) and in Figure 14.4 (a single solid arrow) without involving circular path.

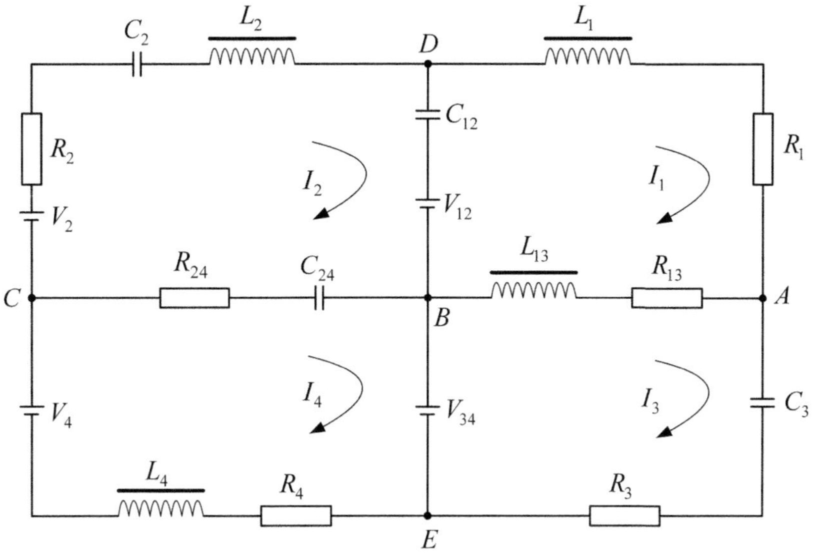

Fig. 14.2. Electrical circuit of four circuit loops with shared edges.

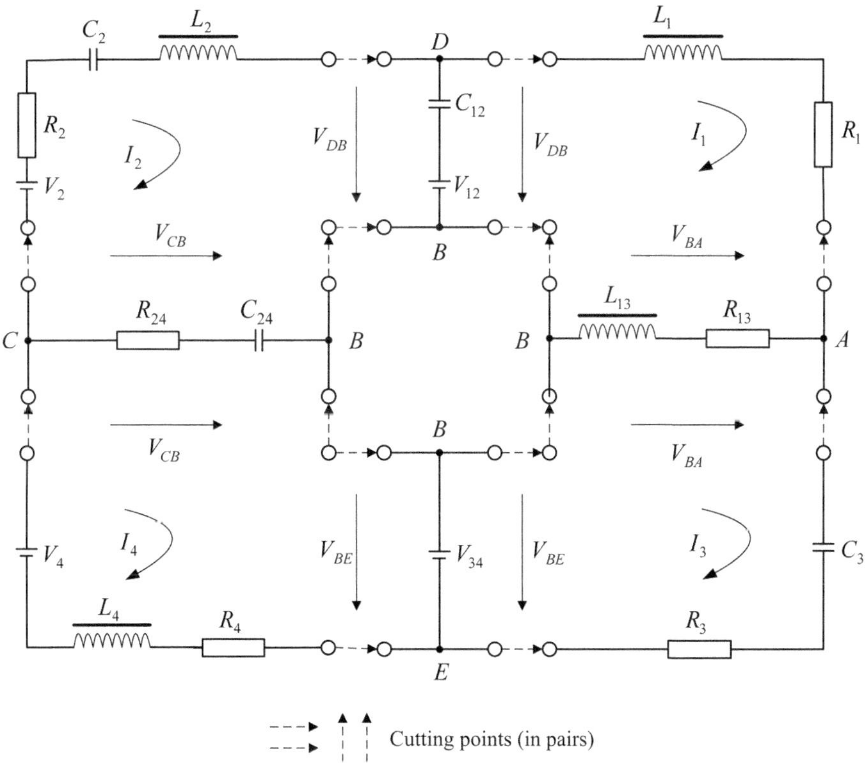

Fig. 14.3. Electrical circuit after virtual decomposition.

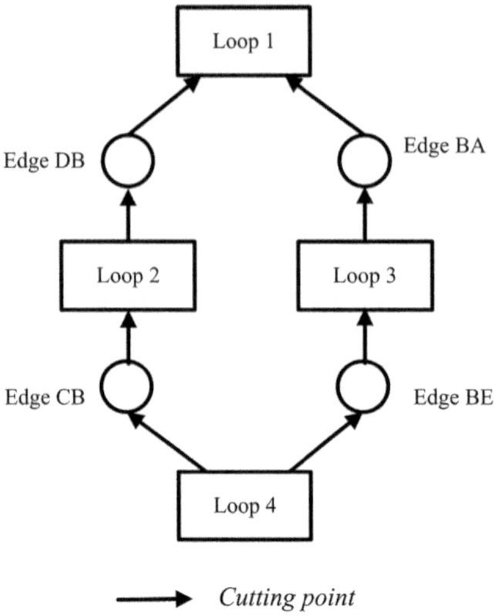

Fig. 14.4. Graphic expression of the electrical circuit.

With the specified reference directions, a *cutting point* is interpreted as a *driven cutting point* by the subsystem to which the reference direction points and is simultaneously interpreted as a *driving cutting point* by the other subsystem from which the reference direction points. No circular path of the reference directions is allowed.

The four circuit loops and the four shared edges are specified below:

- Circuit loop 1 has two *driven cutting points* associated with shared edges DB and BA, respectively.
- Circuit loop 2 has one *driving cutting point* associated with shared edge DB and one *driven cutting point* associated with shared edge CB.
- Circuit loop 3 has one *driving cutting point* associated with shared edge BA and one *driven cutting point* associated with shared edge BE.
- Circuit loop 4 has two *driving cutting points* associated with shared edges CB and BE, respectively.
- Shared edge DB has one *driving cutting point* associated with circuit loop 1 and one *driven cutting point* associated with circuit loop 2.
- Shared edge BA has one *driving cutting point* associated with circuit loop 1 and one *driven cutting point* associated with circuit loop 3.
- Shared edge CB has one *driving cutting point* associated with circuit loop 2 and one *driven cutting point* associated with circuit loop 4.
- Shared edge BE has one *driving cutting point* associated with circuit loop 3 and one *driven cutting point* associated with circuit loop 4.

14.3.2 Circuit Loop 1

The voltage equation along the first circuit loop can be written as

$$L_1\dot{I}_1 + R_1 I_1 = V_{DB} + V_{BA}. \tag{14.1}$$

With known-parameter components, the model-based circuit loop voltage equation of the required variables[2] is

$$L_1\dot{I}_{1r} + R_1 I_{1r} = V_{DBr} + V_{BAr}. \tag{14.2}$$

Theorem 14.1. *The first circuit loop described by (14.1), combined with its control equation (14.2), is virtually stable with its circuit loop current error* $I_{1r} - I_1$ *being a virtual function in both* L_2 *and* L_∞*, in the sense of Definition 2.17.*

Proof: Subtracting (14.1) from (14.2) yields

$$L_1(\dot{I}_{1r} - \dot{I}_1) + R_1(I_{1r} - I_1) = (V_{DBr} - V_{DB}) + (V_{BAr} - V_{BA}). \tag{14.3}$$

Let the non-negative accompanying function be

$$\nu_1 = \frac{1}{2} L_1 (I_{1r} - I_1)^2. \tag{14.4}$$

It follows from (14.3) that

$$\begin{aligned}\dot{\nu}_1 &= L_1(I_{1r} - I_1)(\dot{I}_{1r} - \dot{I}_1) \\ &= -R_1(I_{1r} - I_1)^2 + (I_{1r} - I_1)(V_{DBr} - V_{DB}) + (I_{1r} - I_1)(V_{BAr} - V_{BA}) \\ &= -R_1(I_{1r} - I_1)^2 + p_{1DB} + p_{1BA} \qquad (14.5)\end{aligned}$$

holds with

$$p_{1DB} \overset{def}{=} (I_{1r} - I_1)(V_{DBr} - V_{DB}) \tag{14.6}$$

$$p_{1BA} \overset{def}{=} (I_{1r} - I_1)(V_{BAr} - V_{BA}) \tag{14.7}$$

representing the two *virtual power flows* at the two *cutting points* of the loop.

Consider the fact that circuit loop 1 has two *driven cutting points* associated with shared edges DB and BA. Using (14.4), (14.5), and Definition 2.17 completes the proof. ∎

14.3.3 Circuit Loop 2

The voltage equation along the second circuit loop can be written as

$$L_2\dot{I}_2 + R_2 I_2 + \frac{1}{C_2}\int I_2 dt - V_2 = V_{CB} - V_{DB}. \tag{14.8}$$

With known-parameter components, the model-based circuit loop voltage equation of the required variables is

$$L_2\dot{I}_{2r} + R_2 I_{2r} + \frac{1}{C_2}\int I_{2r} dt - V_2 = V_{CBr} - V_{DBr}. \tag{14.9}$$

[2] Any variable with a post-subscript "r" is considered as a required (design) variable throughout this chapter.

Theorem 14.2. *The second circuit loop described by (14.8), combined with its control equation (14.9), is virtually stable with its circuit loop current error $I_{2r} - I_2$ being a virtual function in both L_2 and L_∞ and its integral error $\int(I_{2r} - I_2)dt$ being a virtual function in L_∞, in the sense of Definition 2.17.*

Proof: Subtracting (14.8) from (14.9) yields

$$\begin{aligned} L_2(\dot{I}_{2r} - \dot{I}_2) + R_2(I_{2r} - I_2) + \frac{1}{C_2}\int(I_{2r} - I_2)dt \\ = (V_{CBr} - V_{CB}) - (V_{DBr} - V_{DB}). \end{aligned} \tag{14.10}$$

Let the non-negative accompanying function be

$$\nu_2 = \frac{1}{2}L_2(I_{2r} - I_2)^2 + \frac{1}{2C_2}\left(\int(I_{2r} - I_2)dt\right)^2. \tag{14.11}$$

It follows from (14.10) that

$$\begin{aligned} \dot{\nu}_2 &= L_2(I_{2r} - I_2)(\dot{I}_{2r} - \dot{I}_2) + \frac{1}{C_2}(I_{2r} - I_2)\int(I_{2r} - I_2)dt \\ &= -R_2(I_{2r} - I_2)^2 + (I_{2r} - I_2)(V_{CBr} - V_{CB}) \\ &\quad -(I_{2r} - I_2)(V_{DBr} - V_{DB}) \\ &= -R_2(I_{2r} - I_2)^2 + p_{2CB} - p_{2DB} \end{aligned} \tag{14.12}$$

holds with

$$p_{2CB} \stackrel{def}{=} (I_{2r} - I_2)(V_{CBr} - V_{CB}) \tag{14.13}$$

$$p_{2DB} \stackrel{def}{=} (I_{2r} - I_2)(V_{DBr} - V_{DB}) \tag{14.14}$$

representing the two *virtual power flows* at the two *cutting points* of the loop.

Consider the fact that circuit loop 2 has one *driving cutting point* associated with shared edge DB and one *driven cutting point* associated with shared edge CB. Using (14.11), (14.12), and Definition 2.17 completes the proof. ∎

14.3.4 Circuit Loop 3

The voltage equation along the third circuit loop can be written as

$$R_3I_3 + \frac{1}{C_3}\int I_3dt = V_{BE} - V_{BA}. \tag{14.15}$$

With known-parameter components, the model-based circuit loop voltage equation of the required variables is

$$R_3I_{3r} + \frac{1}{C_3}\int I_{3r}dt = V_{BEr} - V_{BAr}. \tag{14.16}$$

Theorem 14.3. *The third circuit loop described by (14.15), combined with its control equation (14.16), is virtually stable with its circuit loop current error* $I_{3r} - I_3$ *being a virtual function in* L_2 *and its integral error* $\int(I_{3r} - I_3)dt$ *being a virtual function in* L_∞*, in the sense of Definition 2.17.*

Proof: Subtracting (14.15) from (14.16) yields

$$\begin{aligned} & R_3(I_{3r} - I_3) + \frac{1}{C_3}\int(I_{3r} - I_3)dt \\ & = (V_{BEr} - V_{BE}) - (V_{BAr} - V_{BA}). \end{aligned} \tag{14.17}$$

Let the non-negative accompanying function be

$$\nu_3 = \frac{1}{2C_3}\left(\int(I_{3r} - I_3)dt\right)^2. \tag{14.18}$$

It follows from (14.17) that

$$\begin{aligned} \dot{\nu}_3 &= \frac{1}{C_3}(I_{3r} - I_3)\int(I_{3r} - I_3)dt \\ &= -R_3(I_{3r} - I_3)^2 + (I_{3r} - I_3)(V_{BEr} - V_{BE}) \\ &\quad -(I_{3r} - I_3)(V_{BAr} - V_{BA}) \\ &= -R_3(I_{3r} - I_3)^2 + p_{3BE} - p_{3BA} \end{aligned} \tag{14.19}$$

holds with

$$p_{3BE} \stackrel{def}{=} (I_{3r} - I_3)(V_{BEr} - V_{BE}) \tag{14.20}$$

$$p_{3BA} \stackrel{def}{=} (I_{3r} - I_3)(V_{BAr} - V_{BA}) \tag{14.21}$$

representing the two *virtual power flows* at the two *cutting points* of the loop.

Consider the fact that circuit loop 3 has one *driving cutting point* associated with shared edge BA and one *driven cutting point* associated with shared edge BE. Using (14.18), (14.19), and Definition 2.17 completes the proof. ∎

14.3.5 Circuit Loop 4

The voltage equation along the fourth circuit loop can be written as

$$L_4\dot{I}_4 + R_4 I_4 - V_4 = -V_{BE} - V_{CB}. \tag{14.22}$$

With known-parameter components, the model-based circuit loop voltage equation of the required variables is

$$L_4\dot{I}_{4r} + R_4 I_{4r} - V_4 = -V_{BEr} - V_{CBr}. \tag{14.23}$$

Theorem 14.4. *The fourth circuit loop described by (14.22), combined with its control equation (14.23), is virtually stable with its circuit loop current error* $I_{4r} - I_4$ *being a virtual function in both* L_2 *and* L_∞*, in the sense of Definition 2.17.*

Proof: Subtracting (14.22) from (14.23) yields

$$L_4(\dot{I}_{4r} - \dot{I}_4) + R_4(I_{4r} - I_4) = -(V_{BEr} - V_{BE}) - (V_{CBr} - V_{CB}). \tag{14.24}$$

Let the non-negative accompanying function be

$$\nu_4 = \frac{1}{2}L_4(I_{4r} - I_4)^2. \tag{14.25}$$

It follows from (14.24) that

$$\begin{aligned}\dot{\nu}_4 &= L_4(I_{4r} - I_4)(\dot{I}_{4r} - \dot{I}_4)\\ &= -R_4(I_{4r} - I_4)^2 - (I_{4r} - I_4)(V_{BEr} - V_{BE}) - (I_{4r} - I_4)(V_{CBr} - V_{CB})\\ &= -R_4(I_{4r} - I_4)^2 - p_{4BE} - p_{4CB} \end{aligned} \tag{14.26}$$

holds with

$$p_{4BE} \stackrel{def}{=} (I_{4r} - I_4)(V_{BEr} - V_{BE}) \tag{14.27}$$

$$p_{4CB} \stackrel{def}{=} (I_{4r} - I_4)(V_{CBr} - V_{CB}) \tag{14.28}$$

representing the two *virtual power flows* at the two *cutting points* of the loop.

Consider the fact that circuit loop 4 has two *driving cutting points* associated with shared edges CB and BE. Using (14.25), (14.26), and Definition 2.17 completes the proof. ■

14.3.6 Shared Edge DB

The voltage equation of shared edge DB can be written as

$$V_{DB} = \frac{1}{C_{12}}\int (I_2 - I_1)dt + V_{12}. \tag{14.29}$$

With known-parameter components, the model-based voltage equation of the required variables is

$$V_{DBr} = \frac{1}{C_{12}}\int (I_{2r} - I_{1r})dt + V_{12}. \tag{14.30}$$

Theorem 14.5. *Shared edge DB described by (14.29), combined with its control equation (14.30), is virtually stable with its integral error* $\int [(I_{2r} - I_2) - (I_{1r} - I_1)]\,dt$ *being a virtual function in* L_∞*, in the sense of Definition 2.17.*

Proof: Subtracting (14.29) from (14.30) yields

$$V_{DBr} - V_{DB} = \frac{1}{C_{12}}\int [(I_{2r} - I_2) - (I_{1r} - I_1)]\,dt. \tag{14.31}$$

Let the non-negative accompanying function be

$$\nu_{DB} = \frac{1}{2C_{12}} \left\{ \int [(I_{2r} - I_2) - (I_{1r} - I_1)]\, dt \right\}^2 . \tag{14.32}$$

It follows from (14.31) that

$$\begin{aligned} \dot{\nu}_{DB} &= [(I_{2r} - I_2) - (I_{1r} - I_1)] \frac{1}{C_{12}} \int [(I_{2r} - I_2) - (I_{1r} - I_1)]\, dt \\ &= [(I_{2r} - I_2) - (I_{1r} - I_1)] (V_{DBr} - V_{DB}) \\ &= p_{2DB} - p_{1DB} \end{aligned} \tag{14.33}$$

holds, where p_{2DB} and p_{1DB} are defined by (14.14) and (14.6), respectively.

Consider the fact that shared edge DB has one *driving cutting point* associated with circuit loop 1 and one *driven cutting point* associated with circuit loop 2. Using (14.32), (14.33), and Definition 2.17 completes the proof. ■

14.3.7 Shared Edge BA

The voltage equation of shared edge BA can be written as

$$V_{BA} = L_{13}(\dot{I}_3 - \dot{I}_1) + R_{13}(I_3 - I_1). \tag{14.34}$$

With known-parameter components, the model-based voltage equation of the required variables is

$$V_{BAr} = L_{13}(\dot{I}_{3r} - \dot{I}_{1r}) + R_{13}(I_{3r} - I_{1r}). \tag{14.35}$$

Theorem 14.6. *Shared edge BA described by (14.34), combined with its control equation (14.35), is virtually stable with its circuit loop current error* $(I_{3r} - I_3) - (I_{1r} - I_1)$ *being a virtual function in both* L_2 *and* L_∞*, in the sense of Definition 2.17.*

Proof: Subtracting (14.34) from (14.35) yields

$$V_{BAr} - V_{BA} = L_{13} \left[(\dot{I}_{3r} - \dot{I}_3) - (\dot{I}_{1r} - \dot{I}_1) \right] + R_{13} [(I_{3r} - I_3) - (I_{1r} - I_1)] . \tag{14.36}$$

Let the non-negative accompanying function be

$$\nu_{BA} = \frac{1}{2} L_{13} [(I_{3r} - I_3) - (I_{1r} - I_1)]^2 . \tag{14.37}$$

It follows from (14.36) that

$$\begin{aligned} \dot{\nu}_{BA} &= [(I_{3r} - I_3) - (I_{1r} - I_1)] L_{13} \left[(\dot{I}_{3r} - \dot{I}_3) - (\dot{I}_{1r} - \dot{I}_1) \right] \\ &= -R_{13} [(I_{3r} - I_3) - (I_{1r} - I_1)]^2 \\ &\quad + [(I_{3r} - I_3) - (I_{1r} - I_1)] (V_{BAr} - V_{BA}) \\ &= -R_{13} [(I_{3r} - I_3) - (I_{1r} - I_1)]^2 \\ &\quad + p_{3BA} - p_{1BA} \end{aligned} \tag{14.38}$$

holds, where p_{3BA} and p_{1BA} are defined by (14.21) and (14.7), respectively.

Consider the fact that shared edge BA has one *driving cutting point* associated with circuit loop 1 and one *driven cutting point* associated with circuit loop 3. Using (14.37), (14.38), and Definition 2.17 completes the proof. ■

14.3.8 Shared Edge CB

The voltage equation of shared edge CB can be written as

$$V_{CB} = R_{24}(I_4 - I_2) + \frac{1}{C_{24}} \int (I_4 - I_2) dt. \tag{14.39}$$

With known-parameter components, the model-based voltage equation of the required variables is

$$V_{CBr} = R_{24}(I_{4r} - I_{2r}) + \frac{1}{C_{24}} \int (I_{4r} - I_{2r}) dt. \tag{14.40}$$

Theorem 14.7. *Shared edge CB described by (14.39), combined with its control equation (14.40), is virtually stable with its current error* $(I_{4r} - I_4) - (I_{2r} - I_2)$ *being a virtual function in* L_2 *and its integral error* $\int [(I_{4r} - I_4) - (I_{2r} - I_2)]\, dt$ *being a virtual function in* L_∞*, in the sense of Definition 2.17.*

Proof: Subtracting (14.39) from (14.40) yields

$$\begin{aligned} V_{CBr} - V_{CB} = {} & R_{24} \left[(I_{4r} - I_4) - (I_{2r} - I_2) \right] \\ & + \frac{1}{C_{24}} \int \left[(I_{4r} - I_4) - (I_{2r} - I_2) \right] dt. \end{aligned} \tag{14.41}$$

Let the non-negative accompanying function be

$$\nu_{CB} = \frac{1}{2C_{24}} \left\{ \int \left[(I_{4r} - I_4) - (I_{2r} - I_2) \right] dt \right\}^2. \tag{14.42}$$

It follows from (14.41) that

$$\begin{aligned} \dot{\nu}_{CB} &= \left[(I_{4r} - I_4) - (I_{2r} - I_2) \right] \frac{1}{C_{24}} \int \left[(I_{4r} - I_4) - (I_{2r} - I_2) \right] dt \\ &= -R_{24} \left[(I_{4r} - I_4) - (I_{2r} - I_2) \right]^2 \\ &\quad + \left[(I_{4r} - I_4) - (I_{2r} - I_2) \right] (V_{CBr} - V_{CB}) \\ &= -R_{24} \left[(I_{4r} - I_4) - (I_{2r} - I_2) \right]^2 \\ &\quad + p_{4CB} - p_{2CB} \end{aligned} \tag{14.43}$$

holds, where p_{4CB} and p_{2CB} are defined by (14.28) and (14.13), respectively.

Consider the fact that shared edge CB has one *driving cutting point* associated with circuit loop 2 and one *driven cutting point* associated with circuit loop 4. Using (14.42), (14.43), and Definition 2.17 completes the proof. ■

14.3.9 Shared Edge BE

The voltage equation of shared edge BE can be written as

$$V_{BE} = V_{34}. \tag{14.44}$$

The model-based voltage equation of the required variables is

$$V_{BEr} = V_{34}. \tag{14.45}$$

Theorem 14.8. *Shared edge BE described by (14.44), combined with its control equation (14.45), is virtually stable in the sense of Definition 2.17.*

Proof: Subtracting (14.44) from (14.45) yields

$$V_{BEr} - V_{BE} = 0. \tag{14.46}$$

Let the non-negative accompanying function be *zero*. It follows from (14.46) that

$$\begin{aligned} 0 &= [(I_{4r} - I_4) - (I_{3r} - I_3)]\,(V_{BEr} - V_{BE}) \\ &= p_{4BE} - p_{3BE} \end{aligned} \tag{14.47}$$

holds, where p_{4BE} and p_{3BE} are defined by (14.27) and (14.20), respectively.

Consider the fact that shared edge BE has one *driving cutting point* associated with circuit loop 3 and one *driven cutting point* associated with circuit loop 4. Using Definition 2.17 completes the proof. ∎

Remark 14.1. The differences of the design equations (14.2), (14.9), (14.16), (14.23), (14.30), (14.35), (14.40), and (14.45) from the system equations (14.1), (14.8), (14.15), (14.22), (14.29), (14.34), (14.39), and (14.44) are that all circuit loop currents and shared edge voltages are replaced by their required (design) variables.

14.4 System Stability

The stability of the complete circuit system is presented in this section.

14.4.1 L_2 and L_∞ Stability

Since every subsystem is *virtually stable* in the sense of Definition 2.17, it follows from Theorem 2.1 that the entire circuit system is stable with

$$I_{1r} - I_1 \in L_2 \bigcap L_\infty \tag{14.48}$$

$$I_{2r} - I_2 \in L_2 \bigcap L_\infty \tag{14.49}$$

$$I_{3r} - I_3 \in L_2 \tag{14.50}$$

$$I_{4r} - I_4 \in L_2 \bigcap L_\infty \tag{14.51}$$

$$(I_{3r} - I_3) - (I_{1r} - I_1) \in L_2 \bigcap L_\infty \tag{14.52}$$

$$\int (I_{2r} - I_2)dt \in L_\infty \tag{14.53}$$

$$\int (I_{3r} - I_3)dt \in L_\infty \tag{14.54}$$

$$\int \left[(I_{2r} - I_2) - (I_{1r} - I_1)\right] dt \in L_\infty \tag{14.55}$$

$$\int \left[(I_{4r} - I_4) - (I_{2r} - I_2)\right] dt \in L_\infty. \tag{14.56}$$

Expressions (14.48)–(14.52) lead to

$$I_{1r} - I_1 \in L_2 \bigcap L_\infty \tag{14.57}$$

$$I_{2r} - I_2 \in L_2 \bigcap L_\infty \tag{14.58}$$

$$I_{3r} - I_3 \in L_2 \bigcap L_\infty \tag{14.59}$$

$$I_{4r} - I_4 \in L_2 \bigcap L_\infty. \tag{14.60}$$

14.4.2 Asymptotic Stability

Substituting (14.31), (14.36), (14.41), and (14.46) into (14.3), (14.10), (14.17), and (14.24) yields

$$\begin{aligned}
&(L_1 + L_{13})(\dot{I}_{1r} - \dot{I}_1) - L_{13}(\dot{I}_{3r} - \dot{I}_3) \\
&\quad = -R_1(I_{1r} - I_1) + \frac{1}{C_{12}} \int \left[(I_{2r} - I_2) - (I_{1r} - I_1)\right] dt \\
&\quad\quad + R_{13}\left[(I_{3r} - I_3) - (I_{1r} - I_1)\right]
\end{aligned} \tag{14.61}$$

$$\begin{aligned}
&L_2(\dot{I}_{2r} - \dot{I}_2) \\
&\quad = -R_2(I_{2r} - I_2) - \frac{1}{C_2} \int (I_{2r} - I_2)dt \\
&\quad\quad + R_{24}\left[(I_{4r} - I_4) - (I_{2r} - I_2)\right] \\
&\quad\quad + \frac{1}{C_{24}} \int \left[(I_{4r} - I_4) - (I_{2r} - I_2)\right] dt \\
&\quad\quad - \frac{1}{C_{12}} \int \left[(I_{2r} - I_2) - (I_{1r} - I_1)\right] dt
\end{aligned} \tag{14.62}$$

$$\begin{aligned}
&-L_{13}(\dot{I}_{1r} - \dot{I}_1) + L_{13}(\dot{I}_{3r} - \dot{I}_3) \\
&\quad = -R_3(I_{3r} - I_3) - \frac{1}{C_3} \int (I_{3r} - I_3)dt \\
&\quad\quad - R_{13}\left[(I_{3r} - I_3) - (I_{1r} - I_1)\right]
\end{aligned} \tag{14.63}$$

$$\begin{aligned} L_4(\dot{I}_{4r} - \dot{I}_4) &= -R_4(I_{4r} - I_4) - R_{24}\left[(I_{4r} - I_4) - (I_{2r} - I_2)\right] \\ &\quad -\frac{1}{C_{24}}\int \left[(I_{4r} - I_4) - (I_{2r} - I_2)\right] dt. \end{aligned} \tag{14.64}$$

Expressions (14.48)–(14.56) ensure that the right hand sides of (14.61)–(14.64) are bounded. Therefore

$$\dot{I}_{1r} - \dot{I}_1 \in L_\infty \tag{14.65}$$
$$\dot{I}_{2r} - \dot{I}_2 \in L_\infty \tag{14.66}$$
$$\dot{I}_{3r} - \dot{I}_3 \in L_\infty \tag{14.67}$$
$$\dot{I}_{4r} - \dot{I}_4 \in L_\infty \tag{14.68}$$

can be guaranteed. Finally, it follows from (14.57)–(14.60), (14.65)–(14.68), and Lemma 2.8 that

$$I_{1r} - I_1 \to 0 \tag{14.69}$$
$$I_{2r} - I_2 \to 0 \tag{14.70}$$
$$I_{3r} - I_3 \to 0 \tag{14.71}$$
$$I_{4r} - I_4 \to 0 \tag{14.72}$$

hold.

14.4.3 Control Algorithm

Given the required currents, denoted as I_{1r}, I_{2r}, I_{3r}, and I_{4r}, substituting (14.30), (14.35), (14.40), and (14.45) into (14.2), (14.9), (14.16), and (14.23) gives the solution to the voltage control sources as

$$\begin{aligned} V_{12} &= L_1\dot{I}_{1r} + R_1 I_{1r} - \frac{1}{C_{12}}\int (I_{2r} - I_{1r})dt \\ &\quad -L_{13}(\dot{I}_{3r} - \dot{I}_{1r}) - R_{13}(I_{3r} - I_{1r}) \end{aligned} \tag{14.73}$$

$$\begin{aligned} V_2 &= L_2\dot{I}_{2r} + R_2 I_{2r} + \frac{1}{C_2}\int I_{2r}dt - R_{24}(I_{4r} - I_{2r}) \\ &\quad -\frac{1}{C_{24}}\int (I_{4r} - I_{2r})dt + \frac{1}{C_{12}}\int (I_{2r} - I_{1r})dt + V_{12} \end{aligned} \tag{14.74}$$

$$\begin{aligned} V_{34} &= R_3 I_{3r} + \frac{1}{C_3}\int I_{3r}dt \\ &\quad +L_{13}(\dot{I}_{3r} - \dot{I}_{1r}) + R_{13}(I_{3r} - I_{1r}) \end{aligned} \tag{14.75}$$

$$\begin{aligned} V_4 &= L_4\dot{I}_{4r} + R_4 I_{4r} + V_{34} \\ &\quad +R_{24}(I_{4r} - I_{2r}) + \frac{1}{C_{24}}\int (I_{4r} - I_{2r})dt. \end{aligned} \tag{14.76}$$

14.5 Adaptive Control

When the parameters of all components are unknown, the design equations (14.2), (14.9), (14.16), (14.23), (14.30), (14.35), (14.40), and (14.45) have to be changed to

$$\hat{L}_1\dot{I}_{1r} + \hat{R}_1 I_{1r} = V_{DBr} + V_{BAr} \tag{14.77}$$

$$\hat{L}_2\dot{I}_{2r} + \hat{R}_2 I_{2r} + \widehat{\left(\frac{1}{C_2}\right)} \int I_{2r}dt - V_2 = V_{CBr} - V_{DBr} \tag{14.78}$$

$$\hat{R}_3 I_{3r} + \widehat{\left(\frac{1}{C_3}\right)} \int I_{3r}dt = V_{BEr} - V_{BAr} \tag{14.79}$$

$$\hat{L}_4\dot{I}_{4r} + \hat{R}_4 I_{4r} - V_4 = -V_{BEr} - V_{CBr} \tag{14.80}$$

with

$$V_{DBr} = \widehat{\left(\frac{1}{C_{12}}\right)} \int (I_{2r} - I_{1r})dt + V_{12} \tag{14.81}$$

$$V_{BAr} = \hat{L}_{13}(\dot{I}_{3r} - \dot{I}_{1r}) + \hat{R}_{13}(I_{3r} - I_{1r}) \tag{14.82}$$

$$V_{CBr} = \hat{R}_{24}(I_{4r} - I_{2r}) + \widehat{\left(\frac{1}{C_{24}}\right)} \int (I_{4r} - I_{2r})dt \tag{14.83}$$

$$V_{BEr} = V_{34} \tag{14.84}$$

subject to the following parameter adaptation

$$\hat{L}_1 = \mathcal{P}\left((I_{1r} - I_1)\dot{I}_{1r}, \rho_{L_1}, \underline{L}_1, \overline{L}_1, t\right) \tag{14.85}$$

$$\hat{R}_1 = \mathcal{P}\left((I_{1r} - I_1)I_{1r}, \rho_{R_1}, \underline{R}_1, \overline{R}_1, t\right) \tag{14.86}$$

$$\hat{L}_2 = \mathcal{P}\left((I_{2r} - I_2)\dot{I}_{2r}, \rho_{L_2}, \underline{L}_2, \overline{L}_2, t\right) \tag{14.87}$$

$$\hat{R}_2 = \mathcal{P}\left((I_{2r} - I_2)I_{2r}, \rho_{R_2}, \underline{R}_2, \overline{R}_2, t\right) \tag{14.88}$$

$$\widehat{\left(\frac{1}{C_2}\right)} = \mathcal{P}\left((I_{2r} - I_2)\int I_{2r}dt, \rho_{C_2}, \underline{\left(\frac{1}{C_2}\right)}, \overline{\left(\frac{1}{C_2}\right)}, t\right) \tag{14.89}$$

$$\hat{R}_3 = \mathcal{P}\left((I_{3r} - I_3)I_{3r}, \rho_{R_3}, \underline{R}_3, \overline{R}_3, t\right) \tag{14.90}$$

$$\widehat{\left(\frac{1}{C_3}\right)} = \mathcal{P}\left((I_{3r} - I_3)\int I_{3r}dt, \rho_{C_3}, \underline{\left(\frac{1}{C_3}\right)}, \overline{\left(\frac{1}{C_3}\right)}, t\right) \tag{14.91}$$

$$\hat{L}_4 = \mathcal{P}\left((I_{4r} - I_4)\dot{I}_{4r}, \rho_{L_4}, \underline{L}_4, \overline{L}_4, t\right) \tag{14.92}$$

$$\hat{R}_4 = \mathcal{P}\left((I_{4r} - I_4)I_{4r}, \rho_{R_4}, \underline{R}_4, \overline{R}_4, t\right) \tag{14.93}$$

$$\widehat{\left(\frac{1}{C_{12}}\right)} = \mathcal{P}\left([(I_{2r} - I_2) - (I_{1r} - I_1)]\int (I_{2r} - I_{1r})dt,\right.$$

$$\rho_{C_{12}}, \underline{\left(\frac{1}{C_{12}}\right)}, \overline{\left(\frac{1}{C_{12}}\right)}, t\Bigg) \tag{14.94}$$

$$\hat{L}_{13} = \mathcal{P}\Big(\left[(I_{3r} - I_3) - (I_{1r} - I_1)\right](\dot{I}_{3r} - \dot{I}_{1r}),$$
$$\rho_{L_{13}}, \underline{L}_{13}, \overline{L}_{13}, t\Big) \tag{14.95}$$

$$\hat{R}_{13} = \mathcal{P}\left(\left[(I_{3r} - I_3) - (I_{1r} - I_1)\right](I_{3r} - I_{1r}),\right.$$
$$\left.\rho_{R_{13}}, \underline{R}_{13}, \overline{R}_{13}, t\right) \tag{14.96}$$

$$\hat{R}_{24} = \mathcal{P}\left(\left[(I_{4r} - I_4) - (I_{2r} - I_2)\right](I_{4r} - I_{2r}),\right.$$
$$\left.\rho_{R_{24}}, \underline{R}_{24}, \overline{R}_{24}, t\right) \tag{14.97}$$

$$\widehat{\left(\frac{1}{C_{24}}\right)} = \mathcal{P}\Bigg(\left[(I_{4r} - I_4) - (I_{2r} - I_2)\right]\int(I_{4r} - I_{2r})dt,$$
$$\rho_{C_{24}}, \underline{\left(\frac{1}{C_{24}}\right)}, \overline{\left(\frac{1}{C_{24}}\right)}, t\Bigg). \tag{14.98}$$

Then, subtracting (14.1), (14.8), (14.15), and (14.22) from (14.77)–(14.80) yields

$$L_1(\dot{I}_{1r} - \dot{I}_1) + R_1(I_{1r} - I_1) - (L_1 - \hat{L}_1)\dot{I}_{1r} - (R_1 - \hat{R}_1)I_{1r}$$
$$= (V_{DBr} - V_{DB}) + (V_{BAr} - V_{BA}) \tag{14.99}$$

$$L_2(\dot{I}_{2r} - \dot{I}_2) + R_2(I_{2r} - I_2) + \frac{1}{C_2}\int(I_{2r} - I_2)dt - (L_2 - \hat{L}_2)\dot{I}_{2r}$$
$$-(R_2 - \hat{R}_2)I_{2r} - \left[\frac{1}{C_2} - \widehat{\left(\frac{1}{C_2}\right)}\right]\int I_{2r}dt$$
$$= (V_{CBr} - V_{CB}) - (V_{DBr} - V_{DB}) \tag{14.100}$$

$$R_3(I_{3r} - I_3) + \frac{1}{C_3}\int(I_{3r} - I_3)dt - (R_3 - \hat{R}_3)I_{3r}$$
$$-\left[\frac{1}{C_3} - \widehat{\left(\frac{1}{C_3}\right)}\right]\int I_{3r}dt$$
$$= (V_{BEr} - V_{BE}) - (V_{BAr} - V_{BA}) \tag{14.101}$$

$$L_4(\dot{I}_{4r} - \dot{I}_4) + R_4(I_{4r} - I_4) - (L_4 - \hat{L}_4)\dot{I}_{4r} - (R_4 - \hat{R}_4)I_{4r}$$
$$= -(V_{BEr} - V_{BE}) - (V_{CBr} - V_{CB}). \tag{14.102}$$

Meanwhile, subtracting (14.29), (14.34), (14.39), and (14.44) from (14.81)–(14.84) yields

$$V_{DBr} - V_{DB} = \frac{1}{C_{12}}\int\left[(I_{2r} - I_2) - (I_{1r} - I_1)\right]dt$$
$$-\left[\frac{1}{C_{12}} - \widehat{\left(\frac{1}{C_{12}}\right)}\right]\int(I_{2r} - I_{1r})dt \tag{14.103}$$

$$
\begin{aligned}
V_{BAr} - V_{BA} &= L_{13}\left[(\dot{I}_{3r} - \dot{I}_3) - (\dot{I}_{1r} - \dot{I}_1)\right] \\
&\quad + R_{13}\left[(I_{3r} - I_3) - (I_{1r} - I_1)\right] \\
&\quad - (L_{13} - \hat{L}_{13})(\dot{I}_{3r} - \dot{I}_{1r}) \\
&\quad - (R_{13} - \hat{R}_{13})(I_{3r} - I_{1r}) && (14.104) \\
V_{CBr} - V_{CB} &= R_{24}\left[(I_{4r} - I_4) - (I_{2r} - I_2)\right] \\
&\quad + \frac{1}{C_{24}}\int \left[(I_{4r} - I_4) - (I_{2r} - I_2)\right] dt \\
&\quad - (R_{24} - \hat{R}_{24})(I_{4r} - I_{2r}) \\
&\quad - \left[\frac{1}{C_{24}} - \widehat{\left(\frac{1}{C_{24}}\right)}\right]\int (I_{4r} - I_{2r})dt && (14.105) \\
V_{BEr} - V_{BE} &= 0. && (14.106)
\end{aligned}
$$

The following theorems ensure the *virtual stability* of the four loops and the four shared edges subject to the adaptive control.

Theorem 14.9. *The first circuit loop described by (14.1), combined with its control equation (14.77) and with the parameter adaptation (14.85) and (14.86), is virtually stable with its circuit loop current error $I_{1r} - I_1$ being a virtual function in both L_2 and L_∞, in the sense of Definition 2.17.*

Proof: Let the non-negative accompanying function be

$$
\nu_{1a} = \frac{1}{2}L_1(I_{1r} - I_1)^2 + \frac{1}{2\rho_{L_1}}(L_1 - \hat{L}_1)^2 + \frac{1}{2\rho_{R_1}}(R_1 - \hat{R}_1)^2. \qquad (14.107)
$$

It follows from (14.85), (14.86), (14.99), and Lemma 2.9 that

$$
\begin{aligned}
\dot{\nu}_{1a} &= L_1(I_{1r} - I_1)(\dot{I}_{1r} - \dot{I}_1) - (L_1 - \hat{L}_1)\dot{\hat{L}}_1/\rho_{L_1} \\
&\quad - (R_1 - \hat{R}_1)\dot{\hat{R}}_1/\rho_{R_1} \\
&= (L_1 - \hat{L}_1)\left(\dot{I}_{1r}(I_{1r} - I_1) - \dot{\hat{L}}_1/\rho_{L_1}\right) \\
&\quad + (R_1 - \hat{R}_1)\left(I_{1r}(I_{1r} - I_1) - \dot{\hat{R}}_1/\rho_{R_1}\right) \\
&\quad - R_1(I_{1r} - I_1)^2 + (I_{1r} - I_1)(V_{DBr} - V_{DB}) \\
&\quad + (I_{1r} - I_1)(V_{BAr} - V_{BA}) \\
&\leqslant -R_1(I_{1r} - I_1)^2 + p_{1DB} + p_{1BA} && (14.108)
\end{aligned}
$$

holds, where p_{1DB} and p_{1BA} are defined by (14.6) and (14.7), respectively.

Consider the fact that circuit loop 1 has two *driven cutting points* associated with shared edges DB and BA. Using (14.107), (14.108), and Definition 2.17 completes the proof. ∎

Theorem 14.10. *The second circuit loop described by (14.8), combined with its control equation (14.78) and with the parameter adaptation (14.87)–(14.89), is* *virtually stable with its circuit loop current error $I_{2r} - I_2$ being a virtual function in both L_2 and L_∞ and its integral error $\int(I_{2r} - I_2)dt$ being a virtual function in L_∞, in the sense of Definition 2.17.*

Proof: Let the non-negative accompanying function be

$$\begin{aligned}\nu_{2a} &= \frac{1}{2}L_2(I_{2r} - I_2)^2 + \frac{1}{2C_2}\left(\int(I_{2r} - I_2)dt\right)^2 \\ &+\frac{1}{2\rho_{L_2}}(L_2 - \hat{L}_2)^2 + \frac{1}{2\rho_{R_2}}(R_2 - \hat{R}_2)^2 \\ &+\frac{1}{2\rho_{C_2}}\left[\frac{1}{C_2} - \widehat{\left(\frac{1}{C_2}\right)}\right]^2. \end{aligned} \tag{14.109}$$

It follows from (14.87)–(14.89), (14.100), and Lemma 2.9 that

$$\begin{aligned}\dot{\nu}_{2a} &= L_2(I_{2r} - I_2)(\dot{I}_{2r} - \dot{I}_2) + \frac{1}{C_2}(I_{2r} - I_2)\int(I_{2r} - I_2)dt \\ &-(L_2 - \hat{L}_2)\dot{\hat{L}}_2/\rho_{L_2} - (R_2 - \hat{R}_2)\dot{\hat{R}}_2/\rho_{R_2} \\ &-\left[\frac{1}{C_2} - \widehat{\left(\frac{1}{C_2}\right)}\right]\dot{\widehat{\left(\frac{1}{C_2}\right)}}/\rho_{C_2} \\ &= (L_2 - \hat{L}_2)\left[\dot{I}_{2r}(I_{2r} - I_2) - \dot{\hat{L}}_2/\rho_{L_2}\right] \\ &+(R_2 - \hat{R}_2)\left[I_{2r}(I_{2r} - I_2) - \dot{\hat{R}}_2/\rho_{R_2}\right] \\ &+\left[\frac{1}{C_2} - \widehat{\left(\frac{1}{C_2}\right)}\right]\left[\int I_{2r}dt(I_{2r} - I_2) - \dot{\widehat{\left(\frac{1}{C_2}\right)}}/\rho_{C_2}\right] \\ &-R_2(I_{2r} - I_2)^2 + (I_{2r} - I_2)(V_{CBr} - V_{CB}) \\ &-(I_{2r} - I_2)(V_{DBr} - V_{DB}) \\ &\leqslant -R_2(I_{2r} - I_2)^2 + p_{2CB} - p_{2DB} \end{aligned} \tag{14.110}$$

holds, where p_{2CB} and p_{2DB} are defined by (14.13) and (14.14), respectively.

Consider the fact that circuit loop 2 has one *driving cutting point* associated with shared edge DB and one *driven cutting point* associated with shared edge CB. Using (14.109), (14.110), and Definition 2.17 completes the proof. ■

Theorem 14.11. *The third circuit loop described by (14.15), combined with its control equation (14.79) and with the parameter adaptation (14.90) and (14.91), is virtually stable with its circuit loop current error $I_{3r} - I_3$ being a virtual function in L_2 and its integral error $\int(I_{3r} - I_3)dt$ being a virtual function in L_∞, in the sense of Definition 2.17.*

Proof: Let the non-negative accompanying function be

$$\nu_{3a} = \frac{1}{2C_3}\left(\int (I_{3r} - I_3)dt\right)^2 + \frac{1}{2\rho_{R_3}}(R_3 - \hat{R}_3)^2 \\ + \frac{1}{2\rho_{C_3}}\left[\frac{1}{C_3} - \widehat{\left(\frac{1}{C_3}\right)}\right]^2. \tag{14.111}$$

It follows from (14.90), (14.91), (14.101), and Lemma 2.9 that

$$\begin{aligned}
\dot{\nu}_{3a} &= \frac{1}{C_3}(I_{3r} - I_3)\int (I_{3r} - I_3)dt \\
&\quad -(R_3 - \hat{R}_3)\dot{\hat{R}}_3/\rho_{R_3} - \left[\frac{1}{C_3} - \widehat{\left(\frac{1}{C_3}\right)}\right]\dot{\widehat{\left(\frac{1}{C_3}\right)}}/\rho_{C_3} \\
&= (R_3 - \hat{R}_3)\left[I_{3r}(I_{3r} - I_3) - \dot{\hat{R}}_3/\rho_{R_3}\right] \\
&\quad + \left[\frac{1}{C_3} - \widehat{\left(\frac{1}{C_3}\right)}\right]\left[(I_{3r} - I_3)\int I_{3r}dt - \dot{\widehat{\left(\frac{1}{C_3}\right)}}/\rho_{C_3}\right] \\
&\quad -R_3(I_{3r} - I_3)^2 + (I_{3r} - I_3)(V_{BEr} - V_{BE}) \\
&\quad -(I_{3r} - I_3)(V_{BAr} - V_{BA}) \\
&\leqslant -R_3(I_{3r} - I_3)^2 + p_{3BE} - p_{3BA}
\end{aligned} \tag{14.112}$$

holds, where p_{3BE} and p_{3BA} are defined by (14.20) and (14.21), respectively.

Consider the fact that circuit loop 3 has one *driving cutting point* associated with shared edge BA and one *driven cutting point* associated with shared edge BE. Using (14.111), (14.112), and Definition 2.17 completes the proof. ■

Theorem 14.12. *The fourth circuit loop described by (14.22), combined with its control equation (14.80) and with the parameter adaptation (14.92) and (14.93), is virtually stable with its circuit loop current error $I_{4r} - I_4$ being a virtual function in both L_2 and L_∞, in the sense of Definition 2.17.*

Proof: Let the non-negative accompanying function be

$$\nu_{4a} = \frac{1}{2}L_4(I_{4r} - I_4)^2 + \frac{1}{2\rho_{L_4}}(L_4 - \hat{L}_4)^2 + \frac{1}{2\rho_{R_4}}(R_4 - \hat{R}_4)^2. \tag{14.113}$$

It follows from (14.92), (14.93), (14.102), and Lemma 2.9 that

$$\begin{aligned}
\dot{\nu}_{4a} &= L_4(I_{4r} - I_4)(\dot{I}_{4r} - \dot{I}_4) - (L_4 - \hat{L}_4)\dot{\hat{L}}_4/\rho_{L_4} \\
&\quad -(R_4 - \hat{R}_4)\dot{\hat{R}}_4/\rho_{R_4} \\
&= (L_4 - \hat{L}_4)\left[\dot{I}_{4r}(I_{4r} - I_4) - \dot{\hat{L}}_4/\rho_{L_4}\right] \\
&\quad +(R_4 - \hat{R}_4)\left[I_{4r}(I_{4r} - I_4) - \dot{\hat{R}}_4/\rho_{R_4}\right]
\end{aligned}$$

$$
\begin{aligned}
&-R_4(I_{4r}-I_4)^2-(I_{4r}-I_4)(V_{BEr}-V_{BE})\\
&-(I_{4r}-I_4)(V_{CBr}-V_{CB})\\
&\leqslant -R_4(I_{4r}-I_4)^2-p_{4BE}-p_{4CB}
\end{aligned}
\tag{14.114}
$$

holds, where p_{4BE} and p_{4CB} are defined by (14.27) and (14.28), respectively.

Consider the fact that circuit loop 4 has two *driving cutting points* associated with shared edges CB and BE. Using (14.113), (14.114), and Definition 2.17 completes the proof. ■

Theorem 14.13. *Shared edge DB described by (14.29), combined with its control equation (14.81) and with the parameter adaptation (14.94), is virtually stable with its integral error* $\int[(I_{2r}-I_2)-(I_{1r}-I_1)]\,dt$ *being a virtual function in* L_∞*, in the sense of Definition 2.17.*

Proof: Let the non-negative accompanying function be

$$
\begin{aligned}
\nu_{DBa} &= \frac{1}{2C_{12}}\left\{\int[(I_{2r}-I_2)-(I_{1r}-I_1)]\,dt\right\}^2\\
&\quad+\frac{1}{2\rho_{C_{12}}}\left[\frac{1}{C_{12}}-\widehat{\left(\frac{1}{C_{12}}\right)}\right]^2.
\end{aligned}
\tag{14.115}
$$

It follows from (14.94), (14.103), and Lemma 2.9 that

$$
\begin{aligned}
\dot{\nu}_{DBa} &= [(I_{2r}-I_2)-(I_{1r}-I_1)]\frac{1}{C_{12}}\int[(I_{2r}-I_2)-(I_{1r}-I_1)]\,dt\\
&\quad-\left[\frac{1}{C_{12}}-\widehat{\left(\frac{1}{C_{12}}\right)}\right]\dot{\widehat{\left(\frac{1}{C_{12}}\right)}}/\rho_{C_{12}}\\
&=\left\{[(I_{2r}-I_2)-(I_{1r}-I_1)]\int(I_{2r}-I_{1r})dt-\dot{\widehat{\left(\frac{1}{C_{12}}\right)}}/\rho_{C_{12}}\right\}\\
&\qquad\times\left[\frac{1}{C_{12}}-\widehat{\left(\frac{1}{C_{12}}\right)}\right]\\
&\quad+[(I_{2r}-I_2)-(I_{1r}-I_1)](V_{DBr}-V_{DB})\\
&\leqslant p_{2DB}-p_{1DB}
\end{aligned}
\tag{14.116}
$$

holds, where p_{2DB} and p_{1DB} are defined by (14.14) and (14.6), respectively.

Consider the fact that shared edge DB has one *driving cutting point* associated with circuit loop 1 and one *driven cutting point* associated with circuit loop 2. Using (14.115), (14.116), and Definition 2.17 completes the proof. ■

Theorem 14.14. *Shared edge BA described by (14.34), combined with its control equation (14.82) and with the parameter adaptation (14.95) and (14.96), is virtually stable with its circuit loop current error* $(I_{3r}-I_3)-(I_{1r}-I_1)$ *being a virtual function in both* L_2 *and* L_∞*, in the sense of Definition 2.17.*

Proof: Let the non-negative accompanying function be

$$\nu_{BAa} = \frac{1}{2}L_{13}\left[(I_{3r}-I_3)-(I_{1r}-I_1)\right]^2 + \frac{1}{2\rho_{L_{13}}}(L_{13}-\hat{L}_{13})^2 + \frac{1}{2\rho_{R_{13}}}(R_{13}-\hat{R}_{13})^2. \tag{14.117}$$

It follows from (14.95), (14.96), (14.104), and Lemma 2.9 that

$$\begin{aligned}\dot{\nu}_{BAa} &= \left[(I_{3r}-I_3)-(I_{1r}-I_1)\right]L_{13}\left[(\dot{I}_{3r}-\dot{I}_3)-(\dot{I}_{1r}-\dot{I}_1)\right] \\ &\quad -(L_{13}-\hat{L}_{13})\dot{\hat{L}}_{13}/\rho_{L_{13}} - (R_{13}-\hat{R}_{13})\dot{\hat{R}}_{13}/\rho_{R_{13}} \\ &= (L_{13}-\hat{L}_{13})\left\{\left[(I_{3r}-I_3)-(I_{1r}-I_1)\right](\dot{I}_{3r}-\dot{I}_{1r}) - \dot{\hat{L}}_{13}/\rho_{L_{13}}\right\} \\ &\quad +(R_{13}-\hat{R}_{13})\left\{\left[(I_{3r}-I_3)-(I_{1r}-I_1)\right](I_{3r}-I_{1r}) - \dot{\hat{R}}_{13}/\rho_{R_{13}}\right\} \\ &\quad -R_{13}\left[(I_{3r}-I_3)-(I_{1r}-I_1)\right]^2 \\ &\quad +\left[(I_{3r}-I_3)-(I_{1r}-I_1)\right](V_{BAr}-V_{BA}) \\ &\leqslant -R_{13}\left[(I_{3r}-I_3)-(I_{1r}-I_1)\right]^2 \\ &\quad +p_{3BA}-p_{1BA}\end{aligned} \tag{14.118}$$

holds, where p_{3BA} and p_{1BA} are defined by (14.21) and (14.7), respectively.

Consider the fact that shared edge BA has one *driving cutting point* associated with circuit loop 1 and one *driven cutting point* associated with circuit loop 3. Using (14.117), (14.118), and Definition 2.17 completes the proof. ∎

Theorem 14.15. *Shared edge CB described by (14.39), combined with its control equation (14.83) and with the parameter adaptation (14.97) and (14.98), is virtually stable with its current error* $(I_{4r}-I_4)-(I_{2r}-I_2)$ *being a virtual function in* L_2 *and its integral error* $\int\left[(I_{4r}-I_4)-(I_{2r}-I_2)\right]dt$ *being a virtual function in* L_∞*, in the sense of Definition 2.17.*

Proof: Let the non-negative accompanying function be

$$\nu_{CBa} = \frac{1}{2C_{24}}\left\{\int\left[(I_{4r}-I_4)-(I_{2r}-I_2)\right]dt\right\}^2 + \frac{1}{2\rho_{R_{24}}}(R_{24}-\hat{R}_{24})^2 + \frac{1}{2\rho_{C_{24}}}\left[\frac{1}{C_{24}}-\widehat{\left(\frac{1}{C_{24}}\right)}\right]^2. \tag{14.119}$$

It follows from (14.97), (14.98), (14.105), and Lemma 2.9 that

$$\begin{aligned}\dot{\nu}_{CBa} &= \left[(I_{4r}-I_4)-(I_{2r}-I_2)\right]\frac{1}{C_{24}}\int\left[(I_{4r}-I_4)-(I_{2r}-I_2)\right]dt \\ &\quad -(R_{24}-\hat{R}_{24})\dot{\hat{R}}_{24}/\rho_{R_{24}} - \left[\frac{1}{C_{24}}-\widehat{\left(\frac{1}{C_{24}}\right)}\right]\dot{\widehat{\left(\frac{1}{C_{24}}\right)}}/\rho_{C_{24}}\end{aligned}$$

$$
\begin{aligned}
&= (R_{24} - \hat{R}_{24}) \left\{ [(I_{4r} - I_4) - (I_{2r} - I_2)] (I_{4r} - I_{2r}) - \dot{\hat{R}}_{24}/\rho_{R_{24}} \right\} \\
&\quad + \left\{ [(I_{4r} - I_4) - (I_{2r} - I_2)] \int (I_{4r} - I_{2r})dt - \dot{\widehat{\left(\frac{1}{C_{24}}\right)}} / \rho_{C_{24}} \right\} \\
&\qquad \times \left[\frac{1}{C_{24}} - \widehat{\left(\frac{1}{C_{24}}\right)} \right] \\
&\quad - R_{24} \left[(I_{4r} - I_4) - (I_{2r} - I_2) \right]^2 \\
&\quad + [(I_{4r} - I_4) - (I_{2r} - I_2)] (V_{CBr} - V_{CB}) \\
&\leqslant -R_{24} \left[(I_{4r} - I_4) - (I_{2r} - I_2) \right]^2 \\
&\quad + p_{4CB} - p_{2CB}
\end{aligned} \tag{14.120}
$$

holds, where p_{4CB} and p_{2CB} are defined by (14.28) and (14.13), respectively.

Consider the fact that shared edge CB has one *driving cutting point* associated with circuit loop 2 and one *driven cutting point* associated with circuit loop 4. Using (14.119), (14.120), and Definition 2.17 completes the proof. ■

Finally, since there is no parameter uncertainty associated with shared edge BE, Theorem 14.8 is still valid.

When all four circuit loops and all four shared edges combined with their respective control equations and parameter adaptation are *virtually stable* in the sense of Definition 2.17, it follows from Theorem 2.1 that the entire circuit system is stable such that expressions (14.57)–(14.60) and (14.69)–(14.72) hold indefinitely.

14.6 Concluding Remarks

In this chapter, the VDC approach has been applied to electrical circuits for the first time, based on the duality between mechanical and electrical systems. The control objective is to control the currents in different loops by applying appropriate voltages at different locations. Similar to the previous chapters, the control design is model based, and parameter (inductance, resistance, and capacitance) uncertainties can be accommodated. The stability of the entire circuit system solely depends on the *virtual stability* of each subsystem (a circuit loop or a shared edge) combined with its respective control equations and parameter adaptation. This chapter paves the way, allowing the subsystem based control to be applied to even more complex electrical systems.

A

Regressor Matrix and Parameter Vector of an *Object*

For a rigid body to which frame $\{\mathbf{A}\}$ is attached, the regressor matrix $\mathbf{Y_A} \in \mathbb{R}^{6\times 13}$ and the parameter vector $\boldsymbol{\theta}_\mathbf{A} \in \mathbb{R}^{13}$ as appeared in (2.78) are expressed in this appendix.

The non-zero elements in $\mathbf{Y_A} \in \mathbb{R}^{6\times 13}$ are listed as

$$y_\mathbf{A}(1,1) = \frac{d}{dt}\left({}^\mathbf{A}v_r\right)(1) + {}^\mathbf{A}v(5){}^\mathbf{A}v_r(3) - {}^\mathbf{A}v(6){}^\mathbf{A}v_r(2) + {}^\mathbf{A}g(1) \tag{A.1}$$

$$y_\mathbf{A}(1,2) = -{}^\mathbf{A}v(5){}^\mathbf{A}v_r(5) - {}^\mathbf{A}v(6){}^\mathbf{A}v_r(6) \tag{A.2}$$

$$y_\mathbf{A}(1,3) = -\frac{d}{dt}\left({}^\mathbf{A}v_r\right)(6) + {}^\mathbf{A}v(5){}^\mathbf{A}v_r(4) \tag{A.3}$$

$$y_\mathbf{A}(1,4) = \frac{d}{dt}\left({}^\mathbf{A}v_r\right)(5) + {}^\mathbf{A}v(6){}^\mathbf{A}v_r(4) \tag{A.4}$$

$$y_\mathbf{A}(2,1) = \frac{d}{dt}\left({}^\mathbf{A}v_r\right)(2) + {}^\mathbf{A}v(6){}^\mathbf{A}v_r(1) - {}^\mathbf{A}v(4){}^\mathbf{A}v_r(3) + {}^\mathbf{A}g(2) \tag{A.5}$$

$$y_\mathbf{A}(2,2) = \frac{d}{dt}\left({}^\mathbf{A}v_r\right)(6) + {}^\mathbf{A}v(4){}^\mathbf{A}v_r(5) \tag{A.6}$$

$$y_\mathbf{A}(2,3) = -{}^\mathbf{A}v(4){}^\mathbf{A}v_r(4) - {}^\mathbf{A}v(6){}^\mathbf{A}v_r(6) \tag{A.7}$$

$$y_\mathbf{A}(2,4) = -\frac{d}{dt}\left({}^\mathbf{A}v_r\right)(4) + {}^\mathbf{A}v(6){}^\mathbf{A}v_r(5) \tag{A.8}$$

$$y_\mathbf{A}(3,1) = \frac{d}{dt}\left({}^\mathbf{A}v_r\right)(3) + {}^\mathbf{A}v(4){}^\mathbf{A}v_r(2) - {}^\mathbf{A}v(5){}^\mathbf{A}v_r(1) + {}^\mathbf{A}g(3) \tag{A.9}$$

$$y_\mathbf{A}(3,2) = -\frac{d}{dt}\left({}^\mathbf{A}v_r\right)(5) + {}^\mathbf{A}v(4){}^\mathbf{A}v_r(6) \tag{A.10}$$

$$y_\mathbf{A}(3,3) = \frac{d}{dt}\left({}^\mathbf{A}v_r\right)(4) + {}^\mathbf{A}v(5){}^\mathbf{A}v_r(6) \tag{A.11}$$

$$y_\mathbf{A}(3,4) = -{}^\mathbf{A}v(4){}^\mathbf{A}v_r(4) - {}^\mathbf{A}v(5){}^\mathbf{A}v_r(5) \tag{A.12}$$

$$y_\mathbf{A}(4,3) = y_\mathbf{A}(3,1) \tag{A.13}$$

$$y_\mathbf{A}(4,4) = -y_\mathbf{A}(2,1) \tag{A.14}$$

$$y_\mathbf{A}(4,6) = y_\mathbf{A}(3,3) \tag{A.15}$$

$$y_\mathbf{A}(4,7) = -y_\mathbf{A}(2,4) \tag{A.16}$$

$$y_{\mathbf{A}}(4,8) = y_{\mathbf{A}}(3,2) \tag{A.17}$$
$$y_{\mathbf{A}}(4,9) = -y_{\mathbf{A}}(2,2) \tag{A.18}$$
$$y_{\mathbf{A}}(4,10) = {}^{\mathbf{A}}v(6)^{\mathbf{A}}v_r(6) - {}^{\mathbf{A}}v(5)^{\mathbf{A}}v_r(5) \tag{A.19}$$
$$y_{\mathbf{A}}(4,11) = \frac{d}{dt}\left({}^{\mathbf{A}}v_r\right)(4) + {}^{\mathbf{A}}v(5)^{\mathbf{A}}v_r(6) - {}^{\mathbf{A}}v(6)^{\mathbf{A}}v_r(5) \tag{A.20}$$
$$y_{\mathbf{A}}(4,12) = -{}^{\mathbf{A}}v(6)^{\mathbf{A}}v_r(5) \tag{A.21}$$
$$y_{\mathbf{A}}(4,13) = {}^{\mathbf{A}}v(5)^{\mathbf{A}}v_r(6) \tag{A.22}$$
$$y_{\mathbf{A}}(5,2) = -y_{\mathbf{A}}(3,1) \tag{A.23}$$
$$y_{\mathbf{A}}(5,4) = y_{\mathbf{A}}(1,1) \tag{A.24}$$
$$y_{\mathbf{A}}(5,5) = -y_{\mathbf{A}}(3,2) \tag{A.25}$$
$$y_{\mathbf{A}}(5,7) = y_{\mathbf{A}}(1,4) \tag{A.26}$$
$$y_{\mathbf{A}}(5,8) = -y_{\mathbf{A}}(3,3) \tag{A.27}$$
$$y_{\mathbf{A}}(5,9) = {}^{\mathbf{A}}v(4)^{\mathbf{A}}v_r(4) - {}^{\mathbf{A}}v(6)^{\mathbf{A}}v_r(6) \tag{A.28}$$
$$y_{\mathbf{A}}(5,10) = y_{\mathbf{A}}(1,3) \tag{A.29}$$
$$y_{\mathbf{A}}(5,11) = {}^{\mathbf{A}}v(6)^{\mathbf{A}}v_r(4) \tag{A.30}$$
$$y_{\mathbf{A}}(5,12) = \frac{d}{dt}\left({}^{\mathbf{A}}v_r\right)(5) + {}^{\mathbf{A}}v(6)^{\mathbf{A}}v_r(4) - {}^{\mathbf{A}}v(4)^{\mathbf{A}}v_r(6) \tag{A.31}$$
$$y_{\mathbf{A}}(5,13) = -{}^{\mathbf{A}}v(4)^{\mathbf{A}}v_r(6) \tag{A.32}$$
$$y_{\mathbf{A}}(6,2) = y_{\mathbf{A}}(2,1) \tag{A.33}$$
$$y_{\mathbf{A}}(6,3) = -y_{\mathbf{A}}(1,1) \tag{A.34}$$
$$y_{\mathbf{A}}(6,5) = y_{\mathbf{A}}(2,2) \tag{A.35}$$
$$y_{\mathbf{A}}(6,6) = -y_{\mathbf{A}}(1,3) \tag{A.36}$$
$$y_{\mathbf{A}}(6,8) = {}^{\mathbf{A}}v(5)^{\mathbf{A}}v_r(5) - {}^{\mathbf{A}}v(4)^{\mathbf{A}}v_r(4) \tag{A.37}$$
$$y_{\mathbf{A}}(6,9) = y_{\mathbf{A}}(2,4) \tag{A.38}$$
$$y_{\mathbf{A}}(6,10) = -y_{\mathbf{A}}(1,4) \tag{A.39}$$
$$y_{\mathbf{A}}(6,11) = -{}^{\mathbf{A}}v(5)^{\mathbf{A}}v_r(4) \tag{A.40}$$
$$y_{\mathbf{A}}(6,12) = {}^{\mathbf{A}}v(4)^{\mathbf{A}}v_r(5) \tag{A.41}$$
$$y_{\mathbf{A}}(6,13) = \frac{d}{dt}\left({}^{\mathbf{A}}v_r\right)(6) + {}^{\mathbf{A}}v(4)^{\mathbf{A}}v_r(5) - {}^{\mathbf{A}}v(5)^{\mathbf{A}}v_r(4) \tag{A.42}$$

where $y_{\mathbf{A}}(j,k) \in \mathbb{R}$ denotes an element of $\mathbf{Y_A} \in \mathbb{R}^{6\times 13}$ at row j and column k for $j \in \{1,6\}$ and $k \in \{1,13\}$; the three variables $\frac{d}{dt}\left({}^{\mathbf{A}}v_r\right)(j) \in \mathbb{R}$, ${}^{\mathbf{A}}v(j) \in \mathbb{R}$, and ${}^{\mathbf{A}}v_r(j) \in \mathbb{R}$ denote the jth elements of $\frac{d}{dt}\left({}^{\mathbf{A}}V_r\right) \in \mathbb{R}^6$, ${}^{\mathbf{A}}V \in \mathbb{R}^6$, and ${}^{\mathbf{A}}V_r \in \mathbb{R}^6$, respectively, for all $j \in \{1,6\}$; and ${}^{\mathbf{A}}g(j) \in \mathbb{R}$ denotes the jth element of ${}^{\mathbf{A}}\mathbf{R_I g} \in \mathbb{R}^3$ with $\mathbf{g} = [0,0,9.8]^T \in \mathbb{R}^3$ for all $j \in \{1,3\}$.

The 13 elements of $\boldsymbol{\theta}_{\mathbf{A}} \in \mathbb{R}^{13}$ are listed as

$$\theta_{\mathbf{A}1} = m_{\mathbf{A}} \tag{A.43}$$

$$\theta_{\mathbf{A}2} = m_{\mathbf{A}}{}^{\mathbf{A}}r_{mx} \tag{A.44}$$

$$\theta_{\mathbf{A}3} = m_{\mathbf{A}}{}^{\mathbf{A}}r_{my} \tag{A.45}$$

$$\theta_{\mathbf{A}4} = m_{\mathbf{A}}{}^{\mathbf{A}}r_{mz} \tag{A.46}$$

$$\theta_{\mathbf{A}5} = m_{\mathbf{A}}{}^{\mathbf{A}}r_{mx}^2 \tag{A.47}$$

$$\theta_{\mathbf{A}6} = m_{\mathbf{A}}{}^{\mathbf{A}}r_{my}^2 \tag{A.48}$$

$$\theta_{\mathbf{A}7} = m_{\mathbf{A}}{}^{\mathbf{A}}r_{mx}^2 \tag{A.49}$$

$$\theta_{\mathbf{A}8} = m_{\mathbf{A}}{}^{\mathbf{A}}r_{mx}{}^{\mathbf{A}}r_{my} - I_{\mathbf{A}xy} \tag{A.50}$$

$$\theta_{\mathbf{A}9} = m_{\mathbf{A}}{}^{\mathbf{A}}r_{mx}{}^{\mathbf{A}}r_{mz} - I_{\mathbf{A}xz} \tag{A.51}$$

$$\theta_{\mathbf{A}10} = m_{\mathbf{A}}{}^{\mathbf{A}}r_{my}{}^{\mathbf{A}}r_{mz} - I_{\mathbf{A}yz} \tag{A.52}$$

$$\theta_{\mathbf{A}11} = I_{\mathbf{A}xx} \tag{A.53}$$

$$\theta_{\mathbf{A}12} = I_{\mathbf{A}yy} \tag{A.54}$$

$$\theta_{\mathbf{A}13} = I_{\mathbf{A}zz} \tag{A.55}$$

where $\theta_{\mathbf{A}k}$ denotes the kth element of $\boldsymbol{\theta}_{\mathbf{A}} \in \mathbb{R}^{13}$ for all $k \in \{1, 13\}$; $m_{\mathbf{A}}$ is the mass; ${}^{\mathbf{A}}\mathbf{r}_m = [{}^{\mathbf{A}}r_{mx}, {}^{\mathbf{A}}r_{my}, {}^{\mathbf{A}}r_{mz}]^T \in \mathbb{R}^3$ denotes a vector pointing from the origin of frame $\mathbf{A}$ toward the mass center and expressed in frame $\mathbf{A}$ (it is equivalent to ${}^{\mathbf{A}}\mathbf{r}_{\mathbf{AB}}$ in (2.75)-(2.77)), and $I_{\mathbf{A}xx}$, $I_{\mathbf{A}yy}$, $I_{\mathbf{A}zz}$, $I_{\mathbf{A}xy}$, $I_{\mathbf{A}xz}$, and $I_{\mathbf{A}yz}$ are elements of $\mathbf{I}_{\mathbf{A}}$.

B

Mathematical Derivations

B.1 Proof of Lemma 4.1

Subtracting (4.24) from (4.32) yields

$$
\begin{aligned}
{}^{\mathbf{O}_i}F_r^* - {}^{\mathbf{O}_i}F^* = \mathbf{M}_{\mathbf{O}_i}\left[\frac{d}{dt}({}^{\mathbf{O}_i}V_r) - \frac{d}{dt}({}^{\mathbf{O}_i}V)\right] + \mathbf{C}_{\mathbf{O}_i}({}^{\mathbf{O}_i}\omega)\left({}^{\mathbf{O}_i}V_r - {}^{\mathbf{O}_i}V\right) \\
-\mathbf{Y}_{\mathbf{O}_i}\left(\boldsymbol{\theta}_{\mathbf{O}_i} - \hat{\boldsymbol{\theta}}_{\mathbf{O}_i}\right) + \mathbf{K}_{\mathbf{O}_i}\left({}^{\mathbf{O}_i}V_r - {}^{\mathbf{O}_i}V\right). \qquad \text{(B.1)}
\end{aligned}
$$

In view of (B.1), the skew-symmetric property of $\mathbf{C}_{\mathbf{O}_i}({}^{\mathbf{O}_i}\omega)$, and Lemma 2.9, the time derivative of $\nu_{\mathbf{O}_i}$ in (4.40) can be written as

$$
\begin{aligned}
\dot{\nu}_{\mathbf{O}_i} &= \left({}^{\mathbf{O}_i}V_r - {}^{\mathbf{O}_i}V\right)^T \mathbf{M}_{\mathbf{O}_i}\frac{d}{dt}\left({}^{\mathbf{O}_i}V_r - {}^{\mathbf{O}_i}V\right) \\
&\quad -\sum_{\gamma=1}^{13}\left(\theta_{\mathbf{O}_i\gamma} - \hat{\theta}_{\mathbf{O}_i\gamma}\right)\dot{\hat{\theta}}_{\mathbf{O}_i\gamma}/\rho_{\mathbf{O}_i\gamma} \\
&= -\left({}^{\mathbf{O}_i}V_r - {}^{\mathbf{O}_i}V\right)^T \mathbf{C}_{\mathbf{O}_i}({}^{\mathbf{O}_i}\omega)\left({}^{\mathbf{O}_i}V_r - {}^{\mathbf{O}_i}V\right) \\
&\quad +\left({}^{\mathbf{O}_i}V_r - {}^{\mathbf{O}_i}V\right)^T \mathbf{Y}_{\mathbf{O}_i}\left(\boldsymbol{\theta}_{\mathbf{O}_i} - \hat{\boldsymbol{\theta}}_{\mathbf{O}_i}\right) \\
&\quad -\left({}^{\mathbf{O}_i}V_r - {}^{\mathbf{O}_i}V\right)^T \mathbf{K}_{\mathbf{O}_i}\left({}^{\mathbf{O}_i}V_r - {}^{\mathbf{O}_i}V\right) \\
&\quad +\left({}^{\mathbf{O}_i}V_r - {}^{\mathbf{O}_i}V\right)^T \left({}^{\mathbf{O}_i}F_r^* - {}^{\mathbf{O}_i}F^*\right) \\
&\quad -\sum_{\gamma=1}^{13}\left(\theta_{\mathbf{O}_i\gamma} - \hat{\theta}_{\mathbf{O}_i\gamma}\right)\dot{\hat{\theta}}_{\mathbf{O}_i\gamma}/\rho_{\mathbf{O}_i\gamma} \\
&= -\left({}^{\mathbf{O}_i}V_r - {}^{\mathbf{O}_i}V\right)^T \mathbf{K}_{\mathbf{O}_i}\left({}^{\mathbf{O}_i}V_r - {}^{\mathbf{O}_i}V\right) \\
&\quad +\left({}^{\mathbf{O}_i}V_r - {}^{\mathbf{O}_i}V\right)^T \left({}^{\mathbf{O}_i}F_r^* - {}^{\mathbf{O}_i}F^*\right) \\
&\quad +\sum_{\gamma=1}^{13}\Big\{\left(\theta_{\mathbf{O}_i\gamma} - \hat{\theta}_{\mathbf{O}_i\gamma}\right)
\end{aligned}
$$

$$\left[s_{\mathbf{O}_i\gamma} - \frac{\dot{\hat{\theta}}_{\mathbf{O}_i\gamma}}{\rho_{\mathbf{O}_i\gamma}}\right]\Bigg\}$$

$$\leqslant -\left({}^{\mathbf{O}_i}V_r - {}^{\mathbf{O}_i}V\right)^T \mathbf{K}_{\mathbf{O}_i}\left({}^{\mathbf{O}_i}V_r - {}^{\mathbf{O}_i}V\right)$$
$$+\left({}^{\mathbf{O}_i}V_r - {}^{\mathbf{O}_i}V\right)^T\left({}^{\mathbf{O}_i}F_r^* - {}^{\mathbf{O}_i}F^*\right). \tag{B.2}$$

B.2 Derivation of (4.121)

Premultiplying $\mathcal{F}_b$ in (4.112) by $\mathcal{J}_b^T$ in (4.116) and using (4.25), (4.51)–(4.53), (4.73), (4.112), (4.113), and (4.117)–(4.119) yields

$$\mathcal{J}_b^T\mathcal{F}_b = \sum_{i=1}^{n_o}\left(\mathcal{J}_{\mathbf{O}_i}^T\,{}^{\mathbf{O}_i}F^{*T}\right) + \sum_{j=1}^{n_c}\left(\mathcal{J}_{bj}^T\mathcal{F}_{bj}\right)$$
$$= \sum_{i=1}^{n_o}\mathcal{J}_{\mathbf{O}_i}^T\,{}^{\mathbf{O}_i}F^* + \sum_{j=1}^{n_c}\left(\mathcal{J}_{\mathbf{B}_j}^T\,{}^{\mathbf{B}_j}F^* + \sum_{k=1}^{l_j}\mathcal{J}_{\mathbf{B}_{jk}}^T\,{}^{\mathbf{B}_{jk}}F^*\right)$$
$$= \sum_{i=1}^{n_o}\left(\sum_{\{\mathbf{T}_j\}\in\mathcal{T}_i}\mathcal{J}_{\mathbf{O}_i}^T\,{}^{\mathbf{O}_i}\mathbf{U}_{\mathbf{T}_j}\,{}^{\mathbf{T}_j}F - \sum_{\{\mathbf{B}_j\}\in\mathcal{B}_i}\mathcal{J}_{\mathbf{O}_i}^T\,{}^{\mathbf{O}_i}\mathbf{U}_{\mathbf{B}_j}\,{}^{\mathbf{B}_j}F - \mathcal{J}_{\mathbf{O}_i}^T\,{}^{\mathbf{O}_i}\mathbf{U}_{\mathbf{S}_i}\,{}^{\mathbf{S}_i}F\right)$$
$$+\sum_{j=1}^{n_c}\Bigg[\mathcal{J}_{\mathbf{B}_j}^T\,{}^{\mathbf{B}_j}F - \mathcal{J}_{\mathbf{B}_j}^T\,{}^{\mathbf{B}_j}\mathbf{U}_{\mathbf{T}_{j1}}\,{}^{\mathbf{T}_{j1}}F$$
$$+\sum_{k=1}^{l_j-1}\left(\mathcal{J}_{\mathbf{B}_{jk}}^T\,{}^{\mathbf{B}_{jk}}F - \mathcal{J}_{\mathbf{B}_{jk}}^T\,{}^{\mathbf{B}_{jk}}\mathbf{U}_{\mathbf{T}_{j(k+1)}}\,{}^{\mathbf{T}_{j(k+1)}}F\right)$$
$$+\mathcal{J}_{\mathbf{B}_{jl_j}}^T\,{}^{\mathbf{B}_{jl_j}}F - \mathcal{J}_{\mathbf{B}_{jl_j}}^T\,{}^{\mathbf{B}_{jl_j}}\mathbf{U}_{\mathbf{T}_j}\,{}^{\mathbf{T}_j}F\Bigg]$$
$$= \sum_{i=1}^{n_o}\left(\sum_{\{\mathbf{T}_j\}\in\mathcal{T}_i}\mathcal{J}_{\mathbf{O}_i}^T\,{}^{\mathbf{O}_i}\mathbf{U}_{\mathbf{T}_j}\,{}^{\mathbf{T}_j}F - \sum_{\{\mathbf{B}_j\}\in\mathcal{B}_i}\mathcal{J}_{\mathbf{O}_i}^T\,{}^{\mathbf{O}_i}\mathbf{U}_{\mathbf{B}_j}\,{}^{\mathbf{B}_j}F - \mathcal{J}_{\mathbf{O}_i}^T\,{}^{\mathbf{O}_i}\mathbf{U}_{\mathbf{S}_i}\,{}^{\mathbf{S}_i}F\right)$$
$$+\sum_{j=1}^{n_c}\Bigg[\mathcal{J}_{\mathbf{B}_j}^T\,{}^{\mathbf{B}_j}F - \mathcal{J}_{\mathbf{B}_j}^T\,{}^{\mathbf{B}_j}\mathbf{U}_{\mathbf{B}_{j1}}\,{}^{\mathbf{B}_{j1}}F$$
$$+\sum_{k=1}^{l_j-1}\left(\mathcal{J}_{\mathbf{B}_{jk}}^T\,{}^{\mathbf{B}_{jk}}F - \mathcal{J}_{\mathbf{B}_{jk}}^T\,{}^{\mathbf{B}_{jk}}\mathbf{U}_{\mathbf{B}_{j(k+1)}}\,{}^{\mathbf{B}_{j(k+1)}}F\right)$$
$$+\mathcal{J}_{\mathbf{B}_{jl_j}}^T\,{}^{\mathbf{B}_{jl_j}}F - \mathcal{J}_{\mathbf{B}_{jl_j}}^T\,{}^{\mathbf{B}_{jl_j}}\mathbf{U}_{\mathbf{T}_j}\,{}^{\mathbf{T}_j}F\Bigg]. \tag{B.3}$$

For $\{\mathbf{T}_j\}\in\mathcal{T}_i$, it yields

$${}^{\mathbf{T}_j}V = {}^{\mathbf{O}_i}\mathbf{U}_{\mathbf{T}_j}^T\mathcal{J}_{\mathbf{O}_i}\mathcal{V} \tag{B.4}$$
$${}^{\mathbf{T}_j}V = {}^{\mathbf{B}_{jl_j}}\mathbf{U}_{\mathbf{T}_j}^T\mathcal{J}_{\mathbf{B}_{jl_j}}\mathcal{V} \tag{B.5}$$

from the definition of $\mathcal{J}_b$ in (4.116).

Since $\mathcal{V} \in \mathbb{R}^m$ is an independent vector, it follows from (B.4) and (B.5) that

$$ {}^{\mathbf{O}_i}\mathbf{U}_{\mathbf{T}_j}^T \mathcal{J}_{\mathbf{O}_i} = {}^{\mathbf{B}_{jl_j}}\mathbf{U}_{\mathbf{T}_j}^T \mathcal{J}_{\mathbf{B}_{jl_j}} \tag{B.6} $$

holds or equivalently

$$ \mathcal{J}_{\mathbf{O}_i}^T \, {}^{\mathbf{O}_i}\mathbf{U}_{\mathbf{T}_j} = \mathcal{J}_{\mathbf{B}_{jl_j}}^T \, {}^{\mathbf{B}_{jl_j}}\mathbf{U}_{\mathbf{T}_j} \tag{B.7} $$

holds after performing the transpose operation.

Following the same procedure yields

$$ \mathcal{J}_{\mathbf{O}_i}^T \, {}^{\mathbf{O}_i}\mathbf{U}_{\mathbf{B}_j} = \mathcal{J}_{\mathbf{B}_j}^T \tag{B.8} $$

for $\{\mathbf{B}_j\} \in \mathcal{B}_i$.

Meanwhile, for $k \in \mathcal{C}_{1j}$ and $j \in \{1, n_c\}$, re-write (4.73) as

$$ \begin{aligned} {}^{\mathbf{B}_{jk}}V &= \mathcal{J}_{\mathbf{B}_{jk}} \mathcal{V} \\ &= \mathbf{z}\dot{q}_{jk} + {}^{\mathbf{B}_{j(k-1)}}\mathbf{U}_{\mathbf{B}_{jk}}^T \, {}^{\mathbf{B}_{j(k-1)}}V \\ &= \mathbf{z}\mathcal{J}_{jk}\mathcal{V} + {}^{\mathbf{B}_{j(k-1)}}\mathbf{U}_{\mathbf{B}_{jk}}^T \mathcal{J}_{\mathbf{B}_{j(k-1)}} \mathcal{V}. \end{aligned} \tag{B.9} $$

Since $\mathcal{V} \in \mathbb{R}^m$ is an independent vector, it follows from (B.9) that

$$ \mathcal{J}_{\mathbf{B}_{jk}} = \mathbf{z}\mathcal{J}_{jk} + {}^{\mathbf{B}_{j(k-1)}}\mathbf{U}_{\mathbf{B}_{jk}}^T \mathcal{J}_{\mathbf{B}_{j(k-1)}} \tag{B.10} $$

holds. Performing a transpose operation yields

$$ \mathcal{J}_{\mathbf{B}_{jk}}^T = \mathcal{J}_{jk}^T \mathbf{z}^T + \mathcal{J}_{\mathbf{B}_{j(k-1)}}^T \, {}^{\mathbf{B}_{j(k-1)}}\mathbf{U}_{\mathbf{B}_{jk}} \tag{B.11} $$

with $\mathbf{B}_{j0} = \mathbf{B}_j$.

Accordingly, for $k \in \mathcal{C}_{3j}$ and $j \in \{1, n_c\}$, it follows that

$$ \mathcal{J}_{\mathbf{B}_{jk}}^T = \mathcal{J}_{jk}^T \mathbf{Z}^T + \mathcal{J}_{\mathbf{B}_{j(k-1)}}^T \, {}^{\mathbf{B}_{j(k-1)}}\mathbf{U}_{\mathbf{B}_{jk}} \tag{B.12} $$

holds with $\mathbf{B}_{j0} = \mathbf{B}_j$.

Combining (B.11) and (B.12) validates the following equation

$$ \begin{aligned} &- \sum_{k \in \mathcal{C}_{1j}} \mathcal{J}_{jk}^T \mathbf{z}^T \, {}^{\mathbf{B}_{jk}}F - \sum_{k \in \mathcal{C}_{3j}} \mathcal{J}_{jk}^T \mathbf{Z}^T \, {}^{\mathbf{B}_{jk}}F \\ &+ \mathcal{J}_{\mathbf{B}_j}^T \, {}^{\mathbf{B}_j}F - \mathcal{J}_{\mathbf{B}_j}^T \, {}^{\mathbf{B}_j}\mathbf{U}_{\mathbf{B}_{j1}} \, {}^{\mathbf{B}_{j1}}F \\ &+ \sum_{k=1}^{l_j - 1} \left(\mathcal{J}_{\mathbf{B}_{jk}}^T \, {}^{\mathbf{B}_{jk}}F - \mathcal{J}_{\mathbf{B}_{jk}}^T \, {}^{\mathbf{B}_{jk}}\mathbf{U}_{\mathbf{B}_{j(k+1)}} \, {}^{\mathbf{B}_{j(k+1)}}F \right) \\ &+ \mathcal{J}_{\mathbf{B}_{jl_j}}^T \, {}^{\mathbf{B}_{jl_j}}F - \mathcal{J}_{\mathbf{B}_{jl_j}}^T \, {}^{\mathbf{B}_{jl_j}}\mathbf{U}_{\mathbf{T}_j} \, {}^{\mathbf{T}_j}F \\ = {} & \mathcal{J}_{\mathbf{B}_j}^T \, {}^{\mathbf{B}_j}F + \sum_{k=1}^{l_j} \left[\mathcal{J}_{\mathbf{B}_{jk}}^T - \mathcal{J}_{jk}^T \mathbf{z}_k^T - \mathcal{J}_{\mathbf{B}_{j(k-1)}}^T \, {}^{\mathbf{B}_{j(k-1)}}\mathbf{U}_{\mathbf{B}_{jk}} \right] {}^{\mathbf{B}_{jk}}F \\ &- \mathcal{J}_{\mathbf{B}_{jl_j}}^T \, {}^{\mathbf{B}_{jl_j}}\mathbf{U}_{\mathbf{T}_j} \, {}^{\mathbf{T}_j}F \\ = {} & \mathcal{J}_{\mathbf{B}_j}^T \, {}^{\mathbf{B}_j}F - \mathcal{J}_{\mathbf{B}_{jl_j}}^T \, {}^{\mathbf{B}_{jl_j}}\mathbf{U}_{\mathbf{T}_j} \, {}^{\mathbf{T}_j}F \end{aligned} \tag{B.13} $$

with $\mathbf{B}_{j0} = \mathbf{B}_j$ and

$$\mathbf{z}_k = \begin{cases} \mathbf{z} & \text{if } k \in \mathcal{C}_{1j} \\ \mathbf{Z} & \text{if } k \in \mathcal{C}_{3j} \end{cases} . \tag{B.14}$$

In view of (B.7), (B.8), (4.118), and (4.119), substituting (B.13) into (B.3) yields

$$\begin{aligned}
\mathcal{J}_b^T \mathcal{F}_b = & \sum_{i=1}^{n_o} \left(\sum_{\{\mathbf{T}_j\} \in \mathcal{T}_i} \mathcal{J}_{\mathbf{O}_i}^T {}^{\mathbf{O}_i}\mathbf{U}_{\mathbf{T}_j} {}^{\mathbf{T}_j}F - \sum_{\{\mathbf{B}_j\} \in \mathcal{B}_i} \mathcal{J}_{\mathbf{O}_i}^T {}^{\mathbf{O}_i}\mathbf{U}_{\mathbf{B}_j} {}^{\mathbf{B}_j}F - \mathcal{J}_{\mathbf{O}_i}^T {}^{\mathbf{O}_i}\mathbf{U}_{\mathbf{S}_i} {}^{\mathbf{S}_i}F \right) \\
& + \sum_{j=1}^{n_c} \Bigg(\mathcal{J}_{\mathbf{B}_j}^T {}^{\mathbf{B}_j}F - \mathcal{J}_{\mathbf{B}_{jl_j}}^T {}^{\mathbf{B}_{jl_j}}\mathbf{U}_{\mathbf{T}_j} {}^{\mathbf{T}_j}F \\
& + \sum_{k \in \mathcal{C}_{1j}} \mathcal{J}_{jk}^T \mathbf{z}^{T\mathbf{B}_{jk}}F + \sum_{k \in \mathcal{C}_{3j}} \mathcal{J}_{jk}^T \mathbf{Z}^{T\mathbf{B}_{jk}}F \Bigg) \\
= & \left(\sum_{i=1}^{n_o} \sum_{\{\mathbf{T}_j\} \in \mathcal{T}_i} \mathcal{J}_{\mathbf{O}_i}^T {}^{\mathbf{O}_i}\mathbf{U}_{\mathbf{T}_j} {}^{\mathbf{T}_j}F - \sum_{j=1}^{n_c} \mathcal{J}_{\mathbf{B}_{jl_j}}^T {}^{\mathbf{B}_{jl_j}}\mathbf{U}_{\mathbf{T}_j} {}^{\mathbf{T}_j}F \right) \\
& + \left(- \sum_{i=1}^{n_o} \sum_{\{\mathbf{B}_j\} \in \mathcal{B}_i} \mathcal{J}_{\mathbf{O}_i}^T {}^{\mathbf{O}_i}\mathbf{U}_{\mathbf{B}_j} {}^{\mathbf{B}_j}F + \sum_{j=1}^{n_c} \mathcal{J}_{\mathbf{B}_j}^T {}^{\mathbf{B}_j}F \right) \\
& - \sum_{i=1}^{n_o} \mathcal{J}_{\mathbf{O}_i}^T {}^{\mathbf{O}_i}\mathbf{U}_{\mathbf{S}_i} {}^{\mathbf{S}_i}F \\
& + \sum_{j=1}^{n_c} \left(\sum_{k \in \mathcal{C}_{1j}} \mathcal{J}_{jk}^T \mathbf{z}^{T\mathbf{B}_{jk}}F + \sum_{k \in \mathcal{C}_{3j}} \mathcal{J}_{jk}^T \mathbf{Z}^{T\mathbf{B}_{jk}}F \right) \\
= & - \sum_{i=1}^{n_o} \mathcal{J}_{\mathbf{O}_i}^T {}^{\mathbf{O}_i}\mathbf{U}_{\mathbf{S}_i} {}^{\mathbf{S}_i}F \\
& + \sum_{j=1}^{n_c} \left(\sum_{k \in \mathcal{C}_{1j}} \mathcal{J}_{jk}^T \mathbf{z}^{T\mathbf{B}_{jk}}F + \sum_{k \in \mathcal{C}_{3j}} \mathcal{J}_{jk}^T \mathbf{Z}^{T\mathbf{B}_{jk}}F \right) .
\end{aligned} \tag{B.15}$$

B.3 Proof of Lemma 6.1

With appropriate frame substitutions, it follows from (6.3)–(6.5), (6.16)–(6.24), and Lemma 4.1 that

$$\begin{aligned}
\dot{\nu}_{\mathbf{T}} \leqslant & - \left({}^{\mathbf{T}}V_r - {}^{\mathbf{T}}V \right)^T \mathbf{K}_{\mathbf{T}} \left({}^{\mathbf{T}}V_r - {}^{\mathbf{T}}V \right) \\
& + \left({}^{\mathbf{T}}V_r - {}^{\mathbf{T}}V \right)^T \left({}^{\mathbf{T}}F_r^* - {}^{\mathbf{T}}F^* \right)
\end{aligned} \tag{B.16}$$

$$\dot{\nu}_{\mathbf{A}} \leqslant -\left({}^{\mathbf{A}}V_r - {}^{\mathbf{A}}V\right)^T \mathbf{K}_{\mathbf{A}} \left({}^{\mathbf{A}}V_r - {}^{\mathbf{A}}V\right) + \left({}^{\mathbf{A}}V_r - {}^{\mathbf{A}}V\right)^T \left({}^{\mathbf{A}}F_r^* - {}^{\mathbf{A}}F^*\right) \tag{B.17}$$

$$\dot{\nu}_{\mathbf{B}} \leqslant -\left({}^{\mathbf{B}}V_r - {}^{\mathbf{B}}V\right)^T \mathbf{K}_{\mathbf{B}} \left({}^{\mathbf{B}}V_r - {}^{\mathbf{B}}V\right) + \left({}^{\mathbf{B}}V_r - {}^{\mathbf{B}}V\right)^T \left({}^{\mathbf{B}}F_r^* - {}^{\mathbf{B}}F^*\right) \tag{B.18}$$

hold.

Furthermore, in view of (2.85), (6.1), (6.2), (6.6)–(6.12), (6.14), (6.15), and (6.25)–(6.29), it yields

$$\begin{aligned}
&\left({}^{\mathbf{T}}V_r - {}^{\mathbf{T}}V\right)^T \left({}^{\mathbf{T}}F_r^* - {}^{\mathbf{T}}F^*\right) \\
&= \left({}^{\mathbf{T}}V_r - {}^{\mathbf{T}}V\right)^T \left[-\left({}^{\mathbf{T}}F_r - {}^{\mathbf{T}}F\right) + \left({}^{\mathbf{T}}F_{qr} - {}^{\mathbf{T}}F_q\right)\right] \\
&= -\left({}^{\mathbf{T}}V_r - {}^{\mathbf{T}}V\right)^T \left({}^{\mathbf{T}}F_r - {}^{\mathbf{T}}F\right) \\
&\quad + \left[\mathbf{z}_\tau \left(\dot{q}_r - \dot{q}\right) + {}^{\mathbf{B}}\mathbf{U}_{\mathbf{T}}^T \left({}^{\mathbf{B}}V_r - {}^{\mathbf{B}}V\right)\right]^T \left({}^{\mathbf{T}}F_{qr} - {}^{\mathbf{T}}F_q\right) \\
&= -p_{\mathbf{T}} + \varrho \left(\dot{q}_r - \dot{q}\right)\left(\tau_{td} - \tau_t\right) - \varrho k_{vq} \left(\dot{q}_r - \dot{q}\right)^2 \\
&\quad + \left(\dot{q}_r - \dot{q}\right) \mathbf{Y}_b(\dot{q}) \left(\boldsymbol{\theta}_b - \hat{\boldsymbol{\theta}}_b\right) \\
&\quad + ({}^{\mathbf{B}}V_r - {}^{\mathbf{B}}V)^{T\mathbf{B}}\mathbf{U}_{\mathbf{T}}({}^{\mathbf{T}}F_{qr} - {}^{\mathbf{T}}F_q)
\end{aligned} \tag{B.19}$$

$$\begin{aligned}
&\left({}^{\mathbf{A}}V_r - {}^{\mathbf{A}}V\right)^T \left({}^{\mathbf{A}}F_r^* - {}^{\mathbf{A}}F^*\right) \\
&= \left[\mathbf{z}_\tau \varrho(\dot{\phi}_r - \dot{\phi}) + {}^{\mathbf{B}}\mathbf{U}_{\mathbf{A}}^T({}^{\mathbf{B}}V_r - {}^{\mathbf{B}}V)\right]^T \left({}^{\mathbf{A}}F_{\phi r} - {}^{\mathbf{A}}F_\phi\right) \\
&= \varrho(\dot{\phi}_r - \dot{\phi})(\tau_d - \tau) - \varrho(\dot{\phi}_r - \dot{\phi})(\tau_{td} - \tau_t) \\
&\quad - \varrho k_{v\phi} \left(\dot{\phi}_r - \dot{\phi}\right)^2 + \varrho(\dot{\phi}_r - \dot{\phi})\mathbf{Y}_m(\dot{\phi}) \left(\boldsymbol{\theta}_m - \hat{\boldsymbol{\theta}}_m\right) \\
&\quad + ({}^{\mathbf{B}}V_r - {}^{\mathbf{B}}V)^{T\mathbf{B}}\mathbf{U}_{\mathbf{A}}({}^{\mathbf{A}}F_{\phi r} - {}^{\mathbf{A}}F_\phi)
\end{aligned} \tag{B.20}$$

$$\begin{aligned}
&\left({}^{\mathbf{B}}V_r - {}^{\mathbf{B}}V\right)^T \left({}^{\mathbf{B}}F_r^* - {}^{\mathbf{B}}F^*\right) \\
&= \left({}^{\mathbf{B}}V_r - {}^{\mathbf{B}}V\right)^T \left[({}^{\mathbf{B}}F_r - {}^{\mathbf{B}}F)\right. \\
&\qquad \left. - {}^{\mathbf{B}}\mathbf{U}_{\mathbf{T}}({}^{\mathbf{T}}F_{qr} - {}^{\mathbf{T}}F_q) - {}^{\mathbf{B}}\mathbf{U}_{\mathbf{A}}({}^{\mathbf{A}}F_{\phi r} - {}^{\mathbf{A}}F_\phi)\right] \\
&= p_{\mathbf{B}} - ({}^{\mathbf{B}}V_r - {}^{\mathbf{B}}V)^{T\mathbf{B}}\mathbf{U}_{\mathbf{T}}({}^{\mathbf{T}}F_{qr} - {}^{\mathbf{T}}F_q) \\
&\quad - ({}^{\mathbf{B}}V_r - {}^{\mathbf{B}}V)^{T\mathbf{B}}\mathbf{U}_{\mathbf{A}}({}^{\mathbf{A}}F_{\phi r} - {}^{\mathbf{A}}F_\phi).
\end{aligned} \tag{B.21}$$

Moreover, it follows from (6.35)–(6.38) and Lemma 2.9 that

$$\begin{aligned}
&\frac{d}{dt}\left(\frac{1}{2}\sum_\gamma \left(\theta_{b\gamma} - \hat{\theta}_{b\gamma}\right)^2 / \rho_{b\gamma}\right) + \left(\dot{q}_r - \dot{q}\right) \mathbf{Y}_b(\dot{q}) \left(\boldsymbol{\theta}_b - \hat{\boldsymbol{\theta}}_b\right) \\
&= -\sum_\gamma \left(\theta_{b\gamma} - \hat{\theta}_{b\gamma}\right) \dot{\hat{\theta}}_{b\gamma}/\rho_{b\gamma} + \sum_\gamma s_{b\gamma} \left(\theta_{b\gamma} - \hat{\theta}_{b\gamma}\right) \\
&= \sum_\gamma \left(\theta_{b\gamma} - \hat{\theta}_{b\gamma}\right) \left(s_{b\gamma} - \dot{\hat{\theta}}_{b\gamma}/\rho_{b\gamma}\right) \leqslant 0
\end{aligned} \tag{B.22}$$

$$
\begin{aligned}
&\frac{d}{dt}\left(\frac{1}{2}\sum_{\gamma}\left(\theta_{m\gamma}-\hat{\theta}_{m\gamma}\right)^2/\rho_{m\gamma}\right)+\varrho(\dot{\phi}_r-\dot{\phi})\mathbf{Y}_m(\dot{\phi})\left(\boldsymbol{\theta}_m-\hat{\boldsymbol{\theta}}_m\right)\\
&\quad=-\sum_{\gamma}\left(\theta_{m\gamma}-\hat{\theta}_{m\gamma}\right)\dot{\hat{\theta}}_{m\gamma}/\rho_{m\gamma}+\sum_{\gamma}s_{m\gamma}\left(\theta_{m\gamma}-\hat{\theta}_{m\gamma}\right)\\
&\quad=\sum_{\gamma}\left(\theta_{m\gamma}-\hat{\theta}_{m\gamma}\right)\left(s_{m\gamma}-\dot{\hat{\theta}}_{m\gamma}/\rho_{m\gamma}\right)\leqslant 0
\end{aligned}
\tag{B.23}
$$

hold.

Finally, substituting (B.19)–(B.21) into (B.16)–(B.18) and using (B.22) and (B.23) ensures that the time derivative of ν defined by (6.39) can be expressed as (6.43).

B.4 Proof of Lemma 7.3

When $\dot{x}=0$, re-write (7.137) as

$$
\begin{aligned}
\dot{f}_{ps}=&\ \beta u\left[c_{p1}v\left(p_s-\frac{f_{ps}-xf_p}{s_a l_o}\right)-c_{n2}v\left(\frac{f_{ps}+(l_o-x)f_p}{s_b l_o}-p_r\right)\right]\varepsilon(u)\\
&+\beta u\left[c_{n1}v\left(\frac{f_{ps}-xf_p}{s_a l_o}-p_r\right)-c_{p2}v\left(p_s-\frac{f_{ps}+(l_o-x)f_p}{s_b l_o}\right)\right]\varepsilon(-u).
\end{aligned}
\tag{B.24}
$$

The fact that function $v(x)$ is *monotonically increasing* will be used in the proof below.

With $u>0$, $f_{ps}\leqslant p_r s_b l_o+\gamma_p l_o$, $\delta_c\geqslant 1$, $p_r>0$, and $0<s_a<s_b$, it follows from (7.131), (7.138), (7.139), Assumption 5, and (B.24) that

$$
\begin{aligned}
\dot{f}_{ps}&=\beta u\left[c_{p1}v\left(p_s-\frac{f_{ps}-xf_p}{s_a l_o}\right)-c_{n2}v\left(\frac{f_{ps}+(l_o-x)f_p}{s_b l_o}-p_r\right)\right]\\
&>\beta u\left[c_{p1}v\left(2\delta_c^2\frac{2\gamma_p+s_b p_r}{s_a}-\frac{p_r s_b l_o+\gamma_p l_o-xf_p}{s_a l_o}\right)\right.\\
&\qquad\left.-c_{n2}v\left(\frac{p_r s_b l_o+\gamma_p l_o+(l_o-x)f_p}{s_b l_o}-p_r\right)\right]\\
&>\beta u\left[c_{p1}v\left(\delta_c^2\frac{4\gamma_p+2s_b p_r}{s_a}-\frac{2\gamma_p+p_r s_b}{s_a}\right)-c_{n2}v\left(\frac{2\gamma_p}{s_b}\right)\right]\\
&\geqslant\beta u\left[c_{p1}v\left(\delta_c^2\frac{2\gamma_p+s_b p_r}{s_a}\right)-c_{n2}v\left(\frac{2\gamma_p}{s_b}\right)\right]\\
&=\beta u\left[c_{p1}\delta_c v\left(\frac{2\gamma_p+s_b p_r}{s_a}\right)-c_{n2}v\left(\frac{2\gamma_p}{s_b}\right)\right]\\
&>0
\end{aligned}
\tag{B.25}
$$

holds.

With $u < 0$, $f_{ps} \leqslant p_r s_b l_o + \gamma_p l_o$, $\delta_c \geqslant 1$, $p_r > 0$, and $0 < s_a < s_b$, it follows from (7.131), (7.138), (7.139), Assumption 5, and (B.24) that

$$\begin{aligned}
\dot{f}_{ps} &= \beta u \left[c_{n1} \upsilon \left(\frac{f_{ps} - x f_p}{s_a l_o} - p_r \right) - c_{p2} \upsilon \left(p_s - \frac{f_{ps} + (l_o - x) f_p}{s_b l_o} \right) \right] \\
&> \beta u \left[c_{n1} \upsilon \left(\frac{p_r s_b l_o + \gamma_p l_o - x f_p}{s_a l_o} - p_r \right) \right. \\
&\qquad \left. - c_{p2} \upsilon \left(2\delta_c^2 \frac{2\gamma_p + s_b p_r}{s_a} - \frac{p_r s_b l_o + \gamma_p l_o + (l_o - x) f_p}{s_b l_o} \right) \right] \\
&> \beta u \left[c_{n1} \upsilon \left(\frac{2\gamma_p + p_r s_b}{s_a} \right) - c_{p2} \upsilon \left(\delta_c^2 \frac{4\gamma_p + 2 s_b p_r}{s_a} - \frac{2\gamma_p + p_r s_b}{s_b} \right) \right] \\
&> \beta u \left[c_{n1} \upsilon \left(\frac{2\gamma_p + p_r s_b}{s_a} \right) - c_{p2} \delta_c \upsilon \left(\frac{2\gamma_p + s_b p_r}{s_a} \right) \right] \\
&\geqslant 0
\end{aligned} \tag{B.26}$$

holds. The inequalities (B.25) and (B.26) prove (7.140).

With $u > 0$, $f_{ps} \geqslant p_s s_b l_o - \gamma_p l_o$, $\delta_c \geqslant 1$, $p_r > 0$, and $0 < s_a < s_b$, it follows from (7.131), (7.138), (7.139), Assumption 5, and (B.24) that

$$\begin{aligned}
\dot{f}_{ps} &= \beta u \left[c_{p1} \upsilon \left(p_s - \frac{f_{ps} - x f_p}{s_a l_o} \right) - c_{n2} \upsilon \left(\frac{f_{ps} + (l_o - x) f_p}{s_b l_o} - p_r \right) \right] \\
&\leqslant \beta u \left[c_{p1} \upsilon \left(p_s - \frac{p_s s_b l_o - \gamma_p l_o - x f_p}{s_a l_o} \right) \right. \\
&\qquad \left. - c_{n2} \upsilon \left(\frac{p_s s_b l_o - \gamma_p l_o + (l_o - x) f_p}{s_b l_o} - p_r \right) \right] \\
&< \beta u \left[c_{p1} \upsilon \left(p_s - \frac{p_s s_b - 2\gamma_p}{s_a} \right) - c_{n2} \upsilon \left(p_s - \frac{2\gamma_p}{s_b} - p_r \right) \right] \\
&< \beta u \left[c_{p1} \upsilon \left(\frac{2\gamma_p}{s_a} \right) - c_{n2} \upsilon \left(2\delta_c^2 \frac{2\gamma_p + s_b p_r}{s_a} - \frac{2\gamma_p}{s_b} - p_r \right) \right] \\
&< \beta u \left[c_{p1} \upsilon \left(\frac{2\gamma_p}{s_a} \right) - c_{n2} \delta_c \upsilon \left(\frac{2\gamma_p + s_b p_r}{s_u} \right) \right] \\
&< 0
\end{aligned} \tag{B.27}$$

holds.

With $u < 0$, $f_{ps} \geqslant p_s s_b l_o - \gamma_p l_o$, $\delta_c \geqslant 1$, $p_r > 0$, and $0 < s_a < s_b$, it follows from (7.131), (7.138), (7.139), Assumption 5, and (B.24) that

$$\begin{aligned}
\dot{f}_{ps} &= \beta u \left[c_{n1} \upsilon \left(\frac{f_{ps} - x f_p}{s_a l_o} - p_r \right) - c_{p2} \upsilon \left(p_s - \frac{f_{ps} + (l_o - x) f_p}{s_b l_o} \right) \right] \\
&\leqslant \beta u \left[c_{n1} \upsilon \left(\frac{p_s s_b l_o - \gamma_p l_o - x f_p}{s_a l_o} - p_r \right) \right. \\
&\qquad \left. - c_{p2} \upsilon \left(p_s - \frac{p_s s_b l_o - \gamma_p l_o + (l_o - x) f_p}{s_b l_o} \right) \right]
\end{aligned}$$

$$
\begin{aligned}
&< \beta u \left[c_{n1} v \left(\frac{p_s s_b - 2\gamma_p}{s_a} - p_r \right) - c_{p2} v \left(p_s - \frac{p_s s_b - 2\gamma_p}{s_b} \right) \right] \\
&< \beta u \left[c_{n1} v \left(2\delta_c^2 \frac{2\gamma_p + s_b p_r}{s_a} \frac{s_b}{s_a} - \frac{2\gamma_p}{s_a} - p_r \right) - c_{p2} v \left(\frac{2\gamma_p}{s_b} \right) \right] \\
&< \beta u \left[c_{n1} \delta_c v \left(\frac{2\gamma_p + s_b p_r}{s_a} \frac{s_b}{s_a} \right) - c_{p2} v \left(\frac{2\gamma_p}{s_b} \right) \right] \\
&< 0
\end{aligned} \tag{B.28}
$$

holds. The inequalities (B.27) and (B.28) prove (7.141).

Equation (7.142) is obvious, in view of (B.24) with $u = 0$.

B.5 Proof of Lemma 7.4

Re-write (7.107) as

$$
\begin{aligned}
u &= -\frac{u_{fd}}{\frac{\hat{c}_{p1} v (p_s - p_a)}{l_o - x} + \frac{\hat{c}_{n2} v (p_b - p_r)}{x}} \varepsilon(u) - \frac{u_{fd}}{\frac{\hat{c}_{n1} v (p_a - p_r)}{l_o - x} + \frac{\hat{c}_{p2} v (p_s - p_b)}{x}} \varepsilon(-u) \\
&= -\frac{u_{fd}}{\left[\frac{\hat{c}_{p1} v (p_s - p_a)}{l_o - x} + \frac{\hat{c}_{n2} v (p_b - p_r)}{x} \right] \varepsilon(u) + \left[\frac{\hat{c}_{n1} v (p_a - p_r)}{l_o - x} + \frac{\hat{c}_{p2} v (p_s - p_b)}{x} \right] \varepsilon(-u)}
\end{aligned} \tag{B.29}
$$

where $\varepsilon(u) = \varepsilon(-u_{fd})$ and $\varepsilon(-u) = \varepsilon(u_{fd})$ are used.

Substituting (7.106) into (B.29) yields

$$
u(t) = u_s(t) + k_u(t)\dot{x}(t) \tag{B.30}
$$

with $k_u(t) \in \mathbb{R}$ and $u_s(t) \in \mathbb{R}$ being defined by (7.145) and (7.149), respectively.

The following lemma will be used to complete the proof.

Lemma B.1. *Given three positive numbers, a, b, and c, the inequality*

$$
a \geqslant 2b + 2c^2 \tag{B.31}
$$

implies

$$
\sqrt{a} - \sqrt{b} - c \geqslant 0. \tag{B.32}
$$

Proof: Note that the inequality $(\sqrt{b} - c)^2 \geqslant 0$ leads to $b + c^2 \geqslant 2\sqrt{b}c$. Inequality (B.31) can be rewritten as

$$
\begin{aligned}
a &\geqslant 2b + 2c^2 \\
&= b + c^2 + b + c^2 \\
&\geqslant b + c^2 + 2\sqrt{b}c \\
&= (\sqrt{b} + c)^2.
\end{aligned}
$$

Thus, $a \geqslant 2b + 2c^2$ implies $a \geqslant (\sqrt{b} + c)^2$ and further implies $\sqrt{a} \geqslant \sqrt{b} + c$ and $\sqrt{a} - \sqrt{b} - c \geqslant 0$, since $\sqrt{a}$ is monotonically increasing for $a > 0$. ■

By using (B.30), re-write (7.137) as

$$\begin{aligned}\dot{f}_{ps} &= \beta u \left[c_{p1} \upsilon \left(p_s - \frac{f_{ps} - x f_p}{s_a l_o} \right) - c_{n2} \upsilon \left(\frac{f_{ps} + (l_o - x) f_p}{s_b l_o} - p_r \right) \right] \varepsilon(u) \\ &+ \beta u \left[c_{n1} \upsilon \left(\frac{f_{ps} - x f_p}{s_a l_o} - p_r \right) - c_{p2} \upsilon \left(p_s - \frac{f_{ps} + (l_o - x) f_p}{s_b l_o} \right) \right] \varepsilon(-u) \\ &+ [\beta(s_a - s_b) + f_p] \frac{u - u_s}{k_u}. \end{aligned} \tag{B.33}$$

The fact that function $\upsilon(x)$ is *monotonically increasing* will be used in the proof below.

With $u \geqslant \frac{1}{\alpha_s}|u_s|$, $f_{ps} \leqslant p_r s_b l_o + \gamma_p l_o$, $\delta_c \geqslant 1$, $p_r > 0$, and $0 < s_a < s_b$, it follows from (7.131), (7.138), (7.143), (B.33), Assumption 5, and Lemma B.1 that

$$\begin{aligned}\dot{f}_{ps} &= \beta u \left[c_{p1} \upsilon \left(p_s - \frac{f_{ps} - x f_p}{s_a l_o} \right) - c_{n2} \upsilon \left(\frac{f_{ps} + (l_o - x) f_p}{s_b l_o} - p_r \right) \right] \\ &\quad + [\beta(s_a - s_b) + f_p] \frac{u - u_s}{k_u} \\ &\geqslant \beta u \left[c_{p1} \upsilon \left(p_s - \frac{p_r s_b l_o + \gamma_p l_o - x f_p}{s_a l_o} \right) \right. \\ &\qquad \left. - c_{n2} \upsilon \left(\frac{p_r s_b l_o + \gamma_p l_o + (l_o - x) f_p}{s_b l_o} - p_r \right) \right] \\ &\quad - [\beta(s_b - s_a) + \gamma_p] \frac{|u| + |u_s|}{|k_u|} \\ &> \beta u \left[c_{p1} \upsilon \left(p_s - \frac{p_r s_b + 2\gamma_p}{s_a} \right) - c_{n2} \upsilon \left(\frac{p_r s_b + 2\gamma_p}{s_b} - p_r \right) \right] \\ &\quad - [\beta(s_b - s_a) + \gamma_p] \frac{(1 + \alpha_s)|u|}{|k_u|} \\ &> \beta u \left[c_{p1} \upsilon \left(2\delta_c^2 \frac{2\gamma_p + s_b p_r}{s_a} + 2 \left\{ \frac{[\beta(s_b - s_a) + \gamma_p](1 + \alpha_s)}{|k_u| \beta \min\{c_{p1}, c_{n1}, c_{p2}, c_{n2}\}} \right\}^2 \right) \right. \\ &\qquad \left. - c_{n2} \upsilon \left(\frac{2\gamma_p}{s_b} \right) \right] - [\beta(s_b - s_a) + \gamma_p] \frac{(1 + \alpha_s)|u|}{|k_u|} \\ &= \beta u \left\{ c_{p1} \delta_c \upsilon \left(2 \frac{2\gamma_p + s_b p_r}{s_a} + 2 \left\{ \frac{[\beta(s_b - s_a) + \gamma_p](1 + \alpha_s)}{|k_u| \beta \max\{c_{p1}, c_{n1}, c_{p2}, c_{n2}\}} \right\}^2 \right) \right. \\ &\qquad \left. - c_{n2} \upsilon \left(\frac{2\gamma_p}{s_b} \right) - [\beta(s_b - s_a) + \gamma_p] \frac{(1 + \alpha_s)}{|k_u| \beta} \right\} \\ &\geqslant \beta u c_{p1} \delta_c \left\{ \upsilon \left(2 \frac{2\gamma_p + s_b p_r}{s_a} + 2 \left\{ \frac{[\beta(s_b - s_a) + \gamma_p](1 + \alpha_s)}{|k_u| \beta \max\{c_{p1}, c_{n1}, c_{p2}, c_{n2}\}} \right\}^2 \right) \right. \end{aligned}$$

$$
\begin{aligned}
&\quad -v\left(\frac{2\gamma_p}{s_b}\right) - [\beta(s_b - s_a) + \gamma_p] \frac{(1+\alpha_s)}{|k_u|\beta c_{p1}\delta_c}\Bigg\} \\
&> \beta u c_{p1}\delta_c \Bigg\{ v\left(2\frac{2\gamma_p + s_b p_r}{s_a} + 2\left\{\frac{[\beta(s_b - s_a) + \gamma_p](1+\alpha_s)}{|k_u|\beta\max\{c_{p1}, c_{n1}, c_{p2}, c_{n2}\}}\right\}^2\right) \\
&\quad -v\left(\frac{2\gamma_p + s_b p_r}{s_a}\right) - [\beta(s_b - s_a) + \gamma_p] \frac{(1+\alpha_s)}{|k_u|\beta c_{p1}\delta_c}\Bigg\} \\
&\geqslant 0 \qquad (B.34)
\end{aligned}
$$

holds. The last inequality uses (7.91) and Lemma B.1 after equalizing

$$
a = 2\frac{2\gamma_p + s_b p_r}{s_a} + 2\left\{\frac{[\beta(s_b - s_a) + \gamma_p](1+\alpha_s)}{|k_u|\beta\max\{c_{p1}, c_{n1}, c_{p2}, c_{n2}\}}\right\}^2 \qquad (B.35)
$$

$$
b = \frac{2\gamma_p + s_b p_r}{s_a} \qquad (B.36)
$$

$$
c = [\beta(s_b - s_a) + \gamma_p] \frac{(1+\alpha_s)}{|k_u|\beta c_{p1}\delta_c}. \qquad (B.37)
$$

With $u \leqslant -\frac{1}{\alpha_s}|u_s|$, $f_{ps} \leqslant p_r s_b l_o + \gamma_p l_o$, $\delta_c \geqslant 1$, $p_r > 0$, and $0 < s_a < s_b$, it follows from (7.131), (7.138), (7.143), (B.33), Assumption 5, and Lemma B.1 that

$$
\begin{aligned}
\dot{f}_{ps} &= \beta u \left[c_{n1} v\left(\frac{f_{ps} - x f_p}{s_a l_o} - p_r\right) - c_{p2} v\left(p_s - \frac{f_{ps} + (l_o - x) f_p}{s_b l_o}\right)\right] \\
&\quad + [\beta(s_a - s_b) + f_p] \frac{u - u_s}{k_u} \\
&\geqslant \beta u \Bigg[c_{n1} v\left(\frac{\gamma_p l_o + p_r s_b l_o - x f_p}{s_a l_o} - p_r\right) \\
&\qquad -c_{p2} v\left(p_s - \frac{\gamma_p l_o + p_r s_b l_o + (l_o - x) f_p}{s_b l_o}\right)\Bigg] \\
&\quad - [\beta(s_b - s_a) + \gamma_p] \frac{|u| + |u_s|}{|k_u|} \\
&> \beta u \left[c_{n1} v\left(\frac{2\gamma_p + p_r s_b}{s_a} - p_r\right) - c_{p2} v\left(p_s - \frac{2\gamma_p + p_r s_b}{s_b}\right)\right] \\
&\quad - [\beta(s_b - s_a) + \gamma_p] \frac{(1+\alpha_s)|u|}{|k_u|} \\
&> \beta u \Bigg[c_{n1} v\left(\frac{2\gamma_p + p_r s_b}{s_a} - p_r\right) \\
&\qquad -c_{p2} v\left(2\delta_c^2 \frac{2\gamma_p + s_b p_r}{s_a} + 2\left\{\frac{[\beta(s_b - s_a) + \gamma_p](1+\alpha_s)}{|k_u|\beta\min\{c_{p1}, c_{n1}, c_{p2}, c_{n2}\}}\right\}^2\right)\Bigg] \\
&\quad - [\beta(s_b - s_a) + \gamma_p] \frac{(1+\alpha_s)|u|}{|k_u|}
\end{aligned}
$$

$$
\begin{aligned}
&= -\beta u \left\{ -c_{n1}\upsilon \left(\frac{2\gamma_p + p_r s_b}{s_a} - p_r \right) - \left[\beta(s_b - s_a) + \gamma_p\right] \frac{(1+\alpha_s)}{|k_u|\beta} \right. \\
&\qquad \left. + c_{p2}\delta_c \upsilon \left(2\frac{2\gamma_p + s_b p_r}{s_a} + 2\left\{ \frac{\left[\beta(s_b - s_a) + \gamma_p\right](1+\alpha_s)}{|k_u|\beta \max\{c_{p1}, c_{n1}, c_{p2}, c_{n2}\}} \right\}^2 \right) \right\} \\
&\geqslant -\beta u c_{p2}\delta_c \left\{ -\upsilon \left(\frac{2\gamma_p + p_r s_b}{s_a} - p_r \right) - \left[\beta(s_b - s_a) + \gamma_p\right] \frac{(1+\alpha_s)}{|k_u|\beta c_{p2}\delta_c} \right. \\
&\qquad \left. + \upsilon \left(2\frac{2\gamma_p + s_b p_r}{s_a} + 2\left\{ \frac{\left[\beta(s_b - s_a) + \gamma_p\right](1+\alpha_s)}{|k_u|\beta \max\{c_{p1}, c_{n1}, c_{p2}, c_{n2}\}} \right\}^2 \right) \right\} \\
&> -\beta u c_{p2}\delta_c \left\{ -\upsilon \left(\frac{2\gamma_p + p_r s_b}{s_a} \right) - \left[\beta(s_b - s_a) + \gamma_p\right] \frac{(1+\alpha_s)}{|k_u|\beta c_{p2}\delta_c} \right. \\
&\qquad \left. + \upsilon \left(2\frac{2\gamma_p + s_b p_r}{s_a} + 2\left\{ \frac{\left[\beta(s_b - s_a) + \gamma_p\right](1+\alpha_s)}{|k_u|\beta \max\{c_{p1}, c_{n1}, c_{p2}, c_{n2}\}} \right\}^2 \right) \right\} \\
&\geqslant 0
\end{aligned} \tag{B.38}
$$

holds. The last inequality uses (7.91), (B.35)-(B.37), and Lemma B.1. This proves (7.147).

With $u \geqslant \frac{1}{\alpha_s}|u_s|$, $f_{ps} \geqslant p_s s_b l_o - \gamma_p l_o$, $\delta_c \geqslant 1$, $p_r > 0$, and $0 < s_a < s_b$, it follows from (7.131), (7.138), (7.143), (B.33), Assumption 5, and Lemma B.1 that

$$
\begin{aligned}
\dot{f}_{ps} &= \beta u \left[c_{p1}\upsilon \left(p_s - \frac{f_{ps} - x f_p}{s_a l_o} \right) - c_{n2}\upsilon \left(\frac{f_{ps} + (l_o - x) f_p}{s_b l_o} - p_r \right) \right] \\
&\quad + \left[\beta(s_a - s_b) + f_p\right] \frac{u - u_s}{k_u} \\
&\leqslant \beta u \left[c_{p1}\upsilon \left(p_s - \frac{p_s s_b l_o - \gamma_p l_o - x f_p}{s_a l_o} \right) \right. \\
&\qquad \left. - c_{n2}\upsilon \left(\frac{p_s s_b l_o - \gamma_p l_o + (l_o - x) f_p}{s_b l_o} - p_r \right) \right] \\
&\quad + \left[\beta(s_b - s_a) + \gamma_p\right] \frac{|u| + |u_s|}{|k_u|} \\
&< \beta u \left[c_{p1}\upsilon \left(\frac{2\gamma_p}{s_a} \right) - c_{n2}\upsilon \left(p_s - \frac{2\gamma_p}{s_b} - p_r \right) \right] \\
&\quad + \left[\beta(s_b - s_a) + \gamma_p\right] \frac{(1+\alpha_s)|u|}{|k_u|} \\
&< \beta u \left[c_{p1}\upsilon \left(\frac{2\gamma_p}{s_a} \right) \right. \\
&\qquad \left. - c_{n2}\upsilon \left(2\delta_c^2 \frac{2\gamma_p + s_b p_r}{s_a} + 2\left\{ \frac{\left[\beta(s_b - s_a) + \gamma_p\right](1+\alpha_s)}{|k_u|\beta \min\{c_{p1}, c_{n1}, c_{p2}, c_{n2}\}} \right\}^2 \right) \right] \\
&\quad + \left[\beta(s_b - s_a) + \gamma_p\right] \frac{(1+\alpha_s)|u|}{|k_u|}
\end{aligned}
$$

$$
\begin{aligned}
&= \beta u \left\{ c_{p1} \upsilon \left(\frac{2\gamma_p}{s_a} \right) + [\beta(s_b - s_a) + \gamma_p] \frac{(1+\alpha_s)}{|k_u|\beta} \right. \\
&\quad \left. - c_{n2}\delta_c \upsilon \left(2\frac{2\gamma_p + s_b p_r}{s_a} + 2\left\{ \frac{[\beta(s_b - s_a) + \gamma_p](1+\alpha_s)}{|k_u|\beta \max\{c_{p1}, c_{n1}, c_{p2}, c_{n2}\}} \right\}^2 \right) \right\} \\
&\leqslant -\beta u c_{n2}\delta_c \left\{ -\upsilon \left(\frac{2\gamma_p}{s_a} \right) - [\beta(s_b - s_a) + \gamma_p] \frac{(1+\alpha_s)}{|k_u|\beta c_{n2}\delta_c} \right. \\
&\quad \left. + \upsilon \left(2\frac{2\gamma_p + s_b p_r}{s_a} + 2\left\{ \frac{[\beta(s_b - s_a) + \gamma_p](1+\alpha_s)}{|k_u|\beta \max\{c_{p1}, c_{n1}, c_{p2}, c_{n2}\}} \right\}^2 \right) \right\} \\
&< -\beta u c_{n2}\delta_c \left\{ -\upsilon \left(\frac{2\gamma_p + s_b p_r}{s_a} \right) - [\beta(s_b - s_a) + \gamma_p] \frac{(1+\alpha_s)}{|k_u|\beta c_{n2}\delta_c} \right. \\
&\quad \left. + \upsilon \left(2\frac{2\gamma_p + s_b p_r}{s_a} + 2\left\{ \frac{[\beta(s_b - s_a) + \gamma_p](1+\alpha_s)}{|k_u|\beta \max\{c_{p1}, c_{n1}, c_{p2}, c_{n2}\}} \right\}^2 \right) \right\} \\
&\leqslant 0
\end{aligned} \tag{B.39}
$$

holds. The last inequality uses (7.91), (B.35)–(B.37), and Lemma B.1.

With $u \leqslant -\frac{1}{\alpha_s}|u_s|$, $f_{ps} \geqslant p_s s_b l_o - \gamma_p l_o$, $\delta_c \geqslant 1$, $p_r > 0$, and $0 < s_a < s_b$, it follows from (7.131), (7.138), (7.143), (B.33), Assumption 5, and Lemma B.1 that

$$
\begin{aligned}
\dot{f}_{ps} &= \beta u \left[c_{n1} \upsilon \left(\frac{f_{ps} - x f_p}{s_a l_o} - p_r \right) - c_{p2} \upsilon \left(p_s - \frac{f_{ps} + (l_o - x) f_p}{s_b l_o} \right) \right] \\
&\quad + [\beta(s_a - s_b) + f_p] \frac{u - u_s}{k_u} \\
&\leqslant \beta u \left[c_{n1} \upsilon \left(\frac{p_s s_b l_o - \gamma_p l_o - x f_p}{s_a l_o} - p_r \right) \right. \\
&\quad \left. - c_{p2} \upsilon \left(p_s - \frac{p_s s_b l_o - \gamma_p l_o + (l_o - x) f_p}{s_b l_o} \right) \right] \\
&\quad + [\beta(s_b - s_a) + \gamma_p] \frac{|u| + |u_s|}{|k_u|} \\
&< \beta u \left[c_{n1} \upsilon \left(p_s - \frac{2\gamma_p}{s_a} - p_r \right) - c_{p2} \upsilon \left(\frac{2\gamma_p}{s_b} \right) \right] \\
&\quad + [\beta(s_b - s_a) + \gamma_p] \frac{(1+\alpha_s)|u|}{|k_u|} \\
&< \beta u \left[c_{n1} \upsilon \left(2\delta_c^2 \frac{2\gamma_p + s_b p_r}{s_a} + 2\left\{ \frac{[\beta(s_b - s_a) + \gamma_p](1+\alpha_s)}{|k_u|\beta \min\{c_{p1}, c_{n1}, c_{p2}, c_{n2}\}} \right\}^2 \right) \right. \\
&\quad \left. - c_{p2} \upsilon \left(\frac{2\gamma_p}{s_b} \right) \right] + [\beta(s_b - s_a) + \gamma_p] \frac{(1+\alpha_s)|u|}{|k_u|} \\
&= \beta u \left\{ c_{n1}\delta_c \upsilon \left(2\frac{2\gamma_p + s_b p_r}{s_a} + 2\left\{ \frac{[\beta(s_b - s_a) + \gamma_p](1+\alpha_s)}{|k_u|\beta \max\{c_{p1}, c_{n1}, c_{p2}, c_{n2}\}} \right\}^2 \right) \right.
\end{aligned}
$$

$$
\begin{aligned}
& -c_{p2}\upsilon\left(\frac{2\gamma_p}{s_b}\right) - [\beta(s_b - s_a) + \gamma_p]\,\frac{(1+\alpha_s)}{|k_u|\beta}\Bigg\} \\
\leqslant\ & \beta u c_{n1}\delta_c\Bigg\{\upsilon\left(2\frac{2\gamma_p + s_b p_r}{s_a} + 2\left\{\frac{[\beta(s_b - s_a) + \gamma_p]\,(1+\alpha_s)}{|k_u|\beta\max\{c_{p1}, c_{n1}, c_{p2}, c_{n2}\}}\right\}^2\right) \\
& -\upsilon\left(\frac{2\gamma_p}{s_b}\right) - [\beta(s_b - s_a) + \gamma_p]\,\frac{(1+\alpha_s)}{|k_u|\beta c_{n1}\delta_c}\Bigg\} \\
<\ & \beta u c_{n1}\delta_c\Bigg\{\upsilon\left(2\frac{2\gamma_p + s_b p_r}{s_a} + 2\left\{\frac{[\beta(s_b - s_a) + \gamma_p]\,(1+\alpha_s)}{|k_u|\beta\max\{c_{p1}, c_{n1}, c_{p2}, c_{n2}\}}\right\}^2\right) \\
& -\upsilon\left(\frac{2\gamma_p + s_b p_r}{s_a}\right) - [\beta(s_b - s_a) + \gamma_p]\,\frac{(1+\alpha_s)}{|k_u|\beta c_{n1}\delta_c}\Bigg\} \\
\leqslant\ & 0
\end{aligned} \tag{B.40}
$$

holds. The last inequality uses (7.91), (B.35)-(B.37), and Lemma B.1. This proves (7.148).

B.6 Derivation of (10.72) and (10.73)

In view of (2.85), (10.41)–(10.44), (10.47)–(10.55), and (10.62)–(10.66), it yields

$$
\begin{aligned}
& \left({}^{\mathbf{B}_{LT}}V_r - {}^{\mathbf{B}_{LT}}V\right)^T \left({}^{\mathbf{B}_{LT}}F_r^* - {}^{\mathbf{B}_{LT}}F^*\right) \\
&= \left({}^{\mathbf{B}_{LT}}V_r - {}^{\mathbf{B}_{LT}}V\right)^T \left[\left({}^{\mathbf{B}_{LT}}F_r - {}^{\mathbf{B}_{LT}}F\right) - {}^{\mathbf{B}_{LT}}\mathbf{U}_{\mathbf{B}_T}\left({}^{\mathbf{B}_{T1}}F_r - {}^{\mathbf{B}_{T1}}F\right)\right] \\
&= p_{\mathbf{B}_{LT}} - \left({}^{\mathbf{B}_{LT}}V_r - {}^{\mathbf{B}_{LT}}V\right)^T {}^{\mathbf{B}_{LT}}\mathbf{U}_{\mathbf{B}_T}\left({}^{\mathbf{B}_{T1}}F_r - {}^{\mathbf{B}_{T1}}F\right) \\
&= p_{\mathbf{B}_{LT}} - \left[{}^{\mathbf{B}_{LT}}\mathbf{U}_{\mathbf{B}_T}^T\left({}^{\mathbf{B}_{LT}}V_r - {}^{\mathbf{B}_{LT}}V\right)\right]^T \left({}^{\mathbf{B}_{T1}}F_r - {}^{\mathbf{B}_{T1}}F\right) \\
&= p_{\mathbf{B}_{LT}} - \left[\left({}^{\mathbf{B}_{T1}}V_r - {}^{\mathbf{B}_{T1}}V\right) - \mathbf{z}\left((\dot{q}_{Tr} - \dot{q}_T) - (\dot{q}_{LTr} - \dot{q}_{LT})\right)\right]^T \\
&\qquad \times \left({}^{\mathbf{B}_{T1}}F_r - {}^{\mathbf{B}_{T1}}F\right) \\
&= p_{\mathbf{B}_{LT}} - p_{\mathbf{B}_{T1}} + \left((\dot{q}_{Tr} - \dot{q}_T) - (\dot{q}_{LTr} - \dot{q}_{LT})\right)\mathbf{z}^T\left({}^{\mathbf{B}_{T1}}F_r - {}^{\mathbf{B}_{T1}}F\right) \\
&= p_{\mathbf{B}_{LT}} - p_{\mathbf{B}_{T1}} + \left((\dot{q}_{Tr} - \dot{q}_T) - (\dot{q}_{LTr} - \dot{q}_{LT})\right)\left(\mathbf{z}^{T\,\mathbf{B}_{T1}}F_r - \tau_{LT}\right) \\
&= p_{\mathbf{B}_{LT}} - p_{\mathbf{B}_{T1}} - k_{LT}\left((\dot{q}_{Tr} - \dot{q}_T) - (\dot{q}_{LTr} - \dot{q}_{LT})\right)^2
\end{aligned} \tag{B.41}
$$

$$
\begin{aligned}
& \left({}^{\mathbf{B}_{LS}}V_r - {}^{\mathbf{B}_{LS}}V\right)^T \left({}^{\mathbf{B}_{LS}}F_r^* - {}^{\mathbf{B}_{LS}}F^*\right) \\
&= \left({}^{\mathbf{B}_{LS}}V_r - {}^{\mathbf{B}_{LS}}V\right)^T \left[\left({}^{\mathbf{B}_{LS}}F_r - {}^{\mathbf{B}_{LS}}F\right) - {}^{\mathbf{B}_{LS}}\mathbf{U}_{\mathbf{B}_{LT}}\left({}^{\mathbf{B}_{LT}}F_r - {}^{\mathbf{B}_{LT}}F\right)\right] \\
&= (\dot{q}_{LSr} - \dot{q}_{LS})\,\mathbf{z}^T\left({}^{\mathbf{B}_{LS}}F_r - {}^{\mathbf{B}_{LS}}F\right) \\
&\quad - \left[{}^{\mathbf{B}_{LS}}\mathbf{U}_{\mathbf{B}_{LT}}^{-T}\left(\left({}^{\mathbf{B}_{LT}}V_r - {}^{\mathbf{B}_{LT}}V\right) - \mathbf{z}\left(\dot{q}_{LNr} - \dot{q}_{LN}\right)\right)\right]^T \\
&\qquad \times {}^{\mathbf{B}_{LS}}\mathbf{U}_{\mathbf{B}_{LT}}\left({}^{\mathbf{B}_{LT}}F_r - {}^{\mathbf{B}_{LT}}F\right) \\
&= (\dot{q}_{LSr} - \dot{q}_{LS})\left(\mathbf{z}^{T\,\mathbf{B}_{LS}}F_r - \tau_{LS}\right) \\
&\quad - \left[\left({}^{\mathbf{B}_{LT}}V_r - {}^{\mathbf{B}_{LT}}V\right) - \mathbf{z}\left(\dot{q}_{LNr} - \dot{q}_{LN}\right)\right]^T \left({}^{\mathbf{B}_{LT}}F_r - {}^{\mathbf{B}_{LT}}F\right)
\end{aligned}
$$

$$
\begin{aligned}
&= \left(\dot{q}_{LSr} - \dot{q}_{LS}\right) \mathbf{z}^{T\mathbf{B}_{LS}} F_r - p_{\mathbf{B}_{LT}} \\
&\quad + \left(\dot{q}_{LNr} - \dot{q}_{LN}\right) \mathbf{z}^T \left({}^{\mathbf{B}_{LT}}F_r - {}^{\mathbf{B}_{LT}}F\right) \\
&= -k_{LS}\left(\dot{q}_{LSr} - \dot{q}_{LS}\right)^2 - p_{\mathbf{B}_{LT}} + \left(\dot{q}_{LNr} - \dot{q}_{LN}\right)\left(\mathbf{z}^{T\mathbf{B}_{LT}}F_r - \tau_{LN}\right) \\
&= -p_{\mathbf{B}_{LT}} - k_{LS}\left(\dot{q}_{LSr} - \dot{q}_{LS}\right)^2 - k_{LN}\left(\dot{q}_{LNr} - \dot{q}_{LN}\right)^2 .
\end{aligned} \tag{B.42}
$$

B.7 Derivation of (10.106) and (10.107)

In view of (2.85), (10.75)–(10.78), (10.81)–(10.89), and (10.96)–(10.100), it yields

$$
\begin{aligned}
&\left({}^{\mathbf{B}_{RT}}V_r - {}^{\mathbf{B}_{RT}}V\right)^T \left({}^{\mathbf{B}_{RT}}F_r^* - {}^{\mathbf{B}_{RT}}F^*\right) \\
&= \left({}^{\mathbf{B}_{RT}}V_r - {}^{\mathbf{B}_{RT}}V\right)^T \left[\left({}^{\mathbf{B}_{RT}}F_r - {}^{\mathbf{B}_{RT}}F\right) - {}^{\mathbf{B}_{RT}}\mathbf{U}_{\mathbf{B}_T}\left({}^{\mathbf{B}_{T2}}F_r - {}^{\mathbf{B}_{T2}}F\right)\right] \\
&= p_{\mathbf{B}_{RT}} - \left({}^{\mathbf{B}_{RT}}V_r - {}^{\mathbf{B}_{RT}}V\right)^T {}^{\mathbf{B}_{RT}}\mathbf{U}_{\mathbf{B}_T}\left({}^{\mathbf{B}_{T2}}F_r - {}^{\mathbf{B}_{T2}}F\right) \\
&= p_{\mathbf{B}_{RT}} - \left[{}^{\mathbf{B}_{RT}}\mathbf{U}_{\mathbf{B}_T}^T\left({}^{\mathbf{B}_{RT}}V_r - {}^{\mathbf{B}_{RT}}V\right)\right]^T \left({}^{\mathbf{B}_{T2}}F_r - {}^{\mathbf{B}_{T2}}F\right) \\
&= p_{\mathbf{B}_{RT}} - \left[\left({}^{\mathbf{B}_{T2}}V_r - {}^{\mathbf{B}_{T2}}V\right) - \mathbf{z}\left(\left(\dot{q}_{Tr} - \dot{q}_T\right) - \left(\dot{q}_{RTr} - \dot{q}_{RT}\right)\right)\right]^T \\
&\qquad\qquad \times \left({}^{\mathbf{B}_{T2}}F_r - {}^{\mathbf{B}_{T2}}F\right) \\
&= p_{\mathbf{B}_{RT}} - p_{\mathbf{B}_{T2}} + \left(\left(\dot{q}_{Tr} - \dot{q}_T\right) - \left(\dot{q}_{RTr} - \dot{q}_{RT}\right)\right)\mathbf{z}^T\left({}^{\mathbf{B}_{T2}}F_r - {}^{\mathbf{B}_{T2}}F\right) \\
&= p_{\mathbf{B}_{RT}} - p_{\mathbf{B}_{T2}} + \left(\left(\dot{q}_{Tr} - \dot{q}_T\right) - \left(\dot{q}_{RTr} - \dot{q}_{RT}\right)\right)\left(\mathbf{z}^{T\mathbf{B}_{T2}}F_r - \tau_{RT}\right) \\
&= p_{\mathbf{B}_{RT}} - p_{\mathbf{B}_{T2}} - k_{RT}\left(\left(\dot{q}_{Tr} - \dot{q}_T\right) - \left(\dot{q}_{RTr} - \dot{q}_{RT}\right)\right)^2
\end{aligned} \tag{B.43}
$$

$$
\begin{aligned}
&\left({}^{\mathbf{B}_{RS}}V_r - {}^{\mathbf{B}_{RS}}V\right)^T \left({}^{\mathbf{B}_{RS}}F_r^* - {}^{\mathbf{B}_{RS}}F^*\right) \\
&= \left({}^{\mathbf{B}_{RS}}V_r - {}^{\mathbf{B}_{RS}}V\right)^T \\
&\qquad \times \left[\left({}^{\mathbf{B}_{RS}}F_r - {}^{\mathbf{B}_{RS}}F\right) - {}^{\mathbf{B}_{RS}}\mathbf{U}_{\mathbf{B}_{RT}}\left({}^{\mathbf{B}_{RT}}F_r - {}^{\mathbf{B}_{RT}}F\right)\right] \\
&= \left(\dot{q}_{RSr} - \dot{q}_{RS}\right)\mathbf{z}^T\left({}^{\mathbf{B}_{RS}}F_r - {}^{\mathbf{B}_{RS}}F\right) \\
&\quad - \left[{}^{\mathbf{B}_{RS}}\mathbf{U}_{\mathbf{B}_{RT}}^{-T}\left(\left({}^{\mathbf{B}_{RT}}V_r - {}^{\mathbf{B}_{RT}}V\right) - \mathbf{z}\left(\dot{q}_{RNr} - \dot{q}_{RN}\right)\right)\right]^T \\
&\qquad \times {}^{\mathbf{B}_{RS}}\mathbf{U}_{\mathbf{B}_{RT}}\left({}^{\mathbf{B}_{RT}}F_r - {}^{\mathbf{B}_{RT}}F\right) \\
&= \left(\dot{q}_{RSr} - \dot{q}_{RS}\right)\left(\mathbf{z}^{T\mathbf{B}_{RS}}F_r - \tau_{RS}\right) \\
&\quad - \left[\left({}^{\mathbf{B}_{RT}}V_r - {}^{\mathbf{B}_{RT}}V\right) - \mathbf{z}\left(\dot{q}_{RNr} - \dot{q}_{RN}\right)\right]^T\left({}^{\mathbf{B}_{RT}}F_r - {}^{\mathbf{B}_{RT}}F\right) \\
&= \left(\dot{q}_{RSr} - \dot{q}_{RS}\right)\mathbf{z}^{T\mathbf{B}_{RS}}F_r \\
&\quad - p_{\mathbf{B}_{RT}} + \left(\dot{q}_{RNr} - \dot{q}_{RN}\right)\mathbf{z}^T\left({}^{\mathbf{B}_{RT}}F_r - {}^{\mathbf{B}_{RT}}F\right) \\
&= -k_{RS}\left(\dot{q}_{RSr} - \dot{q}_{RS}\right)^2 - p_{\mathbf{B}_{RT}} + \left(\dot{q}_{RNr} - \dot{q}_{RN}\right)\left(\mathbf{z}^{T\mathbf{B}_{RT}}F_r - \tau_{RN}\right) \\
&= -p_{\mathbf{B}_{RT}} - k_{RS}\left(\dot{q}_{RSr} - \dot{q}_{RS}\right)^2 - k_{RN}\left(\dot{q}_{RNr} - \dot{q}_{RN}\right)^2 .
\end{aligned} \tag{B.44}
$$

B.8 Derivation of (10.211)

In view of (2.85), (10.76), (10.82), (10.84), (10.87), (10.97), and (10.99), it yields

$$
\begin{aligned}
&\left({}^{\mathbf{B}_{RS}}V_r - {}^{\mathbf{B}_{RS}}V\right)^T \left({}^{\mathbf{B}_{RS}}F_r^* - {}^{\mathbf{B}_{RS}}F^*\right) \\
&\quad = \left({}^{\mathbf{B}_{RS}}V_r - {}^{\mathbf{B}_{RS}}V\right)^T \\
&\qquad\qquad \times \left[\left({}^{\mathbf{B}_{RS}}F_r - {}^{\mathbf{B}_{RS}}F\right) - {}^{\mathbf{B}_{RS}}\mathbf{U}_{\mathbf{B}_{RT}} \left({}^{\mathbf{B}_{RT}}F_r - {}^{\mathbf{B}_{RT}}F\right)\right] \\
&\quad = p_{\mathbf{B}_{RS}} - \left[{}^{\mathbf{B}_{RS}}\mathbf{U}_{\mathbf{B}_{RT}}^{-T} \left(\left({}^{\mathbf{B}_{RT}}V_r - {}^{\mathbf{B}_{RT}}V\right) - \mathbf{z}\left(\dot{q}_{RNr} - \dot{q}_{RN}\right)\right)\right]^T \\
&\qquad\qquad \times {}^{\mathbf{B}_{RS}}\mathbf{U}_{\mathbf{B}_{RT}} \left({}^{\mathbf{B}_{RT}}F_r - {}^{\mathbf{B}_{RT}}F\right) \\
&\quad = p_{\mathbf{B}_{RS}} - \left[\left({}^{\mathbf{B}_{RT}}V_r - {}^{\mathbf{B}_{RT}}V\right) - \mathbf{z}\left(\dot{q}_{RNr} - \dot{q}_{RN}\right)\right]^T \left({}^{\mathbf{B}_{RT}}F_r - {}^{\mathbf{B}_{RT}}F\right) \\
&\quad = p_{\mathbf{B}_{RS}} - p_{\mathbf{B}_{RT}} + \left(\dot{q}_{RNr} - \dot{q}_{RN}\right)\mathbf{z}^T \left({}^{\mathbf{B}_{RT}}F_r - {}^{\mathbf{B}_{RT}}F\right) \\
&\quad = p_{\mathbf{B}_{RS}} - p_{\mathbf{B}_{RT}} - k_{RN}\left(\dot{q}_{RNr} - \dot{q}_{RN}\right)^2 . \qquad \text{(B.45)}
\end{aligned}
$$

B.9 Proof of Lemma 11.1

For constant σ_f and σ_c, it follows from (11.1)–(11.6), (11.12)–(11.20), (11.24), and Lemma 2.9 that

$$
\begin{aligned}
\int_0^\infty p_{\mathbf{S}_s}(t)dt &= \int_0^\infty ({}^{\mathbf{S}_s}V_r - {}^{\mathbf{S}_s}V)^T({}^{\mathbf{S}_s}F_r - {}^{\mathbf{S}_s}F)dt \\
&= \int_0^\infty (\mathbf{v}_{sfr} - \mathbf{v}_{sf})^T \sigma_f(\mathbf{f}_{sfd} - \mathbf{f}_{sf})dt \\
&\quad + \int_0^\infty [\mathbf{v}_{scr} - (1-\sigma_c)\mathbf{v}_{sc}]^T \sigma_c(\mathbf{f}_{scd} - \mathbf{f}_{sc})dt \\
&= \int_0^\infty \sigma_f(\mathbf{v}_{sfr} - \mathbf{v}_{sf})^T \mathbf{M}_f(\dot{\mathbf{v}}_{sfr} - \dot{\mathbf{v}}_{sf})dt \\
&\quad - \int_0^\infty \sigma_f(\mathbf{v}_{sfr} - \mathbf{v}_{sf})^T \mathbf{Y}_{sf}(\boldsymbol{\theta}_{sf} - \hat{\boldsymbol{\theta}}_{sf})dt \\
&\quad + \int_0^\infty \sigma_c \mathbf{v}_{scr}^T(\mathbf{f}_{scd} - \mathbf{f}_{sc})dt \\
&= \int_0^\infty \sigma_f(\mathbf{v}_{sfr} - \mathbf{v}_{sf})^T \mathbf{M}_f(\dot{\mathbf{v}}_{sfr} - \dot{\mathbf{v}}_{sf})dt \\
&\quad + \int_0^\infty \sum_\gamma (\theta_{sf\gamma} - \hat{\theta}_{sf\gamma})(-\dot{\hat{\theta}}_{sf\gamma}/\rho_{sf\gamma})dt \\
&\quad - \int_0^\infty \sum_\gamma (\theta_{sf\gamma} - \hat{\theta}_{sf\gamma})(s_{sf\gamma} - \dot{\hat{\theta}}_{sf\gamma}/\rho_{sf\gamma})dt \\
&\quad + \int_0^\infty \sigma_c \left(\mathbf{v}_{scd} - \mathbf{A}_c\tilde{\mathbf{f}}_{sc}\right)^T \\
&\qquad\quad [\mathbf{A}_c^{-1}(\mathbf{C}_c^{-1}\dot{\mathbf{v}}_{scd} + \mathbf{v}_{scd}) - (\mathbf{C}_c^{-1}\dot{\tilde{\mathbf{f}}}_{sc} + \tilde{\mathbf{f}}_{sc})]dt \\
&\geqslant \int_0^\infty \sigma_f(\mathbf{v}_{sfr} \quad \mathbf{v}_{sf})^T \mathbf{M}_f(\dot{\mathbf{v}}_{sfr} \quad \dot{\mathbf{v}}_{sf})dt
\end{aligned}
$$

$$
\begin{aligned}
&+\int_0^\infty \sum_\gamma (\theta_{sf\gamma}-\hat{\theta}_{sf\gamma})(-\dot{\hat{\theta}}_{sf\gamma}/\rho_{sf\gamma})dt \\
&+\int_0^\infty \sigma_c\left(\mathbf{v}_{scd}-\mathbf{A}_c\tilde{\mathbf{f}}_{sc}\right)^T \mathbf{A}_c^{-1}\mathbf{C}_c^{-1}\left(\dot{\mathbf{v}}_{scd}-\mathbf{A}_c\dot{\tilde{\mathbf{f}}}_{sc}\right)dt \\
&+\int_0^\infty \sigma_c\left(\mathbf{v}_{scd}-\mathbf{A}_c\tilde{\mathbf{f}}_{sc}\right)^T \mathbf{A}_c^{-1}\left(\mathbf{v}_{scd}-\mathbf{A}_c\tilde{\mathbf{f}}_{sc}\right)dt \\
&\geqslant -\frac{1}{2}\sigma_f(\mathbf{v}_{sfr}(0)-\mathbf{v}_{sf}(0))^T\mathbf{M}_f(\mathbf{v}_{sfr}(0)-\mathbf{v}_{sf}(0)) \\
&-\frac{1}{2}\sum_\gamma(\theta_{sf\gamma}-\hat{\theta}_{sf\gamma}(0))^2/\rho_{sf\gamma} \\
&-\frac{1}{2}\sigma_c(\mathbf{v}_{scd}(0)-\mathbf{A}_c\tilde{\mathbf{f}}_{sc}(0))^T\mathbf{A}_c^{-1}\mathbf{C}_c^{-1}(\mathbf{v}_{scd}(0)-\mathbf{A}_c\tilde{\mathbf{f}}_{sc}(0))
\end{aligned}
\tag{B.46}
$$

holds. This proves the lemma by equalizing

$$
\begin{aligned}
\gamma_s =& \frac{1}{2}\sigma_f(\mathbf{v}_{sfr}(0)-\mathbf{v}_{sf}(0))^T\mathbf{M}_f(\mathbf{v}_{sfr}(0)-\mathbf{v}_{sf}(0)) \\
&+\frac{1}{2}\sum_\gamma(\theta_{sf\gamma}-\hat{\theta}_{sf\gamma}(0))^2/\rho_{sf\gamma} \\
&+\frac{1}{2}\sigma_c(\mathbf{v}_{scd}(0)-\mathbf{A}_c\tilde{\mathbf{f}}_{sc}(0))^T\mathbf{A}_c^{-1}\mathbf{C}_c^{-1}(\mathbf{v}_{scd}(0)-\mathbf{A}_c\tilde{\mathbf{f}}_{sc}(0)).
\end{aligned}
\tag{B.47}
$$

B.10 Proof of Lemma 11.2

It follows from (11.28)–(11.31), (11.34)–(11.36), (11.41)–(11.46), (11.50), and Lemma 2.9 that

$$
\begin{aligned}
\int_0^\infty p_{\mathbf{S}_m}(t)dt =& \int_0^\infty ({}^{\mathbf{S}_m}V_r - {}^{\mathbf{S}_m}V)^T({}^{\mathbf{S}_m}F_r - {}^{\mathbf{S}_m}F)dt \\
=& \int_0^\infty (\mathbf{v}_{mr}-\mathbf{v}_m)^T(\mathbf{f}_{md}-\mathbf{f}_m)dt \\
=& \int_0^\infty (\mathbf{v}_{mr}-\mathbf{v}_m)^T\mathbf{M}_h(\dot{\mathbf{v}}_{mr}-\dot{\mathbf{v}}_m)dt \\
&-\int_0^\infty (\mathbf{v}_{mr}-\mathbf{v}_m)^T\mathbf{Y}_m(\boldsymbol{\theta}_m-\hat{\boldsymbol{\theta}}_m)dt \\
&+\int_0^\infty (\mathbf{v}_{mr}-\mathbf{v}_m)^T[\alpha_h \mathrm{sign}(\mathbf{v}_{mr}-\mathbf{v}_m)-\mathbf{f}_h^*]dt \\
=& \int_0^\infty (\mathbf{v}_{mr}-\mathbf{v}_m)^T\mathbf{M}_h(\dot{\mathbf{v}}_{mr}-\dot{\mathbf{v}}_m)dt \\
&+\int_0^\infty \sum_\gamma(\theta_{m\gamma}-\hat{\theta}_{m\gamma})(-\dot{\hat{\theta}}_{m\gamma}/\rho_{m\gamma})dt
\end{aligned}
$$

$$
\begin{aligned}
&-\int_0^\infty \sum_\gamma (\boldsymbol{\theta}_m - \hat{\boldsymbol{\theta}}_m)(s_{m\gamma} - \dot{\hat{\theta}}_{m\gamma}/\rho_{m\gamma})dt \\
&+\int_0^\infty (\mathbf{v}_{mr} - \mathbf{v}_m)^T [\alpha_h \text{sign}(\mathbf{v}_{mr} - \mathbf{v}_m) - \mathbf{f}_h^*]dt \\
&\geqslant -\frac{1}{2}(\mathbf{v}_{mr}(0) - \mathbf{v}_m(0))^T \mathbf{M}_h (\mathbf{v}_{mr}(0) - \mathbf{v}_m(0)) \\
&-\frac{1}{2}\sum_\gamma (\theta_{m\gamma} - \hat{\theta}_{m\gamma}(0))^2/\rho_{m\gamma}
\end{aligned} \tag{B.48}
$$

holds. This proves the lemma by equalizing

$$
\begin{aligned}
\gamma_m =& \frac{1}{2}(\mathbf{v}_{mr}(0) - \mathbf{v}_m(0))^T \mathbf{M}_h (\mathbf{v}_{mr}(0) - \mathbf{v}_m(0)) \\
&+\frac{1}{2}\sum_\gamma (\theta_{m\gamma} - \hat{\theta}_{m\gamma}(0))^2/\rho_{m\gamma}.
\end{aligned} \tag{B.49}
$$

B.11 Proof of Theorem 13.2

Subtracting (13.7) from (13.34) yields

$$
\begin{aligned}
&\rho\left[\ddot{y}_r(x,t) - \ddot{y}(x,t)\right] + EI\left[y_r''''(x,t) - y''''(x,t)\right] \\
&+k_v\left[\dot{y}_r(x,t) - \dot{y}(x,t)\right] - \left[\rho - \hat{\rho}(t)\right]\ddot{y}_r(x,t) - \left[EI - \widehat{EI}(t)\right] y_r''''(x,t) = 0.
\end{aligned} \tag{B.50}
$$

With integration by parts, it follows from (B.50) and (13.19) that

$$
\begin{aligned}
\dot{\nu}_K =& \int_0^l \rho\left[\dot{y}_r(x,t) - \dot{y}(x,t)\right]\left[\ddot{y}_r(x,t) - \ddot{y}(x,t)\right]dx \\
=& -\int_0^l \left[\dot{y}_r(x,t) - \dot{y}(x,t)\right] EI\left[y_r''''(x,t) - y''''(x,t)\right]dx \\
&-\int_0^l k_v\left[\dot{y}_r(x,t) - \dot{y}(x,t)\right]^2 dx \\
&+\int_0^l \left[\dot{y}_r(x,t) - \dot{y}(x,t)\right]\left[\rho - \hat{\rho}(t)\right]\ddot{y}_r(x,t)dx \\
&+\int_0^l \left[\dot{y}_r(x,t) - \dot{y}(x,t)\right]\left[EI - \widehat{EI}(t)\right] y_r''''(x,t)dx \\
=& -\left[\dot{y}_r(x,t) - \dot{y}(x,t)\right] EI\left[y_r'''(x,t) - y'''(x,t)\right]\Big|_0^l \\
&+\int_0^l \left[\dot{y}_r'(x,t) - \dot{y}'(x,t)\right] EI\left[y_r'''(x,t) - y'''(x,t)\right]dx \\
&-\int_0^l k_v\left[\dot{y}_r(x,t) - \dot{y}(x,t)\right]^2 dx
\end{aligned}
$$

$$
\begin{aligned}
&+[\rho-\hat{\rho}(t)]\int_0^l[\dot{y}_r(x,t)-\dot{y}(x,t)]\,\ddot{y}_r(x,t)dx\\
&+\left[EI-\widehat{EI}(t)\right]\int_0^l[\dot{y}_r(x,t)-\dot{y}(x,t)]\,y_r''''(x,t)dx\\
=&-[\dot{y}_r(x,t)-\dot{y}(x,t)]\,EI\,[y_r'''(x,t)-y'''(x,t)]\,|_0^l\\
&+[\dot{y}_r'(x,t)-\dot{y}'(x,t)]\,EI\,[y_r''(x,t)-y''(x,t)]\,|_0^l\\
&-\int_0^l[\dot{y}_r''(x,t)-\dot{y}''(x,t)]\,EI\,[y_r''(x,t)-y''(x,t)]\,dx\\
&-\int_0^l k_v\,[\dot{y}_r(x,t)-\dot{y}(x,t)]^2\,dx\\
&+[\rho-\hat{\rho}(t)]\int_0^l[\dot{y}_r(x,t)-\dot{y}(x,t)]\,\ddot{y}_r(x,t)dx\\
&+\left[EI-\widehat{EI}(t)\right]\int_0^l[\dot{y}_r(x,t)-\dot{y}(x,t)]\,y_r''''(x,t)dx\\
=&-[\dot{y}_r(l,t)-\dot{y}(l,t)]\,EI\,[y_r'''(l,t)-y'''(l,t)]\\
&+[\dot{y}_r(0,t)-\dot{y}(0,t)]\,EI\,[y_r'''(0,t)-y'''(0,t)]\\
&+[\dot{y}_r'(l,t)-\dot{y}'(l,t)]\,EI\,[y_r''(l,t)-y''(l,t)]\\
&-[\dot{y}_r'(0,t)-\dot{y}'(0,t)]\,EI\,[y_r''(0,t)-y''(0,t)]\\
&-\int_0^l[\dot{y}_r''(x,t)-\dot{y}''(x,t)]\,EI\,[y_r''(x,t)-y''(x,t)]\,dx\\
&-\int_0^l k_v\,[\dot{y}_r(x,t)-\dot{y}(x,t)]^2\,dx\\
&+[\rho-\hat{\rho}(t)]\int_0^l[\dot{y}_r(x,t)-\dot{y}(x,t)]\,\ddot{y}_r(x,t)dx\\
&+\left[EI-\widehat{EI}(t)\right]\int_0^l[\dot{y}_r(x,t)-\dot{y}(x,t)]\,y_r''''(x,t)dx\\
=&-[\dot{y}_r(l,t)-\dot{y}(l,t)]\left[\widehat{EI}y_r'''(l,t)-EIy'''(l,t)\right]\\
&+[\dot{y}_r(0,t)-\dot{y}(0,t)]\left[\widehat{EI}y_r'''(0,t)-EIy'''(0,t)\right]\\
&+[\dot{y}_r'(l,t)-\dot{y}'(l,t)]\left[\widehat{EI}y_r''(l,t)-EIy''(l,t)\right]\\
&-[\dot{y}_r'(0,t)-\dot{y}'(0,t)]\left[\widehat{EI}y_r''(0,t)-EIy''(0,t)\right]\\
&-\int_0^l[\dot{y}_r''(x,t)-\dot{y}''(x,t)]\,EI\,[y_r''(x,t)-y''(x,t)]\,dx\\
&-\int_0^l k_v\,[\dot{y}_r(x,t)-\dot{y}(x,t)]^2\,dx\\
&+[\rho-\hat{\rho}(t)]\int_0^l[\dot{y}_r(x,t)-\dot{y}(x,t)]\,\ddot{y}_r(x,t)dx
\end{aligned}
$$

$$
\begin{aligned}
&+\left[EI-\widehat{EI}(t)\right]\int_0^l\left[\dot{y}_r(x,t)-\dot{y}(x,t)\right]y_r''''(x,t)dx\\
&-\left[EI-\widehat{EI}(t)\right]\left[\dot{y}_r(l,t)-\dot{y}(l,t)\right]y_r'''(l,t)\\
&+\left[EI-\widehat{EI}(t)\right]\left[\dot{y}_r(0,t)-\dot{y}(0,t)\right]y_r'''(0,t)\\
&+\left[EI-\widehat{EI}(t)\right]\left[\dot{y}_r'(l,t)-\dot{y}'(l,t)\right]y_r''(l,t)\\
&-\left[EI-\widehat{EI}(t)\right]\left[\dot{y}_r'(0,t)-\dot{y}'(0,t)\right]y_r''(0,t)
\end{aligned}
\tag{B.51}
$$

holds.

Re-define the non-negative accompanying function ν as

$$
\nu=\nu_K+\nu_V+\frac{1}{2\rho_\rho}\left[\rho-\hat{\rho}(t)\right]^2+\frac{1}{2\rho_{EI}}\left[EI-\widehat{EI}(t)\right]^2.
\tag{B.52}
$$

Substituting (B.51) and the time derivative of (13.20) into the time derivative of (B.52) and using (13.41), (13.42), and Lemma 2.9 yields

$$
\begin{aligned}
\dot{\nu}=&-\int_0^l k_v\left[\dot{y}_r(x,t)-\dot{y}(x,t)\right]^2dx\\
&+\left[\rho-\hat{\rho}(t)\right]\left[s_\rho-\frac{\dot{\hat{\rho}}(t)}{\rho_\rho}\right]+\left[EI-\widehat{EI}(t)\right]\left[s_{EI}-\frac{\dot{\widehat{EI}}(t)}{\rho_{EI}}\right]\\
&+p_B-p_T\\
\leqslant&-\int_0^l k_v\left[\dot{y}_r(x,t)-\dot{y}(x,t)\right]^2dx\\
&+p_B-p_T
\end{aligned}
\tag{B.53}
$$

with

$$
\begin{aligned}
p_B&=\left[\dot{y}_r(0,t)-\dot{y}(0,t)\right]\left[\widehat{EI}y_r'''(0,t)-EIy'''(0,t)\right]\\
&\quad-\left[\dot{y}_r'(0,t)-\dot{y}'(0,t)\right]\left[\widehat{EI}y_r''(0,t)-EIy''(0,t)\right]\\
&=\left[\dot{y}_r(0,t)-\dot{y}(0,t)\right](f_{Br}-f_B)\\
&\quad+\left[\dot{y}_r'(0,t)-\dot{y}'(0,t)\right](m_{Br}-m_B)
\end{aligned}
\tag{B.54}
$$

$$
\begin{aligned}
p_T&=\left[\dot{y}_r(l,t)-\dot{y}(l,t)\right]\left[\widehat{EI}y_r'''(l,t)-EIy'''(l,t)\right]\\
&\quad-\left[\dot{y}_r'(l,t)-\dot{y}'(l,t)\right]\left[\widehat{EI}y_r''(l,t)-EIy''(l,t)\right]\\
&=\left[\dot{y}_r(l,t)-\dot{y}(l,t)\right](f_{Tr}-f_T)\\
&\quad+\left[\dot{y}_r'(l,t)-\dot{y}'(l,t)\right](m_{Tr}-m_T)
\end{aligned}
\tag{B.55}
$$

in view of (13.8)–(13.11) and (13.35)–(13.38).

Consider the fact that the flexible beam has one *driving cutting point* associated with point T and one *driven cutting point* associated with point B. Using (13.19), (13.20), (B.52), (B.53), and Definition 2.17 completes the proof.

B.12 Derivation of (13.83)–(13.88)

Substituting (13.64), (13.74), and (13.75) into (13.4) yields

$$
\begin{aligned}
&\int_{t_1}^{t_2} (\delta e_K - \delta e_P + \delta \overline{w_E})\, dt \\
&= -\int_0^l \int_{t_1}^{t_2} \rho \left({}^{\mathbf{B}_b}\mathbf{R}_\mathbf{I} \frac{\partial^2}{\partial t^2} \mathbf{r}_b(x,t) \right)^T {}^{\mathbf{B}_b}\mathbf{R}_\mathbf{I} \delta \mathbf{r}_b(x,t) dt dx \\
&\quad - \int_0^l \int_{t_1}^{t_2} \left\{ \frac{\rho}{S_x} \mathrm{diag}(J_x, I_y, I_z) {}^{\mathbf{B}_b}\mathbf{R}_\mathbf{I} \frac{\partial}{\partial t} \boldsymbol{\omega}_b + \frac{\rho}{S_x} \begin{bmatrix} (I_z - I_y) {}^{\mathbf{B}_b}\omega_y {}^{\mathbf{B}_b}\omega_z \\ I_y {}^{\mathbf{B}_b}\omega_x {}^{\mathbf{B}_b}\omega_z \\ -I_z {}^{\mathbf{B}_b}\omega_x {}^{\mathbf{B}_b}\omega_y \end{bmatrix} \right\}^T \\
&\qquad \times {}^{\mathbf{B}_b}\mathbf{R}_\mathbf{I} \delta \mathbf{q}_b dt dx \\
&\quad - \int_{t_1}^{t_2} E S_x \left(\frac{\partial u_x(x,t)}{\partial x} \right) \delta u_x(x,t) \Big|_0^l dt \\
&\quad + \int_{t_1}^{t_2} \int_0^l E S_x \left(\frac{\partial^2 u_x(x,t)}{\partial x^2} \right) \delta u_x(x,t) dx dt \\
&\quad - \int_{t_1}^{t_2} E I_z \left(\frac{\partial^2 y_b(x,t)}{\partial x^2} \right) \delta \left(\frac{\partial y_b(x,t)}{\partial x} \right) \Big|_0^l dt \\
&\quad + \int_{t_1}^{t_2} \int_0^l \alpha_z(x,t) E I_z \left(\frac{\partial^3 y_b(x,t)}{\partial x^3} \right) \delta \left(\frac{\partial y_b(x,t)}{\partial x} \right) dx dt \\
&\quad + \int_{t_1}^{t_2} (1 - \alpha_z(x,t)) E I_z \left(\frac{\partial^3 y_b(x,t)}{\partial x^3} \right) \delta y_b(x,t) \Big|_0^l dt \\
&\quad - \int_{t_1}^{t_2} \int_0^l E I_z \frac{\partial}{\partial x} \left((1 - \alpha_z(x,t)) \frac{\partial^3 y_b(x,t)}{\partial x^3} \right) \delta y_b(x,t) dx dt \\
&\quad - \int_{t_1}^{t_2} E I_y \left(\frac{\partial^2 z_b(x,t)}{\partial x^2} \right) \delta \left(\frac{\partial z_b(x,t)}{\partial x} \right) \Big|_0^l dt \\
&\quad + \int_{t_1}^{t_2} \int_0^l \alpha_y(x,t) E I_y \left(\frac{\partial^3 z_b(x,t)}{\partial x^3} \right) \delta \left(\frac{\partial z_b(x,t)}{\partial x} \right) dx dt \\
&\quad + \int_{t_1}^{t_2} (1 - \alpha_y(x,t)) E I_y \left(\frac{\partial^3 z_b(x,t)}{\partial x^3} \right) (\delta z_b(x,t)) \Big|_0^l dt \\
&\quad - \int_{t_1}^{t_2} \int_0^l E I_y \frac{\partial}{\partial x} \left((1 - \alpha_y(x,t)) \frac{\partial^3 z_b(x,t)}{\partial x^3} \right) \delta z_b(x,t) dx dt \\
&\quad - \int_{t_1}^{t_2} G J_x \left(\frac{\partial \Theta_x(x,t)}{\partial x} \right) \delta \Theta_x(x,t) \Big|_0^l dt \\
&\quad + \int_{t_1}^{t_2} \int_0^l G J_x \left(\frac{\partial^2 \Theta_x(x,t)}{\partial x^2} \right) \delta \Theta_x(x,t) dx dt
\end{aligned}
$$

$$
+\int_{t_1}^{t_2}\left[({}^{\mathbf{B}}\mathbf{R_I}\mathbf{f_B})^T({}^{\mathbf{B}}\mathbf{R_I}\delta\mathbf{r_B})+({}^{\mathbf{B}}\mathbf{R_I}\mathbf{m_B})^T({}^{\mathbf{B}}\mathbf{R_I}\delta\mathbf{q_B})\right.
$$

$$
\left.-({}^{\mathbf{T}}\mathbf{R_I}\mathbf{f_T})^T({}^{\mathbf{T}}\mathbf{R_I}\delta\mathbf{r_T})-({}^{\mathbf{T}}\mathbf{R_I}\mathbf{m_T})^T({}^{\mathbf{T}}\mathbf{R_I}\delta\mathbf{q_T})\right]dt
$$

$$
=-\int_{t_1}^{t_2}\int_0^l\rho\left({}^{\mathbf{B}_b}\mathbf{R_I}\frac{\partial^2}{\partial t^2}\mathbf{r}_b(x,t)\right)^T{}^{\mathbf{B}_b}\mathbf{R_I}\delta\mathbf{r}_b(x,t)dxdt
$$

$$
-\int_{t_1}^{t_2}\int_0^l\left\{\frac{\rho}{S_x}\mathrm{diag}(J_x,I_y,I_z){}^{\mathbf{B}_b}\mathbf{R_I}\frac{\partial}{\partial t}\boldsymbol{\omega}_b+\frac{\rho}{S_x}\begin{bmatrix}(I_z-I_y){}^{\mathbf{B}_b}\omega_y{}^{\mathbf{B}_b}\omega_z\\ I_y{}^{\mathbf{B}_b}\omega_x{}^{\mathbf{B}_b}\omega_z\\ -I_z{}^{\mathbf{B}_b}\omega_x{}^{\mathbf{B}_b}\omega_y\end{bmatrix}\right\}^T
$$

$$
\times{}^{\mathbf{B}_b}\mathbf{R_I}\delta\mathbf{q}_b dxdt
$$

$$
+\int_{t_1}^{t_2}\int_0^l ES_x\left(\frac{\partial^2u_x(x,t)}{\partial x^2}\right)\delta u_x(x,t)dxdt
$$

$$
-\int_{t_1}^{t_2}\int_0^l EI_z\frac{\partial}{\partial x}\left((1-\alpha_z(x,t))\frac{\partial^3y_b(x,t)}{\partial x^3}\right)\delta y_b(x,t)dxdt
$$

$$
-\int_{t_1}^{t_2}\int_0^l EI_y\frac{\partial}{\partial x}\left((1-\alpha_y(x,t))\frac{\partial^3z_b(x,t)}{\partial x^3}\right)\delta z_b(x,t)dxdt
$$

$$
+\int_{t_1}^{t_2}\int_0^l GJ_x\left(\frac{\partial^2\Theta_x(x,t)}{\partial x^2}\right)\delta\Theta_x(x,t)dxdt
$$

$$
+\int_{t_1}^{t_2}\int_0^l \alpha_y(x,t)EI_y\left(\frac{\partial^3z_b(x,t)}{\partial x^3}\right)\delta\left(\frac{\partial z_b(x,t)}{\partial x}\right)dxdt
$$

$$
+\int_{t_1}^{t_2}\int_0^l \alpha_z(x,t)EI_z\left(\frac{\partial^3y_b(x,t)}{\partial x^3}\right)\delta\left(\frac{\partial y_b(x,t)}{\partial x}\right)dxdt
$$

$$
-\int_{t_1}^{t_2}ES_x\left(\frac{\partial u_x(x,t)}{\partial x}\right)\delta u_x(x,t)\bigg|_{x=l}dt
$$

$$
+\int_{t_1}^{t_2}(1-\alpha_z(x,t))EI_z\left(\frac{\partial^3y_b(x,t)}{\partial x^3}\right)\delta y_b(x,t)\bigg|_{x=l}dt
$$

$$
+\int_{t_1}^{t_2}(1-\alpha_y(x,t))EI_y\left(\frac{\partial^3z_b(x,t)}{\partial x^3}\right)\delta z_b(x,t)\bigg|_{x=l}dt
$$

$$
-\int_{t_1}^{t_2}GJ_x\left(\frac{\partial\Theta_x(x,t)}{\partial x}\right)\delta\Theta_x(x,t)\bigg|_{x=l}dt
$$

$$
-\int_{t_1}^{t_2}EI_y\left(\frac{\partial^2z_b(x,t)}{\partial x^2}\right)\delta\left(\frac{\partial z_b(x,t)}{\partial x}\right)\bigg|_{x=l}dt
$$

$$
-\int_{t_1}^{t_2}EI_z\left(\frac{\partial^2y_b(x,t)}{\partial x^2}\right)\delta\left(\frac{\partial y_b(x,t)}{\partial x}\right)\bigg|_{x=l}dt
$$

$$
+\int_{t_1}^{t_2}ES_x\left(\frac{\partial u_x(x,t)}{\partial x}\right)\delta u_x(x,t)\bigg|_{x=0}dt
$$

$$
-\int_{t_1}^{t_2}(1-\alpha_z(x,t))EI_z\left(\frac{\partial^3y_b(x,t)}{\partial x^3}\right)\delta y_b(x,t)\bigg|_{x=0}dt
$$

$$
\begin{aligned}
&- \int_{t_1}^{t_2} (1-\alpha_y(x,t)) EI_y \left(\frac{\partial^3 z_b(x,t)}{\partial x^3}\right) \delta z_b(x,t)\Big|_{x=0} dt \\
&+ \int_{t_1}^{t_2} GJ_x \left(\frac{\partial \Theta_x(x,t)}{\partial x}\right) \delta \Theta_x(x,t)\Big|_{x=0} dt \\
&+ \int_{t_1}^{t_2} EI_y \left(\frac{\partial^2 z_b(x,t)}{\partial x^2}\right) \delta \left(\frac{\partial z_b(x,t)}{\partial x}\right)\Big|_{x=0} dt \\
&+ \int_{t_1}^{t_2} EI_z \left(\frac{\partial^2 y_b(x,t)}{\partial x^2}\right) \delta \left(\frac{\partial y_b(x,t)}{\partial x}\right)\Big|_{x=0} dt \\
&+ \int_{t_1}^{t_2} \left[({}^{\mathbf{B}}\mathbf{R_I}\mathbf{f_B})^T({}^{\mathbf{B}}\mathbf{R_I}\delta\mathbf{r_B}) + ({}^{\mathbf{B}}\mathbf{R_I}\mathbf{m_B})^T({}^{\mathbf{B}}\mathbf{R_I}\delta\mathbf{q_B})\right. \\
&\left. -({}^{\mathbf{T}}\mathbf{R_I}\mathbf{f_T})^T({}^{\mathbf{T}}\mathbf{R_I}\delta\mathbf{r_T}) - ({}^{\mathbf{T}}\mathbf{R_I}\mathbf{m_T})^T({}^{\mathbf{T}}\mathbf{R_I}\delta\mathbf{q_T})\right] dt \\
=& - \int_{t_1}^{t_2}\int_0^l \rho \left({}^{\mathbf{B}_b}\mathbf{R_I}\frac{\partial^2}{\partial t^2}\mathbf{r}_b(x,t)\right)^T {}^{\mathbf{B}_b}\mathbf{R_I}\delta\mathbf{r}_b(x,t) dx dt \\
&- \int_{t_1}^{t_2}\int_0^l \begin{bmatrix} -ES_x\left(\frac{\partial^2 u_x(x,t)}{\partial x^2}\right) \\ EI_z\frac{\partial}{\partial x}\left((1-\alpha_z(x,t))\frac{\partial^3 y_b(x,t)}{\partial x^3}\right) \\ EI_y\frac{\partial}{\partial x}\left((1-\alpha_y(x,t))\frac{\partial^3 z_b(x,t)}{\partial x^3}\right) \end{bmatrix}^T \begin{bmatrix} \delta u_x(x,t) \\ \delta y_b(x,t) \\ \delta z_b(x,t) \end{bmatrix} dx dt \\
&- \int_{t_1}^{t_2}\int_0^l \left\{ \frac{\rho}{S_x}\mathrm{diag}(J_x, I_y, I_z)^{\mathbf{B}_b}\mathbf{R_I}\frac{\partial}{\partial t}\boldsymbol{\omega}_b + \frac{\rho}{S_x}\begin{bmatrix} (I_z - I_y)^{\mathbf{B}_b}\omega_y{}^{\mathbf{B}_b}\omega_z \\ I_y{}^{\mathbf{B}_b}\omega_x{}^{\mathbf{B}_b}\omega_z \\ -I_z{}^{\mathbf{B}_b}\omega_x{}^{\mathbf{B}_b}\omega_y \end{bmatrix}\right\}^T \\
&\quad \times {}^{\mathbf{B}_b}\mathbf{R_I}\delta\mathbf{q}_b dx dt \\
&+ \int_{t_1}^{t_2}\int_0^l \begin{bmatrix} GJ_x\left(\frac{\partial^2 \Theta_x(x,t)}{\partial x^2}\right) \\ -\alpha_y(x,t)EI_y\left(\frac{\partial^3 z_b(x,t)}{\partial x^3}\right) \\ \alpha_z(x,t)EI_z\left(\frac{\partial^3 y_b(x,t)}{\partial x^3}\right) \end{bmatrix}^T \begin{bmatrix} \delta\Theta_x(x,t) \\ -\delta\left(\frac{\partial z_b(x,t)}{\partial x}\right) \\ \delta\left(\frac{\partial y_b(x,t)}{\partial x}\right) \end{bmatrix} dx dt \\
&+ \int_{t_1}^{t_2} \begin{bmatrix} -ES_x\left(\frac{\partial u_x(x,t)}{\partial x}\right)\Big|_{x=l} \\ (1-\alpha_z(x,t))EI_z\left(\frac{\partial^3 y_b(x,t)}{\partial x^3}\right)\Big|_{x=l} \\ (1-\alpha_y(x,t))EI_y\left(\frac{\partial^3 z_b(x,t)}{\partial x^3}\right)\Big|_{x=l} \end{bmatrix}^T \begin{bmatrix} \delta u_x(x,t)|_{x=l} \\ \delta y_b(x,t)|_{x=l} \\ \delta z_b(x,t)|_{x=l} \end{bmatrix} dt \\
&+ \int_{t_1}^{t_2} \begin{bmatrix} -GJ_x\left(\frac{\partial \Theta_x(x,t)}{\partial x}\right)\Big|_{x=l} \\ EI_y\left(\frac{\partial^2 z_b(x,t)}{\partial x^2}\right)\Big|_{x=l} \\ -EI_z\left(\frac{\partial^2 y_b(x,t)}{\partial x^2}\right)\Big|_{x=l} \end{bmatrix}^T \begin{bmatrix} \delta\Theta_x(x,t)|_{x=l} \\ -\delta\left(\frac{\partial z_b(x,t)}{\partial x}\right)|_{x=l} \\ \delta\left(\frac{\partial y_b(x,t)}{\partial x}\right)|_{x=l} \end{bmatrix} dt
\end{aligned}
$$

$$
-\int_{t_1}^{t_2}\begin{bmatrix} -ES_x\left(\frac{\partial u_x(x,t)}{\partial x}\right)\Big|_{x=0} \\ (1-\alpha_z(x,t))EI_z\left(\frac{\partial^3 y_b(x,t)}{\partial x^3}\right)\Big|_{x=0} \\ (1-\alpha_y(x,t))EI_y\left(\frac{\partial^3 z_b(x,t)}{\partial x^3}\right)\Big|_{x=0}\end{bmatrix}^T \begin{bmatrix}\delta u_x(x,t)|_{x=0} \\ \delta y_b(x,t)|_{x=0} \\ \delta z_b(x,t)|_{x=0}\end{bmatrix} dt
$$

$$
-\int_{t_1}^{t_2}\begin{bmatrix} -GJ_x\left(\frac{\partial \Theta_x(x,t)}{\partial x}\right)\Big|_{x=0} \\ EI_y\left(\frac{\partial^2 z_b(x,t)}{\partial x^2}\right)\Big|_{x=0} \\ -EI_z\left(\frac{\partial^2 y_b(x,t)}{\partial x^2}\right)\Big|_{x=0}\end{bmatrix}^T \begin{bmatrix}\delta \Theta_x(x,t)|_{x=0} \\ -\delta\left(\frac{\partial z_b(x,t)}{\partial x}\right)\Big|_{x=0} \\ \delta\left(\frac{\partial y_b(x,t)}{\partial x}\right)\Big|_{x=0}\end{bmatrix} dt
$$

$$
+\int_{t_1}^{t_2}\left[({}^{\mathbf{B}}\mathbf{R_I}\mathbf{f_B})^T({}^{\mathbf{B}}\mathbf{R_I}\delta\mathbf{r_B}) + ({}^{\mathbf{B}}\mathbf{R_I}\mathbf{m_B})^T({}^{\mathbf{B}}\mathbf{R_I}\delta\mathbf{q_B})\right.
$$

$$
\left. -({}^{\mathbf{T}}\mathbf{R_I}\mathbf{f_T})^T({}^{\mathbf{T}}\mathbf{R_I}\delta\mathbf{r_T}) - ({}^{\mathbf{T}}\mathbf{R_I}\mathbf{m_T})^T({}^{\mathbf{T}}\mathbf{R_I}\delta\mathbf{q_T})\right] dt
$$

$$
= 0. \tag{B.56}
$$

It leads to (13.83)–(13.88) from given (13.77)–(13.82).

B.13 Proof of Theorem 13.3

A non-negative accompany function

$$
\nu = \nu_a + \nu_a + \nu_c \tag{B.57}
$$

with

$$
\nu_a = \frac{1}{2}\int_0^l \rho \frac{\partial}{\partial t}\left(\mathbf{r}_{br}(x,t) - \mathbf{r}_b(x,t)\right)^T \frac{\partial}{\partial t}\left(\mathbf{r}_{br}(x,t) - \mathbf{r}_b(x,t)\right) dx \tag{B.58}
$$

$$
\begin{aligned}
\nu_b = \frac{1}{2}\int_0^l \frac{\rho}{S_x}&\left[{}^{\mathbf{B}_b}\mathbf{R_I}\left(\boldsymbol{\omega}_{br} - \boldsymbol{\omega}_b\right)\right]^T \mathrm{diag}(J_x, I_y, I_z) \\
&\times \left[{}^{\mathbf{B}_b}\mathbf{R_I}\left(\boldsymbol{\omega}_{br} - \boldsymbol{\omega}_b\right)\right] dx
\end{aligned} \tag{B.59}
$$

$$
\begin{aligned}
\nu_c = &\frac{1}{2}\int_0^l ES_x\left[\frac{\partial}{\partial x}\left(u_{xr}(x,t) - u_x(x,t)\right)\right]^2 dx \\
&+\frac{1}{2}\int_0^l EI_z\left[\frac{\partial^2}{\partial x^2}\left(y_{br}(x,t) - y_b(x,t)\right)\right]^2 dx \\
&+\frac{1}{2}\int_0^l EI_y\left[\frac{\partial^2}{\partial x^2}\left(z_{br}(x,t) - z_b(x,t)\right)\right]^2 dx \\
&+\frac{1}{2}\int_0^l GJ_x\left[\frac{\partial}{\partial x}\left(\Theta_{xr}(x,t) - \Theta_x(x,t)\right)\right]^2 dx
\end{aligned} \tag{B.60}
$$

is assigned.

Subtracting (13.83) and (13.84) from (13.96) and (13.97) yields

$$\rho^{\mathbf{B}_b}\mathbf{R}_\mathbf{I}\frac{\partial^2}{\partial t^2}\left(\mathbf{r}_{br}(x,t)-\mathbf{r}_b(x,t)\right)$$
$$=-\begin{bmatrix} -ES_x\frac{\partial^2}{\partial x^2}\left(u_{xr}(x,t)-u_x(x,t)\right) \\ EI_z\frac{\partial}{\partial x}\left((1-\alpha_{zr}(x,t))\frac{\partial^3 y_{br}(x,t)}{\partial x^3}-(1-\alpha_z(x,t))\frac{\partial^3 y_b(x,t)}{\partial x^3}\right) \\ EI_y\frac{\partial}{\partial x}\left((1-\alpha_{yr}(x,t))\frac{\partial^3 z_{br}(x,t)}{\partial x^3}-(1-\alpha_y(x,t))\frac{\partial^3 z_b(x,t)}{\partial x^3}\right)\end{bmatrix}$$
$$-\mathbf{K}_p{}^{\mathbf{B}_b}\mathbf{R}_\mathbf{I}\left(\frac{\partial}{\partial t}\mathbf{r}_{br}(x,t)-\frac{\partial}{\partial t}\mathbf{r}_b(x,t)\right) \tag{B.61}$$

$$\frac{\rho}{S_x}\mathrm{diag}(J_x,I_y,I_z)^{\mathbf{B}_b}\mathbf{R}_\mathbf{I}\left[\left(\frac{\partial}{\partial t}\boldsymbol{\omega}_{br}-\frac{\partial}{\partial t}\boldsymbol{\omega}_b\right)-(\boldsymbol{\omega}_b\times)(\boldsymbol{\omega}_{br}-\boldsymbol{\omega}_b)\right]$$
$$=-\frac{\rho}{S_x}\begin{bmatrix} I_z{}^{\mathbf{B}_b}\omega_y\left({}^{\mathbf{B}_b}\omega_{zr}-{}^{\mathbf{B}_b}\omega_z\right)-I_y{}^{\mathbf{B}_b}\omega_z\left({}^{\mathbf{B}_b}\omega_{yr}-{}^{\mathbf{B}_b}\omega_y\right) \\ I_y{}^{\mathbf{B}_b}\omega_z\left({}^{\mathbf{B}_b}\omega_{xr}-{}^{\mathbf{B}_b}\omega_x\right) \\ -I_z{}^{\mathbf{B}_b}\omega_y\left({}^{\mathbf{B}_b}\omega_{xr}-{}^{\mathbf{B}_b}\omega_x\right)\end{bmatrix}$$
$$+\begin{bmatrix} GJ_x\frac{\partial^2}{\partial x^2}\left(\Theta_{xr}(x,t)-\Theta_x(x,t)\right) \\ -EI_y\left(\alpha_{yr}(x,t)\frac{\partial^3 z_{br}(x,t)}{\partial x^3}-\alpha_y(x,t)\frac{\partial^3 z_b(x,t)}{\partial x^3}\right) \\ EI_z\left(\alpha_{zr}(x,t)\frac{\partial^3 y_{br}(x,t)}{\partial x^3}-\alpha_z(x,t)\frac{\partial^3 y_b(x,t)}{\partial x^3}\right)\end{bmatrix}$$
$$-\mathbf{K}_\omega{}^{\mathbf{B}_b}\mathbf{R}_\mathbf{I}\left(\boldsymbol{\omega}_{br}-\boldsymbol{\omega}_b\right). \tag{B.62}$$

In view of (13.49), (13.92), (B.61), and

$$\frac{d}{dt}\left({}^{\mathbf{B}_b}\mathbf{R}_\mathbf{I}\right)=-{}^{\mathbf{B}_b}\mathbf{R}_\mathbf{I}(\boldsymbol{\omega}_b\times) \tag{B.63}$$

the time-derivative of (B.58) can be expressed as

$$\dot{\nu}_a=\int_0^l\rho\frac{\partial}{\partial t}\left(\mathbf{r}_{br}(x,t)-\mathbf{r}_b(x,t)\right)^T\frac{\partial^2}{\partial t^2}\left(\mathbf{r}_{br}(x,t)-\mathbf{r}_b(x,t)\right)dx$$
$$=\int_0^l\rho\left[{}^{\mathbf{B}_b}\mathbf{R}_\mathbf{I}\frac{\partial}{\partial t}\left(\mathbf{r}_{br}(x,t)-\mathbf{r}_b(x,t)\right)\right]^T$$
$$\times\left[{}^{\mathbf{B}_b}\mathbf{R}_\mathbf{I}\frac{\partial^2}{\partial t^2}\left(\mathbf{r}_{br}(x,t)-\mathbf{r}_b(x,t)\right)\right]dx$$
$$=-\int_0^l\begin{bmatrix}{}^{\mathbf{B}_b}v_{bxr}-{}^{\mathbf{B}_b}v_{bx} \\ {}^{\mathbf{B}_b}v_{byr}-{}^{\mathbf{B}_b}v_{by} \\ {}^{\mathbf{B}_b}v_{bzr}-{}^{\mathbf{B}_b}v_{bz}\end{bmatrix}^T$$
$$\times\begin{bmatrix} -ES_x\frac{\partial^2}{\partial x^2}\left(u_{xr}(x,t)-u_x(x,t)\right) \\ EI_z\frac{\partial}{\partial x}\left((1-\alpha_{zr}(x,t))\frac{\partial^3 y_{br}(x,t)}{\partial x^3}-(1-\alpha_z(x,t))\frac{\partial^3 y_b(x,t)}{\partial x^3}\right) \\ EI_y\frac{\partial}{\partial x}\left((1-\alpha_{yr}(x,t))\frac{\partial^3 z_{br}(x,t)}{\partial x^3}-(1-\alpha_y(x,t))\frac{\partial^3 z_b(x,t)}{\partial x^3}\right)\end{bmatrix}dx$$

$$
\begin{aligned}
&- \int_0^l \left[{}^{\mathbf{B}_b}\mathbf{R}_{\mathbf{I}} \frac{\partial}{\partial t} \left(\mathbf{r}_{br}(x,t) - \mathbf{r}_b(x,t) \right) \right]^T \\
&\qquad \times \mathbf{K}_p {}^{\mathbf{B}_b}\mathbf{R}_{\mathbf{I}} \frac{\partial}{\partial t} \left(\mathbf{r}_{br}(x,t) - \mathbf{r}_b(x,t) \right) dx \\
&= \int_0^l \left({}^{\mathbf{B}_b}v_{bxr} - {}^{\mathbf{B}_b}v_{bx} \right) ES_x \frac{\partial^2}{\partial x^2} \left(u_{xr}(x,t) - u_x(x,t) \right) dx \\
&- \int_0^l \left({}^{\mathbf{B}_b}v_{byr} - {}^{\mathbf{B}_b}v_{by} \right) EI_z \\
&\qquad \times \frac{\partial}{\partial x} \left[(1 - \alpha_{zr}(x,t)) \frac{\partial^3 y_{br}(x,t)}{\partial x^3} - (1 - \alpha_z(x,t)) \frac{\partial^3 y_b(x,t)}{\partial x^3} \right] dx \\
&- \int_0^l \left({}^{\mathbf{B}_b}v_{bzr} - {}^{\mathbf{B}_b}v_{bz} \right) EI_y \\
&\qquad \times \frac{\partial}{\partial x} \left[(1 - \alpha_{yr}(x,t)) \frac{\partial^3 z_{br}(x,t)}{\partial x^3} - (1 - \alpha_y(x,t)) \frac{\partial^3 z_b(x,t)}{\partial x^3} \right] dx \\
&- \int_0^l \frac{\partial}{\partial t} \left(\mathbf{r}_{br}(x,t) - \mathbf{r}_b(x,t) \right)^T \left({}^{\mathbf{B}_b}\mathbf{R}_{\mathbf{I}}^T \mathbf{K}_p {}^{\mathbf{B}_b}\mathbf{R}_{\mathbf{I}} \right) \\
&\qquad \times \frac{\partial}{\partial t} \left(\mathbf{r}_{br}(x,t) - \mathbf{r}_b(x,t) \right) dx.
\end{aligned}
\tag{B.64}
$$

The first to third terms in the right hand side of (B.64) can be further expressed by

$$
\begin{aligned}
&\int_0^l \left({}^{\mathbf{B}_b}v_{bxr} - {}^{\mathbf{B}_b}v_{bx} \right) ES_x \frac{\partial^2}{\partial x^2} \left(u_{xr}(x,t) - u_x(x,t) \right) dx \\
&\quad = \left({}^{\mathbf{B}_b}v_{bxr} - {}^{\mathbf{B}_b}v_{bx} \right) ES_x \frac{\partial}{\partial x} \left(u_{xr}(x,t) - u_x(x,t) \right) \Big|_0^l \\
&\qquad - \int_0^l \frac{\partial}{\partial x} \left({}^{\mathbf{B}_b}v_{bxr} - {}^{\mathbf{B}_b}v_{bx} \right) ES_x \frac{\partial}{\partial x} \left(u_{xr}(x,t) - u_x(x,t) \right) dx
\end{aligned}
\tag{B.65}
$$

$$
\begin{aligned}
&- \int_0^l \left({}^{\mathbf{B}_b}v_{byr} - {}^{\mathbf{B}_b}v_{by} \right) EI_z \\
&\qquad \times \frac{\partial}{\partial x} \left[(1 - \alpha_{zr}(x,t)) \frac{\partial^3 y_{br}(x,t)}{\partial x^3} - (1 - \alpha_z(x,t)) \frac{\partial^3 y_b(x,t)}{\partial x^3} \right] dx \\
&\quad = \left({}^{\mathbf{B}_b}v_{byr} \quad {}^{\mathbf{B}_b}v_{by} \right) EI_z \\
&\qquad\quad \times \left[(1 - \alpha_{zr}(x,t)) \frac{\partial^3 y_{br}(x,t)}{\partial x^3} - (1 - \alpha_z(x,t)) \frac{\partial^3 y_b(x,t)}{\partial x^3} \right] \Bigg|_0^l \\
&\qquad + \int_0^l \frac{\partial}{\partial x} \left({}^{\mathbf{B}_b}v_{byr} - {}^{\mathbf{B}_b}v_{by} \right) EI_z \\
&\qquad\quad \times \left[(1 - \alpha_{zr}(x,t)) \frac{\partial^3 y_{br}(x,t)}{\partial x^3} - (1 - \alpha_z(x,t)) \frac{\partial^3 y_b(x,t)}{\partial x^3} \right] dx
\end{aligned}
\tag{B.66}
$$

$$
\begin{aligned}
&-\int_0^l \left({}^{\mathbf{B}_b}v_{bzr} - {}^{\mathbf{B}_b}v_{bz}\right) EI_y \\
&\qquad \times \frac{\partial}{\partial x}\left[(1-\alpha_{yr}(x,t))\frac{\partial^3 z_{br}(x,t)}{\partial x^3} - (1-\alpha_y(x,t))\frac{\partial^3 z_b(x,t)}{\partial x^3}\right] dx \\
&= -\left({}^{\mathbf{B}_b}v_{bzr} - {}^{\mathbf{B}_b}v_{bz}\right) EI_y \\
&\qquad \times \left[(1-\alpha_{yr}(x,t))\frac{\partial^3 z_{br}(x,t)}{\partial x^3} - (1-\alpha_y(x,t))\frac{\partial^3 z_b(x,t)}{\partial x^3}\right]\Bigg|_0^l \\
&\quad + \int_0^l \frac{\partial}{\partial x}\left({}^{\mathbf{B}_b}v_{bzr} - {}^{\mathbf{B}_b}v_{bz}\right) EI_y \\
&\qquad \times \left[(1-\alpha_{yr}(x,t))\frac{\partial^3 z_{br}(x,t)}{\partial x^3} - (1-\alpha_y(x,t))\frac{\partial^3 z_b(x,t)}{\partial x^3}\right] dx.
\end{aligned}
\tag{B.67}
$$

Then, in view of (13.48), (13.91), (B.62), and (B.63), the time-derivative of (B.59) can be expressed by

$$
\begin{aligned}
\dot{\nu}_b &= \int_0^l \frac{\rho}{S_x}\left[{}^{\mathbf{B}_b}\mathbf{R}_\mathbf{I}(\boldsymbol{\omega}_{br}-\boldsymbol{\omega}_b)\right]^T \mathrm{diag}(J_x, I_y, I_z) \\
&\qquad \times \left[{}^{\mathbf{B}_b}\mathbf{R}_\mathbf{I}\left(\frac{\partial}{\partial t}\boldsymbol{\omega}_{br} - \frac{\partial}{\partial t}\boldsymbol{\omega}_b\right) - {}^{\mathbf{B}_b}\mathbf{R}_\mathbf{I}(\boldsymbol{\omega}_b\times)(\boldsymbol{\omega}_{br}-\boldsymbol{\omega}_b)\right] dx \\
&= -\int_0^l \frac{\rho}{S_x}\begin{bmatrix} {}^{\mathbf{B}_b}\omega_{xr} - {}^{\mathbf{B}_b}\omega_x \\ {}^{\mathbf{B}_b}\omega_{yr} - {}^{\mathbf{B}_b}\omega_y \\ {}^{\mathbf{B}_b}\omega_{zr} - {}^{\mathbf{B}_b}\omega_z \end{bmatrix}^T \\
&\qquad \times \begin{bmatrix} (I_z {}^{\mathbf{B}_b}\omega_y\left({}^{\mathbf{B}_b}\omega_{zr} - {}^{\mathbf{B}_b}\omega_z\right) - I_y {}^{\mathbf{B}_b}\omega_z\left({}^{\mathbf{B}_b}\omega_{yr} - {}^{\mathbf{B}_b}\omega_y\right) \\ I_y {}^{\mathbf{B}_b}\omega_z\left({}^{\mathbf{B}_b}\omega_{xr} - {}^{\mathbf{B}_b}\omega_x\right) \\ -I_z {}^{\mathbf{B}_b}\omega_y\left({}^{\mathbf{B}_b}\omega_{xr} - {}^{\mathbf{B}_b}\omega_x\right) \end{bmatrix} dx \\
&\quad + \int_0^l \begin{bmatrix} {}^{\mathbf{B}_b}\omega_{xr} - {}^{\mathbf{B}_b}\omega_x \\ {}^{\mathbf{B}_b}\omega_{yr} - {}^{\mathbf{B}_b}\omega_y \\ {}^{\mathbf{B}_b}\omega_{zr} - {}^{\mathbf{B}_b}\omega_z \end{bmatrix}^T \\
&\qquad \times \begin{bmatrix} GJ_x \frac{\partial^2}{\partial x^2}\left(\Theta_{xr}(x,t) - \Theta_x(x,t)\right) \\ -EI_y\left(\alpha_{yr}(x,t)\frac{\partial^3 z_{br}(x,t)}{\partial x^3} - \alpha_y(x,t)\frac{\partial^3 z_b(x,t)}{\partial x^3}\right) \\ EI_z\left(\alpha_{zr}(x,t)\frac{\partial^3 y_{br}(x,t)}{\partial x^3} - \alpha_z(x,t)\frac{\partial^3 y_b(x,t)}{\partial x^3}\right) \end{bmatrix} dx \\
&\quad - \int_0^l \left[{}^{\mathbf{B}_b}\mathbf{R}_\mathbf{I}(\boldsymbol{\omega}_{br}-\boldsymbol{\omega}_b)\right]^T \mathbf{K}_\omega \left[{}^{\mathbf{B}_b}\mathbf{R}_\mathbf{I}(\boldsymbol{\omega}_{br}-\boldsymbol{\omega}_b)\right] dx \\
&= \int_0^l \left({}^{\mathbf{B}_b}\omega_{xr} - {}^{\mathbf{B}_b}\omega_x\right) GJ_x \frac{\partial^2}{\partial x^2}\left(\Theta_{xr}(x,t) - \Theta_x(x,t)\right) dx \\
&\quad - \int_0^l \left({}^{\mathbf{B}_b}\omega_{yr} - {}^{\mathbf{B}_b}\omega_y\right) EI_y
\end{aligned}
$$

$$\times\left(\alpha_{yr}(x,t)\frac{\partial^3 z_{br}(x,t)}{\partial x^3}-\alpha_y(x,t)\frac{\partial^3 z_b(x,t)}{\partial x^3}\right)dx$$
$$+\int_0^l\left({}^{\mathbf{B}_b}\omega_{zr}-{}^{\mathbf{B}_b}\omega_z\right)EI_z$$
$$\times\left(\alpha_{zr}(x,t)\frac{\partial^3 y_{br}(x,t)}{\partial x^3}-\alpha_z(x,t)\frac{\partial^3 y_b(x,t)}{\partial x^3}\right)dx$$
$$-\int_0^l(\boldsymbol{\omega}_{br}-\boldsymbol{\omega}_b)^T\left({}^{\mathbf{B}_b}\mathbf{R}_{\mathbf{I}}^T\mathbf{K}_\omega{}^{\mathbf{B}_b}\mathbf{R}_{\mathbf{I}}\right)(\boldsymbol{\omega}_{br}-\boldsymbol{\omega}_b)\,dx. \tag{B.68}$$

Subtracting (13.50) and (13.51) from (13.93) and (13.94), respectively, yields

$${}^{\mathbf{B}_b}\omega_{yr}-{}^{\mathbf{B}_b}\omega_y=-\frac{\partial}{\partial x}\left({}^{\mathbf{B}_b}v_{bzr}-{}^{\mathbf{B}_b}v_{bz}\right) \tag{B.69}$$
$${}^{\mathbf{B}_b}\omega_{zr}-{}^{\mathbf{B}_b}\omega_z=\frac{\partial}{\partial x}\left({}^{\mathbf{B}_b}v_{byr}-{}^{\mathbf{B}_b}v_{by}\right). \tag{B.70}$$

By using (13.48), (13.49), (13.91), (13.92), (B.69), and (B.70), it follows from (13.85)–(13.88), (13.98)–(13.101), and (B.64)–(B.68) that

$$\dot{\nu}_a+\dot{\nu}_b=\left({}^{\mathbf{B}_b}v_{bxr}-{}^{\mathbf{B}_b}v_{bx}\right)ES_x\frac{\partial}{\partial x}\left(u_{xr}(x,t)-u_x(x,t)\right)\Big|_0^l$$
$$-\int_0^l\frac{\partial}{\partial x}\left({}^{\mathbf{B}_b}v_{bxr}-{}^{\mathbf{B}_b}v_{bx}\right)ES_x\frac{\partial}{\partial x}\left(u_{xr}(x,t)-u_x(x,t)\right)dx$$
$$-\left({}^{\mathbf{B}_b}v_{byr}-{}^{\mathbf{B}_b}v_{by}\right)EI_z$$
$$\times\left[(1-\alpha_{zr}(x,t))\frac{\partial^3 y_{br}(x,t)}{\partial x^3}-(1-\alpha_z(x,t))\frac{\partial^3 y_b(x,t)}{\partial x^3}\right]\Big|_0^l$$
$$+\int_0^l\frac{\partial}{\partial x}\left({}^{\mathbf{B}_b}v_{byr}-{}^{\mathbf{B}_b}v_{by}\right)EI_z\left(\frac{\partial^3 y_{br}(x,t)}{\partial x^3}-\frac{\partial^3 y_b(x,t)}{\partial x^3}\right)dx$$
$$-\left({}^{\mathbf{B}_b}v_{bzr}-{}^{\mathbf{B}_b}v_{bz}\right)EI_y$$
$$\times\left[(1-\alpha_{yr}(x,t))\frac{\partial^3 z_{br}(x,t)}{\partial x^3}-(1-\alpha_y(x,t))\frac{\partial^3 z_b(x,t)}{\partial x^3}\right]\Big|_0^l$$
$$+\int_0^l\frac{\partial}{\partial x}\left({}^{\mathbf{B}_b}v_{bzr}-{}^{\mathbf{B}_b}v_{bz}\right)EI_y\left(\frac{\partial^3 z_{br}(x,t)}{\partial x^3}-\frac{\partial^3 z_b(x,t)}{\partial x^3}\right)dx$$
$$-\int_0^l\frac{\partial}{\partial t}\left(\mathbf{r}_{br}(x,t)-\mathbf{r}_b(x,t)\right)^T\left({}^{\mathbf{B}_b}\mathbf{R}_{\mathbf{I}}^T\mathbf{K}_p{}^{\mathbf{B}_b}\mathbf{R}_{\mathbf{I}}\right)$$
$$\times\frac{\partial}{\partial t}\left(\mathbf{r}_{br}(x,t)-\mathbf{r}_b(x,t)\right)dx$$
$$+\int_0^l\left({}^{\mathbf{B}_b}\omega_{xr}-{}^{\mathbf{B}_b}\omega_x\right)GJ_x\frac{\partial^2}{\partial x^2}\left(\Theta_{xr}(x,t)-\Theta_x(x,t)\right)dx$$
$$-\int_0^l(\boldsymbol{\omega}_{br}-\boldsymbol{\omega}_b)^T\left({}^{\mathbf{B}_b}\mathbf{R}_{\mathbf{I}}^T\mathbf{K}_\omega{}^{\mathbf{B}_b}\mathbf{R}_{\mathbf{I}}\right)(\boldsymbol{\omega}_{br}-\boldsymbol{\omega}_b)\,dx$$

$$
\begin{aligned}
&= -\int_0^l \frac{\partial}{\partial t}\left(\mathbf{r}_{br}(x,t) - \mathbf{r}_b(x,t)\right)^T \left({}^{\mathbf{B}_b}\mathbf{R}_{\mathbf{I}}^T \mathbf{K}_p {}^{\mathbf{B}_b}\mathbf{R}_{\mathbf{I}}\right) \\
&\qquad \times \frac{\partial}{\partial t}\left(\mathbf{r}_{br}(x,t) - \mathbf{r}_b(x,t)\right) dx \\
&\quad - \int_0^l \left(\boldsymbol{\omega}_{br} - \boldsymbol{\omega}_b\right)^T \left({}^{\mathbf{B}_b}\mathbf{R}_{\mathbf{I}}^T \mathbf{K}_\omega {}^{\mathbf{B}_b}\mathbf{R}_{\mathbf{I}}\right)\left(\boldsymbol{\omega}_{br} - \boldsymbol{\omega}_b\right) dx \\
&\quad + \left({}^{\mathbf{B}_b}v_{bxr} - {}^{\mathbf{B}_b}v_{bx}\right) ES_x \frac{\partial}{\partial x}\left(u_{xr}(x,t) - u_x(x,t)\right)\Big|_0^l \\
&\quad - \left({}^{\mathbf{B}_b}v_{byr} - {}^{\mathbf{B}_b}v_{by}\right) EI_z \\
&\qquad \times \left[(1-\alpha_{zr}(x,t))\frac{\partial^3 y_{br}(x,t)}{\partial x^3} - (1-\alpha_z(x,t))\frac{\partial^3 y_b(x,t)}{\partial x^3}\right]\Bigg|_0^l \\
&\quad - \left({}^{\mathbf{B}_b}v_{bzr} - {}^{\mathbf{B}_b}v_{bz}\right) EI_y \\
&\qquad \times \left[(1-\alpha_{yr}(x,t))\frac{\partial^3 z_{br}(x,t)}{\partial x^3} - (1-\alpha_y(x,t))\frac{\partial^3 z_b(x,t)}{\partial x^3}\right]\Bigg|_0^l \\
&\quad + \left({}^{\mathbf{B}_b}\omega_{xr} - {}^{\mathbf{B}_b}\omega_x\right) GJ_x \frac{\partial}{\partial x}\left(\Theta_{xr}(x,t) - \Theta_x(x,t)\right)\Big|_0^l \\
&\quad - \frac{\partial}{\partial x}\left({}^{\mathbf{B}_b}v_{bzr} - {}^{\mathbf{B}_b}v_{bz}\right)\left[-EI_y\left(\frac{\partial^2 z_{br}(x,t)}{\partial x^2} - \frac{\partial^2 z_b(x,t)}{\partial x^2}\right)\right]\Bigg|_0^l \\
&\quad + \frac{\partial}{\partial x}\left({}^{\mathbf{B}_b}v_{byr} - {}^{\mathbf{B}_b}v_{by}\right) EI_z\left(\frac{\partial^2 y_{br}(x,t)}{\partial x^2} - \frac{\partial^2 y_b(x,t)}{\partial x^2}\right)\Bigg|_0^l \\
&\quad - \int_0^l \frac{\partial}{\partial x}\left({}^{\mathbf{B}_b}v_{bxr} - {}^{\mathbf{B}_b}v_{bx}\right) ES_x \frac{\partial}{\partial x}\left(u_{xr}(x,t) - u_x(x,t)\right) dx \\
&\quad - \int_0^l \frac{\partial^2}{\partial x^2}\left({}^{\mathbf{B}_b}v_{byr} - {}^{\mathbf{B}_b}v_{by}\right) EI_z \frac{\partial^2}{\partial x^2}\left(y_{br}(x,t) - y_b(x,t)\right) dx \\
&\quad - \int_0^l \frac{\partial^2}{\partial x^2}\left({}^{\mathbf{B}_b}v_{bzr} - {}^{\mathbf{B}_b}v_{bz}\right) EI_y \frac{\partial^2}{\partial x^2}\left(z_{br}(x,t) - z_b(x,t)\right) dx \\
&\quad - \int_0^l \frac{\partial}{\partial x}\left({}^{\mathbf{B}_b}\omega_{xr} - {}^{\mathbf{B}_b}\omega_x\right) GJ_x \frac{\partial}{\partial x}\left(\Theta_{xr}(x,t) - \Theta_x(x,t)\right) dx \\
&= -\int_0^l \frac{\partial}{\partial t}\left(\mathbf{r}_{br}(x,t) - \mathbf{r}_b(x,t)\right)^T \left({}^{\mathbf{B}_b}\mathbf{R}_{\mathbf{I}}^T \mathbf{K}_p {}^{\mathbf{B}_b}\mathbf{R}_{\mathbf{I}}\right) \\
&\qquad \times \frac{\partial}{\partial t}\left(\mathbf{r}_{br}(x,t) - \mathbf{r}_b(x,t)\right) dx \\
&\quad - \int_0^l \left(\boldsymbol{\omega}_{br} - \boldsymbol{\omega}_b\right)^T \left({}^{\mathbf{B}_b}\mathbf{R}_{\mathbf{I}}^T \mathbf{K}_\omega {}^{\mathbf{B}_b}\mathbf{R}_{\mathbf{I}}\right)\left(\boldsymbol{\omega}_{br} - \boldsymbol{\omega}_b\right) dx \\
&\quad + \left[{}^{\mathbf{B}}\mathbf{R}_{\mathbf{I}}\left(\dot{\mathbf{r}}_{\mathbf{B}r} - \dot{\mathbf{r}}_{\mathbf{B}}\right)\right]^T {}^{\mathbf{B}}\mathbf{R}_{\mathbf{I}}\left(\mathbf{f}_{\mathbf{B}r} - \mathbf{f}_{\mathbf{B}}\right) \\
&\quad + \left[{}^{\mathbf{B}}\mathbf{R}_{\mathbf{I}}\left(\boldsymbol{\omega}_{\mathbf{B}r} - \boldsymbol{\omega}_{\mathbf{B}}\right)\right]^T {}^{\mathbf{B}}\mathbf{R}_{\mathbf{I}}\left(\mathbf{m}_{\mathbf{Br}} - \mathbf{m}_{\mathbf{B}}\right)
\end{aligned}
$$

$$
\begin{aligned}
&- \left[{}^{\mathbf{T}}\mathbf{R}_{\mathbf{I}} \left(\dot{\mathbf{r}}_{\mathbf{T}r} - \dot{\mathbf{r}}_{\mathbf{T}}\right)\right]^T {}^{\mathbf{T}}\mathbf{R}_{\mathbf{I}} \left(\mathbf{f}_{\mathbf{T}r} - \mathbf{f}_{\mathbf{T}}\right) \\
&- \left[{}^{\mathbf{T}}\mathbf{R}_{\mathbf{I}} \left(\boldsymbol{\omega}_{\mathbf{T}r} - \boldsymbol{\omega}_{\mathbf{T}}\right)\right]^T {}^{\mathbf{T}}\mathbf{R}_{\mathbf{I}} \left(\mathbf{m}_{\mathbf{T}r} - \mathbf{m}_{\mathbf{T}}\right) \\
&- \int_0^l \frac{\partial}{\partial x} \left({}^{\mathbf{B}_b}v_{bxr} - {}^{\mathbf{B}_b}v_{bx}\right) ES_x \frac{\partial}{\partial x} \left(u_{xr}(x,t) - u_x(x,t)\right) dx \\
&- \int_0^l \frac{\partial^2}{\partial x^2} \left({}^{\mathbf{B}_b}v_{byr} - {}^{\mathbf{B}_b}v_{by}\right) EI_z \frac{\partial^2}{\partial x^2} \left(y_{br}(x,t) - y_b(x,t)\right) dx \\
&- \int_0^l \frac{\partial^2}{\partial x^2} \left({}^{\mathbf{B}_b}v_{bzr} - {}^{\mathbf{B}_b}v_{bz}\right) EI_y \frac{\partial^2}{\partial x^2} \left(z_{br}(x,t) - z_b(x,t)\right) dx \\
&- \int_0^l \frac{\partial}{\partial x} \left({}^{\mathbf{B}_b}\omega_{xr} - {}^{\mathbf{B}_b}\omega_{x}\right) GJ_x \frac{\partial}{\partial x} \left(\Theta_{xr}(x,t) - \Theta_x(x,t)\right) dx
\end{aligned}
\qquad (B.71)
$$

holds with

$$
\begin{aligned}
\dot{\mathbf{r}}_{\mathbf{B}r} &= \left.\frac{\partial}{\partial t}\mathbf{r}_{br}(x,t)\right|_{x=0} \\
\dot{\mathbf{r}}_{\mathbf{B}} &= \left.\frac{\partial}{\partial t}\mathbf{r}_{b}(x,t)\right|_{x=0} \\
\boldsymbol{\omega}_{\mathbf{B}r} &= \left.\boldsymbol{\omega}_{br}(x,t)\right|_{x=0} \\
\boldsymbol{\omega}_{\mathbf{B}} &= \left.\boldsymbol{\omega}_{b}(x,t)\right|_{x=0} \\
\dot{\mathbf{r}}_{\mathbf{T}r} &= \left.\frac{\partial}{\partial t}\mathbf{r}_{br}(x,t)\right|_{x=l} \\
\dot{\mathbf{r}}_{\mathbf{T}} &= \left.\frac{\partial}{\partial t}\mathbf{r}_{b}(x,t)\right|_{x=l} \\
\boldsymbol{\omega}_{\mathbf{T}r} &= \left.\boldsymbol{\omega}_{br}(x,t)\right|_{x=l} \\
\boldsymbol{\omega}_{\mathbf{T}} &= \left.\boldsymbol{\omega}_{b}(x,t)\right|_{x=l} .
\end{aligned}
$$

Substituting (13.95) into (B.71) and using (13.89), (13.90), and (B.60) yields

$$
\begin{aligned}
\dot{\nu} &= \dot{\nu}_a + \dot{\nu}_b + \dot{\nu}_c \\
&= - \int_0^l \frac{\partial}{\partial t} \left[\mathbf{r}_{br}(x,t) - \mathbf{r}_b(x,t)\right]^T \left({}^{\mathbf{B}_b}\mathbf{R}_{\mathbf{I}}^T \mathbf{K}_p {}^{\mathbf{B}_b}\mathbf{R}_{\mathbf{I}}\right) \\
&\qquad \times \frac{\partial}{\partial t} \left[\mathbf{r}_{br}(x,t) - \mathbf{r}_b(x,t)\right] dx \\
&\quad - \int_0^l \left(\boldsymbol{\omega}_{br} - \boldsymbol{\omega}_b\right)^T \left({}^{\mathbf{B}_b}\mathbf{R}_{\mathbf{I}}^T \mathbf{K}_\omega {}^{\mathbf{B}_b}\mathbf{R}_{\mathbf{I}}\right) \left(\boldsymbol{\omega}_{br} - \boldsymbol{\omega}_b\right) dx \\
&\quad - ES_x\lambda_x \int_0^l \left[\frac{\partial}{\partial x} \left(u_{xr}(x,t) - u_x(x,t)\right)\right]^2 dx \\
&\quad - EI_z\lambda_y \int_0^l \left[\frac{\partial^2}{\partial x^2} \left(y_{br}(x,t) - y_b(x,t)\right)\right]^2 dx \\
&\quad - EI_y\lambda_z \int_0^l \left[\frac{\partial^2}{\partial x^2} \left(z_{br}(x,t) - z_b(x,t)\right)\right]^2 dx
\end{aligned}
$$

$$-GJ_x\lambda_\Theta \int_0^l \left[\frac{\partial}{\partial x}\left(\Theta_{xr}(x,t) - \Theta_x(x,t)\right)\right]^2 dx$$
$$+p_{\mathbf{B}} - p_{\mathbf{T}} \tag{B.72}$$

where

$$\begin{aligned}
p_{\mathbf{B}} &= \left[{}^{\mathbf{B}}\mathbf{R_I}\left(\dot{\mathbf{r}}_{\mathbf{B}r} - \dot{\mathbf{r}}_{\mathbf{B}}\right)\right]^T {}^{\mathbf{B}}\mathbf{R_I}\left(\mathbf{f}_{\mathbf{B}r} - \mathbf{f}_{\mathbf{B}}\right)\\
&\quad + \left[{}^{\mathbf{B}}\mathbf{R_I}\left(\boldsymbol{\omega}_{\mathbf{B}r} - \boldsymbol{\omega}_{\mathbf{B}}\right)\right]^T {}^{\mathbf{B}}\mathbf{R_I}\left(\mathbf{m}_{\mathbf{B}r} - \mathbf{m}_{\mathbf{B}}\right)\\
&= -\left({}^{\mathbf{B}_b}v_{bxr} - {}^{\mathbf{B}_b}v_{bx}\right) ES_x \frac{\partial}{\partial x}\left(u_{xr}(x,t) - u_x(x,t)\right)\Big|_{x=0}\\
&\quad + \left({}^{\mathbf{B}_b}v_{byr} - {}^{\mathbf{B}_b}v_{by}\right) EI_z\\
&\qquad \times \left[(1-\alpha_{zr}(x,t))\frac{\partial^3 y_{br}(x,t)}{\partial x^3} - (1-\alpha_z(x,t))\frac{\partial^3 y_b(x,t)}{\partial x^3}\right]\Bigg|_{x=0}\\
&\quad + \left({}^{\mathbf{B}_b}v_{bzr} - {}^{\mathbf{B}_b}v_{bz}\right) EI_y\\
&\qquad \times \left[(1-\alpha_{yr}(x,t))\frac{\partial^3 z_{br}(x,t)}{\partial x^3} - (1-\alpha_y(x,t))\frac{\partial^3 z_b(x,t)}{\partial x^3}\right]\Bigg|_{x=0}\\
&\quad + \left({}^{\mathbf{B}_b}\omega_{xr} - {}^{\mathbf{B}_b}\omega_x\right)\left[-GJ_x\frac{\partial}{\partial x}\left(\Theta_{xr}(x,t) - \Theta_x(x,t)\right)\right]\Bigg|_{x=0}\\
&\quad - \frac{\partial}{\partial x}\left({}^{\mathbf{B}_b}v_{bzr} - {}^{\mathbf{B}_b}v_{bz}\right)\left[EI_y\left(\frac{\partial^2 z_{br}(x,t)}{\partial x^2} - \frac{\partial^2 z_b(x,t)}{\partial x^2}\right)\right]\Bigg|_{x=0}\\
&\quad + \frac{\partial}{\partial x}\left({}^{\mathbf{B}_b}v_{byr} - {}^{\mathbf{B}_b}v_{by}\right)\left[-EI_z\left(\frac{\partial^2 y_{br}(x,t)}{\partial x^2} - \frac{\partial^2 y_b(x,t)}{\partial x^2}\right)\right]\Bigg|_{x=0}
\end{aligned} \tag{B.73}$$

$$\begin{aligned}
p_{\mathbf{T}} &= \left[{}^{\mathbf{T}}\mathbf{R_I}\left(\dot{\mathbf{r}}_{\mathbf{T}r} - \dot{\mathbf{r}}_{\mathbf{T}}\right)\right]^T {}^{\mathbf{T}}\mathbf{R_I}\left(\mathbf{f}_{\mathbf{T}r} - \mathbf{f}_{\mathbf{T}}\right)\\
&\quad + \left[{}^{\mathbf{T}}\mathbf{R_I}\left(\boldsymbol{\omega}_{\mathbf{T}r} - \boldsymbol{\omega}_{\mathbf{T}}\right)\right]^T {}^{\mathbf{T}}\mathbf{R_I}\left(\mathbf{m}_{\mathbf{T}r} - \mathbf{m}_{\mathbf{T}}\right)\\
&= -\left({}^{\mathbf{B}_b}v_{bxr} - {}^{\mathbf{B}_b}v_{bx}\right) ES_x \frac{\partial}{\partial x}\left(u_{xr}(x,t) - u_x(x,t)\right)\Big|_{x=l}\\
&\quad + \left({}^{\mathbf{B}_b}v_{byr} - {}^{\mathbf{B}_b}v_{by}\right) EI_z\\
&\qquad \times \left[(1-\alpha_{zr}(x,t))\frac{\partial^3 y_{br}(x,t)}{\partial x^3} - (1-\alpha_z(x,t))\frac{\partial^3 y_b(x,t)}{\partial x^3}\right]\Bigg|_{x=l}\\
&\quad + \left({}^{\mathbf{B}_b}v_{bzr} - {}^{\mathbf{B}_b}v_{bz}\right) EI_y\\
&\qquad \times \left[(1-\alpha_{yr}(x,t))\frac{\partial^3 z_{br}(x,t)}{\partial x^3} - (1-\alpha_y(x,t))\frac{\partial^3 z_b(x,t)}{\partial x^3}\right]\Bigg|_{x=l}\\
&\quad + \left({}^{\mathbf{B}_b}\omega_{xr} - {}^{\mathbf{B}_b}\omega_x\right)\left[-GJ_x\frac{\partial}{\partial x}\left(\Theta_{xr}(x,t) - \Theta_x(x,t)\right)\right]\Bigg|_{x=l}\\
&\quad - \frac{\partial}{\partial x}\left({}^{\mathbf{B}_b}v_{bzr} - {}^{\mathbf{B}_b}v_{bz}\right)\left[EI_y\left(\frac{\partial^2 z_{br}(x,t)}{\partial x^2} - \frac{\partial^2 z_b(x,t)}{\partial x^2}\right)\right]\Bigg|_{x=l}\\
&\quad + \frac{\partial}{\partial x}\left({}^{\mathbf{B}_b}v_{byr} - {}^{\mathbf{B}_b}v_{by}\right)\left[-EI_z\left(\frac{\partial^2 y_{br}(x,t)}{\partial x^2} - \frac{\partial^2 y_b(x,t)}{\partial x^2}\right)\right]\Bigg|_{x=l}
\end{aligned} \tag{B.74}$$

denote the *virtual power flows* at the two ends of the flexible beam.

B.14 Proof of Theorem 13.4

Subtracting (13.83) and (13.84) from (13.107) and (13.108) yields

$$\begin{aligned}
&\rho^{\mathbf{B}_b}\mathbf{R}_{\mathbf{I}}\frac{\partial^2}{\partial t^2}\left(\mathbf{r}_{br}(x,t)-\mathbf{r}_b(x,t)\right)\\
&\quad=-\mathbf{K}_p{}^{\mathbf{B}_b}\mathbf{R}_{\mathbf{I}}\left(\frac{\partial}{\partial t}\mathbf{r}_{br}(x,t)-\frac{\partial}{\partial t}\mathbf{r}_b(x,t)\right)\\
&\qquad-\begin{bmatrix}-ES_x\frac{\partial^2}{\partial x^2}\left(u_{xr}(x,t)-u_x(x,t)\right)\\ EI_z\frac{\partial}{\partial x}\left((1-\alpha_{zr}(x,t))\frac{\partial^3 y_{br}(x,t)}{\partial x^3}-(1-\alpha_z(x,t))\frac{\partial^3 y_b(x,t)}{\partial x^3}\right)\\ EI_y\frac{\partial}{\partial x}\left((1-\alpha_{yr}(x,t))\frac{\partial^3 z_{br}(x,t)}{\partial x^3}-(1-\alpha_y(x,t))\frac{\partial^3 z_b(x,t)}{\partial x^3}\right)\end{bmatrix}\\
&\qquad+(\rho-\hat{\rho})\,{}^{\mathbf{B}_b}\mathbf{R}_{\mathbf{I}}\frac{\partial^2}{\partial t^2}\mathbf{r}_{br}(x,t)\\
&\qquad+\begin{bmatrix}-\left(ES_x-\widehat{ES_x}\right)\frac{\partial^2}{\partial x^2}u_{xr}(x,t)\\ \left(EI_z-\widehat{EI_z}\right)\frac{\partial}{\partial x}\left((1-\alpha_{zr}(x,t))\frac{\partial^3 y_{br}(x,t)}{\partial x^3}\right)\\ \left(EI_y-\widehat{EI_y}\right)\frac{\partial}{\partial x}\left((1-\alpha_{yr}(x,t))\frac{\partial^3 z_{br}(x,t)}{\partial x^3}\right)\end{bmatrix}
\end{aligned}\tag{B.75}$$

$$\begin{aligned}
&\frac{\rho}{S_x}\mathrm{diag}(J_x,I_y,I_z)^{\mathbf{B}_b}\mathbf{R}_{\mathbf{I}}\left[\left(\frac{\partial}{\partial t}\boldsymbol{\omega}_{br}-\frac{\partial}{\partial t}\boldsymbol{\omega}_b\right)-(\boldsymbol{\omega}_b\times)(\boldsymbol{\omega}_{br}-\boldsymbol{\omega}_b)\right]\\
&\quad=-\frac{\rho}{S_x}\begin{bmatrix}(I_z{}^{\mathbf{B}_b}\omega_y\left({}^{\mathbf{B}_b}\omega_{zr}-{}^{\mathbf{B}_b}\omega_z\right)-I_y{}^{\mathbf{B}_b}\omega_z\left({}^{\mathbf{B}_b}\omega_{yr}-{}^{\mathbf{B}_b}\omega_y\right)\\ I_y{}^{\mathbf{B}_b}\omega_z\left({}^{\mathbf{B}_b}\omega_{xr}-{}^{\mathbf{B}_b}\omega_x\right)\\ -I_z{}^{\mathbf{B}_b}\omega_y\left({}^{\mathbf{B}_b}\omega_{xr}-{}^{\mathbf{B}_b}\omega_x\right)\end{bmatrix}\\
&\qquad-\mathbf{K}_\omega{}^{\mathbf{B}_b}\mathbf{R}_{\mathbf{I}}\left(\boldsymbol{\omega}_{br}-\boldsymbol{\omega}_b\right)\\
&\qquad+\begin{bmatrix}GJ_x\frac{\partial^2}{\partial x^2}\left(\Theta_{xr}(x,t)-\Theta_x(x,t)\right)\\ -EI_y\left(\alpha_{yr}(x,t)\frac{\partial^3 z_{br}(x,t)}{\partial x^3}-\alpha_y(x,t)\frac{\partial^3 z_b(x,t)}{\partial x^3}\right)\\ EI_z\left(\alpha_{zr}(x,t)\frac{\partial^3 y_{br}(x,t)}{\partial x^3}-\alpha_z(x,t)\frac{\partial^3 y_b(x,t)}{\partial x^3}\right)\end{bmatrix}\\
&\qquad+\mathrm{diag}\left(\left(\frac{\rho}{S_x}J_x-\widehat{\frac{\rho}{S_x}J_x}\right),\left(\frac{\rho}{S_x}I_y-\widehat{\frac{\rho}{S_x}I_y}\right),\left(\frac{\rho}{S_x}I_z-\widehat{\frac{\rho}{S_x}I_z}\right)\right)\\
&\qquad\quad\times\left({}^{\mathbf{B}_b}\mathbf{R}_{\mathbf{I}}\frac{\partial}{\partial t}\boldsymbol{\omega}_{br}-{}^{\mathbf{B}_b}\mathbf{R}_{\mathbf{I}}(\boldsymbol{\omega}_b\times)\boldsymbol{\omega}_{br}\right)\\
&\qquad+\begin{bmatrix}\left(\frac{\rho}{S_x}I_z-\widehat{\frac{\rho}{S_x}I_z}\right){}^{\mathbf{B}_b}\omega_y{}^{\mathbf{B}_b}\omega_{zr}-\left(\frac{\rho}{S_x}I_y-\widehat{\frac{\rho}{S_x}I_y}\right){}^{\mathbf{B}_b}\omega_z{}^{\mathbf{B}_b}\omega_{yr}\\ \left(\frac{\rho}{S_x}I_y-\widehat{\frac{\rho}{S_x}I_y}\right){}^{\mathbf{B}_b}\omega_z{}^{\mathbf{B}_b}\omega_{xr}\\ -\left(\frac{\rho}{S_x}I_z-\widehat{\frac{\rho}{S_x}I_z}\right){}^{\mathbf{B}_b}\omega_y{}^{\mathbf{B}_b}\omega_{xr}\end{bmatrix}
\end{aligned}$$

$$-\begin{bmatrix} \left(GJ_x-\widehat{GJ_x}\right)\frac{\partial^2}{\partial x^2}\Theta_{xr}(x,t) \\ -\left(EI_y-\widehat{EI_y}\right)\alpha_{yr}(x,t)\frac{\partial^3 z_{br}(x,t)}{\partial x^3} \\ \left(EI_z-\widehat{EI_z}\right)\alpha_{zr}(x,t)\frac{\partial^3 y_{br}(x,t)}{\partial x^3} \end{bmatrix}. \tag{B.76}$$

With the same non-negative accompanying function defined by (B.57)–(B.60), replacing (B.61) and (B.62) by (B.75) and (B.76) and using (13.48), (13.49), (13.91), (13.92), and (B.72)–(B.74) yields

$$\begin{aligned}
\dot{\nu} = &-\int_0^l \frac{\partial}{\partial t}\left[\mathbf{r}_{br}(x,t)-\mathbf{r}_b(x,t)\right]^T \left({}^{\mathbf{B}_b}\mathbf{R}_\mathbf{I}^T\mathbf{K}_p{}^{\mathbf{B}_b}\mathbf{R}_\mathbf{I}\right) \\
&\quad\times\frac{\partial}{\partial t}\left[\mathbf{r}_{br}(x,t)-\mathbf{r}_b(x,t)\right]dx \\
&-\int_0^l \left(\boldsymbol{\omega}_{br}-\boldsymbol{\omega}_b\right)^T\left({}^{\mathbf{B}_b}\mathbf{R}_\mathbf{I}^T\mathbf{K}_\omega{}^{\mathbf{B}_b}\mathbf{R}_\mathbf{I}\right)\left(\boldsymbol{\omega}_{br}-\boldsymbol{\omega}_b\right)dx \\
&-ES_x\lambda_x\int_0^l\left[\frac{\partial}{\partial x}\left(u_{xr}(x,t)-u_x(x,t)\right)\right]^2 dx \\
&-EI_z\lambda_y\int_0^l\left[\frac{\partial^2}{\partial x^2}\left(y_{br}(x,t)-y_b(x,t)\right)\right]^2 dx \\
&-EI_y\lambda_z\int_0^l\left[\frac{\partial^2}{\partial x^2}\left(z_{br}(x,t)-z_b(x,t)\right)\right]^2 dx \\
&-GJ_x\lambda_\Theta\int_0^l\left[\frac{\partial}{\partial x}\left(\Theta_{xr}(x,t)-\Theta_x(x,t)\right)\right]^2 dx \\
&+\left({}^{\mathbf{B}_b}v_{bxr}-{}^{\mathbf{B}_b}v_{bx}\right)ES_x\frac{\partial}{\partial x}\left(u_{xr}(x,t)-u_x(x,t)\right)\Big|_0^l \\
&-\left({}^{\mathbf{B}_b}v_{byr}-{}^{\mathbf{B}_b}v_{by}\right)EI_z \\
&\quad\times\left[\left(1-\alpha_{zr}(x,t)\right)\frac{\partial^3 y_{br}(x,t)}{\partial x^3}-\left(1-\alpha_z(x,t)\right)\frac{\partial^3 y_b(x,t)}{\partial x^3}\right]\Big|_0^l \\
&-\left({}^{\mathbf{B}_b}v_{bzr}-{}^{\mathbf{B}_b}v_{bz}\right)EI_y \\
&\quad\times\left[\left(1-\alpha_{yr}(x,t)\right)\frac{\partial^3 z_{br}(x,t)}{\partial x^3}-\left(1-\alpha_y(x,t)\right)\frac{\partial^3 z_b(x,t)}{\partial x^3}\right]\Big|_0^l \\
&+\left({}^{\mathbf{B}_b}\omega_{xr}-{}^{\mathbf{B}_b}\omega_x\right)GJ_x\frac{\partial}{\partial x}\left(\Theta_{xr}(x,t)-\Theta_x(x,t)\right)\Big|_0^l \\
&-\frac{\partial}{\partial x}\left({}^{\mathbf{B}_b}v_{bzr}-{}^{\mathbf{B}_b}v_{bz}\right)\left[-EI_y\left(\frac{\partial^2 z_{br}(x,t)}{\partial x^2}-\frac{\partial^2 z_b(x,t)}{\partial x^2}\right)\right]\Big|_0^l \\
&+\frac{\partial}{\partial x}\left({}^{\mathbf{B}_b}v_{byr}-{}^{\mathbf{B}_b}v_{by}\right)EI_z\left(\frac{\partial^2 y_{br}(x,t)}{\partial x^2}-\frac{\partial^2 y_b(x,t)}{\partial x^2}\right)\Big|_0^l
\end{aligned}$$

$$+\int_0^l \left[{}^{\mathbf{B}_b}\mathbf{R}_\mathbf{I}\frac{\partial}{\partial t}\left(\mathbf{r}_{br}(x,t)-\mathbf{r}_b(x,t)\right)\right]^T(\rho-\hat{\rho})\,{}^{\mathbf{B}_b}\mathbf{R}_\mathbf{I}\frac{\partial^2}{\partial t^2}\mathbf{r}_{br}(x,t)dx$$

$$+\int_0^l \begin{bmatrix} {}^{\mathbf{B}_b}v_{bxr}-{}^{\mathbf{B}_b}v_{bx} \\ {}^{\mathbf{B}_b}v_{byr}-{}^{\mathbf{B}_b}v_{by} \\ {}^{\mathbf{B}_b}v_{bzr}-{}^{\mathbf{B}_b}v_{bz} \end{bmatrix}^T$$

$$\times\begin{bmatrix} -\left(ES_x-\widehat{ES_x}\right)\frac{\partial^2}{\partial x^2}u_{xr}(x,t) \\ \left(EI_z-\widehat{EI_z}\right)\frac{\partial}{\partial x}\left((1-\alpha_{zr}(x,t))\frac{\partial^3 y_{br}(x,t)}{\partial x^3}\right) \\ \left(EI_y-\widehat{EI_y}\right)\frac{\partial}{\partial x}\left((1-\alpha_{yr}(x,t))\frac{\partial^3 z_{br}(x,t)}{\partial x^3}\right) \end{bmatrix}dx$$

$$+\int_0^l \begin{bmatrix} \left({}^{\mathbf{B}_b}\omega_{xr}-{}^{\mathbf{B}_b}\omega_x\right)\left(\frac{\rho}{S_x}J_x-\widehat{\frac{\rho}{S_x}J_x}\right) \\ \left({}^{\mathbf{B}_b}\omega_{yr}-{}^{\mathbf{B}_b}\omega_y\right)\left(\frac{\rho}{S_x}I_y-\widehat{\frac{\rho}{S_x}I_y}\right) \\ \left({}^{\mathbf{B}_b}\omega_{zr}-{}^{\mathbf{B}_b}\omega_z\right)\left(\frac{\rho}{S_x}I_z-\widehat{\frac{\rho}{S_x}I_z}\right) \end{bmatrix}^T$$

$$\times\left({}^{\mathbf{B}_b}\mathbf{R}_\mathbf{I}\frac{\partial}{\partial t}\boldsymbol{\omega}_{br}-{}^{\mathbf{B}_b}\mathbf{R}_\mathbf{I}(\boldsymbol{\omega}_b\times)\boldsymbol{\omega}_{br}\right)dx$$

$$+\int_0^l \begin{bmatrix} {}^{\mathbf{B}_b}\omega_{xr}-{}^{\mathbf{B}_b}\omega_x \\ {}^{\mathbf{B}_b}\omega_{yr}-{}^{\mathbf{B}_b}\omega_y \\ {}^{\mathbf{B}_b}\omega_{zr}-{}^{\mathbf{B}_b}\omega_z \end{bmatrix}^T$$

$$\times\begin{bmatrix} \left(\frac{\rho}{S_x}I_z-\widehat{\frac{\rho}{S_x}I_z}\right){}^{\mathbf{B}_b}\omega_y{}^{\mathbf{B}_b}\omega_{zr}-\left(\frac{\rho}{S_x}I_y-\widehat{\frac{\rho}{S_x}I_y}\right){}^{\mathbf{B}_b}\omega_z{}^{\mathbf{B}_b}\omega_{yr} \\ \left(\frac{\rho}{S_x}I_y-\widehat{\frac{\rho}{S_x}I_y}\right){}^{\mathbf{B}_b}\omega_z{}^{\mathbf{B}_b}\omega_{xr} \\ -\left(\frac{\rho}{S_x}I_z-\widehat{\frac{\rho}{S_x}I_z}\right){}^{\mathbf{B}_b}\omega_y{}^{\mathbf{B}_b}\omega_{xr} \end{bmatrix}dx$$

$$-\int_0^l \begin{bmatrix} {}^{\mathbf{B}_b}\omega_{xr}-{}^{\mathbf{B}_b}\omega_x \\ {}^{\mathbf{B}_b}\omega_{yr}-{}^{\mathbf{B}_b}\omega_y \\ {}^{\mathbf{B}_b}\omega_{zr}-{}^{\mathbf{B}_b}\omega_z \end{bmatrix}^T \begin{bmatrix} \left(GJ_x-\widehat{GJ_x}\right)\frac{\partial^2}{\partial x^2}\Theta_{xr}(x,t) \\ -\left(EI_y-\widehat{EI_y}\right)\alpha_{yr}(x,t)\frac{\partial^3 z_{br}(x,t)}{\partial x^3} \\ \left(EI_z-\widehat{EI_z}\right)\alpha_{zr}(x,t)\frac{\partial^3 y_{br}(x,t)}{\partial x^3} \end{bmatrix}dx$$

$$=-\int_0^l \frac{\partial}{\partial t}\left[\mathbf{r}_{br}(x,t)-\mathbf{r}_b(x,t)\right]^T\left({}^{\mathbf{B}_b}\mathbf{R}_\mathbf{I}^T\mathbf{K}_p{}^{\mathbf{B}_b}\mathbf{R}_\mathbf{I}\right)$$

$$\times\frac{\partial}{\partial t}\left[\mathbf{r}_{br}(x,t)-\mathbf{r}_b(x,t)\right]dx$$

$$-\int_0^l(\boldsymbol{\omega}_{br}-\boldsymbol{\omega}_b)^T\left({}^{\mathbf{B}_b}\mathbf{R}_\mathbf{I}^T\mathbf{K}_\omega{}^{\mathbf{B}_b}\mathbf{R}_\mathbf{I}\right)(\boldsymbol{\omega}_{br}-\boldsymbol{\omega}_b)\,dx$$

$$-ES_x\lambda_x\int_0^l\left[\frac{\partial}{\partial x}\left(u_{xr}(x,t)-u_x(x,t)\right)\right]^2dx$$

$$-EI_z\lambda_y\int_0^l\left[\frac{\partial^2}{\partial x^2}\left(y_{br}(x,t)-y_b(x,t)\right)\right]^2dx$$

$$
\begin{aligned}
&-EI_y\lambda_z \int_0^l \left[\frac{\partial^2}{\partial x^2}\left(z_{br}(x,t) - z_b(x,t)\right)\right]^2 dx \\
&-GJ_x\lambda_\Theta \int_0^l \left[\frac{\partial}{\partial x}\left(\Theta_{xr}(x,t) - \Theta_x(x,t)\right)\right]^2 dx \\
&+ \left({}^{\mathbf{B}_b}v_{bxr} - {}^{\mathbf{B}_b}v_{bx}\right)\left(\widehat{ES_x}\frac{\partial}{\partial x}u_{xr}(x,t) - ES_x\frac{\partial}{\partial x}u_x(x,t)\right)\Bigg|_0^l \\
&- \left({}^{\mathbf{B}_b}v_{byr} - {}^{\mathbf{B}_b}v_{by}\right)\left[\widehat{EI_z}(1-\alpha_{zr}(x,t))\frac{\partial^3 y_{br}(x,t)}{\partial x^3}\right. \\
&\qquad \left. -EI_z(1-\alpha_z(x,t))\frac{\partial^3 y_b(x,t)}{\partial x^3}\right]\Bigg|_0^l \\
&- \left({}^{\mathbf{B}_b}v_{bzr} - {}^{\mathbf{B}_b}v_{bz}\right)\left[\widehat{EI_y}(1-\alpha_{yr}(x,t))\frac{\partial^3 z_{br}(x,t)}{\partial x^3}\right. \\
&\qquad \left. -EI_y(1-\alpha_y(x,t))\frac{\partial^3 z_b(x,t)}{\partial x^3}\right]\Bigg|_0^l \\
&+ \left({}^{\mathbf{B}_b}\omega_{xr} - {}^{\mathbf{B}_b}\omega_x\right)\left(\widehat{GJ_x}\frac{\partial}{\partial x}\Theta_{xr}(x,t) - GJ_x\frac{\partial}{\partial x}\Theta_x(x,t)\right)\Bigg|_0^l \\
&- \frac{\partial}{\partial x}\left({}^{\mathbf{B}_b}v_{bzr} - {}^{\mathbf{B}_b}v_{bz}\right)\left(-\widehat{EI_y}\frac{\partial^2 z_{br}(x,t)}{\partial x^2} - (-EI_y)\frac{\partial^2 z_b(x,t)}{\partial x^2}\right)\Bigg|_0^l \\
&+ \frac{\partial}{\partial x}\left({}^{\mathbf{B}_b}v_{byr} - {}^{\mathbf{B}_b}v_{by}\right)\left(\widehat{EI_z}\frac{\partial^2 y_{br}(x,t)}{\partial x^2} - EI_z\frac{\partial^2 y_b(x,t)}{\partial x^2}\right)\Bigg|_0^l \\
&+ \left(ES_x - \widehat{ES_x}\right)\left({}^{\mathbf{B}_b}v_{bxr} - {}^{\mathbf{B}_b}v_{bx}\right)\frac{\partial}{\partial x}u_{xr}(x,t)\Bigg|_0^l \\
&- \left(EI_z - \widehat{EI_z}\right)\left({}^{\mathbf{B}_b}v_{byr} - {}^{\mathbf{B}_b}v_{by}\right)(1-\alpha_{zr}(x,t))\frac{\partial^3 y_{br}(x,t)}{\partial x^3}\Bigg|_0^l \\
&- \left(EI_y - \widehat{EI_y}\right)\left({}^{\mathbf{B}_b}v_{bzr} - {}^{\mathbf{B}_b}v_{bz}\right)(1-\alpha_{yr}(x,t))\frac{\partial^3 z_{br}(x,t)}{\partial x^3}\Bigg|_0^l \\
&+ \left(GJ_x - \widehat{GJ_x}\right)\left({}^{\mathbf{B}_b}\omega_{xr} - {}^{\mathbf{B}_b}\omega_x\right)\frac{\partial}{\partial x}\Theta_{xr}(x,t)\Bigg|_0^l \\
&+ \left(EI_y - \widehat{EI_y}\right)\frac{\partial}{\partial x}\left({}^{\mathbf{B}_b}v_{bzr} - {}^{\mathbf{B}_b}v_{bz}\right)\frac{\partial^2 z_{br}(x,t)}{\partial x^2}\Bigg|_0^l \\
&+ \left(EI_z - \widehat{EI_z}\right)\frac{\partial}{\partial x}\left({}^{\mathbf{B}_b}v_{byr} - {}^{\mathbf{B}_b}v_{by}\right)\frac{\partial^2 y_{br}(x,t)}{\partial x^2}\Bigg|_0^l \\
&+ \int_0^l \left[{}^{\mathbf{B}_b}\mathbf{R}_{\mathbf{I}}\frac{\partial}{\partial t}\left(\mathbf{r}_{br}(x,t) - \mathbf{r}_b(x,t)\right)\right]^T (\rho - \hat{\rho})\, {}^{\mathbf{B}_b}\mathbf{R}_{\mathbf{I}}\frac{\partial^2}{\partial t^2}\mathbf{r}_{br}(x,t)dx
\end{aligned}
$$

$$+\int_0^l \begin{bmatrix} {}^{\mathbf{B}_b}v_{bxr} - {}^{\mathbf{B}_b}v_{bx} \\ {}^{\mathbf{B}_b}v_{byr} - {}^{\mathbf{B}_b}v_{by} \\ {}^{\mathbf{B}_b}v_{bzr} - {}^{\mathbf{B}_b}v_{bz} \end{bmatrix}^T \times \begin{bmatrix} -\left(ES_x - \widehat{ES_x}\right)\frac{\partial^2}{\partial x^2}u_{xr}(x,t) \\ \left(EI_z - \widehat{EI_z}\right)\frac{\partial}{\partial x}\left((1-\alpha_{zr}(x,t))\frac{\partial^3 y_{br}(x,t)}{\partial x^3}\right) \\ \left(EI_y - \widehat{EI_y}\right)\frac{\partial}{\partial x}\left((1-\alpha_{yr}(x,t))\frac{\partial^3 z_{br}(x,t)}{\partial x^3}\right) \end{bmatrix} dx$$

$$+\int_0^l \begin{bmatrix} \left({}^{\mathbf{B}_b}\omega_{xr} - {}^{\mathbf{B}_b}\omega_x\right)\left(\frac{\rho}{S_x}J_x - \widehat{\frac{\rho}{S_x}J_x}\right) \\ \left({}^{\mathbf{B}_b}\omega_{yr} - {}^{\mathbf{B}_b}\omega_y\right)\left(\frac{\rho}{S_x}I_y - \widehat{\frac{\rho}{S_x}I_y}\right) \\ \left({}^{\mathbf{B}_b}\omega_{zr} - {}^{\mathbf{B}_b}\omega_z\right)\left(\frac{\rho}{S_x}I_z - \widehat{\frac{\rho}{S_x}I_z}\right) \end{bmatrix}^T \times \left({}^{\mathbf{B}_b}\mathbf{R}_\mathbf{I}\frac{\partial}{\partial t}\boldsymbol{\omega}_{br} - {}^{\mathbf{B}_b}\mathbf{R}_\mathbf{I}(\boldsymbol{\omega}_b\times)\boldsymbol{\omega}_{br}\right) dx$$

$$+\int_0^l \begin{bmatrix} {}^{\mathbf{B}_b}\omega_{xr} - {}^{\mathbf{B}_b}\omega_x \\ {}^{\mathbf{B}_b}\omega_{yr} - {}^{\mathbf{B}_b}\omega_y \\ {}^{\mathbf{B}_b}\omega_{zr} - {}^{\mathbf{B}_b}\omega_z \end{bmatrix}^T \times \begin{bmatrix} \left(\frac{\rho}{S_x}I_z - \widehat{\frac{\rho}{S_x}I_z}\right){}^{\mathbf{B}_b}\omega_y{}^{\mathbf{B}_b}\omega_{zr} - \left(\frac{\rho}{S_x}I_y - \widehat{\frac{\rho}{S_x}I_y}\right){}^{\mathbf{B}_b}\omega_z{}^{\mathbf{B}_b}\omega_{yr} \\ \left(\frac{\rho}{S_x}I_y - \widehat{\frac{\rho}{S_x}I_y}\right){}^{\mathbf{B}_b}\omega_z{}^{\mathbf{B}_b}\omega_{xr} \\ -\left(\frac{\rho}{S_x}I_z - \widehat{\frac{\rho}{S_x}I_z}\right){}^{\mathbf{B}_b}\omega_y{}^{\mathbf{B}_b}\omega_{xr} \end{bmatrix} dx$$

$$-\int_0^l \begin{bmatrix} {}^{\mathbf{B}_b}\omega_{xr} - {}^{\mathbf{B}_b}\omega_x \\ {}^{\mathbf{B}_b}\omega_{yr} - {}^{\mathbf{B}_b}\omega_y \\ {}^{\mathbf{B}_b}\omega_{zr} - {}^{\mathbf{B}_b}\omega_z \end{bmatrix}^T \begin{bmatrix} \left(GJ_x - \widehat{GJ_x}\right)\frac{\partial^2}{\partial x^2}\Theta_{xr}(x,t) \\ -\left(EI_y - \widehat{EI_y}\right)\alpha_{yr}(x,t)\frac{\partial^3 z_{br}(x,t)}{\partial x^3} \\ \left(EI_z - \widehat{EI_z}\right)\alpha_{zr}(x,t)\frac{\partial^3 y_{br}(x,t)}{\partial x^3} \end{bmatrix} dx. \tag{B.77}$$

Finally, define a non-negative accompanying function for the three-dimensional flexible beam with adaptive control as

$$\begin{aligned}
\nu_p = \nu &+ \frac{1}{2}\rho_\rho(\rho - \hat{\rho}(t))^2 + \frac{1}{2}\rho_{ES_x}(ES_x - \widehat{ES_x}(t))^2 \\
&+\frac{1}{2}\rho_{EI_z}(EI_z - \widehat{EI_z}(t))^2 + \frac{1}{2}\rho_{EI_y}(EI_y - \widehat{EI_y}(t))^2 \\
&+\frac{1}{2}\rho_{\frac{\rho}{S_x}J_x}\left(\frac{\rho}{S_x}J_x - \widehat{\frac{\rho}{S_x}J_x}(t)\right)^2 + \frac{1}{2}\rho_{\frac{\rho}{S_x}I_z}\left(\frac{\rho}{S_x}I_z - \widehat{\frac{\rho}{S_x}I_z}(t)\right)^2 \\
&+\frac{1}{2}\rho_{\frac{\rho}{S_x}I_y}\left(\frac{\rho}{S_x}I_y - \widehat{\frac{\rho}{S_x}I_y}(t)\right)^2 + \frac{1}{2}\rho_{GJ_x}(GJ_x - \widehat{GJ_x}(t))^2.
\end{aligned} \tag{B.78}$$

It follows from (13.113)–(13.128), (B.77), (B.78), and Lemma 2.9 that

$$
\begin{aligned}
\dot{\nu}_p \leqslant & -\int_0^l \frac{\partial}{\partial t}\left[\mathbf{r}_{br}(x,t)-\mathbf{r}_b(x,t)\right]^T\left({}^{\mathbf{B}_b}\mathbf{R}_{\mathbf{I}}^T\mathbf{K}_p{}^{\mathbf{B}_b}\mathbf{R}_{\mathbf{I}}\right) \\
& \qquad\times\frac{\partial}{\partial t}\left[\mathbf{r}_{br}(x,t)-\mathbf{r}_b(x,t)\right]dx \\
& -\int_0^l \left(\boldsymbol{\omega}_{br}-\boldsymbol{\omega}_b\right)^T\left({}^{\mathbf{B}_b}\mathbf{R}_{\mathbf{I}}^T\mathbf{K}_\omega{}^{\mathbf{B}_b}\mathbf{R}_{\mathbf{I}}\right)\left(\boldsymbol{\omega}_{br}-\boldsymbol{\omega}_b\right)dx \\
& -ES_x\lambda_x\int_0^l\left[\frac{\partial}{\partial x}\left(u_{xr}(x,t)-u_x(x,t)\right)\right]^2dx \\
& -EI_z\lambda_y\int_0^l\left[\frac{\partial^2}{\partial x^2}\left(y_{br}(x,t)-y_b(x,t)\right)\right]^2dx \\
& -EI_y\lambda_z\int_0^l\left[\frac{\partial^2}{\partial x^2}\left(z_{br}(x,t)-z_b(x,t)\right)\right]^2dx \\
& -GJ_x\lambda_\Theta\int_0^l\left[\frac{\partial}{\partial x}\left(\Theta_{xr}(x,t)-\Theta_x(x,t)\right)\right]^2dx \\
& +p_{\mathbf{B}}-p_{\mathbf{T}}
\end{aligned}
\tag{B.79}
$$

holds with

$$
\begin{aligned}
p_{\mathbf{B}} = & \left[{}^{\mathbf{B}}\mathbf{R}_{\mathbf{I}}\left(\dot{\mathbf{r}}_{\mathbf{B}r}-\dot{\mathbf{r}}_{\mathbf{B}}\right)\right]^T{}^{\mathbf{B}}\mathbf{R}_{\mathbf{I}}\left(\mathbf{f}_{\mathbf{B}r}-\mathbf{f}_{\mathbf{B}}\right) \\
& +\left[{}^{\mathbf{B}}\mathbf{R}_{\mathbf{I}}\left(\boldsymbol{\omega}_{\mathbf{B}r}-\boldsymbol{\omega}_{\mathbf{B}}\right)\right]^T{}^{\mathbf{B}}\mathbf{R}_{\mathbf{I}}\left(\mathbf{m}_{\mathbf{Br}}-\mathbf{m}_{\mathbf{B}}\right) \\
= & -\left({}^{\mathbf{B}_b}v_{bxr}-{}^{\mathbf{B}_b}v_{bx}\right)\left(\widehat{ES}_x\frac{\partial}{\partial x}u_{xr}(x,t)-ES_x\frac{\partial}{\partial x}u_x(x,t)\right)\Bigg|_{x=0} \\
& +\left({}^{\mathbf{B}_b}v_{byr}-{}^{\mathbf{B}_b}v_{by}\right)\Bigg[\widehat{EI}_z(1-\alpha_{zr}(x,t))\frac{\partial^3 y_{br}(x,t)}{\partial x^3} \\
& \qquad -EI_z(1-\alpha_z(x,t))\frac{\partial^3 y_b(x,t)}{\partial x^3}\Bigg]\Bigg|_{x=0} \\
& +\left({}^{\mathbf{B}_b}v_{bzr}-{}^{\mathbf{B}_b}v_{bz}\right)\Bigg[\widehat{EI}_y(1-\alpha_{yr}(x,t))\frac{\partial^3 z_{br}(x,t)}{\partial x^3} \\
& \qquad -EI_y(1-\alpha_y(x,t))\frac{\partial^3 z_b(x,t)}{\partial x^3}\Bigg]\Bigg|_{x=0} \\
& +\left({}^{\mathbf{B}_b}\omega_{xr}-{}^{\mathbf{B}_b}\omega_x\right)\left[-\left(\widehat{GJ}_x\frac{\partial}{\partial x}\Theta_{xr}(x,t)-GJ_x\frac{\partial}{\partial x}\Theta_x(x,t)\right)\right]\Bigg|_{x=0} \\
& -\frac{\partial}{\partial x}\left({}^{\mathbf{B}_b}v_{bzr}-{}^{\mathbf{B}_b}v_{bz}\right)\left(\widehat{EI}_y\frac{\partial^2 z_{br}(x,t)}{\partial x^2}-EI_y\frac{\partial^2 z_b(x,t)}{\partial x^2}\right)\Bigg|_{x=0} \\
& +\frac{\partial}{\partial x}\left({}^{\mathbf{B}_b}v_{byr}-{}^{\mathbf{B}_b}v_{by}\right)\left[-\left(\widehat{EI}_z\frac{\partial^2 y_{br}(x,t)}{\partial x^2}-EI_z\frac{\partial^2 y_b(x,t)}{\partial x^2}\right)\right]\Bigg|_{x=0}
\end{aligned}
\tag{B.80}
$$

$$
\begin{aligned}
p_{\mathbf{T}} = & \left[{}^{\mathbf{T}}\mathbf{R}_{\mathbf{I}}\left(\dot{\mathbf{r}}_{\mathbf{T}r}-\dot{\mathbf{r}}_{\mathbf{T}}\right)\right]^T{}^{\mathbf{T}}\mathbf{R}_{\mathbf{I}}\left(\mathbf{f}_{\mathbf{T}r}-\mathbf{f}_{\mathbf{T}}\right) \\
& +\left[{}^{\mathbf{T}}\mathbf{R}_{\mathbf{I}}\left(\boldsymbol{\omega}_{\mathbf{T}r}-\boldsymbol{\omega}_{\mathbf{T}}\right)\right]^T{}^{\mathbf{T}}\mathbf{R}_{\mathbf{I}}\left(\mathbf{m}_{\mathbf{T}r}-\mathbf{m}_{\mathbf{T}}\right)
\end{aligned}
$$

$$
\begin{aligned}
= & -\left({}^{\mathbf{B}_b}v_{bxr} - {}^{\mathbf{B}_b}v_{bx}\right)\left(\widehat{ES_x}\frac{\partial}{\partial x}u_{xr}(x,t) - ES_x\frac{\partial}{\partial x}u_x(x,t)\right)\Big|_{x=l} \\
& + \left({}^{\mathbf{B}_b}v_{byr} - {}^{\mathbf{B}_b}v_{by}\right)\left[\widehat{EI_z}(1-\alpha_{zr}(x,t))\frac{\partial^3 y_{br}(x,t)}{\partial x^3}\right. \\
& \qquad \left. - EI_z(1-\alpha_z(x,t))\frac{\partial^3 y_b(x,t)}{\partial x^3}\right]\Big|_{x=l} \\
& + \left({}^{\mathbf{B}_b}v_{bzr} - {}^{\mathbf{B}_b}v_{bz}\right)\left[\widehat{EI_y}(1-\alpha_{yr}(x,t))\frac{\partial^3 z_{br}(x,t)}{\partial x^3}\right. \\
& \qquad \left. - EI_y(1-\alpha_y(x,t))\frac{\partial^3 z_b(x,t)}{\partial x^3}\right]\Big|_{x=l} \\
& + \left({}^{\mathbf{B}_b}\omega_{xr} - {}^{\mathbf{B}_b}\omega_{x}\right)\left[-\left(\widehat{GJ_x}\frac{\partial}{\partial x}\Theta_{xr}(x,t) - GJ_x\frac{\partial}{\partial x}\Theta_x(x,t)\right)\right]\Big|_{x=l} \\
& - \frac{\partial}{\partial x}\left({}^{\mathbf{B}_b}v_{bzr} - {}^{\mathbf{B}_b}v_{bz}\right)\left(\widehat{EI_y}\frac{\partial^2 z_{br}(x,t)}{\partial x^2} - EI_y\frac{\partial^2 z_b(x,t)}{\partial x^2}\right)\Big|_{x=l} \\
& + \frac{\partial}{\partial x}\left({}^{\mathbf{B}_b}v_{byr} - {}^{\mathbf{B}_b}v_{by}\right)\left[-\left(\widehat{EI_z}\frac{\partial^2 y_{br}(x,t)}{\partial x^2} - EI_z\frac{\partial^2 y_b(x,t)}{\partial x^2}\right)\right]\Big|_{x=l} .
\end{aligned}
\tag{B.81}
$$

Consider the fact that the flexible beam has one *driving cutting point* associated with frame $\{\mathbf{T}\}$ and one *driven cutting point* associated with frame $\{\mathbf{B}\}$. Using (B.78), (B.79), and Definition 2.17 completes the proof.

References

1. Abbati-Marescotti, A., Bonivento, C., Melchiorri, C.: On the invariance of the hybrid position/force control. J. Intell. Robot. Syst. 3, 233–250 (1990)
2. Alleyne, A., Liu, R.: On the limitations of force tracking control for hydraulic servosystems. ASME J. Dynamic Syst. Measure. Contr. 121(2), 184–190 (1999)
3. Albu-Schaffer, A., Hirzinger, G.: Parameter identification and passivity based joint control for a 7 DOF torque controlled light weight robot. In: Proc. of 2001 IEEE Int. Conf. Robot. Automat., Seoul, Korea, pp. 2852–2858 (2001)
4. An, C.H., Atkeson, C.G., Hollerbach, J.: Model-based control of a robot manipulator. MIT Press, Cambridge (1988)
5. Anderson, R.J., Spong, M.W.: Bilateral control of teleoperators with time delay. IEEE Trans. Automat. Contr. 34(5), 494–501 (1989)
6. Anderson, R.J., Spong, M.W.: Asymptotic stability for force reflecting teleoperators with time delay. Int. J. Robot. Res. 11(2), 135–149 (1992)
7. Angeles, J.: Fundamentals of robotic mechanical systems: theory, methods, and algorithms. Springer, Heidelberg (2003)
8. Antonelli, G.: Underwater robots: motion and force control of vehicle-manipulator. Springer, Heidelberg (2003)
9. Arimoto, S., Miyazaki, F., Kawamura, S.: Cooperative motion control of multiple robot arms or fingers. In: Proc. of 1987 IEEE Int. Conf. Robot. Automat., pp. 1407–1412 (1987)
10. Arimoto, S.: Fundamental problems of robot control: part 1: innovation in the realm of robot servo-loops, and part 2: a nonlinear circuit theory toward an understanding of dexterous motions. Robotica 13, 19–27, 111–122 (1995)
11. Asada, H., Slotine, J.J.E.: Robot analysis and control. John Wiley & Sons, Chichester (1986)
12. Bejczy, A.K.: Teleoperators. IEEE Industrial Electronics Society Newsletter, 4–12 (1996)
13. Bessonnet, G., Chesse, S., Sardain, P.: Optimal gait synthesis of a seven-link planar biped. Int. J. Robot. Res. 23(10-11), 1059–1073 (2004)
14. Bonitz, R.G., Hsia, T.C.: Internal force-based impedance control for cooperating manipulators. IEEE Trans. Robot. Automat. 12(1), 78–89 (1996)
15. Bridges, M.M., Dawson, D.M., Qu, Z., Martindale, S.C.: Robust control of rigid-link flexible-joint robots with redundant joint actuators. IEEE Trans. Syst. Man Cybern. 24(7), 961–970 (1994)

16. Brogliato, B., Ortega, R., Lozano, R.: Global tracking controllers for flexible-joint manipulators: a comparative study. Automatica 31(7), 941–956 (1995)
17. Bu, F., Yao, B.: Desired compensation adaptive robust control of single-rod electro-hydraulic actuator. In: Proc. of 2001 American Control Conference, Washington, DC, pp. 3926–3931 (2001)
18. Butler, Z., Kotay, K., Rus, D., Tomita, K.: Generic decentralized control for a class of self-reconfigurable robots. In: Proc. of 2002 IEEE Int. Conf. Robot. Automat., pp. 809–816 (2002)
19. Caballero, R., Armada, M.A., Akinfiev, T.: Robust cascade controller for nonlinearly actuated biped robots: experimental evaluation. Int. J. Robot. Res. 23(10-11), 1075–1095 (2004)
20. Caccavale, F., Chiacchio, P., Chiaverini, S.: Stability analysis of a joint space control law for a two-manipulator system. IEEE Trans. Automat. Contr. 44(1), 85–88 (1999)
21. Canudas de Wit, C., Siciliano, B., Bastin, G.: Theory of robot control. Springer, Berlin (1996)
22. Canudas de Wit, C., Brogliato, B.: Direct adaptive impedance control including transition phases. Automatica 33(4), 643–649 (1997)
23. Chang, Y., Daniel, R.W.: On the adaptive control of flexible joint robots. Automatica 28(5), 969–974 (1992)
24. Chartrand, G.: Directed graphs as mathematical models. In: Introductory graph theory, pp. 16–19. Courier Dover Publications, New York (1985)
25. Chen, I.M., Burdick, J.W.: Determining task optimal modular robot assembly configurations. In: Proc. of 1995 IEEE Int. Conf. Robot. Automat., pp. 132–137 (1995)
26. Cheng, F.T., Orin, D.: Optimal force distribution in multiple-chain robotic systems. IEEE Trans. Syst. Man Cybern. 21, 13–24 (1991)
27. Chevallereau, C., Aoustin, Y.: Optimal reference trajectories for walking and running a biped robot. Robotica 19(5), 557–569 (2001)
28. Chevallereau, C., Abba, G., Aoustin, Y., Plestan, F., Westervelt, E.R., Canudas de Wit, C., Grizzle, J.W.: RABBIT: a testbed for advanced control theory. IEEE Control Systems Magazine 23, 57–79 (2003)
29. Chevallereau, C.: Time-scaling control for an underactuated biped robot. IEEE Trans. Robot. Automat. 19(2), 362–368 (2003)
30. Chiacchio, P., Chiaverini, S., Sciavicco, L., Siciliano, B.: Global task space manipulability ellipsoids for multiple-arm systems. IEEE Trans. Robot. Automat. 7, 678–685 (1991)
31. Chiaverini, S., Sciavicco, L.: The parallel approach to force/position control of robotic manipulators. IEEE Trans. Robot. Automat. 9(4), 361–373 (1993)
32. Chou, J.C.: Quaternion kinematic and dynamic differential equations. IEEE Trans. Robot. Automat. 8(1), 53–64 (1992)
33. Chirikjian, G.S., Zhou, Y., Suthakorn, J.: Self-replicating robots for lunar development. IEEE Trans. Mechatronics 7(4), 462–472 (2002)
34. Christensen, D.J., Stoy, K.: Selecting a meta-module to shape-change the ATRON self-reconfigurable robot. In: Proc. of 2006 Int. Conf. Robot. Automat., Orlando, FL, pp. 2532–2538 (2006)
35. Chung, W.K., Han, J., Youm, Y., Kim, S.H.: Task based design of modular robot manipulator using efficient genetic algorithm. In: Proc. of 1997 IEEE Int. Conf. Robot. Automat., Albuquerque, NM, pp. 507–512 (1997)
36. Colgate, J.E.: Robust impedance shaping telemanipulation. IEEE Trans. Robot. Automat. 9(4), 374–384 (1993)

37. Collins, S.H., Wisse, M., Ruina, A.: A three-dimensional passive-dynamic walking robot with two legs and knees. Int. J. Robot. Res. 20(7), 607–615 (2001)
38. Cooke, J.D.: Dependence of human arm movements on limb mechanical properties. Brain Res. 165, 366–369 (1979)
39. Craig, J.J.: Introduction to robotics: mechanics and control. Addison-Wesley, Reading (1986)
40. Dauchez, P., Fournier, A., Jourdan, R.: Hybrid control of a two-arm robot for complex tasks. Robot. Autonomous Syst. 5, 323–332 (1989)
41. Dawson, D.M., Qu, Z., Carrol, J.J.: Tracking control of rigid-link electrically-driven robot manipulators. Int. J. Control. 56, 991–1006 (1992)
42. De Luca, A., Siciliano, B.: Inversion-based nonlinear control of robot arms with flexible links. AIAA J. Guidance Control and Dynamics 16(6), 1169–1176 (1993)
43. De Luca, A., Manes, C.: Modeling of robots in contact with a dynamic environment. IEEE Trans. Robot. Automat. 10(3), 542–548 (1994)
44. De Luca, A., Farina, R., Lucibello, P.: On the control of robots with visco-elastic joints. In: Proc. of 2005 IEEE Int. Conf. Robot. Automat., Barcelona, Spain, pp. 4308–4313 (2005)
45. De Queiroz, M.S., Dawson, D.M., Agarwal, M., Zhang, F.: Adaptive nonlinear boundary control of a flexible link robot arm. IEEE Trans. Robot. Automat. 15(4), 779–787 (1999)
46. De Schutter, J., Van Brussel, H.: Compliant robot motion. Int. J. Robot. Res. 7(4), 3–33 (1988)
47. De Schutter, J., Bruyninckx, H., Demey, S., Dutre, S., De Geeter, J., Katupitiya, J.: Force control. In: Melchiorri, C., Tornambe, A. (eds.) Modelling and control of mechanisms and robots, pp. 227–264. World Scientific, Singapore (1996)
48. De Schutter, J., Torfs, D., Bruyninckx, H., Dutre, S.: Invariant hybrid force/position control of a velocity controlled robot with compliant end effector using modal decoupling. Int. J. Robot. Res. 16(3), 340–356 (1997)
49. Dhaouadi, R., Ghorbel, F.H., Gandhi, P.S.: A new dynamic model of hysteresis in harmonic drives. IEEE Trans. Ind. Electron. 50(6), 1165–1171 (2003)
50. Dickson, W.C., Cannon Jr., R.H.: Experimental results of two free-flying robots capturing and manipulating a free-flying object. In: Proc. of IEEE/IROS 1995, pp. 51–58 (1995)
51. Dubowsky, S., Papadopoulos, E.: The kinematics, dynamics, and control of free-flying and free-floating space robotic systems. IEEE Trans. Robot. Automat. 9(5), 531–543 (1993)
52. Egeland, O., Sagli, J.R.: Coordination of motion in a spacecraft/manipulator system. Int. J. Rob. Res. 12(4), 366–379 (1993)
53. Eryilmaz, B., Wilson, B.H.: Improved tracking control of hydraulic systems. ASME J. Dynamic Syst. Measure. Contr. 123(3), 457–462 (2001)
54. Fu, K.S., Gonzalez, R.C., Lee, C.S.G.: Robotics: control, sensing, vision, and intelligence. McGraw-Hill, New York (1987)
55. Fukuda, T., Kawauchi, Y.: Cellular robotic system (CEBOT) as one of the realization of self-organizing intelligent universal manipulator. In: Proc. of 1990 IEEE Int. Conf. Robot. Automat., pp. 662–667 (1990)
56. Funda, J., Taylor, R.H., Eldridge, B., Gomory, S., Gruben, K.G.: Constrained Cartesian motion control for teleoperated surgical robots. IEEE Trans. Robot. Automat. 12(3), 453–465 (1996)
57. Godler, I., Horiuchi, M., Hashimoto, M., Ninomiya, T.: Accuracy improvement of built-in torque sensing for harmonic drives. IEEE/ASME Trans. Mechatronics 5(4), 360–366 (2000)

58. Goldenberg, A.A., Wiercienski, J., Kuzan, P., Szymczyk, C., Fenton, R.G., Shaver, B.: A remote manipulator for forestry operation. IEEE Trans. Robot. Automat. 11(2), 185–197 (1995)
59. Good, M.C., Sweet, L.M., Strbel, K.L.: Dynamic models for control system design of integrated robot and drive systems. ASME J. Dynamic Syst. Measure. Contr. 107(1), 53–59 (1985)
60. Gorinevsky, D.M., Formalsky, A.M., Schneider, A.Y.: Force control of robotics systems. CRC Press, Boca Raton (1997)
61. Gray, A.: Euclidean spaces. In: Modern differential geometry of curves and surfaces with mathematica, 2nd edn., pp. 2–5. CRC Press, Boca Raton (1997)
62. Grizzle, J.W., Abba, G., Plestan, F.: Asymptotically stable walking for biped robots: analysis via systems with impulse effects. IEEE Trans. Automat. Contr. 46(1), 51–64 (2001)
63. Gupta, K.C.: Mechanics and control of robots. Springer, New York (1997)
64. Hannaford, B.: A design framework for teleoperators with kinesthetic feedback. IEEE Trans. Robot. Automat. 5(4), 426–434 (1989)
65. Hannaford, B., Wood, L., McAffee, D.A., Zak, H.: Performance evaluation of a six-axis generalized force-reflecting teleoperator. IEEE Trans. Syst. Man Cybern. 21(3), 620–633 (1991)
66. Hashimoto, M.: Robot motion control based on joint torque sensing. In: Proc. of 1989 IEEE Int. Conf. Robot. Automat., pp. 256–261 (1989)
67. Hashimoto, M., Kiyosawa, Y., Paul, R.P.: A torque sensing technique for robots with harmonic drives. IEEE Trans. Robot. Automat. 9(1), 108–116 (1993)
68. Hayati, S.: Hybrid position/force control of multi-arm cooperating robots. In: Proc. of 1986 IEEE Int. Conf. Robot. Automat., pp. 82–89 (1986)
69. Hirai, K., Hirose, M., Haikawa, Y., Takenaka, T.: The development of Honda humanoid robot. In: Proc. of 1998 IEEE Int. Conf. Robot. Automat., pp. 1321–1326 (1998)
70. Hirzinger, G., Brunner, B., Dietrich, J., Heindl, J.: Sensor-based space robotics – ROTEX and its telerobotic features. IEEE Trans. Robot. Automat. 9(5), 649–663 (1993)
71. Hogan, N.: Impedance control: an approach to manipulation, parts 1-3. ASME J. Dynamic Syst. Measure. Contr. 107, 1–24 (1985)
72. Honegger, M., Corke, P.: Model-based control of hydraulically actuated manipulators. In: Proc. of 2001 IEEE Int. Conf. Robot. Automat., Seoul, Korea, pp. 2553–2559 (2001)
73. Hu, Y.R., Goldenberg, A.A.: Dynamic control of coordinated redundant robots with torque optimization. Automatica 29(6), 1411–1424 (1993)
74. Hu, Y.R., Goldenberg, A.A.: An adaptive approach to motion and force control of multiple coordinated robots. ASME J. Dynamic Syst. Measure. Contr. 115(1), 60–69 (1993)
75. Hsu, P.: Coordinated control of multiple manipulator systems. IEEE Trans. Robot. Automat. 9(4), 400–410 (1993)
76. Inaba, N., Oda, M.: Autonomous satellite capture by a space robot. In: Proc. of 2000 Int. Conf. Robot. Automat., San Francisco, CA, pp. 1169–1174 (2000)
77. Jain, A., Rodriguez, G.: An analysis of the kinematics and dynamics of underactuated manipulators. IEEE Trans. Robot. Automat. 9(5), 411–422 (1993)
78. Jankowski, K.P., Van Brussel, H.: An approach to discrete inverse dynamics control of flexible joint robots. IEEE Trans. Robot. Automat. 8(5), 651–658 (1992)
79. Jean, J.H., Fu, L.C.: Adaptive hybrid control strategies for constrained robots. IEEE Trans. Automat. Contr. 38(4), 598–603 (1993)

80. Jean, J.H., Fu, L.C.: An adaptive control scheme for coordinated multi-manipulator systems. IEEE Trans. Robot. Automat. 9, 226–231 (1993)
81. Junkins, J., Kim, Y.: Introduction to dynamics and control of flexible structure. In: AIAA Education Series, Washington, DC (1993)
82. Kanoh, H., Tzafestas, S., Lee, H.G., Kalat, J.: Modelling and control of flexible robot arms. In: Proc. of 25th IEEE Conference on Decision and Control, pp. 1866–1870 (1986)
83. Kazerooni, H., Tsay, T.I., Hollerbach, K.: A controller design framework for telerobotic systems. IEEE Trans. Control Systems Technology 1(1), 50–62 (1993)
84. Kazerooni, H., Her, M.G.: The dynamics and control of a haptic interface device. IEEE Trans. Robot. Automat. 10(4), 453–464 (1994)
85. Kelly, R., Ortega, R., Ailon, A., Loria, A.: Global regulation of flexible joint robots using approximate differentiation. IEEE Trans. Automat. Contr. 39(6), 1222–1224 (1994)
86. Khadem, S.E., Pirmohammadi, A.A.: Analytical development of dynamic equations of motion for a three-dimensional flexible link manipulator with revolute and prismatic joints. IEEE Trans. Syst. Man Cybern. B 33(2), 237–249 (2003)
87. Khalil, H.K.: Nonlinear systems, 2nd edn. Prentice Hall, Englewood Cliffs (1996)
88. Khatib, O.: A unified approach for motion and force control of robot manipulators: the operational space formulation. IEEE J. of Robot. Automat. 3(1), 43–53 (1987)
89. Khorasani, K.: Adaptive control of flexible joint robots. In: Proc. of 1991 IEEE Int. Conf. Robot. Automat., pp. 2127–2134 (1991)
90. Kircanski, N.M., Goldenberg, A.A.: An experimental study of nonlinear stiffness, hysteresis, and friction effects in robot joints with harmonic drives and torque sensors. Int. J. Robot. Res. 16(2), 214–239 (1997)
91. Koga, M., Kosuge, K., Furuta, K., Nosaki, K.: Coordinated motion control of robot arms based on the virtual internal model. IEEE Trans. Robot. Automat. 8, 77–85 (1992)
92. Koivo, A.J.: Fundamentals for control of robot manipulators. John Wiley & Sons, Chichester (1989)
93. Koivo, A.J., Unseren, M.A.: Reduced order model and decoupled control architecture for two manipulators holding a rigid object. ASME J. Dynamic Syst. Measure. Contr. 113, 646–654 (1991)
94. Kosuge, K., Ishikawa, J., Furuta, K., Sakai, M.: Control of single-master multi-slave manipulator system using VIM. In: Proc. of 1990 IEEE Int. Conf. Robot. Automat., pp. 1172–1177 (1990)
95. Kosuge, K., Takeuchi, H., Furuta, K.: Motion control of a robot arm using joint torque sensors. IEEE Trans. Robot. Automat. 6(2), 258–263 (1990)
96. Kreutz, K., Lokshin, A.: Load balancing and closed chain multiple arm control. In: Proc. of 1988 American Contr. Conf., pp. 2148–2155 (1988)
97. Kuo, A.D.: Stabilization of lateral motion in passive dynamic walking. Int. J. Robot. Res. 18(9), 917–930 (1999)
98. Lawrence, D.A.: Stability and transparency in bilateral teleoperation. IEEE Trans. Robot. Automat. 9(5), 624–637 (1993)
99. Lee, S., Lee, H.S.: Modeling, design, and evaluation of advanced teleoperator control systems with short time delay. IEEE Trans. Robot. Automat. 9(5), 607–623 (1993)
100. Lewis, F.L., Dawson, D.M., Abdallah, C.T.: Robot manipulator control: theory and practice. Marcel Dekker Publishing Company, New York (2004)

101. Leung, G.M.H., Francis, B.A., Apkarian, J.: Bilateral controller for teleoperators with time delay via μ-synthesis. IEEE Trans. Robot. Automat. 11(1), 105–116 (1995)
102. Liu, G., Goldenberg, A.A.: Robust control of robot manipulators incorporating motor dynamics. In: Proc. of IEEE/IROS 1993, pp. 68–75 (1993)
103. Liu, G., Abdul, S., Goldenberg, A.A.: Distributed control of modular and reconfigurable robot with torque sensing. Robotica 26(1), 75–84 (2008)
104. Liu, R., Alleyne, A.: Nonlinear force/pressure tracking of an electro-hydraulic actuator. ASME J. Dynamic Syst. Measure. Contr. 122(1), 232–237 (2000)
105. Liu, Y.H., Arimoto, S., Kitagaki, K.: Adaptive control for holonomically constrained robots: time-invariant and time-variant cases. In: Proc. of 1995 IEEE Int. Conf. Robot. Automat., pp. 905–912 (1995)
106. Liu, Y.H., Xu, Y., Bergerman, M.: Cooperation control of multiple manipulators with passive joints. IEEE Trans. Robot. Automat. 15(2), 258–267 (1999)
107. Lohmeier, S., Buschmann, T., Ulbrich, H., Pfeiffer, F.: Modular joint design for performance enhanced humanoid robot LOLA. In: Proc. of 2006 IEEE Int. Conf. Robot. Automat., Orlando, FL, pp. 88–93 (2006)
108. Longman, R.W., Lindberg, R.E., Zedd, M.F.: Satellite-mounted robot manipulators - new kinematics and reaction moment compensation. Int. J. Robot. Res. 6(3), 87–103 (1987)
109. Lozano, R., Brogliato, B.: Adaptive control of robot manipulators with flexible joints. IEEE Trans. Automat. Contr. 37(2), 174–181 (1992)
110. Luh, J.Y.S., Walker, M.W., Paul, R.P.C.: On-line computational scheme for mechanical manipulators. ASME J. Dynamic Syst. Measure. Contr. 102, 69–76 (1980)
111. Luh, J.Y.S., Zheng, Y.F.: Constrained relations between two coordinated industrial robots for motion control. Int. J. Robot. Res. 6(3), 60–70 (1987)
112. Luo, Z.H.: Direct strain feedback control of flexible robot arms: new theoretical and experimental results. IEEE Trans. Automat. Contr. 38(11), 1610–1622 (1993)
113. Luo, Z.H., Guo, B.Z., Morgul, O.: Stability and stabilization of infinite dimensional systems with applications. Springer, London (1999)
114. Macchelli, A., Melchiorri, C., Stramigioli, S.: Port-based modeling of a flexible link. IEEE Trans. Robot. Automat. 23(4), 650–660 (2007)
115. Marino, R., Nicosia, S.: Singular perturbation techniques in the adaptive control of elastic robots. In: Proc. of First IFAC Symp. Robot Control (1985)
116. Mattila, J., Virvalo, T.: Energy-efficient motion control of a hydraulic manipulator. In: Proc. 2000 IEEE Int. Conf. Robot. Automat., San Francisco, CA, pp. 3000–3006 (2000)
117. McClamroch, N.H., Wang, D.W.: Feedback stabilization and tracking of constrained robots. IEEE Trans. Automat. Contr. 33, 419–426 (1988)
118. McGeer, T.: Passive dynamic walking. Int. J. Robot. Res. 9(2), 62–82 (1990)
119. Meirovitch, L.: Principles and techniques of vibrations. Prentice-Hall, Englewood Cliffs (1997)
120. Merritt, H.E.: Hydraulic control systems. Wiley, New York (1976)
121. Miles, E.S., Cannon Jr., R.H.: Utilizing human vision and computer vision to direct a robot in a semi-structured environment via task-level commands. In: Proc. of IEEE/IROS 1995, pp. 366–371 (1995)
122. Mrad, F., Ahmad, S.: Control of flexible joint robots. In: Proc. of 1991 IEEE Int. Conf. Robot. Automat., pp. 2832–2837 (1991)

123. Murata, S., Yoshida, E., Kamimura, A., Kurokawa, H., Tomita, K., Kokaji, S.: M-TRAN: self-reconfigurable modular robotic system. IEEE/ASME Trans. Mechatronics 7(4), 431–441 (2002)
124. Murphy, S.H., Wen, J.T., Saridis, G.N.: Simulation of cooperating robot manipulators on a mobile platform. IEEE Trans. Robot. Automat. 7(4), 468–477 (1991)
125. Nakamura, Y., Nagai, K., Yoshikawa, T.: Dynamics and stability in coordination of multiple robotic mechanisms. Int. J. Robot. Res. 8(2), 44–61 (1989)
126. Nakamura, Y., Mukherjee, R.: Exploiting nonholonomic redundancy of free-flying space robots. IEEE Trans. Robot. Automat. 9(4), 499–506 (1993)
127. Namvar, M., Aghili, F.: A combined scheme for identification and robust torque control of hydraulic actuators. ASME J. Dynamic Syst. Measure. Contr. 125(4), 595–606 (2003)
128. Natale, C.: Interaction control of robot manipulators - six-degrees-of-freedom tasks. Springer, Heidelberg (2003)
129. Nicosia, S., Tomei, P.: A tracking controller for flexible joint robots using only link position feedback. IEEE Trans. Automat. Contr. 40(5), 885–890 (1995)
130. Niksefat, N., Sepehri, N.: Robust force controller design for a hydraulic actuator based on experimental input-output data. In: Proc. of 1999 American Contr. Conf., San Diego, CA, pp. 3718–3722 (1999)
131. Niksefat, N., Wu, C.Q., Sepehri, N.: Design of a Lyapunov controller for an electro-hydraulic actuator during contact tasks. ASME J. Dynamic Syst. Measure. Contr. 123(2), 299–307 (2001)
132. Nof, S.Y. (ed.): Handbook of industrial robotics, 2nd edn. John Wiley & Sons, Chichester (1999)
133. Oda, M.: Coordinated control of spacecraft attitude and its manipulator. In: Proc. of 1996 IEEE Int. Conf. Robot. Automat., pp. 732–738 (1996)
134. Orin, D.E., Oh, S.Y.: Control of force distribution in robotic mechanisms containing closed kinematic chains. ASME J. Dynamic Syst. Measure. Contr. 102, 134–141 (1981)
135. Papadopoulos, E., Dubowsky, S.: On the nature of control algorithms for free-floating space manipulators. IEEE Trans. Robot. Automat. 7(6), 750–758 (1991)
136. Papadopoulos, E., Dubowsky, S.: Dynamic singularities in free-floating space manipulators. ASME J. Dynamic Syst. Measure. Contr. 115, 44–52 (1993)
137. Paredis, C.J.J., Khosla, P.K.: Kinematic design of serial link manipulators from task specifications. Int. J. Robot. Res. 12(3), 274–287 (1993)
138. Paul, R.P.: Robot manipulator: mathematics, programming, and control. MIT Press, Cambridge (1981)
139. Qu, Z.: Input-output robust tracking control design for flexible joint robots. IEEE Trans. Automat. Contr. 40(1), 78–83 (1995)
140. Qu, Z., Dawson, D.M.: Robust tracking control of robot manipulators. IEEE Press, Los Alamitos (1996)
141. Raibert, M.H., Craig, J.J.: Hybrid position/force control of manipulators. ASME J. Dynamic Syst. Measure. Contr. 102(2), 126–133 (1981)
142. Royden, H.: Real analysis. Prentice-Hall, Englewood Cliffs (1988)
143. Rus, D., Vona, M.: Crystalline robots: self-reconfiguration with compressible unit modules. Autonomous Robots 10(1), 107–124 (2001)
144. Schneider, S.A., Cannon Jr., R.H.: Object impedance control for cooperative manipulation: theory and experimental. IEEE Trans. Robot. Automat. 8, 383–394 (1992)
145. Sciavicco, L., Siciliano, B.: Modelling and control of robot manipulators. Springer, Heidelberg (2000)

146. Shen, W.M., Salemi, B., Will, P.: Hormone-inspired adaptive communication and distributed control for CONRO self-reconfigurable robots. IEEE Trans. Robot. Automat. 18(5), 700–712 (2002)
147. Seraji, H., Colbaugh, R.: Force tracking in impedance control. Int. J. Robot. Res. 16(1), 97–117 (1997)
148. Sheridan, T.B.: Telerobotics. Automatica 25(4), 487–507 (1989)
149. Siciliano, B., Villani, L.: Robot force control. Kluwer Academic Publishers, Dordrecht (2000)
150. Siciliano, B., Khatib, O.: Springer handbook of robotics. Springer, Heidelberg (2008)
151. Sirouspour, M.R., Salcudean, S.E.: Nonlinear control of hydraulic robots. IEEE Trans. Robot. Automat. 17(2), 173–182 (2001)
152. Skogestad, S., Postlethwaite, I.: Multivariable feedback control - analysis and design. John Wiley & Sons, Chichester (1996)
153. Slotine, J.J.E., Li, W.P.: On the adaptive control of robot manipulators. Int. J. Robot. Res. 6(3), 49–59 (1987)
154. Slotine, J.J.E., Li, W.P.: Adaptive manipulator control: a case study. IEEE Trans. Automat. Contr. 33(11), 995–1003 (1988)
155. Slotine, J.J.E., Li, W.P.: Applied nonlinear control. Prentice Hall, Englewood Cliffs (1991)
156. Sohl, G.A., Bobrow, J.E.: Experiments and simulations on the nonlinear control of a hydraulic servosystem. IEEE Trans. Control Systems Technology 7(2), 238–247 (1999)
157. Somolinos, J.A., Feliu, V., Sanchez, L.: Design, dynamic modelling and experimental validation of a new three-degree-of-freedom flexible arm. Mechatronics 12(7), 919–948 (2002)
158. Spofford, J.R., Akin, D.L.: Redundancy control of a free-flying telerobot. AIAA J. Guidance Control and Dynamics 12(3), 515–523 (1990)
159. Spong, M.W.: Modeling and control of elastic joint robots. ASME J. Dynamic Syst. Measure. Contr. 109, 310–319 (1987)
160. Spong, M.W., Khorasani, K., Kokotovic, P.V.: An integral manifold approach to the feedback control of flexible joint robots. IEEE Trans. Robot. Automat. 3(4), 291–300 (1987)
161. Spong, M.W., Vidyasagar, M.: Robot dynamics and control. John Wiley & Sons, Chichester (1989)
162. Spong, M.W.: On the force control problem for flexible joint manipulators. IEEE Trans. Automat. Contr. 34(1), 107–111 (1989)
163. Su, C.Y., Stepanenko, Y.: Hybrid adaptive/robust motion control of rigid-link electrically-driven robot manipulators. IEEE Trans. Robot. Automat. 11(3), 426–432 (1995)
164. Sun, Z., Tsao, T.C.: Adaptive control with asymptotic tracking performance and its application to an electro-hydraulic servo system. ASME J. Dynamic Syst. Measure. Contr. 122(1), 188–195 (2000)
165. Tafazoli, S., Lawrence, P.D., Salcudean, S.E.: Identification of inertial and friction parameters for excavator arms. IEEE Trans. Robot. Automat. 15(5), 966–971 (1999)
166. Takegaki, M., Arimoto, S.: A new feedback method for dynamic control of manipulators. ASME J. Dynamic Syst. Measure. Contr. 103, 119–125 (1981)
167. Tao, J.M., Luh, J.Y.S., Zheng, Y.F.: Compliant coordination control of two moving industrial robots. IEEE Trans. Robot. Automat. 6, 322–330 (1990)

168. Tao, G.: A simple alternative to the Barbălat lemma. IEEE Trans. Automat. Contr. 42(5), 698 (1997)
169. Tarn, T.J., Bejczy, A.K., Yun, X.: Design of dynamic control of two cooperating robot arms: closed chain formulation. In: Proc. of 1987 IEEE Int. Conf. Robot. Automat., pp. 7–13 (1987)
170. Tarn, T.J., Bejczy, A.K., Yun, X., Li, Z.: Effect of motor dynamics on nonlinear feedback robot arm control. IEEE Trans. Robot. Automat. 7(1), 114–122 (1991)
171. Tomei, P.: A simple PD controller for robots with elastic joints. IEEE Trans. Automat. Contr. 36(10), 1208–1213 (1991)
172. Tomei, P.: Tracking control of flexible joint robots with uncertain parameters and disturbances. IEEE Trans. Automat. Contr. 39(5), 1067–1072 (1994)
173. Turetta, A., Casalino, G., Sorbara, A.: Distributed control architecture for self-reconfigurable manipulators. Int. J. Robot. Res. 27(3-4), 481–504 (2008)
174. Tuttle, T.D., Seering, W.P.: A nonlinear model of a harmonic drive gear transmission. IEEE Trans. Robot. Automat. 12(3), 368–374 (1996)
175. Uchiyama, M., Dauchez, P.: A symmetric hybrid position/force control scheme for the coordination of two robots. In: Proc. of 1988 IEEE Int. Conf. Robot. Automat., pp. 350–356 (1988)
176. Umetani, Y., Yoshida, K.: Resolved motion rate control of space manipulators with generalized Jacobian matrix. IEEE Trans. Robot. Automat. 5(3), 303–314 (1989)
177. Unseren, M.A.: A rigid body model and decoupled control architecture for two manipulators holding a complex object. Robot. Autonomous Syst. 10, 115–131 (1992)
178. Van der Schaft, A.: L_2-gain and passivity techniques in nonlinear control. Springer, Heidelberg (2000)
179. Vukobratovic, M., Borovac, B., Surla, D., Stokic, D.: Biped locomotion - dynamics, stability, control and application. Springer, Heidelberg (1990)
180. Walker, I.D., Freeman, R.A., Marcus, S.I.: Internal object loading for multiple cooperating robot manipulators. In: Proc. of 1989 IEEE Int. Conf. Robot. Automat., pp. 606–611 (1989)
181. Walker, M.W., Kim, D., Dionise, J.: Adaptive coordinated motion control of two manipulator arms. In: Proc. of 1989 IEEE Int. Conf. Robot. Automat., pp. 1084–1090 (1989)
182. Walker, M.W.: Adaptive control of manipulators containing closed kinematic loops. IEEE Trans. Robot. Automat. 6, 10–19 (1990)
183. Walker, M.W., Wee, L.B.: Adaptive control of space-based robot manipulators. IEEE Trans. Robot. Automat. 7(6), 828–835 (1991)
184. Wang, D., Vidyasagar, M.: Modeling a class of multilink manipulators with the last link flexible. IEEE Trans. Robot. Automat. 8(1), 33–41 (1992)
185. Wen, J.T., Kreutz-Delgado, K.: The attitude control problem. IEEE Trans. Automat. Contr. 36(10), 1148–1162 (1991)
186. Wen, J.T., Kreutz-Delgado, K.: Motion and force control of multiple robotic manipulators. Automatica 28(4), 729–743 (1992)
187. Westervelt, E.R., Grizzle, J.W., Koditschek, D.E.: Hybrid zero dynamics of planar biped walkers. IEEE Trans. Automat. Contr. 48(1), 42–56 (2003)
188. Xi, N., Tarn, T.J., Bejczy, A.K.: Intelligent planning and control for multirobot coordination: an event-based approach. IEEE Trans. Robot. Automat. 12(3), 439–452 (1996)
189. Xu, Y., Kanade, T.: Space robotics: dynamics and control. Kluwer Academic Publishers, Dordrecht (1992)

190. Xu, Y., Shum, H.Y., Kanade, T., Lee, J.J.: Parameterization and adaptive control of space robot systems. IEEE Trans. Aerospace Electron. Syst. 30(2), 435–451 (1994)
191. Yabuta, T.: Nonlinear basic stability concept of the hybrid position/force control scheme for robot manipulators. IEEE Trans. Robot. Automat. 8(5), 663–670 (1992)
192. Yan, J., Salcudean, S.E.: Teleoperation controller design using H_∞-optimization with application to motion-scaling. IEEE Trans. Contr. Syst. Technology 4(3), 244–258 (1996)
193. Yao, B., Gao, W.B., Chan, S.P., Cheng, M.: VSC coordinated control of two manipulator arms in the presence of environmental constraints. IEEE Trans. Automat. Contr. 37, 1806–1812 (1992)
194. Yao, B., Bu, F., Reedy, J., Chiu, G.: Adaptive robust motion control of single-rod hydraulic actuators: theory and experiments. IEEE/ASME Trans. Mechatronics 5(1), 79–91 (2000)
195. Yim, M., Duff, D.G., Roufas, K.D.: Polybot: a modular reconfigurable robot. In: Proc. of 2000 IEEE Int. Conf. Robot. Automat., San Francisco, CA, pp. 514–520 (2000)
196. Yim, M., Shen, W.M., Salemi, B., Rus, D., Moll, M., Lipson, H., Klavins, E., Chirikjian, G.S.: Modular self-reconfigurable robot systems. IEEE Robot. Automat. Magazine 14(1), 43–52 (2007)
197. Yokokohji, Y., Toyoshima, T., Yoshikawa, T.: Efficient computational algorithm for trajectory control of free-flying space robots with multiple arms. IEEE Trans. Robot. Automat. 9(5), 571–579 (1993)
198. Yokokohji, Y., Yoshikawa, T.: Bilateral control of master-slave manipulators for ideal kinesthetic coupling – formulation and experiment. IEEE Trans. Robot. Automat. 10(5), 605–620 (1994)
199. Yoshida, K., Kurazume, R., Umetani, Y.: Dual arm coordination in space free-flying robot. In: Proc. of 1991 IEEE Int. Conf. Robot. Automat., pp. 2516–2521 (1991)
200. Yoshikawa, T.: Foundations of robotics: analysis and control. MIT Press, Cambridge (1990)
201. Yoshikawa, T., Zheng, X.Z.: Coordinated dynamic hybrid position/force control for multiple robot manipulators handling one constrained object. Int. J. Robot. Res. 12(3), 219–230 (1993)
202. Yuan, J., Stepanenko, Y.: Composite adaptive control of flexible joint robots. Automatica 29(3), 609–619 (1993)
203. Yuan, J.: Composite adaptive control of constrained robots. IEEE Trans. Robot. Automat. 12(4), 640–645 (1996)
204. Yun, X.P., Kumar, V.R.: An approach to simultaneous control of trajectory and interaction forces in dual-arm configurations. IEEE Trans. Robot. Automat. 7(5), 618–625 (1991)
205. Yun, X.P.: Modeling and control of two constrained manipulators. J. Intell. Robot. Syst. 4, 363–377 (1991)
206. Zhen, R.R., Goldenberg, A.A.: An adaptive approach to constrained robot motion control. In: Proc. of 1995 IEEE Int. Conf. Robot. Automat., pp. 1833–1838 (1995)
207. Zheng, Y.F., Hemami, H.: Impact effects of biped contact with the environment. IEEE Trans. Syst. Man Cybern. 14(3), 437–443 (1984)
208. Zheng, Y.F., Luh, J.Y.S.: Control of two coordinated robots in motion. In: Proc. of 1985 IEEE Int. Conf. Robot. Automat., pp. 1761–1766 (1985)

209. Zhu, W.H., Xi, Y.G., Zhang, Z.J.: Coordinative control of two spacerobots based on virtual decomposition. In: Proc. of 3rd IEEE Conf. Control Application., Glasgow, Scotland, pp. 327–332 (1994)
210. Zhu, M., Salcudean, S.E.: Achieving transparency for teleoperator systems under position and rate control. In: Proc. of IEEE/IROS 1995, vol. 2, pp. 7–12 (1995)
211. Zhu, W.H., Xi, Y.G., Zhang, Z.J., Bien, Z., De Schutter, J.: Virtual decomposition based control for generalized high dimensional robotic systems with complicated structure. IEEE Trans. Robot. Automat. 13(3), 411–436 (1997)
212. Zhu, W.H., Bien, Z., De Schutter, J.: Adaptive motion/force control of coordinated multiple manipulators with joint flexibility based on virtual decomposition. IEEE Trans. Automat. Contr. 43(1), 46–60 (1998)
213. Zhu, W.H., De Schutter, J.: Experiment with two industrial robot manipulators rigidly holding an egg. In: Proc. of 1998 IEEE Int. Conf. Robot. Automat., Leuven, Belgium, pp. 1534–1539 (1998)
214. Zhu, W.H., De Schutter, J.: Adaptive control of mixed rigid/flexible joint robot manipulators based on virtual decomposition. IEEE Trans. Robot. Automat. 15(2), 310–317 (1999)
215. Zhu, W.H., De Schutter, J.: Control of two industrial manipulators rigidly holding an egg. IEEE Control Systems Magazine 19(2), 24–30 (1999)
216. Zhu, W.H., De Schutter, J.: Adaptive control of electrically driven space robots based on virtual decomposition. AIAA J. Guidance Control and Dynamics 22(2), 329–339 (1999)
217. Zhu, W.H., Salcudean, S.E.: Stability guaranteed teleoperation: an adaptive motion/force control approach. IEEE Trans. Automat. Contr. 45(11), 1951–1969 (2000)
218. Zhu, W.H., De Schutter, J.: Experimental verifications of virtual decomposition based motion/force control. IEEE Trans. Robot. Automat. 18(3), 379–386 (2002)
219. Zhu, W.H., Piedboeuf, J.C., Gonthier, Y.: Emulation of a space robot using a hydraulic manipulator on ground. In: Proc. of 2002 IEEE Int. Conf. Robot. Automat., Washington, DC, pp. 2315–2320 (2002)
220. Zhu, W.H., De Schutter, J.: Virtual decomposition based motion/force control of an industrial manipulator KUKA361. In: Preprint of 15th IFAC World Congress, Barcelona, Spain, pp. 1902–1907 (2002)
221. Zhu, W.H., Doyon, M.: Adaptive control of harmonic drives. In: Proc. of 43rd IEEE Conference on Decision and Control, The Atlantis, Bahamas, pp. 2602–2608 (2004)
222. Zhu, W.H., Piedboeuf, J.C.: Adaptive output force tracking control of hydraulic cylinders with applications to robot manipulators. ASME J. Dynamic Syst. Measure. Contr. 127(2), 206–217 (2005)
223. Zhu, W.H.: On adaptive synchronization control of coordinated multirobots with flexible/rigid constraints. IEEE Trans. Robotics 21(3), 520–525 (2005)
224. Zhu, W.H., Dupuis, E., Doyon, M., Piedboeuf, J.C.: Adaptive control of harmonic drives based on virtual decomposition. IEEE/ASME Trans. Mechatronics 11(5), 604–614 (2006)
225. Zhu, W.H., Dupuis, E., Doyon, M.: Adaptive control of harmonic drives. ASME J. Dynamic Syst. Measure. Contr. 129, 182–193 (2007)
226. Zhu, W.H., Lamarche, T.: Modular robot manipulators based on virtual decomposition control. In: Proc. of 2007 IEEE Int. Conf. Robot. Automat., pp. 2235–2240 (2007)
227. Zhu, W.H.: Dynamics of general constrained robots derived from rigid bodies. ASME J. Applied Mechanics 75(3) (2008)

228. Zhu, W.H., Lange, C., Callot, M.: Virtual decomposition control of a planar flexible-link robot. In: Preprint of 17th IFAC World Congress, Seoul, Korea, pp. 1697–1702 (2008)
229. Ziaei, K., Sepehri, N.: Design of a nonlinear adaptive controller for an electro-hydraulic actuator. ASME J. Dynamic Syst. Measure. Contr. 123(3), 449–456 (2001)

Index

Springer Tracts in Advanced Robotics

Edited by B. Siciliano, O. Khatib and F. Groen

Further volumes of this series can be found on our homepage: springer.com

Vol. 59: Otake, M.
Electroactive Polymer Gel Robots – Modelling and Control of Artificial Muscles
238 p. 2010 [978-3-540-23955-0]

Vol. 58: Kröger, T.
On-Line Trajectory Generation in Robotic Systems – Basic Concepts for Instantaneous Reactions to Unforeseen (Sensor) Events
230 p. 2010 [978-3-642-05174-6]

Vol. 57: Chirikjian, G.S.; Choset, H.; Morales, M., Murphey, T. (Eds.)
Algorithmic Foundations of Robotics VIII – Selected Contributions of the Eighth International Workshop on the Algorithmic Foundations of Robotics
680 p. 2010 [978-3-642-00311-0]

Vol. 56: Buehler, M.; Iagnemma, K.; Singh S. (Eds.)
The DARPA Urban Challenge – Autonomous Vehicles in City Traffic
625 p. 2009 [978-3-642-03990-4]

Vol. 55: Stachniss, C.
Robotic Mapping and Exploration
196 p. 2009 [978-3-642-01096-5]

Vol. 54: Khatib, O.; Kumar, V.; Pappas, G.J. (Eds.)
Experimental Robotics: The Eleventh International Symposium
579 p. 2009 [978-3-642-00195-6]

Vol. 53: Duindam, V.; Stramigioli, S.
Modeling and Control for Efficient Bipedal Walking Robots
211 p. 2009 [978-3-540-89917-4]

Vol. 52: Nüchter, A.
3D Robotic Mapping
201 p. 2009 [978-3-540-89883-2]

Vol. 51: Song, D.
Sharing a Vision
186 p. 2009 [978-3-540-88064-6]

Vol. 50: Alterovitz, R.; Goldberg, K.
Motion Planning in Medicine: Optimization and Simulation Algorithms for Image-Guided Procedures
153 p. 2008 [978-3-540-69257-7]

Vol. 49: Ott, C.
Cartesian Impedance Control of Redundant and Flexible-Joint Robots
190 p. 2008 [978-3-540-69253-9]

Vol. 48: Wolter, D.
Spatial Representation and Reasoning for Robot Mapping
185 p. 2008 [978-3-540-69011-5]

Vol. 47: Akella, S.; Amato, N.; Huang, W.; Mishra, B.; (Eds.)
Algorithmic Foundation of Robotics VII
524 p. 2008 [978-3-540-68404-6]

Vol. 46: Bessière, P.; Laugier, C.; Siegwart R. (Eds.)
Probabilistic Reasoning and Decision Making in Sensory-Motor Systems
375 p. 2008 [978-3-540-79006-8]

Vol. 45: Bicchi, A.; Buss, M.; Ernst, M.O.; Peer A. (Eds.)
The Sense of Touch and Its Rendering
281 p. 2008 [978-3-540-79034-1]

Vol. 44: Bruyninckx, H.; Přeučil, L.; Kulich, M. (Eds.)
European Robotics Symposium 2008
356 p. 2008 [978-3-540-78315-2]

Vol. 43: Lamon, P.
3D-Position Tracking and Control for All-Terrain Robots
105 p. 2008 [978-3-540-78286-5]

Vol. 42: Laugier, C.; Siegwart, R. (Eds.)
Field and Service Robotics
597 p. 2008 [978-3-540-75403-9]

Vol. 41: Milford, M.J.
Robot Navigation from Nature
194 p. 2008 [978-3-540-77519-5]

Vol. 40: Birglen, L.; Laliberté, T.; Gosselin, C.
Underactuated Robotic Hands
241 p. 2008 [978-3-540-77458-7]

Vol. 39: Khatib, O.; Kumar, V.; Rus, D. (Eds.)
Experimental Robotics
563 p. 2008 [978-3-540-77456-3]

Vol. 38: Jefferies, M.E.; Yeap, W.-K. (Eds.)
Robotics and Cognitive Approaches to
Spatial Mapping
328 p. 2008 [978-3-540-75386-5]

Vol. 37: Ollero, A.; Maza, I. (Eds.)
Multiple Heterogeneous Unmanned Aerial
Vehicles
233 p. 2007 [978-3-540-73957-9]

Vol. 36: Buehler, M.; Iagnemma, K.;
Singh, S. (Eds.)
The 2005 DARPA Grand Challenge – The Great
Robot Race
520 p. 2007 [978-3-540-73428-4]

Vol. 35: Laugier, C.; Chatila, R. (Eds.)
Autonomous Navigation in Dynamic
Environments
169 p. 2007 [978-3-540-73421-5]

Vol. 34: Wisse, M.; van der Linde, R.Q.
Delft Pneumatic Bipeds
136 p. 2007 [978-3-540-72807-8]

Vol. 33: Kong, X.; Gosselin, C.
Type Synthesis of Parallel
Mechanisms
272 p. 2007 [978-3-540-71989-2]

Vol. 32: Milutinović, D.; Lima, P.
Cells and Robots – Modeling and Control of
Large-Size Agent Populations
130 p. 2007 [978-3-540-71981-6]

Vol. 31: Ferre, M.; Buss, M.; Aracil, R.;
Melchiorri, C.; Balaguer C. (Eds.)
Advances in Telerobotics
500 p. 2007 [978-3-540-71363-0]

Vol. 30: Brugali, D. (Ed.)
Software Engineering for Experimental Robotics
490 p. 2007 [978-3-540-68949-2]

Vol. 29: Secchi, C.; Stramigioli, S.; Fantuzzi, C.
Control of Interactive Robotic Interfaces – A
Port-Hamiltonian Approach
225 p. 2007 [978-3-540-49712-7]

Vol. 28: Thrun, S.; Brooks, R.;
Durrant-Whyte, H. (Eds.)
Robotics Research – Results of the 12th
International Symposium ISRR
602 p. 2007 [978-3-540-48110-2]

Vol. 27: Montemerlo, M.; Thrun, S.
FastSLAM – A Scalable Method for the
Simultaneous Localization and Mapping
Problem in Robotics
120 p. 2007 [978-3-540-46399-3]

Vol. 26: Taylor, G.; Kleeman, L.
Visual Perception and Robotic Manipulation – 3D
Object Recognition, Tracking and Hand-Eye
Coordination
218 p. 2007 [978-3-540-33454-5]

Vol. 25: Corke, P.; Sukkarieh, S. (Eds.)
Field and Service Robotics – Results of the 5th
International Conference
580 p. 2006 [978-3-540-33452-1]

Vol. 24: Yuta, S.; Asama, H.; Thrun, S.;
Prassler, E.; Tsubouchi, T. (Eds.)
Field and Service Robotics – Recent Advances in
Research and Applications
550 p. 2006 [978-3-540-32801-8]

Vol. 23: Andrade-Cetto, J,; Sanfeliu, A.
Environment Learning for Indoor Mobile Robots
– A Stochastic State Estimation Approach
to Simultaneous Localization and Map Building
130 p. 2006 [978-3-540-32795-0]

Vol. 22: Christensen, H.I. (Ed.)
European Robotics Symposium 2006
209 p. 2006 [978-3-540-32688-5]

Vol. 21: Ang Jr., H.; Khatib, O. (Eds.)
Experimental Robotics IX – The 9th International
Symposium on Experimental Robotics
618 p. 2006 [978-3-540-28816-9]

Vol. 20: Xu, Y.; Ou, Y.
Control of Single Wheel Robots
188 p. 2005 [978-3-540-28184-9]

Vol. 19: Lefebvre, T.; Bruyninckx, H.;
De Schutter, J. Nonlinear Kalman Filtering
for Force-Controlled Robot Tasks
280 p. 2005 [978-3-540-28023-1]

Vol. 18: Barbagli, F.; Prattichizzo, D.;
Salisbury, K. (Eds.)
Multi-point Interaction with Real
and Virtual Objects
281 p. 2005 [978-3-540-26036-3]

Batch number: 10614741

Printed by Printforce, the Netherlands